CENTROID LOCATIONS FOR A FEW COMMON LINE SEGMENTS AND AREAS

Circular arc

$L = 2r\alpha$

$x_C = \dfrac{r \sin \alpha}{\alpha}$

$y_C = 0$

Circular sector

$A = r^2\alpha$

$x_C = \dfrac{2r \sin \alpha}{3\alpha}$

$y_C = 0$

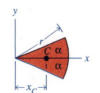

Quarter circular arc

$L = \dfrac{\pi r}{2}$

$x_C = \dfrac{2r}{\pi}$

$y_C = \dfrac{2r}{\pi}$

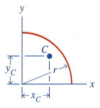

Quadrant of a circle

$A = \dfrac{\pi r^2}{4}$

$x_C = \dfrac{4r}{3\pi}$

$y_C = \dfrac{4r}{3\pi}$

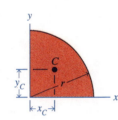

Semicircular arc

$L = \pi r$

$x_C = r$

$y_C = \dfrac{2r}{\pi}$

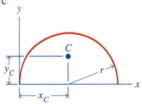

Semicircular area

$A = \dfrac{\pi r^2}{2}$

$x_C = r$

$y_C = \dfrac{4r}{3\pi}$

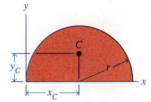

Rectangular area

$A = bh$

$x_C = \dfrac{b}{2}$

$y_C = \dfrac{h}{2}$

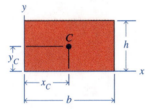

Quadrant of an ellipse

$A = \dfrac{\pi ab}{4}$

$x_C = \dfrac{4a}{3\pi}$

$y_C = \dfrac{4b}{3\pi}$

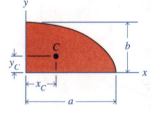

Triangular area

$A = \dfrac{bh}{2}$

$x_C = \dfrac{2b}{3}$

$y_C = \dfrac{h}{3}$

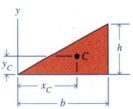

Parabolic spandrel

$A = \dfrac{bh}{3}$

$x_C = \dfrac{3b}{4}$

$y_C = \dfrac{3h}{10}$

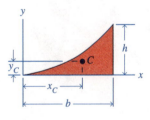

Triangular area

$A = \dfrac{bh}{2}$

$x_C = \dfrac{a + b}{3}$

$y_C = \dfrac{h}{3}$

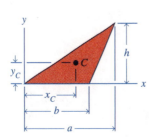

Quadrant of a parabola

$A = \dfrac{2bh}{3}$

$x_C = \dfrac{5b}{8}$

$y_C = \dfrac{2h}{5}$

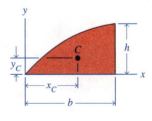

STATICS AND MECHANICS OF MATERIALS: AN INTEGRATED APPROACH

STATICS AND MECHANICS OF MATERIALS: AN INTEGRATED APPROACH

William F. Riley
Professor Emeritus
Iowa State University
Leroy D. Sturges
Aerospace Engineering and Engineering Mechanics
Iowa State University
Don H. Morris
Engineering Science and Mechanics
Virginia Polytechnic Institute and State University

JOHN WILEY & SONS, INC.
New York   Chichester   Brisbane   Toronto   Singapore

ACQUISITIONS EDITOR   Charity Robey

MARKETING MANAGER   Debra Riegert

PRODUCTION MANAGER   Charlotte Hyland

DESIGNER   Dawn L. Stanley

MANUFACTURING MANAGER   Mark Cirillo

ILLUSTRATION COORDINATOR   Jaime Perea

This book was set in Times New Roman by York Graphic Services and
printed and bound by von Hoffman. The cover was printed by Phoenix Color.

Library of Congress Cataloging in Publication Data:
Riley, William F. (William Franklin), 1925–
    Statics and mechanics of materials: an integrated approach/
   William F. Riley, Leroy D. Sturges, Don H. Morris.
        p.   cm.
      Includes index.
      ISBN 0-471-01334-X (cloth: acid-free paper)
      1. Strength of materials.   2. Statics.   I. Sturges, Leroy D.
   II. Morris, Don H.   III. Title.
   TA405.R56   1996   95–4072
   620.1′12—dc20   CIP

Printed in the United States of America

10 9 8 7 6 5 4 3 2

PREFACE

The purpose of courses in engineering mechanics is to describe the effects that forces have on bodies and structures. The traditional introduction to mechanics consists of a course in statics followed by a course in mechanics of materials. The principles of statics are used to determine the forces that act on or in a structure, assuming that the structure is perfectly rigid and does not deform. These forces, along with the theory developed in mechanics of materials, are used to determine how the material deforms or reacts.

This book approaches the teaching of mechanics using the *just-in-time* approach. As soon as the student has studied equilibrium of concurrent force systems, he or she is ready to calculate stretches of wires and rods using a one-dimensional Hooke's law. After studying rigid body equilibrium, the student is ready to calculate stresses and deformation in members such as shafts and beams. When the two subjects are integrated in a unified course in this manner, students can immediately see the use of the principles of statics, they can see the inter-relationship of statics and mechanics of materials.

This text would normally be used to teach a sophomore-level course of the same name—Statics and Mechanics of Materials—or a course titled Statics and Strength of Materials, or Introduction to Mechanics, or possibly Essentials of Mechanics. The course could be taught out of various engineering departments, including Aerospace Engineering, Civil Engineering, Engineering Fundamentals, Engineering Mechanics, Mechanical Engineering, and General Engineering. We envision a single-semester course of at least four or five credits. Prerequisites for the course would be a year of calculus, however a semester of physics, while not necessary, would be desirable.

After a brief introduction to mechanics in Chapter One, Chapter Two describes the characteristics of forces and develops the mathematics necessary to work with concurrent forces. These concepts are immediately used in Chapter Three to calculate the forces acting on a particle in equilibrium. The basic discussion is completed in Chapter Four where stress, strain, and the relationship between loads and deformation is presented. Chapter Five continues the description of forces, develops the concept of equivalent force–couple systems, and explores the effects of forces and couples on rigid bodies. Chapter Six presents the equilibrium of rigid bodies and its use in several structural applications. The final five chapters consist of standard topics of Mechanics of Materials—torsion of circular shafts (Chapter 7), flexural stresses in beams (Chapter 8), deflections of beams (Chapter 9), combined loadings (Chapter 10), and columns and other compressive members (Chapter 11). Second moments of areas are introduced and developed where needed in Chapters 7 and 8.

Every chapter opens with a brief introduction; all principles are illustrated by one or more Example Problems and several Homework Problems. The Homework Problems are graded in difficulty and are separated into a section of Standard problems followed by a set of more Challenging problems.

U.S. customary and SI units are used equally throughout the book. To help the instructor who wants to assign problems of one type or the other, odd-numbered homework problems are in U.S. customary units and even-numbered homework problems are in SI units. Answers to about half of the homework problems are in the back of the book. Since the convenient designation of problems for which answers are provided is of great value to those who make up assignment sheets, the problems for which answers are provided are indicated by means of an asterisk (∗) after the number.

Most chapters conclude with a section on Design Problems, including Example Problems and followed by a set of Homework Problems. Every chapter ends with a Summary of important concepts covered in the chapter, followed by a set of Review Problems. Since the Review Problems are not tied to a specific section, they often combine principles from several different sections of the chapter. Most chapters end with a set of Computer Problems that require students to analyze how the solution depends on some parameter of the problem.

This book is designed to emphasize the required fundamental principles, with numerous applications to demonstrate and develop logical, orderly methods of procedure. Instead of deriving numerous formulas for all types of problems, we have stressed the use of free-body diagrams and the equations of equilibrium, together with the geometry of the deformed body and the observed relations between stress and strain, for the analysis of the force system acting on a body.

The emphasis is always on keeping the material understandable to the student. Clarity is never sacrificed for the sake of mathematical elegance. Calculus and vector methods are used where necessary and where appropriate. However, if scalar methods are more appropriate and/or are commonly used by practicing engineers, then these methods are generally used in the Example Problems.

We are grateful for comments and suggestions received from reviewers and have incorporated those recommended changes we considered important.

WILLIAM F. RILEY
LEROY D. STURGES
DON H. MORRIS

CONTENTS

CHAPTER 6: EQUILIBRIUM: RIGID BODIES

CHAPTER 10: COMBINED STATIC LOADING

CHAPTER 11: COLUMNS

APPENDIX A : VECTOR OPERATIONS

APPENDIX B: TABLES OF PROPERTIES

GENERAL PRINCIPLES

1

1-1 INTRODUCTION

Mechanics is the branch of the physical sciences that deals with the response of bodies to the action of forces. The subject matter of this field constitutes a large part of our knowledge of the laws governing the behavior of solid bodies as well as the laws governing the behavior of gases and liquids. The laws of mechanics find application in most machines and structures involved in engineering practice. For convenience, the study of mechanics is divided into three parts: namely, the mechanics of rigid bodies, the mechanics of deformable bodies, and the mechanics of fluids.

A study of the mechanics of rigid bodies can be further subdivided into three main divisions: statics, kinematics, and kinetics. *Statics* deals with bodies that are acted upon by balanced forces and hence are at rest or have uniform motion. Such bodies are said to be in "equilibrium." Statics is an important part of the study of mechanics because it provides methods for determining support reactions and relationships between internal force distributions and external loads for stationary structures. Many practical engineering problems involving loads carried by structural components can be solved using relationships developed in statics. The relationships developed in statics between internal force distributions and external loads play an important role in the development of deformable body mechanics.

Kinematics deals with the motion of bodies without considering the manner in which the motion is produced. Kinematics is sometimes referred to as the "geometry of motion." Kinematics forms an important part of the study of mechanics, not only because of its application to problems in which forces are involved, but also because of its application to problems that involve only motions of parts of a machine. For many motion problems, the principles of kinematics, alone, are sufficient for solution of the problem. Such problems are discussed in "Kinematics of Machinery" books, in which the motion of machine elements such as cam shafts, gears, connecting rods, and quick-return mechanisms is considered.

Kinetics deals with bodies that are acted upon by unbalanced forces; hence, they have nonuniform or accelerated motions. A study of kinetics is an important part of the study of mechanics because it provides relationships between the motion of a body and the forces and moments acting upon the body. Kinetic relationships may be obtained by direct application of Newton's laws of motion or by using integrated forms of the equations of motion that result in the principles of work–energy or impulse–momentum. Frequently, the term *dynamics* is used in the technical literature to denote the subdivisions of mechanics with which the idea of motion is most closely associated—namely, kinematics and kinetics.

The branch of mechanics that deals with internal force distributions and the deformations developed in actual engineering structures and machine components when they are subjected to systems of forces is known as *mechanics of deformable bodies.* Books covering this part of mechanics commonly have the title "Mechanics of Materials," "Strength of Materials," or "Mechanics of Deformable Bodies."

The branch of mechanics that deals with liquids and gases at rest or in motion is known as *fluid mechanics.* Fluids can be classified as compressible or incompressible. A fluid is said to be compressible if the density of the fluid varies with temperature and pressure. If the volume of a fluid remains constant during a change in pressure, the fluid is said to be incompressible. Liquids are considered incompressible for most engineering applications. A subdivision of fluid mechanics that deals with incompressible liquids is commonly known as *hydraulics.*

This integrated book on statics and mechanics of materials will cover all topics normally treated in books dealing with statics of rigid and deformable bodies. The book will provide the foundation required for follow-on courses in many fields of engineering.

1-2 FUNDAMENTAL QUANTITIES OF MECHANICS

The fundamental quantities of mechanics are space, time, mass, and force. Three of the quantities—space, time, and mass—are absolute quantities, meaning that they are independent of each other and cannot be expressed in terms of the other quantities or in simpler terms. The quantity known as a force is not independent of the other three quantities but is related to the mass of the body and to the manner in which the velocity of the body varies with time. The following paragraphs provide a brief description of these and other important concepts.

Space is the geometric region in which the physical events of interest in mechanics occur. The region extends without limit in all directions. The measure used to describe the size of a physical system is known as a *length.* The position of a point in space can be determined relative to some reference point by using linear and angular measurements with respect to a coordinate system whose origin is at the reference point. The basic reference system used as an aid in solving mechanics problems is one that is considered fixed in space. Measurements relative to this system are called *absolute.*

Time can be defined as the interval between two events. Measurements of this interval are made by making comparisons with some reproducible event such as the time required for the earth to orbit the sun or the time required for the earth to rotate on its axis. Solar time is earth rotation time measured with respect to the sun and is used for navigation on Earth and for daily living purposes.

Any device that is used to indicate the passage of time is referred to as a *clock*. Reproducible events commonly used as sensing mechanisms for clocks include the swing of a pendulum, oscillation of a spiral spring and balance wheel, and oscillation of a piezoelectric crystal. The time required for one of these devices to complete one cycle of motion is known as the *period*. The frequency of the motion is the number of cycles occurring in a given unit of time.

Matter is any substance that occupies space. A *body* is matter bounded by a closed surface. The property of a body that causes it to resist any change in motion is known as *inertia*. *Mass* is a quantitative measure of inertia. The resistance a body offers to a change in translational motion is independent of the size and shape of the body. It depends only on the mass of the body. The resistance a body offers to a change in rotational motion depends on the distribution of the body's mass. Mass is also a factor in the gravitational attraction between two bodies.

A *force* can be defined as the action of one body upon another body. Our concept of force comes mainly from personal experiences in which we are one of the bodies and tension or compression of our muscles results when we try to pull or push the second body. This is an example of force resulting from direct contact between bodies. A force can also be exerted between bodies that are physically separated. Gravitational forces exerted by the earth on the moon and on artificial satellites to keep them in earth orbit are examples. Since a body cannot exert a force on a second body unless the second body offers a resistance, a force never exists alone. Forces always occur in pairs and the two forces have equal magnitude and opposite sense. Although a single force never exists, it is usually convenient in mechanics problems to think only of the actions of other bodies on the body in question without taking into account the reactions of the body in question. The external effect of a force on a body is either acceleration of the body or development of resisting forces (reactions) on the body.

A *particle* has mass but no size or shape. When a body (large or small) in a mechanics problem can be treated as a particle, the analysis is greatly simplified since the mass can be assumed to be concentrated at a point and the concept of rotation is not involved in the solution of the problem.

A *rigid body* can be represented as a collection of particles. The size and shape of the body remain constant at all times and under all conditions of loading. The rigid-body concept represents an idealization of the true situation since all real bodies will change shape to a certain extent when subjected to a system of forces. Such changes are small for most structural elements and machine parts encountered in engineering practice; therefore, they have only a negligible effect upon the reactions required to maintain equilibrium of the body.

1-2-1 Newton's Laws

The foundations for studies in engineering mechanics are the laws formulated and published by Sir Isaac Newton in 1687. In a treatise called *The Principia*, Newton stated the basic laws governing the motion of a particle as[1]:

---

[1]As stated in Dr. Ernst Mach, *The Science of Mechanics*, 9th ed. The Open Court Publishing Company, LaSalle, Ill., 1942.

NEWTON'S LAWS OF MOTION

Law 1: "Every body perseveres in its state of rest or of uniform motion in a straight line, except insofar as it is compelled to change that state by impressed forces."

Law 2: "Change of motion is proportional to the moving force impressed, and takes place in the direction of the straight line in which such force is impressed."

Law 3: "Reaction is always equal and opposite to action; that is to say, the actions of two bodies upon each other are always equal and directly opposite."

These laws, which have come to be known as "Newton's Laws of Motion," are commonly expressed today as:

Law 1. In the absence of external forces, a particle originally at rest or moving with a constant velocity will remain at rest or continue to move with a constant velocity along a straight line.

Law 2. If an external force acts on a particle, the particle will be accelerated in the direction of the force and the magnitude of the acceleration will be directly proportional to the force and inversely proportional to the mass of the particle.

Law 3. For every action there is an equal and opposite reaction. The forces of action and reaction between bodies are equal in magnitude, opposite in direction, and collinear.

Newton's three laws were developed from a study of planetary motion (the motion of particles); therefore, they apply only to the motion of particles. During the eighteenth century, Leonhard Euler (1707–1783) extended Newton's work on particles to rigid-body systems.

The first law of motion is a special case of the second law and covers the case in which the particle is in equilibrium. Thus, the first law provides the foundation for the study of statics. The second law of motion provides the foundation for the study of dynamics. The mathematical statement of the second law, which is widely used in dynamics, is

$$\mathbf{F} = m\mathbf{a}$$

(1-1)

where $\mathbf{F}$ is the external force acting on the particle, m is the mass of the particle, and $\mathbf{a}$ is the acceleration of the particle. The third law of motion provides the foundation for an understanding of the concept of a force, since in practical engineering applications the word "action" is taken to mean force. Thus, if one body exerts a force on a second body, the second body exerts an equal and opposite force on the first.

The law that governs the mutual attraction between two bodies was also formulated by Newton and is known as the "Law of Gravitation." This law can be expressed mathematically as

$$F = G\frac{m_1 m_2}{r^2} \tag{1-2}$$

where F is the magnitude of the mutual force of attraction between the two bodies, G is the universal gravitational constant, m_1 is the mass of one of the bodies, m_2 is the mass of the other body, and r is the distance between the centers of mass of the two bodies.

Approximate values of the universal gravitational constant that are suitable for most engineering computations are

$G = 3.439(10^{-8})$ ft³/(slug · s²) (in the U.S. customary system of units)

$G = 6.673(10^{-11})$ m³/(kg · s²) (in the SI system of units)

The mutual forces of attraction between the two bodies represent the action of one body on the other; therefore, they obey Newton's third law, which requires that they be equal in magnitude, opposite in direction, and collinear (lie along the line joining the centers of mass of the two bodies).

1-2-2 Mass and Weight

The mass m of a body is an absolute quantity that is independent of the position of the body and independent of the surroundings in which the body is placed. The weight W of a body is the gravitational attraction exerted on the body by the planet Earth or by any other massive body, such as the moon. Therefore, the weight of a body depends on the position of the body relative to some other body. Thus, for Eq. (1-2), at the surface of the earth,

$$W = G\frac{m_e m}{r_e^2} = mg \tag{1-3}$$

where m_e is the mass of the earth, r_e is the mean radius of the earth, and $g = Gm_e/r_e^2$ is the gravitational acceleration.

Approximate values for the gravitational acceleration that are suitable for most engineering computations are

$g = 32.17$ ft/s²

$g = 9.807$ m/s²

A source of some confusion arises because the pound is sometimes used as a unit of mass and the kilogram is sometimes used as a unit of force. In grocery stores in Europe, weights of packages are marked in kilograms. In the United States, weights of packages are often marked in both pounds and kilograms. Similarly, a unit of mass called the *pound* or the *pound mass,* which is the mass whose weight is one pound under standard gravitational conditions, is sometimes used.

Throughout this book, without exception, the pound (lb) will be used as the unit of force and the slug will be used as the unit of mass for problems and examples when the U.S. customary system of units is used. Similarly, the newton (N) will be used as the unit of force and the kilogram (kg) will be used as the unit of mass for problems and examples when the SI system of units is used.

Example Problem 1-1

A body weighs 250 lb at the earth's surface. Determine

(a) The mass of the body.
(b) The weight of the body 500 miles above the earth's surface.
(c) The weight of the body on the moon's surface.

SOLUTION

(a) The weight of the body at the earth's surface is given by Eq. (1-3) as

$$W = mg$$

Thus,

$$m = \frac{W}{g} = \frac{250}{32.17} = 7.77 \frac{\text{lb} \cdot \text{s}^2}{\text{ft}} = 7.77 \text{ slug} \qquad \textbf{Ans.}$$

(b) The force of attraction between two bodies is given by Eq. (1-2) as

$$W = F = G\frac{m_1 m_2}{r^2} \qquad \text{or} \qquad Wr^2 = Gm_1 m_2 = \text{Constant}$$

The mean radius of the earth is

$$r_e = 3960 \text{ mi}$$

Thus, for the two positions of the body

$$Wr_e^2 = W_{500}(r_e + 500)^2 = Gm_1 m_2 = \text{Constant}$$

$$W_{500} = \frac{Wr_e^2}{(r_e + 500)^2} = \frac{250(3960)^2}{(3960 + 500)^2} = 197.1 \text{ lb} \qquad \textbf{Ans.}$$

(c) The mean radius and mass of the moon are

$$r_m = 1080 \text{ mi} = 5.702(10^6) \text{ ft}$$
$$m_m = 5.037(10^{21}) \text{ slug}$$

Also,

$$G = 3.439(10^{-8}) \text{ ft}^3/(\text{slug} \cdot \text{s}^2)$$

On the moon's surface, the weight of the body is given by Eq. (1-2) as

$$W = \frac{Gmm_m}{r_m^2} = 3.439(10^{-8})\frac{7.77(5.037)(10^{21})}{[5.702(10^6)]^2} = 41.4 \text{ lb} \ \blacksquare \qquad \textbf{Ans.}$$

PROBLEMS

Standard Problems

1-1* Calculate the mass *m* of a body that weighs 500 lb at the surface of the earth.

1-2* Calculate the weight *W* of a body at the surface of the earth if it has a mass *m* of 575 kg.

1-3* If a man weighs 180 lb at sea level, determine the weight *W* of the man
  (a) At the top of Mt. McKinley (20,320 ft above sea level).
  (b) At the top of Mt. Everest (29,028 ft above sea level).

1-4* Calculate the weight *W* of a navigation satellite at a distance of 20,200 km above the earth's surface if the satellite weighs 9750 N at the earth's surface.

1-5 Compute the gravitational force acting between two spheres that are touching each other if each sphere weighs 1125 lb and has a diameter of 20 in.

1-6 Two spherical bodies have masses of 60 kg and 80 kg, respectively. Determine the force of gravity acting between them if the distance from center to center of the bodies is 500 mm.

Challenging Problems

1-7 At what distance from the surface of the earth, in miles, is the weight of a body equal to one half of its weight on the earth's surface?

1-8 At what distance, in kilometers, from the surface of the earth on a line from center to center would the gravitational force of the earth on a body be exactly balanced by the gravitational force of the moon on the body?

1-3 UNITS OF MEASUREMENT

The building blocks of mechanics are the physical quantities used to express the laws of mechanics. Some of these quantities are mass, length, force, and time. Physical quantities are often divided into fundamental quantities and derived quantities. *Fundamental quantities* cannot be defined in terms of other physical quantities. The number of quantities regarded as fundamental is the minimum number needed to give a consistent and complete description of all the physical quantities ordinarily encountered in the subject area. Length and time are examples of quantities viewed as fundamental in mechanics. *Derived quantities* are those whose defining operations are based on measurements of other physical quantities. Area, volume, velocity, and acceleration are examples of derived quantities in mechanics. Some quantities may be viewed as either fundamental or derived; mass and force are examples of such quantities. In the SI system of units, mass is regarded as a fundamental quantity and force is a derived quantity. In the U.S. customary system of units, force is regarded as a fundamental quantity and mass is a derived quantity.

The magnitude of each of the fundamental quantities is defined by an arbitrarily chosen unit or "Standard." The familiar yard, foot, and inch, for example, come from the practice of using the human arm, foot, and thumb as length standards. However, for any type of precise calculations, such units of length are unsatisfactory. The first truly international standard of length was a bar of platinum–iridium alloy, called the *standard meter*,[2] which was kept at the International Bureau of Weights and Measures in Sèvres, France. The distance between two fine lines engraved on gold plugs near the ends of the bar is defined to be

---

[2]The United States has accepted the meter as a standard of length since 1893.

one meter. Historically, the meter was intended to be one ten-millionth of the distance from the pole to the equator along the meridian line through Paris. Accurate measurements made after the standard meter bar was constructed show that it differs from its intended value by approximately 0.023 percent.

In 1961 an international atomic standard of length was adopted. The wavelength in vacuum of the orange-red line from the spectrum of krypton 86 was chosen. One meter (m) is now defined to be 1,650,763.73 wavelengths of this light. The choice of an atomic standard offers advantages other than increased precision in length measurements. Krypton 86 can be obtained relatively easily and inexpensively everywhere, and all atoms of the material are identical and emit light of the same wavelength. The particular wavelength chosen is uniquely characteristic of krypton 86 and is very sharply defined. The definition of the yard, by international agreement, is 1 yard = 0.9144 m, exactly.[3] Thus, 1 inch = 25.4 mm, exactly; and 1 foot = 0.3048 m, exactly.

Similarly, time can be measured in a number of ways. Since the earliest times, the length of a day has been an accepted standard of time measurement. The internationally accepted standard unit of time—the *second* (s)—has been defined in the past as 1/86,400 of a mean solar day or 1/31,557,700 of a mean solar year. Time defined in terms of the rotation of the earth must be determined by astronomical observations. Since these observations require at least several weeks, a good secondary terrestrial measure, calibrated by astronomical observations, is used. Quartz crystal clocks, based on the electrically sustained natural periodic vibrations of a quartz wafer, have been used as secondary time standards. The best of these quartz clocks have kept time for a year with a maximum error of 0.02 second.

To meet the need for an even better time standard, an atomic clock has been developed that uses the periodic atomic vibrations of isotope cesium 133. The second based on this cesium clock is defined as the duration of 9,192,631,770 cycles of vibration. The cesium clock provides an improvement over the accuracy associated with astronomical methods by a factor of 200. Two cesium clocks will differ by no more than one second after running 3000 years.

The standard unit of mass, the *kilogram* (kg), is defined by a bar of platinum–iridium alloy that is kept at the International Bureau of Weights and Measures in Sèvres, France.

1-3-1 The U.S. Customary System of Units

Most engineers in the United States use the U.S. customary system of units (USCS) (sometimes called the "British gravitational system"), in which the base units are foot (ft) for length, pound (lb) for force, and second (s) for time. In this system, a foot is defined as 0.3048 m, exactly. The *pound* is defined as the weight at sea level and at a latitude of 45 degrees of a platinum standard that is kept at the Bureau of Standards in Washington, D.C. This platinum standard has a mass of 0.453,592,43 kg. The second is defined in the same manner as in the SI system.

In the U.S. customary system, the unit of mass is derived and is called a *slug.* One slug is the mass that is accelerated one foot per second squared by a force of one pound; that is, 1 slug equals 1 lb · s²/ft. Since the weight of the plat-

---

[3]Guide for the Use of the International System of Units, National Institute of Standards and Technology (NIST) Special Publication 811, September 1991.

inum standard depends on the gravitational attraction of the earth, the U.S. customary system is a gravitational system of units rather than an absolute system of units.

1-3-2 The International System of Units

The original metric system provided a set of units for the measurement of length, area, volume, capacity, and mass based on two fundamental units: the meter and the kilogram. With the addition of a unit of time, practical measurements began to be based on the meter–kilogram–second (MKS) system of units. In 1960, the Eleventh General Conference on Weights and Measures adopted as the international standard the Système International d'Unités (International System of Units), for which the abbreviation is SI in all languages.

The International System of Units adopted by the conference includes three classes of units: (1) base units, (2) supplementary units, and (3) derived units. The system is founded on the seven base units listed in Table 1-1.

Certain units of the International System have not been classified under either base units or derived units. These units, listed in Table 1-2, are called *supplementary units* and may be regarded as either base or derived units.

Derived units are expressed algebraically in terms of base units and/or supplementary units. Their symbols are obtained by means of the mathematical signs of multiplication and division. For example, the SI unit for velocity is meter per second (m/s) and the SI unit for angular velocity is radian per second (rad/s). In the SI system, the unit of force is derived and is called a *newton.* One newton is the force required to give one kilogram of mass an acceleration of one meter per second squared. Thus, $1 \text{ N} = 1 \text{ kg} \cdot \text{m/s}^2$. For some of the derived units, special names and symbols exist; those of interest in mechanics are listed in Table 1-3.

Prefixes are used to form names and symbols for decimal multiples and submultiples of SI names. The multiple usually should be chosen so that numerical values of the quantity will be between 0.1 and 1000. Only one prefix should be used in forming a multiple of a compound unit, and prefixes in the de-

TABLE 1-1 Base Units and Their Symbols

Quantity	Name of Unit	Symbol
Length	meter	m
Mass	kilogram	kg
Time	second	s
Electric current	ampere	A
Thermodynamic temperature	kelvin	K
Amount of substance	mole	mol
Luminous intensity	candela	cd

TABLE 1-2 Supplementary Units and Their Symbols

Quantity	Name of Unit	Symbol
Plane angle	radian	rad
Solid angle	steradian	sr

TABLE 1-3 Derived Units and Their Symbols and Special Names

Quantity	Derived SI Unit	Symbol	Special Name
Area	square meter	m^2	
Volume	cubic meter	m^3	
Linear velocity	meter per second	m/s	
Angular velocity	radian per second	rad/s	
Linear acceleration	meter per second squared	m/s^2	
Frequency	(cycle) per second	Hz	hertz
Density	kilogram per cubic meter	kg/m^3	
Force	kilogram · meter per second squared	N	newton
Moment of force	newton · meter	N · m	
Pressure	newton per meter squared	Pa	pascal
Stress	newton per meter squared	Pa	pascal
Work	newton · meter	J	joule
Energy	newton · meter	J	joule
Power	joule per second	W	watt

nominator should be avoided. Approved prefixes with their names and symbols are listed in Table 1-4.

As the use of the SI system becomes more commonplace in the United States, engineers will be required to be familiar with both the SI system and the U.S. customary system in common use today. As an aid to interpreting the physical significance of answers in SI units for those more accustomed to the U.S. customary system, some conversion factors for the quantities normally encountered in mechanics are provided in Table 1-5.

TABLE 1-4 Multiples of SI Units

Factor by Which Unit is Multiplied	PREFIX	
	Name	Symbol
10^{18}	exa	E
10^{15}	peta	P
10^{12}	tera	T
10^{9}	giga	G
10^{6}	mega	M
10^{3}	kilo	k
10^{2}	hecto*	h
10	deca*	da
10^{-1}	deci*	d
10^{-2}	centi*	c
10^{-3}	milli	m
10^{-6}	micro	μ
10^{-9}	nano	n
10^{-12}	pico	p
10^{-15}	femto	f
10^{-18}	atto	a

*To be avoided when possible.

TABLE 1-5 Conversion Factors between the SI and U.S. Customary Systems

Quantity	U.S. Customary to SI	SI to U.S. Customary
Length	1 in. = 25.40 mm	1 m = 39.37 in.
	1 ft = 0.3048 m	1 m = 3.281 ft
	1 mi = 1.609 km	1 km = 0.6214 mi
Area	1 in.^{2} = 645.2 mm^2	1 m^2 = 1550 in.^{2}
	1 ft^2 = 0.0929 m^2	1 m^2 = 10.76 ft^2
Volume	1 in.^{3} = 16.39(10^3) mm^3	1 mm^3 = 61.02(10^{-6}) in.^{3}
	1 ft^3 = 0.02832 m^3	1 m^3 = 35.31 ft^3
	1 gal = 3.785 L*	1 L = 0.2642 gal
Velocity	1 in./s = 0.0254 m/s	1 m/s = 39.37 in./s
	1 ft/s = 0.3048 m/s	1 m/s = 3.281 ft/s
	1 mi/h = 1.609 km/h	1 km/h = 0.6214 mi/h
Acceleration	1 in./s^2 = 0.0254 m/s^2	1 m/s^2 = 39.37 in./s^2
	1 ft/s^2 = 0.3048 m/s^2	1 m/s^2 = 3.281 ft/s^2
Mass	1 slug = 14.59 kg	1 kg = 0.06854 slug
Second moment of area	1 in.^{4} = 0.4162(10^6) mm^4	1 mm^4 = 2.402(10^{-6}) in.^{4}
Force	1 lb = 4.448 N	1 N = 0.2248 lb
Distributed load	1 lb/ft = 14.59 N/m	1 kN/m = 68.54 lb/ft
Pressure or stress	1 psi = 6.895 kPa	1 kPa = 0.1450 psi
	1 ksi = 6.895 MPa	1 MPa = 145.0 psi
Bending moment or torque	1 ft · lb = 1.356 N · m	1 N · m = 0.7376 ft · lb
Work or energy	1 ft · lb = 1.356 J	1 J = 0.7376 ft · lb
Power	1 ft · lb/s = 1.356 W	1 W = 0.7376 ft · lb/s
	1 hp = 745.7 W	1 kW = 1.341 hp

* Both L and l are accepted symbols for liter. Because the letter "l" can easily be confused with the numeral "1," the symbol "L" is recommended for United States use by the National Institute of Standards and Technology (see NITS Special Publication 811, September 1991).

For the foreseeable future, engineers in the United States will be required to work with both the U.S. customary and SI systems of units; therefore, we have used both sets of units in examples and problems in this book.

Example Problem 1-2

A manufacturer lists the fuel consumption for a new automobile as 15 kilometers per liter. Determine the fuel consumption in miles per gallon.

SOLUTION

One accepted procedure for converting units is to write the associated units in abbreviated form with each of the numerical values used in the conversion. Like-unit symbols can then be canceled in the same manner as algebraic symbols. The conversion factors (see Table 1-5) needed for this example are

$$1 \text{ km} = 0.6214 \text{ mi}$$

$$1 \text{ gal} = 3.785 \text{ L}$$

Thus

$$15\frac{km}{L} \times 0.6214\frac{mi}{km} \times 3.785\frac{L}{gal} = 35.3 \text{ mi/gal} \quad \blacksquare \qquad \textbf{Ans.}$$

Example Problem 1-3

The value of G (universal gravitational constant) used for engineering computations in the U.S. customary system of units is $G = 3.439(10^{-8})$ ft^3/(slug $\cdot$ s^2). Use the conversion factors listed in Table 1-5 to determine a value of G with units of m^3/(kg $\cdot$ s^2) suitable for computations in the SI system of units.

SOLUTION

The conversion factors (see Table 1-5) needed for this example are:

$$1 \text{ ft}^3 = 0.02832 \text{ m}^3$$

$$1 \text{ kg} = 0.06854 \text{ slug}$$

Thus

$$G = 3.439(10^{-8})\,\frac{\text{ft}^3}{\text{slug} \cdot \text{s}^2} \times 0.02832\frac{\text{m}^3}{\text{ft}^3} \times 0.06854\frac{\text{slug}}{\text{kg}}$$

$$= 6.675(10^{-11})\,\frac{\text{m}^3}{\text{kg} \cdot \text{s}^2} \quad \blacksquare \qquad \textbf{Ans.}$$

PROBLEMS

Standard Problems

1-9* Determine the weight W, in U.S. customary units, of a 75-kg steel bar under standard conditions (sea level at a latitude of 45 degrees).

1-10* Determine the mass m, in SI units, for a 500-lb steel beam under standard conditions (sea level at a latitude of 45 degrees).

1-11 An automobile has a 440 cubic inch engine displacement. The engine displacement, in liters, is?

1-12 The viscosity of crude oil under conditions of standard temperature and pressure is $7.13(10^{-3})$ N $\cdot$ s/m^2. The viscosity of crude oil in U.S. customary units is?

Challenging Problems

1-13* Express the density, in SI units, of a specimen of material that has a specific weight of 0.025 lb/in.^{3}.

1-14* Express the specific weight, in U.S. customary units, of a specimen of material that has a density of 8.86 Mg/m^3.

1-15 One acre equals 43,560 ft^2. One gallon equals 231 in.^{3}. The number of liters of water in 2000 acre $\cdot$ ft of water is?

1-16 The specific heat of air under standard atmospheric pressure, in SI units, is 1003 N $\cdot$ m/kg $\cdot$ K. The specific heat of air under standard atmospheric pressure, in U.S. customary units (ft $\cdot$ lb/slug $\cdot$ °R), is?

1-4 DIMENSIONAL CONSIDERATIONS

All the physical quantities encountered in mechanics can be expressed dimensionally in terms of the three fundamental quantities: mass, length, and time, denoted, respectively, by M, L, and T. The dimensions of quantities other than the fundamental quantities follow from definitions or from physical laws. For example, the dimension of velocity, L/T, follows from the definition of velocity: rate of change of position with time. Similarly, acceleration is defined as the rate of change of velocity with time and has the dimension L/T^2. From Newton's second law (Eq. 1-1), force is defined as the product of mass and acceleration; therefore, force has the dimension ML/T^2. The dimensions of a number of other physical quantities commonly encountered in mechanics are given in Table 1-6.

1-4-1 Dimensional Homogeneity

An equation is said to be dimensionally homogeneous if the form of the equation does not depend on the units of measurement. For example, the equation describing the distance h a body released from rest has fallen is $h = gt^2/2$, where h is the distance traveled, t is the time since release, and g is the gravitational ac-

TABLE 1-6 Dimensions of the Physical Quantities of Mechanics

Physical Quantity	Dimension	COMMON UNITS	
		SI System	U.S. Customary System
Length	L	m, mm	in., ft
Area	L^2	m^2, mm^2	in.^{2}, ft^2
Volume	L^3	m^3, mm^3	in.^{3}, ft^3
Angle	$1\,(L/L)$	rad, degree	rad, degree
Time	T	s	s
Linear velocity	L/T	m/s	ft/s
Linear acceleration	L/T^2	m/s^2	ft/s^2
Angular velocity	$1/T$	rad/s	rad/s
Angular acceleration	$1/T^2$	rad/s^2	rad/s^2
Mass	M	kg	slug
Force	ML/T^2	N	lb
Moment of a force	ML^2/T^2	N $\cdot$ m	ft $\cdot$ lb
Pressure	M/LT^2	Pa, kPa	psi, ksi
Stress	M/LT^2	Pa, MPa	psi, ksi
Energy	ML^2/T^2	J	ft $\cdot$ lb
Work	ML^2/T^2	J	ft $\cdot$ lb
Power	ML^2/T^3	W	hp
Linear impulse	ML/T	N $\cdot$ s	lb $\cdot$ s
Momentum	ML/T	N $\cdot$ s	lb $\cdot$ s
Specific weight	M/L^2T^2	N/m^3	lb/ft^3
Density	M/L^3	kg/m^3	slug/ft^3
Second moment of area	L^4	m^4, mm^4	in.^{4}, ft^4
Moment of inertia	ML^2	kg $\cdot$ m^2	slug $\cdot$ ft^2

celeration. This equation is valid whether length is measured in feet, meters, or inches and whether time is measured in hours, years, or seconds, provided g is measured in the same units of length and time as h and t. Therefore, by definition, this equation is dimensionally homogeneous.

If the value $g = 32.2$ ft/s^2 is substituted in the previous equation, the equation obtained is $h = 16.1t^2$ ft/s^2. This equation is not dimensionally homogeneous since the equation applies only if length is measured in feet and time is measured in seconds. Dimensionally homogeneous equations are usually preferred in order to eliminate any uncertainty regarding units associated with constants appearing in dimensionally nonhomogeneous equations.

All like dimensions in a given equation should be measured with the same unit. For example, if the length dimension of a beam is measured in feet and the cross-sectional dimensions are measured in inches, all measurements should be converted to either feet or inches before they are used in a given equation. If this is done, the terms of the equation can be combined after numerical values are substituted for the variables.

Example Problem 1-4

Determine the dimensions of I, R, w, M, and C in the dimensionally homogeneous equation

$$EIy = Rx^3 - P(x - a)^3 - wx^4 + Mx^2 + C$$

in which x and y are lengths, P is a force, and E is a force per unit area.

SOLUTION

The equation can be written dimensionally as

$$\frac{F}{L^2}(I)(L) = R(L^3) - F(L - a)^3 - w(L^4) + M(L^2) + C$$

For this equation to be dimensionally homogeneous, a must be a length; hence, all terms must have the dimensions FL^3. Thus,

$$(I)\frac{F}{L} = (R)L^3 = (w)L^4 = (M)L^2 = C = FL^3$$

The dimensions for each of the unknown quantities are obtained as follows:

$$I = \frac{L}{F}(FL^3) = L^4 \qquad \text{Ans.}$$

$$R = \frac{1}{L^3}(FL^3) = F \qquad \text{Ans.}$$

$$w = \frac{1}{L^4}(FL^3) = \frac{F}{L} \qquad \text{Ans.}$$

$$M = \frac{1}{L^2}(FL^3) = FL \qquad \text{Ans.}$$

$$C = FL^3 \quad \blacksquare \qquad \text{Ans.}$$

PROBLEMS

1-17* The angle of twist for a circular shaft subjected to a twisting moment is given by the equation $\theta = TL/GJ$. What are the dimensions of J if θ is an angle in radians, T is the moment of a force, L is a length, and G is a force per unit area?

1-18* The elongation of a bar of uniform cross section subjected to an axial force is given by the equation $\delta = PL/EA$. What are the dimensions of E if δ and L are lengths, P is a force, and A is an area?

1-19 The period of oscillation of a simple pendulum is given by the equation $T = k(L/g)^{1/2}$, where T is in seconds, L is in feet, g is the acceleration due to gravity,

and k is a constant. What are the dimensions of k for dimensional homogeneity?

1-20 The equation $x = Ae^{-t/b} \sin (at + \alpha)$ is dimensionally homogeneous. If A is a length and t is time, determine the dimensions of x, a, b, and α.

1-21* In the dimensionally homogeneous equation $w = x^3 + ax^2 + bx + a^2b/x$, if x is a length, what are the dimensions of a, b, and w?

1-22 In the dimensionally homogeneous equation $d^5 = Ad^4 + Bd^3 + Cd^2 + D/d^2$, if d is a length, what are the dimensions of A, B, C, and D?

1-5 METHOD OF PROBLEM SOLVING

The principles of mechanics are few and relatively simple; however, the applications are infinite in number, variety, and complexity. Success in engineering mechanics depends to a large degree on a well-disciplined method of problem solving. Experience has shown that the development of good problem-solving methods and skills results from solving a large variety of problems. Professional problem solving consists of three phases: problem definition and identification, model development and simplification, and mathematical solution and result interpretation. The problem-solving method outlined in this section will prove useful for the engineering mechanics courses that follow and for most situations encountered later in engineering practice.

Problems in engineering mechanics (statics, dynamics, and mechanics of deformable bodies) are concerned with the external and internal effects of a system of forces on a physical body. The approach usually used in solving a problem requires identification of all external forces acting on the "body of interest." A carefully prepared drawing that shows the "body of interest" separated from all other interacting bodies and with all external forces applied is known as a *free-body diagram* (FBD).

> *Most engineers consider an appropriate free-body diagram to be the single most important tool for the solution of mechanics problems.*

Given that the relationships between the external forces applied to a body and the motions or deformations that they produce are stated in mathematical form, the true physical situation must be represented by a mathematical model to obtain the required solution. Often, to simplify the solution, it is necessary to make assumptions or approximations in setting up this model. The most common approximation is to treat most of the bodies in statics and dynamics problems as rigid bodies. No real body is absolutely rigid; however, the changes in shape of a real body usually have a negligible effect upon the acceleration produced by a force system or upon the reactions required to maintain equilibrium of the body. Considerations of changes in shape under these circumstances would be an unnecessary complication of the problem. Similarly, the weights of many members can be neglected since they are small with respect to the applied loads. A distributed force, which acts over a small area, can often be considered to be concentrated at a point.

Most actual physical problems cannot be solved exactly or completely. However, even in complicated problems, a simplified model can provide good qualitative results. Appropriate interpretation of such results can lead to approximate predictions of physical behavior or can be used to verify the "reasonableness" of more sophisticated analytical, numerical, or experimental results. An engineer must always be aware of the actual physical problem under consideration and of any limitations associated with the mathematical model used. Assumptions must be continually evaluated to ensure that the mathematical problem solved provides an adequate representation of the physical process or device of interest.

As stated previously, the most effective way to learn the material contained in engineering mechanics courses is to solve a variety of problems. To become an effective engineer, the student must develop the ability to reduce complicated problems to simple parts that can be easily analyzed and to present the results of the work in a clear, logical, and neat manner. This can be accomplished by using the following sequence of steps.

1. Read the problem carefully.
2. Identify the result requested.
3. Identify the principles to be used to obtain the result.
4. Prepare a scaled sketch and tabulate the information provided.
5. Draw the appropriate free-body diagrams.
6. Apply the appropriate principles and equations.
7. Report the answer with the appropriate number of significant figures and the appropriate units.
8. Study the answer and determine whether it is reasonable.

The development of an ability to apply an orderly approach to problem solving constitutes a significant part of an engineering education. Also, the problem-identification, model-simplification, and result-interpretation phases of engineering problem solving are often more important than the mathematical-solution phase.

1-6 SIGNIFICANCE OF NUMERICAL RESULTS

The accuracy of solutions to real engineering problems depends on three factors:

1. The accuracy of the known physical data.
2. The accuracy of the physical model.
3. The accuracy of the computations performed.

An accuracy greater than 0.2 percent is seldom possible for practical engineering problems since physical data are seldom known with any greater accuracy. A practical rule for "rounding off" the final numbers obtained in the computations involved in engineering analysis, which provides answers to approximately this degree of accuracy, is to retain four significant figures for numbers beginning with the figure "1" and to retain three significant figures for numbers beginning with any figure from "2" through "9."

Pocket electronic calculators are widely used to perform the numerical computations required to solve engineering problems. The number of significant figures obtained when these calculators are used, however, should not be taken as an indication of the accuracy of the solution. As noted previously, engineering data are seldom known to an accuracy greater than 0.2 percent; therefore, calculated results should always be rounded off to the number of significant figures which will yield the same degree of accuracy as the data on which they are based. Three significant figures are used for most of the data provided in this book for example and homework problems.

For closed-form analytical solutions, the accuracy of the data and the adequacy of the model determine the accuracy of the results. For numerical solutions, the computational accuracy of the algorithms used also influences the accuracy of the results.

An error can be defined as the difference between two quantities. The difference, for example, might be between an experimentally measured value and a computed theoretical value. An error may also result from the rounding off of numbers during a calculation. One method for describing an error is to state a percent difference (%D). Thus, for two numbers A and B, if it is desired to compare number A with number B, the percent difference between the two numbers is defined as

$$\%D = \frac{A - B}{B}(100)$$

In this equation, B is the reference value with which A is to be compared. The percent difference resulting from round-off error is illustrated in the following example.

Example Problem 1-5

Round off the number 12345 to two, three, and four significant figures. Find the percent difference between the rounded-off numbers and the original number by using the original number as the reference.

SOLUTION

Rounding off the number 12345 to two, three, and four significant figures yields 12000, 12300, and 12350. The percent difference for each of these numbers is

$$\%D = \frac{A - B}{B}(100)$$

For 12,000:

$$\%D = \frac{12000 - 12345}{12345}(100) = -2.79\% \qquad \textbf{Ans.}$$

For 12,300:

$$\%D = \frac{12300 - 12345}{12345}(100) = -0.36\% \qquad \textbf{Ans.}$$

For 12,350:

$$\%D = \frac{12350 - 12345}{12345}(100) = +0.041\% \qquad \textbf{Ans.}$$

The minus signs associated with the above percent differences indicate that the rounded-off numbers are smaller than the reference number. Similarly, a positive percent difference indicates that the rounded-off number is larger than the reference number. ■

PROBLEMS

Round off the numbers in the following problems to two significant figures. Find the percent difference between each rounded-off number and the original number by using the original number as the reference.

1-23* (a) 0.015362   (b) 0.034739   (c) 0.056623

1-24   (a) 0.837482   (b) 0.472916   (c) 0.664473

Round off the numbers in the following problems to three significant figures. Find the percent difference between each rounded-off number and the original number by using the original number as the reference.

1-25* (a) 26.39473   (b) 74.82917   (c) 55.33682

1-26   (a) 374.9371   (b) 826.4836   (c) 349.3378

Round off the numbers in the following problems to four significant figures. Find the percent difference between each rounded-off number and the original number by using the original number as the reference.

1-27* (a) 63746.27   (b) 27382.84   (c) 55129.92

1-28   (a) 937284.9   (b) 274918.2   (c) 339872.8

1-7 SUMMARY

The foundations for studies in mechanics are the laws formulated by Sir Isaac Newton in 1687. The first law deals with conditions for equilibrium of a particle; therefore, it provides the foundation for the study of statics. The second law, which establishes a relationship between the force acting upon a particle and the motion of the particle, provides the foundation for the study of dynamics. The third law provides the foundation for understanding the concept of a force. In addition to the basic laws of motion, Newton also formulated the law of gravitation, which governs the mutual attraction between two bodies.

Physical quantities used to express the laws of mechanics can be divided into fundamental quantities and derived quantities. The magnitude of each fundamental quantity is defined by an arbitrarily chosen unit or "Standard." The units used in the SI system are the meter (m) for length, the kilogram (kg) for mass, and the second (s) for time. The unit of force is a derived unit called a newton (N). In the U.S. customary system of units, the units used are the foot (ft) for length, the pound (lb) for force, and the second (s) for time. The unit of mass is a derived unit called a slug.

The terms of an equation used to describe a physical process should not depend on the units of measurement (that is, they should be dimensionally homogeneous). If an equation is dimensionally homogeneous, the equation is valid for use with any system of units provided all quantities in the equation are measured in the same system. Use of dimensionally homogeneous equations eliminates the need for unit conversion factors.

Success in engineering depends to a large degree on a well-disciplined method of problem solving. Professional problem solving consists of three phases:

1. Problem definition and identification.
2. Model development and simplification.
3. Mathematical solution and result interpretation.

Problems in mechanics are concerned primarily with the effects of force systems on physical bodies. As a result, an extremely important part of the solution of any problem involves identification of the external forces acting upon the body. This is accomplished efficiently and accurately by using a free-body diagram. In obtaining a solution to most problems, the true physical situation must be represented by a mathematical model. A common approximation made in setting up this model is to treat the body as a rigid body. Even though no real body is absolutely rigid, changes in shape usually have a negligible effect upon accelerations produced by a force system or upon reactions required to maintain equilibrium of the body. Solution of some mechanics of materials problems, however, requires use of deformation characteristics of the body. Whenever a mathematical model is used in solving a problem, care must be exercised to ensure that the model and the associated mathematical problem being solved provide an adequate representation of the physical process or device that they represent.

The accuracy of solutions to real engineering problems depends on three factors:

1. The accuracy of the known physical data.
2. The accuracy of the physical model.
3. The accuracy of the computations performed.

An accuracy greater than 0.2 percent is seldom possible. Calculated results should always be "rounded off" to the number of significant figures that will yield the same degree of accuracy as the data on which they are based. In this book, numbers beginning with "1" will be rounded off to four significant figures and numbers beginning with "2" through "9" will be rounded off to three significant figures when presenting final results. One additional figure will be retained in all intermediate results to maintain this degree of accuracy in the final results.

REVIEW PROBLEMS

1-29* On the surface of the earth the weight of a body is 150 lb. At what distance from the center of the earth would the weight of the body be (a) 100 lb? (b) 50 lb?

1-30* At what distance from the center of the earth would the force of attraction between two 1 m–diameter spheres in contact equal the force of attraction of the earth on one of the spheres? The mass of each sphere is 100 kg.

1-31* Convert 640 acres (1 square mile) to hectares if 1 acre equals 4840 yd^2 and 1 hectare equals 10^4 m^2.

1-32* A fluid has a dynamic viscosity of $1.2(10^{-3})$ N $\cdot$ s/m^2. Express its dynamic viscosity in U.S. customary units.

COMPUTER PROBLEMS

C1-33 A common practice in rounding answers is to report numbers whose leading digit is 1 to an accuracy of four significant figures and all other numbers to an accuracy of three significant figures. Although this practice probably started with the accuracy with which slide rules could be read, it also reflects that fact that an accuracy of greater than 0.2 percent is seldom possible. This project will examine the error introduced by this and some other rounding schemes. For each of the rounding schemes below,

1. Generate 20,000 random numbers between 1 and 10.
2. Round each number to the specified number of significant figures. (Note that three significant figures is equivalent to two decimal places, four significant figures is equivalent to three decimal places, etc., since all numbers are between 1 and 10.)
3. Calculate the percent relative error for each number.

$$\text{PercentRelError} = \left| \frac{\text{Number} - \text{RoundNumber}}{\text{Number}} \right| \cdot 100$$

4. Plot PercentRelError versus Number.
5. Comment on the maximum round-off error and the distribution of round-off error.

(a) Round all numbers to an accuracy of three significant figures.
(b) Round numbers less than 2 to an accuracy of four significant figures and numbers greater than 2 to an accuracy of three significant figures.
(c) Round numbers less than 3 to an accuracy of four significant figures and numbers greater than 3 to an accuracy of three significant figures.
(d) Round numbers less than 5 to an accuracy of four significant figures and numbers greater than 5 to an accuracy of three significant figures.

C1-34 When engineers deal with angles, they are usually more interested in the sine or cosine of the angle than they are with the angle itself. Given that

$$\sin 5° = \cos 85° = \sin 175° = \sin 1085° = \cdots = 0.08716$$

the rounding of angles requires a different scheme than that described in Problem C1-33. That is, angles should be rounded to a specified number of decimal places rather than a specified number of significant figures. This project will examine the error introduced by rounding angles to various numbers of decimal places. For each of the cases below,

1. Generate 20,000 random angles between 1° and 89°. (Use a random number generator that produces decimal numbers and not just integers.) Calculate the sine and cosine of each angle.

2. Round each angle to the specified number of decimal places and calculate the sine and cosine of the rounded angle.
3. Calculate the percent relative error for each angle.

$$\text{PercentRelError} = \left| \frac{\sin(\text{Angle}) - \sin(\text{RoundAngle})}{\sin(\text{Angle})} \right| \cdot 100$$

or

$$\text{PercentRelError} = \left| \frac{\cos(\text{Angle}) - \cos(\text{RoundAngle})}{\cos(\text{Angle})} \right| \cdot 100$$

4. Plot PercentRelError versus Angle.
5. Comment on the maximum round-off error and the distribution of round-off error.

(a) Round all angles to an accuracy of one decimal place.
(b) Round all angles to an accuracy of two decimal places.
(c) Round angles less than 10° to an accuracy of three decimal places and angles greater than 10° to an accuracy of two decimal places.

C1-35 When two numbers are added or multiplied, the result is always less accurate than the original numbers. This project will examine the error introduced by rounding two numbers before they are multiplied.

(a) Generate 80 random numbers between 4.51 and 5.49 (that is, $5 \pm 0.5 \cdot \text{RND}$). If any pair of these numbers are rounded to the nearest integer (5) and then multiplied, the result will be 25. How does this result compare with the correct product obtained by multiplying the original two numbers? Is the result accurate to the nearest integer? Is the result accurate to less than 10 percent?
(b) Repeat part (a) for numbers between 49.51 and 50.49 ($50 \pm 0.5 \cdot \text{RND}$). Is the result accurate to the nearest integer? Is the result accurate to less than 1 percent?
(c) Generate 20,000 random integers between 1 and 49. For each integer N, generate two random numbers which will round to that integer.

$$N1 = N \pm 0.5 \cdot \text{RND} \qquad N2 = N \pm 0.5 \cdot \text{RND}$$

Plot the percent relative difference in the products

$$\text{ProdDiff} = \left| \frac{N1 \cdot N2 - N \cdot N}{N \cdot N} \right| \cdot 100$$

versus N. Compare this with the percent relative difference in the original numbers.

$$\text{NumDiff} = \left| \frac{N1 - N}{N} \right| \cdot 100$$

CONCURRENT FORCE SYSTEMS

2

2-1 INTRODUCTION

A force was defined in Section 1-2 as the action of one body on another. The action may be the result of direct physical contact between the bodies or it may be the result of gravitational, electrical, or magnetic effects for bodies that are separated.

A force exerted on a body has two effects on the body: (1) the external effect, which is the tendency to change the motion of the body or to develop resisting forces (reactions) on the body, and (2) the internal effect, which is the tendency to deform the body. Both external and internal effects are discussed in this book. In many problems, the external effect is significant but the internal effect is not of interest. This is the case in many statics and dynamics problems when the body is assumed to be rigid. In other problems, when the body cannot be assumed to be rigid, the internal effects are important. Problems of this type are usually discussed in textbooks on "Mechanics of Materials" or "Mechanics of Deformable Bodies."

When a number of forces are treated as a group, they are referred to as a *force system.* If a force system acting upon a body produces no external effect, the forces are said to be in balance and the body is said to be in equilibrium. If a body is acted upon by a force system that is not in balance, a change in motion of the body must occur. Such a force system is said to be unbalanced or to have a resultant.

Two force systems are said to be equivalent if they produce the same external effect when applied in turn to a given body. The resultant of a force system is the simplest equivalent system to which the original system will reduce. The process of reducing a force system to a simpler equivalent force system is called *reduction.* The process of expanding a force or a force system into a less simple equivalent system is called *resolution.* A component of a force is one of the two or more forces into which a given force may be resolved.

2-2 FORCES AND THEIR CHARACTERISTICS

The properties needed to describe a force are called the *characteristics of the force.* The characteristics of a force are

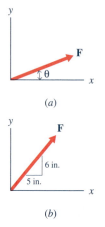

(a)

(b)

Figure 2-1

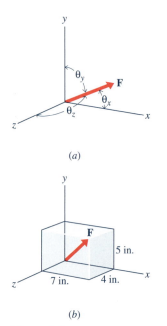

(a)

(b)

Figure 2-2

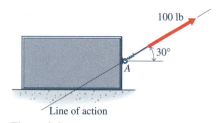

Figure 2-3

1. Its magnitude,
2. Its direction (line of action and sense), and
3. Its point of application.

The magnitude (positive numerical value) of a force is the amount or size of the force. In this book, the magnitude of a force will be expressed in newtons (N) or kilonewtons (kN) when the SI system of units is used and in pounds (lb) or kilopounds (kip) when the U.S. customary system of units is used.

The direction of a force is the slope and sense of the line segment used to represent the force. In a two-dimensional problem, the slope can be specified by providing an angle, as shown in Fig. 2-1a, or by providing two dimensions, as shown in Fig. 2-1b. In a three-dimensional problem, the slope can be specified by providing three angles, as shown in Fig. 2-2a, or by providing three dimensions, as shown in Fig. 2-2b. The sense of the force can be specified by placing an arrowhead on the appropriate end of the line segment used to represent the force. Alternatively, a plus or minus sign can be used with the magnitude of a force to indicate the sense of the force.

The point of application of a force is the point of contact between the two bodies. A straight line extending through the point of application in the direction of the force is called its *line of action.*

The three characteristics of a force are illustrated on the sketch of a block shown in Fig. 2-3. In this case, the force applied to the block can be described as a 100-lb (magnitude) force acting 30° upward and to the right (direction—slope and sense) through point *A* (point of application). A discussion of the manner in which these characteristics influence the reactions developed in holding a body at rest forms an important part of the study of statics. In a similar manner, a discussion of the manner in which these characteristics influence the change in motion of a body forms an important part of the study of kinetics.

2-2-1 Scalar and Vector Quantities

Scalar quantities can be completely described with a magnitude (number). Examples of scalar quantities in mechanics are mass, density, length, area, volume, speed, energy, time, and temperature. In mathematical operations, scalars follow the rules of elementary algebra.

A vector quantity has both a magnitude and a direction (line of action and sense) and obeys the parallelogram law of addition (see Appendix A). Examples of vector quantities in mechanics are force, moment, displacement, velocity, acceleration, impulse, and momentum. Vectors can be classified into three types: free, sliding, or fixed.

1. A free vector has a specific magnitude, slope, and sense but its line of action does not pass through a unique point in space.
2. A sliding vector has a specific magnitude, slope, and sense and its line of action passes through a unique point in space. The point of application of a sliding vector can be anywhere along its line of action.
3. A fixed vector has a specific magnitude, slope, and sense and its line of action passes through a unique point in space. The point of application of a fixed vector is confined to a fixed point on its line of action.

The results shown in Fig. 2-9 indicate that the resultant **R** does not depend on the order in which forces **F**₁ and **F**₂ are selected. Figure 2-9 is a graphical illustration of the commutative law of vector addition.

Graphical methods for determining the resultant of two forces require an accurate scaled drawing if accurate results are to be obtained. In practice, numerical results are obtained by using trigonometric methods based on the law of sines and the law of cosines in conjunction with sketches of the force system. For example, consider the triangle shown in Fig. 2-10, which is similar to the force triangles illustrated in Figs. 2-8 and 2-9. For this general triangle, the law of sines is

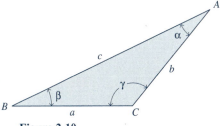
Figure 2-10

$$\frac{a}{\sin \alpha} = \frac{b}{\sin \beta} = \frac{c}{\sin \gamma}$$

and the law of cosines is

$$c^2 = a^2 + b^2 - 2ab \cos \gamma$$

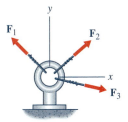
Figure 2-11

The method can easily be extended to cover three or more forces. As an example, consider the case of three coplanar, concurrent forces acting on a bracket, as shown in Fig. 2-11. Application of the parallelogram law to forces **F**₁ and **F**₂, as shown graphically in Fig. 2-12, yields resultant **R**₁₂. Then, combining resultant **R**₁₂ with force **F**₃, through a second graphical application of the parallelogram law, yields resultant **R**₁₂₃, which is the vector sum of the three forces.

In practice, numerical results for specific problems involving three or more forces are obtained algebraically by using the law of sines and the law of cosines in conjunction with sketches of the force system similar to those shown in Fig. 2-13. The sketches shown in Fig. 2-13 are known as *force polygons*. The order in which the forces are added can be arbitrary, as shown in Figs. 2-13a and b, where the forces are added in the order **F**₁, **F**₂, **F**₃ in Fig. 2-13a and in the order **F**₃, **F**₁, **F**₂ in Fig. 2-13b. Although the shape of the polygon changes, the resultant force remains the same. The fact that the sum of the three vectors is the same, regardless of the order in which they are added, illustrates the associative law of vector addition.

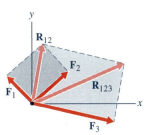
Figure 2-12

If there are more than three forces (as an example, Fig. 2-14 shows the bracket of Fig. 2-11 with four forces), the process of adding additional forces can be continued, as shown in Fig. 2-15, until all the forces are joined in head-to-tail fashion. The closing side of the polygon, drawn from the tail of the first vector to the head of the last vector, is the resultant of the force system.

Application of the parallelogram law to more than three forces requires extensive geometric and trigonometric calculation. Therefore, problems of this type are usually solved by using the rectangular-component method, which is developed in Section 2-6 of this book.

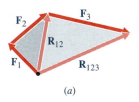
(a)

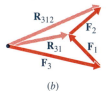
(b)
Figure 2-13

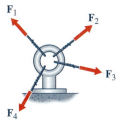
Figure 2-14

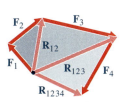
Figure 2-15

The procedure for determining the resultant **R** of a force system by using the law of sines and the law of cosines is demonstrated in the following example.

Example Problem 2-1

Two forces are applied to a bracket as shown in Fig. 2-16a. Determine the magnitude of the resultant **R** of the two forces and the angle θ between the x axis and the line of action of the resultant.

SOLUTION

The two forces, the resultant **R,** and the angle θ are shown in Fig. 2-16b. The triangle law for the addition of the two forces can be applied, as shown in Fig. 2-16c. Applying the law of cosines to the triangle yields

$$R^2 = 900^2 + 600^2 - 2(900)(600) \cos (180° - 40°)$$

where R, or $|\mathbf{R}|$, is the magnitude of the resultant. Thus,

$$R = |\mathbf{R}| = 1413.3 \cong 1413 \text{ lb} \qquad \textbf{Ans.}$$

Applying the law of sines to the triangle yields

$$\sin \alpha = \frac{600}{1413.3} \sin (180° - 40°) = 0.2729$$

from which

$$\alpha = 15.84°$$

Thus,

$$\theta = 15.84° + 35° = 50.84° \cong 50.8° \ \blacksquare \qquad \textbf{Ans.}$$

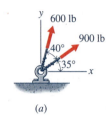

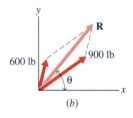

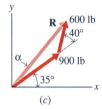

Figure 2-16

PROBLEMS

Use the law of sines and the law of cosines, in conjunction with sketches of the force triangles, to solve the following problems. Determine the magnitude of the resultant **R** and the angle θ between the x-axis and the line of action of the resultant for the following problems.

Standard Problems

2-1* The two forces shown in Fig. P2-1.

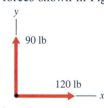

Figure P2-1

2-2* The two forces shown in Fig. P2-2.

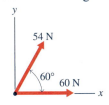

Figure P2-2

2-3* The two forces shown in Fig. P2-3.

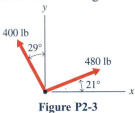

Figure P2-3

2-4* The two forces shown in Fig. P2-4.

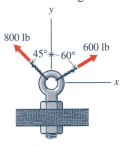

Figure P2-4

2-5 The two forces shown in Fig. P2-5.

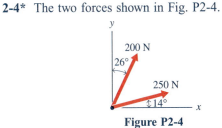

Figure P2-5

2-6 The two forces shown in Fig. P2-6.

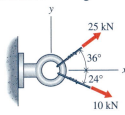

Figure P2-6

Challenging Problems

2-7* The three forces shown in Fig. P2-7.

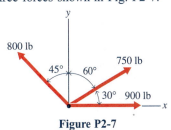

Figure P2-7

2-8* The three forces shown in Fig. P2-8.

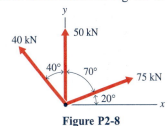

Figure P2-8

2-9 The three forces shown in Fig. P2-9.

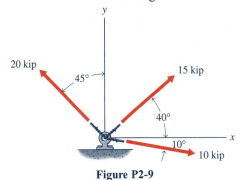

Figure P2-9

2-10 The three forces shown in Fig. P2-10.

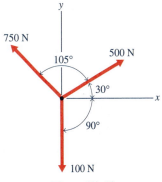

Figure P2-10

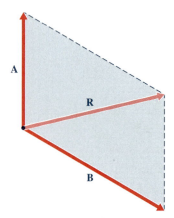

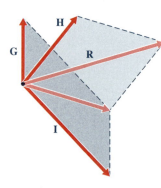

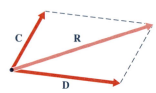

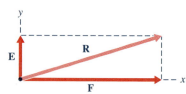

Figure 2-17

2-4 RESOLUTION OF A FORCE INTO COMPONENTS

In the previous section of this chapter, use of the parallelogram and triangle laws to determine the resultant **R** of two concurrent forces $\mathbf{F}_1$ and $\mathbf{F}_2$ or three or more concurrent forces $\mathbf{F}_1$, $\mathbf{F}_2$, $\cdots$, $\mathbf{F}_n$ was discussed. In a similar manner, a single force **F** can be replaced by a system of two or more forces $\mathbf{F}_a$, $\mathbf{F}_b$, $\cdots$, $\mathbf{F}_n$. Forces $\mathbf{F}_a$, $\mathbf{F}_b$, $\cdots$, $\mathbf{F}_n$ are called *components of the original force.* In the most general case, the components of a force can be any system of forces that can be combined by the parallelogram law to produce the original force. Such components are not required to be concurrent or coplanar. Normally, however, the term "component" is used to specify either one of two coplanar concurrent forces or one of three noncoplanar concurrent forces that can be combined vectorially to produce the original force. The point of concurrency must be on the line of action of the original force. The process of replacing a force by two or more forces is called *resolution.*

The process of resolution does not produce a unique set of vector components. For example, consider the four coplanar sketches shown in Fig. 2-17. It is obvious from these sketches that

$$\mathbf{A} + \mathbf{B} = \mathbf{R} \qquad \mathbf{G} + \mathbf{H} + \mathbf{I} = \mathbf{R}$$

$$\mathbf{C} + \mathbf{D} = \mathbf{R} \qquad \mathbf{E} + \mathbf{F} = \mathbf{R}$$

where **R** is the same vector in each expression. Thus, an infinite number of sets of components exist for any vector.

The following examples illustrate use of the parallelogram and triangle laws to resolve a force into components along any two oblique lines of action.

Example Problem 2-2

Determine the magnitudes of the *u*- and *v*-components of the 900-N force shown in Fig. 2-18*a*.

SOLUTION

The magnitude and direction of the 900-N force are shown in Fig. 2-18*b*. The components $\mathbf{F}_u$ and $\mathbf{F}_v$ along the *u*- and *v*-axes can be determined by drawing lines parallel to the *u*- and *v*-axes through the head and tail of the vector used

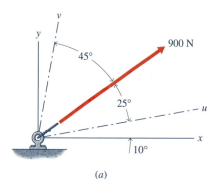

(a)

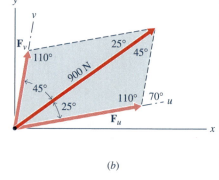

(b)

Figure 2-18

to represent the 900-N force. From the parallelogram thus produced, the law of sines can be applied to determine the forces $\mathbf{F}_u$ and $\mathbf{F}_v$, since all the angles for the two triangles which form the parallelogram are known. Thus,

$$\frac{F_u}{\sin 45°} = \frac{F_v}{\sin 25°} = \frac{900}{\sin 110°}$$

from which

$$F_u = |\mathbf{F}_u| = \frac{900 \sin 45°}{\sin 110°} = 677 \text{ N} \qquad \textbf{Ans.}$$

$$F_v = |\mathbf{F}_v| = \frac{900 \sin 25°}{\sin 110°} = 405 \text{ N} \enspace \blacksquare \qquad \textbf{Ans.}$$

Example Problem 2-3

Two forces are applied to an eye bracket as shown in Fig. 2-19a. The resultant $\mathbf{R}$ of the two forces has a magnitude of 1000 lb and its line of action is directed along the x-axis. If the force $\mathbf{F}_1$ has a magnitude of 250 lb, determine

(a) The magnitude of force $\mathbf{F}_2$.

(b) The angle α between the x-axis and the line of action of the force $\mathbf{F}_2$.

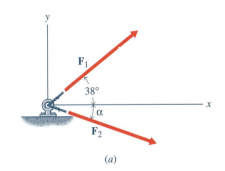

(a)

SOLUTION

The two forces $\mathbf{F}_1$ and $\mathbf{F}_2$, the resultant $\mathbf{R}$, and the angle α are shown in Fig. 2-19b. The force triangle was drawn by using $\mathbf{F}_1$, $\mathbf{R}_1$, and the 38° angle. Completing the parallelogram identifies force $\mathbf{F}_2$ and the angle α.

(a) Applying the law of cosines to the top triangle of Fig. 2-19b yields

$$F_2^2 = 250^2 + 1000^2 - 2(250)(1000) \cos 38°$$

from which

$$F_2 = |\mathbf{F}_2| = 817.6 \cong 818 \text{ lb} \qquad \textbf{Ans.}$$

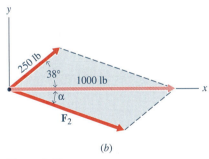

(b)

Figure 2-19

(b) Applying the law of sines to the top triangle yields

$$\frac{250}{\sin \alpha} = \frac{817.6}{\sin 38°}$$

Thus,

$$\sin \alpha = \frac{250}{817.6} \sin 38° = 0.18825$$

from which

$$\alpha = 10.85° \enspace \blacksquare \qquad \textbf{Ans.}$$

Use the law of sines and the law of cosines in conjunction with sketches of the force triangles to solve the following problems.

Standard Problems

2-11* Determine the magnitudes of the u- and v-components of the 1000-lb force shown in Fig. P2-11.

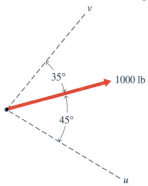

Figure P2-11

2-12* Determine the magnitudes of the u- and v-components of the 750-N force shown in Fig. P2-12.

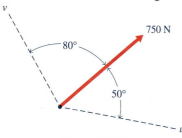

Figure P2-12

2-13* Determine the magnitudes of the u- and v-components of the 650-lb force shown in Fig. P2-13.

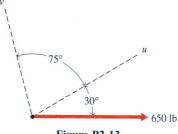

Figure P2-13

2-14* Determine the magnitudes of the u- and v-components of the 25-kN force shown in Fig. P2-14.

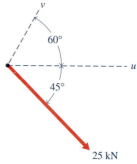

25 kN

Figure P2-14

2-15 Two cables are used to support a stop light as shown in Fig. P2-15. The resultant **R** of the cable forces $\mathbf{F}_u$ and $\mathbf{F}_v$ has a magnitude of 300 lb and its line of action is vertical. Determine the magnitudes of forces $\mathbf{F}_u$ and $\mathbf{F}_v$.

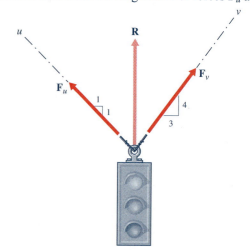

Figure P2-15

2-16 Two ropes are used to tow a boat upstream as shown in Fig. P2-16. The resultant **R** of the rope forces $\mathbf{F}_u$ and $\mathbf{F}_v$ has a magnitude of 1500 N and its line of action is directed along the axis of the boat. Determine the magnitudes of forces $\mathbf{F}_u$ and $\mathbf{F}_v$.

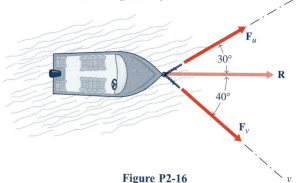

Figure P2-16

Challenging Problems

2-17* A 25-kip force is resisted by an eye bar and a pipe strut as shown in Fig. P2-17. Determine the component F_u of the force along the axis of eye bar AB and the component F_v of the force along the axis of strut BC.

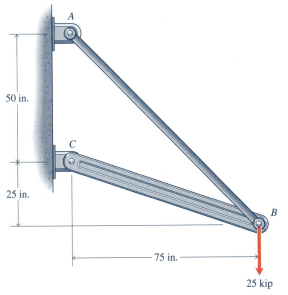

Figure P2-17

2-18* A 100-kN force is resisted by an eye bar and a pipe strut as shown in Fig. P2-18. Determine the component F_u of the force along the axis of eye bar AB and the component F_v of the force along the axis of strut AC.

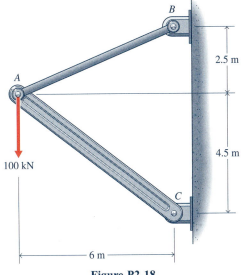

Figure P2-18

2-19 Three forces are applied to a bracket as shown in Fig. P2-19. The magnitude of the resultant R of the three forces is 50 kip. If the force F_1 has a magnitude of 30 kip, determine the magnitudes of forces F_2 and F_3.

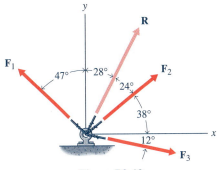

Figure P2-19

2-20 A gusset plate is used to transfer forces from three bars to a beam as shown in Fig. P2-20. The magnitude of the resultant R of the three forces is 100 kN. If the force F_1 has a magnitude of 20 kN, determine the magnitudes of forces F_2 and F_3.

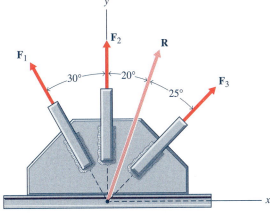

Figure P2-20

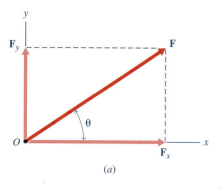

(a)

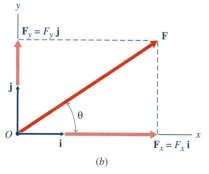

(b)

Figure 2-20

2-5 RECTANGULAR COMPONENTS OF A FORCE

General oblique components of a force are not widely used for solving most practical engineering problems. Mutually perpendicular (rectangular) components, on the other hand, find wide usage. The process for obtaining rectangular components is greatly simplified since the parallelogram used to represent the force and its components reduces to a rectangle and the law of cosines used to obtain numerical values of the components reduces to the Pythagorean theorem.

A force $\mathbf{F}$ can be resolved into a rectangular component $\mathbf{F}_x$ along the x-axis and a rectangular component $\mathbf{F}_y$ along the y-axis as shown in Fig. 2-20a. The forces $\mathbf{F}_x$ and $\mathbf{F}_y$ are the vector components of the force $\mathbf{F}$. The x- and y-axis are usually chosen horizontal and vertical, as shown in Fig. 2-20a; however, they may be chosen in any two perpendicular directions. The choice is usually indicated by the geometry of the problem.[1]

The force $\mathbf{F}$ and its two-dimensional vector components $\mathbf{F}_x$ and $\mathbf{F}_y$ can be written in Cartesian vector form by using unit vectors $\mathbf{i}$ and $\mathbf{j}$ directed along the positive x- and y-coordinate axes, as shown in Fig. 2-20b. Thus,

$$\mathbf{F} = \mathbf{F}_x + \mathbf{F}_y = F_x\,\mathbf{i} + F_y\,\mathbf{j} \tag{2-1}$$

where the scalars F_x and F_y are the x and y scalar components of the force $\mathbf{F}$. The scalar components F_x and F_y are related to the magnitude of the force $|\mathbf{F}| = F$ and the angle of inclination θ (direction) of the force $\mathbf{F}$ by the following expressions

$$
\begin{aligned}
F_x &= F\cos\theta \qquad F = \sqrt{F_x^2 + F_y^2} \\[2mm]
F_y &= F\sin\theta \qquad \theta = \tan^{-1}\frac{F_y}{F_x}
\end{aligned}
\tag{2-2}
$$

The scalar components F_x and F_y of the force $\mathbf{F}$ can be positive or negative, depending on the sense of the vector components $\mathbf{F}_x$ and $\mathbf{F}_y$. A scalar component is positive when the vector component has the same sense as the unit vector with which it is associated, and negative when the vector component has the opposite sense.

Similarly, for problems requiring analysis in three dimensions, a force $\mathbf{F}$ in space can be resolved into three mutually perpendicular rectangular components $\mathbf{F}_x$, $\mathbf{F}_y$, and $\mathbf{F}_z$ along the x-, y-, and z-coordinate axes, as shown in Fig. 2-21. The force $\mathbf{F}$ and its three-dimensional vector components $\mathbf{F}_x$, $\mathbf{F}_y$, and $\mathbf{F}_z$ can also be written in Cartesian vector form by using unit vectors $\mathbf{i}$, $\mathbf{j}$, and $\mathbf{k}$ directed along the positive x-, y-, and z-coordinate axes, as shown in Fig. 2-22. Thus,

$$
\begin{aligned}
\mathbf{F} &= \mathbf{F}_x + \mathbf{F}_y + \mathbf{F}_z \\
&= F_x\,\mathbf{i} + F_y\,\mathbf{j} + F_z\,\mathbf{k} \\
&= F\cos\theta_x\,\mathbf{i} + F\cos\theta_y\,\mathbf{j} + F\cos\theta_z\,\mathbf{k}
\end{aligned}
\tag{2-3}
$$

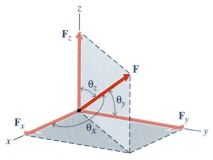

Figure 2-21

---

[1]Coordinate systems and axes are tools that may be used to advantage by the analyst. Machine components and structural elements do not come inscribed with x- and y-axes; therefore, the analyst is free to select directions that are convenient for his or her work.

Thus, the scalar components F_x, F_y, and F_z are related to the magnitude F and the direction θ of the force $\mathbf{F}$ by the following expressions:

$$F_x = F \cos \theta_x \qquad F_y = F \cos \theta_y \qquad F_z = F \cos \theta_z$$

$$\theta_x = \cos^{-1}\frac{F_x}{F} \qquad \theta_y = \cos^{-1}\frac{F_y}{F} \qquad \theta_z = \cos^{-1}\frac{F_z}{F} \quad (2\text{-}4)$$

$$F = \sqrt{F_x^2 + F_y^2 + F_z^2}$$

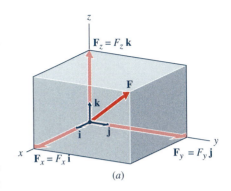

The angles θ_x, θ_y, and θ_z are the angles ($0 \le \theta \le 180°$) between the force $\mathbf{F}$ and the positive coordinate axes. The cosines of these angles, called *direction cosines,* must satisfy the equation

$$\cos^2 \theta_x + \cos^2 \theta_y + \cos^2 \theta_z = 1$$

If an angle is greater than 90°, the cosine is negative, indicating that the sense of the component is opposite to the positive direction of the coordinate axis. Thus, Eqs. (2-4) provide the sign as well as the magnitude of the scalar components of the force and hold for any value of the angle.

The scalar component of a force $\mathbf{F}$ along an arbitrary direction n can be obtained by using the vector operation known as the *dot product* or *scalar product* (see Appendix A). For example, the scalar component F_x of a force $\mathbf{F}$ is obtained as

$$\begin{aligned} F_x &= \mathbf{F} \cdot \mathbf{i} = (F_x\,\mathbf{i} + F_y\,\mathbf{j} + F_z\,\mathbf{k}) \cdot \mathbf{i} \\ &= F_x(\mathbf{i} \cdot \mathbf{i}) + F_y(\mathbf{j} \cdot \mathbf{i}) + F_z(\mathbf{k} \cdot \mathbf{i}) = F_x \end{aligned}$$

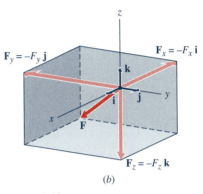

Figure 2-22

since

$$\mathbf{i} \cdot \mathbf{i} = \mathbf{j} \cdot \mathbf{j} = \mathbf{k} \cdot \mathbf{k} = 1$$

and

$$\mathbf{i} \cdot \mathbf{j} = \mathbf{j} \cdot \mathbf{i} = \mathbf{i} \cdot \mathbf{k} = \mathbf{k} \cdot \mathbf{i} = \mathbf{j} \cdot \mathbf{k} = \mathbf{k} \cdot \mathbf{j} = 0$$

In more general terms, if $\mathbf{e}_n$ is a unit vector in a specified direction n, then the scalar component F_n of the force $\mathbf{F}$ is

$$F_n = \mathbf{F} \cdot \mathbf{e}_n = (F_x\,\mathbf{i} + F_y\,\mathbf{j} + F_z\,\mathbf{k}) \cdot \mathbf{e}_n$$

Since the angles between the direction n and the x-, y-, and z-axes are θ_x', θ_y', and θ_z', as shown in Fig. 2-23, the direction cosines for the unit vector $\mathbf{e}_n$ are $\cos \theta_x'$, $\cos \theta_y'$, and $\cos \theta_z'$. Therefore, the unit vector $\mathbf{e}_n$ can be written in Cartesian vector form as

$$\mathbf{e}_n = \cos \theta_x'\,\mathbf{i} + \cos \theta_y'\,\mathbf{j} + \cos \theta_z'\,\mathbf{k}$$

Thus, the force F_n is

$$\begin{aligned} F_n &= \mathbf{F} \cdot \mathbf{e}_n = (F_x\,\mathbf{i} + F_y\,\mathbf{j} + F_z\,\mathbf{k}) \cdot (\cos \theta_x'\,\mathbf{i} + \cos \theta_y'\,\mathbf{j} + \cos \theta_z'\,\mathbf{k}) \\ &= F_x \cos \theta_x' + F_y \cos \theta_y' + F_z \cos \theta_z' \end{aligned} \quad (2\text{-}5)$$

Substituting Eqs. (2-4) into Eq. (2-5) yields an expression for the scalar component F_n in terms of F and the direction cosines associated with $\mathbf{F}$ and n. Thus,

$$F_n = \mathbf{F} \cdot \mathbf{e}_n = F\,(\cos \theta_x \cos \theta_x' + \cos \theta_y \cos \theta_y' + \cos \theta_z \cos \theta_z') \quad (2\text{-}6)$$

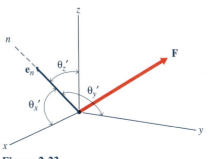

Figure 2-23

The scalar component F_n of the force $\mathbf{F}$ can be expressed in Cartesian vector form by multiplying the scalar component F_n by the unit vector $\mathbf{e}_n$. Thus,

$$\mathbf{F}_n = (\mathbf{F} \cdot \mathbf{e}_n) \, \mathbf{e}_n = F_n \, \mathbf{e}_n$$
$$= F_n \, (\cos \theta_x' \, \mathbf{i} + \cos \theta_y' \, \mathbf{j} + \cos \theta_z' \, \mathbf{k}) \tag{2-7}$$

The angle α between the line of action of the force $\mathbf{F}$ and the direction n can be determined by using the dot-product relationship and the definition of a scalar component of a force. Thus

$$F_n = F \cos \alpha = \mathbf{F} \cdot \mathbf{e}_n$$

therefore,

$$\alpha = \cos^{-1} \frac{\mathbf{F} \cdot \mathbf{e}_n}{F} = \cos^{-1} \frac{F_n}{F} \tag{2-8}$$

Equation (2-8) can be used to determine the angle α between any two vectors $\mathbf{A}$ and $\mathbf{B}$ or between any two lines by using the unit vectors $\mathbf{e}_1$ and $\mathbf{e}_2$ associated with the lines. Thus,

$$\alpha = \cos^{-1} \frac{\mathbf{A} \cdot \mathbf{B}}{AB} \tag{2-9}$$

or

$$\alpha = \cos^{-1} (\mathbf{e}_1 \cdot \mathbf{e}_2) \tag{2-10}$$

Example Problem 2-4

A force $\mathbf{F}$ is applied at a point in a body as shown in Fig. 2-24.

(a) Determine the x and y scalar components of the force.
(b) Determine the x' and y' scalar components of the force.
(c) Express the force $\mathbf{F}$ in Cartesian vector form for the xy- and $x'y'$-axes.

SOLUTION

(a) The magnitude F of the force is 450 N. The angle θ_x between the x-axis and the line of action of the force is

$$\theta_x = 90° - 28° = 62°$$

Thus,

$$F_x = F \cos \theta_x = 450 \cos 62° = +211 \text{ N} \qquad \textbf{Ans.}$$
$$F_y = F \sin \theta_x = 450 \sin 62° = +397 \text{ N} \qquad \textbf{Ans.}$$

(b) The magnitude F of the force is 450 N. The angle $\theta_{x'}$ between the x'-axis and the line of action of the force is

$$\theta_{x'} = \theta_x - 30° = 62° - 30° = 32°$$

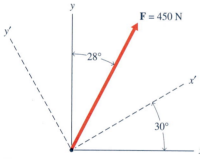

Figure 2-24

Thus,

$$F_{x'} = F \cos \theta_{x'} = 450 \cos 32° = +382 \text{ N} \qquad \textbf{Ans.}$$

$$F_{y'} = F \sin \theta_{x'} = 450 \sin 32° = +238 \text{ N} \qquad \textbf{Ans.}$$

As a check, note that

$$F = \sqrt{F_x^2 + F_y^2} = \sqrt{F_{x'}^2 + F_{y'}^2}$$
$$= \sqrt{211^2 + 397^2} = \sqrt{382^2 + 238^2} = 450 \text{ N}$$

(c) The force **F** expressed in Cartesian vector form for the xy- and $x'y'$-axes are

$$\textbf{F} = 211 \, \textbf{i} + 397 \, \textbf{j} \text{ N} \qquad \textbf{Ans.}$$

$$\textbf{F} = 382 \, \textbf{e}_{x'} + 238 \, \textbf{e}_{y'} \text{ N} \ \blacksquare \qquad \textbf{Ans.}$$

Example Problem 2-5

A force **F** is applied at a point in a body as shown in Fig. 2-25.
(a) Determine the x, y, and z scalar components of the force.
(b) Express the force in Cartesian vector form.

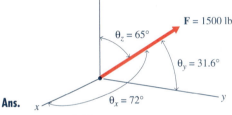

Figure 2-25

SOLUTION

(a) The magnitude F of the force is 1500 lb. Thus,

$$F_x = F \cos \theta_x = 1500 \cos 72.0° = +464 \text{ lb} \qquad \textbf{Ans.}$$

$$F_y = F \cos \theta_y = 1500 \cos 31.6° = +1278 \text{ lb} \qquad \textbf{Ans.}$$

$$F_z = F \cos \theta_z = 1500 \cos 65.0° = +634 \text{ lb} \qquad \textbf{Ans.}$$

As a check, note that

$$F = \sqrt{F_x^2 + F_y^2 + F_z^2}$$
$$= \sqrt{464^2 + 1278^2 + 634^2} = 1500 \text{ lb}$$

(b) The force **F** expressed in Cartesian vector form is

$$\textbf{F} = F_x \, \textbf{i} + F_y \, \textbf{j} + F_z \, \textbf{k}$$
$$= 464 \, \textbf{i} + 1278 \, \textbf{j} + 634 \, \textbf{k} \text{ lb} \ \blacksquare \qquad \textbf{Ans.}$$

Example Problem 2-6

A force **F** is applied at a point in a body as shown in Fig. 2-26. Determine

(a) The angles θ_x, θ_y, and θ_z.
(b) The x, y, and z scalar components of the force.
(c) The rectangular component F_n of the force along line OA.
(d) The rectangular component F_t of the force perpendicular to line OA.

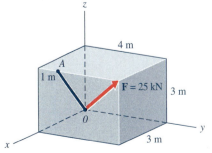

Figure 2-26

SOLUTION

(a) The angles θ_x, θ_y, and θ_z can be determined from the geometry of the box shown in Fig. 2-26. The length of diagonal d of the box is

$$d = \sqrt{x^2 + y^2 + z^2} = \sqrt{3^2 + 4^2 + 3^2} = 5.831 \text{ m}$$

Thus,

$$\theta_x = \cos^{-1}\frac{x}{d} = \cos^{-1}\frac{3}{5.831} = 59.0° \qquad \textbf{Ans.}$$

$$\theta_y = \cos^{-1}\frac{y}{d} = \cos^{-1}\frac{4}{5.831} = 46.7° \qquad \textbf{Ans.}$$

$$\theta_z = \cos^{-1}\frac{z}{d} = \cos^{-1}\frac{3}{5.831} = 59.0° \qquad \textbf{Ans.}$$

(b) The magnitude F of the force is 25 kN. Thus,

$$F_x = F \cos \theta_x = 25\left(\frac{3}{5.831}\right) = +12.86 \text{ kN} \qquad \textbf{Ans.}$$

$$F_y = F \cos \theta_y = 25\left(\frac{4}{5.831}\right) = +17.15 \text{ kN} \qquad \textbf{Ans.}$$

$$F_z = F \cos \theta_z = 25\left(\frac{3}{5.831}\right) = +12.86 \text{ kN} \qquad \textbf{Ans.}$$

(c) The angles θ'_x, θ'_y, and θ'_z between the n direction (along OA) and the x-, y-, and z-axes can also be determined from the geometry of the box shown in Fig. 2-26. The length of diagonal d' of the line from O to A is

$$d' = \sqrt{(x')^2 + (y')^2 + (z')^2} = \sqrt{3^2 + 1^2 + 3^2} = 4.359 \text{ m}$$

Thus,

$$\theta'_x = \cos^{-1}\frac{x'}{d'} = \cos^{-1}\frac{3}{4.359} = 46.5°$$

$$\theta'_y = \cos^{-1}\frac{y'}{d'} = \cos^{-1}\frac{1}{4.359} = 76.7°$$

$$\theta'_z = \cos^{-1}\frac{z'}{d'} = \cos^{-1}\frac{3}{4.359} = 46.5°$$

The unit vector $\mathbf{e}_n$ along line OA is

$$\mathbf{e}_n = \cos \theta'_x \, \mathbf{i} + \cos \theta'_y \, \mathbf{j} + \cos \theta'_z \, \mathbf{k}$$
$$= 0.6882 \, \mathbf{i} + 0.2294 \, \mathbf{j} + 0.6882 \, \mathbf{k}$$

The force F expressed in Cartesian vector form is

$$\mathbf{F} = 12.86 \, \mathbf{i} + 17.15 \, \mathbf{j} + 12.86 \, \mathbf{k} \text{ kN}$$

Therefore,

$F_n = \mathbf{F} \cdot \mathbf{e}_n$

$\quad = (12.86\ \mathbf{i} + 17.15\ \mathbf{j} + 12.86\ \mathbf{k}) \cdot (0.6882\ \mathbf{i} + 0.2294\ \mathbf{j} + 0.6882\ \mathbf{k})$

$\quad = 12.86(0.6882) + 17.15(0.2294) + 12.86(0.6882)$

$\quad = 21.64 \cong 21.6\ \text{kN}$ **Ans.**

(d) The force **F** can be resolved into components $\mathbf{F}_n$ along OA and $\mathbf{F}_t$ perpendicular to OA. Thus

$$\mathbf{F} = \mathbf{F}_n + \mathbf{F}_t \qquad \text{and} \qquad F = \sqrt{F_n^2 + F_t^2}$$

Therefore

$$F_t = \sqrt{F^2 - F_n^2} = \sqrt{25^2 - 21.64^2} = 12.52\ \text{kN} \quad \blacksquare \qquad \textbf{Ans.}$$

PROBLEMS

Standard Problems

2-21* Determine the x- and y-components of the force shown in Fig. P2-21.

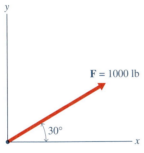

Figure P2-21

2-22* Determine the x- and y-components of the force shown in Fig. P2-22.

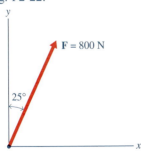

Figure P2-22

2-23 Two forces are applied at a point on a body as shown in Fig. P2-23.
(a) Determine the x- and y-components of each force
(b) Determine the x'- and y'-components of each force.

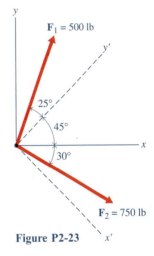

Figure P2-23

2-24 Two forces are applied at a point on a body as shown in Fig. P2-24.
(a) Determine the x- and y-components of each force.
(b) Determine the x'- and y'-components of each force.

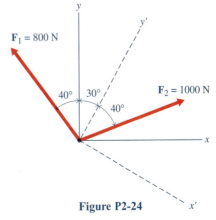

Figure P2-24

2-25* Four forces are applied to a bracket as shown in Fig. P2-25. Determine the x- and y-components of each force.

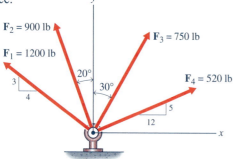

Figure P2-25

2-26* Four forces are applied to a plate as shown in Fig. P2-26. Determine the x- and y-components of each force.

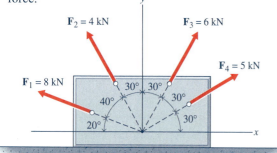

Figure P2-26

2-27 A force of 800 lb is applied to a bracket as shown in Fig. P2-27.
(a) Determine the angles θ_x, θ_y, and θ_z.
(b) Determine the x-, y-, and z-components of the force.
(c) Express the force in Cartesian vector form.

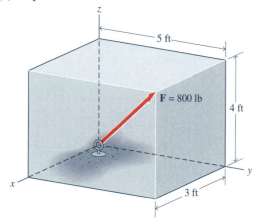

Figure P2-27

2-28 A force of 50 kN is applied to a bracket as shown in Fig. P2-28.
(a) Determine the angles θ_x, θ_y, and θ_z.
(b) Determine the x-, y-, and z-components of the force.
(c) Express the force in Cartesian vector form.

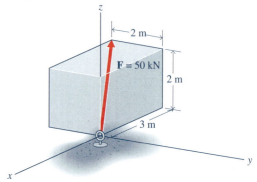

Figure P2-28

Challenging Problems

2-29* Two forces are applied to an eye bolt as shown in Fig. P2-29.
(a) Determine the x-, y-, and z-components of force $\mathbf{F}_1$.
(b) Express force $\mathbf{F}_1$ in Cartesian vector form.
(c) Determine the magnitude of the rectangular component of force $\mathbf{F}_1$ along the line of action of force $\mathbf{F}_2$.
(d) Determine the angle α between forces $\mathbf{F}_1$ and $\mathbf{F}_2$.

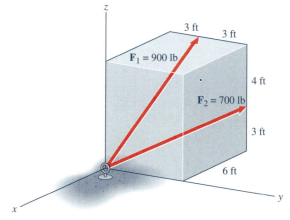

Figure P2-29

2-30* Two forces are applied to an eye bolt as shown in Fig. P2-30.
(a) Determine the x-, y-, and z-components of force $\mathbf{F}_1$.
(b) Express force $\mathbf{F}_1$ in Cartesian vector form.
(c) Determine the magnitude of the rectangular component of force $\mathbf{F}_1$ along the line of action of force $\mathbf{F}_2$.
(d) Determine the angle α between forces $\mathbf{F}_1$ and $\mathbf{F}_2$.

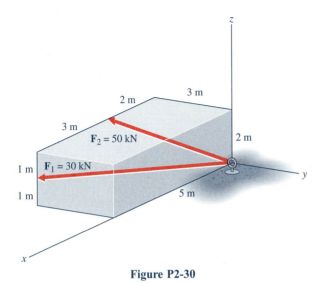

Figure P2-30

2-31 Two forces are applied to an eye bolt as shown in Fig. P2-29.
(a) Determine the x-, y-, and z-components of force $\mathbf{F}_2$.
(b) Express force $\mathbf{F}_2$ in Cartesian vector form.
(c) Determine the magnitude of the rectangular component of force $\mathbf{F}_2$ along the line of action of force $\mathbf{F}_1$.

2-32 Two forces are applied to an eye bolt as shown in Fig. P2-30.
(a) Determine the x-, y-, and z-components of force $\mathbf{F}_2$.
(b) Express force $\mathbf{F}_2$ in Cartesian vector form.
(c) Determine the magnitude of the rectangular component of force $\mathbf{F}_2$ along the line of action of force $\mathbf{F}_1$.

2-6 RESULTANTS BY RECTANGULAR COMPONENTS

Previous sections of this chapter discussed the use of parallelogram and triangle laws to determine the resultant $\mathbf{R}$ of two or more concurrent coplanar forces $\mathbf{F}_1$, $\mathbf{F}_2$, $\mathbf{F}_3$, $\cdots$, $\mathbf{F}_n$. Using the parallelogram law to add more than two forces is time-consuming and tedious since the procedure requires extensive geometric and trigonometric calculation to determine the magnitude and locate the line of action of the resultant $\mathbf{R}$. Problems of this type, however, are easily solved using the rectangular components of a force discussed in Section 2-5.

For any system of coplanar concurrent forces, such as the three shown in Fig. 2-27a, rectangular components $\mathbf{F}_{1x}$ and $\mathbf{F}_{1y}$, $\mathbf{F}_{2x}$ and $\mathbf{F}_{2y}$, $\mathbf{F}_{3x}$ and $\mathbf{F}_{3y}$, $\cdots$, and $\mathbf{F}_{nx}$ and $\mathbf{F}_{ny}$ can be determined as shown in Fig. 2-27b. Adding the respective x- and y-components yields

$$\mathbf{R}_x = \Sigma \mathbf{F}_x = \mathbf{F}_{1x} + \mathbf{F}_{2x} + \mathbf{F}_{3x} + \cdots + \mathbf{F}_{nx}$$
$$= (F_{1x} + F_{2x} + F_{3x} + \cdots + F_{nx})\, \mathbf{i} = R_x\, \mathbf{i}$$

$$\mathbf{R}_y = \Sigma \mathbf{F}_y = \mathbf{F}_{1y} + \mathbf{F}_{2y} + \mathbf{F}_{3y} + \cdots + \mathbf{F}_{ny}$$
$$= (F_{1y} + F_{2y} + F_{3y} + \cdots + F_{ny})\, \mathbf{j} = R_y\, \mathbf{j}$$

By the parallelogram law

$$\mathbf{R} = \mathbf{R}_x + \mathbf{R}_y = R_x\, \mathbf{i} + R_y\, \mathbf{j}$$

The magnitude R of the resultant can be determined from the Pythagorean theorem. Thus,

$$R = \sqrt{R_x^2 + R_y^2} = \sqrt{(\Sigma F_x)^2 + (\Sigma F_y)^2}$$

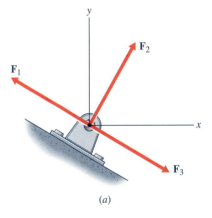

(a)

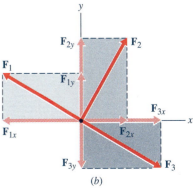

(b)

Figure 2-27

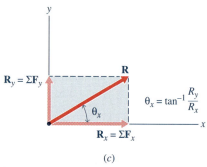

(c)

Figure 2-27

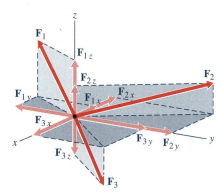

Figure 2-28

The angle θ_x between the x-axis and the line of action of the resultant **R,** as shown in Fig. 2-27c, is

$$\theta_x = \tan^{-1}\frac{R_y}{R_x} = \tan^{-1}\frac{\Sigma F_y}{\Sigma F_x}$$

The angle θ_x can also be determined, if it is more convenient, from the equations

$$\theta_x = \cos^{-1}\frac{\Sigma F_x}{R} \quad \text{or} \quad \theta_x = \sin^{-1}\frac{\Sigma F_y}{R}$$

The sense of each component must be designated in the summations by using a plus sign if the component acts in the positive x- or y-direction, and a minus sign if the component acts in the negative x- or y-direction.

In the general case of three or more concurrent forces in space, such as the three shown in Fig. 2-28, rectangular components $\mathbf{F}_{1x}$, $\mathbf{F}_{1y}$, and $\mathbf{F}_{1z}$; $\mathbf{F}_{2x}$, $\mathbf{F}_{2y}$, and $\mathbf{F}_{2z}$; $\mathbf{F}_{3x}$, $\mathbf{F}_{3y}$, and $\mathbf{F}_{3z}$; $\cdots$; and $\mathbf{F}_{nx}$, $\mathbf{F}_{ny}$, and $\mathbf{F}_{nz}$ can be determined. Adding the respective x-, y-, and z-components yields

$$\mathbf{R}_x = \Sigma \mathbf{F}_x = \mathbf{F}_{1x} + \mathbf{F}_{2x} + \mathbf{F}_{3x} + \cdots + \mathbf{F}_{nx}$$
$$= (F_{1x} + F_{2x} + F_{3x} + \cdots + F_{nx})\,\mathbf{i} = R_x\,\mathbf{i}$$

$$\mathbf{R}_y = \Sigma \mathbf{F}_y = \mathbf{F}_{1y} + \mathbf{F}_{2y} + \mathbf{F}_{3y} + \cdots + \mathbf{F}_{ny}$$
$$= (F_{1y} + F_{2y} + F_{3y} + \cdots + F_{ny})\,\mathbf{j} = R_y\,\mathbf{j}$$

$$\mathbf{R}_z = \Sigma \mathbf{F}_z = \mathbf{F}_{1z} + \mathbf{F}_{2z} + \mathbf{F}_{3z} + \cdots + \mathbf{F}_{nz}$$
$$= (F_{1z} + F_{2z} + F_{3z} + \cdots + F_{nz})\,\mathbf{k} = R_z\,\mathbf{k}$$

The resultant **R** is then obtained from the expression

$$\mathbf{R} = \mathbf{R}_x + \mathbf{R}_y + \mathbf{R}_z = R_x\,\mathbf{i} + R_y\,\mathbf{j} + R_z\,\mathbf{k}$$

Once the scalar components R_x, R_y, and R_z are known, the magnitude R of the resultant and the angles θ_x, θ_y, and θ_z between the line of action of the resultant and the positive coordinate axes can be obtained from the expressions

$$R = \sqrt{R_x^2 + R_y^2 + R_z^2}$$

and

$$\theta_x = \cos^{-1}\frac{R_x}{R} \qquad \theta_y = \cos^{-1}\frac{R_y}{R} \qquad \theta_z = \cos^{-1}\frac{R_z}{R}$$

Example Problem 2-7

Determine the magnitude R of the resultant of the four forces shown in Fig. 2-29a and the angle θ_x between the x-axis and the line of action of the resultant.

SOLUTION

The magnitude R of the resultant will be determined by using the rectangular components F_x and F_y of each of the forces. Thus

$F_{1x} = 80 \cos 140° = -61.28$ lb $F_{1y} = 80 \sin 140° = +51.42$ lb

$F_{2x} = 60 \cos 110° = -20.52$ lb $F_{2y} = 60 \sin 110° = +56.38$ lb

$F_{3x} = 75 \cos 45° = +53.03$ lb $F_{3y} = 75 \sin 45° = +53.03$ lb

$F_{4x} = 90 \cos 17° = +86.07$ lb $F_{4y} = 90 \sin 17° = +26.31$ lb

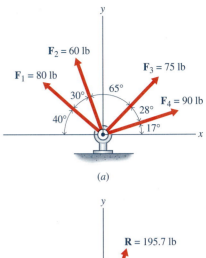

Once the rectangular components of the forces are known, the components R_x and R_y of the resultant are obtained from the expressions

$$R_x = \Sigma F_x = F_{1x} + F_{2x} + F_{3x} + F_{4x}$$
$$= -61.28 - 20.52 + 53.03 + 86.07 = +57.30 \text{ lb}$$

$$R_y = \Sigma F_y = F_{1y} + F_{2y} + F_{3y} + F_{4y}$$
$$= +51.42 + 56.38 + 53.03 + 26.31 = +187.14 \text{ lb}$$

The magnitude R of the resultant is

$$R = \sqrt{R_x^2 + R_y^2} = \sqrt{(57.30)^2 + (187.14)^2} = 195.7 \text{ lb} \qquad \textbf{Ans.}$$

The angle θ_x is obtained from the expression

$$\theta_x = \tan^{-1}\frac{R_y}{R_x} = \tan^{-1}\frac{+187.14}{+57.30} = 73.0° \qquad \textbf{Ans.}$$

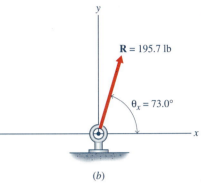

Figure 2-29

The resultant **R** of the four forces of Fig. 2-29a is shown in Fig. 2-29b. ∎

Example Problem 2-8

Three forces are applied at a point on a body as shown in Fig. 2-30. Determine the magnitude R of the resultant of the three forces and the angles θ_x, θ_y, and θ_z between the line of action of the resultant and the positive x-, y-, and z-coordinate axes.

SOLUTION

The magnitude R of the resultant will first be determined by using the rectangular components F_x, F_y, and F_z of each of the forces. Thus

$$F_{1x} = 25 \cos 26° \cos 120° = -11.235 \text{ kN}$$

$$F_{1y} = 25 \cos 26° \sin 120° = +19.459 \text{ kN}$$

$$F_{1z} = 25 \sin 26° = +10.959 \text{ kN}$$

$$F_{2x} = 10 \cos 60° \cos (-60°) = +2.500 \text{ kN}$$

$$F_{2y} = 10 \cos 60° \sin (-60°) = -4.330 \text{ kN}$$

$$F_{2z} = 10 \sin 60° = +8.660 \text{ kN}$$

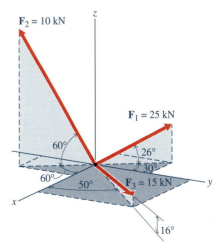

Figure 2-30

$$F_{3x} = 15 \cos 16° \cos 50° = +9.268 \text{ kN}$$

$$F_{3y} = 15 \cos 16° \sin 50° = +11.046 \text{ kN}$$

$$F_{3z} = 15 \sin 16° = +4.135 \text{ kN}$$

Once the rectangular components of the forces are known, the components R_x, R_y, and R_z of the resultant are obtained from the expressions

$$R_x = \Sigma F_x = F_{1x} + F_{2x} + F_{3x}$$
$$= -11.235 + 2.500 + 9.268 = +0.533 \text{ kN}$$

$$R_y = \Sigma F_y = F_{1y} + F_{2y} + F_{3y}$$
$$= +19.459 - 4.330 + 11.046 = +26.175 \text{ kN}$$

$$R_z = \Sigma F_z = F_{1z} + F_{2z} + F_{3z}$$
$$= +10.959 + 8.660 + 4.135 = +23.754 \text{ kN}$$

The magnitude R of the resultant is

$$R = \sqrt{R_x^2 + R_y^2 + R_z^2}$$
$$= \sqrt{(0.533)^2 + (26.175)^2 + (23.754)^2} = 35.35 \cong 35.4 \text{ kN} \qquad \textbf{Ans.}$$

The angles θ_x, θ_y, and θ_z are obtained from the expressions

$$\theta_x = \cos^{-1}\frac{R_x}{R} = \cos^{-1}\frac{+0.533}{35.35} = 89.1° \qquad \textbf{Ans.}$$

$$\theta_y = \cos^{-1}\frac{R_y}{R} = \cos^{-1}\frac{+26.175}{35.35} = 42.2° \qquad \textbf{Ans.}$$

$$\theta_z = \cos^{-1}\frac{R_z}{R} = \cos^{-1}\frac{+23.754}{35.35} = 47.8° \qquad \textbf{Ans.}$$

Alternatively, the solution can be obtained using vector methods. Unit vectors $\mathbf{e}_1$, $\mathbf{e}_2$, and $\mathbf{e}_3$, along the lines of action of forces $\mathbf{F}_1$, $\mathbf{F}_2$, and $\mathbf{F}_3$, respectively, are

$$\mathbf{e}_1 = (\cos 26° \cos 120°) \mathbf{i} + (\cos 26° \cos 30°) \mathbf{j} + (\cos 64°) \mathbf{k}$$
$$= -0.4494 \mathbf{i} + 0.7784 \mathbf{j} + 0.4384 \mathbf{k}$$

$$\mathbf{e}_2 = (\cos 60° \cos 60°) \mathbf{i} + (\cos 60° \cos 150°) \mathbf{j} + (\cos 30°) \mathbf{k}$$
$$= 0.2500 \mathbf{i} - 0.4330 \mathbf{j} + 0.8660 \mathbf{k}$$

$$\mathbf{e}_3 = (\cos 16° \cos 50°) \mathbf{i} + (\cos 16° \cos 40°) \mathbf{j} + (\cos 74°) \mathbf{k}$$
$$= 0.6179 \mathbf{i} + 0.7364 \mathbf{j} + 0.2756 \mathbf{k}$$

The forces are then written in Cartesian vector form as

$$\mathbf{F}_1 = F_1\mathbf{e}_1 = 25(-0.4494 \mathbf{i} + 0.7784 \mathbf{j} + 0.4384 \mathbf{k})$$
$$= -11.235 \mathbf{i} + 19.460 \mathbf{j} + 10.960 \mathbf{k} \text{ kN}$$
$$\mathbf{F}_2 = F_2\mathbf{e}_2 = 10(0.2500 \mathbf{i} - 0.4330 \mathbf{j} + 0.8660 \mathbf{k})$$
$$= 2.500 \mathbf{i} - 4.330 \mathbf{j} + 8.660 \mathbf{k} \text{ kN}$$

$$\mathbf{F}_3 = F_3\mathbf{e}_3 = 15(0.6179\ \mathbf{i} + 0.7364\ \mathbf{j} + 0.2756\ \mathbf{k})$$
$$= 9.269\ \mathbf{i} + 11.046\ \mathbf{j} + 4.134\ \mathbf{k}\ \text{kN}$$

The resultant **R** of the three forces is

$$\mathbf{R} = \mathbf{F}_1 + \mathbf{F}_2 + \mathbf{F}_3 = \Sigma F_x\ \mathbf{i} + \Sigma F_y\ \mathbf{j} + \Sigma F_z\ \mathbf{k}$$
$$= 0.533\ \mathbf{i} + 26.175\ \mathbf{j} + 23.754\ \mathbf{k}\ \text{kN}$$

Once the scalar components of the resultant are determined, the magnitude of the resultant and the direction angles are determined as in the first part of the example. ∎

Example Problem 2-9

Determine the magnitude R of the resultant of the three forces shown in Fig. 2-31 and the angles θ_x, θ_y, and θ_z between the line of action of the resultant and the positive x-, y-, and z-coordinate axes.

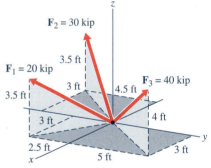

Figure 2-31

SOLUTION

In addition to the origin of coordinates, the lines of action of forces $\mathbf{F}_1$, $\mathbf{F}_2$, and $\mathbf{F}_3$ pass through points $(3, -2.5, 3.5)$, $(-3, -4.5, 3.5)$, and $(3, 5, 4)$, respectively. Since the coordinates of these points in space are known, it is easy to determine the unit vectors associated with each of the forces. Thus,

$$\mathbf{e}_1 = \frac{3\ \mathbf{i} - 2.5\ \mathbf{j} + 3.5\ \mathbf{k}}{\sqrt{(3)^2 + (-2.5)^2 + (3.5)^2}} = +0.5721\ \mathbf{i} - 0.4767\ \mathbf{j} + 0.6674\ \mathbf{k}$$

$$\mathbf{e}_2 = \frac{-3\ \mathbf{i} - 4.5\ \mathbf{j} + 3.5\ \mathbf{k}}{\sqrt{(-3)^2 + (-4.5)^2 + (3.5)^2}} = -0.4657\ \mathbf{i} - 0.6985\ \mathbf{j} + 0.5433\ \mathbf{k}$$

$$\mathbf{e}_3 = \frac{3\ \mathbf{i} + 5\ \mathbf{j} + 4\ \mathbf{k}}{\sqrt{(3)^2 + (5)^2 + (4)^2}} = +0.4243\ \mathbf{i} + 0.7071\ \mathbf{j} + 0.5657\ \mathbf{k}$$

Once the unit vectors $\mathbf{e}_1$, $\mathbf{e}_2$, and $\mathbf{e}_3$ are known, the three forces can be expressed in Cartesian vector form as

$$\mathbf{F}_1 = F_1\mathbf{e}_1 = 20(+0.5721\ \mathbf{i} - 0.4767\ \mathbf{j} + 0.6674\ \mathbf{k})$$
$$= +11.442\ \mathbf{i} - 9.534\ \mathbf{j} + 13.348\ \mathbf{k}\ \text{kip}$$

$$\mathbf{F}_2 = F_2\mathbf{e}_2 = 30(-0.4657\ \mathbf{i} - 0.6985\ \mathbf{j} + 0.5433\ \mathbf{k})$$
$$= -13.971\ \mathbf{i} - 20.955\ \mathbf{j} + 16.299\ \mathbf{k}\ \text{kip}$$

$$\mathbf{F}_3 = F_3\mathbf{e}_3 = 40(+0.4243\ \mathbf{i} + 0.7071\ \mathbf{j} + 0.5657\ \mathbf{k})$$
$$= +16.972\ \mathbf{i} + 28.284\ \mathbf{j} + 22.628\ \mathbf{k}\ \text{kip}$$

The resultant **R** of the three forces is

$$\mathbf{R} = \mathbf{F}_1 + \mathbf{F}_2 + \mathbf{F}_3 = R_x\ \mathbf{i} + R_y\ \mathbf{j} + R_z\ \mathbf{k}\ \text{kip}$$

where

$$R_x = \Sigma F_x = F_{1x} + F_{2x} + F_{3x}$$
$$= +11.442 - 13.971 + 16.972 = +14.443 \text{ kip}$$

$$R_y = \Sigma F_y = F_{1y} + F_{2y} + F_{3y}$$
$$= -9.534 - 20.955 + 28.284 = -2.205 \text{ kip}$$

$$R_z = \Sigma F_z = F_{1z} + F_{2z} + F_{3z}$$
$$= +13.348 + 16.299 + 22.628 = +52.28 \text{ kip}$$

Thus

$$\mathbf{R} = +14.443 \, \mathbf{i} - 2.205 \, \mathbf{j} + 52.28 \, \mathbf{k} \text{ kip}$$

The magnitude R of the resultant is

$$R = \sqrt{R_x^2 + R_y^2 + R_z^2}$$
$$= \sqrt{(+14.443)^2 + (-2.205)^2 + (+52.28)^2} = 54.28 \cong 54.3 \text{ kip} \quad \textbf{Ans.}$$

The angles θ_x, θ_y, and θ_z are obtained from the expressions

$$\theta_x = \cos^{-1}\frac{R_x}{R} = \cos^{-1}\frac{+14.443}{54.28} = 74.6° \qquad \textbf{Ans.}$$

$$\theta_y = \cos^{-1}\frac{R_y}{R} = \cos^{-1}\frac{-2.205}{54.28} = 92.3° \qquad \textbf{Ans.}$$

$$\theta_z = \cos^{-1}\frac{R_z}{R} = \cos^{-1}\frac{+52.28}{54.28} = 15.60° \quad ■ \qquad \textbf{Ans.}$$

PROBLEMS

Use the rectangular-component method to solve the following problems. Determine the magnitude R of the resultant and the angle θ_x between the line of action of the resultant and the x-axis for:

Standard Problems

2-33* The three forces shown in Fig. P2-33.

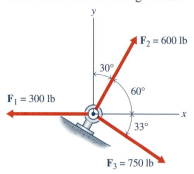

Figure P2-33

2-34* The three forces shown in Fig. P2-34.

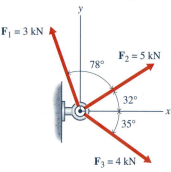

Figure P2-34

2-35* The three forces shown in Fig. P2-35.

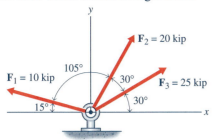

Figure P2-35

2-36* The three forces shown in Fig. P2-36.

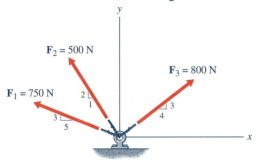

Figure P2-36

2-37 The four forces shown in Fig. P2-37.

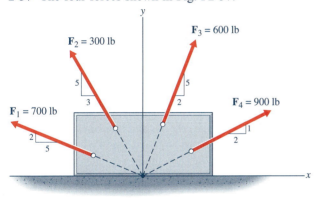

Figure P2-37

2-38 The four forces shown in Fig. P2-38.

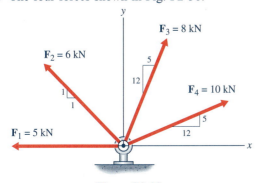

Figure P2-38

Challenging Problems

Use the rectangular-component method to solve the following problems. Determine the magnitude R of the resultant and the angles θ_x, θ_y, and θ_z between the line of action of the resultant and the positive x-, y-, and z-coordinate axes for:

2-39* The three forces shown in Fig. P2-39.

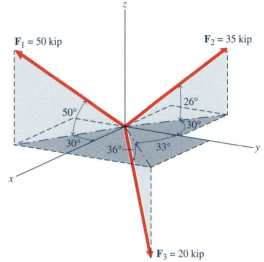

Figure P2-39

2-40* The three forces shown in Fig. P2-40.

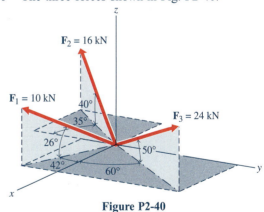

Figure P2-40

2-41 The three forces shown in Fig. P2-41.

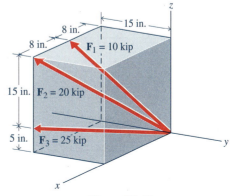

Figure P2-41

2-42 The three forces shown in Fig. P2-42.

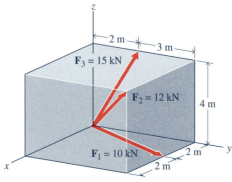

Figure P2-42

2-7 SUMMARY

A force is defined as the action of one physical body upon another. Since the interaction can occur when the bodies are in contact or when the bodies are physically separated, forces are classified as either surface forces (a push or a pull produced by mechanical means) or body forces (the gravitational pull of the earth). The characteristics of a force are its magnitude, its direction (line of action and sense), and its point of application. Since both a magnitude and a direction are needed to characterize a force and forces add according to the parallelogram law of addition, forces are vector quantities. A number of forces treated as a group constitute a force system.

The components of a force are any system of forces that can be combined by the parallelogram law to produce the original force. For most engineering problems, rectangular (mutually perpendicular) components of a force are more useful than general oblique components. The three rectangular components of a force $\mathbf{F}$ in space are $\mathbf{F}_x$, $\mathbf{F}_y$, and $\mathbf{F}_z$ along the x-, y-, and z-coordinate axes, respectively. The force $\mathbf{F}$ and its scalar components F_x, F_y, and F_z can be written in Cartesian vector form by using unit vectors $\mathbf{i}$, $\mathbf{j}$, and $\mathbf{k}$ directed along the positive x-, y-, and z-coordinate axes as

$$\mathbf{F} = \mathbf{F}_x + \mathbf{F}_y + \mathbf{F}_z = F_x\,\mathbf{i} + F_y\,\mathbf{j} + F_z\,\mathbf{k}$$
$$= F\cos\theta_x\,\mathbf{i} + F\cos\theta_y\,\mathbf{j} + F\cos\theta_z\,\mathbf{k} \qquad (2\text{-}3)$$

where the scalar components F_x, F_y, and F_z are related to the magnitude F and the direction θ of the force $\mathbf{F}$ by the expressions

$$F_x = F\cos\theta_x \qquad F_y = F\cos\theta_y \qquad F_z = F\cos\theta_z$$
$$F = \sqrt{F_x^2 + F_y^2 + F_z^2} \qquad\qquad (2\text{-}4)$$

$$\theta_x = \cos^{-1}\frac{F_x}{F} \qquad \theta_y = \cos^{-1}\frac{F_y}{F} \qquad \theta_z = \cos^{-1}\frac{F_z}{F}$$

and

$$\cos^2\theta_x + \cos^2\theta_y + \cos^2\theta_z = 1$$

The rectangular component of a force **F** along an arbitrary direction n can be obtained by using the vector dot product. Thus, if $\mathbf{e}_n$ is a unit vector in the specified direction n, the magnitude of the rectangular component $\mathbf{F}_n$ of the force **F** is

$$F_n = \mathbf{F} \cdot \mathbf{e}_n = (F_x \mathbf{i} + F_y \mathbf{j} + F_z \mathbf{k}) \cdot \mathbf{e}_n$$

If the angles between the direction n and the x-, y-, and z-axes are θ'_x, θ'_y, and θ'_z, the unit vector $\mathbf{e}_n$ can be written in Cartesian vector form as

$$\mathbf{e}_n = \cos \theta'_x \mathbf{i} + \cos \theta'_y \mathbf{j} + \cos \theta'_z \mathbf{k}$$

Thus, the magnitude of the force $\mathbf{F}_n$ is

$$F_n = \mathbf{F} \cdot \mathbf{e}_n = F_x \cos \theta'_x + F_y \cos \theta'_y + F_z \cos \theta'_z \qquad (2\text{-}5)$$

The rectangular component $\mathbf{F}_n$ of the force **F** is expressed in Cartesian vector form as

$$\mathbf{F}_n = (\mathbf{F} \cdot \mathbf{e}_n) \mathbf{e}_n = F_n \mathbf{e}_n = F_n(\cos \theta'_x \mathbf{i} + \cos \theta'_y \mathbf{j} + \cos \theta'_z \mathbf{k}) \qquad (2\text{-}7)$$

The angle α between the line of action of the force **F** and the direction n is determined by using the definition of a rectangular component of a force ($F_n = F \cos \alpha = \mathbf{F} \cdot \mathbf{e}_n$). Thus

$$\alpha = \cos^{-1}\frac{\mathbf{F} \cdot \mathbf{e}_n}{F} = \cos^{-1}\frac{F_n}{F} \qquad (2\text{-}8)$$

A single force, called the resultant **R,** will produce the same effect on a body as a system of concurrent forces. The resultant can be determined by adding the forces using the parallelogram law; however, this procedure is time-consuming and tedious when the system comprises more than two forces. Resultants are easily obtained, however, by using rectangular components of the forces. For the general case of two or more concurrent forces in space,

$$\mathbf{R}_x = \Sigma \mathbf{F}_x = R_x \mathbf{i} \qquad \mathbf{R}_y = \Sigma \mathbf{F}_y = R_y \mathbf{j} \qquad \mathbf{R}_z = \Sigma \mathbf{F}_z = R_z \mathbf{k}$$

The magnitude R of the resultant and the angles θ_x, θ_y, and θ_z between the line of action of the resultant and the positive coordinate axes are

$$R = \sqrt{R_x^2 + R_y^2 + R_z^2}$$

$$\theta_x = \cos^{-1}\frac{R_x}{R} \qquad \theta_y = \cos^{-1}\frac{R_y}{R} \qquad \theta_z = \cos^{-1}\frac{R_z}{R}$$

REVIEW PROBLEMS

2-43* A 500-lb force is applied to the post shown in Fig. P2-43. Determine
(a) The magnitudes of the x- and y-components of the force.
(b) The magnitudes of the u- and v-components of the force.

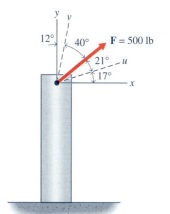

Figure P2-43

2-44* Two forces $\mathbf{F}_1$ and $\mathbf{F}_2$ are applied to an eye bolt as shown in Fig. P2-44. Determine
(a) The magnitude and direction (angle θ_x) of the resultant $\mathbf{R}$ of the two forces.
(b) The magnitudes of two other forces $\mathbf{F}_u$ and $\mathbf{F}_v$ that would have the same resultant.

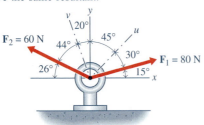

Figure P2-44

2-45 Three forces are applied to an eye bolt as shown in Fig. P2-45. Determine the magnitude R of the resultant of the forces and the angle θ_x between the line of action of the resultant and the x-axis.

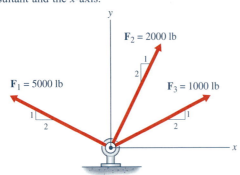

Figure P2-45

2-46 Four forces act on a small airplane in flight, as shown in Fig. P2-46: its weight, the thrust provided by the engine, the lift provided by the wings, and the drag resulting from its motion through the air. Determine the resultant of the four forces and its line of action with respect to the axis of the plane.

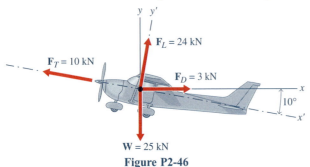

Figure P2-46

2-47* Three forces are applied at a point on a body as shown in Fig. P2-47. Determine the resultant $\mathbf{R}$ of the three forces and the angles θ_x, θ_y, and θ_z between the line of action of the resultant and the positive x-, y-, and z-coordinate axes.

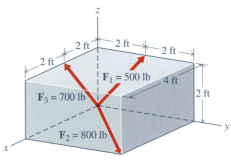

Figure P2-47

2-48* Three forces are applied at a point on a body as shown in Fig. P2-48. Determine the resultant $\mathbf{R}$ of the three forces and the angles θ_x, θ_y, and θ_z between the line of action of the resultant and the positive x-, y-, and z-coordinate axes.

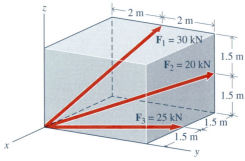

Figure P2-48

2-49* Three forces are applied to a stalled automobile as shown in Fig. P2-49. Determine the magnitude of the force $\mathbf{F}_3$ and the magnitude of the resultant $\mathbf{R}$ if the line of action of the resultant is along the x-axis.

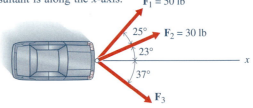

Figure P2-49

2-50 Three cables are used to drag a heavy crate on a horizontal surface as shown in Fig. P2-50. The resultant $\mathbf{R}$ of the forces has a magnitude of 2800 N and its line of action is directed along the x-axis. Determine the magnitudes of forces $\mathbf{F}_1$ and $\mathbf{F}_3$.

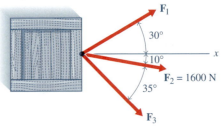

Figure P2-50

2-51 Two forces are applied at a point in a body as shown in Fig. P2-51. Determine

(a) The magnitude and direction (angles θ_x, θ_y, and θ_z) of the resultant **R** of the two forces.

(b) The magnitude of the rectangular component of force $\mathbf{F}_1$ along the line of action of force $\mathbf{F}_2$.

(c) The angle α between forces $\mathbf{F}_1$ and $\mathbf{F}_2$.

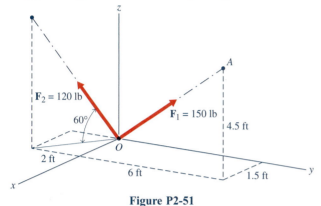

Figure P2-51

2-52 Three forces are applied with cables to the anchor block shown in Fig. P2-52. Determine

(a) The magnitude and direction (angles θ_x, θ_y, and θ_z) of the resultant **R** of the three forces.

(b) The magnitude of the rectangular component of force $\mathbf{F}_1$ along the line of action of force $\mathbf{F}_2$.

(c) The angle α between forces $\mathbf{F}_1$ and $\mathbf{F}_3$.

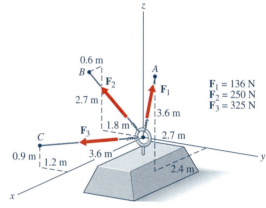

Figure P2-52

COMPUTER PROBLEMS

C2-53 Two forces **A** and **B** are applied to an eye bolt as shown in Fig. P2-53. If the magnitudes of the two forces are $A = 50$ lb and $B = 100$ lb, calculate and plot the magnitude of the resultant **R** as a function of the angle θ_A ($0° < \theta_A < 180°$). Also calculate and plot the angle θ_R that the resultant makes with the force **B** as a function of the angle θ_A. When is the resultant a maximum? When is the resultant a minimum? When is the angle θ_R a maximum? Repeat for $A = 100$ lb and $B = 50$ lb.

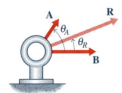

Figure P2-53

C2-54 Two forces **A** and **B** are applied to an eye bolt using ropes as shown in Fig. P2-54. The resultant **R** of the two forces has a magnitude $R = 4$ kN and makes an angle of $30°$ with the force **A** as shown. If both the forces pull on the eye bolt as shown (ropes cannot push on the eye bolt), what is the range of angles ($\theta_{min} < \theta_B < \theta_{max}$) for which this problem has a solution? Calculate and plot the required magnitudes A and B as functions of the angle θ_B ($\theta_{min} < \theta_B < \theta_{max}$). Why is the magnitude of **B** a minimum when $\theta_B = 90°$?

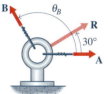

Figure P2-54

C2-55 Three forces **A**, **B**, and **C** are applied to an eye bolt using ropes as shown in Fig. P2-55. Force **A** has a magnitude $A = 50$ lb, and the resultant of the three forces is zero. If all the forces pull on the eye bolt as shown (ropes cannot push on the eye bolt), what is the range of angles ($\theta_{min} < \theta_C < \theta_{max}$) for which this problem has a solution? Calculate and plot the required magnitudes B and C as functions of the angle θ_C ($\theta_{min} < \theta_C < \theta_{max}$). Why is the magnitude of **C** a minimum when $\theta_C = 90°$?

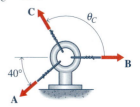

Figure P2-55

3

EQUILIBRIUM: CONCURRENT FORCE SYSTEMS

3-1 INTRODUCTION

Statics was defined in Chapter 1 as the branch of rigid-body mechanics concerned with bodies that are acted upon by a balanced system of forces (the resultant of all forces acting on the body is zero) and hence are at rest or moving with a constant velocity in a straight line.

A body with negligible dimensions is commonly referred to as a *particle.* In mechanics, either large bodies or small bodies can be referred to as a particle when the size and shape of the body have no effect on the response of the body to a system of forces. Under these conditions, the mass of the body can be assumed to be concentrated at a point. For example, the earth can be modeled as a particle for orbital motion studies since the size of the earth is insignificant when compared with the size of its orbit, and the shape of the earth does not influence the description of its position or the action of forces applied to it.

Since it is assumed that the mass of a particle is concentrated at a point and that the size and shape of a particle can be neglected, a particle can be subjected only to a system of concurrent forces. Newton's first law of motion states that "in the absence of external forces ($\mathbf{R} = \mathbf{0}$), a particle originally at rest or moving with a constant velocity (in equilibrium) will remain at rest or continue to move with a constant velocity along a straight line." Thus, a necessary condition for equilibrium of a particle is

$$\mathbf{R} = \Sigma \mathbf{F} = \mathbf{0} \tag{3-1}$$

A particle in equilibrium must also satisfy Newton's second law of motion, which can be expressed in equation form [Eq. (1-1)] as

$$\mathbf{R} = \Sigma \mathbf{F} = m\mathbf{a} \tag{1-1}$$

In order to satisfy both Eqs. (1-1) and (3-1),

$$m\mathbf{a} = \mathbf{0}$$

Since the mass of the particle is not zero, the acceleration of a particle in equilibrium is zero ($\mathbf{a} = \mathbf{0}$). Thus, a particle initially at rest will remain at rest and a particle moving with a constant velocity will maintain that velocity. Therefore, Eq. (3-1) is both a necessary condition and a sufficient condition for equilibrium.

The particle assumption is valid for many practical applications and thus provides a means for introducing the student to some interesting engineering problems early in a statics course. For this reason, this short chapter on statics (equilibrium) of a particle has been introduced before consideration of the more difficult problems associated with equilibrium of a rigid body (which involves the concepts of moments and distributed loads).

The force system acting on a body in a typical statics problem consists of known forces and unknown forces. Both must be clearly identified before a solution to a specific problem is attempted. A method commonly used to identify all forces acting on a body in a given situation is described in the following section.

3-2 FREE-BODY DIAGRAMS

A carefully prepared drawing or sketch that shows a "body of interest" separated from all interacting bodies is known as a *free-body diagram* (FBD). Once the body of interest is selected, the forces exerted by all other bodies on the one being considered must be determined and shown on the diagram. It is important that "all" forces acting "on" the body of interest be shown. Recall also that "a force cannot exist unless there is a body to exert the force." Frequently, the student will overlook and omit a force from the free-body diagram or show a force on the free-body diagram when there is no body present to exert the force.

The number of forces on a free-body diagram is determined by noting the number of bodies that exert forces on the body of interest. These forces may be either forces of contact or body forces. An important body force is the earth-pull on (or weight of) a body.

Each known force should be shown on a free-body diagram with its correct magnitude, slope, and sense. Letter symbols are used for the magnitudes of unknown forces. If a force has a known line of action but an unknown magnitude and sense, the sense of the force can be assumed. The correct sense will become apparent after solving for the unknown magnitude. By definition, the magnitude of a force is always positive; therefore, if the solution yields a negative magnitude, the minus sign indicates that the sense of the force is opposite to that assumed on the free-body diagram.

If both the magnitude and direction of a force acting on the body of interest are unknown (such as a pin reaction in a pin-connected structure), it is frequently convenient to show the two rectangular components of the force on the free-body diagram instead of the actual force. In this way, one deals with two forces of unknown magnitude but known direction. After the two rectangular components of the force are determined, the magnitude and direction of the actual force can easily be found. However, do not show both the force of unknown magnitude and its rectangular components on the same diagram.

The word "free" in the name "free-body diagram" emphasizes the idea that all bodies exerting forces on the body of interest are removed or withdrawn and are replaced by the forces they exert. Do not show both the bodies removed and the forces exerted by them on the free-body diagram. Sometimes it may be convenient to indicate, by light-weight dotted lines, the faint outlines of the bodies removed, in order to visualize the geometry and specify dimensions required for solution of the problem.

In drawing a free-body diagram of a given body, certain assumptions are made regarding the nature of the forces (reactions) exerted by other bodies on the body of interest. Two common assumptions are the following:

1. If a surface of contact at which a force is applied by one body to another body has only a small degree of roughness, it may be assumed to be smooth (frictionless), and hence, the action (or reaction) of the one body on the other is directed normal to the surface of contact.
2. A body that possesses only a small degree of bending stiffness (resistance to bending), such as a cord, rope, or chain, may be considered to be perfectly flexible, and hence the pull of such a body on any other body is directed along the axis of the flexible body.

A large number of additional assumptions will be discussed in Section 6-2 in the chapter on equilibrium of rigid bodies, where the topic of idealization of supports and connections is considered.

The term "body of interest" used in the definition of a free-body diagram may mean any definite part of a structure or machine, such as an eye bar in a bridge truss or a connecting rod in an automobile engine. The body of interest may also be taken as a group of physical bodies joined together (considered as one body), such as an entire bridge or a complete engine.

The importance of drawing a free-body diagram before attempting to solve a mechanics problem cannot be overemphasized. A procedure that can be followed to construct a complete and correct free-body diagram contains the following four steps.

CONSTRUCTING A FREE-BODY DIAGRAM

Step 1: Decide which body or combination of bodies is to be shown on the free-body diagram.

Step 2: Prepare a drawing or sketch of the outline of this isolated or free body.

Step 3: Carefully trace around the boundary of the free body and identify all the forces exerted by contacting or attracting bodies that were removed during the isolation process.

Step 4: Choose the set of coordinate axes to be used in solving the problem and indicate these directions on the free-body diagram. Place on the diagram any dimensions required for solution of the problem.

Application of these four steps to any mechanics problem should produce a complete and correct free-body diagram, which is an essential first step for the solution of any problem.

The free-body diagram is the "road map" for writing the equations of equilibrium. Every equation of equilibrium must be supported by a properly drawn, complete, free-body diagram. The symbols used in the equations of equilibrium must match the symbols used on the free-body diagram. For example, use A cos $30°$ rather than A_x if A is used on the free-body diagram to represent a force of known direction ($30°$ with respect to the x axis).

3-3 EQUILIBRIUM OF A PARTICLE

The term "particle" is used in statics to describe a body when the size and shape of the body will not significantly affect the solution of the problem being considered and when the mass of the body can be assumed to be concentrated at a point. As a result, a particle can be subjected only to a system of concurrent forces, and the necessary and sufficient conditions for equilibrium can be expressed mathematically as

$$\mathbf{R} = \Sigma\mathbf{F} = \mathbf{0} \tag{3-1}$$

where $\Sigma\mathbf{F}$ is the vector sum of all forces acting on the particle.

3-3-1 Two-Dimensional Problems

For a system of coplanar (say the xy-plane), concurrent forces, Eq. (3-1) can be written as

$$\begin{aligned}\mathbf{R} &= \mathbf{R}_x + \mathbf{R}_y = \mathbf{R}_n + \mathbf{R}_t \\ &= R_x\,\mathbf{i} + R_y\,\mathbf{j} = R_n\mathbf{e}_n + R_t\mathbf{e}_t \\ &= \Sigma F_x\,\mathbf{i} + \Sigma F_y\,\mathbf{j} = \Sigma F_n\mathbf{e}_n + \Sigma F_t\mathbf{e}_t = \mathbf{0}\end{aligned} \tag{3-2}$$

Equation (3-2) is satisfied only if

$$\mathbf{R}_x = R_x\,\mathbf{i} = \Sigma F_x\,\mathbf{i} = \mathbf{0} \qquad \mathbf{R}_n = R_n\mathbf{e}_n = \Sigma F_n\mathbf{e}_n = \mathbf{0}$$
$$\text{or}$$
$$\mathbf{R}_y = R_y\,\mathbf{j} = \Sigma F_y\,\mathbf{j} = \mathbf{0} \qquad \mathbf{R}_t = R_t\mathbf{e}_t = \Sigma F_t\mathbf{e}_t = \mathbf{0}$$

In scalar form, these equations become

$$R_x = \Sigma F_x = 0 \qquad R_n = \Sigma F_n = 0$$
$$\text{or} \qquad\qquad\qquad\qquad \tag{3-3}$$
$$R_y = \Sigma F_y = 0 \qquad R_t = \Sigma F_t = 0$$

That is, the sum of the rectangular components of the forces in any direction must be zero. While this would appear to give an infinite number of equations, no more than two of the equations are independent. The remaining equations can be obtained from combinations of the two independent equations. It is sometimes convenient to use $\Sigma F_x = 0$ and $\Sigma F_n = 0$ as the two independent equations rather than $\Sigma F_x = 0$ and $\Sigma F_y = 0$ (see Example 3-1). Equations (3-3) can be used to determine two unknown quantities (two magnitudes, two slopes, or a magnitude and a slope).

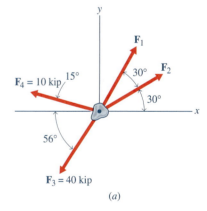

(a)

▌ Example Problem 3-1

A free-body diagram of a particle subjected to the action of four forces is shown in Fig. 3-1a. Determine the magnitudes of forces $\mathbf{F}_1$ and $\mathbf{F}_2$ so that the particle is in equilibrium.

SOLUTION

The particle is subjected to a system of coplanar, concurrent forces. The necessary and sufficient conditions for equilibrium are given by Eqs. (3-3) as

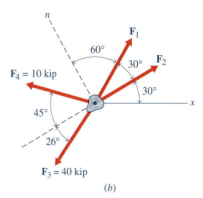

(b)

Figure 3-1

$\Sigma F_x = 0$ and $\Sigma F_y = 0$. Applying these equations, by using the free-body diagram shown in Fig. 3-1a, yields

$$+ \rightarrow \Sigma F_x = F_{1x} + F_{2x} + F_{3x} + F_{4x}$$
$$= F_1 \cos 60° + F_2 \cos 30° - 40 \cos 56° - 10 \cos 15°$$
$$= 0.5000F_1 + 0.8660F_2 - 22.37 - 9.659 = 0$$

from which

$$F_1 + 1.732F_2 = 64.06 \qquad\qquad (a)$$

$$+ \uparrow \Sigma F_y = F_{1y} + F_{2y} + F_{3y} + F_{4y}$$
$$= F_1 \sin 60° + F_2 \sin 30° - 40 \sin 56° + 10 \sin 15°$$
$$= 0.8660F_1 + 0.5000F_2 - 33.16 + 2.588 = 0$$

from which

$$F_1 + 0.5774F_2 = 35.30 \qquad\qquad (b)$$

Solving Eqs. (a) and (b) simultaneously yields

$$F_1 = 20.9 \text{ kip} \qquad\qquad \textbf{Ans.}$$
$$F_2 = 24.9 \text{ kip} \qquad\qquad \textbf{Ans.}$$

Alternatively, summing forces in a direction n perpendicular to the line of action of force $\mathbf{F}_2$, as shown in Fig. 3-1b, yields

$$+ \nwarrow \Sigma F_n = F_{1n} + F_{3n} + F_{4n}$$
$$= F_1 \sin 30° - 40 \sin 26° + 10 \sin 45° = 0$$

from which

$$F_1 = 20.93 \cong 20.9 \text{ kip} \qquad\qquad \textbf{Ans.}$$

Once force $\mathbf{F}_1$ is known, forces can be summed in any other direction to obtain force $\mathbf{F}_2$. Thus, summing forces in the x-direction yields

$$+ \rightarrow \Sigma F_x = F_{1x} + F_{2x} + F_{3x} + F_{4x}$$
$$= 20.93 \cos 60° + F_2 \cos 30° - 40 \cos 56° - 10 \cos 15° = 0$$

from which

$$F_2 = 24.90 \cong 24.9 \text{ kip} \qquad\qquad \textbf{Ans.}$$

Summing forces in a direction perpendicular to one of the unknown forces eliminates the need to solve simultaneous equations in two-dimensional problems. ∎

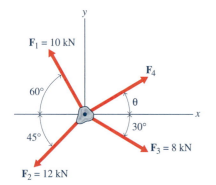

Figure 3-2

Example Problem 3-2

A free-body diagram of a particle subjected to the action of four forces is shown in Fig. 3-2. Determine the magnitude and direction (angle θ) of force $\mathbf{F}_4$ so that the particle is in equilibrium.

SOLUTION

The particle is subjected to a system of coplanar, concurrent forces. The necessary and sufficient conditions for equilibrium are given by Eqs. (3-3) as $\Sigma F_x = 0$ and $\Sigma F_y = 0$. Applying these equations, by using the free-body diagram shown in Fig. 3-2, yields

$$+ \rightarrow \Sigma F_x = F_{1x} + F_{2x} + F_{3x} + F_{4x}$$
$$= -10 \cos 60° - 12 \cos 45° + 8 \cos 30° + F_4 \cos \theta = 0$$

from which

$$F_{4x} = F_4 \cos \theta = 6.557 \text{ kN} \qquad \text{(a)}$$

$$+ \uparrow \Sigma F_y = F_{1y} + F_{2y} + F_{3y} + F_{4y}$$
$$= 10 \sin 60° - 12 \sin 45° - 8 \sin 30° + F_4 \sin \theta = 0$$

from which

$$F_{4y} = F_4 \sin \theta = 3.825 \text{ kN} \qquad \text{(b)}$$

Once the components F_{4x} and F_{4y} are known, Eqs. (2-2) are used to determine F_4 and θ. Thus,

$$F_4 = \sqrt{F_{4x}^2 + F_{4y}^2} = \sqrt{(6.557)^2 + (3.825)^2} = 7.59 \text{ kN} \qquad \textbf{Ans.}$$

$$\theta = \tan^{-1} \frac{F_{4y}}{F_{4x}} = \tan^{-1} \frac{3.825}{6.557} = 30.3° \quad \blacksquare \qquad \textbf{Ans.}$$

3-3-2 Three-Dimensional Problems

For a three-dimensional system of concurrent forces, Eq. (3-1) can be written as

$$\mathbf{R} = \Sigma \mathbf{F} = \mathbf{R}_x + \mathbf{R}_y + \mathbf{R}_z$$
$$= R_x \mathbf{i} + R_y \mathbf{j} + R_z \mathbf{k}$$
$$= \Sigma F_x \mathbf{i} + \Sigma F_y \mathbf{j} + \Sigma F_z \mathbf{k} = \mathbf{0} \qquad \text{(3-4)}$$

Equation (3-4) is satisfied only if

$$\mathbf{R}_x = R_x \mathbf{i} = \Sigma F_x \mathbf{i} = \mathbf{0}$$
$$\mathbf{R}_y = R_y \mathbf{j} = \Sigma F_y \mathbf{j} = \mathbf{0} \qquad \text{(3-5)}$$
$$\mathbf{R}_z = R_z \mathbf{k} = \Sigma F_z \mathbf{k} = \mathbf{0}$$

In scalar form, these equations become

$$R_x = \Sigma F_x = 0$$
$$R_y = \Sigma F_y = 0 \qquad \text{(3-6)}$$
$$R_z = \Sigma F_z = 0$$

Equations (3-5) and (3-6) can be used to determine three unknown quantities (three magnitudes, three slopes, or any combination of three magnitudes and slopes). The procedure is illustrated in Examples 3-3 and 3-4. Example 3-3 illustrates the scalar method of solution for a three-dimensional problem. Example 3-4 illustrates the vector method of solution for a similar problem.

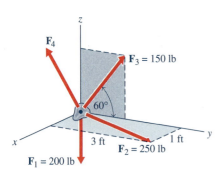

Figure 3-3

Example Problem 3-3

A free-body diagram of a particle subjected to the action of four forces is shown in Fig. 3-3. Determine the magnitude and the coordinate direction angles of the unknown force $\mathbf{F}_4$ so that the particle is in equilibrium.

SOLUTION

The necessary and sufficient conditions for equilibrium of a particle subjected to a three-dimensional system of concurrent forces are given by Eqs. (3-6) as

$$R_x = \Sigma F_x = 0,$$
$$R_y = \Sigma F_y = 0,$$
$$R_z = \Sigma F_z = 0$$

The scalar components for each of the known forces F_1, F_2, and F_3 shown on the free-body diagram of Fig. 3-3 are:

For $\mathbf{F}_1$:

$$F_{1x} = 0$$
$$F_{1y} = 0$$
$$F_{1z} = -200 \text{ lb}$$

For $\mathbf{F}_2$:

$$F_{2x} = (1/\sqrt{10})(250) = 79.06 \text{ lb}$$
$$F_{2y} = (3/\sqrt{10})(250) = 237.2 \text{ lb}$$
$$F_{2z} = 0$$

For $\mathbf{F}_3$:

$$F_{3x} = 0$$
$$F_{3y} = 150 \cos 60° = 75.00 \text{ lb}$$
$$F_{3z} = 150 \sin 60° = 129.90 \text{ lb}$$

Thus

$$R_x = \Sigma F_x = F_{1x} + F_{2x} + F_{3x} + F_{4x}$$
$$= 0 + 79.06 + 0 + F_{4x} = 0 \qquad F_{4x} = -79.06 \text{ lb}$$

$$R_y = \Sigma F_y = F_{1y} + F_{2y} + F_{3y} + F_{4y}$$
$$= 0 + 237.2 + 75.00 + F_{4y} = 0 \qquad F_{4y} = -312.2 \text{ lb}$$

$$R_z = \Sigma F_z = F_{1z} + F_{2z} + F_{3z} + F_{4z}$$
$$= -200 + 0 + 129.90 + F_{4z} = 0 \qquad F_{4z} = 70.10 \text{ lb}$$

Once the rectangular components of the force $\mathbf{F}_4$ are known, Eqs. (2-4) can be used to determine its magnitude and coordinate direction angles. Thus

$$F_4 = \sqrt{F_{4x}^2 + F_{4y}^2 + F_{4z}^2}$$
$$= \sqrt{(-79.06)^2 + (-312.2)^2 + (70.10)^2}$$
$$= 329.6 \cong 330 \text{ lb} \qquad \qquad \textbf{Ans.}$$

$$\theta_x = \cos^{-1}\frac{F_{4x}}{F_4} = \cos^{-1}\frac{-79.06}{329.6} = 103.9° \qquad \textbf{Ans.}$$

$$\theta_y = \cos^{-1}\frac{F_{4y}}{F_4} = \cos^{-1}\frac{-312.2}{329.6} = 161.3° \qquad \textbf{Ans.}$$

$$\theta_z = \cos^{-1}\frac{F_{4z}}{F_4} = \cos^{-1}\frac{70.10}{329.6} = 77.7° \; \blacksquare \qquad \textbf{Ans.}$$

Example Problem 3-4

A block is supported by a system of cables as shown in Fig. 3-4a. The weight of the block is 500 N. Determine the tensions in cables A, B, and C.

SOLUTION

A free-body diagram for joint D of the cable system is shown in Fig. 3-4b. This diagram shows that joint D is subjected to a three-dimensional system of concurrent forces with three unknown cable tensions T_A, T_B, and T_C. The coordinates of the support points for each of the cables is shown on the free-body diagram in (x, y, z) format as an aid in writing vector equations for the cable tensions. The necessary and sufficient conditions for equilibrium of the joint are given by Eq. (3-4) as

$$\mathbf{R} = \Sigma\mathbf{F} = \mathbf{T}_A + \mathbf{T}_B + \mathbf{T}_C + \mathbf{W} = \mathbf{0}$$

The cable tensions and the weight of the block can be expressed in Cartesian vector form as

$$\mathbf{T}_A = T_A\left[\frac{1.2\,\mathbf{i} - 1.5\,\mathbf{j} + 2.4\,\mathbf{k}}{\sqrt{9.45}}\right] = T_A(0.3904\,\mathbf{i} - 0.4880\,\mathbf{j} + 0.7807\,\mathbf{k})$$

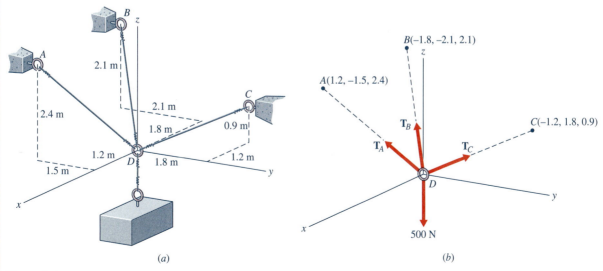

(a)   (b)

Figure 3-4

$$\mathbf{T}_B = T_B\left[\frac{-1.8\,\mathbf{i} - 2.1\,\mathbf{j} + 2.1\,\mathbf{k}}{\sqrt{12.06}}\right] = T_B(-0.5183\,\mathbf{i} - 0.6047\,\mathbf{j} + 0.6047\,\mathbf{k}) \tag{a}$$

$$\mathbf{T}_C = T_C\left[\frac{-1.2\,\mathbf{i} + 1.8\,\mathbf{j} + 0.9\,\mathbf{k}}{\sqrt{5.49}}\right] = T_C(-0.5121\,\mathbf{i} + 0.7682\,\mathbf{j} + 0.3841\,\mathbf{k})$$

$$\mathbf{W} = -500\,\mathbf{k}$$

Substituting Eqs. (a) into Eq. (3-4) yields

$$\begin{aligned}\mathbf{R} = \ &(0.3904T_A - 0.5183T_B - 0.5121T_C)\,\mathbf{i} \\ &+ (-0.4880T_A - 0.6047T_B + 0.7682T_C)\,\mathbf{j} \\ &+ (0.7807T_A + 0.6047T_B + 0.3841T_C - 500)\,\mathbf{k} = \mathbf{0}\end{aligned} \tag{b}$$

Since each of the components of Eq. (b) must equal zero if the resultant $\mathbf{R}$ is to be zero, the following equations must be satisfied.

$$0.3904T_A - 0.5183T_B - 0.5121T_C = 0$$

$$-0.4880T_A - 0.6047T_B + 0.7682T_C = 0$$

$$0.7807T_A + 0.6047T_B + 0.3841T_C = 500$$

The simultaneous solution of these three linear equations gives

$$T_A = 459 \text{ N} \qquad\qquad\qquad \textbf{Ans.}$$

$$T_B = 32.4 \text{ N} \qquad\qquad\qquad \textbf{Ans.}$$

$$T_C = 317 \text{ N} \ \blacksquare \qquad\qquad\qquad \textbf{Ans.}$$

PROBLEMS

Standard Problems

3-1* Determine the magnitude of forces $\mathbf{F}_2$ and $\mathbf{F}_3$ so that the particle shown in Fig. P3-1 is in equilibrium.

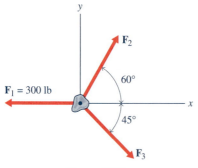

Figure P3-1

3-2* Determine the magnitudes of forces $\mathbf{F}_1$ and $\mathbf{F}_2$ so that the particle shown in Fig. P3-2 is in equilibrium.

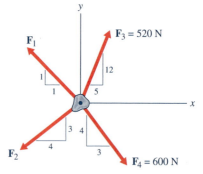

Figure P3-2

3-14* The particle shown in Fig. P3-14 is in equilibrium under the action of the four forces shown on the free-body diagram. Determine the magnitude and the coordinate direction angles of the unknown force $\mathbf{F}_4$.

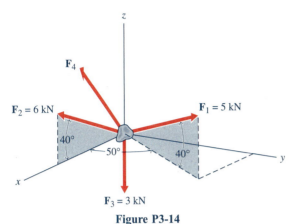

Figure P3-14

3-15* The particle shown in Fig. P3-15 is in equilibrium under the action of the four forces shown on the free-body diagram. Determine the magnitude and the coordinate direction angles of the unknown force $\mathbf{F}_4$.

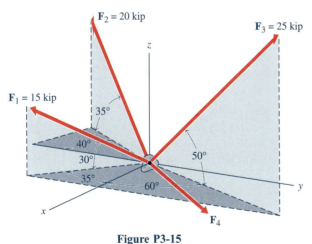

Figure P3-15

3-16 The particle shown in Fig. P3-16 is in equilibrium under the action of the four forces shown on the free-body diagram. Determine the magnitude and the coordinate direction angles of the unknown force $\mathbf{F}_4$.

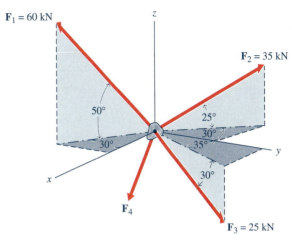

Figure P3-16

3-17 A 3000-lb cylinder is supported by a system of cables as shown in Fig. P3-17. Determine the tensions in cables A, B, and C.

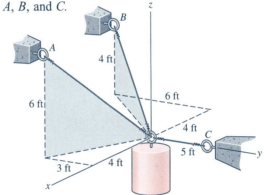

Figure P3-17

3-18 The traffic light shown in Fig. P3-18 is supported by a system of cables. Determine the tensions in cables A, B, and C if the traffic light has a mass of 75 kg.

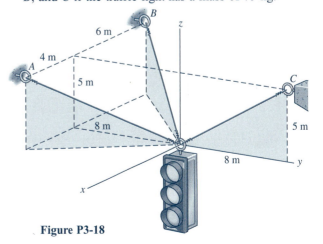

Figure P3-18

3-4 SUMMARY

The term "particle" is used in statics to describe a body when the size and shape of the body will not significantly affect the solution of the problem being considered and when the mass of the body can be assumed to be concentrated at a point. As a result, a particle can be subjected only to a system of concurrent forces, and the necessary and sufficient conditions for equilibrium can be expressed mathematically as

$$\mathbf{R} = \Sigma\mathbf{F} = \mathbf{0} \qquad (3\text{-}1)$$

For a three-dimensional system of concurrent forces, Eq. (3-1) can be written as

$$\begin{aligned}\mathbf{R} = \Sigma\mathbf{F} &= \mathbf{R}_x + \mathbf{R}_y + \mathbf{R}_z \\ &= R_x\,\mathbf{i} + R_y\,\mathbf{j} + R_z\,\mathbf{k} \\ &= \Sigma F_x\,\mathbf{i} + \Sigma F_y\,\mathbf{j} + \Sigma F_z\,\mathbf{k} = \mathbf{0}\end{aligned} \qquad (3\text{-}4)$$

Equation (3-4) is satisfied only if

$$\Sigma F_x = 0 \qquad \Sigma F_y = 0 \qquad \Sigma F_z = 0 \qquad (3\text{-}6)$$

Equations (3-6) can be used to determine three unknown quantities (three magnitudes, three slopes, or any combination of three magnitudes and slopes).

REVIEW PROBLEMS

3-19* Determine the magnitudes of forces $\mathbf{F}_1$ and $\mathbf{F}_2$ so that the particle shown in Fig. P3-19 is in equilibrium.

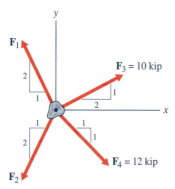

Figure P3-19

3-20* Determine the magnitude and direction angle θ of force $\mathbf{F}_4$ so that the particle shown in Fig. P3-20 is in equilibrium.

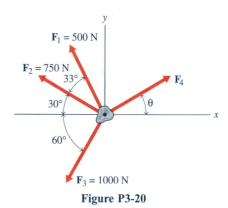

Figure P3-20

3-21* Two 10-in.-diameter pipes and a 6-in.-diameter pipe are supported in a pipe rack as shown in Fig. P3-21. The 10-in.-diameter pipes each weigh 300 lb and the 6-in.-diameter pipe weighs 175 lb. Determine the forces exerted on the pipes by the supports at contact surfaces A, B, and C. Assume all surfaces to be smooth.

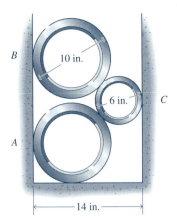

Figure P3-21

3-22* The masses of cylinders A and B shown in Fig. P3-22 are 40 kg and 90 kg, respectively. Determine the forces exerted on the cylinders by the inclined surfaces and the magnitude and direction of the force exerted by cylinder A on cylinder B when the cylinders are in equilibrium. Assume that all surfaces are smooth.

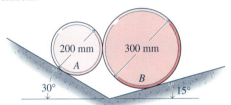

Figure P3-22

3-23 A continuous cable is used to support blocks A and B as shown in Fig. P3-23. The tension in the cable does not change as it passes around the small frictionless pulleys. Block A is supported by a small wheel that is free to roll on the cable. Determine the displacement y of block A for equilibrium if the weights of blocks A and B are 50 lb and 75 lb, respectively.

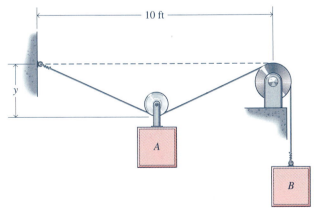

Figure P3-23

3-24 The mass of block A shown in Fig. P3-24 is 200 kg. Block A is supported by a small wheel that is free to roll on the continuous cable between supports B and C. The length of the cable is 43 m. Determine the distance x and the tension T in the cable when the system is in equilibrium. The tension in the cable does not change as it passes around the small frictionless pulley.

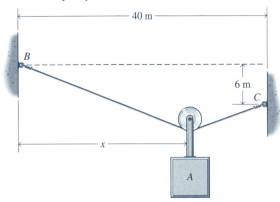

Figure P3-24

3-25* The circular disk shown in Fig. P3-25 weighs 500 lb. Determine the tensions in cables A, B, and C.

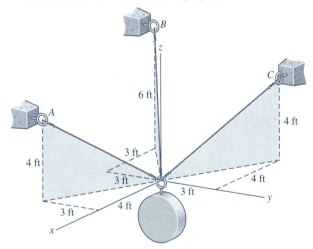

Figure P3-25

3-26* The particle shown in Fig. P3-26 is in equilibrium under the action of the four forces shown on the free-body diagram. Determine the magnitudes of forces F_1, F_2, and F_3.

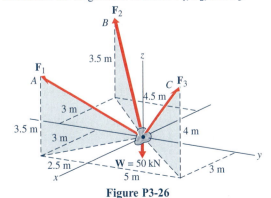

Figure P3-26

3-27 The hot-air balloon shown in Fig. P3-27 is tethered with three mooring cables. If the net lift of the balloon is 750 lb, determine the force exerted on the balloon by each of the cables.

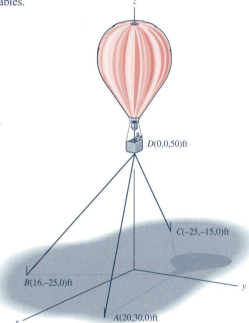

Figure P3-27

3-28 The crate shown in Fig. P3-28 has a mass of 500 kg. Determine the tensions in cables A, B, and C used to support the crate.

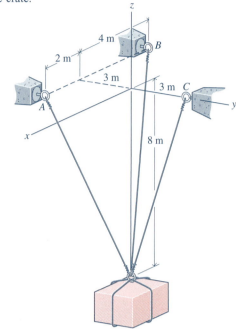

Figure P3-28

COMPUTER PROBLEMS

C3-29 A 75-lb stop light is suspended between two poles as shown in Fig. P3-29. Neglect the weight of the flexible cables and plot the tension in both cables as a function of the sag distance d ($0 \le d \le 8$ ft.) Determine the minimum sag d for which both tensions are less than
(a) 100 lb   (b) 250 lb   (c) 500 lb

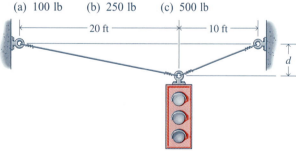

Figure P3-29

C3-30 A 50-kg load is suspended from a pulley as shown in Fig. P3-30a. The tension in the flexible cable does not change as it passes around the small frictionless pulleys, and the weight of the cable may be neglected. Plot the force P required for equilibrium as a function of the sag distance d ($0 \le d \le 1$ m). Determine the minimum sag d for which P is less than
(a) Twice the weight of the load.
(b) Four times the weight of the load.
(c) Eight times the weight of the load.

Repeat the problem if the load is securely fastened to the hoisting rope as shown in Fig. P3-30b. (Plot both the force P and the tension T_{AB} in segment AB of the cable on the same graph.)

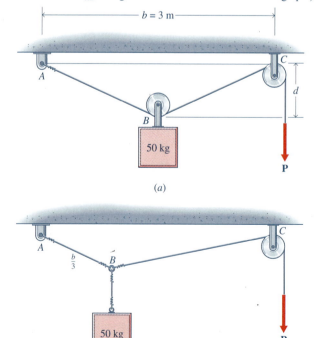

Figure P3-30   (b)

C3-31 Three identical steel disks are stacked in a box as shown in Fig. P3-31. The weight and diameter of the smooth disks are 50 lb and 12 in., respectively. Plot the three forces exerted on disk C (by disk A, by the side wall, and by the floor) as a function of the distance b between the walls of the box (24 in. $\leq b \leq$ 36 in.). Determine the range of b for which
(a) The force at the floor is larger than the other two forces.
(b) None of the three forces exceeds 100 lb.
(c) None of the three forces exceeds 200 lb.

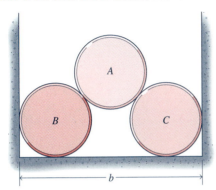

Figure P3-31

C3-32 A pair of steel pipes is stacked in a box as shown in Fig. P3-32. The masses and diameters of the smooth pipes are $m_A = 5$ kg, $m_B = 20$ kg, $d_A = 100$ mm, and $d_B = 200$ mm. Plot the two forces exerted on pipe A (by pipe B and by the side wall) as a function of the distance b between the walls of the box (200 mm $\leq b \leq$ 300 mm). Determine the range of b for which
(a) The force at the side wall is less than W_A, the weight of pipe A.
(b) None of the two forces exceed $2W_A$.
(c) None of the two forces exceed $4W_A$.

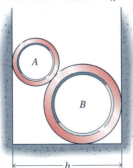

Figure P3-32

C3-33 Two small blocks are connected by a flexible cord as shown in Fig. P3-33. The radius of the smooth cylindrical surface is 18 in., and the length of the rope is such that the angle between the two blocks is 90°. Plot the angle θ (between block 2 and the vertical) as a function of the weight W_2 ($W_2 \leq$ 150 lb). Determine the weight W_2 for which
(a) $\theta = 10°$ (b) $\theta = 80°$
Do you think the equilibrium positions of parts a and b are stable? (That is, if the blocks were disturbed slightly, do you think they would return to the equilibrium position or do you think they would slide off the cylindrical surface?)

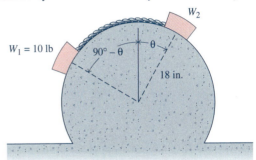

Figure P3-33

C3-34 Two small wheels are connected by a light-weight rigid rod as shown in Fig. P3-34. Plot the angle θ (between the rod and the horizontal) as a function of the weight W_1 ($W_1 \leq 10$ W_2). Determine the weight W_1 for which
(a) $\theta = -50°$ (b) $\theta = 10°$ (c) $\theta = 25°$
Do you think the equilibrium positions of parts a, b, and c are stable? (That is, if the wheels were disturbed slightly, do you think they would return to the equilibrium position or do you think they would slide off the triangular surface?)

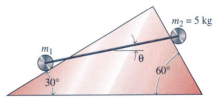

Figure P3-34

4

STRESS, STRAIN, AND DEFORMATION: AXIAL LOADING

4-1 INTRODUCTION

Application of the equations of equilibrium is usually just the first step in solving engineering problems. Using these equations, an engineer can determine the forces exerted on a structure by its supports, the forces on bolts and rivets that connect parts of a machine together, the internal forces in cables or rods that either support the structure or are a part of the structure, and so on. A second and equally important step is determining the effect of the forces on the structure or machine. It is important, therefore, that all engineers understand the behavior of materials under the action of forces.

Safety and economy in a design are two considerations for which an engineer must accept responsibility. He or she must be able to calculate the intensity of the internal forces to which each part of a machine or structure is subjected and the deformation that each part experiences during the performance of its intended function. Then, by knowing the properties of the material from which the parts will be made, the engineer establishes the most effective size and shape of the individual parts, and the appropriate means of connecting them. Problems of this type are considered in courses commonly known as "Mechanics of Materials."

In this chapter, the behavior of engineering materials subjected to uniaxial loading situations will be described. Several terms used to described the material behavior are introduced and defined. An experimental setup used to determine the material properties is described. Finally, simple problems involving axially loaded members are solved.

4-2 AXIALLY LOADED MEMBERS—INTERNAL FORCES

Axial loading is produced by two or more collinear forces acting along the axis of a long slender member, such as the eye bar shown in Fig. 4-1. This type of loading occurs in many engineering elements including struts and connecting rods of engines, and in the individual members that make up bridge and building trusses. When a structural member or machine component is subjected to a system of external loads (applied loads and support reactions), a system of internal resisting forces develops within the member to balance the external forces.

For example, consider the eye bar shown in Fig. 4-1, which is subjected to a system of balanced, external, axial forces F_1, F_2, F_3, . . . , F_n. These forces tend to either crush the bar (compression) or pull it apart (tension). In either case, internal resisting forces develop within the bar to resist the crushing or pulling apart of the bar.

The internal forces, **R**, that develop on plane a-a of Fig. 4-1 are shown in Fig. 4-2. These internal forces are the result of the mutual attraction (or repulsion) of the molecules on one side of the plane a-a for the molecules on the other side of the plane a-a and are distributed over the entire surface of the cutting plane. Since the bar is in equilibrium, both parts of the bar with plane a-a as one of its bounding surfaces must also be in equilibrium under the action of the external forces and the internal forces that develop on the plane. Thus, the resultant **R** of the internal forces on plane a-a can be determined by using either the left or right part of the bar. The intensities of these internal forces (force per unit area) are called *stresses*. In general, forces acting over the small elements of area dA which make up the total cross-sectional area A of the bar at plane a-a are not uniformly distributed; therefore, the stress distribution on plane a-a does not have to be uniform.

If the cutting plane a-a is perpendicular to the axis of the bar, as shown in Fig. 4-2, then the internal stresses, the internal forces, and the resultant of the internal forces are all perpendicular to the cutting plane (along the axis of the bar). If the cutting plane a-a cuts the bar at an arbitrary angle, as shown in Fig. 4-3, the resultant of the internal forces is still along the axis of the bar. However, experimental studies indicate that materials respond differently to forces that tend to pull surfaces apart than to forces that tend to slide surfaces relative to each other. Therefore, as shown in Fig. 4-3, the resultant **R** is usually resolved into a component R_n perpendicular to plane a-a (a normal force from which a normal stress is determined) and a component R_t tangent to plane a-a (a shear force from which a shear stress is determined).

4-2-1 Normal Stress Under Axial Loading

In the simplest qualitative terms, stress is the intensity of internal force. A body must be able to withstand the intensity of internal force, or else the body may rupture or deform excessively. Force intensity (stress) is force divided by the area over which the force is distributed. Thus,

$$\text{Stress} = \frac{\text{Force}}{\text{Area}} \qquad (4\text{-}1)$$

The forces shown in Fig. 4-4 are collinear with the axis of the eyebar and produce an axial tensile loading of the bar. When the eye bar is cut by a transverse plane, such as plane a-a of Fig. 4-4, a free-body diagram of the bottom half of the bar can be drawn as shown in Fig. 4-5. Equilibrium of this portion of the bar is obtained with a distribution of internal force that develops on the exposed cross section. This distribution of internal force has a resultant **F** that is normal to the

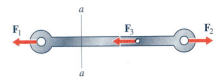

Figure 4-1

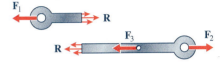

Figure 4-2

Figure 4-3

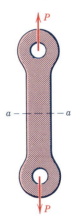

Figure 4-4

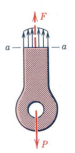

Figure 4-5

exposed surface, is equal in magnitude to **P**, and has a line of action that is collinear with the line of action of **P**. An average intensity of internal force, which is also known as the average normal stress σ_{avg} on the cross section, can be computed as:

$$\sigma_{avg} = \frac{F}{A}$$ (4-2)

where F is the magnitude of the force **F** and A is the cross-sectional area of the eye bar.

The Greek letter sigma (σ) will be used to denote a normal stress in this book. A positive sign will be used to indicate a tensile normal stress (member in tension), and a negative sign will be used to indicate a compressive normal stress (member in compression). This sign convention is independent of the selection of a coordinate system.

Consider now a small area ΔA on the exposed cross section of the bar and let ΔF represent the magnitude of the resultant of the internal forces transmitted by this small area as shown in Fig. 4-6. The average intensity of internal force being transmitted by area ΔA is obtained by dividing ΔF by ΔA. If the internal forces transmitted across the section are assumed to be continuously distributed, the area ΔA can be made smaller and smaller and will approach a point on the exposed surface in the limit. The corresponding force ΔF also becomes smaller and smaller. The stress at the point on the cross section to which ΔA converges is defined as

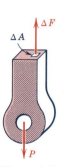

Figure 4-6

$$\sigma = \lim_{\Delta A \to 0} \frac{\Delta F}{\Delta A}$$ (4-3)

In general, the stress σ at a given point on a transverse cross section of an axially loaded bar will not be the same as the average stress computed by dividing the force F by the cross-sectional area A. For long, slender, axially loaded members such as those found in trusses and similar structures, however, it is generally assumed that the normal stresses are uniformly distributed except in the vicinity of the points of application of the loads. The subject of nonuniform stress distributions under axial loading will be discussed in a later chapter of this book.

4-2-2 Shearing Stress in Connections

Loads applied to a structure or machine are generally transmitted to the individual members through connections that use rivets, bolts, pins, nails, or welds. In all of these connections, one of the most significant stresses induced is a shearing stress. The bolted and pinned connection shown in Fig. 4-7 will be used to introduce the concept of a shearing stress.

The method by which loads are transferred from one member of the connection to another is by means of a distribution of (internal) shearing force on a transverse cross section of the bolt or pin used to effect the connection. A free-body diagram of the left member of the connection of Fig. 4-7 is shown in Fig. 4-8. In this diagram, a transverse cut has been made through the bolt, and the lower portion of the bolt remains in contact with the left member. The distribution of shearing force on the transverse cross section of the bolt has been replaced by a resultant shear force V. Since only one transverse cross section of the

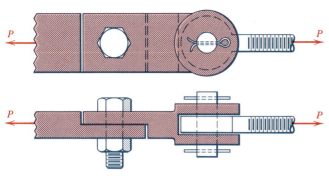

Figure 4-7

bolt is used to effect load transfer between the members, the bolt is said to be in single shear; therefore, equilibrium requires that the resultant shear force V equal the applied load P. A free-body diagram for the threaded eye bar at the right end of the connection of Fig. 4-7 is shown in Fig. 4-9. In this diagram, two transverse cuts have been made through the bolt, and the middle portion of the bolt remains in contact with the eye bar. In this case, two transverse cross sections of the pin are used to effect load transfer between members of the connection and the pin is said to be in double shear. As a result, equilibrium requires that the resultant shear force V on each cross section of the pin equals one half the applied load P.

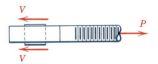

Figure 4-8

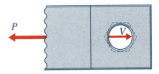

Figure 4-9

From the definition of stress given by Eq. (4-1), an average shearing stress on the transverse cross section of the bolt or pin can be computed as

$$\tau_{\text{avg}} = \frac{V}{A_s} \qquad (4\text{-}4)$$

where V is the magnitude of the shear force $\mathbf{V}$ and A_s is the cross-sectional area of the bolt or pin.

The Greek letter tau (τ) will be used to denote shearing stress in this book. A sign convention for shearing stress will be presented in a later section of the book.

The stress at a point on the transverse cross section of the bolt or pin can be obtained by using the same type of limit process that was used to obtain Eq. (4-3) for the normal stress at a point. Thus,

$$\tau = \lim_{\Delta A_s \to 0} \frac{\Delta V}{\Delta A_s} \qquad (4\text{-}5)$$

Unlike the normal stress in long slender members, it can be shown that the shear stress τ cannot be uniformly distributed over the area. Therefore, the actual shear stress at any particular point and the maximum shear stress on a cross section will generally be different than the average shear stress calculated using Eq. (4-4). However, the design of simple connections is usually based on average stress considerations and this procedure will be followed in this book.

Another type of shear loading is termed "punching shear." Examples of this type of loading include the action of a punch in forming rivet holes in a metal plate, the tendency of building columns to punch through footings, and the tendency of a tensile axial load on a bolt to pull the shank of the bolt through the head (Fig. 4-10). Under a punching shear load, the significant stress is the average shearing stress on the surface described by the periphery of the punching member and the thickness of the punched member; for example, the shaded cylindrical area $A_s = \pi dt$ shown extending through the head of the bolt in Fig. 4-10b.

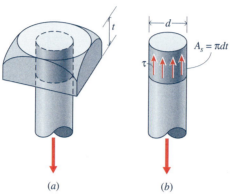

(a)   (b)

Figure 4-10

4-2-3 Bearing Stress

Bearing stresses (compressive normal stresses) occur on the surface of contact between two interacting members. For the case of the connection shown in Fig. 4-7, bearing stresses occur on the surfaces of contact between the head of the bolt and the top plate and between the nut and the bottom plate. The force producing the stress is the axial tensile internal force $\mathbf{F}$ developed in the shank of the bolt as the nut is tightened. The area of interest for bearing stress calculations is the annular area $A_b = \dfrac{\pi}{4}(d_o^2 - d_i^2)$ of the bolt head or nut (see Fig. 4-11a) that is in contact with the plate. Thus, the average bearing stress σ_b is expressed as

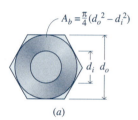

$A_b = \frac{\pi}{4}(d_o^2 - d_i^2)$

(a)

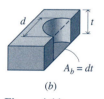

$A_b = dt$

(b)

Figure 4-11

$$\sigma_b = \frac{F}{A_b} \tag{4-6}$$

Bearing stresses also develop on surfaces of contact where the shanks of bolts and pins are pressed against the sides of the hole through which they pass. Since the distribution of these forces is quite complicated, an average bearing stress σ_b is often used for design purposes. This stress is computed by dividing the force F transmitted across the surface of contact by the projected area $A_b = dt$ shown in Fig. 4-11b, instead of the actual contact area.

Stress, being the intensity of internal force, has the dimensions of force per unit area (FL^{-2}). Until recently, the commonly used unit for stress in the United States was the pound per square inch, abbreviated as psi. Since metals can sustain stresses of several thousand pounds per square inch, the unit ksi (kip per square inch) is also frequently used (1 ksi = 1000 psi). With the advent of the International System of Units (SI units), units of stress based on the international system are beginning to be used in the United States and will undoubtedly come into wider use in the future. During the transition, both systems will be encountered by engineers; therefore, approximately one half of the example problems and homework problems in this book are given using the U.S. customary system (pounds and inches) and the other half are given in SI units (newtons and meters). For problems with SI units, forces will be given in newtons (N) or kilo-newtons (kN), dimensions in meters (m) or millimeters (mm), and masses in kilo-grams (kg). The SI unit for stress is a newton per square meter (N/m^2), also known as a pascal (Pa). Stress magnitudes normally encountered in engineering applications are expressed in meganewtons per square meter (MN/m^2), or mega-pascals (MPa).

Example Problem 4-1

A flat steel bar has axial loads applied at points A, B, C, and D as shown in Fig. 4-12a. If the bar has a cross-sectional area of 3.00 in.^{2}, determine the ax-ial stress in the bar

(a) On a cross section 20 in. to the right of point A.
(b) On a cross section 20 in. to the right of point B.
(c) On a cross section 20 in. to the right of point C.

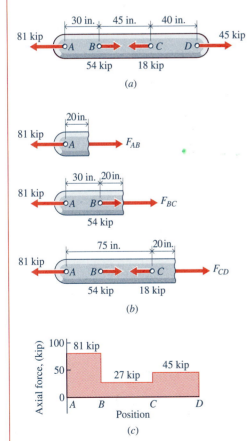

Figure 4-12

SOLUTION

The forces transmitted by cross sections in intervals AB, BC, and CD of the bar shown in Fig. 4-12a are obtained by using the free-body diagrams shown in Fig. 4-12b. Summing forces along the axis of the bar yields

$$+ \rightarrow \Sigma F = F_{AB} - 81 = 0 \qquad\qquad F_{AB} = 81 \text{ kip}$$

$$+ \rightarrow \Sigma F = F_{BC} + 54 - 81 = 0 \qquad\qquad F_{BC} = 27 \text{ kip}$$

$$+ \rightarrow \Sigma F = F_{CD} - 18 + 54 - 81 = 0 \qquad F_{CD} = 45 \text{ kip}$$

A pictorial representation of the distribution of axial force (internal) in bar $ABCD$ is shown in Fig. 4-12c. This type of representation is known as an axial-force diagram and has been shown to be useful in solving problems involving axial-force distributions. The required stresses from Eq. (4-2) are

(a) $$\sigma_{AB} = \frac{F_{AB}}{3} = \frac{81}{3} = 27.0 \text{ ksi T}$$ **Ans.**

(b) $$\sigma_{BC} = \frac{F_{BC}}{3} = \frac{27}{3} = 9.00 \text{ ksi T}$$ **Ans.**

(c) $$\sigma_{CD} = \frac{F_{CD}}{3} = \frac{45}{3} = 15.00 \text{ ksi T}$$ **Ans.**

where T indicates that the normal stress is tensile. ■

Example Problem 4-2

The round bar shown in Fig. 4-13a has steel, brass, and aluminum sections. Axial loads are applied at cross sections A, B, C, and D. If the allowable axial normal stresses are 125 MPa in the steel, 70 MPa in the brass, and 85 MPa in the aluminum, determine the diameters required for each of the sections.

SOLUTION

The forces transmitted by cross sections in intervals AB, BC, and CD of the bar shown in Fig. 4-13a are obtained by using the free-body diagrams shown in Fig. 4-13b. Summing forces along the axis of the bar yields

$$+ \rightarrow \Sigma F = F_s + 270 = 0 \qquad\qquad F_s = -270 \text{ kN} = 270 \text{ kN C}$$

$$+ \rightarrow \Sigma F = F_b + 270 - 245 = 0 \qquad\qquad F_b = -25 \text{ kN} = 25 \text{ kN C}$$

$$+ \rightarrow \Sigma F = F_a + 270 - 245 + 200 = 0 \qquad F_a = -225 \text{ kN} = 225 \text{ kN C}$$

where C indicates that the normal stress is compressive.

An axial-force diagram for bar $ABCD$ is shown in Fig. 4-13c. The cross-sectional areas of the bar required to limit the stresses to the specified values are obtained from Eq. (4-2). Thus,

(a) $$A_s = \frac{\pi}{4}d_s^2 = \frac{F_s}{\sigma_s} = \frac{270(10^3)}{125(10^6)} \qquad d_s = 52.4(10^{-3}) \text{ m} = 52.4 \text{ mm}$$ **Ans.**

(b) $$A_b = \frac{\pi}{4}d_b^2 = \frac{F_b}{\sigma_b} = \frac{25(10^3)}{70(10^6)} \qquad d_b = 21.3(10^{-3}) \text{ m} = 21.3 \text{ mm}$$ **Ans.**

(c) $$A_a = \frac{\pi}{4}d_a^2 = \frac{F_a}{\sigma_a} = \frac{225(10^3)}{85(10^6)} \qquad d_a = 58.1(10^{-3}) \text{ m} = 58.1 \text{ mm}$$ ■ **Ans.**

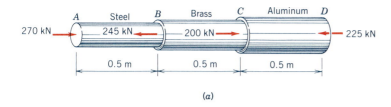

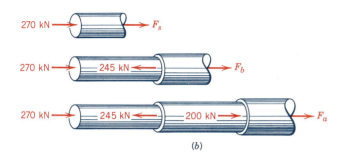

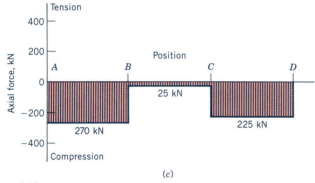

(c)

Figure 4-13

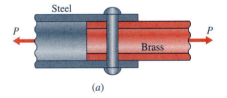

(a)

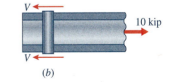

(b)

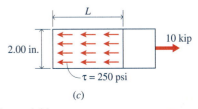

(c)

Figure 4-14

Example Problem 4-3

A brass tube with an outside diameter of 2.00 in. and a wall thickness of 0.375 in. is connected to a steel tube with an inside diameter of 2.00 in. and a wall thickness of 0.250 in. by using a 0.750 in.–diameter pin as shown in Fig. 4-14a. Determine

(a) The shearing stress in the pin when the joint is carrying an axial load of 10 kip.
(b) The length of joint required if the pin is replaced by a glued joint and the shearing stress in the glue must be limited to 250 psi.

SOLUTION

(a) A free-body diagram of the brass tube and pin is shown in Fig. 4-14b.

Since the pin is in double shear $A_s = 2A = 2\left(\dfrac{\pi}{4}\right)(0.750)^2 = 0.8836 \text{ in.}^2$.

Thus, from Eq. (4-4),

$$\tau = \frac{P}{A_s} = \frac{10}{0.8836} = 11.32 \text{ ksi} \qquad \textbf{Ans.}$$

(b) A free-body diagram of the brass tube and glued joint is shown in Fig. 4-14c. For the glued joint $A_s = \pi dL = \pi(2.00)L = 2.00\pi L$ in.^{2}. Thus, from Eq. (4-4),

$$\tau = \frac{P}{A_s} = \frac{10{,}000}{2.00\pi L} = 250$$

From which

$$L = 6.37 \text{ in.} \ \blacksquare \qquad \textbf{Ans.}$$

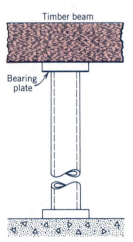

Timber beam

Bearing plate

Figure 4-15

Example Problem 4-4

The steel pipe column shown in Fig. 4-15 has an outside diameter of 150 mm and a wall thickness of 15 mm. The load imposed on the column by the timber beam is 150 kN. Determine

(a) The average bearing stress at the surface between the steel pipe column and the steel bearing plate.

(b) The diameter of a circular bearing plate if the average bearing stress between the steel bearing plate and the wood beam is not to exceed 3.25 MPa.

SOLUTION

(a) The annular area between the steel column and the bearing plate is

$$A_b = \frac{\pi}{4}(d_o^2 - d_i^2) = \frac{\pi}{4}[(150)^2 - (120)^2] = 6362 \text{ mm}^2$$

Thus, from Eq. (4-6),

$$\sigma_b = \frac{F}{A_b} = \frac{150(10^3)}{6362(10^{-6})} = 23.6(10^6) \text{ N/m}^2 = 23.6 \text{ MPa} \qquad \textbf{Ans.}$$

(b) The circular area between the bearing plate and the timber beam is

$$A_b = \frac{\pi}{4}d^2$$

Thus, from Eq. (4-6),

$$\sigma_b = \frac{F}{A_b} = \frac{150(10^3)}{(\pi/4)}d^2 = 3.25(10^6)$$

From which

$$d = 242(10^{-3}) \text{ m} = 242 \text{ mm} \ \blacksquare \qquad \textbf{Ans.}$$

PROBLEMS

Standard Problems

4-1* Axial loads are applied with rigid bearing plates to the steel pipes shown in Fig. P4-1. The cross-sectional areas are 16 in.^{2} for *AB*, 4 in.^{2} for *BC*, and 12 in.^{2} for *CD*. Determine the axial stress in the pipe
(a) On a cross section in the interval *AB*.
(b) On a cross section in the interval *BC*.
(c) On a cross section in the interval *CD*.

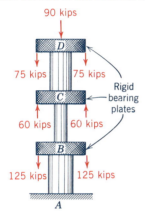

Figure P4-1

4-2* An aluminum bar is loaded and supported as shown in Fig. P4-2. If the axial stress in the bar must not exceed 150 MPa T, determine the cross-sectional areas required for each of the sections.

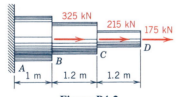

Figure P4-2

4-3 An aluminum tube with an outside diameter of 1.000 in. will be used to support a 10-kip load. If the axial stress in the member must be limited to 30 ksi T or C, determine the wall thickness required for the tube.

4-4 A stainless steel tube with an outside diameter of 50 mm and a wall thickness of 5 mm will be used as a compression member. If the axial stress in the member must be limited to 500 MPa, determine the maximum load that the member can support.

4-5* A coupling is used to connect 2 in.–diameter plastic rods to 1.5 in.–diameter rods as shown in Fig. P4-5. If the average shearing stress in the adhesive must be limited to 500 psi, determine the minimum lengths L_1 and L_2 required for the joint if the applied axial load *P* is 8000 lb.

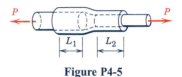

Figure P4-5

4-6* Two strips of a plastic material are bonded together as shown in Fig. P4-6. The average shearing stress in the glue must be limited to 950 kPa. What length *L* of splice plate is needed if the axial load carried by the joint is 50 kN?

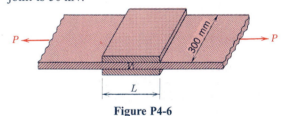

Figure P4-6

4-7 A 100-ton hydraulic punch press is used to punch holes in a 0.50 in.–thick steel plate as illustrated schematically in Fig. P4-7. If the average punching shear resistance of the steel plate is 40 ksi, determine the maximum diameter hole that can be punched.

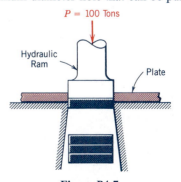

FIG. P1-9 **Figure P4-7**

4-8 A device for determining the shearing strength of wood is shown in Fig. P4-8. The dimensions of the wood specimen are 150 mm wide by 200 mm high by 50 mm thick. If a load of 75 kN is required to fail the specimen, determine the shearing strength of the wood.

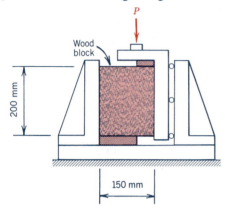

Figure P4-8

4-9* The tie rod shown in Fig. P4-9 has a diameter of 1.50 in. and is used to resist the lateral pressure against the walls of a grain bin. If the tensile stress in the rod is 10 ksi, determine

(a) The diameter of the washer that must be used if the bearing stress on the wall is not to exceed 350 psi.

(b) The thickness of the head of the rod if the punching shear stress in the head is not to exceed 5 ksi.

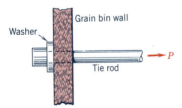

Figure P4-9

4-10* A vertical shaft is supported by a thrust collar and bearing plate, as shown in Fig. P4-10. Determine the maximum axial load that can be applied to the shaft if the average punching shear stress in the collar and the average bearing stress between the collar and the plate are limited to 75 and 100 MPa, respectively.

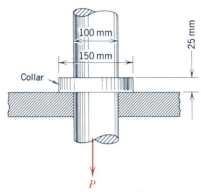

Figure P4-10

4-11 A brass tube with an outside diameter of 2.00 in. and a wall thickness of 0.375 in. is connected to a steel tube with an inside diameter of 2.00 in. and a wall thickness of 0.250 in. by using a 0.750-in. diameter pin as shown in Fig. P4-11. When the joint is carrying an axial load of 12 kip, determine

(a) The average bearing stress between the pin and the steel tube.

(b) The average bearing stress between the pin and the brass tube.

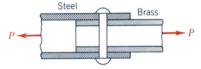

Figure P4-11

4-12 The tension member shown in Fig. P4-12 consists of steel pipe A, which has an outside diameter of 150 mm and an inside diameter of 120 mm, and a solid aluminum alloy bar B, which has an outside diameter of 100 mm. Determine the average bearing stress between the collar on bar B and the flange on pipe A.

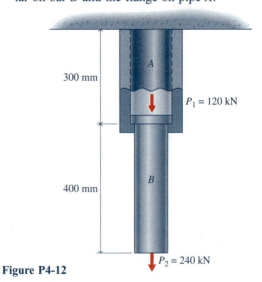

Figure P4-12

Challenging Problems

4-13* The inclined member AB of a timber truss is framed into a 4×6-in. bottom chord as shown in Fig. P4-13. Determine the axial compressive force in member AB when the average shearing stress parallel to the grain at the end of the bottom chord is 225 psi.

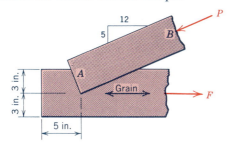

Figure P4-13

4-14* Three plates are joined with a 12 mm–diameter pin as shown in Fig. P4-14. Determine the maximum load P that can be transmitted by the joint if

(a) The maximum axial stress on a cross section at the pin must be limited to 350 MPa.
(b) The maximum bearing stress between a plate and the pin must be limited to 650 MPa.
(c) The maximum shearing stress on a cross section of the pin must be limited to 240 MPa.
(d) The punching shear resistance of the material in the top and bottom plates is 300 MPa.

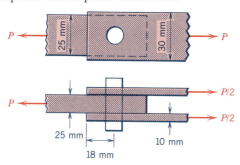

Figure P4-14

4-15 The turnbuckles in Fig. P4-15 are tightened until the compression block D exerts a force of 28,000 lb on the beam AC at B. Member D is a hollow tube with an inside diameter of 1.00 in. and an outside diameter of 2.00 in. Determine

(a) The axial stress in member D.
(b) The minimum diameter for rod AE if the axial stress is limited to 15 ksi.
(c) The minimum diameter for the pin at joint A if the shearing stress is limited to 12 ksi.

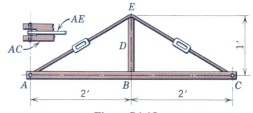

Figure P4-15

4-16 A flat steel bar 100 mm wide by 25 mm thick has axial loads applied with 40 mm–diameter pins in double shear at points A, B, C, and D as shown in Fig. P4-16. Determine

(a) The axial stress in the bar on a cross section at pin B.
(b) The average bearing stress on the bar at pin B.
(c) The shearing stress on the pin at A.

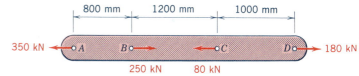

4-3 STRESSES ON AN INCLINED PLANE IN AN AXIALLY LOADED MEMBER

In Section 4-2, normal, shear, and bearing stresses for axially loaded members were introduced. Stresses on planes inclined to the axis of axially loaded bars will now be considered. When the eye bar shown in Fig. 4-4 is cut by an inclined plane, a free-body diagram of the upper portion of the bar would appear as shown in Fig. 4-16. Equilibrium of the upper portion of the bar is established by plac-

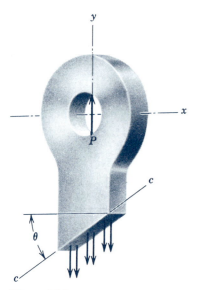

Figure 4-16

ing a distribution of internal force on the cut section as shown in Fig. 4-17. The resultant F of this distribution of internal force is equal in magnitude to the applied load P and has a line of action that is coincident with the axis of the bar as shown in Fig. 4-17. An average total stress S_{avg} on the inclined surface can be computed by using Eq. (4-2). This total stress conveys very little information that is useful for design purposes. The resultant F, however, can be replaced by normal and tangential components N and V, as shown in Fig. 4-18. These components can be used to compute normal and shear stresses on the inclined surface by using Eqs. (4-2) and (4-4). The area A_n of the inclined surface equals $A/\cos \theta$, where A is the cross-sectional area of the axially loaded member, the normal force is $N = P \cos \theta$, and the shear force is $V = P \sin \theta$. Therefore,

$$\sigma_n = \frac{N}{A_n} = \frac{P \cos \theta}{A/\cos \theta} = \frac{P}{A} \cos^2 \theta = \frac{P}{2A} (1 + \cos 2\theta) \qquad (4\text{-}7)$$

$$\tau_n = \frac{V}{A_n} = \frac{P \sin \theta}{A/\cos \theta} = \frac{P}{A} \sin \theta \cos \theta = \frac{P}{2A} \sin 2\theta \qquad (4\text{-}8)$$

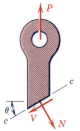

Figure 4-17

In the preceding discussion, the assumption was made that the stress is uniformly distributed over the inclined surface. Nonuniform stress distribution under axial loading will be discussed later in this book.

Both the area of the inclined surface A_n and the values for the normal and shear forces N and V on the surface depend on the angle θ of the inclined plane; therefore, the normal and shear stresses σ_n and τ_n on the inclined plane also depend on the angle θ. This dependence of stress on both force and area means that stress is not a vector quantity; therefore, the laws of vector addition do not apply to stresses that act on different planes. This need not be cause for concern if, in the application of the equations of equilibrium (or motion), one always replaces a stress with a total force (stress multiplied by the appropriate area), thus reducing the problem to one involving ordinary force vectors. However, stresses that act on a single particular plane can be treated as vectors, since they all are associated with the same area.

A graph showing the magnitudes of σ_n and τ_n as a function of θ is shown in Fig. 4-19. These results indicate that σ_n is maximum when θ is 0° or 180°, that τ_n is maximum when θ is 45° or 135°, and also that $\tau_{max} = \sigma_{max}/2$. Therefore, the maximum normal and shearing stresses for axial tensile or compressive loading are

$$\sigma_{max} = P/A \qquad (4\text{-}9)$$

$$\tau_{max} = P/2A \qquad (4\text{-}10)$$

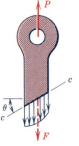

Figure 4-18

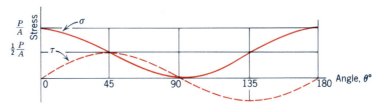

Figure 4-19

Note that the normal stress is either maximum or minimum on planes for which the shearing stress is zero. It can be shown that the shearing stress is always zero on the planes of maximum or minimum normal stress. The concepts of maximum and minimum normal stress and maximum shearing stress for more general cases will be treated in later sections of this book.

Laboratory experiments indicate that both normal and shearing stresses under axial loading are important, since a brittle material loaded in tension will fail in tension on a transverse plane, whereas a ductile material loaded in tension will fail in shear on a 45° plane.

The plot of normal and shear stresses for axial loading, shown in Fig. 4-19, indicates that the sign of the shearing stress changes when θ is greater than 90°. The magnitude of the shearing stress for any angle θ, however, is the same as that for 90° + θ. The sign change merely indicates that the shear force V changes sense, being directed down to the right on plane 90° + θ instead of down to the left on plane θ as shown in Fig. 4-18. Later in the book it will be shown that if a shearing stress exists at a point on any plane, a shearing stress of the same magnitude must also exist at this point on an orthogonal plane in order to maintain equilibrium of the body.

Example Problem 4-5

A plastic bar with a circular cross section will be used to support an axial load of 1000 lb as shown in Fig. 4-20a. The allowable normal and shearing stresses in the adhesive joint used to connect the two parts of the bar are 675 psi and 350 psi, respectively. Determine the required diameter d for the bar.

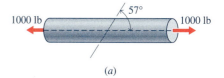

(a)

SOLUTION

A free-body diagram for the portion of the bar to the left of the joint is shown in Fig. 4-20b. From a summation of forces normal and tangent to the inclined surface,

(b)

Figure 4-20

$$+\searrow \Sigma F_n = N - 1000 \cos 33° = 0 \qquad N = 838.7 \text{ lb}$$

$$+\nearrow \Sigma F_t = V - 1000 \sin 33° = 0 \qquad V = 544.6 \text{ lb}$$

$$A = \frac{\pi}{4}d^2 \qquad A_n = A/\cos\theta = \frac{\pi}{4}d^2/\cos 33° = 0.9365d^2$$

From Eq. (4-1):

$$\sigma_n = \frac{N}{A_n} = \frac{838.7}{0.9365d^2} = 675 \qquad d = 1.1519 \text{ in.}$$

$$\tau_n = \frac{V}{A_n} = \frac{544.6}{0.9365d^2} = 350 \qquad d = 1.2890 \text{ in.}$$

Therefore,

$$d_{min} = 1.289 \text{ in.} \qquad\qquad \textbf{Ans.}$$

Alternatively, from Eqs. (4-7) and (4-8):

$$\sigma_n = \frac{P}{2A}(1 + \cos 2\theta) = \frac{1000}{2(\pi d^2/4)}(1 + \cos 66°) = 675 \qquad d = 1.1518 \text{ in.}$$

$$\tau_n = \frac{P}{2A}(\sin 2\theta) = \frac{1000}{2(\pi d^2/4)}(\sin 66°) = 350 \qquad d = 1.2891 \text{ in.}$$

Therefore,

$$d_{min} = 1.289 \text{ in.} \quad \blacksquare \qquad \text{Ans.}$$

Example Problem 4-6

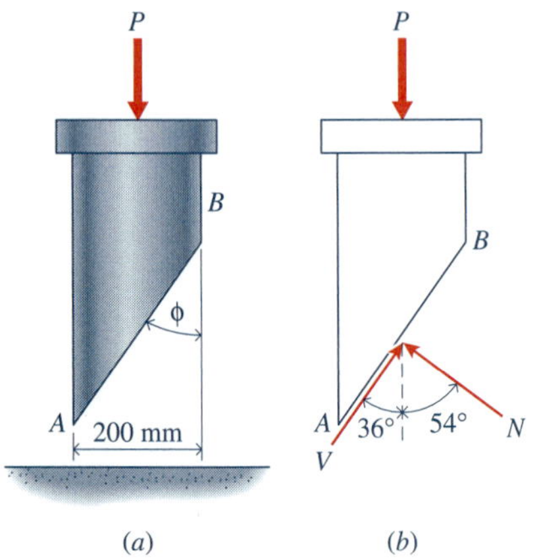

The block shown in Fig. 4-21a has a 200 × 100-mm rectangular cross section. The normal stress on plane AB is 12.00 MPa C when the load P is applied. If angle ϕ is 36°, determine

(a) The load P.

(b) The shearing stress on plane AB.

(c) The maximum normal and shearing stresses in the block.

SOLUTION

$$A = 200(100) = 20,000 \text{ mm}^2 = 0.0200 \text{ m}^2$$

$$A_n = A/\cos \theta = 0.020/\cos 54° = 0.03403 \text{ m}^2$$

(a) A free-body diagram for the portion of the bar above plane AB is shown in Fig. 4-21b. From Eq. (4-1):

$$N = \sigma_n A_n = 12(10^6)(0.03403) = 408.4(10^3) \text{ N} = 408.4 \text{ kN}$$

From a summation of forces normal to the plane,

$$+ \nwarrow \Sigma F_n = 408.4 - P \cos 54° = 0$$

$$P = 694.8 \text{ kN} \cong 695 \text{ kN C} \qquad \text{Ans.}$$

(b) From Eq. (4-8):

$$\tau_n = \frac{P}{2A}(\sin 2\theta) = \frac{694.8(10^3)}{2(0.0200)}(\sin 108°)$$

$$= 16.520(10^6) \text{ N/m}^2 = 16.52 \text{ MPa} \qquad \text{Ans.}$$

(c) From Eqs. (4-9) and (4-10):

$$\sigma_{max} = \frac{P}{A} = \frac{694.8(10^3)}{0.0200} = 34.74(10^6) \text{ N/m}^2 \cong 34.7 \text{ MPa C} \qquad \text{Ans.}$$

$$\tau_{max} = \frac{P}{2A} = \frac{694.8(10^3)}{2(0.0200)} = 17.37(10^6) \text{ N/m}^2 \cong 17.37 \text{ MPa} \quad \blacksquare \qquad \text{Ans.}$$

(a)   (b)

Figure 4-21

PROBLEMS

Standard Problems

4-17* A steel rod of circular cross section will be used to carry an axial tensile load of 50 kip. The maximum stresses in the rod must be limited to 25 ksi in tension and 15 ksi in shear. Determine the required diameter d for the rod.

4-18* A concrete cylinder 75 mm in diameter and 150 mm high failed along a plane making an angle of 57° with the horizontal when subjected to an axial vertical compressive load of 80 kN. Determine the normal and shearing stresses on the failure plane.

4-19 A structural steel eye bar with a 2 × 6-in. rectangular cross section is subjected to an axial tensile load of 270 kip. Determine
(a) The normal and shearing stresses on a plane through the bar that makes an angle of 40° with the direction of the load.
(b) The maximum normal and shearing stresses in the bar.

4-20 A structural steel bar 25 mm in diameter is subjected to an axial tensile load of 55 kN. Determine
(a) The normal and shearing stresses on a plane through the bar that makes an angle of 30° with the direction of the load.
(b) The maximum normal and shearing stresses in the bar.

4-21* Specifications for the 3 × 3 × 21-in. block shown in Fig. P4-21 require that the normal and shearing stresses on plane *A-A* not exceed 800 psi and 500 psi, respectively. Determine the maximum load P that can be applied without exceeding the specifications.

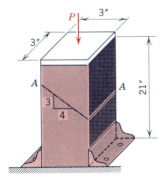

Figure P4-21

4-22* Determine the maximum axial load P that can be applied to the wood compression block shown in Fig. P4-22 if specifications require that the shearing stress parallel to the grain not exceed 5.25 MPa, the compressive stress perpendicular to the grain not exceed 13.60 MPa, and the maximum shearing stress in the block not exceed 8.75 MPa.

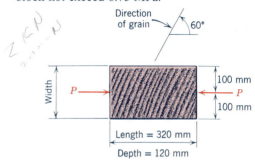

Figure P4-22

4-23 A timber block with a square cross section will be used to support a compressive load of 32 kip as shown in Fig. P4-23. Determine the size of the block required if the shearing stress parallel to the grain is not to exceed 200 psi and if the normal stress parallel to the grain is not to exceed 2000 psi.

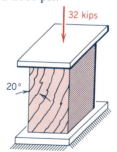

Figure P4-23

4-24 A steel bar with a butt-welded joint, as shown in Fig. P4-24, will be used to carry an axial tensile load of 400 kN. If the normal and shearing stresses on the plane of the butt weld must be limited to 70 MPa and 45 MPa, respectively, determine the minimum thickness required for the bar.

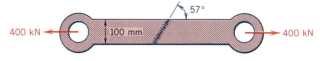

Figure P4-24

Challenging Problems

4-25* The two parts of the eye bar shown in Fig. P4-25 are connected with $\frac{1}{2}$ in.–diameter bolts (one on each side). Specifications for the bolts require that the axial tensile stress not exceed 12.0 ksi and the shearing stress not exceed 8.0 ksi. Determine the maximum load P that can be applied to the eye bar without exceeding either specification.

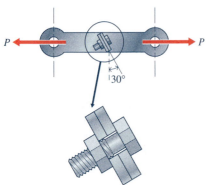

Figure P4-25

4-26* A wood tension member with a 50 × 100-mm rectangular cross section will be fabricated with an inclined glued joint ($45° \leq \phi \leq 90°$) at its midsection as shown in Fig. P4-26. If the allowable stresses for the glue are 5 MPa in tension and 3 MPa in shear, determine
(a) The optimum angle ϕ for the joint.
(b) The maximum safe load P for the member.

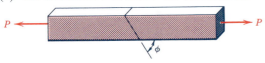

Figure P4-26

4-27 A steel eye bar with a 4 × 1-in. rectangular cross section has been designed to transmit an axial tensile load. The length of the eye bar must be increased by welding a new center section in the bar ($45° \leq \phi \leq 90°$) as shown in Fig. P4-27. The stresses in the weld material must be limited to 12,000 psi in tension and 9000 psi in shear. Determine
(a) The optimum angle ϕ for the joint.
(b) The maximum safe load P for the redesigned member.

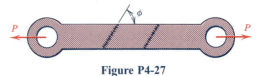

Figure P4-27

4-28 The bar shown in Fig. P4-28 has a 200 × 100-mm rectangular cross section. Determine
(a) The normal and shearing stresses on plane *a-a*.
(b) The maximum normal and shearing stresses in the bar.

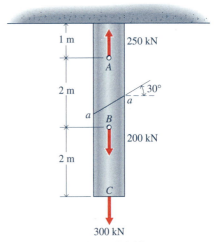

Figure P4-28

✂ 4-4 DISPLACEMENT, DEFORMATION, AND STRAIN

Relationships were developed in Sections 4-2 and 4-3 between forces and stresses and between stresses on planes having different orientations at a point using equilibrium considerations. No assumptions involving deformations or materials used in fabricating the body were made; therefore, the results are valid for an idealized rigid body or for a real deformable body. In the design of structural elements or machine components, the deformations experienced by the body, as a result of the applied loads, often represent as important a design consideration as the stresses. For this reason, the nature of the deformations experienced by a real deformable body as a result of internal force or stress distributions will be studied, and methods to measure or compute deformations will be established.

Displacement. When a system of loads is applied to a machine component or structural element, individual points of the body generally move. This movement of a point with respect to some convenient reference system of axes is a vector quantity known as a *displacement*. In some instances displacements are associated with a translation and/or rotation of the body as a whole and neither the size nor the shape of the body is changed. The study of displacements in which neither the size nor the shape of the body is changed is the concern of courses in rigid-body mechanics. When displacements induced by applied loads cause the size and/or shape of a body to be altered, individual points of the body move relative to one another. The change in any dimension associated with these relative displacements is known as a *deformation* and will be designated by the Greek letter delta (δ).

Deformation. Deformation is not uniquely related to force or stress, however. Two rods of identical material and identical cross-sectional area subjected to different loads (Fig. 4-22a), can have the same deformation if the second rod is half as long as the first. Similarly, if two rods of identical material and identical cross-sectional area are subjected to identical loads (Fig. 4-22b), the deformation in a 2 m–long rod will be twice as large as the deformation in a 1 m–long rod. Therefore, a quantitative measure of the intensity of the deformation is needed, just as stress is used to measure the intensity of an internal force (force per unit area).

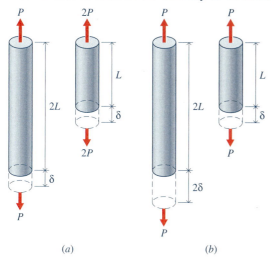

(a)   (b)

Figure 4-22

Strain. *Strain* (deformation per unit length) is the quantity used to measure the intensity of a deformation just as stress (force per unit area) is used to measure the intensity of an internal force. In Section 4-2, two types of stresses were defined: normal stresses and shearing stresses. This same classification is used for strains. *Normal strain*, designated by the Greek letter epsilon (ϵ), measures the change in size (elongation or contraction of an arbitrary line segment) of a body during deformation. *Shearing strain*, designated by the Greek letter gamma (γ), measures the change in shape (change in angle between two lines that are orthogonal in the undeformed state) of a body during deformation. The deformation or strain may be the result of a stress, of a change in temperature, or of other physical phenomena such as grain growth or shrinkage. In this book only strains resulting from changes in temperature or stress are considered.

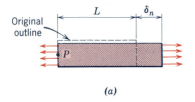

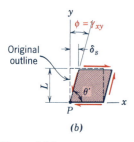

Figure 4-23

Average Axial Strain. The change in length (or width) of a simple bar under an axial load (Fig. 4-23a) can be used to illustrate the idea of a normal strain. The average axial strain (a normal strain, hereafter called axial strain) ϵ_{avg} over the length of the bar is obtained by dividing the axial deformation δ_n by the original length of the bar (the length L)

$$\epsilon_{avg} = \frac{\delta_n}{L} \qquad (4\text{-}11)$$

That is, the axial strain is the deformation δ_n in the direction of the length divided by the length L.

True Axial Strain. In cases in which the deformation is nonuniform along the length of the bar (a long bar hanging under its own weight, for example), the average axial strain given by Eq. (4-11) may be significantly different from the true axial strain at an arbitrary point P along the bar. The true axial strain at a point can be determined by making the length over which the axial deformation is measured smaller and smaller. In the limit as $\Delta L \to 0$, a quantity defined as the axial strain at the point, $\epsilon(P)$, is obtained. This limit process is indicated by the expression

$$\epsilon(P) = \lim_{\Delta L \to 0} \frac{\Delta \delta_n}{\Delta L} = \frac{d\delta_n}{dL} \qquad (4\text{-}12)$$

Shearing Strain. In a similar manner a deformation involving a change in shape can be used to illustrate a shearing strain. The average shearing strain γ_{avg} is obtained by dividing the deformation δ_s in a direction normal to the length L by the length (Fig. 4-23b)

$$\gamma_{avg} = \frac{\delta_s}{L} = \tan \phi \qquad (4\text{-}13)$$

Since δ_s/L is usually very small (typically $\delta_s/L < 0.001$), $\sin \phi \cong \tan \phi \cong \phi$ where ϕ is measured in radians. Therefore $\gamma_{avg} = \phi = \delta_s/L$ is the decrease in the angle between two reference lines that are orthogonal in the undeformed state. Again, for those cases in which the deformation is nonuniform, the shearing strain at a point, $\gamma_{xy}(P)$, associated with two orthogonal reference lines x and y, is obtained by measuring the shearing deformation as the size of the element is made smaller and smaller. In the limit as $\Delta L \to 0$

$$\gamma_{xy}(P) = \lim_{\Delta L \to 0} \frac{\Delta \delta_s}{\Delta L} = \frac{d\delta_s}{dL} \qquad (4\text{-}14a)$$

The angle γ_{xy} is difficult to observe and even more difficult to measure. An equivalent expression for shearing strain that is sometimes useful for calculations is

$$\gamma_{xy}(P) = \frac{\pi}{2} - \theta' \qquad (4\text{-}14b)$$

In this expression θ' is the angle in the deformed state between the two initially orthogonal reference lines.

Units of Strain. Equations (4-11) through (4-14) indicate that both normal and shearing strains are dimensionless quantities; however, normal strains are frequently expressed in units of in./in. or micro-in./in., while shearing strains are expressed in radians or micro-radians. The symbol μ is frequently used to indicate micro- (10^{-6}).

Tensile/Compressive Strains. From the definition of normal strain given by Eq. (4-11) or (4-12) it is evident that normal strain is positive when a line elongates and negative when the line contracts. In general, if the axial stress is tensile, the axial deformation will be an elongation. Therefore, positive normal strains are referred to as *tensile strains*. The reverse will be true for compressive axial stresses; therefore, negative normal strains are referred to as *compressive strains*. From Eq. (4-14b) it is evident that shearing strains will be positive if the angle between reference lines decreases. If the angle increases, the shearing strain is negative. Positive and negative shearing strains are not given special names. Normal and shearing strains for most engineering materials in the elastic range (see Section 4-5) seldom exceed values of 0.2 percent (0.002 in./in. or 0.002 rad).

Example Problem 4-7

A 1.00 in.–diameter steel bar is 8 ft long. The diameter is reduced to $\frac{1}{2}$ in. in a 2-ft central portion of the bar. When an axial load is applied to the ends of the bar, the axial strain in the central portion of the bar is 960 μin./in., and the total elongation of the bar is 0.04032 in. Determine

(a) The elongation of the central portion of the bar.
(b) The axial strain in the end portions of the bar.

SOLUTION

(a) The elongation of the central portion of the bar, δC, is obtained by using Eq. (4-11). Thus,

$$\delta_C = \epsilon_{avg}L = 960(10^{-6})(2)(12)$$
$$= 0.02304 \text{ in.} \cong 0.0230 \text{ in.} \qquad \textbf{Ans.}$$

(b) The elongation of the end portions of the bar, δ_E, is

$f_{total} = \delta_E + \delta_C$

$$\delta_E = \delta_{total} - \delta_C = 0.04032 - 0.02304 = 0.01728 \text{ in.}$$

The axial strain in the end portions of the bar is obtained by using Eq. (4-11):

$$\epsilon_E = \frac{\delta_E}{L} = \frac{0.01728}{6(12)} = 240(10^{-6}) = 240 \ \mu\text{in./in.} \ \blacksquare \qquad \textbf{Ans.}$$

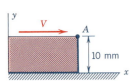

Figure 4-24

Example Problem 4-8

The shear force V shown in Fig. 4-24 produces an average shearing strain γ_{avg} of 1000 μm/m in the block of material. Determine the horizontal movement of point A resulting from application of the shear force V.

SOLUTION

The horizontal movement of point A is obtained by using Eq. (4-13). Thus,

$$\delta_A = \gamma_{avg}L = 1000(10^{-6})(10)$$
$$= 0.0100 \text{ mm} = 10.00 \ \mu\text{m} \ \blacksquare \qquad \textbf{Ans.}$$

PROBLEMS

Standard Problems

4-29* A 50-ft length of steel wire is subjected to a tensile load that produces a change in length of 1.25 in. Determine the axial strain in the wire.

4-30* Compression tests of concrete indicate that concrete fails when the axial compressive strain is 1200 μm/m. Determine the maximum change in length that a 200 mm–diameter by 400 mm–long concrete test specimen can tolerate before failure occurs.

4-31 A rigid steel plate A is supported by three rods as shown in Fig. P4-31. There is no strain in the rods before the load P is applied. After load P is applied, the axial strain in rod C is 900 μin./in. Determine
(a) The axial strain in rods B.
(b) The axial strain in rods B if there is a 0.006-in. clearance in the connections between A and B before the load is applied.

4-32 A structural steel bar was loaded in tension to fracture. A 200 mm–length of the bar was marked off in 25-mm lengths before loading. After the rod broke, the 25-mm segments were found to have lengthened to 30.0, 30.5, 31.5, 34.0, 44.5, 32.0, 31.0, and 30.0 mm, consecutively. Determine
(a) The average strain over the 200-mm length.
(b) The maximum average strain over any 50-mm length.

4-33* Mutually perpendicular axes in an unstressed member were found to be oriented at 89.92° when the member was stressed. Determine the shearing strain associated with these axes in the stressed member.

4-34* A thin triangular plate is uniformly deformed as shown in Fig. P4-34. Determine the shearing strain at P associated with the two edges (PQ and PR) that were orthogonal in the undeformed plate.

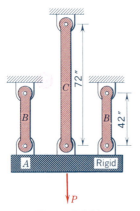

Figure P4-31

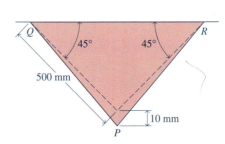

Figure P4-34

4-35 The $0.5 \times 2.0 \times 4.0$-in. rubber mounts shown in Fig. P4-35 are used to isolate the vibrational motion of a machine from its supports. Determine the average shearing strain in the rubber mounts if the rigid frame displaces 0.01 in. vertically relative to the support.

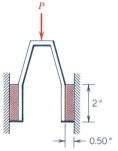

Figure P4-35

4-36 A thin rectangular plate is uniformly deformed as shown by $PRSQ$ in Fig. P4-36. Determine the shearing strain γ_{xy} at P.

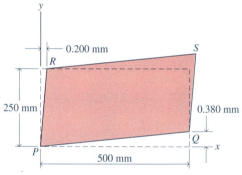

Figure P4-36

Challenging Problems

4-37* The axial strain in a vertically suspended bar of material of varying cross section due to its own weight is given by the expression $(\gamma y)/(3E)$ where γ is the specific weight of the material, y is the distance from the free (bottom) end of the bar, and E is a material constant. Determine, in terms of γ, L, and E,
(a) The change in length of the bar due to its own weight.
(b) The average axial strain over the length L of the bar.
(c) The maximum axial strain in the bar.

4-38* A steel rod is subjected to a nonuniform heating that produces an extensional (axial) strain that is proportional to the square of the distance from the unheated end ($\epsilon = kx^2$). If the strain is 1250 μm/m at the midpoint of a 3.00-m rod, determine
(a) The change in length of the rod.
(b) The average axial strain over the length L of the rod.
(c) The maximum axial strain in the rod.

4-39 A steel cable is used to support an elevator cage at the bottom of a 2000 ft–deep mine shaft. A uniform axial strain of 250 μin./in. is produced in the cable by the weight of the cage. At each point the weight of the cable produces an additional axial strain that is proportional to the length of the cable below the point. If the total axial strain in the cable at the cable drum (upper end of the cable) is 700 μin./in., determine
(a) The strain in the cable at a depth of 500 ft.
(b) The total elongation of the cable.

4-40 A steel cable is used to tether an observation balloon. The force exerted on the cable by the balloon is sufficient to produce a uniform strain of 500 μm/m in the cable. In addition, at each point in the cable, the weight of the cable reduces the axial strain by an amount that is proportional to the length of the cable between the balloon and the point. When the balloon is directly overhead at an elevation of 300 m, the axial strain at the midlength of the cable is 350 μm/m. Determine
(a) The total elongation of the cable.
(b) The maximum height that the balloon could achieve.

4-5 STRESS–STRAIN–TEMPERATURE RELATIONSHIPS

The satisfactory performance of a structure frequently is determined by the amount of deformation or distortion that can be permitted. A deformation of a few thousandths of an inch might make a boring machine useless, whereas the hook on a drag line might deform several inches without impairing its usefulness. It is often necessary to relate loads and temperature changes on a structure to the deformations produced by the loads and temperature changes. Experience has shown that the deformations caused by loads and by temperature effects are essentially independent of each other. The deformations due to the two effects may be computed separately and added together to get the total deformation.

4-5-1 Stress–Strain Diagrams

The relationship between loads and deformation in a structure can be obtained by plotting diagrams showing loads and deformations for each member and each type of loading in a structure. However, the relationship between load and deformation depends on the dimensions of the members as well as on the type of material from which the members are made. For example, the graph of Fig. 4-25a shows the relationship between the force required to stretch three bars of the same material but of different lengths and cross-sectional areas. It is not clear from this graph that these three curves all describe the same material behavior. However, if these curves are redrawn plotting stress versus deformation, as in Fig. 4-25b, the data for the first and third bars form a single line. If the curves are redrawn plotting stress versus strain as in Fig. 4-25c, the data for all three curves forms a single line.

That is, curves showing the relationship between stress and strain (such as Fig. 4-25c) are independent of the size and shape of the member and depend only on the type of material from which the members are made. Such diagrams are called *stress–strain diagrams*.

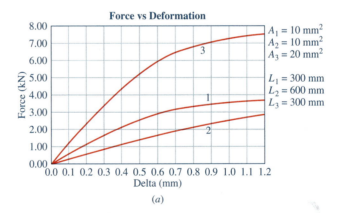

(a)

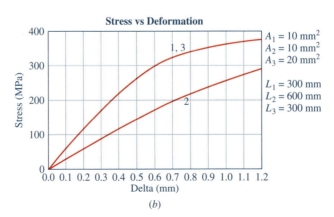

(b)

Figure 4-25

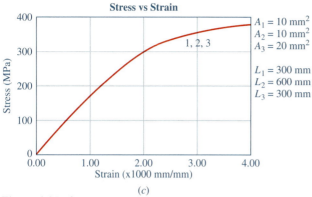

$A_1 = 10 \text{ mm}^2$
$A_2 = 10 \text{ mm}^2$
$A_3 = 20 \text{ mm}^2$

$L_1 = 300 \text{ mm}$
$L_2 = 600 \text{ mm}$
$L_3 = 300 \text{ mm}$

(c)

Figure 4-25 (Cont.)

The Tensile Test. Data for stress–strain diagrams are obtained by applying an axial load to a test specimen and measuring the load and deformation simultaneously. A testing machine (Fig. 4-26) is used to deform the specimen and to measure the load required to produce the deformation. The stress is obtained by dividing the load by the initial cross-sectional area of the specimen. The area will change somewhat during the loading, and the stress obtained using the initial area is obviously not the exact stress occurring at higher loads. However, it is the stress most commonly used in designing structures. Stress obtained by dividing the load by the actual area is frequently called the "true" stress and is useful in explaining the fundamental behavior of materials.

Figure 4-26 Hydraulic testing machine set up for a tension test (Courtesy of MTS Systems Corporation).

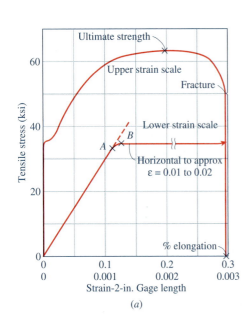

(a)

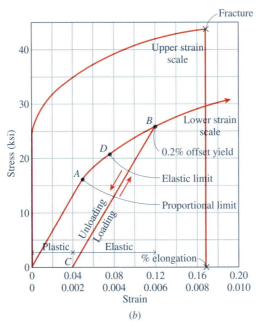

(b)

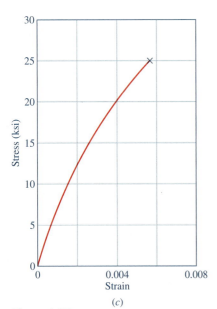

(c)

Figure 4-27

Strain Measurement. Strains are small in materials used in engineering structures, often less than 0.001, and their accurate determination requires special measuring equipment. Normal strain is obtained by measuring the deformation δ in a length L and dividing δ by L. Instruments for measuring the deformation δ are called *strain gages* or *extensometers*, and they obtain the desired accuracy by multiplying levers, dial indicators, beams of light, or other means. The electrical resistance strain gage is widely used for this type of measurement.

True strain, like true stress, is computed on the basis of the actual length of the test specimen during the test and is used primarily to study the fundamental properties of materials. The difference between nominal stress and strain, computed from initial dimensions of the specimen, and true stress and strain is negligible for stresses usually encountered in engineering structures, but sometimes the difference becomes important with larger stresses and strains. A more complete discussion of the experimental determination of stress and strain will be found in various books on experimental stress analysis.[1]

Example Stress–Strain Diagrams. Figures 4-27a, b, and c show tensile stress–strain diagrams for structural steel (a low-carbon steel), for a magnesium alloy, and for a gray cast iron, respectively. These diagrams will be used to explain a number of properties useful in the study of mechanics of materials. Although some of the relationships that follow have a basis in theory (for example, the linear relationship between stress and strain for small strain), others are purely empirical fits to experimental data. In either case, the values of specific constants for various materials must be experimentally determined.

Modulus of Elasticity. The initial portion of the stress–strain diagram for most materials used in engineering structures (see Figs. 4-27a and b) is a straight

[1]Experimental Stress Analysis, J. W. Dally and W. F. Riley, McGraw-Hill, New York, 1992.

line. The stress–strain diagrams for some materials, such as gray cast iron (see Fig. 4-27c) and concrete, show a slight curve even at very small stresses, but it is common practice to draw a straight line to average the data for the first part of the diagram and neglect the curvature. The proportionality of load to deflection was first recorded by Robert Hooke, who observed in 1678, "ut tensio sic vis" (as the stretch so the force); this is frequently referred to as Hooke's law

$$\sigma = \frac{\sigma}{\epsilon} \quad \epsilon \qquad \sigma = E\epsilon \qquad \sigma = \frac{\sigma}{\epsilon} \tag{4-15a}$$

where E is the slope of the straight-line portion of the stress–strain diagram. It is important to realize that Hooke's law describes only the initial linear portion of the stress–strain diagram and is valid only for bars loaded in uniaxial extension, as in the testing machine of Fig. 4-26. A modified version of Hooke's law valid for materials being stretched in two or three directions at the same time will be derived in Chapter 10.

Thomas Young, in 1807, suggested what amounts to using the ratio of stress to strain to measure the stiffness of a material. This ratio is called *Young's modulus* or the modulus of elasticity, and is the slope of the straight line portion of the stress–strain diagram. Young's modulus is written as

$$E = \sigma/\epsilon \qquad \text{(normal stress–strain)} \tag{4-15b}$$

$$G = \tau/\gamma \qquad \text{(shear stress–strain)} \tag{4-15c}$$

where E is the modulus used for normal stress σ and normal strain ϵ and G (sometimes called the shear modulus or the modulus of rigidity) is the modulus used for shearing stress τ and shearing strain γ. The maximum stress for which stress and strain are proportional is called the *proportional limit* and is indicated by the ordinates at points A on Fig. 4-27a or b. The exact point of the proportional limit is difficult to determine from the stress–strain curve.

For points on the stress–strain curve beyond the proportional limit (such as point C on Fig. 4-28), other quantities such as the tangent modulus and the secant modulus are used as measures of the stiffness of a material. The tangent modulus E_t is defined as the slope of the stress–strain diagram at a particular stress level. Thus, the tangent modulus is a function of the stress (or strain) for stresses greater than the proportional limit. For stresses less than the proportional limit, the tangent modulus is the same as Young's modulus. The secant modulus E_s is the ratio of the stress to the strain at any point on the diagram. Young's modulus E, the tangent modulus E_t, and the secant modulus E_s are all illustrated in Fig. 4-28.

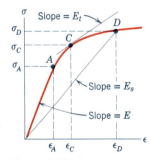

Figure 4-28

Elastic Limit. The action is said to be elastic if the strain resulting from loading disappears when the load is removed. The *elastic limit* is the maximum stress for which the material acts elastically. For most materials it is found that the stress–strain diagram for unloading (see line BC in Fig. 4-27b) is approximately parallel to the loading portion (see line OA in Fig. 4-27b). If the specimen is reloaded, the stress–strain diagram will usually follow the unloading curve until it reaches a stress a little less than the maximum stress attained during the initial loading, at which time it will start to curve in the direction of the initial loading curve. As indicated in Fig. 4-27b, the proportional limit for the second loading (point B) is greater than that for the initial loading (point A). This phenomenon is called *strain hardening* or *work hardening*.

When the stress exceeds the elastic limit (or proportional limit for practical purposes), it is found that a portion of the deformation remains after the load is removed. The deformation remaining after an applied load is removed is called *plastic deformation*. Plastic deformation independent of the time duration of the applied load is known as *slip*. *Creep* is plastic deformation that continues to increase under a constant stress. In many instances creep continues until fracture occurs; however, in other instances the rate of creep decreases and approaches zero as a limit. Some materials are much more susceptible to creep than are others, but most materials used in engineering exhibit creep at elevated temperatures. The total strain is thus made up of elastic strain, possibly combined with plastic strain that results from slip, creep, or both. When the load is removed, the elastic portion of the strain is recovered, but the plastic part (slip and creep) remains as permanent set.[2]

Yield Point. A precise value for the proportional limit is difficult to obtain when the transition of the stress–strain diagram from a straight line to a curve is gradual. For this reason, other measures of stress that can be used as a practical elastic limit are required. The yield point and the yield strength for a specified offset are used for this purpose.

The *yield point* is the stress at which there is an appreciable increase in strain with no increase in stress, with the limitation that, if straining is continued, the stress will again increase. This latter specification indicates that there is a kink or "knee" in the stress–strain diagram, as indicated in Fig. 4-27*a*. The yield point is easily determined without the aid of strain-measuring equipment because the load indicated by the testing machine ceases to rise or drops at the yield point. Unfortunately, few materials possess this property, the most common examples being low-carbon steels.

Yield Strength. The *yield strength* is defined as the stress that will induce a specified permanent set, usually 0.05 to 0.3 percent (which is equivalent to a strain of 0.0005 to 0.003), with 0.2 percent being the most commonly used value. The yield strength can be conveniently determined from a stress–strain diagram by laying off the specified offset (permanent set) on the strain axis (*OC* in Fig. 4-27*b*) and drawing a line *CB* parallel to *OA*. The stress indicated by the intersection of *CB* and the stress–strain diagram is the yield strength for the specified offset.

Ultimate Strength. The maximum stress (based on the original area) developed in a material before rupture is called the *ultimate strength* of the material (Fig. 4-27*a*), and the term may be modified as the ultimate tensile, compressive, or shearing strength of the material. Ductile materials undergo considerable plastic tensile or shearing deformation before rupture. When the ultimate strength of a ductile material is reached, the cross-sectional area of the test specimen starts to decrease or neck down (see Fig. 4-29), and the resultant load that can be carried by the specimen decreases. Thus, the stress based on the original area decreases beyond the ultimate strength of the material (Fig. 4-27*a*), although the true stress continues to increase until rupture.

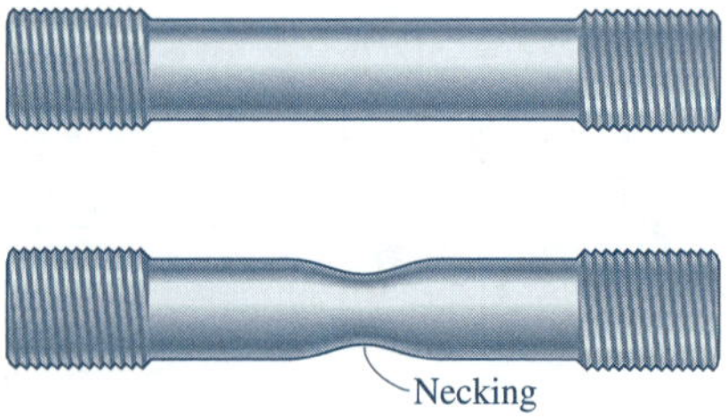

Necking

Figure 4-29

---

[2]In some instances a portion of the strain that remains immediately after the stress is removed may disappear after a period of time. This reduction of strain is sometimes called *recovery*.

Elastoplastic Materials. Most engineering structures are designed so that the stresses are less than the proportional limit; therefore, Young's modulus provides a simple and convenient relationship between stress and strain. When the stress exceeds the proportional limit, no simple relation exists between stress and strain. Various empirical equations have been proposed relating the stress and strain beyond the proportional limit. A stress–strain diagram similar to the one shown in Fig. 4-30 (elastoplastic) is frequently assumed for mild steel or other materials with similar properties in order to simplify calculations.

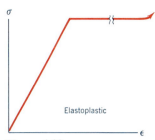

Elastoplastic

Figure 4-30

Ductility. Strength and stiffness are not the only properties of interest to a design engineer. Another important property is *ductility*, defined as the capacity for plastic deformation in tension or shear. This property controls the amount of cold forming to which a material may be subjected. The forming of automobile bodies and the manufacture of fencing and other wire products all require ductile materials. Ductility is also an important property of materials used for fabricated structures. Under static loading, the presence of large stresses in the region of rivet holes or welds may be ignored, since ductility permits considerable plastic action to take place in the region of high stress, with a resulting redistribution of stress and the establishment of equilibrium. Two commonly used quantitative indices of ductility are the ultimate elongation (expressed as a percent elongation of length at rupture) and the reduction of cross-sectional area at the section where rupture occurs (expressed as a percentage of the original area).

Creep Limit. The property indicating the resistance of a material to failure by creep is known as the *creep limit* and is defined as the maximum stress for which the plastic strain will not exceed a specified amount during a specified time interval at a specified temperature. The creep limit is important when designing parts to be fabricated with polymeric materials (commonly known as plastics) and when designing metal parts that will be subjected to high temperatures and sustained loads (for example, the turbine blades in a turbojet engine).

Poisson's Ratio. A material loaded in one direction will undergo strains perpendicular to the direction of the load in addition to those parallel to the load. The ratio of the lateral or perpendicular strain (ϵ_{lat} or ϵ_t) to the longitudinal or axial strain (ϵ_{long} or ϵ_a) is called Poisson's ratio after Siméon D. Poisson, who identified the constant in 1811. Poisson's ratio is a constant for stresses below the proportional limit and has a value between $\frac{1}{4}$ and $\frac{1}{3}$ for most metals. The symbol v is used for Poisson's ratio, which is given by the equation

$$v = -\frac{\epsilon_{\text{lat}}}{\epsilon_{\text{long}}} = -\frac{\epsilon_t}{\epsilon_a} \qquad (4\text{-}16)$$

The ratio $v = -\epsilon_t/\epsilon_a$ is valid only for a uniaxial state of stress. It will be shown later in this book that Poisson's ratio is related to E and G by the formula

$$E = 2(1 + v)G \qquad (4\text{-}17)$$

The properties discussed in this section are primarily concerned with static or continuous loading or with slowly varying loading.

4-5-2 Temperature Strain

Most engineering materials when unrestrained expand when heated and contract when cooled. The thermal strain due to a one degree (1°) change in temperature is designated by α and is known as the *coefficient of thermal expansion*. The thermal strain due to a temperature change of ΔT degrees is

$$\epsilon_T = \alpha \, \Delta T \tag{4-18}$$

Like the constants described in the last section, the value of α for various materials must be determined experimentally. The coefficient of thermal expansion is approximately constant for a large range of temperatures (in general, the coefficient increases with an increase of temperature). For a homogeneous, isotropic material,[3] the coefficient applies to all dimensions (all directions). Values of the coefficient of expansion for several materials are included in Appendix B.

Total Strains. Strains caused by temperature changes and strains caused by applied loads are essentially independent. The total normal strain in a body acted on by both temperature changes and applied loads[4] is given by

$$\epsilon_{\text{total}} = \epsilon_\sigma + \epsilon_T = \frac{\sigma}{E} + \alpha \, \Delta T \tag{4-19}$$

Since homogeneous, isotropic materials, when unrestrained, expand uniformly in all directions when heated (and contract uniformly when cooled), neither the shape of the body nor the shearing stresses and shearing strains are affected by temperature changes.

Example Problem 4-9

A 100-kip axial load is applied to a $1 \times 4 \times 90$-in. rectangular bar. When loaded, the 4-in. side measures 3.9986 in., and the length has increased 0.09 in. Determine Poisson's ratio, Young's modulus, and the modulus of rigidity of the material.

SOLUTION

The lateral and longitudinal strains and the axial stress for the bar are

$$\delta_{\text{lat}} = 3.9986 - 4 = -0.0014 \text{ in.}$$

$$\epsilon_{\text{lat}} = \frac{\delta_{\text{lat}}}{L} = \frac{-0.0014}{4} = -0.00035$$

---

[3]In a homogeneous material, material properties such as modulus of elasticity and Poisson's ratio do not vary from point to point. Examples of nonhomogeneous materials are concrete (which consists of sand and rocks held together by cement) and particle board (which consists of sawdust and wood chips held together by glue). In an isotropic material, material properties such as modulus of elasticity and Poisson's ratio are independent of direction within the material. Examples of nonisotropic materials are fiber-reinforced composites and many crystalline materials.

[4]Assuming the deformation remains in the linearly elastic range so that Hooke's law (Eq. 4-15) applies.

$$\epsilon_{long} = \frac{\delta_{long}}{L} = \frac{0.09}{90} = 0.00100$$

$$\sigma = \frac{P}{A} = \frac{100}{4(1)} = 25 \text{ ksi}$$

Poisson's ratio is obtained by using Eq. (4-16):

$$v = -\frac{\epsilon_{lat}}{\epsilon_{long}} = \frac{0.00035}{0.00100} = 0.35 \qquad \text{Ans.}$$

Young's modulus is obtained by using Eq. (4-15b):

$$E = \frac{\sigma}{\epsilon} = \frac{25}{0.00100} = 25,000 \text{ ksi} \qquad \text{Ans.}$$

The modulus of rigidity is obtained by using Eq. (4-17):

$$G = \frac{E}{2(1 + v)} = \frac{25,000}{2(1 + 0.35)} = 9260 \text{ ksi} \; \blacksquare \qquad \text{Ans.}$$

PROBLEMS

Standard Problems

4-41* At the proportional limit, an 8-in. gage length of a $\frac{1}{2}$ in.–diameter alloy bar has elongated 0.012 in., and the diameter has been reduced 0.00025 in. The total axial load carried was 4800 lb. Determine the modulus of elasticity, Poisson's ratio, and the proportional limit for the material.

4-42* At the proportional limit, a 200 mm–gage length of a 15 mm–diameter alloy bar has elongated 0.90 mm and the diameter has been reduced 0.022 mm. The total axial load carried was 62.6 kN. Determine the modulus of elasticity, Poisson's ratio, and the proportional limit for the material.

4-43 A 1.50 in.–diameter rod 20 ft long elongates 0.48 in. under a load of 53 kip. The diameter of the rod decreases 0.001 in. during the loading. Determine the modulus of elasticity, Poisson's ratio, and the modulus of rigidity for the material.

4-44 A 6 × 50-mm flat alloy bar elongates 1.22 mm in a length of 1.50 m under a total axial load of 35.0 kN. The proportional limit of the material is 300 MPa. Determine the axial stress in the bar, the modulus of elasticity of the material, and the change in the two lateral dimensions if Poisson's ratio for the material is 0.32.

Challenging Problems

4-45 A tensile test specimen having a diameter of 0.505 in. and a gage length of 2.000 in. was tested to fracture. Load and deformation data obtained during the test were as follows:

Load (lb)	Change in Length (in.)	Load (lb)	Change in Length (in.)
0	0	12,600	0.0600
2,200	0.0008	13,200	0.0800
4,300	0.0016	13,900	0.1200
6,400	0.0024	14,300	0.1600
8,200	0.0032	14,500	0.2000
8,600	0.0040	14,600	0.2400
8,800	0.0048	14,500	0.2800
9,200	0.0064	14,400	0.3200
9,500	0.0080	14,300	0.3600
9,600	0.0096	13,800	0.4000
10,600	0.0200	13,000	Fracture
11,800	0.0400		

Determine
(a) The modulus of elasticity.
(b) The proportional limit.
(c) The ultimate strength.
(d) The yield strength (0.05-percent offset).

(e) The yield strength (0.20-percent offset).
(f) The fracture stress.
(g) The true fracture stress if the final diameter of the specimen at the location of the fracture was 0.425 in.
(h) The tangent modulus at a stress level of 46,000 psi.
(i) The secant modulus at a stress level of 46,000 psi.

4-46 A tensile test specimen having a diameter of 11.28 mm and a gage length of 50 mm was tested to fracture. Load and deformation data obtained during the test were as follows:

Load (kN)	Change in Length (mm)	Load (kN)	Change in Length (mm)
0	0	43.8	1.50
7.6	0.02	45.8	2.00
14.9	0.04	48.3	3.00
22.2	0.06	49.7	4.00
28.5	0.08	50.4	5.00
29.9	0.10	50.7	6.00
30.6	0.12	50.4	7.00
32.0	0.16	50.0	8.00
33.0	0.20	49.7	9.00
33.3	0.24	47.9	10.00
36.8	0.50	45.1	Fracture
41.0	1.00		

Determine
(a) The modulus of elasticity.
(b) The proportional limit.
(c) The ultimate strength.
(d) The yield strength (0.05-percent offset).
(e) The yield strength (0.20-percent offset).
(f) The fracture stress.
(g) The true fracture stress if the final diameter of the specimen at the location of the fracture was 9.50 mm.
(h) The tangent modulus at a stress level of 315 MPa.
(i) The secant modulus at a stress level of 315 MPa.

4-6 DEFORMATION OF AXIALLY LOADED MEMBERS

Uniform Member. When a straight bar of uniform cross section is axially loaded by forces applied at the ends, the axial strain along the length of the bar is assumed to have a constant value,[5] and the elongation (or contraction) of the bar resulting from the axial load **P** may be expressed as $\delta = \epsilon L$ (by the definition of average axial strain). If Hooke's law (Eq. 4-15a) applies, the axial deformation may be expressed in terms of either stress or load as

$$\delta = \epsilon L = \frac{\sigma L}{E} \qquad (4\text{-}20a)$$

or

$$\delta = \frac{PL}{EA} \qquad (4\text{-}20b)$$

---

[5] In Section 6-3-1 it will be shown that the forces at the ends of such members must be equal in magnitude, opposite in direction, and directed along the axis of the member. Furthermore, the internal forces at any position along the member must be the same as the forces at the ends of the member and also must act along the axis of the member.

The first form will be convenient in elastic problems in which the limiting axial stress and axial deformation are both specified and either the maximum allowable load or the required size (cross-sectional area) of the member are to be determined. The stress corresponding to the specified deformation can be obtained from Eq. (4-20a) and compared to the working stress (the maximum allowable stress). The smaller of the two values can then be used to compute the allowable load or the required cross-sectional area. In general, Eq. (4-20a) is preferred when the problem involves the determination or comparison of stresses.

Multiple Loads/Sizes. Equation (4-20b), which gives the elongation (or contraction) δ occurring over some length L, applies only to uniform members for which P, A, and E are constant over the entire length L. If a bar is subjected to a number of axial loads at different points along the bar, or if the bar consists of parts having different cross-sectional areas or of parts composed of different materials (Fig. 4-31a), then the change in length of each part can be computed by using Eq. (4-20b). The changes in length of the various parts of the bar can then be added algebraically to give the total change in length of the complete bar:

$$\delta = \sum_{i=1}^{n} \delta_i = \sum_{i=1}^{n} \frac{P_i L_i}{E_i A_i} \qquad (4\text{-}21)$$

where A_i, L_i, and E_i are all constant on segment i and the force P_i is the internal force in segment i of the bar and is usually different than the forces applied at the ends of the segment. These forces must be calculated from equilibrium of the segment and are often shown on an axial force diagram such as Fig. 4-31b.

Nonuniform Deformation. For cases in which the axial force or the cross-sectional area varies continuously along the length of the bar (Fig. 4-32), Eq. (4-20b) is not valid. The axial strain at a point for the case of nonuniform deformation was defined in Section 4-4 as $\epsilon = d\delta/dL$. Thus, the increment of deformation associated with a differential element of length $dL = dx$ may be expressed as $d\delta = \epsilon\, dx$. If Hooke's law applies, the strain may be again expressed as $\epsilon = \sigma/E$, where $\sigma = P_x/A_x$. The subscripts indicate that both the applied load P_x and the cross-sectional area A_x may be functions of position x along the bar. Thus,

$$d\delta = \frac{P_x}{A_x E}\, dx \qquad (a)$$

Integrating Eq. (a) yields the following expression for the total elongation (or contraction) of the bar:

$$\delta = \int_0^L d\delta = \int_0^L \frac{P_x}{A_x E}\, dx \qquad (4\text{-}22)$$

Equation (4-22) gives acceptable results for tapered bars, provided the angle between the sides of the bar does not exceed 20°.

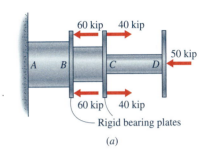

(a)

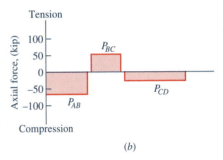

(b)

Figure 4-31

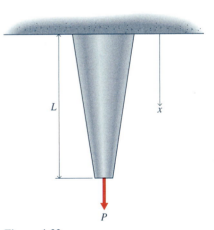

Figure 4-32

Example Problem 4-10

The compression member shown in Fig. 4-33a consists of a solid aluminum bar A, which has an outside diameter of 100 mm; a brass tube B, which has an out-

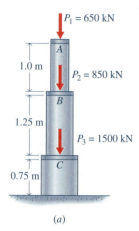

$P_1 = 650$ kN

A

1.0 m

$P_2 = 850$ kN

B

1.25 m

$P_3 = 1500$ kN

C

0.75 m

(a)

Figure 4-33

side diameter of 150 mm and an inside diameter of 100 mm; and a steel pipe C, which has an outside diameter of 200 mm and an inside diameter of 125 mm. The moduli of elasticity of the aluminum, brass, and steel are 73, 100, and 210 GPa, respectively. Determine the overall shortening of the member under the action of the indicated loads.

SOLUTION

The forces transmitted by cross sections in parts A, B, and C of the member shown in Fig. 4-33a are obtained by using the free-body diagrams shown in Fig. 4-33b. Summing forces along the axis of the bar yields

$$+\uparrow \Sigma F = -P_A - 650 = 0 \qquad\qquad P_A = -650 \text{ kN} = 650 \text{ kN C}$$

$$+\uparrow \Sigma F = -P_B - 650 - 850 = 0 \qquad P_B = -1500 \text{ kN} = 1500 \text{ kN C}$$

$$+\uparrow \Sigma F = -P_C - 650 - 850 - 1500 = 0 \quad P_C = -3000 \text{ kN} = 3000 \text{ kN C}$$

A pictorial representation of the distribution of axial, or internal force in the member is shown in Fig. 4-33c. The cross-sectional areas of the aluminum, brass, and steel are

$$A_A = \frac{\pi}{4}d^2 = \frac{\pi}{4}(100)^2 = 7854 \text{ mm}^2 = 0.007854 \text{ m}^2$$

$$A_B = \frac{\pi}{4}(d_o^2 - d_i^2) = \frac{\pi}{4}(150^2 - 100^2) = 9817 \text{ mm}^2 = 0.009817 \text{ m}^2$$

$$A_C = \frac{\pi}{4}(d_o^2 - d_i^2) = \frac{\pi}{4}(200^2 - 125^2) = 19,144 \text{ mm}^2 = 0.019144 \text{ m}^2$$

The changes in length of the different parts are obtained by using Eq. (4-20b).

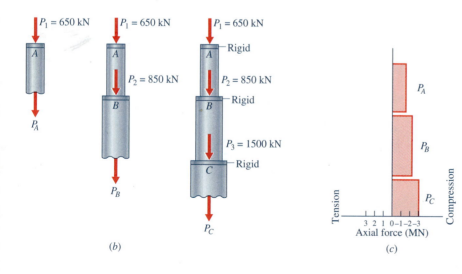

$P_1 = 650$ kN

A

P_A

$P_1 = 650$ kN

A

$P_2 = 850$ kN

B

P_B

$P_1 = 650$ kN

A — Rigid

$P_2 = 850$ kN

B — Rigid

$P_3 = 1500$ kN

C — Rigid

P_C

(b)

Tension — Compression

3 2 1 0 –1 –2 –3
Axial force (MN)

P_A

P_B

P_C

(c)

Thus,

$$\delta_A = \frac{P_A L_A}{E_A A_A} = \frac{-650(10^3)(1.0)}{0.007854(73)(10^9)} = -1.1337(10^{-3})\ \text{m} = -1.1337\ \text{mm}$$

$$\delta_B = \frac{P_B L_B}{E_B A_B} = \frac{-1500(10^3)(1.25)}{0.009817(100)(10^9)} = -1.9100(10^{-3})\ \text{m} = -1.9100\ \text{mm}$$

$$\delta_C = \frac{P_C L_C}{E_C A_C} = \frac{-3000(10^3)(0.75)}{0.019144(210)(10^9)} = -0.5597(10^{-3})\ \text{m} = -0.5597\ \text{mm}$$

The total change in length of the complete bar is given by Eq. (4-21) as

$$\delta_{\text{total}} = \delta_A + \delta_B + \delta_C$$
$$= -1.1337 - 1.9100 - 0.5597 = -3.6034\ \text{mm} \cong -3.60\ \text{mm} \quad \blacksquare \quad \textbf{Ans.}$$

PROBLEMS

Standard Problems

4-47* Specifications for a steel ($E = 30,000$ ksi) machine part require that the axial stress not exceed 20 ksi and that the total elongation not exceed 0.015 in. If the part has a 2-in. diameter and is 3 ft long, determine the maximum axial load to which it may be subjected.

4-48* A structural tension member of aluminum alloy ($E = 70$ GPa) has a rectangular cross section 25×75 mm and is 2 m long. Determine the maximum axial load that may be applied if the axial stress is not to exceed 100 MPa and the total elongation is not to exceed 4 mm.

4-49 A steel ($E = 30,000$ ksi) bar 10 ft long has a 2-in. diameter over one half its length and a 1.5-in. diameter over the other half. If a tensile load of 30 kip is applied to the bar, determine
(a) The elongation of the bar.
(b) The elongation of a bar of uniform cross section having the same weight as the bar of part (a).

4-50 A steel ($E_s = 200$ GPa) rod, which has a diameter of 25 mm and a length of 1.50 m, is attached to the end of a bronze ($E_b = 100$ GPa) tube, which has an internal diameter of 25 mm, a wall thickness of 10 mm, and a length of 3.00 m. Determine the load required to stretch the assembly 3.00 mm.

4-51* The tension member of Fig. P4-51 consists of a steel ($E_s = 30,000$ ksi) pipe A, which has an outside diameter of 6 in. and an inside diameter of 4.5 in., and a solid aluminum alloy ($E_a = 10,600$ ksi) bar B, which

has an outside diameter of 4 in. Determine the overall elongation of the member.

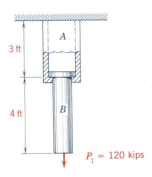

4-52* An aluminum alloy ($E = 73$ GPa) bar is loaded and supported as shown in Fig. P4-52. The diameters of the top and bottom sections of the bar are 25 mm and 15 mm, respectively. Determine
(a) The deflection of cross section a-a.
(b) The deflection of cross section b-b.

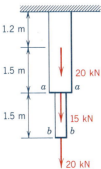

Figure P4-52

4-53* A hollow brass ($E_b = 15{,}000$ ksi) tube A with a 4-in. outside diameter and a 2-in. inside diameter is fastened to a solid 2 in.–diameter steel ($E_s = 30{,}000$ ksi) rod B as shown in Fig. P4-53. Determine
(a) The deflection of cross section a-a.
(b) The deflection of cross section b-b.

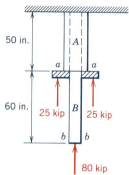

Figure P4-53

4-54 The rigid yokes B and C of Fig. P4-54 are securely fastened to the 50-mm square steel ($E = 210$ MPa) bar AD. Determine
(a) The maximum normal stress in the bar.
(b) The change in length of the complete bar.

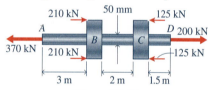

Figure P4-54

4-55 A hollow steel ($E_s = 30{,}000$ ksi) tube A with an outside diameter of 2.5 in. and an inside diameter of 2 in. is fastened to an aluminum ($E_a = 10{,}000$ ksi) bar B that has a 2-in. diameter over one half its length and a 1-in. diameter over the other half. The bar is loaded and supported as shown in Fig. P4-55. Determine
(a) The change in length of the steel tube.
(b) The overall elongation of the member.

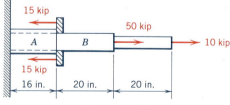

Figure P4-55

4-56 An aluminum alloy ($E_a = 73$ GPa) tube A with an outside diameter of 75 mm is used to support a 25-mm diameter steel ($E_s = 200$ GPa) rod B as shown in Fig.

P4-56. Determine the minimum thickness required for the tube if the maximum deflection of the loaded end of the rod must be limited to 0.40 mm.

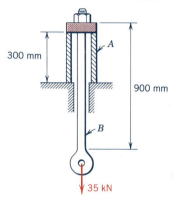

Figure P4-56

Challenging Problems

4-57* A steel ($E = 30{,}000$ ksi) bar of rectangular cross section consists of uniform and tapered sections as shown in Fig. P4-57. The width of the tapered section varies linearly from 2 in. at the bottom to 5 in. at the top. The bar has a constant thickness of $\frac{1}{2}$ in. Determine the elongation of the bar resulting from application of the 30-kip load. Neglect the weight of the bar.

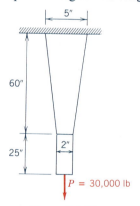

Figure P4-57

4-58 A homogeneous bar of uniform cross section hangs vertically while suspended from one end. Determine the elongation of the bar due to its own weight W in terms of W, L, A, and E.

4-59 Determine the elongation of the homogeneous conical bar of Fig. P4-59 due to its own weight. Express the results in terms of L, E, and the specific weight γ

of the material. The taper of the bar is slight enough for the assumption of a uniform axial stress distribution over a cross section to be valid.

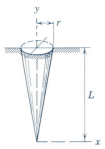

Figure P4-59

4-60 Determine the elongation, due to its own weight, of the homogeneous bar of Fig. P4-60. Express the results

in terms of L, E, and the specific weight γ of the material. The taper of the bar is slight enough for the assumption of a uniform axial stress distribution over a cross section to be valid.

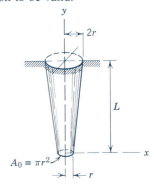

$A_0 = \pi r^2$

Figure P4-60

4-6-1 Statically Indeterminate Axially Loaded Members

If it is possible to find the internal forces in any member of a structure by means of the equations of equilibrium alone, then the structure is called *statically determinate*. If the number of unknown forces in any structure exceeds the number of independent equations of equilibrium, the structure is said to be *statically indeterminate* and it is necessary to write additional equations involving deformations to solve the problem. As an aid in solving problems, it is recommended that free-body diagrams be drawn showing forces and displacement diagrams be drawn showing deformations to assist in obtaining the correct equilibrium and deformation equations. Displacement diagrams should be simple line drawings with the deformations indicated with exaggerated magnitudes. Equilibrium equations and the corresponding deformation equations must be compatible; that is, when a tensile force is assumed for a member on the free-body diagram, a tensile deformation must be indicated for the same member on the deformation diagram. If the diagrams are compatible, a negative result will indicate that an assumption was wrong; however, the magnitude of the result will be correct.

Example Problem 4-11

A rigid plate C is used to transfer a 20-kip load P to a steel ($E_s = 30{,}000$ ksi) rod A and to an aluminum alloy ($E_a = 10{,}000$ ksi) pipe B, as shown in Fig. 4-34a. The supports at the top of the rod and bottom of the pipe are rigid and there are no stresses in the rod or pipe before the load P is applied. The cross-sectional areas of rod A and pipe B are 0.800 in.^{2} and 3.00 in.^{2}, respectively. Determine

(a) The axial stresses in rod A and pipe B.
(b) The displacement of plate C.

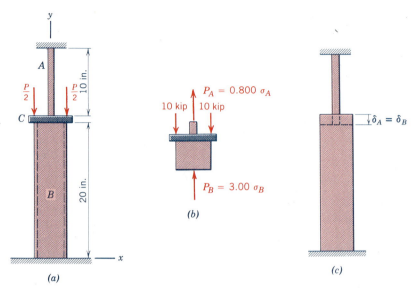

Figure 4-34

SOLUTION

(a) A free-body diagram of plate C and portions of rod A and pipe B is shown in Fig. 4-34b. The free-body diagram contains two unknown forces, P_A and P_B. Since only one equation of equilibrium, $\Sigma F_y = 0$, is available, the problem is statically indeterminate. The additional equation needed to obtain a solution to the problem is obtained from the deformation diagram shown in Fig. 4-34c. As the load P is applied to plate C, it moves downward an amount δ, which represents the magnitude of the deflection experienced by both rod A and pipe B. The relationship between stress and deflection for axial loading is given by Eq. (4-20a). Thus, the two equations needed to solve the problem are as follows:

Equilibrium equation:

$$+\uparrow \Sigma F_y = P_A + P_B - 20 = 0$$

Or in terms of stresses:

$$0.800\ \sigma_A + 3.00\ \sigma_B = 20 \qquad (a)$$

Deformation equation:

$$\delta_A = \delta_B = \frac{\sigma_A L_A}{E_A} = \frac{\sigma_B L_B}{E_B} = \frac{\sigma_A(10)}{30,000} = \frac{\sigma_B(20)}{10,000}$$

from which

$$\sigma_A = 6\sigma_B \qquad (b)$$

Solving Eqs. (a) and (b) simultaneously yields

$$\sigma_A = 15.38 \text{ ksi T} \quad \text{and} \quad \sigma_B = 2.56 \text{ ksi C} \qquad \textbf{Ans.}$$

(b) The displacement of plate C is the same as the magnitude of the deflection of rod A or pipe B. Thus,

$$\delta_C = \delta_A = \delta_B = \frac{\sigma_A(10)}{30,000}$$

$$= \frac{15.38(10)}{30,000} = 0.00513 \text{ in.} \quad \blacksquare \qquad \textbf{Ans.}$$

PROBLEMS

Standard Problems

4-61* Nine 1 in.–diameter steel ($E_s = 30{,}000$ ksi) reinforcing bars are used in the short concrete ($E_c = 3000$ ksi) pier shown in Fig. P4-61. A load P of 200 kip is applied to the pier through a rigid capping plate. Determine
(a) The stresses in the concrete and in the steel parts.
(b) The shortening of the pier.

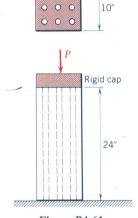

Figure P4-61

4-62* The $150 \times 150 \times 500$-mm oak ($E_w = 12$ GPa) block shown in Fig. P4-62 was reinforced by bolting two $50 \times 150 \times 500$-mm steel ($E_s = 200$ GPa) plates to opposite sides of the block. If stresses in the wood and steel are to be limited to 32 MPa and 150 MPa, respectively, determine
(a) The maximum compressive load that can be applied to the reinforced block.
(b) The shortening of the block when the load of part (a) is applied.

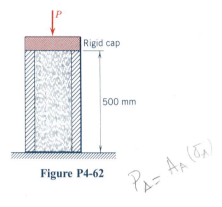

Figure P4-62

$P_A = A_A (\sigma_A)$

4-63 The steel ($E = 30{,}000$ ksi) pipe shown in Fig. P4-63 has a cross-sectional area of 3.00 in.^{2}. When the loading yoke on the pipe is subjected to the two 100-kip loads, determine
(a) The axial stresses in the pipe above and below the loading yoke.
(b) The vertical displacement of the loading yoke.

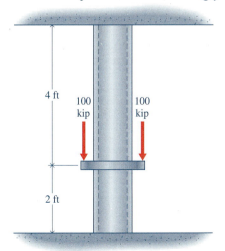

Figure P4-63

4-64 A 3 mm–diameter cord (E_c = 7 GPa) that is covered with a 0.5 mm–thick plastic sheath (E_s = 14 GPa) is subjected to an axial tensile load. The elastic strengths for the cord and sheath are 15 MPa and 56 MPa, respectively. Determine the maximum allowable load.

4-65* The assembly shown in Fig. P4-65 consists of a steel (E_s = 30,000 ksi and A_s = 1.25 in.²) bar A, a rigid bearing plate C that is securely fastened to bar A, and a bronze (E_b = 15,000 ksi and A_b = 3.75 in.²) bar B. The elastic strengths of the steel and bronze are 62 ksi and 75 ksi, respectively. A clearance of 0.015 in. exists between the bearing plate C and bar B before the assembly is loaded. After a load of 95 kip is applied to the bearing plate, determine
(a) The axial stresses in bars A and B.
(b) The vertical displacement of bearing plate C.

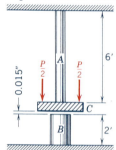

Figure P4-65

4-66* A 150 mm–diameter by 200 mm–long polymer (E_p = 2.10 GPa) cylinder is attached to a 45 mm–diameter by 400 mm–long brass (E_b = 100 GPa) rod by using the flange type of connection shown in Fig. P4-66. A 0.15-mm clearance exists between the parts as a result of a machining error. If the bolts are inserted and tightened, determine
(a) The axial stresses produced in each of the members.
(b) The final position of the flange–polymer interface after assembly, with respect to the left support.

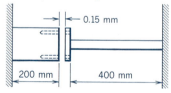

Figure P4-66

4-67 A hollow steel (E_s = 30,000 ksi) tube A with an outside diameter of 2.5 in. and an inside diameter of 2 in. is fastened to an aluminum (E_a = 10,000 ksi) bar B that has a 2-in. diameter over one half its length and a 1.5-in. diameter over the other half as shown in Fig. P4-67.

The assembly is attached to rigid supports at the left and right ends. Determine
(a) The stresses in all parts of the bar.
(b) The deflection of cross-section a-a.

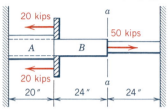

Figure P4-67

4-68 The assembly shown in Fig. P4-68 consists of a steel (E_s = 210 GPa) cylinder A, a rigid bearing plate C, and an aluminum alloy (E_a = 71 GPa) bar B. The elastic strengths of the steel and aluminum are 430 MPa and 170 MPa, respectively. An axial load P of 1200 kN is applied. Determine
(a) The minimum cross-sectional area for cylinder A for elastic action if bar B has a cross-sectional area of 2500 mm².
(b) The displacement of plate C when the area of part (a) is used.

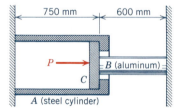

Figure P4-68

Challenging Problems

4-69 A high-strength steel (E_s = 30,000 ksi) bolt passes through a brass (E_b = 15,000 ksi) sleeve as shown in Fig. P4-69. After the unit is assembled, the nut is tightened one-quarter turn (moves 0.03125 in.). The cross-sectional areas are 0.785 in.² for the bolt and 1.767 in.² for the sleeve. Determine
(a) The axial stresses in the bolt and in the sleeve.
(b) The change in length of the brass sleeve.

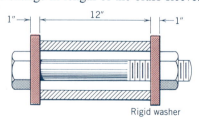

Rigid washer

Figure P4-69

4-70 The two faces of the clamp shown in Fig. P4-70 are 250 mm apart when the two stainless-steel ($E_s = 190$ GPa) bolts connecting them are unstretched. A force P is applied to separate the faces of the clamp so that an aluminum alloy ($E_a = 73$ GPa) bar with a length of 251 mm can be inserted as shown. Each of the bolts has a cross-sectional area of 120 mm^2 and the bar has a cross-sectional area of 625 mm^2. After the load P is removed, determine

(a) The axial stresses in the bolts and in the bar.
(b) The change in length of the aluminum alloy bar.

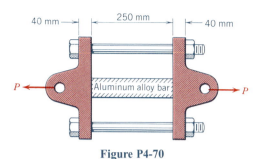

Figure P4-70

4-6-2 Temperature Effects

Thermal Stresses. When a temperature change takes place while a member is restrained (free movement restricted or prevented), stresses (referred to as "thermal stresses") are induced in the member. For example, the bar AB of Fig. 4-35a is securely fastened to rigid supports at the ends and is subjected to a temperature change. Since the ends of the bar are fixed, the total deformation of the bar must be zero.

$$\delta_{total} = \delta_T + \delta_\sigma = \epsilon_T L + \epsilon_\sigma L$$

$$0 = \alpha\,\Delta T\,L + \frac{\sigma}{E}L$$

If the temperature of the bar increases (ΔT positive), then the induced stress must be negative and the wall must push on the ends of the rod. If the temperature of the bar decreases (ΔT negative), then the induced stress must be positive and the wall must pull on the ends of the rod.

That is, if end B were not attached to the wall and the temperature drops, end B would move to B', a distance $|\delta_T| = |\epsilon_T L| = |\alpha\,\Delta T\,L|$, as indicated in Fig. 4-35b. Therefore, for the total deformation of the bar to be zero, the wall at B must apply a force $P = \sigma A$ (Fig. 4-35c) of sufficient magnitude to move end B

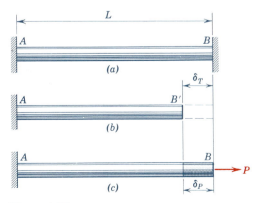

Figure 4-35

through a distance $\delta_P = \epsilon_\sigma L = (\sigma/E)L$ so that the length of the bar is again L, the distance between the walls. Since the walls do not move, $|\delta_T| = \delta_P$, or

$$\delta_P - |\delta_T| = \delta_P + \delta_T = 0$$

and thus the total deformation of the bar is zero.

Example Problem 4-12

A 10-m section of steel [$E = 200$ GPa and $\alpha = 11.9(10^{-6})/°C$] rail has a cross-sectional area of 7500 mm². Both ends of the rail are tight against adjacent rails that, for this problem, can be assumed to be rigid. The rail is supported against lateral movement. For an increase in temperature of 50°C, determine

(a) The normal stress in the rail.

(b) The internal force on a cross section of the rail.

SOLUTION

(a) The change in length of the rail resulting from the temperature change is given by modifying Eq. (4-18) as

$$\delta = \epsilon_T L = \alpha L \, \Delta T = 11.9(10^{-6})(10)(50) = 5.95(10^{-3}) \text{ m} = 5.95 \text{ mm}$$

The stress required to resist a change in length of 5.95 mm is given by Eq. (4-20a) as

$$\sigma = \frac{E\delta}{L} = \frac{200(10^9)(5.95)(10^{-3})}{10} = 119.0(10^6) \text{ N/m}^2 = 119.0 \text{ MPa}$$

(b) The internal force on a cross section of the rail is

$$F = \sigma A = 119.0(10^6)(7500)(10^{-6}) = 892.5(10^3) \text{ N} \cong 893 \text{ kN} \quad \blacksquare$$

PROBLEMS

Standard Problems

4-71* A bridge has a span of 4800 ft. Determine the change in length of a steel [$E = 30{,}000$ ksi and $\alpha = 6.5(10^{-6})/°F$] longitudinal floor beam (which must be provided for with expansion joints) due to a seasonal temperature variation from $-30°F$ to $100°F$.

4-72* An airplane has a wing span of 40 m. Determine the change in length of the aluminum alloy [$E = 73$ GPa and $\alpha = 22.5(10^{-6})/°C$] wing spar if the plane leaves the ground at a temperature of 40°C and climbs to an altitude where the temperature is $-40°C$.

4-73 A steel surveyor's tape $\frac{1}{2}$-in. wide by $\frac{1}{64}$-in. thick is exactly 100 ft long at 70°F and under a pull of 10 lb. What correction should be introduced if the tape is used to make a 100-ft measurement at a temperature of 100°F and under a pull of 20 lb? Use $E = 30{,}000$ ksi and $\alpha = 6.5(10^{-6})/°F$.

4-74 A 25-mm diameter aluminum [$\alpha = 22.5(10^{-6})/°C$, $E = 73$ GPa, and $v = 0.33$] rod hangs vertically while suspended from one end. A 2500-kg mass is attached at the other end. After the load is applied, the temperature decreases 50°C. Determine

(a) The axial stress in the rod.
(b) The axial strain in the rod.
(c) The change in diameter of the rod.

4-75* Determine the movement of the pointer of Fig. P4-75 with respect to the scale 0 when the temperature increases 100°F. The coefficients of thermal expansion are $6.6(10^{-6})/°F$ for the steel and $12.5(10^{-6})/°F$ for the aluminum.

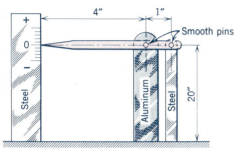

Figure P4-75

4-76* Determine the horizontal movement of point A in Fig. P4-76 due to a temperature drop of 45°C. Assume that member AE has an insignificant coefficient of thermal expansion. The coefficients of thermal expansion are $11.9(10^{-6})/°C$ for the steel and $22.5(10^{-6})/°C$ for the aluminum alloy.

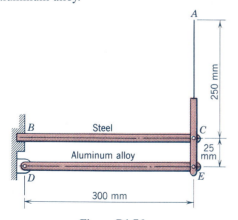

Figure P4-76

4-77 A 3 in.–diameter by 80 in.–long aluminum alloy [$E = 10,600$ ksi and $\alpha = 12.5(10^{-6})/°F$] bar is stress free when it is attached to rigid supports as shown in Fig. P4-77. Determine the stress in the bar after the temperature drops 100°F.

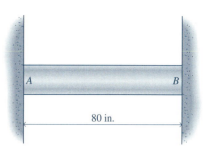

Figure P4-77

4-78 A steel tie rod containing a turnbuckle has its ends attached to rigid walls. During the summer when the temperature is 30°C, the turnbuckle is tightened to produce a stress in the rod of 15 MPa. Determine the stress in the rod in the winter when the temperature is −10°C. Use $\alpha = 11.9(10^{-6})/°C$ and $E = 200$ GPa.

4-79* A bar consists of 3 in.–diameter aluminum alloy [$E = 10,600$ ksi and $\alpha = 12.5(10^{-6})/°F$] and 4 in.–diameter steel [$E = 30,000$ ksi and $\alpha = 6.6(10^{-6})/°F$] parts as shown in Fig. P4-79. If end supports are rigid and the bar is stress free at 0°F, determine
(a) The axial stress in both parts of the bar at 80°F.
(b) The change in length of the steel part of the bar resulting from the 80°F temperature increase.

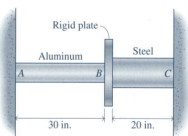

Figure P4-79

4-80 A 6 m–long by 50 mm–diameter rod of aluminum alloy ($E = 70$ GPa, $v = 0.346$, and $\alpha = 22.5(10^{-6})/°C$) is attached at its ends to supports that yield to permit a change in length of 1.00 mm in the rod when stressed. When the temperature is 35°C, there is no stress in the rod. After the temperature of the rod drops to −20°C, determine
(a) The normal stress in the rod.
(b) The change in diameter of the rod.

Challenging Problems

4-81* A load P will be supported by a structure consisting of a rigid bar A, two aluminum alloy bars B ($E = 10,600$ ksi), and a stainless steel bar C ($E = 28,000$ ksi), as shown in Fig. P4-81. The bars are unstressed when the structure is assembled at room temperature (72°F). Each bar has a cross-sectional area of 2.00 in.^{2}. The coefficients of thermal expansion are $12.5(10^{-6})$/°F for the aluminum and $9.6(10^{-6})$/°F for the stainless steel. Determine the axial stresses in the bars after a 40-kip load is applied and the temperature is increased to 250°F.

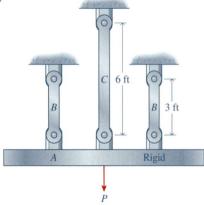

Figure P4-81

4-82* A prismatic bar $[E = 70\ \text{GPa}$ and $\alpha = 22.5(10^{-6})$/°C], free of stress at room temperature, is fastened to rigid walls at its ends. One end of the bar is heated to 100°C above room temperature while the other end is maintained at room temperature. The change in temperature ΔT along the bar is proportional to the square of the distance from the unheated end. Determine the axial stress in the bar after the change in temperature.

4-83 A high-strength steel bolt passes through a brass sleeve as shown in Fig. P4-69 (p. 104). After the unit is assembled at 40°F, the temperature is increased to 100°F. If the unit is free of stress at 40°F, determine the axial stresses in the bolt and in the sleeve at 100°F. The following data apply:

Steel: $[E_s = 30{,}000\ \text{ksi}, A_s = 0.785\ \text{in.}^2$, and
$\alpha_s = 6.6(10^{-6})$/°F]

Brass: $[E_b = 15{,}000\ \text{ksi}, A_b = 1.767\ \text{in.}^2$, and
$\alpha_b = 9.8(10^{-6})$/°F]

4-84 The two faces of the clamp of Fig. P4-70 (p. 105) are 250 mm apart when the two stainless-steel bolts connecting them are unstretched. A force P is applied to separate the faces of the clamp so that an aluminum alloy bar with a length of 250.50 mm can be inserted as shown. After the load P is removed, the temperature is raised 100°C. Determine the axial stresses in the bolts and in the bar, and the distance between the faces of the clamps. The following data apply:

Steel: $[E_s = 190\ \text{GPa}, A_s = 115\ \text{mm}^2$ (each), and
$\alpha_s = 17.3(10^{-6})$/°C]

Aluminum: $[E_a = 73\ \text{GPa}, A_a = 625\ \text{mm}^2$, and
$\alpha_a = 22.5(10^{-6})$/°C]

4-7 DESIGN LOADS, WORKING STRESSES, AND FACTOR OF SAFETY

As mentioned earlier, a designer is required to select a material and properly proportion a member to perform a specified function without failure. Failure is defined as the state or condition in which a member or structure no longer functions as intended. Several types of failure warrant discussion at this point. *Elastic failure* occurs as a result of excessive elastic deformation—a bridge may deflect elastically under traffic to an extent that may result in discomfort to vehicle passengers. Similarly, a close-fitting machine part may deform sufficiently under load to prevent proper operation of the machine. When a structure is designed to avoid elastic failure, the stiffness of the material, indicated by Young's modulus, is the significant property. *Slip failure* is characterized by excessive plastic deformation. Yield strength, yield point, and proportional limit are used as indices of strength with respect to failure by slip for members subjected to static loads. *Creep failure* occurs as a result of excessive plastic deformation over a long time under constant stress. For machines or structures that are to be subjected to relatively high stress, high temperature, or both over a long time, creep is a major design consideration, and the creep limit is the strength index to be used. The creep limit normally decreases as the temperature increases. *Failure by fracture* is a complete separation of the material. The ultimate strength of a material is the index of resistance to failure by fracture under static loads in which creep is not involved.

Most design problems involve many unknown variables. The load that the structure or machine must carry is usually estimated. The actual load may vary considerably from the estimate, especially when loads at some future time must be considered. Since testing usually damages a material, the properties of a material used in a structure cannot be evaluated directly but are normally determined by testing specimens of a similar material. Furthermore, the actual stresses that will exist in a structure are unknown because the calculations are based on assumptions about the distribution of stresses in the material. Because of these and other unknown variables, it is customary to design for the load required to produce failure, which is larger than the estimated actual load, or to use a working (or design) stress below the stress required to produce failure.

A *working stress* or *allowable stress* is defined as the maximum stress permitted in the design computation. The *factor of safety* may be defined as the ratio of a failure-producing load to the estimated actual load. Factor of safety may also be defined as the ratio of the strength of a material to the maximum computed stress in the material. The latter factor of safety may be used to determine an allowable (working) stress by dividing the strength of a material by the factor of safety.

In this book, the term "factor of safety" will denote the ratio of the strength of the material to the maximum computed stress. For a given design, the factor of safety based on the ultimate strength is, in general, quite different from that based on the elastic strength (proportional limit, yield point, or yield strength); hence, the term "strength" is ambiguous without a qualifying statement regarding the pertinent mode of failure. The following example illustrates the use of a factor of safety.

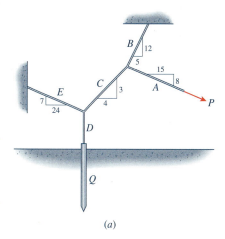

(a)

Example Problem 4-13

The improvised system of cables in Fig. 4-36a is used to remove the pile Q. A tractor with a maximum drawbar pull of 16 kip is available to apply the force P. Cable A is a 1 in.–diameter iron hoisting cable that has a breaking load of 29.0 kip. Cables B, C, and D are parts of a continuous $\frac{3}{4}$ in.–diameter carbon steel aircraft cable with a breaking strength of 49.6 kip. Cable E is a $\frac{7}{8}$ in.–diameter steel hoisting cable with a nominal area of 0.60 in.^{2} and an ultimate tensile strength, based on the nominal area, of 71.6 ksi. The minimum factor of safety for each cable, based on fracture, is to be 2.0. Determine

(a) The maximum allowable force that can be exerted on the pile.
(b) The actual factor of safety for each of the five cables when the load in part (a) is applied.

SOLUTION

(a) The two free-body diagrams needed to solve the problem are shown in Fig. 4-36b and c. The maximum allowable tension in each cable can be obtained from the strength of the cable and the factor of safety, because for axial loading the factor of safety based on load is the same as that based on stress.

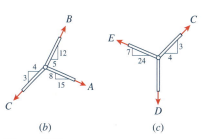

(b)

(c)

Figure 4-36

The allowable tensions are

$$A = 29.0/2 = 14.50 \text{ kip}$$
$$B = C = D = 49.6/2 = 24.8 \text{ kip}$$
$$E = 0.60(71.6)/2 = 43.0/2 = 21.5 \text{ kip}$$

The maximum tractor pull (16.0 kip) is greater than the allowable force in cable A; therefore, the maximum tractor pull cannot be used without exceeding the specifications. Assuming that the force in A is 14.50 kip, the equations of equilibrium applied to Fig. 4-36b give

$$\frac{5}{13}B + \frac{15}{17}(14.50) - \frac{4}{5}C = 0$$

$$\frac{12}{13}B - \frac{8}{17}(14.50) - \frac{3}{5}C = 0$$

from which $B = 25.9$ kip and $C = 28.4$ kip. Since the maximum allowable load on cables B and C is only 24.8 kip, it is apparent that the force in A must be less than the allowable load. The force in B is less than that in C; therefore, cable C is the controlling member of Fig. 4-36b. The equations of equilibrium for this free-body diagram with $C = 24.8$ kip give $A = 12.65$ kip and $B = 22.6$ kip.

The equations of equilibrium for Fig. 4-36c are

$$\frac{4}{5}C - \frac{24}{25}E = 0 \quad \text{and} \quad \frac{3}{5}C + \frac{7}{25}E - D = 0$$

When $C = 24.8$ kip is substituted in these equations, the results are

$$D = E = 20.7 \text{ kip}$$

Both of these forces are less than the maximum allowable loads in the corresponding cables and are therefore safe loads. The strength of cable C is thus the limiting factor. The maximum pull on the pile is

$$D = 20.7 \text{ kip} \qquad \qquad \textbf{Ans.}$$

(b) The actual factor of safety, *FS,* in each of the cables is

$$(FS)_A = 29.0/12.65 = 2.29 \qquad \textbf{Ans.}$$
$$(FS)_B = 49.6/22.6 = 2.19 \qquad \textbf{Ans.}$$
$$(FS)_C = 49.6/24.8 = 2.00 \qquad \textbf{Ans.}$$
$$(FS)_D = 49.6/20.7 = 2.40 \qquad \textbf{Ans.}$$
$$(FS)_E = 43.0/20.7 = 2.08 \qquad \textbf{Ans.}$$

As noted previously, failure may be due to excessive elastic deformation. To guard against such failures, maximum deformations may be specified, and for many designs the limiting specification is the maximum deformation rather than the maximum stress. ■

PROBLEMS

Standard Problems

4-85* A 1 in.-diameter structural steel (see Appendix B for properties) bar is supporting an axial tensile load of 8 kip. Determine
(a) The factor of safety with respect to failure by slip.
(b) The factor of safety with respect to failure by fracture.

4-86* A structural steel (see Appendix B for properties) bar with a 25×50-mm rectangular cross section is supporting an axial tensile load of 125 kN. Determine
(a) The factor of safety with respect to failure by slip.
(b) The factor of safety with respect to failure by fracture.

4-87 A 10,000-lb load is supported by two tie rods as shown in Fig. P4-87. Rod AC is made of annealed red brass (see Appendix B for properties), has a cross-sectional area of 1.00 in.^{2}, and has a working stress specification of 10 ksi. Rod BC is made of annealed bronze, has a cross sectional area of 1.25 in.^{2}, and has a working stress specification of 14 ksi. Determine
(a) The factor of safety with respect to failure by slip for each of the rods based on the working stress specifications.
(b) The actual factor of safety with respect to failure by fracture for each of the rods.

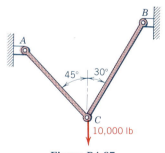

Figure P4-87

4-88 Two aluminum alloy plates are connected by using a lap joint with 3 aluminum alloy rivets. The yield and ultimate strengths of the rivets in shear are 190 MPa and 280 MPa, respectively. The plate will be used to transmit an axial tensile load of 100 kN. Determine the minimum diameter for the rivets if

(a) A factor of safety of 2.5 with respect to failure by slip is specified.
(b) A factor of safety of 2 with respect to failure by fracture is specified.

4-89* The load W of Fig. P4-89 is supported by two cables. Cable A has a breaking strength of 3.5 kip. Cable B has a breaking strength of 3 kip. Determine the maximum safe load that can be supported if a factor of safety of 2.5 with respect to failure by fracture is specified.

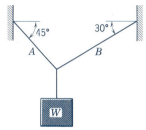

Figure P4-89

Challenging Problems

4-90 The load P of Fig. P4-90 is supported by a continuous, carbon-steel aircraft cable having a length L and a breaking strength of 36 kN. Determine
(a) The equilibrium position of load P (distance a in terms of cable length L).
(b) The maximum load P that can be supported if a factor of safety of 2 with respect to failure by fracture is specified.

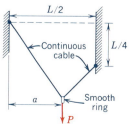

Figure P4-90

4-8 DESIGN

According to the Accreditation Board for Engineering and Technology[6] (ABET)

> Engineering design is the process of devising a system, component, or process to meet desired needs. It is a decision-making process (often iterative), in which the basic sciences and mathematics and engineering sciences are applied to convert resources optimally to meet a stated objective. Among the fundamental elements of the design process are the establishment of objectives and criteria, synthesis, analysis, construction, testing, and evaluation . . . .

The ABET definition of design is comprehensive, and is beyond the scope of this book. Thus, in the following example and homework problems, design is limited to selecting a material and properly proportioning a member to perform a specified function without failure. To accomplish these tasks, a designer must anticipate the type of failure that may occur. Several types of failure were discussed in Section 4-7; in this book's context of design attention will be focused on elastic failure, slip failure, and failure by fracture.

Once the type of failure has been determined, the significant material property that controls failure is established; this property is then divided by the factor of safety to determine the allowable or working stress (deformation) when carrying out the design computations. These computations are accomplished using mathematical relationships between load and stress or load and deformation. The following examples illustrate the use of design principles.

Example Problem 4-14

An axially loaded circular bar made of structural steel has a constant cross-sectional area and is subjected to the forces shown in Fig. 4-37a. The factor of safety, based on slip failure, is to be 1.8. Determine the minimum diameter of the bar required to support the loads.

SOLUTION

The forces transmitted by sections AB and BC are obtained from free-body diagrams of portions of the bar isolated by using cutting planes to the right of pin A and to the left of pin C and drawing the axial force diagram shown in Fig. 4-37b. Thus, the maximum load transmitted by any cross section is $F_{AB} =$ 36 kN. Since the criteria for failure is slip, the significant property of the material is the yield strength. From Table B-18,

$$\sigma_{\text{yield}} = 250 \text{ MPa}$$

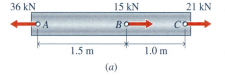

(a)

(b)

Figure 4-37

---

[6]Engineering Accreditation Commission, Accreditation Board for Engineering and Technology, Inc., "Criteria for Accrediting Programs in Engineering in the United States, Effective for Evaluations During the 1993–94 Accreditation Cycle," New York, New York, 1993, p. 7.

Since the factor of safety is 1.8,

$$\sigma_{allowable} = \frac{250}{1.8} = 138.89 \text{ MPa}$$

Therefore,

$$\sigma_{allowable} = \frac{F_{AB}}{A} = \frac{36(10^3)}{(\pi/4)d^2} = 138.89 \text{ MPa}$$

$$d = d_{min} = 18.166(10^{-3}) \text{ m} \cong 18.17 \text{ mm} \quad \blacksquare \qquad \textbf{Ans.}$$

Example Problem 4-15

A 40-lb light is supported at the midpoint of a 10-ft length of wire that is made of 0.2%C hardened steel, as shown in Fig. 4-38a. For reasons of safety, a factor of safety of 3 based on the yield strength of the wire is specified. Spools of wire are available with diameters of 10, 20, 30, 40, and 50 mil (1 mil = 0.001 in.). What spool size would you select for suspending the light?

SOLUTION

The forces transmitted by cables AB and AC are obtained from the free-body diagram of joint A of the cable system shown in Fig. 4-38b:

$$\theta = \cos^{-1}\frac{4}{5} = 36.87°$$

$$+ \rightarrow \Sigma F_x = T_{AC} \cos \theta - T_{AB} \cos \theta = 0$$

From which

$$T_{AC} = T_{AB}$$
$$+ \uparrow \Sigma F_y = T_{AC} \sin 36.87° + T_{AB} \sin 36.87° - 40 = 0$$
$$T_{AC} = T_{AB} = 33.33 \text{ lb}$$

For 0.2% C hardened steel (see Table B-17):

$$\sigma_{yield} = 62 \text{ ksi}$$

Therefore

$$\sigma_{allowable} = \frac{62}{3} = 20.67 \text{ ksi}$$

$$\sigma_{allowable} = \frac{T_{AB}}{A} = \frac{33.33}{(\pi/4)d^2} = 20.67(10^3) \text{ psi}$$

$$d = d_{min} = 0.04531 \text{ in.} \cong 0.0453 \text{ in.}$$

Use spool size 50 (50-mil wire). $\blacksquare$ **Ans.**

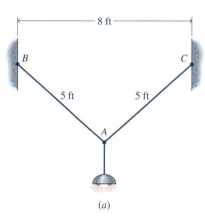

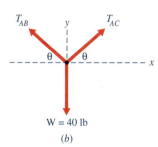

Figure 4-38

Example Problem 4-16

A short post made of air-dried Douglas fir is subjected to a centric compressive load of magnitude $P = 50$ kN, as shown in Fig. 4-39a. The load is supplied to the post through a rigid steel plate. The grain of the wood makes an angle of 30° with the horizontal. If failure is by fracture, and the factor of safety is 2.0, determine the smallest nominal size timber that may be used.

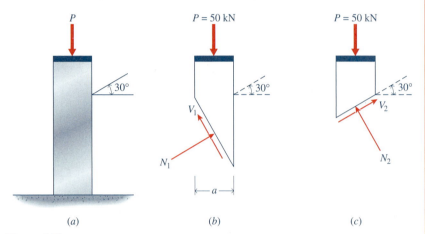

(a) (b) (c)

Figure 4-39

SOLUTION

Internal forces parallel and perpendicular to the grain of the wood are obtained by using the free-body diagrams shown in Figs. 4-39b and 4-39c. From the free-body diagram of Fig. 4-39b:

$$+\nearrow \Sigma F = N_1 - 50 \sin 30° = 0 \qquad N_1 = 25.00 \text{ kN}$$

$$+\nwarrow \Sigma F = V_1 - 50 \cos 30° = 0 \qquad V_1 = 43.30 \text{ kN}$$

From the free-body diagram of Fig. 4-39c:

$$+\nwarrow \Sigma F = N_2 - 50 \cos 30° = 0 \qquad N_2 = 43.30 \text{ kN}$$

$$+\nearrow \Sigma F = V_2 - 50 \sin 30° = 0 \qquad V_2 = 25.00 \text{ kN}$$

Since the criterion for failure is fracture, the significant property is the ultimate strengths (parallel to the grain) of the material in compression and in shear. From Table B-18:

$$\sigma_{\text{ultimate}} = 51 \text{ MPa C}$$

$$\tau_{\text{ultimate}} = 7.6 \text{ MPa}$$

Since the factor of safety is 2,

$$\sigma_{\text{allowable}} = \frac{51}{2} = 25.5 \text{ MPa}$$

$$\tau_{\text{allowable}} = \frac{7.6}{2} = 3.8 \text{ MPa}$$

Therefore,

$$\sigma_{\text{allowable}} = \frac{N_1}{A_1'} = \frac{25.00(10^3)}{A_1'} = 25.5(10^6) \text{ N/m}^2$$

$$A_1' = 980.4(10^{-6}) \text{ m}^2 = 980.4 \text{ mm}^2$$

$$A = A_1' \cos 60° = 980.4 \cos 60° = 490.2 \text{ mm}^2$$

$$\tau_{\text{allowable}} = \frac{V_2}{A_2'} = \frac{25.00(10^3)}{A_2'} = 3.8(10^6) \text{ N/m}^2$$

$$A_2' = 6579(10^{-6}) \text{ m}^2 = 6579 \text{ mm}^2$$

$$A = A_2' \cos 30° = 6579 \cos 30° = 5698 \text{ mm}^2$$

From Table B-16:

Use a 51 × 152 structural timber. ■ **Ans.**

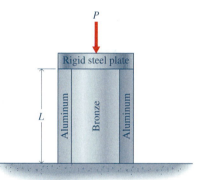

 Example Problem 4-17

The short post shown in Fig. 4-40a is subjected to an axial compressive load $P = 150$ kip. The load is applied to the post through a rigid steel plate. The core of the post is annealed bronze, and the outer segment of the post is composed of two symmetrically placed plates of 2024-T4 aluminum. If failure by slip is to be avoided, determine the minimum thickness t required for each of the aluminum plates.

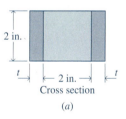

(a)

SOLUTION

Since the criterion for failure is slip the significant property for each material is the yield strength. For the bronze and the aluminum (see Table B-17):

Bronze	Aluminum
$E_B = 15,000$ ksi	$E_A = 10,600$ ksi
$\sigma_{\text{yield}} = 20$ ksi	$\sigma_{\text{yield}} = 48$ ksi

From the deformation equation $\delta_A = \delta_B$:

$$\frac{\sigma_A L_A}{E_A} = \frac{\sigma_B L_B}{E_B} = \frac{\sigma_A L}{10,600} = \frac{\sigma_B L}{15,000}$$

$$\sigma_A = 0.7067 \, \sigma_B$$

Therefore, when $\sigma_B = \sigma_{\text{yield}} = 20$ ksi:

$$\sigma_A = 0.7067\sigma_B = 0.7067(20) = 14.134 \text{ ksi} < 48 \text{ ksi}$$

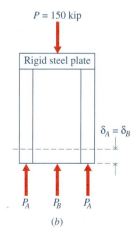

(b)

Figure 4-40

Forces in the bronze and aluminum parts of the post are obtained by using the free-body diagram shown in Fig. 4-40b. Thus

$$+ \uparrow \Sigma F = 2P_A + P_B - 150 = 0$$
$$2P_A + P_B = 150$$
$$2\sigma_A A_A + \sigma_B A_B = 150$$
$$2(14.134)(2t) + 20(4) = 150$$
$$t = 1.238 \text{ in. } \blacksquare \qquad \textbf{Ans.}$$

PROBLEMS

Standard Problems

4-91 An axially loaded circular bar is 4 ft long, and is made of SAE 4340 heat-treated steel. The bar is subjected to a tensile load of 50 kip. If the allowable normal stress is 95 ksi, and the allowable deformation is 0.1 in., determine the minimum diameter of the bar so that neither allowable value is exceeded.

4-92 A short column made of structural steel is used to support the floor of a building, as shown in Fig. P4-92. Each floor beam (A and B) transmits a force of 200 kN to the column. The column has the shape of a wide-flange *(W)* section (see Table B-2). The factor of safety based on slip failure is 3.0. Select the lightest wide-flange section that will support the applied loads.

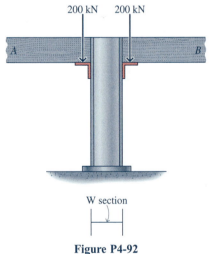

Figure P4-92

4-93 A standard-weight steel pipe (see Table B-13) is used to support an axial compressive load of 25 kip. If the allowable compressive stress is 10 ksi, select the proper size steel pipe to support the load.

4-94 The two structural steel rods A and B shown in Fig. P4-94 are used to support a mass $m = 2000$ kg. If failure is by slip and a factor of safety of 1.75 is specified, determine the diameters of the rods (to the nearest millimeter) that must be used to support the mass.

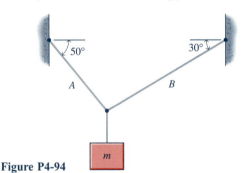

Figure P4-94

4-95 The two parts of the eye bar shown in Fig. P4-95 are connected by bolts (one on each side of the eye bar). The bolts are ASTM A325 steel with an allowable tensile stress of 44 ksi and an allowable shearing stress of 17.5 ksi. The eye bar is subjected to the forces $P = 20$ kip. Determine the minimum bolt diameter required to safely support the forces.

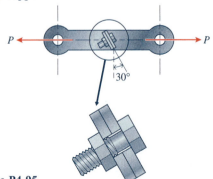

Figure P4-95

4-96 A square-headed structural steel rod is subjected to a tensile load $P = 70$ kN, as shown in Fig. P4-96. The rod passes through a hole (that is slightly larger than the diameter of the rod) in a heavy steel plate. The assembly may fail in one of three ways: (1) tension failure of the rod (allowable normal stress is 138 MPa), (2) punching failure when the rod pulls through the head of the rod (allowable shearing stress is 110 MPa), or (3) bearing failure where the head of the assembly makes contact with the heavy steel plate (allowable bearing stress is 205 MPa). Determine

(a) The diameter d of the rod.
(b) The thickness h of the head.
(c) The dimension a of the head of the assembly.

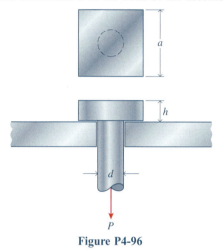

P

Figure P4-96

4-97 An arrangement to support the floor beams of a building is shown in Fig. P4-97. Post A is air-dried Douglas fir, and is used to support the 8 × 16-in. air-dried red oak floor beams B and C. Each floor beam transmits 12,000 lb to post A. The bearing area between the post and floor beams is increased by adding a section of steel channel. The bearing load between the post and concrete footing is distributed over a larger area by means of a rigid steel base plate; likewise for the bearing force between the concrete footing and the ground. The allowable bearing stresses are 300 psi perpendicular to the grain of the red oak floor beams, 500 psi on the concrete footing, and 60 psi on the ground. The allowable stress parallel to the grain of the Douglas fir post is 1000 psi. The weights of all members may be neglected. Determine

(a) The minimum length of channel if the channel fits snugly on the floor beams.
(b) The dimensions of the square post to the nearest $\frac{1}{2}$ in.

(c) The dimensions of the square base plate to the nearest $\frac{1}{2}$ in.
(d) The dimensions of the concrete footing to the nearest 1 in.

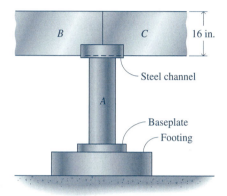

Steel channel

Baseplate
Footing

Figure P4-97

4-98 Four axial forces are applied to the 25 mm–thick structural steel bar with 40 mm–diameter pins as shown in Fig. P4-98. If the maximum allowable tensile normal stress in the bar is 135 MPa and the maximum allowable deformation (extension or contraction) of the bar is 1.25 mm, determine the minimum width w of the bar.

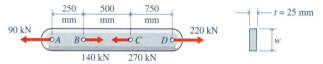

Figure P4-98

4-99 The solid circular annealed red brass bar shown in Fig. P4-99 has a diameter of 1.50 in. The brass bar fits inside a structural steel tube that has an inside diameter of 1.70 in. and an outside diameter of 2.00 in. Initially the tube and bar are at 70°F. A pin is placed through each end of the assembly, and the temperature is raised to 100°F. The pins are made of 6061-T6 aluminum. Determine the required diameters of the pins if failure is by slip, and a factor of safety of 2.0 is specified.

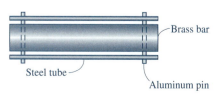

Brass bar

Steel tube

Aluminum pin

Figure P4-99

4-100 An axial load $P = 1000$ kN is applied to the rigid steel bearing plate on the top of the short column shown in Fig. P4-100. The outside segment of the column is made of structural steel with an allowable stress of 175 MPa, and the inside core is made of concrete with an allowable stress of 20 MPa and a modulus of elasticity of 16 GPa. If the area of the concrete is to be 10 times the area of the steel, determine the required dimensions if both segments are square.

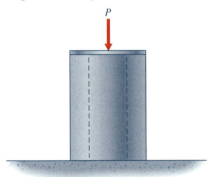

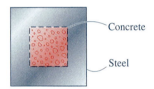

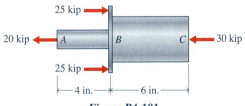

Figure P4-100

Challenging Problems

4-101 The machine component shown in Fig. P4-101 is made of hot-rolled Monel. The forces at B are applied to the component with a rigid collar that is firmly attached to the component. If a maximum normal stress of 30 ksi and a maximum overall change in length (extension or contraction) of 0.01 in. is specified, determine the minimum diameters permitted for the two segments of the component.

Figure P4-101

4-102 The member shown in Fig. P4-102 is made from a piece of standard-weight structural steel pipe. The forces at B and C are applied to rigid collars that are firmly attached to the pipe. If a maximum stress of 140 MPa and a maximum overall change in length (extension or contraction) of 1.00 mm is specified, select the lightest standard steel pipe (see Table B-14) that can be used.

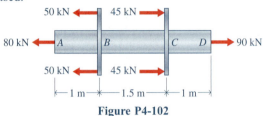

Figure P4-102

4-103 A structural connection is shown in Fig. P4-103. The main straps (center plates) are connected to the splice plates (outer plates) by two bolts. The connection is subjected to a load of 30 kip. The plates are made of ASTM A36 steel and the bolts are ASTM A325 high-strength steel. Failure may occur in several ways: shearing failure of the bolts (allowable shearing stress is 17.5 ksi); bearing failure of the plates (allowable bearing stress is 43.5 ksi); tension failure of a plate on a section through the center of a hole (allowable normal stress is 22 ksi); punching shear tearout of a bolt through an end of a plate (allowable shearing stress is 11 ksi). The thickness t of the main straps is to be twice that of a splice plate, and the width w is the same for all plates. Determine
(a) The required diameter for the bolts.
(b) The required thickness t and width w of the plates.
(c) The margin required to prevent tearout (distance e).

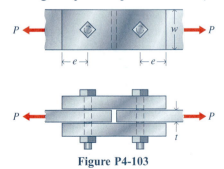

Figure P4-103

4-104 The two solid rods shown in Fig. P4-104 are pin-connected at the ends and support a mass of 5000 kg. The rods are made of SAE 4340 heat-treated steel. The factor of safety for failure by slip is to be 1.5. For a minimum weight of rod design, determine

(a) The optimum angle θ.
(b) The required diameter for the rods.
(c) The weight of each rod. Is it reasonable to neglect the weight of the rods in the design?

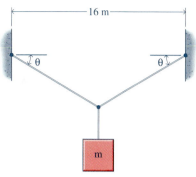

Figure P4-104

4-9 SUMMARY

When a structural member or machine component is subjected to a system of external loads (applied loads and support reactions), a system of internal resisting forces develops within the member to balance the external forces. These forces tend to either crush the member (compression) or pull it apart (tension). The intensities of these internal forces (forces per unit area) are called stresses.

For long, slender, axially loaded members, it is generally assumed that the normal stresses on a transverse plane are uniformly distributed except near the points of load application. In these members an average normal stress on the cross section can be computed as

$$\sigma_{avg} = \frac{F}{A} \qquad (4\text{-}2)$$

where F is the magnitude of the force transmitted by the cross section and A is the cross-sectional area of the member. A positive sign indicates a tensile normal stress and a negative sign indicates a compressive normal stress. This sign convention is independent of the selection of a coordinate system.

Experimental studies indicate that materials respond differently to forces that tend to pull surfaces apart than to forces that tend to slide surfaces relative to each other. Stresses on planes inclined to the axes of axially loaded bars are obtained by resolving the internal forces on the inclined plane into a component perpendicular to the plane (a normal force N from which a normal stress σ is determined) and a component tangent to the plane (a shear force V from which a shear stress τ is determined). Both normal and shearing stresses under axial loading are important since a brittle material loaded in tension will fail in tension on a transverse plane (where the normal stress is a maximum):

$$\sigma_{max} = \frac{P}{A} \qquad (4\text{-}9)$$

whereas a ductile material loaded in tension will fail in shear on a 45° plane (where the shear stress is a maximum):

$$\tau_{\max} = \frac{P}{2A} \tag{4-10}$$

Loads applied to a structure or machine are generally transmitted to the individual members through connections that use rivets, bolts, or pins. In all of these connections, one of the most significant stresses induced is a shearing stress. From the definition of stress, an average shearing stress on the transverse cross section of the rivet, bolt, or pin can be computed as

$$\tau_{\text{avg}} = \frac{V}{A_s} \tag{4-4}$$

where V is the magnitude of the shear force transmitted by the rivet, bolt, or pin and A_s is the cross-sectional area of the rivet, bolt, or pin.

Deformation is a measure of the change in size or shape of a body. Strain (deformation per unit length) is the quantity used to measure the intensity of a deformation, just as stress (force per unit area) is used to measure the intensity of an internal force. Normal strain,

$$\epsilon_{\text{avg}} = \frac{\delta_n}{L} \tag{4-11}$$

measures the change in size (elongation or contraction of an arbitrary line segment) of a body during deformation. Normal strain is positive when the line elongates and negative when the line contracts. In general, if the axial stress is tensile, the axial deformation will be an elongation. Therefore, positive normal strains are referred to as tensile strains and negative normal strains are referred to as compressive strains.

Shearing strain,

$$\gamma_{\text{avg}} = \frac{\delta_s}{L} = \tan \phi \tag{4-13}$$

$$\gamma_{xy}(P) = \frac{\pi}{2} - \theta' \tag{4-14b}$$

measures the change in shape (change in angle between two lines that are orthogonal in the undeformed state) of a body during a deformation. Shearing strains will be positive if the angle between reference lines decreases and negative if the angle increases. Normal and shearing strains for most engineering materials in the elastic range seldom exceed values of 0.2 percent (0.002 in./in. or 0.002 rad). Curves showing the relationship between stress and strain (called stress–strain diagrams) are independent of the size and shape of the member and depend only on the type of material from which the member is made. Data for stress–strain diagrams are obtained by applying an axial load to a test specimen and measuring the load and deformation simultaneously. The initial portion of the stress–strain diagram for most materials used in engineering structures is a straight line and is represented by Hooke's law

$$\sigma = E\epsilon \tag{4-15a}$$

where the constant of proportionality (Young's modulus, E) must be determined from the experimental data. Although the stress–strain diagram for some materials such as gray cast iron and concrete show a slight curve even at very small stresses, it is common practice to draw a straight line to average the data for the first part of the diagram and neglect the curvature. It is important to realize that Hooke's Law only describes the initial linear portion of the stress–strain diagram and is valid only for uniaxally loaded bars.

A body loaded in one direction will undergo strains perpendicular to the direction of the load in addition to those parallel to the load. The ratio of the lateral or perpendicular strain to the longitudinal or axial strain is called Poisson's ratio:

$$v = -\frac{\epsilon_{\text{lat}}}{\epsilon_{\text{long}}} = -\frac{\epsilon_t}{\epsilon_a} \tag{4-16}$$

and is a constant for stresses below the proportional limit. Poisson's ratio is related to E and G by

$$E = 2(1 + v)G \tag{4-17}$$

and has a value between $\frac{1}{4}$ and $\frac{1}{3}$ for most metals. The ratio $v = -\epsilon_t/\epsilon_a$ is valid only for a uniaxial state of stress.

When unrestrained, most engineering materials expand when heated and contract when cooled. The thermal strain of an unrestrained body due to a temperature change of ΔT degrees is

$$\epsilon_T = \alpha\, \Delta T \tag{4-18}$$

where α is known as the coefficient of thermal expansion.

Strains caused by temperature changes and strains caused by applied loads are essentially independent. The total normal strain in a body acted upon by both temperature changes and axially applied loads is given by

$$\epsilon_{\text{total}} = \epsilon_\sigma + \epsilon_T = \frac{\sigma}{E} + \alpha\, \Delta T \tag{4-19}$$

Since homogeneous, isotropic materials expand uniformly in all directions when heated (and contract uniformly when cooled), neither the shape of the body nor the shearing stresses and shearing strains are affected by temperature changes, if the body is unrestrained.

When a straight bar of uniform cross section is axially loaded by forces applied at the ends, the axial strain along the length of the bar is assumed to have a constant value, and the elongation (or contraction) of the bar resulting from the axial load P may be expressed as $\delta = \epsilon L$ (by the definition of average axial strain). If Hooke's law (Eq. 4-15a) applies, the axial deformation may be expressed in terms of either stress or load as

$$\delta = \epsilon L = \frac{\sigma L}{E} \tag{4-20a}$$

or

$$\delta = \frac{PL}{EA} \tag{4-20b}$$

where the stress in the member $\sigma = P/A$, the cross-sectional area of the member A, and Young's modulus E are all constant over the length L. If a bar is subjected to a number of axial loads at different points along the bar, or if the bar consists of parts having different cross-sectional areas or of parts composed of different materials, then the change in length of each part can be computed using Eq. (4-20b). The changes in length of the various parts of the bar can then be added algebraically to give the total change in length of the complete bar

$$\delta = \sum_{i=1}^{n} \delta_i = \sum_{i=1}^{n} \frac{P_i L_i}{E_i A_i} \tag{4-21}$$

where A_i and E_i are constant on segment i of length L_i and the force P_i is the internal force in segment i of the bar. Force P_i is usually different than the forces applied at the ends of the segment. These forces must be calculated from equilibrium of the segment and are often shown on an axial force diagram.

A designer is required to select a material and properly proportion a member to perform a specified function without failure. Failure is defined as the state or condition in which a member or structure no longer functions as intended. Elastic failure occurs as a result of excessive elastic deformation. When a structure is designed to avoid elastic failure, the stiffness of the material, indicated by Young's modulus, is the significant property. Slip failure is characterized by excessive plastic deformation. Yield strength, yield point, and proportional limit are used as indices of strength with respect to failure by slip for members subjected to static loads. Failure by fracture is a complete separation of the material. The ultimate strength of a material is the index of resistance to failure by fracture under static loads in which creep is not involved.

A working stress or allowable stress is defined as the maximum stress permitted in the design computation. The factor of safety may be defined as the ratio of a failure-producing load to the estimated actual load. Factor of safety may also be defined as the ratio of the strength of a material to the maximum computed stress in the material. The latter factor of safety may be used to determine an allowable (working) stress by dividing the strength of a material by the factor of safety.

REVIEW PROBLEMS

Standard Problems

4-105* A control rod in a mechanism must stretch 0.200 in. when a load of 1000 lb is applied. The rod is made from an aluminum alloy with an elastic strength of 48 ksi and a modulus of elasticity of 10,600 ksi. If a factor of safety of 2 with respect to failure by slip is specified, determine the diameter and length of the lightest rod that can be used.

4-106* A 2024-T4 aluminum alloy bar (see Appendix B for properties) 3 m long has a 25 mm–square cross section over

1 m of its length and a 25 mm–diameter circular cross section over the other 2 m of its length. The bar is supporting an axial tensile load of 50 kN. Determine
(a) The elongation of the bar.
(b) The factor of safety with respect to failure by fracture.

4-107 The joint shown in Fig. P4-107 is used in a steel tension member that has a 2×1-in. rectangular cross section. If the allowable normal, bearing, and punching-shear stresses in the joint are 13.5 ksi, 18.0 ksi, and 6.50 ksi, respectively, determine the maximum load P that can be carried by the joint.

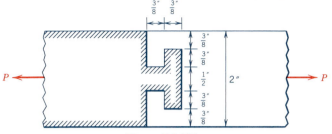

Figure P4-107

4-108 A steel sleeve is connected to a steel shaft with a flexible rubber insert as shown in Fig. P4-108. The insert has an inside diameter of 85 mm and an outside diameter of 110 mm. When the unit is subjected to a torque T, the shaft rotates 1.5° with respect to the sleeve. Assume that radial lines in the unstressed state remain straight as the rubber deforms. Determine the shearing strain $\gamma_{r\theta}$ in the rubber insert
(a) At the inside surface.
(b) At the outside surface.

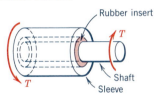

Figure P4-108

4-109* The sanding-drum mandrel shown in Fig. P4-109 is made for use with a hand drill. The mandrel is made from a rubber-like material that expands when the nut is tightened to secure the sanding drum placed over the outside surface. If the diameter of the mandrel increases from 2.00 in. to 2.20 in. as the nut is tightened, determine
(a) The average normal strain along a diameter of the mandrel.
(b) The circumferential strain at the outside surface of the mandrel.

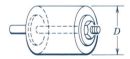

Figure P4-109

4-110* A $25 \times 25 \times 200$ mm–long cold-rolled bronze ($E = 100$ GPa and $G = 45$ GPa) bar is subjected to an axial tensile load of 120 kN. Determine
(a) The axial strain in the bar.
(b) The lateral strain in the bar.
(c) The change in volume of the bar.

4-111 A 1 in.–diameter steel [$\alpha = 6.5(10^{-6})$/°F, $E = 30{,}000$ ksi, and $\upsilon = 0.30$] bar is subjected to a temperature decrease of 150°F. The ends of the bar are supported by two walls that displace a small amount during the temperature change. If the measured strain in the bar is -600 µin./in. after the temperature change, determine the load being transmitted to the walls.

4-112 At a temperature of 25°C, a cold-rolled red brass [$E = 100$ GPa and $\alpha = 17.6(10^{-6})$/°C] sleeve has an inside diameter of 299.75 mm and an outside diameter of 310 mm. The sleeve is to be placed on a steel shaft with an outside diameter of 300 mm. Determine
(a) The temperature at which the sleeve will slip over the shaft with a clearance of 0.05 mm.
(b) The stress in the sleeve after it has cooled to 25°C if the shaft is assumed to be rigid.

Challenging Problems

4-113* An elastomer ($G = 100$ ksi) is bonded to a rigid circular shaft and to a rigid circular sleeve as shown in Fig. P4-113 to form a spring. An axial load P of 10 kip is applied to the shaft. Determine
(a) The shearing strain at the outer surface of the elastomer.
(b) The shearing strain at the inner surface of the elastomer.
(c) The deflection of the end of the shaft with respect to its no-load position.

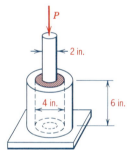

Figure P4-113

4-114* A hollow brass ($E = 100$ GPa) tube A with an outside diameter of 100 mm and an inside diameter of 50 mm is fastened to a 50 mm–diameter steel ($E = 200$ GPa) rod B as shown in Fig. P4-114. The supports at the top and bottom of the assembly and the collar C used to apply the 500-kN load are rigid. Determine
(a) The stresses in each of the members.
(b) The deflection of collar C.

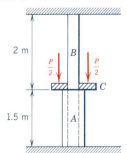

Figure P4-114

4-115 The three bars shown in Fig. P4-115 will be used to support a load P of 150 kip. Bar A will be made of aluminum alloy (E = 10,500 ksi) and will have a circular cross section. Bars B will be made of structural steel (E = 29,000 ksi) and will have 2.5 × 1.0-in. rectangular cross sections. If the stress in bar A must be limited to 15 ksi, determine the minimum satisfactory diameter for bar A.

4-116 Solve Problem 4-114 if the top support in Fig. P4-114 yields and displaces 0.75 mm as the load P is applied.

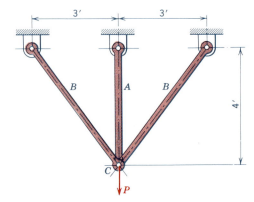

Figure P4-115

COMPUTER PROBLEMS

C4-117 An aluminum tube with an outside diameter d_o will be used to support a 10-kip load. If the axial stress in the tube must be limited to 30 ksi T or C, calculate and plot the required wall thickness t as a function of d_o (0.75 in. < d_o < 4 in.). What diameter would be required for a solid aluminum shaft?

C4-118 The steel pipe column shown in Fig. P4-118a has an outside diameter of 150 mm and a wall thickness of 15 mm. The load imposed on the column by the timber beam is 150 kN. If the bearing stress between the circular steel bearing plate and the timber beam is not to exceed 3.25 MPa, determine the minimum diameter bearing plate that must be used between the column and the beam. Assume that the bearing stress is uniformly distributed over the surface of the plate.

If the bearing plate is not rigid, the stress between the bearing plate and the timber beam will not be uniform. If the stress varies as shown in Fig. P4-118b (a uniform value of σ_{max} above the column and decreasing linearly to $\sigma_{max}/5$ at the outside edge r_P of the bearing plate), calculate and plot σ_{max} versus the radius r_P of the bearing plate (75 mm < r_P < 500 mm). Now what minimum diameter bearing plate must be used if the bearing stress must not exceed 3.25 MPa? What is the percent decrease in σ_{max} for a 400 mm–diameter bearing plate compared with a 150 mm–diameter bearing plate? For a 600 mm–diameter bearing plate compared with a 150 mm–diameter bearing plate?

C4-119 A vertical shaft is supported by a thrust collar and bearing plate as shown in Fig. P4-119a. The force imposed on the bearing plate by the collar is 50 kip. If the bearing stress between the collar and the bearing plate must not exceed 10 ksi, determine the minimum diameter collar that must be used. Assume that the bearing stress is uniformly distributed over the surface of the washer.

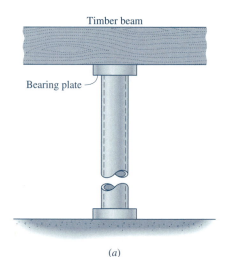

Timber beam

Bearing plate

(a)

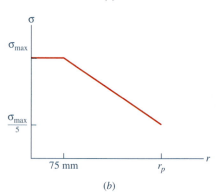

(b)

Figure P4-118

If the collar is not rigid, the stress between the collar and the bearing plate will not be uniform. If the stress varies as shown in Fig. P4-119b (decreasing linearly from σ_{max} at the edge of the shaft to $\sigma_{max}/2$ at $r = 3$ in.), calculate and plot σ_{max} versus the radius r_c of the collar (1 in. $< r_c <$ 2.5 in.). Now what minimum diameter collar must be used if the bearing stress must not exceed 10 ksi? What is the percent decrease in σ_{max} for a 3.2 in.–diameter collar compared with a 2.4 in.–diameter collar? For a 4.0 in.–diameter collar? For a 5.0 in.–diameter collar?

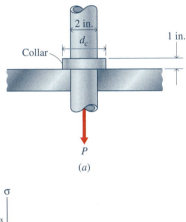

(a)

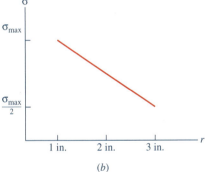

(b)

Figure P4-119

C4-120 The tie rod shown in Fig. P4-120a has a diameter of 40 mm and is used to resist the lateral pressure against the walls of a grain bin. The force imposed on the wall by the rod is 80 kN. If the bearing stress between the washer and the wall must not exceed 2.8 MPa, determine the minimum diameter washer that must be used between the head of the bolt and the grain bin wall. Assume that the bearing stress is uniformly distributed over the surface of the washer.

If the washer is not rigid, the stress between the washer and the wall will not be uniform. If the stress varies as shown in Fig. P4-120b (a uniform value of σ_{max} under the 60 mm–diameter restraining nut and decreasing as $1/r$ to the outside edge r_w of the washer), calculate and plot σ_{max} versus the radius r_w of the washer (30 mm $< r_w <$ 200 mm). Now what minimum diameter washer must be used if the bearing stress must not exceed 2.8 MPa? What is the percent decrease in σ_{max} for a 200 mm–diameter washer compared with no washer? For a 300 mm–diameter washer?

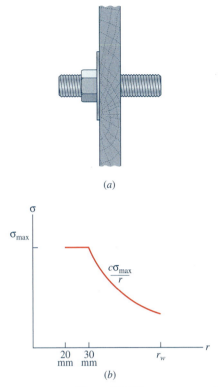

(a)

(b)

Figure P4-120

C4-121 Specifications for the rectangular ($3 \times 3 \times 21$-in.) block shown in Fig. P4-121 require that the normal and shearing stresses on plane A-A not exceed 800 psi and 500 psi, respectively. If the plane A-A makes an angle $\theta = 37°$ with the horizontal, calculate and plot the ratios σ/σ_{max} and τ/τ_{max} as a function of the load P ($0 < P < 13$ kip). What is the maximum load P_{max} that can be applied to the block? Which condition controls what the maximum load can be? Repeat for $\theta = 25°$. For what angle θ will the normal stress and the shear stress both reach their limiting values at the same time?

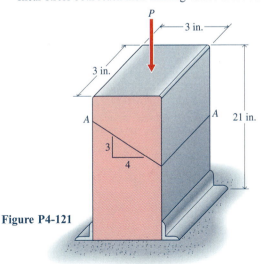

Figure P4-121

C4-122 Specifications for the rectangular (200 × 320 × 120-mm) wooden block shown in Fig. P4-122 require that the shearing stress parallel to the grain not exceed 5.25 MPa and that the normal stress parallel to the grain not exceed 52.5 MPa. If the grain makes an angle $\theta = 60°$ with the horizontal, calculate and plot the ratios σ/σ_{max} and τ/τ_{max} as a function of the load P ($0 < P < 500$ kN). What is the maximum load P_{max} that can be applied to the block? Which condition controls what the maximum load can be? Repeat for $\theta = 30°$.

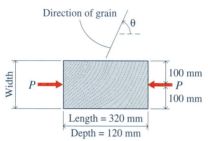

Direction of grain

θ

Width

P ———→   ←——— P

100 mm
100 mm

Length = 320 mm
Depth = 120 mm

Figure P4-122

C4-123 A high-strength steel bolt ($E_s = 30,000$ ksi and $A_s = 0.785$ in.²) passes through a brass sleeve ($E_b = 15,000$ ksi and $A_b = 1.767$ in.²) as shown in Fig. P4-123. As the nut is tightened, it advances a distance of 0.125-in. along the bolt for each complete turn of the nut. Compute and plot

(a) The axial stress σ_s in the steel bolt and the axial stress σ_b in the brass sleeve as functions of the angle of rotation θ of the nut ($0° < \theta < 180°$).

(b) The elongation δ_s of the steel bolt and the elongation δ_b of the brass sleeve as functions of the angle of rotation θ of the nut ($0° < \theta < 180°$).

(c) The distance L between the two washers as a function of the angle of rotation θ of the nut ($0° < \theta < 180°$).

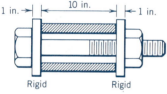

1 in.   10 in.   1 in.

Rigid   Rigid

Figure P4-123

C4-124 The two faces of the clamp shown in Fig. P4-124 are 200 mm apart when the two steel bolts [$E_s = 190$ GPa, $A_s = 115$ mm² each, and $\alpha_s = 17.3(10^{-6})/°C$] connecting them are unstretched. A force P is applied to separate the faces of the clamp so that a brass bar [$E_b = 100$ GPa, $A_b = 625$ mm², and

$\alpha_b = 17.6(10^{-6})/°C$] with a length of 200.50 mm can be inserted as shown. After the load P is removed, the temperature of the system is slowly raised. Compute and plot

(a) The axial stress σ_s in the steel bolt and the axial stress σ_b in the brass bar as functions of the temperature rise ΔT ($0°C < \Delta T < 100°C$).

(b) The elongation δ_s of the steel bolts and the elongation δ_b of the brass bar as functions of the temperature rise ΔT ($0°C < \Delta T < 100°C$).

(c) The distance L between the faces of the clamp as a function of the temperature rise ΔT ($0°C < \Delta T < 100°C$).

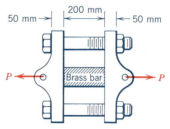

50 mm →   200 mm   ← 50 mm

P ←——— Brass bar ———→ P

Figure P4-124

C4-125 The 10-in. diameter, structural steel ($E = 30,000$ ksi; $\gamma = 0.284$ lb/in.³) pile shown in Fig. P4-125 is being extracted from the ground with a vertical force F. Assume that the entire 30-ft length of the pile is in the ground and that the horizontal normal stresses σ_n and the vertical shearing stresses τ (which are a function of the type of soil surrounding the pile) are uniformly distributed over the surface of the cylinder. The magnitudes of the normal and shearing stresses are 40 psi and 20 psi, respectively. Compute and plot

(a) The axial stress $\sigma_x(x)$ in the pile as a function of the distance x from the ground surface.

(b) The axial deformation δ of the segment of the pile from any value of x to the bottom as a function of the distance x from the ground surface.

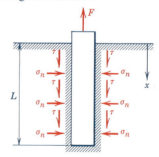

F

L

σ_n τ σ_n
x
σ_n τ σ_n
σ_n τ σ_n

Figure P4-125

EQUIVALENT FORCE/ MOMENT SYSTEMS

5-1 INTRODUCTION

The resultant force $\mathbf{R}$ of a system of two or more concurrent forces $\mathbf{F}_1$, $\mathbf{F}_2$, . . . , $\mathbf{F}_n$ was defined in Chapter 2 as the single force that will produce the same effect on a body as the original system of forces. When the resultant force $\mathbf{R}$ of a concurrent force system is zero, the body on which the system of forces acts is in equilibrium and the force system is said to be balanced. Methods to determine resultants of concurrent force systems were discussed in Chapter 2 and applied to equilibrium of a particle in Chapter 3.

For the case of a three-dimensional body that has a definite size and shape, the particle idealization discussed in Chapter 3 is no longer valid, in general, since the forces acting on the body are usually not concurrent. For these more general force systems, the condition $\mathbf{R} = \mathbf{0}$ is a necessary but not a sufficient condition for equilibrium of the body. A second restriction related to the tendency of a force to produce rotation of a body must also be satisfied and gives rise to the concept of a moment. In this chapter, the moment of a force about a point and the moment of a force about a line (axis) will be defined and methods will be developed for finding the resultant forces and the resultant moments for force systems that are not concurrent.

5-2 MOMENTS AND THEIR CHARACTERISTICS

The moment of a force about a point or axis is a measure of the tendency of the force to rotate a body about that point or axis. For example, the moment of force $\mathbf{F}$ about point O in Fig. 5-1a is a measure of the tendency of the force to rotate the body about line A-A. Line A-A is perpendicular to the plane containing force $\mathbf{F}$ and point O.

A moment has both a magnitude and a direction, and adds according to the parallelogram law of addition; therefore, it is a vector quantity. The magnitude of a moment $|\mathbf{M}|$ is defined as the product of the magnitude of a force $|\mathbf{F}|$ and the perpendicular distance d from the line of action of the force to the axis. Thus,

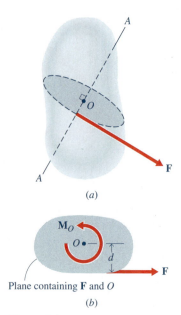

(a)

M_O

Plane containing **F** and O

(b)

Figure 5-1

in Fig. 5-1*b*, the magnitude of the moment of the force **F** about point O (actually about axis *A-A*, which is perpendicular to the page and passes through point O) is

$$M_O = |\mathbf{M}_O| = |\mathbf{F}|d \tag{5-1}$$

Point O is called the *moment center*, distance d is called the *moment arm*, and line *A-A* is called the *axis of the moment*.

   The direction (sense) of a moment in a two-dimensional problem can be specified by using a small curved arrow about the point, as shown in Fig. 5-1*b*. If the force tends to produce a counterclockwise rotation, the moment is assumed to be positive. In a similar manner, if the force tends to produce a clockwise rotation, the moment is negative.

   Since the magnitude of a moment of a force is the product of a force and a length, the dimensional expression for a moment is *FL*. In the U.S. customary system, the units commonly used for moments are lb · ft and lb · in. or in. · lb, ft · lb, and ft · kip. In the SI system, the units commonly used for moments are N · m, kN · m, and so on. It is immaterial whether the unit of force or the unit of length is stated first.

5-2-1 Principle of Moments—Varignon's Theorem

A concept often used in solving mechanics (statics, dynamics, mechanics of materials) problems is the principle of moments. This principle, when applied to a system of forces, states that the moment **M** of the resultant **R** of a system of forces with respect to any axis or point is equal to the vector sum of the moments of the individual forces of the system with respect to the same axis or point. Application of this principle to a pair of concurrent forces is known as Varignon's theorem. Varignon's theorem can be illustrated by using the concurrent force system, shown in Fig. 5-2, where **R** is the resultant of forces **A** and **B** that lie in the *xy*-plane. The point of concurrency A and the moment center O have been arbitrarily selected to lie on the *y*-axis. The distances d, a, and b are the perpendicular distances from the moment center O to the lines of action of forces **R, A,** and **B,** respectively. The angles γ, α, and β (measured from the *x*-axis) locate the forces **R, A,** and **B,** respectively.

   The magnitudes of the moments produced by the resultant **R** and by the two forces **A** and **B** with respect to point O are

$$M_R = Rd = R(h \cos \gamma)$$
$$M_A = Aa = A(h \cos \alpha) \tag{a}$$
$$M_B = Bb = B(h \cos \beta)$$

From Fig. 5-2 note also that

$$R \cos \gamma = A \cos \alpha + B \cos \beta \tag{b}$$

Substituting Eqs. (a) into Eq. (b) and multiplying both sides of the equation by h yields

$$M_R = M_A + M_B \tag{5-2}$$

Equation (5-2) indicates that the moment of the resultant **R** with respect to a point O is equal to the sum of the moments of the forces **A** and **B** with respect to the same point O.

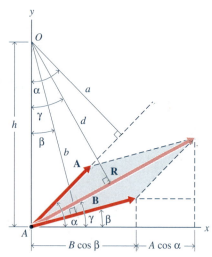

Figure 5-2

Example Problem 5-1

Three forces are applied to a triangular plate as shown in Fig. 5-3. Determine

(a) The moment of force $\mathbf{F}_3$ about point C.
(b) The moment of force $\mathbf{F}_2$ about point B.
(c) The moment of force $\mathbf{F}_1$ about point B.
(d) The moment of force $\mathbf{F}_3$ about point E.

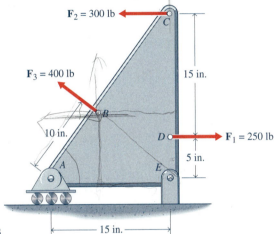

Figure 5-3

SOLUTION

The angle of inclination ϕ of the inclined edge of the plate with respect to a horizontal line is

$$\phi = \tan^{-1}\frac{20}{15} = 53.13°$$

$$L_{AC} = \sqrt{20^2 + 15^2} = 25 \text{ in.}$$

From Eq. (5-1):

$$M = |\mathbf{F}|\, d$$

(a) $M_C = F_3 d_3 = 400(15) = 6000 \text{ in.} \cdot \text{lb}$

$$\mathbf{M}_C = 6000 \text{ in.} \cdot \text{lb} \downarrow \quad \textbf{Ans.}$$

(b) $M_B = F_2 d_2 = 300(15 \sin 53.13°) = 3600 \text{ in.} \cdot \text{lb}$

$$\mathbf{M}_B = 3600 \text{ in.} \cdot \text{lb} \uparrow \quad \textbf{Ans.}$$

(c) $M_B = F_1 d_1 = 250(15 - 15 \sin 53.13°) = 750 \text{ in.} \cdot \text{lb}$

$$\mathbf{M}_B = 750 \text{ in.} \cdot \text{lb} \uparrow \quad \textbf{Ans.}$$

(d) $M_E = F_3 d_3 = 400(10 - 15 \cos 53.13°) = 400 \text{ in.} \cdot \text{lb}$

$$\mathbf{M}_E = 400 \text{ in.} \cdot \text{lb} \uparrow \; \blacksquare \quad \textbf{Ans.}$$

Example Problem 5-2

Use the principle of moments to determine the moment about point B of the 300-N force shown in Fig. 5-4a.

SOLUTION

The magnitudes of the rectangular components of the 300-N force are

$$F_x = F \cos 30° = 300 \cos 30° = 259.8 \text{ N}$$

$$F_y = F \sin 30° = 300 \sin 30° = 150.0 \text{ N}$$

Once the forces F_x and F_y are known, the moment M_B is

$$+\!\!\downarrow M_B = -F_x(0.250) - F_y(0.200)$$
$$= -259.8(0.250) - 150.0(0.200) = -95.0 \text{ N} \cdot \text{m}$$

$$\mathbf{M}_B = 95.0 \text{ N} \cdot \text{m} \,\downarrow \qquad \textbf{Ans.}$$

The moment about point B can also be determined by moving the force $\mathbf{F}$ along its line of action (principle of transmissibility) to points C or D, as shown in Fig. 5-4b. For both of these points, one component of the force produces no moment about point B. Thus,
For point C:

$$d_2 = 250 + 200 \tan 30° = 365.5 \text{ mm}$$

$$+\!\!\downarrow M_B = -F_x d_2 = -259.8(0.3655) = -95.0 \text{ N} \cdot \text{m}$$

$$\mathbf{M}_B = 95.0 \text{ N} \cdot \text{m} \,\downarrow \qquad \textbf{Ans.}$$

For point D:

$$d_3 = 200 + 250 \cot 30° = 633.0 \text{ mm}$$

$$+\!\!\downarrow M_B = -F_y d_3 = -150.0(0.633) = -95.0 \text{ N} \cdot \text{m}$$

$$\mathbf{M}_B = 95.0 \text{ N} \cdot \text{m} \,\downarrow \;\blacksquare \qquad \textbf{Ans.}$$

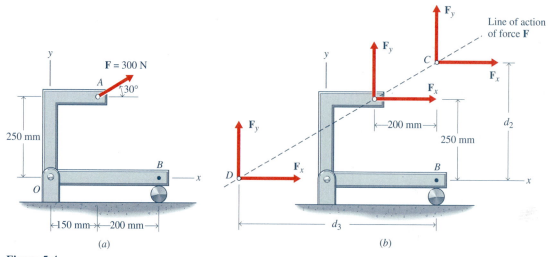

Figure 5-4

PROBLEMS

Standard Problems

5-1* Three forces are applied to the bar shown in Fig. P5-1. Determine
(a) The moment of force $\mathbf{F}_A$ about point E.
(b) The moment of force $\mathbf{F}_E$ about point A.
(c) The moment of force $\mathbf{F}_D$ about point B.

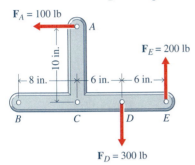

Figure P5-1

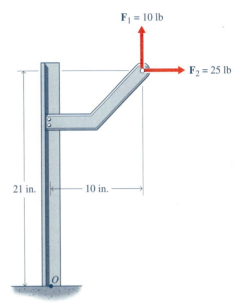

Figure P5-3

5-2* Four forces are applied to a plate as shown in Fig. P5-2. Determine
(a) The moment of force $\mathbf{F}_B$ about point A.
(b) The moment of force $\mathbf{F}_C$ about point B.
(c) The moment of force $\mathbf{F}_C$ about point A.

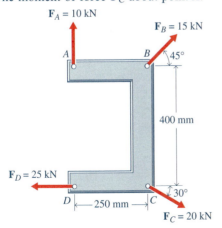

Figure P5-2

5-3 Two forces are applied to a bracket as shown in Fig. P5-3. Determine
(a) The moment of force $\mathbf{F}_1$ about point O.
(b) The moment of force $\mathbf{F}_2$ about point O.

5-4 Three forces are applied to a plate as shown in Fig. P5-4. Determine
(a) The moment of force $\mathbf{F}_1$ about point B.
(b) The moment of force $\mathbf{F}_3$ about point A.
(c) The moment of force $\mathbf{F}_2$ about point B.

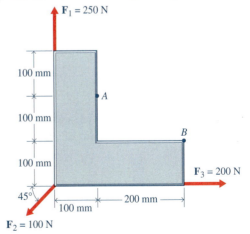

Figure P5-4

5-5* Three forces are applied to a circular plate as shown in Fig. P5-5. Determine
(a) The moment of force $\mathbf{F}_1$ about point O.
(b) The moment of force $\mathbf{F}_3$ about point O.
(c) The moment of force $\mathbf{F}_2$ about point A.

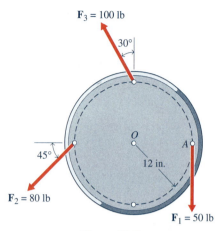

Figure P5-5

5-6* Three forces are applied to a bracket as shown in Fig. P5-6. Determine
(a) The moment of force $\mathbf{F}_1$ about point B.
(b) The moment of force $\mathbf{F}_2$ about point A.
(c) The moment of force $\mathbf{F}_3$ about point C.
(d) The moment of force $\mathbf{F}_3$ about point E.

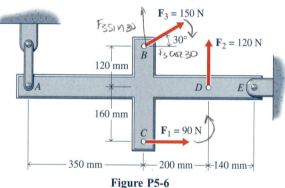

Figure P5-6

5-7 A 300-lb force is applied to a bracket as shown in Fig. P5-7. Determine the moment of the force about point A.

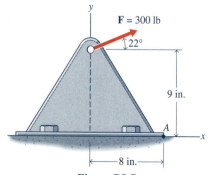

Figure P5-7

5-8 A 250-N force is applied to a bracket as shown in Fig. P5-8. Determine the moment of the force about point A.

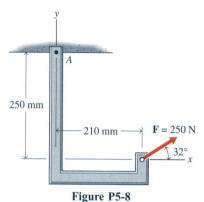

Figure P5-8

Challenging Problems

5-9* Two forces $\mathbf{F}_1$ and $\mathbf{F}_2$ are applied to a plate as shown in Fig. P5-9. Determine
(a) The moment of force $\mathbf{F}_1$ about point A.
(b) The moment of force $\mathbf{F}_2$ about point B.

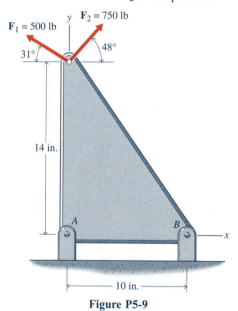

Figure P5-9

5-10* Two forces are applied to a ring as shown in Fig. P5-10. Determine
(a) The moment of force $\mathbf{F}_1$ about point A.
(b) The moment of force $\mathbf{F}_2$ about point A.

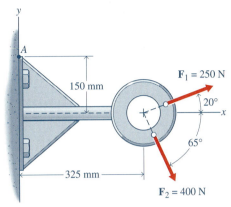

Figure P5-10

5-11 Two forces $\mathbf{F}_1$ and $\mathbf{F}_2$ are applied to a bracket as shown in Fig. P5-11. Determine
(a) The moment of force $\mathbf{F}_1$ about point B.
(b) The moment of force $\mathbf{F}_2$ about point A.

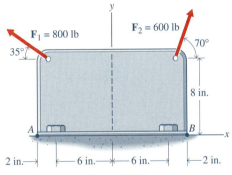

Figure P5-11

5-12 Two forces $\mathbf{F}_1$ and $\mathbf{F}_2$ are applied to a beam as shown in Fig. P5-12. Determine
(a) The moment of force $\mathbf{F}_1$ about point A
(b) The moment of force $\mathbf{F}_2$ about point B.

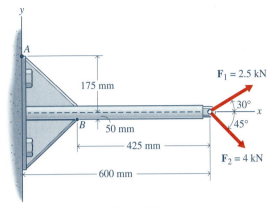

Figure P5-12

5-3 VECTOR REPRESENTATION OF A MOMENT

For some two-dimensional problems and for most three-dimensional problems, use of Eq. (5-1) for moment determinations is not convenient owing to difficulties in determining the perpendicular distance d between the line of action of the force and the moment center O. For these types of problems, a vector approach simplifies moment calculations.

In Fig. 5-1, the moment of the force $\mathbf{F}$ about point O can be represented by the expression

$$\mathbf{M}_O = \mathbf{r} \times \mathbf{F} \qquad (5\text{-}3)$$

where $\mathbf{r}$ is a position vector (see Appendix A) from point O to a point A on the line of action of the force $\mathbf{F}$, as shown in Fig. 5-5. By definition, the cross product

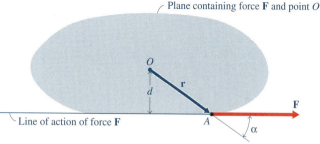

Plane containing force **F** and point O

Line of action of force **F**

Figure 5-5

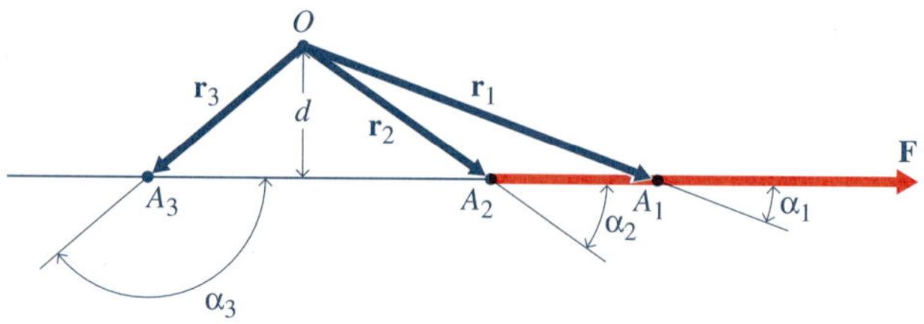

Figure 5-6

(see Appendix A) of the two intersecting vectors **r** and **F** is

$$\mathbf{M}_O = \mathbf{r} \times \mathbf{F}$$
$$= |\mathbf{r}||\mathbf{F}| \sin \alpha \ \mathbf{e} \tag{5-4}$$

where α is the angle ($0 \le \alpha \le 180°$) between the two intersecting vectors **r** and **F** and **e** is a unit vector perpendicular to the plane containing vectors **r** and **F**. It is obvious from Fig. 5-5 that the term $|\mathbf{r}| \sin \alpha$ equals the perpendicular distance d from the line of action of the force to the moment center O. Note also from Fig. 5-6 that the distance d is independent of the position A along the line of action of the force since

$$|\mathbf{r}_1| \sin \alpha_1 = |\mathbf{r}_2| \sin \alpha_2 = |\mathbf{r}_3| \sin \alpha_3 = d$$

Thus, Eq. (5-4) can be written

$$\mathbf{M}_O = |\mathbf{F}|d \ \mathbf{e} = Fd \ \mathbf{e} = M_O \ \mathbf{e} \tag{5-5}$$

In Eq. (5-5), the direction of the unit vector **e** is determined (see Fig. 5-7) by using the right-hand rule (fingers of the right hand curl from positive **r** to positive **F** and the thumb points in the direction of positive $\mathbf{M}_O$). Thus, Eq. (5-3) yields both the magnitude M_O and the direction **e** of the moment $\mathbf{M}_O$. It is important to note that the sequence **r** × **F** must be maintained in calculating moments since the sequence **F** × **r** will produce a moment with the opposite sense. The vector product is *not* a commutative operation.

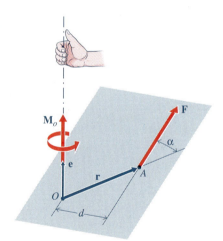

Figure 5-7

5-3-1 Moment of a Force About a Point

The vector **r** from the point about which the moment is to be determined (say, point B) to any point on the line of action of the force **F** (say, point A) can be expressed in terms of the unit vectors **i**, **j**, and **k**, and the coordinates (x_A, y_A, z_A) and (x_B, y_B, z_B) of points A and B, respectively. Thus, as illustrated in Fig. 5-8,

$$\mathbf{r} = \mathbf{r}_{A/B} = \mathbf{r}_A - \mathbf{r}_B = (x_A - x_B) \ \mathbf{i} + (y_A - y_B) \ \mathbf{j} + (z_A - z_B) \ \mathbf{k} \tag{5-6}$$

where the subscript A/B indicates A with respect to B.

Equation (5-3) is applicable for both the two-dimensional case (forces in, say, the xy-plane) and the three-dimensional case (forces with arbitrary space orientations).

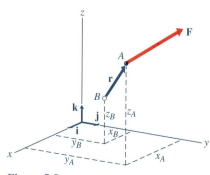

Figure 5-8

The Two-Dimensional Case. Consider first the moment $\mathbf{M}_O$ about the origin of coordinates (see Fig. 5-9a) produced by a force $\mathbf{F}$ in the xy-plane. The line of action of the force passes through point A. For this special case (see Fig. 5-9b),

$$\mathbf{F} = F_x \mathbf{i} + F_y \mathbf{j}$$

and the position vector $\mathbf{r}$ from the origin O to point A is

$$\mathbf{r} = r_x \mathbf{i} + r_y \mathbf{j}$$

The vector product $\mathbf{r} \times \mathbf{F}$ for this two-dimensional case can be written in determinant form as

$$\mathbf{M}_O = \mathbf{r} \times \mathbf{F} = \begin{vmatrix} \mathbf{i} & \mathbf{j} & \mathbf{k} \\ r_x & r_y & 0 \\ F_x & F_y & 0 \end{vmatrix} = (r_x F_y - r_y F_x)\mathbf{k} = M_z \mathbf{k} \qquad (5\text{-}7)$$

Thus, for the two-dimensional case, the moment $\mathbf{M}_O$ about point O due to a force $\mathbf{F}$ in the xy-plane is perpendicular to the plane (directed along the z-axis). The moment is completely defined by the scalar quantity

$$M_O = M_{Oz} = r_x F_y - r_y F_x \qquad (5\text{-}8)$$

since a positive value for M_O indicates a tendency to rotate the body in a counterclockwise direction, which, by the right-hand rule, is along the positive z-axis. Similarly, a negative value indicates a tendency to rotate the body in a clockwise direction, which requires a moment in the negative z-direction.

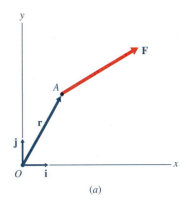

(a)

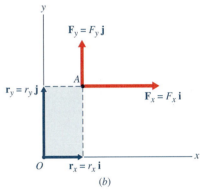

(b)

Figure 5-9

Example Problem 5-3

A 1000-lb force is applied to a beam cross section as shown in Fig. 5-10. Determine

(a) The moment of the force about point O.
(b) The perpendicular distance d from point B to the line of action of the force.

SOLUTION
Scalar Analysis
(a) The scalar components of the force $\mathbf{F}$ are

$$F_x = \frac{4}{5}F = \frac{4}{5}(1000) = 800 \text{ lb}$$

$$F_y = \frac{3}{5}F = \frac{3}{5}(1000) = 600 \text{ lb}$$

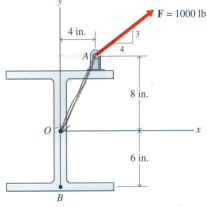

Figure 5-10

The moment of the force **F** about point O is

$$+\!\!\uparrow \Sigma M_O = F_y(4) - F_x(8)$$
$$= 600(4) - 800(8)$$
$$= -4000 \text{ in.} \cdot \text{lb} = 4000 \text{ in.} \cdot \text{lb } \downarrow \qquad \textbf{Ans.}$$

(b) The moment of the force **F** about point B is

$$+\!\!\uparrow \Sigma M_B = F_y(4) - F_x(14)$$
$$= 600(4) - 800(14)$$
$$= -8800 \text{ in.} \cdot \text{lb} = 8800 \text{ in.} \cdot \text{lb } \downarrow$$

$$d = \frac{M_B}{F} = \frac{8800}{1000} = 8.80 \text{ in.} \qquad \textbf{Ans.}$$

 Vector Analysis

(a) The force **F** and the position vector **r** from point O to point A can be expressed in Cartesian vector form as

$$\textbf{F} = 1000(0.80 \textbf{ i} + 0.60 \textbf{ j}) = (800 \textbf{ i} + 600 \textbf{ j}) \text{ lb}$$

$$\textbf{r} = \textbf{r}_{A/O} = (4 \textbf{ i} + 8 \textbf{ j}) \text{ in.}$$

From Eq. (5-7),

$$\textbf{M}_O = \textbf{r} \times \textbf{F} = \begin{vmatrix} \textbf{i} & \textbf{j} & \textbf{k} \\ r_x & r_y & 0 \\ F_x & F_y & 0 \end{vmatrix} = (r_x F_y - r_y F_x)\textbf{k} = M_{Oz}\textbf{k}$$

$$\textbf{M}_O = (r_x F_y - r_y F_x)\textbf{ k} = [(4)(600) - (8)(800)]\textbf{ k}$$
$$= -4000 \textbf{ k} \text{ in.} \cdot \text{lb} = 4000 \text{ in.} \cdot \text{lb } \downarrow \quad \textbf{Ans.}$$

(b) The position vector **r** from point B to point A is

$$\textbf{r} = \textbf{r}_{A/B} = (4 \textbf{ i} + 14 \textbf{ j}) \text{ in.}$$

$$\textbf{M}_B = (r_x F_y - r_y F_x)\textbf{ k} = [(4)(600) - (14)(800)]\textbf{ k}$$
$$= -8800 \textbf{ k} \text{ in.} \cdot \text{lb} = 8800 \text{ in.} \cdot \text{lb } \downarrow$$

$$d = \frac{|\textbf{M}_B|}{|\textbf{F}|} = \frac{8800}{1000} = 8.80 \text{ in.} \quad \blacksquare \qquad \textbf{Ans.}$$

■ PROBLEMS

Use the vector definition $\textbf{M} = \textbf{r} \times \textbf{F}$ in the solution of the following problems.

Standard Problems

5-13* A 100-lb force is applied to a curved bar as shown in Fig. P5-13. Determine the moment of the force about point B.

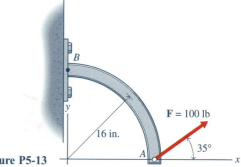

Figure P5-13

5-14* A 450-N force is applied to a bracket as shown in Fig. P5-14. Determine the moment of the force
(a) about point B. (b) about point C.

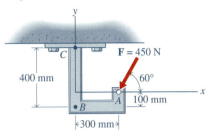

Figure P5-14

5-15 A 583-lb force is applied to a bracket as shown in Fig. P5-15. Determine the moment of the force
(a) about point D. (b) about point E.

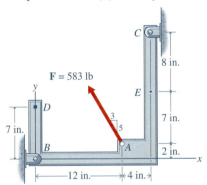

Figure P5-15

5-16 A 650-N force is applied to a bracket as shown in Fig. P5-16. Determine the moment of the force
(a) about point D. (b) about point E.

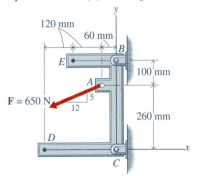

Figure P5-16

5-17* Two forces F_1 and F_2 are applied to a gusset plate as shown in Fig. P5-17. Determine
(a) The moment of force F_1 about point A.
(b) The moment of force F_2 about point B.

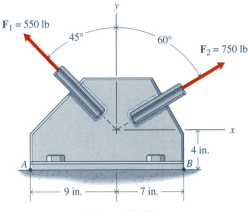

Figure P5-17

5-18* Two forces F_1 and F_2 are applied to a bracket as shown in Fig. P5-18. Determine
(a) The moment of force F_1 about point O.
(b) The moment of force F_2 about point A.

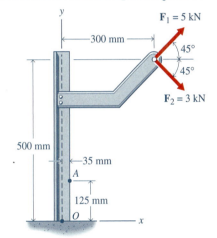

Figure P5-18

5-19 Three forces F_A, F_B, and F_C are applied to a beam as shown in Fig. P5-19. Determine
(a) The moments of forces F_A and F_C about point O.
(b) The moment of force F_B about point D.

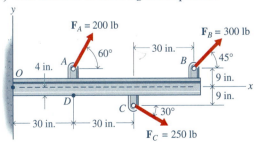

Figure P5-19

5-20 Three forces $\mathbf{F}_1$, $\mathbf{F}_2$, and $\mathbf{F}_3$ are applied to a bracket as shown in Fig. P5-20. Determine the moments of each of the forces about point B.

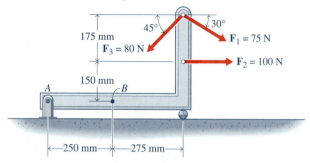

Figure P5-20

Challenging Problems

5-21* A 50-lb force is applied to a triangular plate as shown in Fig. P5-21. Determine the moment of the force
(a) about point A.   (b) about point B.

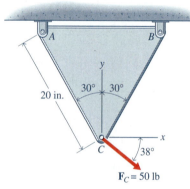

Figure P5-21

5-22* A 300-N force is applied to a triangular bracket as shown in Fig. P5-22. Determine the moment of the force
(a) about point A.   (b) about point B.

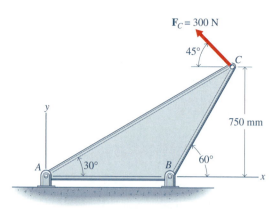

Figure P5-22

5-23 A 300-lb force is applied to a bent bar as shown in Fig. P5-23. Determine the moment of the force
(a) about point A.   (b) about point B.

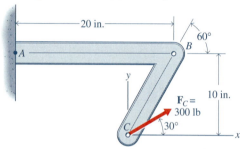

Figure P5-23

5-24 A 450-N force is applied to a bracket as shown in Fig. P5-24. Determine the moment of the force
(a) about point A.   (b) about point B.

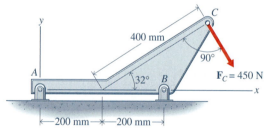

Figure P5-24

The Three-Dimensional Case. The moment $\mathbf{M}_O$ about the origin of coordinates O produced by a force $\mathbf{F}$ with a space (three-dimensional) orientation can also be determined by using Eq. (5-3). For this general case (see Fig. 5-11), the force $\mathbf{F}$ can be expressed in Cartesian vector form as

$$\mathbf{F} = F_x\,\mathbf{i} + F_y\,\mathbf{j} + F_z\,\mathbf{k}$$

and the position vector $\mathbf{r}$ from the origin O to an arbitrary point A on the line of action of the force as

$$\mathbf{r} = r_x\,\mathbf{i} + r_y\,\mathbf{j} + r_z\,\mathbf{k}$$

The vector product $\mathbf{r} \times \mathbf{F}$ for this three-dimensional case can be written in determinant form as

$$\mathbf{M}_O = \mathbf{r} \times \mathbf{F} = \begin{vmatrix} \mathbf{i} & \mathbf{j} & \mathbf{k} \\ r_x & r_y & r_z \\ F_x & F_y & F_z \end{vmatrix}$$

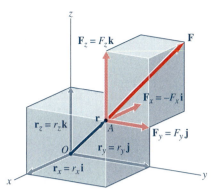

$$= (r_y F_z - r_z F_y) \mathbf{i} + (r_z F_x - r_x F_z) \mathbf{j} + (r_x F_y - r_y F_x) \mathbf{k}$$
$$= M_{Ox} \mathbf{i} + M_{Oy} \mathbf{j} + M_{Oz} \mathbf{k} \qquad (5\text{-}9)$$

where

$$\begin{aligned} M_{Ox} &= r_y F_z - r_z F_y \\ M_{Oy} &= r_z F_x - r_x F_z \\ M_{Oz} &= r_x F_y - r_y F_x \end{aligned} \qquad (5\text{-}10)$$

Figure 5-11

are the three scalar components of the moment of force $\mathbf{F}$ about point O. The magnitude of the moment $|\mathbf{M}_O|$ (see Fig. 5-12) is

$$|\mathbf{M}_O| = \sqrt{M_{Ox}^2 + M_{Oy}^2 + M_{Oz}^2} \qquad (5\text{-}11)$$

Alternatively, the moment $\mathbf{M}_O$ can be written as

$$\mathbf{M}_O = M_O \, \mathbf{e} \qquad (5\text{-}12)$$

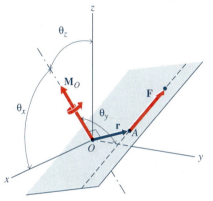

where

$$\mathbf{e} = \cos \theta_x \, \mathbf{i} + \cos \theta_y \, \mathbf{j} + \cos \theta_z \, \mathbf{k} \qquad (5\text{-}13)$$

The direction cosines associated with the unit vector $\mathbf{e}$ are

$$\cos \theta_x = \frac{M_{Ox}}{|\mathbf{M}_O|} \qquad \cos \theta_y = \frac{M_{Oy}}{|\mathbf{M}_O|} \qquad \cos \theta_z = \frac{M_{Oz}}{|\mathbf{M}_O|} \qquad (5\text{-}14)$$

Figure 5-12

A moment obeys all the rules of vector combination and can be considered a sliding vector with a line of action coinciding with the moment axis.

The principle of moments discussed in Section 5-2-1 is not restricted to two concurrent forces but may be extended to any force system. The proof for an arbitrary number of concurrent forces follows from the distributive property of the vector product. Thus,

$$\mathbf{M}_O = \mathbf{r} \times \mathbf{R}$$

where

$$\mathbf{R} = \mathbf{F}_1 + \mathbf{F}_2 + \cdots + \mathbf{F}_n$$

therefore

$$\begin{aligned} \mathbf{M}_O &= \mathbf{r} \times (\mathbf{F}_1 + \mathbf{F}_2 + \cdots + \mathbf{F}_n) \\ &= (\mathbf{r} \times \mathbf{F}_1) + (\mathbf{r} \times \mathbf{F}_2) + \cdots + (\mathbf{r} \times \mathbf{F}_n) \end{aligned}$$

Thus

$$\mathbf{M}_O = \mathbf{M}_{OR} = \mathbf{M}_{O1} + \mathbf{M}_{O2} + \cdots + \mathbf{M}_{On} \qquad (5\text{-}15)$$

Equation (5-15) indicates that the moment of the resultant of any number of forces, $\mathbf{M}_{OR}$, is equal to the sum of the moments of the individual forces.

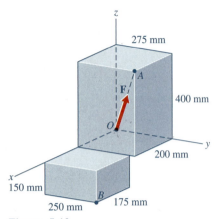

Figure 5-13

Example Problem 5-4

A force **F** of magnitude 840 N acts at a point as shown in Fig. 5-13. Determine

(a) The moment of the force about point B.

(b) The direction angles associated with the unit vector **e** along the axis of the moment.

(c) The perpendicular distance d from point B to the line of action of the force.

SOLUTION

(a) Scalar Analysis

$$d_{OA} = \sqrt{(x_A)^2 + (y_A)^2 + (z_A)^2}$$
$$= \sqrt{(200)^2 + (275)^2 + (400)^2} = 525 \text{ mm}$$

$$F_x = \frac{200}{525}(F) = \frac{200}{525}(840) = 320 \text{ N}$$

$$F_y = \frac{275}{525}(F) = \frac{275}{525}(840) = 440 \text{ N}$$

$$F_z = \frac{400}{525}(F) = \frac{400}{525}(840) = 640 \text{ N}$$

At point B (positive moments counterclockwise when viewing in the negative coordinate direction):

$$M_x = -F_y z_B - F_z y_B = -440(0.150) - 640(0.250) = -226 \text{ N} \cdot \text{m}$$
$$M_y = +F_z x_B + F_x z_B = +640(0.375) + 320(0.150) = +288 \text{ N} \cdot \text{m}$$
$$M_z = +F_x y_B - F_y x_B = +320(0.250) - 440(0.375) = -85.0 \text{ N} \cdot \text{m}$$
$$M_B = \sqrt{M_x^2 + M_y^2 + M_z^2}$$
$$= \sqrt{(-226)^2 + (288)^2 + (-85.0)^2} = 376 \text{ N} \cdot \text{m} \qquad \textbf{Ans.}$$

(a) Vector Analysis

The force **F** and the position vector **r** from point B to point A can be written in Cartesian vector form as

$$\mathbf{F} = 840[(200/525)\,\mathbf{i} + (275/525)\,\mathbf{j} + (400/525)\,\mathbf{k}]$$
$$= (320\,\mathbf{i} + 440\,\mathbf{j} + 640\,\mathbf{k}) \text{ N}$$

$$\mathbf{r}_{A/B} = (-0.175\,\mathbf{i} + 0.025\,\mathbf{j} + 0.550\,\mathbf{k}) \text{ m}$$

The moment $\mathbf{M}_B$ is given by Eq. (5-9) as

$$\mathbf{M}_B = \mathbf{r}_{A/B} \times \mathbf{F} = \begin{vmatrix} \mathbf{i} & \mathbf{j} & \mathbf{k} \\ r_x & r_y & r_z \\ F_x & F_y & F_z \end{vmatrix} = \begin{vmatrix} \mathbf{i} & \mathbf{j} & \mathbf{k} \\ -0.175 & 0.025 & 0.550 \\ 320 & 440 & 640 \end{vmatrix}$$

$$= (-226\,\mathbf{i} + 288\,\mathbf{j} - 85.0\,\mathbf{k}) \text{ N} \cdot \text{m} \qquad \textbf{Ans.}$$

The position vector **r** can also be written from point B to point O as

$$\mathbf{r}_{O/B} = (-0.375 \ \mathbf{i} - 0.250 \ \mathbf{j} + 0.150 \ \mathbf{k}) \ \text{m}$$

The moment $\mathbf{M}_B$ is then given by Eq. (5-9) as

$$\mathbf{M}_B = \mathbf{r}_{O/B} \times \mathbf{F} = \begin{vmatrix} \mathbf{i} & \mathbf{j} & \mathbf{k} \\ r_x & r_y & r_z \\ F_x & F_y & F_z \end{vmatrix} = \begin{vmatrix} \mathbf{i} & \mathbf{j} & \mathbf{k} \\ -0.375 & -0.250 & 0.150 \\ 320 & 440 & 640 \end{vmatrix}$$

$$= (-226 \ \mathbf{i} + 288 \ \mathbf{j} - 85.0 \ \mathbf{k}) \ \text{N} \cdot \text{m} \qquad \textbf{Ans.}$$

(b) The magnitude of moment $\mathbf{M}_B$ is obtained by using Eq. (5-11). Thus

$$|\mathbf{M}_B| = \sqrt{M_x^2 + M_y^2 + M_z^2} = \sqrt{(-226)^2 + (288)^2 + (-85.0)^2} = 375.8 \ \text{N} \cdot \text{m}$$

The direction angles are obtained by using Eqs. (5-14). Thus

$$\theta_x = \cos^{-1}\frac{M_x}{|\mathbf{M}_B|} = \cos^{-1}\frac{-226}{375.8} = 127.0° \qquad \textbf{Ans.}$$

$$\theta_y = \cos^{-1}\frac{M_y}{|\mathbf{M}_B|} = \cos^{-1}\frac{288}{375.8} = 40.0° \qquad \textbf{Ans.}$$

$$\theta_z = \cos^{-1}\frac{M_z}{|\mathbf{M}_B|} = \cos^{-1}\frac{-85.0}{375.8} = 103.1° \qquad \textbf{Ans.}$$

(c) The distance d is obtained by using the definition of a moment. Thus

$$d = \frac{M}{F} = \frac{375.8}{840} = 0.447 \ \text{m} = 447 \ \text{mm} \quad \blacksquare \qquad \textbf{Ans.}$$

PROBLEMS

Standard Problems

5-25* A force with a magnitude of 600 lb acts at a point in a body as shown in Fig. P5-25. Determine the moment of the force about point B.

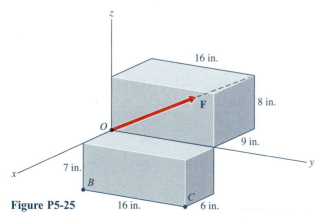

Figure P5-25

5-26* A force with a magnitude of 850 N acts at a point in a body as shown in Fig. P5-26. Determine the moment of the force about point B.

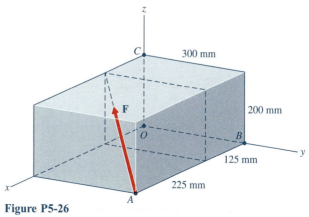

Figure P5-26

5-27 A force with a magnitude of 480 lb acts at a point in a body as shown in Fig. P5-25. Determine the moment of the force about point C.

5-28 A force with a magnitude of 680 N acts at a point in a body as shown in Fig. P5-26. Determine the moment of the force about point C.

5-29* A force with a magnitude of 580 lb acts at a point in a body as shown in Fig. P5-29. Determine
(a) The moment of the force about point B.
(b) The direction angles associated with the unit vector **e** along the axis of the moment.

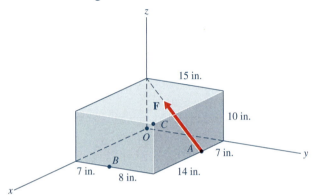

Figure P5-29

5-30* A force with a magnitude of 730 N acts at a point in a body as shown in Fig. P5-30. Determine
(a) The moment of the force about point B.
(b) The direction angles associated with the unit vector **e** along the axis of the moment.

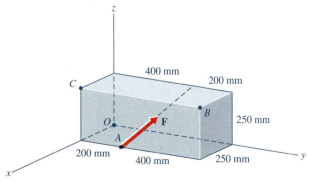

Figure P5-30

5-31 A force with a magnitude of 870 lb acts at a point in a body as shown in Fig. P5-29. Determine

(a) The moment of the force about point C.
(b) The direction angles associated with the unit vector **e** along the axis of the moment.

5-32 A force with a magnitude of 585 N acts at a point in a body as shown in Fig. P5-30. Determine
(a) The moment of the force about point C.
(b) The direction angles associated with the unit vector **e** along the axis of the moment.

Challenging Problems

5-33* Determine the moment of the 580-lb force shown in Fig. P5-33 about point B.

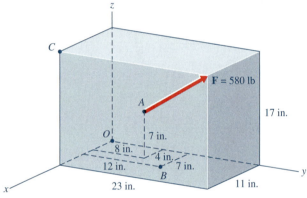

Figure P5-33

5-34* Determine the moment of the 760-N force shown in Fig. P5-34 about point B.

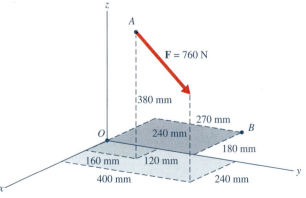

Figure P5-34

5-35 A pipe bracket is loaded as shown in Fig. P5-35. Determine the moment of the force **F** about point *B*.

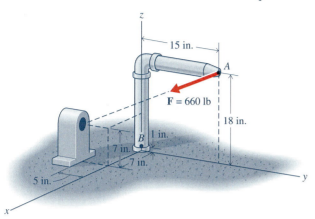

Figure P5-35

5-36 A pipe bracket is loaded as shown in Fig. P5-36. Determine the moment of the force **F** about point *O*.

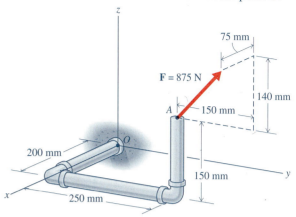

Figure P5-36

5-3-2 Moment of a Force About a Line (Axis)

The moment $\mathbf{M}_O$ of a force **F** about a point *O* was defined as the vector product

$$\mathbf{M}_O = \mathbf{r} \times \mathbf{F} \tag{5-3}$$

While it is possible, mathematically, to define the moment of a force about a point, the quantity has no physical significance in mechanics since bodies rotate about axes (as illustrated in Fig. 5-1) and not points. The vector definition of a moment about a point [Eq. (5-3)] is only an intermediate step in a process that allows us to find the moment about an axis that passes through the point.

The moment $\mathbf{M}_{OB}$ of a force with respect to a line (say line *OB* in Fig. 5-14) can be determined by first calculating the moment $\mathbf{M}_O$ about point *O* on the line (or about any other point on the line). Then, the moment vector $\mathbf{M}_O$ can be resolved into components that are parallel $\mathbf{M}_\parallel$ and perpendicular $\mathbf{M}_\perp$ to the line *OB,* as shown in Fig. 5-15. If $\mathbf{e}_n$ is a unit vector in the *n*-direction along line *OB,* as shown in Fig. 5-14, then

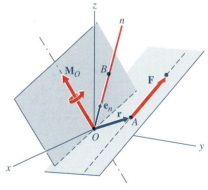

Figure 5-14

$$\mathbf{M}_{OB} = \mathbf{M}_\parallel = (\mathbf{M}_O \cdot \mathbf{e}_n) \, \mathbf{e}_n$$
$$= [(\mathbf{r} \times \mathbf{F}) \cdot \mathbf{e}_n] \, \mathbf{e}_n = M_{OB} \, \mathbf{e}_n \tag{5-16}$$

These two operations, the vector product $\mathbf{r} \times \mathbf{F}$ of the position vector **r** and the force **F** to obtain the moment $\mathbf{M}_O$ about point *O*, followed by the scalar product $\mathbf{M}_O \cdot \mathbf{e}_n$ of the moment $\mathbf{M}_O$ about point *O* and the unit vector $\mathbf{e}_n$ along the desired moment axis, to obtain the moment M_{OB} can be performed in sequence or combined into one operation. The quantity inside the brackets of Eq. (5-16) is called the *triple scalar product.* The triple scalar product can be written in determinant form as

$$M_{OB} = \mathbf{M}_O \cdot \mathbf{e}_n = (\mathbf{r} \times \mathbf{F}) \cdot \mathbf{e}_n = \begin{vmatrix} \mathbf{i} & \mathbf{j} & \mathbf{k} \\ r_x & r_y & r_z \\ F_x & F_y & F_z \end{vmatrix} \cdot \mathbf{e}_n \tag{5-17}$$

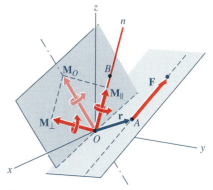

Figure 5-15

or alternatively as

$$M_{OB} = \mathbf{M}_O \cdot \mathbf{e}_n = (\mathbf{r} \times \mathbf{F}) \cdot \mathbf{e}_n = \begin{vmatrix} e_{nx} & e_{ny} & e_{nz} \\ r_x & r_y & r_z \\ F_x & F_y & F_z \end{vmatrix} \quad (5\text{-}18)$$

where e_{nx}, e_{ny}, and e_{nz} are the Cartesian components (direction cosines) of the unit vector $\mathbf{e}_n$. The unit vector $\mathbf{e}_n$ is usually selected in the direction from O toward B. A positive coefficient of $\mathbf{e}_n$ in the expression $\mathbf{M}_{OB} = M_{OB} \mathbf{e}_n$ means that the moment vector has the same sense as that selected for $\mathbf{e}_n$, whereas a negative sign indicates that $\mathbf{M}_{OB}$ is opposite to the sense of $\mathbf{e}_n$.

Example Problem 5-5

The force $\mathbf{F}$ in Fig. 5-16 has a magnitude of 440 lb. Determine

(a) The moment $\mathbf{M}_B$ of the force about point B.
(b) The component of moment $\mathbf{M}_B$ parallel to line BC.
(c) The component of moment $\mathbf{M}_B$ perpendicular to line BC.
(d) The unit vector associated with the component of moment $\mathbf{M}_B$ perpendicular to line BC.

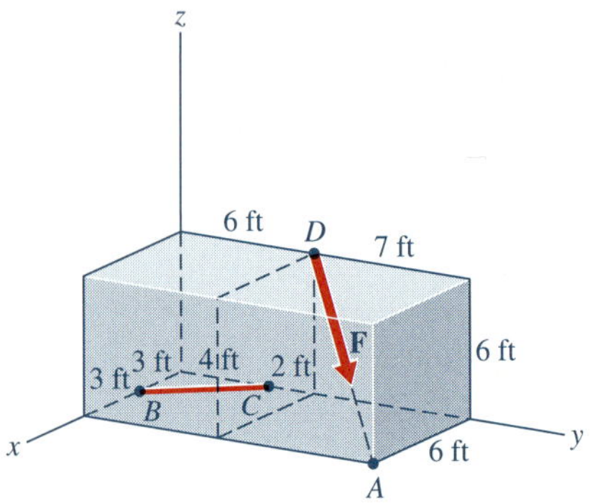

Figure 5-16

SOLUTION

(a) The force $\mathbf{F}$ and the position vector $\mathbf{r}$ from point B to point A can be written in Cartesian vector form as

$$\mathbf{F} = 440[(6/11)\,\mathbf{i} + (7/11)\,\mathbf{j} - (6/11)\,\mathbf{k}] = (240\,\mathbf{i} + 280\,\mathbf{j} - 240\,\mathbf{k}) \text{ lb}$$

$$\mathbf{r}_{A/B} = (3\,\mathbf{i} + 13\,\mathbf{j}) \text{ ft}$$

The moment $\mathbf{M}_B$ is given by Eq. (5-9) as

$$\mathbf{M}_B = \mathbf{r}_{A/B} \times \mathbf{F} = \begin{vmatrix} \mathbf{i} & \mathbf{j} & \mathbf{k} \\ r_x & r_y & r_z \\ F_x & F_y & F_z \end{vmatrix} = \begin{vmatrix} \mathbf{i} & \mathbf{j} & \mathbf{k} \\ 3 & 13 & 0 \\ 240 & 280 & -240 \end{vmatrix}$$

$$= (-3120\,\mathbf{i} + 720\,\mathbf{j} - 2280\,\mathbf{k}) \text{ ft} \cdot \text{lb} \qquad \textbf{Ans.}$$

(b) The unit vector $\mathbf{e}_{BC}$ associated with line BC is

$$\mathbf{e}_{BC} = -0.60\,\mathbf{i} + 0.80\,\mathbf{j}$$

The component of moment $\mathbf{M}_B$ parallel to line BC is given by Eq. (5-16) as

$$\mathbf{M}_{BC} = \mathbf{M}_\| = (\mathbf{M}_B \cdot \mathbf{e}_{BC})\,\mathbf{e}_{BC} = M_{BC}\,\mathbf{e}_{BC}$$

$$\begin{aligned} M_{BC} &= \mathbf{M}_B \cdot \mathbf{e}_{BC} \\ &= (-3120\,\mathbf{i} + 720\,\mathbf{j} - 2280\,\mathbf{k}) \cdot (-0.60\,\mathbf{i} + 0.80\,\mathbf{j}) \\ &= (-3120)(-0.60) + (720)(0.80) = 2448 \text{ ft} \cdot \text{lb} \end{aligned}$$

$$\begin{aligned} \mathbf{M}_{BC} &= M_{BC}\,\mathbf{e}_{BC} \\ &= 2448(-0.60\,\mathbf{i} + 0.80\,\mathbf{j}) \\ &= (-1469\,\mathbf{i} + 1958\,\mathbf{j}) \text{ ft} \cdot \text{lb} \qquad \textbf{Ans.} \end{aligned}$$

The moment M_{BC} can be determined in a single operation by using Eq. (5-18). A different point on the line can also be used. For example, consider point C and the position vector $\mathbf{r}_{D/C}$.

$$\mathbf{r}_{D/C} = (2\,\mathbf{j} + 6\,\mathbf{k}) \text{ ft}$$

From Eq. (5-18):

$$M_{BC} = \begin{vmatrix} e_{nx} & e_{ny} & e_{nz} \\ r_x & r_y & r_z \\ F_x & F_y & F_z \end{vmatrix} = \begin{vmatrix} -0.60 & 0.80 & 0 \\ 0 & 2 & 6 \\ 240 & 280 & -240 \end{vmatrix}$$

$$= (-0.60)(-2160) - (0.80)(-1440) = 2448 \text{ ft} \cdot \text{lb}$$

(c) The moment $\mathbf{M}_\perp$ is obtained as the difference between $\mathbf{M}_B$ and $\mathbf{M}_\parallel$ since $\mathbf{M}_\parallel$ and $\mathbf{M}_\perp$ are the two rectangular components of $\mathbf{M}_B$. Thus,

$$\begin{aligned} \mathbf{M}_\perp &= \mathbf{M}_B - \mathbf{M}_\parallel = \mathbf{M}_B - \mathbf{M}_{BC} \\ &= (-3120\,\mathbf{i} + 720\,\mathbf{j} - 2280\,\mathbf{k}) - (-1469\,\mathbf{i} + 1958\,\mathbf{j}) \\ &= (-1651\,\mathbf{i} - 1238\,\mathbf{j} - 2280\,\mathbf{k}) \text{ ft} \cdot \text{lb} \qquad \textbf{Ans.} \end{aligned}$$

(d) The magnitude of moment $\mathbf{M}_\perp$ is

$$|\mathbf{M}_\perp| = \sqrt{(-1651)^2 + (-1238)^2 + (-2280)^2} = 3075 \text{ ft} \cdot \text{lb}$$

Therefore

$$\begin{aligned} \mathbf{e}_\perp &= (-1651/3075)\,\mathbf{i} + (-1238/3075)\,\mathbf{j} + (-2280/3075)\,\mathbf{k} \\ &= -0.537\,\mathbf{i} - 0.403\,\mathbf{j} - 0.741\,\mathbf{k} \qquad \textbf{Ans.} \end{aligned}$$

As a check:

$$\begin{aligned} \mathbf{e}_\parallel \cdot \mathbf{e}_\perp &= (-0.60\,\mathbf{i} + 0.80\,\mathbf{j}) \cdot (-0.5369\,\mathbf{i} - 0.4026\,\mathbf{j} - 0.7415\,\mathbf{k}) \\ &= 0.00006 \cong 0 \end{aligned}$$

which verifies, except for round-off in the expression for $\mathbf{e}_\perp$, that the moment components $\mathbf{M}_\parallel$ and $\mathbf{M}_\perp$ are perpendicular. ∎

PROBLEMS

Standard Problems

5-37* The force $\mathbf{F}$ in Fig. P5-37 can be expressed in Cartesian vector form as $\mathbf{F} = (60\,\mathbf{i} + 100\,\mathbf{j} + 120\,\mathbf{k})$ lb. Determine the scalar component of the moment at point B about line BC.

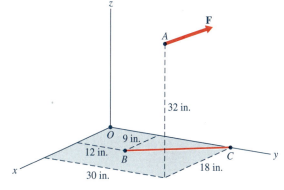

Figure P5-37

5-38* The magnitude of the force **F** in Fig. P5-38 is 735 N. Determine the moment of the force about line *CD*. Express the result in Cartesian vector form.

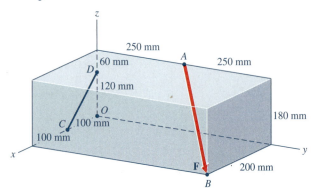

Figure P5-38

5-39 The magnitude of the force **F** in Fig. P5-39 is 450 lb. Determine the moment of the force about line *BC*. Express the result in Cartesian vector form.

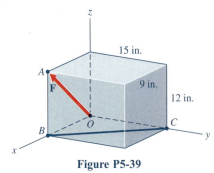

Figure P5-39

5-40 The magnitude of the force **F** in Fig. P5-40 is 635 N. Determine

(a) The scalar component of the moment at point *O* about line *OC*.
(b) The scalar component of the moment at point *D* about line *DE*.

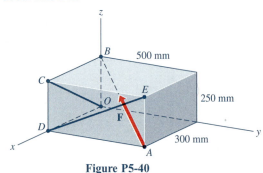

Figure P5-40

5-41* The magnitude of the force **F** in Fig. P5-41 is 680 lb. Determine

(a) The scalar component of the moment at point *O* about line *OC*.
(b) The scalar component of the moment at point *D* about line *DE*.

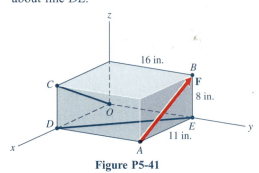

Figure P5-41

5-42 The force **F** in Fig. P5-42 has a magnitude of 721 N. Determine

(a) The moment M_{CD} of the force about line *CD*.
(b) The moment M_{CE} of the force about line *CE*.

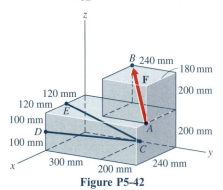

Figure P5-42

5-43* The magnitude of the force **F** in Fig. P5-43 is 107 lb. Determine the scalar component of the moment at point *O* about line *OC*.

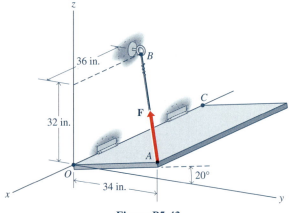

Figure P5-43

5-44* A bracket is subjected to an 825-N force as shown in Fig. P5-44. Determine the scalar component of the moment of the force about line *OB*.

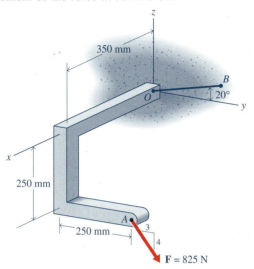

Figure P5-44

5-45 A 200-lb force is applied to a lever-shaft assembly as shown in Fig. P5-45. Determine the moment of the force about line *OC*. Express the results in Cartesian vector form.

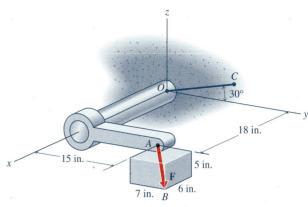

Figure P5-45

5-46 A bracket is subjected to a 384-N force as shown in Fig. P5-46. Determine the moment of the force about line *OC*. Express the results in Cartesian vector form.

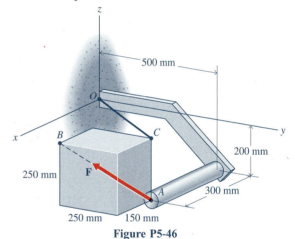

Figure P5-46

5-4 COUPLES

Two equal, noncollinear, parallel forces of opposite sense (see Fig. 5-17) are called a *couple*. Since the two forces are equal, parallel, and of opposite sense, the sum of the forces in any direction is zero. Therefore, a couple tends only to rotate the body on which it acts. The moment of a couple is defined as the sum of the moments of the pair of forces that comprise the couple. For the two forces $\mathbf{F}_1$ and $\mathbf{F}_2$ shown in Fig. 5-17, the moments of the couple about points A and B in the plane of the couple are

$$M_A = \left| \mathbf{F}_2 \right| d \qquad M_B = \left| \mathbf{F}_1 \right| d$$

However,

$$\left| \mathbf{F}_1 \right| = \left| \mathbf{F}_2 \right| = F$$

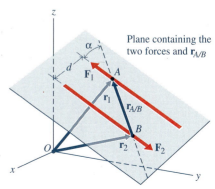

Figure 5-17

therefore,

$$M_A = M_B = Fd$$

which indicates that the magnitude of the moment of a couple about a point in the plane of the couple is equal to the magnitude of one of the forces multiplied by the perpendicular distance between the forces. The point A or B does not have to lie on one of the forces but must lie in the plane of the forces.

Other characteristics of a couple can be determined by considering two parallel forces in space such as those shown in Fig. 5-17. The sum of the moments of the two forces about any point O is

$$\mathbf{M}_O = \mathbf{r}_1 \times \mathbf{F}_1 + \mathbf{r}_2 \times \mathbf{F}_2$$

or since $\mathbf{F}_2$ equals $-\mathbf{F}_1$

$$\begin{aligned}\mathbf{M}_O &= \mathbf{r}_1 \times \mathbf{F}_1 + \mathbf{r}_2 \times (-\mathbf{F}_1) \\ &= (\mathbf{r}_1 - \mathbf{r}_2) \times \mathbf{F}_1 = \mathbf{r}_{A/B} \times \mathbf{F}_1\end{aligned}$$

where $\mathbf{r}_{A/B}$ is the position vector from any point B on the line of action of force $\mathbf{F}_2$ to any point A on the line of action of force $\mathbf{F}_1$. Therefore, from the definition of the vector cross product,

$$\begin{aligned}\mathbf{M}_O &= \mathbf{r}_{A/B} \times \mathbf{F}_1 \\ &= \left|\mathbf{r}_{A/B}\right| \left|\mathbf{F}_1\right| \sin \alpha \, \mathbf{e} = F_1 d \, \mathbf{e}\end{aligned} \qquad (5\text{-}19)$$

where d is the perpendicular distance between the forces of the couple and $\mathbf{e}$ is a unit vector perpendicular to the plane of the couple with its sense in the direction specified for moments by the right-hand rule. It is obvious from Eq. (5-19) that the moment of a couple does not depend upon the location of the moment center O. Thus, the moment of a couple is a free vector.

The characteristics of a couple, which control its "external effect" (overall tendency to translate and to rotate) on a rigid body, are

1. The magnitude of the moment of the couple.
2. The sense (direction of rotation) of the couple.
3. The aspect of the plane of the couple; that is, the direction or slope of the plane (not its location) as defined by a normal n to the plane.

Equation (5-19) indicates that several transformations of a couple can be made without changing any of the external effects of the couple on the body. For example,

1. A couple can be translated to a parallel position in its plane or to any parallel plane [since position vectors $\mathbf{r}_1$ and $\mathbf{r}_2$ do not appear in Eq. (5-19)].
2. A couple can be rotated in its plane.
3. The magnitude of the two forces of a couple and the distance between them can be changed provided the product Fd remains constant.

For two-dimensional problems, a couple is frequently represented by a curved arrow on a sketch of the body. The magnitude of the moment of the couple $|\mathbf{M}|$ $(M = Fd)$ is provided and the curved arrow indicates the sense of the couple.

Any number of couples $\mathbf{C}_1, \mathbf{C}_2, \ldots, \mathbf{C}_n$ in a plane can be combined to yield a resultant couple $\mathbf{C}$ equal to the algebraic sum of the individual couples. A system of couples in space (see Fig. 5-18a) can be combined into a single resultant couple $\mathbf{C}$, by representing each couple of the system (since a couple is a free vector) by a vector, drawn for convenience, from the origin of a set of rectangular axes. Each couple can then be resolved into components $\mathbf{C}_x$, $\mathbf{C}_y$, and $\mathbf{C}_z$ along the coordinate axes. These vector components represent couples lying in the yz-, zx-, and xy-planes, respectively. The x-, y-, and z-components of the resultant couple $\mathbf{C}$ are obtained as $\Sigma \mathbf{C}_x$, $\Sigma \mathbf{C}_y$, and $\Sigma \mathbf{C}_z$, as shown in Fig. 5-18b. The original system of couples is thus reduced to three couples lying in the coordinate planes. The resultant couple $\mathbf{C}$ for the system (see Fig. 5-18c) can be written in vector form as

$$\mathbf{C} = \Sigma \mathbf{C}_x + \Sigma \mathbf{C}_y + \Sigma \mathbf{C}_z = \Sigma C_x\,\mathbf{i} + \Sigma C_y\,\mathbf{j} + \Sigma C_z\,\mathbf{k} \qquad (5\text{-}20)$$

The magnitude of the couple $\mathbf{C}$ is

$$|\mathbf{C}| = \sqrt{(\Sigma C_x)^2 + (\Sigma C_y)^2 + (\Sigma C_z)^2} \qquad (5\text{-}21)$$

Alternatively, the couple $\mathbf{C}$ can be written as

$$\mathbf{C} = |\mathbf{C}|\,\mathbf{e} = C\,\mathbf{e} \qquad (5\text{-}22)$$

where

$$\mathbf{e} = \cos\theta_x\,\mathbf{i} + \cos\theta_y\,\mathbf{j} + \cos\theta_z\,\mathbf{k}$$

The direction cosines associated with the unit vector $\mathbf{e}$ are

$$\theta_x = \cos^{-1}\frac{\Sigma C_x}{|\mathbf{C}|} \qquad \theta_y = \cos^{-1}\frac{\Sigma C_y}{|\mathbf{C}|} \qquad \theta_z = \cos^{-1}\frac{\Sigma C_z}{|\mathbf{C}|} \qquad (5\text{-}23)$$

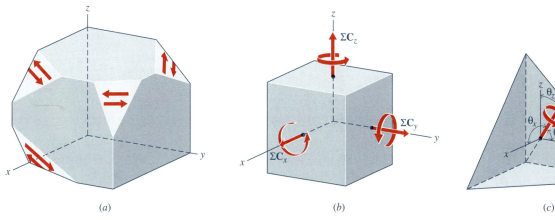

(a)   (b)   (c)

Figure 5-18

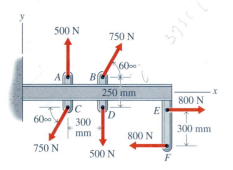

Figure 5-19

Example Problem 5-6

A beam is loaded with a system of forces as shown in Fig. 5-19. Express the resultant of the force system in Cartesian vector form.

SOLUTION

An examination of Fig. 5-19 indicates that the force system consists of a system of three couples in the same plane. A scalar analysis yields:
For forces A and D:

$$M_1 = F_A d_1 = 500(0.300) = 150.0 \text{ N} \cdot \text{m} \downarrow = -150.0 \text{ N} \cdot \text{m}$$

For forces B and C:

The perpendicular distance between forces B and C is not obvious from Fig. 5-19; therefore, components of the forces will be used.

$$F_{Bx} = F_B \cos 60° = 750 \cos 60° = 375 \text{ N}$$

$$F_{By} = F_B \sin 60° = 750 \sin 60° = 649.5 \text{ N}$$

$$M_2 = F_{Bx} d_2 = 375(0.250) = 93.75 \text{ N} \cdot \text{m} \downarrow = -93.75 \text{ N} \cdot \text{m}$$

$$M_3 = F_{By} d_3 = 649.5(0.300) = 194.85 \text{ N} \cdot \text{m} \uparrow = +194.85 \text{ N} \cdot \text{m}$$

For forces E and F:

$$M_4 = F_E d_4 = 800(0.300) = 240 \text{ N} \cdot \text{m} \downarrow = -240 \text{ N} \cdot \text{m}$$

For any number of couples in a plane, the resultant couple $\mathbf{C}$ is equal to the algebraic sum of the individual couples. Thus,

$$C = \Sigma M = -150.0 - 93.75 + 194.85 - 240 = -288.9 \text{ N} \cdot \text{m} \cong -289 \text{ N} \cdot \text{m}$$

$$\mathbf{C} = -289 \text{ k N} \cdot \text{m} \qquad \text{Ans.}$$

Alternatively, a vector analysis yields

$$\mathbf{C} = (\mathbf{r}_{D/A} \times \mathbf{F}_D) + (\mathbf{r}_{B/C} \times \mathbf{F}_B) + (\mathbf{r}_{E/F} \times \mathbf{F}_E)$$
$$= [(0.300 \ \mathbf{i} - 0.250 \ \mathbf{j}) \times (-500 \ \mathbf{j})]$$
$$+ [(0.300 \ \mathbf{i} + 0.250 \ \mathbf{j}) \times (375 \ \mathbf{i} + 649.5 \ \mathbf{j})] + [(0.300 \ \mathbf{j}) \times (800 \ \mathbf{i})]$$
$$= -288.9 \ \mathbf{k} \text{ N} \cdot \text{m} \cong -289 \ \mathbf{k} \text{ N} \cdot \text{m} \qquad \blacksquare \qquad \text{Ans.}$$

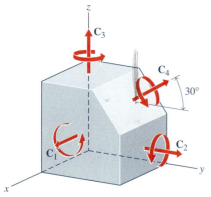

Figure 5-20

Example Problem 5-7

The magnitudes of the four couples applied to the block shown in Fig. 5-20 are $|\mathbf{C}_1| = 75$ ft · lb, $|\mathbf{C}_2| = 50$ ft · lb, $|\mathbf{C}_3| = 60$ ft · lb, and $|\mathbf{C}_4| = 90$ ft · lb. Determine the magnitude of the resultant couple $\mathbf{C}$ and the direction angles associated with the unit vector $\mathbf{e}$ used to describe the normal to the plane of the resultant couple $\mathbf{C}$.

SOLUTION

The x-, y-, and z-components of the resultant couple $\mathbf{C}$ are

$$\Sigma \mathbf{C}_x = \mathbf{C}_1 = 75.0\,\mathbf{i}\ \text{ft}\cdot\text{lb}$$

$$\Sigma \mathbf{C}_y = \mathbf{C}_2 + \mathbf{C}_4 \cos 30° = 50\,\mathbf{j} + 90 \cos 30°\,\mathbf{j} = 127.9\,\mathbf{j}\ \text{ft}\cdot\text{lb}$$

$$\Sigma \mathbf{C}_z = \mathbf{C}_3 + \mathbf{C}_4 \sin 30° = 60\,\mathbf{k} + 90 \sin 30°\,\mathbf{k} = 105.0\,\mathbf{k}\ \text{ft}\cdot\text{lb}$$

The resultant couple $\mathbf{C}$ for the system can be written in vector form as

$$\mathbf{C} = \Sigma \mathbf{C}_x + \Sigma \mathbf{C}_y + \Sigma \mathbf{C}_z = (75.0\,\mathbf{i} + 127.9\,\mathbf{j} + 105.0\,\mathbf{k})\ \text{ft}\cdot\text{lb}$$

The magnitude of the couple $\mathbf{C}$ is

$$|\mathbf{C}| = \sqrt{(\Sigma C_x)^2 + (\Sigma C_y)^2 + (\Sigma C_z)^2}$$
$$= \sqrt{(75.0)^2 + (127.9)^2 + (105.0)^2} = 181.7\ \text{ft}\cdot\text{lb} \qquad \textbf{Ans.}$$

The direction angles are

$$\theta_x = \cos^{-1}\frac{\Sigma C_x}{|\mathbf{C}|} = \cos^{-1}\frac{75.0}{181.7} = 65.6° \qquad \textbf{Ans.}$$

$$\theta_y = \cos^{-1}\frac{\Sigma C_y}{|\mathbf{C}|} = \cos^{-1}\frac{127.9}{181.7} = 45.3° \qquad \textbf{Ans.}$$

$$\theta_z = \cos^{-1}\frac{\Sigma C_z}{|\mathbf{C}|} = \cos^{-1}\frac{105.0}{181.7} = 54.7° \ \blacksquare \qquad \textbf{Ans.}$$

PROBLEMS

Standard Problems

5-47* Two parallel forces of opposite sense $\mathbf{F}_1 = (-70\,\mathbf{i} - 120\,\mathbf{j} - 80\,\mathbf{k})$ lb and $\mathbf{F}_2 = (70\,\mathbf{i} + 120\,\mathbf{j} + 80\,\mathbf{k})$ lb act at points B and A of a body as shown in Fig. P5-47. Determine the moment of the couple and the perpendicular distance between the two forces.

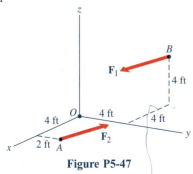

Figure P5-47

5-48* Two parallel forces of opposite sense $\mathbf{F}_1 = (125\,\mathbf{i} + 200\,\mathbf{j} + 250\,\mathbf{k})$ N and $\mathbf{F}_2 = (-125\,\mathbf{i} - 200\,\mathbf{j} - 250\,\mathbf{k})$ N act at points A and B of a body as shown in Fig. P5-48. Determine the moment of the couple and the perpendicular distance between the two forces.

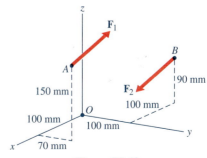

Figure P5-48

5-49 A bracket is loaded with a system of forces as shown in Fig. P5-49. Express the resultant of the force system in Cartesian vector form.

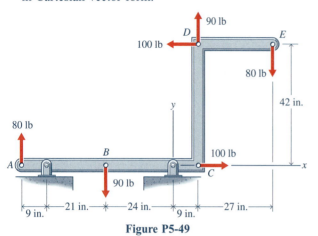

Figure P5-49

5-50 A bracket is loaded with a system of forces as shown in Fig. P5-50. Express the resultant of the force system in Cartesian vector form.

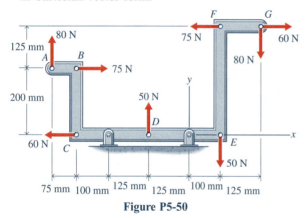

Figure P5-50

Challenging Problems

5-51* Three couples are applied to a bent bar as shown in Fig. P5-51. Determine the magnitude of the resultant couple **C** and the direction angles associated with the unit vector **e** used to describe the normal to the plane of the resultant couple **C**.

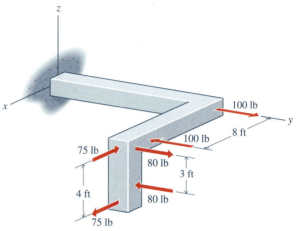

Figure P5-51

5-52* Three couples are applied to a rectangular block as shown in Fig. P5-52. Determine the magnitude of the resultant couple **C** and the direction angles associated with the unit vector **e** used to describe the normal to the plane of the resultant couple **C**.

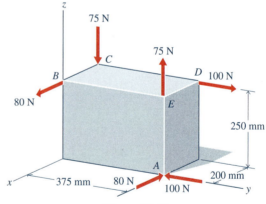

Figure P5-52

5-5 EQUIVALENT FORCE/COUPLE SYSTEMS

In many problems in mechanics it is convenient to resolve a force **F** into a parallel force **F** and a couple **C** (called an *equivalent force-couple*). In Fig. 5-21*a*, let **F** represent a force acting on a body at point *A*. An arbitrary point *O* in the body and the plane containing both force **F** and point *O* are shown shaded in Fig.

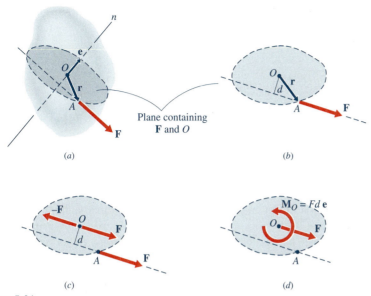

Plane containing
F and O

$M_O = Fd$ **e**

Figure 5-21

5-21a. The aspect of the shaded plane in the body can be described by using its outer normal n and a unit vector **e** along the outer normal. A two-dimensional representation of the shaded plane is shown in Fig. 5-21b. If two equal and opposite collinear forces **F,** parallel to the original force, are introduced at point O, as shown in Fig. 5-21c, the three forces have the same external effect on the body as the original force **F,** since the effects of the two equal, opposite, and collinear forces cancel. The new three-force system can be considered to be a force **F** acting at point O (parallel to the original force and of the same magnitude and sense), and a couple **C,** which has the same moment as the moment of the original force about point O, as shown in Fig. 5-21d. The magnitudes and action lines of the forces of the couple can be changed in accordance with the transformations of a couple discussed in Section 5-4.

Since a force can be resolved into a force and a couple lying in the same plane, it follows conversely that a force and a couple lying in the same plane can be combined into a single force in the plane by reversing the procedure. Thus, the sole external effect of combining a couple with a force is to move the action line of the force into a parallel position. The magnitude and sense of the force remain unchanged.

Two different force systems are equivalent if they produce the same external effect when applied to a rigid body. The "resultant" of any force system is the simplest equivalent system to which the given system will reduce. For some systems, the resultant is a single force. For other systems, the simplest equivalent system is a couple. Still other force systems reduce to a force and a couple as the simplest equivalent system. As will be shown in Chapter 6, for a rigid body to be in equilibrium, the resultant must vanish.

5-5-1 Coplanar Force Systems

The resultant of a system of coplanar forces $\mathbf{F}_1$, $\mathbf{F}_2$, $\mathbf{F}_3$, . . . , $\mathbf{F}_n$ can be determined by using the rectangular components of the forces in any two convenient perpendicular directions. Thus, for a system of forces in the xy-plane, such as the

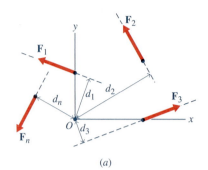

(a)

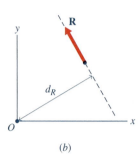

(b)

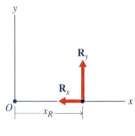

(c)

Figure 5-22

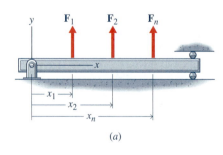

(a)

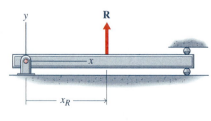

(b)

Figure 5-23

one shown in Fig. 5-22a, the resultant force **R** can be expressed as

$$\mathbf{R} = \mathbf{R}_x + \mathbf{R}_y = R_x\,\mathbf{i} + R_y\,\mathbf{j} = R\,\mathbf{e}$$

where

$$R_x = \Sigma F_x \qquad R_y = \Sigma F_y$$
$$R = |\mathbf{R}| = \sqrt{(\Sigma F_x)^2 + (\Sigma F_y)^2}$$
$$\mathbf{e} = \cos\theta_x\,\mathbf{i} + \cos\theta_y\,\mathbf{j} \qquad (5\text{-}24)$$
$$\cos\theta_x = \frac{\Sigma F_x}{R} \qquad \cos\theta_y = \frac{\Sigma F_y}{R}$$

The location of the line of action of **R** with respect to an arbitrary point O (say the origin of the xy coordinate system) can be computed by using the principle of moments. The moment of **R** about point O must equal the sum of the moments of the forces $\mathbf{F}_1$, $\mathbf{F}_2$, $\mathbf{F}_3$, . . . , $\mathbf{F}_n$ of the original system about the same point O, as shown in Fig. 5-22b. Thus,

$$Rd_R = F_1d_1 + F_2d_2 + F_3d_3 + \cdots + F_nd_n = \Sigma M_O$$

Therefore,

$$d_R = \frac{\Sigma M_O}{R} \qquad (5\text{-}25)$$

where ΣM_O stands for the algebraic sum of the moments of the forces of the original system about point O. The direction of d_R is selected so that the product Rd_R produces a moment about point O with the same sense (clockwise or counterclockwise) as the algebraic sum of the moments of the forces of the original system about point O.

The location of the line of action of **R** with respect to point O can also be specified by determining the intercept of the line of action of the force with one of the coordinate axes. For example, in Fig. 5-22c, the intercept x_R is determined from the equation

$$x_R = \frac{\Sigma M_O}{R_y} \qquad (5\text{-}26)$$

since the component R_x of the resultant force **R** does not produce a moment about point O. The special case for a system of coplanar parallel forces is illustrated in Fig. 5-23a and b.

In the event that the resultant force **R** of a system of coplanar forces $\mathbf{F}_1$, $\mathbf{F}_2$, . . . , $\mathbf{F}_n$ is zero but the moment ΣM_O is not zero, the resultant is a couple **C** whose vector is perpendicular to the plane containing the forces (the xy-plane for the case being discussed). Thus, the resultant of a coplanar system of forces may be either a force **R** or a couple **C**.

5-5-2 Noncoplanar Parallel Force Systems

If all forces of a three-dimensional system are parallel, the resultant force **R** is the algebraic sum of the forces of the system. The line of action of the resultant is determined by using the principle of moments. Thus, as shown in Fig. 5-24a, for a system of forces $\mathbf{F}_1$, $\mathbf{F}_2$, . . . , $\mathbf{F}_n$, perpendicular to the xy-plane,

$$\mathbf{R} = \mathbf{F}_1 + \mathbf{F}_2 + \cdots + \mathbf{F}_n = R\,\mathbf{k} = \Sigma F\,\mathbf{k} \qquad (5\text{-}27)$$

$$\mathbf{M}_O = \mathbf{r} \times \mathbf{R} = \mathbf{r}_1 \times \mathbf{F}_1 + \mathbf{r}_2 \times \mathbf{F}_2 + \mathbf{r}_3 \times \mathbf{F}_3 + \cdots + \mathbf{r}_n \times \mathbf{F}_n \qquad (5\text{-}28)$$

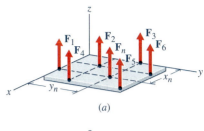

(a)

The intersection of the line of action of the resultant force $\mathbf{R}$ with the xy-plane (see Fig. 5-24b) is located by equating the moments of $\mathbf{R}$ about the x- and y-axes to the sums of the moments of the forces of the system $\mathbf{F}_1$, $\mathbf{F}_2$, . . . , $\mathbf{F}_n$ about the x- and y-axes. It is important to note from Eqs. (5-10) that $M_x = F_z r_y$, while $M_y = -F_z r_x$ (since $F_x = F_y = r_z = 0$). Therefore,

$$Rx_R = F_1 x_1 + F_2 x_2 + \cdots + F_n x_n = -\Sigma M_y$$

$$Ry_R = F_1 y_1 + F_2 y_2 + \cdots + F_n y_n = \Sigma M_x$$

which gives

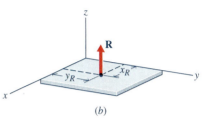

(b)

Figure 5-24

$$x_R = -\frac{\Sigma M_y}{R} \qquad y_R = \frac{\Sigma M_x}{R} \qquad (5\text{-}29)$$

In the event that the resultant force $\mathbf{R}$ of the system of parallel forces is zero but the moments ΣM_x and ΣM_y are not zero, the resultant is a couple $\mathbf{C}$ whose vector lies in a plane perpendicular to the forces (in the xy-plane for the system of forces illustrated in Fig. 5-24). Thus, the resultant of a noncoplanar system of parallel forces may be either a force $\mathbf{R}$ or a couple $\mathbf{C}$.

5-5-3 General Force Systems

The resultant $\mathbf{R}$ of a general, three-dimensional system of forces $\mathbf{F}_1$, $\mathbf{F}_2$, $\mathbf{F}_3$, . . . , $\mathbf{F}_n$, such as the one shown in Fig. 5-25a, can be determined by resolving each force of the system into an equal parallel force through any point (taken for convenience as the origin O of a system of coordinate axes) and a couple. Figure 5-25b shows the resolution for force F_1. Thus, as shown in Fig. 5-25c, the given system is replaced by two systems:

1. A system of noncoplanar, concurrent forces through the origin O that have the same magnitudes and directions as the forces of the original system.
2. A system of noncoplanar couples.

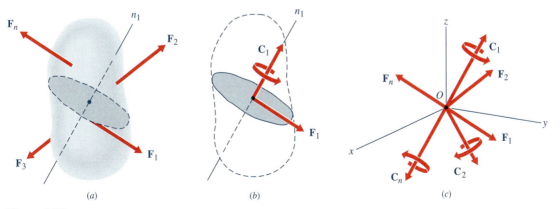

(a)   (b)   (c)

Figure 5-25

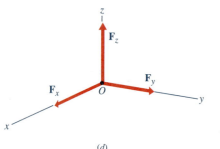

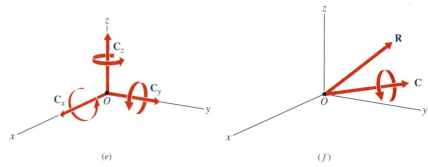

(d)   (e)   (f)

Figure 5-25 (Cont.)

Each force and couple of the two systems can be resolved into components along the coordinate axes as shown in Figs. 5-25d and e. The resultant of the concurrent force system is a force through the origin O, which can be expressed as

$$\mathbf{R} = \mathbf{R}_x + \mathbf{R}_y + \mathbf{R}_z = R_x\,\mathbf{i} + R_y\,\mathbf{j} + R_z\,\mathbf{k} = R\,\mathbf{e} \qquad (5\text{-}30)$$

where

$$R_x = \Sigma F_x \qquad R_y = \Sigma F_y \qquad R_z = \Sigma F_z$$

$$R = |\mathbf{R}| = \sqrt{(\Sigma F_x)^2 + (\Sigma F_y)^2 + (\Sigma F_z)^2}$$

$$\mathbf{e} = \cos\theta_x\,\mathbf{i} + \cos\theta_y\,\mathbf{j} + \cos\theta_z\,\mathbf{k}$$

$$\cos\theta_x = \frac{\Sigma F_x}{|\mathbf{R}|} \qquad \cos\theta_y = \frac{\Sigma F_y}{|\mathbf{R}|} \qquad \cos\theta_z = \frac{\Sigma F_z}{|\mathbf{R}|}$$

The resultant of the system of noncoplanar couples is a couple $\mathbf{C}$ that can be expressed as

$$\mathbf{C} = \Sigma\mathbf{C}_x + \Sigma\mathbf{C}_y + \Sigma\mathbf{C}_z = \Sigma C_x\,\mathbf{i} + \Sigma C_y\,\mathbf{j} + \Sigma C_z\,\mathbf{k} = C\,\mathbf{e} \qquad (5\text{-}31)$$

where

$$C = |\mathbf{C}| = \sqrt{(\Sigma C_x)^2 + (\Sigma C_y)^2 + (\Sigma C_z)^2}$$

$$\mathbf{e} = \cos\theta_x\,\mathbf{i} + \cos\theta_y\,\mathbf{j} + \cos\theta_z\,\mathbf{k}$$

$$\cos\theta_x = \frac{\Sigma C_x}{|\mathbf{C}|} \qquad \cos\theta_y = \frac{\Sigma C_y}{|\mathbf{C}|} \qquad \cos\theta_z = \frac{\Sigma C_z}{|\mathbf{C}|}$$

The resultant force $\mathbf{R}$ and the resultant couple $\mathbf{C}$ shown in Fig. 5-25f together constitute the resultant of the system with respect to point O. In special cases, the resultant couple $\mathbf{C}$ may vanish, leaving the force $\mathbf{R}$ as the resultant of the system. Again in special cases, the resultant force $\mathbf{R}$ may vanish, leaving the couple $\mathbf{C}$ as the resultant of the system. If the resultant force $\mathbf{R}$ and the resultant couple $\mathbf{C}$ both vanish, the resultant of the system is zero and the system is in equilibrium. Thus, the resultant of a general force system may be a force $\mathbf{R}$, a couple $\mathbf{C}$, or both a force $\mathbf{R}$ and a couple $\mathbf{C}$.

When the couple $\mathbf{C}$ is perpendicular to the resultant force $\mathbf{R}$, as shown in Fig. 5-26, the two can be combined to form a single force $\mathbf{R}$ whose line of action is a distance $d = C/R$ from point O in a direction that makes the direction of the moment of R about O the same as that of $\mathbf{C}$.

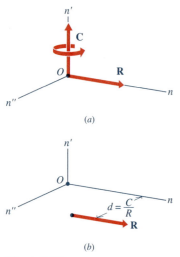

(a)

(b)

Figure 5-26

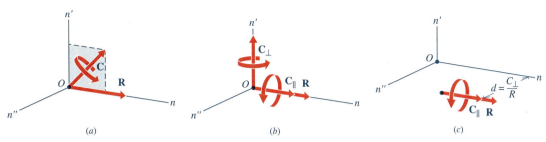

Figure 5-27

Another transformation of the resultant (**R** and **C**) of a general force system is illustrated in Figs. 5-27 and 5-28. In this case, the couple **C** is resolved into components parallel and perpendicular to the resultant force **R**, as shown in Fig. 5-27b. The resultant force **R** and the perpendicular component of the couple $C_\perp$ can be combined as illustrated in Fig. 5-26. In addition, the parallel component of the couple $C_\parallel$ can be translated to coincide with the line of action of the resultant force **R**, as shown in Fig. 5-27c. The combination of couple $C_\parallel$ and resultant force **R** is known as a *wrench*. The action may be described as a push (or pull) and a twist about an axis parallel to the push (or pull). When the force and moment vectors have the same sense, as shown in Fig. 5-27c, the wrench is positive. When the vectors have the opposite sense, as shown in Fig. 5-28c, the wrench is negative. A screwdriver is an example of a wrench.

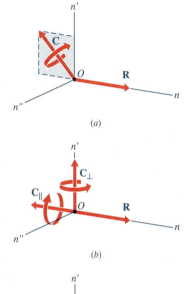

(a)

(b)

(c)

Figure 5-28

Example Problem 5-8

An 800-N force is applied to a bracket as shown in Fig. 5-29. Replace the force by a force at point B and a couple. Express your answer in Cartesian vector form.

SOLUTION
Scalar Analysis

The force **F** of magnitude 800 N can be resolved into the components

$$F_x = \frac{4}{5}F = \frac{4}{5}(800) = 640 \text{ N}$$

$$F_y = -\frac{3}{5}F = -\frac{3}{5}(800) = -480 \text{ N}$$

The moment of the force about point B is

$$
\begin{aligned}
M_B &= -F_x(0.240) + F_y(0.200) \\
&= -640(0.240) + (-480)(0.200) \\
&= -249.6 \text{ N} \cdot \text{m} \cong 250 \text{ N} \cdot \text{m} \downarrow
\end{aligned}
$$

The force and moment in Cartesian vector form are

$$\mathbf{F} = 640\,\mathbf{i} - 480\,\mathbf{j} \text{ N} \qquad \textbf{Ans.}$$

$$\mathbf{C} = 250 \text{ N} \cdot \text{m} \downarrow = -250\,\mathbf{k} \text{ N} \cdot \text{m} \qquad \textbf{Ans.}$$

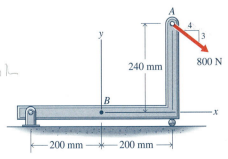

Figure 5-29

Vector Analysis

The force **F** can be expressed in Cartesian vector form as

$$\mathbf{F} = 800\left[\frac{4}{5}\mathbf{i} - \frac{3}{5}\mathbf{j}\right] = 640\,\mathbf{i} - 480\,\mathbf{j}\ \text{N}$$ **Ans.**

The position vector $\mathbf{r}_{A/B}$ is expressed as

$$\mathbf{r}_{A/B} = 0.200\,\mathbf{i} + 0.240\,\mathbf{j}\ \text{m}$$

From Eq. (5-19):

$$\mathbf{C} = \mathbf{r}_{A/B} \times \mathbf{F} = [0.200\,\mathbf{i} + 0.240\,\mathbf{j}] \times [640\,\mathbf{i} - 480\,\mathbf{j}]$$
$$= -249.6\,\mathbf{k}\ \text{N} \cdot \text{m} \cong -250\,\mathbf{k}\ \text{N} \cdot \text{m} \ \blacksquare$$ **Ans.**

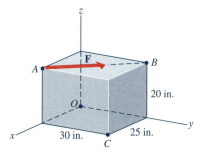

Figure 5-30

Example Problem 5-9

The force **F** shown in Fig. 5-30 has a magnitude of 780 lb. Replace the force **F** by a force $\mathbf{F}_O$ at point O and a couple **C**.

(a) Express the force $\mathbf{F}_O$ and the couple **C** in Cartesian vector form.
(b) Determine the direction angles associated with the unit vector **e** that describes the aspect of the plane of the couple.

SOLUTION

(a)
$$\mathbf{F}_O = \mathbf{F} = 780\left[\frac{-25\,\mathbf{i} + 30\,\mathbf{j}}{\sqrt{1525}}\right]$$
$$= -499.3\,\mathbf{i} + 599.2\,\mathbf{j}\ \text{lb} \cong -499\,\mathbf{i} + 599\,\mathbf{j}\ \text{lb}$$ **Ans.**

$$\mathbf{r}_{A/O} = 25\,\mathbf{i} + 20\,\mathbf{k}\ \text{in.}$$

From Eq. (5-19):

$$\mathbf{C} = \mathbf{M}_O = \mathbf{r}_{A/O} \times \mathbf{F}$$
$$= [25\,\mathbf{i} + 20\,\mathbf{k}] \times [-499.3\,\mathbf{i} + 599.2\,\mathbf{j}]$$
$$= -11984\,\mathbf{i} - 9986\,\mathbf{j} + 14980\,\mathbf{k}\ \text{in.} \cdot \text{lb}$$
$$\cong -11.98\,\mathbf{i} - 9.99\,\mathbf{j} + 14.98\,\mathbf{k}\ \text{in.} \cdot \text{kip}$$ **Ans.**

(b) From Eq. (5-21):

$$C = \sqrt{(-11{,}984)^2 + (-9986)^2 + (14{,}980)^2} = 21{,}627\ \text{in.} \cdot \text{lb}$$

From Eq. (5-23):

$$\theta_x = \cos^{-1}\frac{C_x}{C} = \cos^{-1}\frac{-11{,}984}{21{,}627} = 123.7°$$ **Ans.**

$$\theta_y = \cos^{-1}\frac{C_y}{C} = \cos^{-1}\frac{-9986}{21{,}627} = 117.5°$$ **Ans.**

$$\theta_z = \cos^{-1}\frac{C_z}{C} = \cos^{-1}\frac{14{,}980}{21{,}627} = 46.2° \ \blacksquare$$ **Ans.**

Example Problem 5-10

Three forces and a couple are applied to a bracket as shown in Fig. 5-31a. Determine

(a) The magnitude and direction of the resultant.
(b) The perpendicular distance d_R from point O to the line of action of the resultant.
(c) The distance x_R from point O to the intercept of the line of action of the resultant with the x-axis.

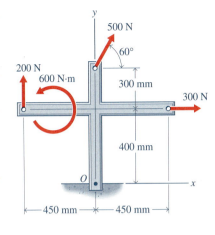

(a)

SOLUTION

(a) The resultant of a system of coplanar forces is either a force **R** or a couple **C**. The resultant force is obtained by using Eqs. (5-24). Thus,

$$R_x = \Sigma F_x = 500 \cos 60° + 300 = 550 \text{ N}$$

$$R_y = \Sigma F_y = 500 \sin 60° + 200 = 633 \text{ N}$$

$$R = |\mathbf{R}| = \sqrt{(\Sigma F_x)^2 + (\Sigma F_y)^2} = \sqrt{(550)^2 + (633)^2} = 838.6 \text{ N} \cong 839 \text{ N} \quad \textbf{Ans.}$$

$$\theta_x = \cos^{-1} \frac{\Sigma F_x}{R} = \cos^{-1} \frac{550}{838.6} = 49.0° \quad \textbf{Ans.}$$

(b) From Eq. (5-25):

$$+\!\downarrow \Sigma M_O = -300(0.400) - 500 \cos 60° (0.700)$$
$$-200(0.450) + 600 = 215 \text{ N} \cdot \text{m}$$

$$d_R = \frac{\Sigma M_O}{R} = \frac{215}{838.6} = 0.2564 \text{ m} \cong 256 \text{ mm} \quad \textbf{Ans.}$$

(c) From Eq. (5-26):

$$x_R = \frac{\Sigma M_O}{R_y} = \frac{215}{633} = 0.3397 \text{ m} \cong 340 \text{ mm} \quad \textbf{Ans.}$$

The results are illustrated in Fig. 5-31b. ■

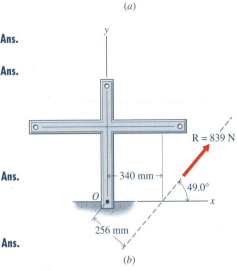

(b)

Figure 5-31

Example Problem 5-11

Determine the resultant of the parallel force system shown in Fig. 5-32 and locate the intersection of the line of action of the resultant with the xy-plane.

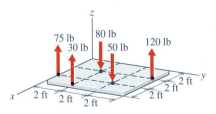

Figure 5-32

SOLUTION

The resultant **R** of the force system is determined by using Eq. (5-27). Thus,

$$\mathbf{R} = \Sigma F \mathbf{k}$$
$$= (75 + 30 - 80 - 50 + 120) \mathbf{k} = 95 \mathbf{k} \text{ lb} \quad \textbf{Ans.}$$

The intersection (x_R, y_R) of the line of action of the resultant with the xy-plane is determined by using Eqs. (5-29). With positive moments defined by using the right-hand rule,

$$\Sigma M_y = F_1 x_1 + F_2 x_2 + \cdots + F_n x_n$$
$$= -75(4) - 30(6) + 80(2) + 50(4) - 120(2) = -360 \text{ ft} \cdot \text{lb}$$

$$\Sigma M_x = F_1 y_1 + F_2 y_2 + \cdots + F_n y_n$$
$$= 75(0) + 30(2) - 80(2) - 50(4) + 120(6) = 420 \text{ ft} \cdot \text{lb}$$

Therefore,

$$x_R = -\frac{\Sigma M_y}{R} = -\frac{-360}{95} = 3.79 \text{ ft}$$ **Ans.**

$$y_R = \frac{\Sigma M_x}{R} = \frac{420}{95} = 4.42 \text{ ft}$$ **Ans.**

The coordinates (x_R, y_R) can also be determined by using a vector analysis. Thus,

$$\mathbf{M}_O = \mathbf{r} \times \mathbf{R} = (\mathbf{r}_1 \times \mathbf{F}_1) + (\mathbf{r}_2 \times \mathbf{F}_2) + (\mathbf{r}_3 \times \mathbf{F}_3) + (\mathbf{r}_4 \times \mathbf{F}_4) + (\mathbf{r}_5 \times \mathbf{F}_5)$$
$$= [(x_R \mathbf{i} + y_R \mathbf{j}) \times (95 \mathbf{k})]$$
$$= [(4 \mathbf{i}) \times (75 \mathbf{k})] + [(6 \mathbf{i} + 2 \mathbf{j}) \times (30 \mathbf{k})] + [(2 \mathbf{i} + 2 \mathbf{j}) \times (-80 \mathbf{k})]$$
$$\quad + [(4 \mathbf{i} + 4 \mathbf{j}) \times (-50 \mathbf{k})] + [(2 \mathbf{i} + 6 \mathbf{j}) \times (120 \mathbf{k})]$$
$$= -95 x_R \mathbf{j} + 95 y_R \mathbf{i} = 420 \mathbf{i} - 360 \mathbf{j}$$

Solving for x_R and y_R yields

$$x_R = \frac{360}{95} = 3.79 \text{ ft}$$ **Ans.**

$$y_R = \frac{420}{95} = 4.42 \text{ ft}$$ ■ **Ans.**

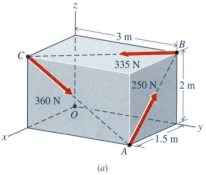

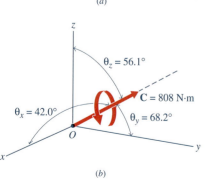

Figure 5-33

Example Problem 5-12

Replace the force system shown in Fig. 5-33a with a force **R** through point O and a couple **C**.

SOLUTION

The three forces and their positions with respect to point O can be written in Cartesian vector form as

$$\mathbf{F}_A = 250 \left[-\frac{1.5}{2.5} \mathbf{i} + \frac{2}{2.5} \mathbf{k} \right] = (-150 \mathbf{i} + 200 \mathbf{k}) \text{ N} \qquad \mathbf{r}_{A/O} = (1.5 \mathbf{i} + 3 \mathbf{j}) \text{ m}$$

$$\mathbf{F}_B = 335 \left[\frac{1.5}{3.354} \mathbf{i} - \frac{3}{3.354} \mathbf{j} \right] = (150 \mathbf{i} - 300 \mathbf{j}) \text{ N} \qquad \mathbf{r}_{B/O} = (3 \mathbf{j} + 2 \mathbf{k}) \text{ m}$$

$$\mathbf{F}_C = 360 \left[\frac{3}{3.606} \mathbf{j} - \frac{2}{3.606} \mathbf{k} \right] = (300 \mathbf{j} - 200 \mathbf{k}) \text{ N} \qquad \mathbf{r}_{C/O} = (1.5 \mathbf{i} + 2 \mathbf{k}) \text{ m}$$

Each of the three forces can be replaced by an equal force through point O and a couple. The vector sum of the concurrent forces is

$$\mathbf{R} = \mathbf{F}_A + \mathbf{F}_B + \mathbf{F}_C$$
$$= -150\,\mathbf{i} + 200\,\mathbf{k} + 150\,\mathbf{i} - 300\,\mathbf{j} + 300\,\mathbf{j} - 200\,\mathbf{k} = 0 \qquad \textbf{Ans.}$$

The moment $\mathbf{M}_O$ of the resultant couple is

$$\mathbf{C} = \mathbf{M}_O = (\mathbf{r}_{A/O} \times \mathbf{F}_A) + (\mathbf{r}_{B/O} \times \mathbf{F}_B) + (\mathbf{r}_{C/O} \times \mathbf{F}_C)$$

$$= \begin{vmatrix} \mathbf{i} & \mathbf{j} & \mathbf{k} \\ 1.5 & 3 & 0 \\ -150 & 0 & 200 \end{vmatrix} + \begin{vmatrix} \mathbf{i} & \mathbf{j} & \mathbf{k} \\ 0 & 3 & 2 \\ 150 & -300 & 0 \end{vmatrix} + \begin{vmatrix} \mathbf{i} & \mathbf{j} & \mathbf{k} \\ 1.5 & 0 & 2 \\ 0 & 300 & -200 \end{vmatrix}$$

$$= (600\,\mathbf{i} + 300\,\mathbf{j} + 450\,\mathbf{k})\ \text{N} \cdot \text{m} \qquad \textbf{Ans.}$$

The magnitude of the resultant couple is

$$|\mathbf{C}| = \sqrt{C_x^2 + C_y^2 + C_z^2}$$
$$= \sqrt{(600)^2 + (300)^2 + (450)^2} = 807.8 \cong 808\ \text{N} \cdot \text{m} \qquad \textbf{Ans.}$$

Finally, the direction angles that locate the axis of the couple are

$$\theta_x = \cos^{-1}\frac{C_x}{|\mathbf{C}|} = \cos^{-1}\frac{600}{807.8} = 42.0° \qquad \textbf{Ans.}$$

$$\theta_y = \cos^{-1}\frac{C_y}{|\mathbf{C}|} = \cos^{-1}\frac{300}{807.8} = 68.2° \qquad \textbf{Ans.}$$

$$\theta_z = \cos^{-1}\frac{C_z}{|\mathbf{C}|} = \cos^{-1}\frac{450}{807.8} = 56.1° \qquad \textbf{Ans.}$$

The resultant of the force system is the couple shown in Fig. 5-33*b*. ■

PROBLEMS

Standard Problems

5-53* A 250-lb force is applied to a beam as shown in Fig. P5-53. Replace the force by a force at point A and a couple. Express your answer in Cartesian vector form.

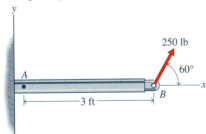

Figure P5-53

5-54* A 500-N force is applied to a beam as shown in Fig. P5-54. Replace the force by a force at point C and a couple. Express your answer in Cartesian vector form.

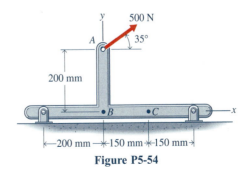

Figure P5-54

5-55 The force $\mathbf{F}$ shown in Fig. P5-55 has a magnitude of 928 lb. Replace the force $\mathbf{F}$ by a force $\mathbf{F}_O$ at point O and a couple $\mathbf{C}$.

(a) Express the force $\mathbf{F}_O$ and the couple $\mathbf{C}$ in Cartesian vector form.

(b) Determine the direction angles associated with the unit vector **e** that describes the aspect of the plane of the couple.

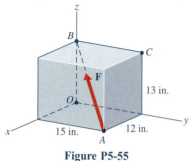

Figure P5-55

5-56 The force **F** shown in Fig. P5-56 has a magnitude of 595 N. Replace the force **F** by a force $\mathbf{F}_O$ at point O and a couple **C**.
(a) Express the force $\mathbf{F}_O$ and the couple **C** in Cartesian vector form.
(b) Determine the direction angles associated with the unit vector **e** that describes the aspect of the plane of the couple.

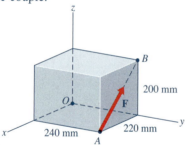

Figure P5-56

5-57* Three forces are applied to a beam as shown in Fig. P5-57. Replace the three forces by an equivalent force-couple system at point C.

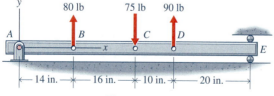

Figure P5-57

5-58* Three forces are applied to an angle bracket as shown in Fig. P5-58. Determine the magnitude and direction of the resultant of the three forces and the perpendicular distance d_R from point O to the line of action of the resultant.

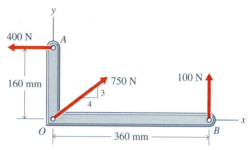

Figure P5-58

5-59 Four forces are applied to the cross section of a beam as shown in Fig. P5-59. Determine the magnitude and direction of the resultant of the four forces and the distance x_R from point O to the intercept of the line of action of the resultant with the x-axis.

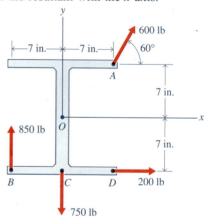

Figure P5-59

5-60 Four forces and a couple are applied to a rectangular plate as shown in Fig. P5-60. Determine the magnitude and direction of the resultant of the force-couple system and the distance x_R from point O to the intercept of the line of action of the resultant with the x-axis.

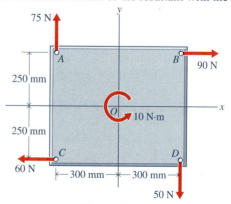

Figure P5-60

5-61* Four forces are applied to a truss as shown in Fig. P5-61. Determine
(a) The magnitude and direction of the resultant.
(b) The perpendicular distance d_R from support A to the line of action of the resultant.

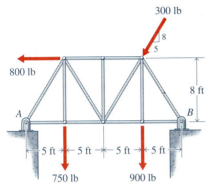

Figure P5-61

5-62* Four forces are applied to a truss as shown in Fig. P5-62. Determine
(a) The magnitude and direction of the resultant.
(b) The perpendicular distance d_R from point A to the line of action of the resultant.

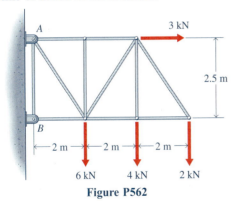

Figure P562

5-63 Determine the resultant of the parallel force system shown in Fig. P5-63 and locate the intersection of the line of action of the resultant with the xy-plane.

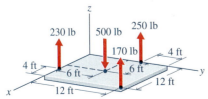

Figure P5-63

5-64 Determine the resultant of the parallel force system shown in Fig. P5-64 and locate the intersection of the line of action of the resultant with the xz-plane.

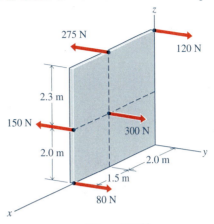

Figure P5-64

5-65* Determine the resultant of the parallel force system shown in Fig. P5-65 and locate the intersection of the line of action of the resultant with the yz-plane.

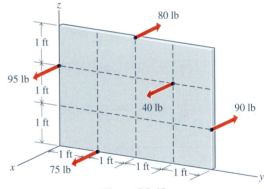

Figure P5-65

5-66* Determine the resultant of the parallel force system shown in Fig. P5-66 and locate the intersection of the line of action of the resultant with the xz-plane.

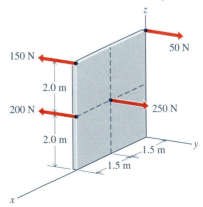

Figure P5-66

Challenging Problems

5-67* Replace the force system shown in Fig. P5-67 with a force **R** through point *D* and a couple **C.**

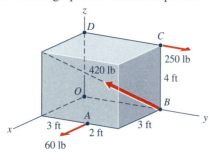

Figure P5-67

5-68* Replace the force system shown in Fig. P5-68 with a force **R** through point *C* and a couple **C.**

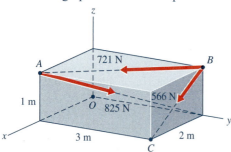

Figure P5-68

5-69 Replace the force system shown in Fig. P5-69 with a force **R** through point *O* and a couple **C.**

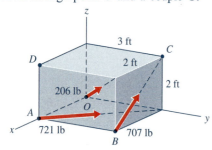

Figure P5-69

5-70 Replace the force system shown in Fig. P5-70 with a force **R** through point *O* and a couple **C.**

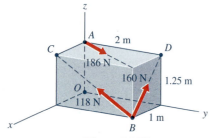

Figure P5-70

5-71* Forces are applied at points *A, B,* and *C* of the bar shown in Fig. P5-71. Replace this system of forces with a force **R** through point *O* and a couple **C.**

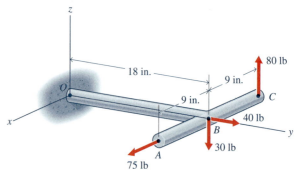

Figure P5-71

5-72* Forces are applied at points *A, B,* and *C* of the bar shown in Fig. P5-72. Replace this system of forces with a force **R** through point *O* and a couple **C.**

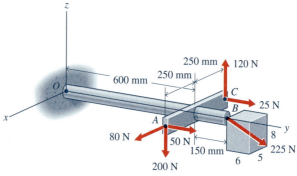

Figure P5-72

5-73 Reduce the forces shown in Fig. P5-73 to a wrench and locate the intersection of the wrench with the *xy*-plane.

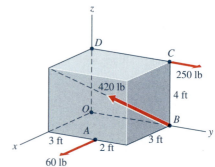

Figure P5-73

5-74 Reduce the forces shown in Fig. P5-74 to a wrench and locate the intersection of the wrench with the *xy*-plane.

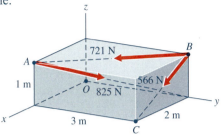

Figure P5-74

5-6 CENTER OF GRAVITY AND CENTER OF MASS

5-6-1 Center of Gravity

An immediate application of equivalent force-couples is the replacement of distributed forces such as gravitational forces with a single equivalent force. For example, consider the set of *n* particles shown in Fig. 5-34a. Approximating the weights of the *n* particles as a parallel force system (Fig. 5-34b),[1] the external effect of the *n* weights is the same as that of a single equivalent force **W** acting at point *G* as shown in Fig. 5-34c. The equivalent force **W** is called the *weight of the body* and the point *G* through which the force acts is called the *center of gravity*. As in Section 5-5-2, the direction of the weight force **W** is parallel to all of the individual weights **W**$_i$, and its magnitude is the sum of the magnitudes of the individual weights (so that the sum of forces is the same on Figs. 5-34b and c)

$$W = \sum_{i=1}^{n} W_i \qquad (5\text{-}32a)$$

and the location of *G* (so that the sum of moments is the same on Figs. 5-34b and c)[2] is

$$M_{yz} = Wx_G = \sum_{i=1}^{n} W_i x_i \quad \text{or} \quad x_G = \frac{1}{W}\sum_{i=1}^{n} W_i x_i \qquad (5\text{-}32b)$$

$$M_{zx} = Wy_G = \sum_{i=1}^{n} W_i y_i \quad \text{or} \quad y_G = \frac{1}{W}\sum_{i=1}^{n} W_i y_i \qquad (5\text{-}32c)$$

$$M_{xy} = Wz_G = \sum_{i=1}^{n} W_i z_i \quad \text{or} \quad z_G = \frac{1}{W}\sum_{i=1}^{n} W_i z_i \qquad (5\text{-}32d)$$

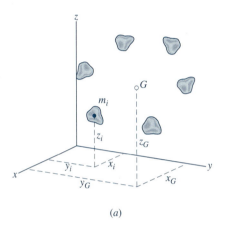

(a)

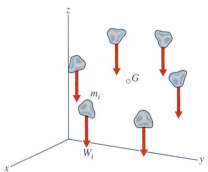

(b)

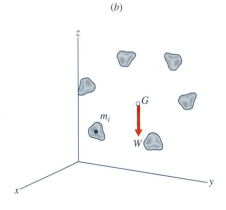

(c)

Figure 5-34

[1]The weight forces are actually concurrent at the center of the earth. However, most engineering structures are small compared to the radius of the earth, and the error introduced by assuming the forces are parallel is very small.

[2]Equation (5-32d) is not directly obtainable from Figs. 5-34b and c. By the Principle of Transmissibility, the force **W** can act anywhere along the vertical line through *G* without affecting its moment. Equation (5-32d) is obtained by rotating the figures, both the masses and the coordinate system, so that gravity acts along the *y*-axis instead of the *z*-axis. Then the sum of moments gives Eq. (5-32d).

where x_i, y_i, z_i, and x_G, y_G, z_G are shown on Fig. 5-34a. The moments M_{yz}, M_{zx}, and M_{xy} are called *first moments of the weight forces* relative to the *yz*-, *zx*-, and *xy*-planes, respectively.

If the particles form a continuous body, as shown in Fig. 5-35a, the summations must be replaced by integrals over the mass of the body, giving

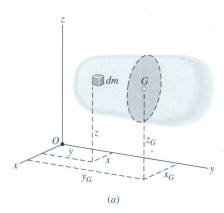

(a)

$$W = \int dW \qquad (5\text{-}33a)$$

$$M_{yz} = Wx_G = \int x\, dW \quad \text{or} \quad x_G = \frac{1}{W}\int x\, dW \qquad (5\text{-}33b)$$

$$M_{zx} = Wy_G = \int y\, dW \quad \text{or} \quad y_G = \frac{1}{W}\int y\, dW \qquad (5\text{-}33c)$$

$$M_{xy} = Wz_G = \int z\, dW \quad \text{or} \quad z_G = \frac{1}{W}\int z\, dW \qquad (5\text{-}33d)$$

Equations (5-32b), (5-32c), and (5-32d) can be combined into a single vector equation by multiplying the first, second, and third equations by **i**, **j**, and **k**, respectively, and adding. Thus,

$$Wx_G\, \mathbf{i} + Wy_G\, \mathbf{j} + Wz_G\, \mathbf{k} = \sum_{i=1}^{n} W_i x_i\, \mathbf{i} + \sum_{i=1}^{n} W_i y_i\, \mathbf{j} + \sum_{i=1}^{n} W_i z_i\, \mathbf{k}$$

from which

$$W(x_G\, \mathbf{i} + y_G\, \mathbf{j} + z_G\, \mathbf{k}) = \sum_{i=1}^{n} W_i (x_i\, \mathbf{i} + y_i\, \mathbf{j} + z_i\, \mathbf{k})$$

which reduces to

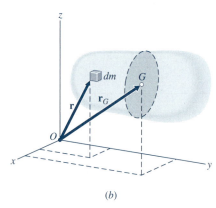

(b)

Figure 5-35

$$\mathbf{M}_O = W\mathbf{r}_G = \sum_{i=1}^{n} W_i \mathbf{r}_i \quad \text{or} \quad \mathbf{r}_G = \frac{1}{W}\sum_{i=1}^{n} W_i \mathbf{r}_i \qquad (5\text{-}34)$$

where $\mathbf{r}_i = x_i\, \mathbf{i} + y_i\, \mathbf{j} + z_i\, \mathbf{k}$ is the position vector from the origin O to the *i*th particle (Fig. 5-35b) and $\mathbf{r}_G = x_G\, \mathbf{i} + y_G\, \mathbf{j} + z_G\, \mathbf{k}$ is the position vector from the origin to the center of gravity. Similarly for Eqs. (5-33) if the particles form a continuous body:

$$W\mathbf{r}_G = \int \mathbf{r}\, dW \quad \text{or} \quad \mathbf{r}_G = \frac{1}{W}\int \mathbf{r}\, dW \qquad (5\text{-}35)$$

The center of gravity G located by Eqs. (5-32) through (5-35) represents the point at which all of the weight of the body could be concentrated without changing the external effects on the body.

5-6-2 Center of Mass

For practical engineering work in which the size of a body is small in comparison to the size of the earth, all of the particles that make up the body can be assumed to be at the same distance from the center of the earth; therefore, they experience the same gravitational acceleration g. Dividing Eqs. (5-33) by the gravitational acceleration gives

$$m = \int dm \qquad (5\text{-}36a)$$

$$mx_G = \int x\, dm \quad \text{or} \quad x_G = \frac{1}{m}\int x\, dm \qquad (5\text{-}36b)$$

$$my_G = \int y \, dm \quad \text{or} \quad y_G = \frac{1}{m}\int y \, dm \qquad (5\text{-}36c)$$

$$mz_G = \int z \, dm \quad \text{or} \quad z_G = \frac{1}{m}\int z \, dm \qquad (5\text{-}36d)$$

The location of the center of mass (x_G, y_G, z_G) is the same as the center of gravity, and both points will be labeled G. However, the center of mass is defined equally well in the weightlessness of space as it is on the surface of the earth.

The moments M_{yz}, M_{zx}, and M_{xy} in Eqs. (5-32) and (5-33) are called *first moments of the weight forces* relative to the yz-, zx-, and xy-planes, respectively. The integrand in each case is the first power of the distance to the respective plane. Similarly, the integrals $\int x \, dm$, $\int y \, dm$, and $\int z \, dm$ are called the *first moments of the mass* (and integrals of the form $\int x \, dA$ are called *first moments of area*) although they are not strictly moments since dm (and dA) are not actually forces. Later, integrals of the form $\int x^2 \, dm$ (and $\int x^2 \, dA$) will be introduced. Such integrals are called *second moments of mass* (and *second moments of area*) since the second power of the distance appears in these expressions.

The following example illustrates the procedure used to locate the "center of gravity" or the "center of mass" of a system of particles.

Example Problem 5-13

Four bodies A, B, C, and D (which can be treated as particles) are attached to a shaft as shown in Fig. 5-36. The masses of the bodies are 0.2, 0.4, 0.6, and 0.8 slug, respectively, and the distances from the z-axis of the shaft to their mass centers (end view) are 1.50, 2.50, 2.00, and 1.25 ft, respectively. Find the mass center for the four bodies.

SOLUTION

A typical equation from Eqs. (5-32) that is used to locate the mass center of a system of particles is

$$x_G = \frac{1}{m}\sum_{i=1}^{n} m_i x_i \qquad \text{where} \qquad m = \sum_{i=1}^{n} m_i$$

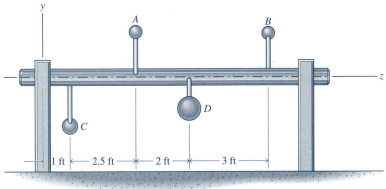

Front View   End View

Figure 5-36

Thus, for the system of four bodies shown in Fig. 5-36:

$$\Sigma m_i = 0.2 + 0.4 + 0.6 + 0.8 = 2.0 \text{ slug}$$

$$
\begin{aligned}
\Sigma m_i x_i &= m_A x_A + m_B x_B + m_C x_C + m_D x_D \\
&= 0.2(-1.50 \cos 60°) + 0.4(2.50 \cos 30°) + 0.6(2.00 \cos 45°) \\
&\quad + 0.8(-1.25 \cos 45°) = 0.8574 \text{ slug} \cdot \text{ft}
\end{aligned}
$$

$$
\begin{aligned}
\Sigma m_i y_i &= m_A y_A + m_B y_B + m_C y_C + m_D y_D \\
&= 0.2(1.50 \sin 60°) + 0.4(2.50 \sin 30°) + 0.6(-2.00 \sin 45°) \\
&\quad + 0.8(-1.25 \sin 45°) = -0.7958 \text{ slug} \cdot \text{ft}
\end{aligned}
$$

$$
\begin{aligned}
\Sigma m_i z_i &= m_A z_A + m_B z_B + m_C z_C + m_D z_D \\
&= 0.2(3.5) + 0.4(8.5) + 0.6(1.0) + 0.8(5.5) = 9.10 \text{ slug} \cdot \text{ft}
\end{aligned}
$$

$$x_G = \frac{\Sigma m_i x_i}{m} = \frac{0.8574}{2.00} = 0.429 \text{ ft} \qquad \textbf{Ans.}$$

$$y_G = \frac{\Sigma m_i y_i}{m} = \frac{-0.7958}{2.00} = -0.398 \text{ ft} \qquad \textbf{Ans.}$$

$$z_G = \frac{\Sigma m_i z_i}{m} = \frac{9.10}{2.00} = 4.55 \text{ ft} \quad\blacksquare \qquad \textbf{Ans.}$$

PROBLEMS

Standard Problems

5-75* Locate the center of gravity for the four particles shown in Fig. P5-75 if $W_A = 20$ lb, $W_B = 25$ lb, $W_C = 30$ lb, and $W_D = 40$ lb.

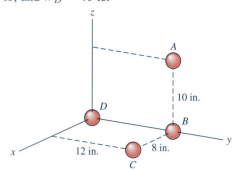

Figure P5-75

5-76* Locate the center of mass for the four particles shown in Fig. P5-76 if $m_A = 16$ kg, $m_B = 24$ kg, $m_C = 14$ kg, and $m_D = 36$ kg.

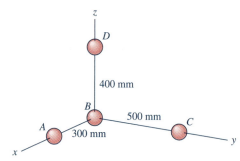

Figure P5-76

5-77 Locate the center of gravity for the five particles shown in Fig. P5-77 if $W_A = 25$ lb, $W_B = 35$ lb, $W_C = 15$ lb, $W_D = 28$ lb, and $W_E = 16$ lb.

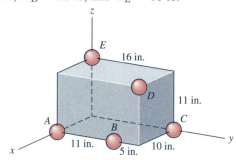

Figure P5-77

5-78 Locate the center of mass for the five particles shown in Fig. P5-78 if $m_A = 2$ kg, $m_B = 3$ kg, $m_C = 4$ kg, $m_D = 3$ kg, and $m_E = 2$ kg.

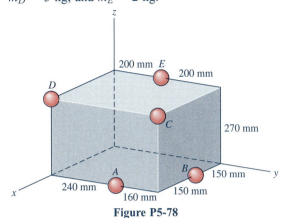

Figure P5-78

5-79* Locate the center of gravity for the five particles shown in Fig. P5-77 if $W_A = 15$ lb, $W_B = 24$ lb, $W_C = 35$ lb, $W_D = 18$ lb, and $W_E = 26$ lb.

5-80* Locate the center of mass for the five particles shown in Fig. P5-78 if $m_A = 6$ kg, $m_B = 9$ kg, $m_C = 5$ kg, $m_D = 7$ kg, and $m_E = 4$ kg.

5-81 Three bodies with masses of 2, 4, and 6 slugs are located at points (2, 3, 4), (3, −4, 5), and (−3, 4, 6), respectively. Locate the mass center of the system if the distances are measured in feet.

5-82 Three bodies with masses of 3, 6, and 7 kg are located at points (4, −3, 1), (−1, 3, 2), and (2, 2, −4), respectively. Locate the mass center of the system if the distances are measured in meters.

5-7 CENTROIDS OF VOLUMES, AREAS, AND LINES

5-7-1 Centroids of Volumes

If the density ρ (mass per unit volume) of the material that makes up a body is constant (the same for every particle that makes up the body), then $m = \rho V$, $dm = \rho\, dV$, and Eqs. (5-36) can be divided by the density ρ to get

$$x_C = \frac{1}{V}\int x\, dV \qquad y_C = \frac{1}{V}\int y\, dV \qquad z_C = \frac{1}{V}\int z\, dV \qquad (5\text{-}37)$$

where V is the volume of the body. The coordinates x_C, y_C, and z_C, defined by Eq. (5-37), depend only on the geometry of the body and are independent of the physical properties. The point located by such a set of coordinates is known as the *centroid C* of the volume of the body. The term "centroid" is usually used in connection with geometrical figures (volumes, areas, and lines), whereas the terms "center of mass" and "center of gravity" are used in connection with physical bodies.

Note that the centroid C of a body can be defined whether the body is homogeneous (made of a uniform material; constant density) or not. If the body is homogeneous, the centroid will have the same position as the center of mass and the center of gravity. If the density of the material varies from point to point within the body, the center of gravity of the body and the centroid of the volume occupied by the body will usually be at different points, as indicated in Fig. 5-37. Since the density of the lower portion of the cone in Fig. 5-37 is greater than the density of the upper portion of the cone, the center of gravity G (which depends on the weights of the two parts) will be below the centroid C (which depends only on the volume of the two parts).

5-7-2 Centroids of Areas

If the body consists of a homogeneous, thin plate of uniform thickness t and surface area A, then $m = \rho t A$, $dm = \rho t\, dA$, and Eqs. (5-36) can be divided by ρt to get

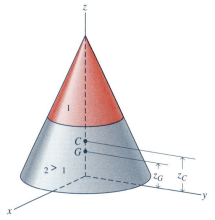

Figure 5-37

$$x_C = \frac{1}{A} \int x \, dA \qquad y_C = \frac{1}{A} \int y \, dA \qquad z_C = \frac{1}{A} \int z \, dA \qquad (5\text{-}38)$$

For a thin three-dimensional shell (such as the dome covering a football stadium), three coordinates x_C, y_C, and z_C are required to specify the location of the centroid C of the shell. The location of the centroid in this case often does not lie on the shell. For example, the centroid of a basketball is not located on the rubber shell of the ball but is located at the center of the hollow cavity inside the shell. For a flat plate in the xy-plane, only two coordinates in the plane of the plate x_C and y_C are required to specify the location of the centroid C of the plate. Even in this case, the centroid may lie inside or outside the plane area.

5-7-3 Centroid of Lines

If the body consists of a homogeneous curved wire with a small uniform cross-sectional area A and length L, then $m = \rho A L$, $dm = \rho A \, dL$, and Eqs. (5-36) can be divided by ρA to get

$$x_C = \frac{1}{L} \int x \, dL \qquad y_C = \frac{1}{L} \int y \, dL \qquad z_C = \frac{1}{L} \int z \, dL \qquad (5\text{-}39)$$

Two or three coordinates, depending on the shape, are required to specify the location of the centroid of the line defining the shape of the wire. Unless the line is straight, the centroid will usually not lie on the line.

5-7-4 Centroid, Center of Mass, or Center of Gravity by Integration

The procedure involved in the determination, by integration, of the coordinates of the centroid, center of mass, or center of gravity of a body can be summarized as follows:

1. Prepare a sketch of the body approximately to scale.
2. Establish a coordinate system. Rectangular coordinates are used with most shapes that have flat planes for boundaries. Polar coordinates are usually used for shapes with circular boundaries. Whenever a line or plane of symmetry exists in a body, a coordinate axis or plane should be chosen to coincide with this line or plane. The centroid, center of mass, or center of gravity will always lie on such a line or plane since the moments of symmetrically located pairs of elements (one with a positive coordinate and the other with an equal negative coordinate) will always cancel.
3. Select an element of volume, area, or length. For center of mass or center of gravity determinations, determine the mass or weight of the element by using the appropriate expression (constant or variable) for the density or specific weight. The element can frequently be selected so that only single integration is required for the complete body or for the several parts into which the body can be divided. Sometimes, however, it may be necessary to use double integration or perhaps triple integration for some shapes. If possible, the element should be chosen so that all parts are the same distance from the reference axis or plane. This distance will be the moment arm for first-moment determinations. When the parts of the element are at different distances from the reference axis or plane, the location of the cen-

troid, center of mass, or center of gravity of the element must be known in order to establish the moment arm for moment calculations. Integrate the expression to determine the volume, area, length, mass, or weight of the body.

4. Write an expression for the first moment of the element with respect to one of the reference axes or planes. Integrate the expression to determine the first moment with respect to the reference axis or plane.

5. Use the appropriate equation [Eqs. (5-32), (5-33), etc.] to obtain the coordinate of the centroid, center of mass, or center of gravity with respect to the reference axis or plane.

6. Repeat steps 3 to 5, using different reference axes or planes for the other coordinates of the centroid, center of mass, or center of gravity.

7. Locate the centroid, center of mass, or center of gravity on the sketch. Gross errors are often detected by using this last step.

The following examples illustrate the procedures for locating centroids (of lines, areas, and volumes) and centers of mass or centers of gravity of bodies by integration.

Example Problem 5-14

Locate the centroid of the rectangular area shown in Fig. 5-38a.

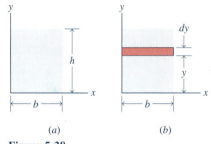

(a)   (b)

Figure 5-38

SOLUTION

Symmetry considerations require that the centroid of a rectangular area be located at the center of the rectangle. Thus, for the rectangular area shown in Fig. 5-38a, $x_C = b/2$ and $y_C = h/2$. These results are established by integration in the following fashion. For the differential element of area shown in Fig. 5-38b, $dA = b\,dy$. The element dA is located a distance y from the x-axis; therefore, the moment of the area about the x-axis is

$$M_x = \int_A y \, dA = \int_0^h y \, (b \, dy) = b \left[\frac{y^2}{2} \right]_0^h = \frac{bh^2}{2}$$

From Eq. (5-38):

$$y_C = \frac{M_x}{A} = \frac{bh^2/2}{bh} = \frac{h}{2} \qquad \textbf{Ans.}$$

In a similar manner by using an element of area $dA = h \, dx$, the moment of the area about the y-axis is

$$M_y = \int_A x \, dA = \int_0^b x \, (h \, dx) = h \left[\frac{x^2}{2} \right]_0^b = \frac{hb^2}{2}$$

From Eq. (5-38):

$$x_C = \frac{M_y}{A} = \frac{hb^2/2}{bh} = \frac{b}{2} \qquad \textbf{Ans.}$$

The element of area $dA = b\,dy$, used to calculate M_x, was not used to calculate M_y since all parts of the horizontal strip are located at different distances x from the y-axis. As a result of this example, it is now known that $x_C = b/2$ for the element of area $dA = b\,dy$ shown in Fig. 5-38b. This result will be used frequently in later examples to simplify the integrations. ■

Example Problem 5-15

Locate the y-coordinate of the centroid of the area of the quarter circle shown in Fig. 5-39a.

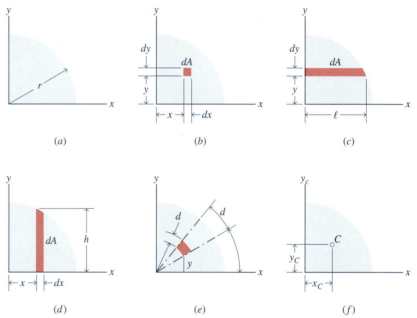

(a)   (b)   (c)

(d)   (e)   (f)

Figure 5-39

SOLUTION

Four different elements will be used to solve this problem.

Element 1: Double Integral in Rectangular Coordinates

For the differential element shown in Fig. 5-39b, $dA = dy\,dx$. The element dA is at a distance y from the x-axis; therefore, the moment of the area about the x-axis is

$$M_x = \int_A y\,dA = \int_0^r \int_0^{\sqrt{r^2 - x^2}} y\,dy\,dx$$

$$= \int_0^r \left[\frac{y^2}{2}\right]_0^{\sqrt{r^2 - x^2}} dx = \int_0^r \frac{r^2 - x^2}{2}\,dx = \left[\frac{r^2 x}{2} - \frac{x^3}{6}\right]_0^r = \frac{r^3}{3}$$

From Eq. (5-38):

$$y_C = \frac{M_x}{A} = \frac{r^3/3}{\pi r^2/4} = \frac{4r}{3\pi} \qquad \textbf{Ans.}$$

Element 2: Single Integral Using a Horizontal Strip

Alternatively, the element of area can be selected as shown in Fig. 5-39c. For this element, which is located a distance y from the x-axis, $dA = \ell\, dy = \sqrt{r^2 - y^2}\, dy$. Therefore, the moment of the area about the x-axis is

$$M_x = \int_A y\, dA = \int_0^r y\sqrt{r^2 - y^2}\, dy = \left[-\frac{(r^2 - y^2)^{3/2}}{3} \right]_0^r = \frac{r^3}{3}$$

From Eq. (5-38):

$$y_C = \frac{M_x}{A} = \frac{r^3/3}{\pi r^2/4} = \frac{4r}{3\pi} \qquad \textbf{Ans.}$$

Element 3: Single Integral Using a Vertical Strip

The element of area could also be selected as shown in Fig. 5-39d. For this element, $dA = h\, dx = \sqrt{r^2 - x^2}\, dx$; however, all parts of the element are at different distances y from the x-axis. For this type of element, the results of Example Problem 5-14 can be used to compute a moment dM_x that can be integrated to yield moment M_x. Thus

$$dM_x = \frac{h}{2}\, dA = \frac{h}{2} h\, dx = \frac{h^2}{2}\, dx = \frac{r^2 - x^2}{2}\, dx$$

$$M_x = \int_A dM_x = \int_0^r \frac{r^2 - x^2}{2}\, dx = \left[\frac{r^2 x}{2} - \frac{x^3}{6} \right]_0^r = \frac{r^3}{3}$$

From Eq. (5-38):

$$y_C = \frac{M_x}{A} = \frac{r^3/3}{\pi r^2/4} = \frac{4r}{3\pi} \qquad \textbf{Ans.}$$

Element 4: Double Integral Using Polar Coordinates

Finally, polar coordinates can be used to locate the centroid of the quarter circle. With polar coordinates, the element of area is $dA = \rho\, d\theta\, d\rho$ and the distance from the x-axis to the element is $y = \rho \sin \theta$, as shown in Fig. 5-39e. Thus

$$M_x = \int_A y\, dA = \int_0^r \int_0^{\pi/2} \rho^2 \sin \theta\, d\theta\, d\rho$$

$$= \int_0^r \rho^2 \left[-\cos \theta \right]_0^{\pi/2} d\rho = \int_0^r \rho^2\, d\rho = \left[\frac{\rho^3}{3} \right]_0^r = \frac{r^3}{3}$$

From Eq. (5-38):

$$y_C = \frac{M_x}{A} = \frac{r^3/3}{\pi r^2/4} = \frac{4r}{3\pi} \qquad \textbf{Ans.}$$

In a completely similar manner, the x-coordinate of the centroid is obtained as

$$x_C = \frac{M_y}{A} = \frac{r^3/3}{\pi r^2/4} = \frac{4r}{3\pi}$$

The results are illustrated in Fig. 5-39f. ■

Example Problem 5-16

A circular arc of thin homogeneous wire is shown in Fig. 5-40a.

(a) Locate the x- and y-coordinates of the mass center.
(b) Use the results of part (a) to determine the coordinates of the mass center for a half circle.

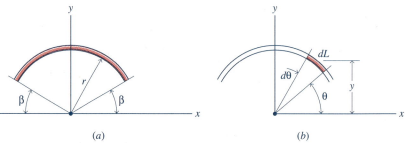

Figure 5-40

SOLUTION

(a) The wire can be assumed to consist of a large number of differential elements of length dL, as shown in Fig. 5-40b. The mass of each of these elements is

$$dm = \rho \, dV = \rho A \, dL = \rho A(r \, d\theta)$$

Therefore, the total mass of the wire is

$$m = \int dm = \int_{\beta}^{\pi-\beta} \rho A r \, d\theta = \rho A r \int_{\beta}^{\pi-\beta} d\theta = \rho A r(\pi - 2\beta)$$

The distance y from the x-axis to the element dm of mass is $y = r \sin \theta$. Thus,

$$my_G = \int y \, dm = \int_{\beta}^{\pi-\beta} (r \sin \theta)(\rho A r \, d\theta)$$

$$= \rho A r^2 \int_{\beta}^{\pi-\beta} \sin \theta \, d\theta = \rho A r^2 (2 \cos \beta)$$

Therefore,

$$y_G = \frac{2\rho A r^2 \cos \beta}{\rho A r(\pi - 2\beta)} = \frac{2r \cos \beta}{\pi - 2\beta} \qquad \textbf{Ans.}$$

Since the length of wire is symmetric about the y-axis,

$$x_G = 0 \qquad \textbf{Ans.}$$

(b) For the half circle, $\beta = 0$

$$y_G = \frac{2r}{\pi}$$ **Ans.**

$$x_G = 0$$ ■   **Ans.**

Example Problem 5-17

Locate the center of gravity G of the homogeneous right circular cone shown in Fig. 5-41a, which has an altitude h, radius r, and is made of a material with a specific weight γ.

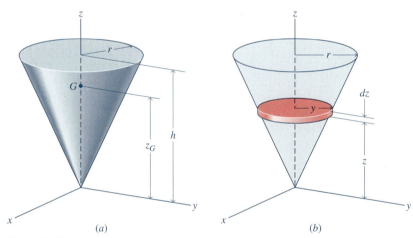

Figure 5-41

SOLUTION

From symmetry, it is obvious that $x_G = y_G = 0$. The coordinate z_G of the center of gravity G of the cone can be located by using the differential element of volume shown in Fig. 5-41b. The weight dW of the differential element is

$$dW = \gamma \, dV = \gamma \, (\pi y^2) \, dz = \gamma \pi \left[\frac{rz}{h}\right]^2 dz = \frac{\gamma \pi r^2}{h^2} z^2 dz$$

From Eq. (5-33):

$$z_G = \frac{1}{W} \int z \, dW$$

Thus,

$$W z_G = \int z \, dW = \int_0^h \frac{\gamma \pi r^2}{h^2} z^3 dz = \frac{1}{4} \gamma \pi r^2 h^2$$

The weight of the cone is

$$W = \int_V \gamma \, dV = \int_0^h \frac{\gamma\pi r^2}{h^2} z^2 dz = \frac{\gamma\pi r^2}{h^2}\left[\frac{z^3}{3}\right]_0^h = \frac{1}{3}\gamma\pi r^2 h$$

Therefore

$$z_G = \frac{\gamma\pi r^2 h^2/4}{\gamma\pi r^2 h/3} = \frac{3h}{4}$$ **Ans.**

Since the xz-plane and the yz-plane are planes of symmetry,

$$x_G = y_G = 0 \quad \blacksquare$$ **Ans.**

PROBLEMS

Standard Problems

5-83* Locate the centroid of the shaded triangular area shown in Fig. P5-83 if $b = 12$ in. and $h = 8$ in.

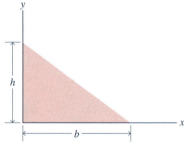

Figure P5-83

5-84* Locate the centroid of the shaded triangular area shown in Fig. P5-84 if $b = 200$ mm and $h = 300$ mm.

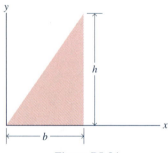

Figure P5-84

5-85 Locate the centroid of the shaded area shown in Fig. P5-85.

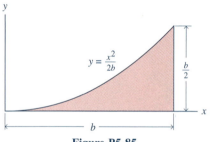

Figure P5-85

5-86 Locate the centroid of the shaded area shown in Fig. P5-86.

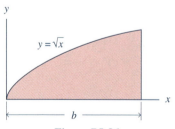

Figure P5-86

5-87* Locate the centroid of the shaded area shown in Fig. P5-87.

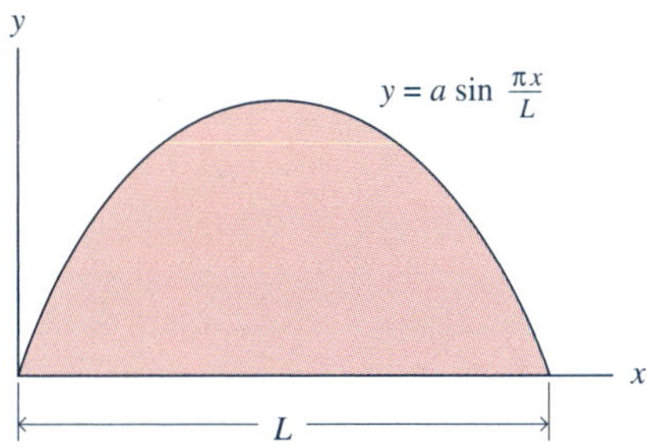

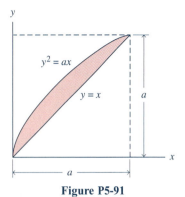

$$y = a \sin \frac{\pi x}{L}$$

Figure P5-87

5-88* Locate the centroid of the shaded area shown in Fig. P5-88.

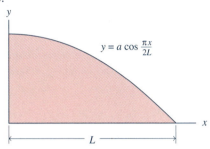

$$y = a \cos \frac{\pi x}{2L}$$

Figure P5-88

5-89 Locate the centroid of the shaded area shown in Fig. P5-89.

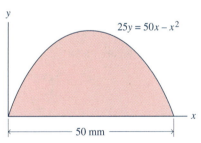

$$25y = 50x - x^2$$

50 mm

Figure P5-89

5-90 Locate the centroid of the shaded area shown in Fig. P5-90.

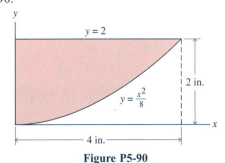

$y = 2$

2 in.

$$y = \frac{x^2}{8}$$

4 in.

Figure P5-90

Challenging Problems

5-91* Locate the centroid of the shaded area shown in Fig. P5-91.

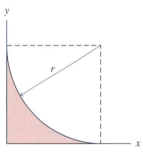

$$y^2 = ax$$

$$y = x$$

a

a

Figure P5-91

5-92* Locate the centroid of the shaded area shown in Fig. P5-92.

r

Figure P5-92

5-93 Locate the centroid of the curved slender rod shown in Fig. P5-93.

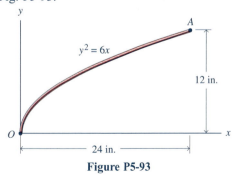

A

$$y^2 = 6x$$

12 in.

O

24 in.

Figure P5-93

5-94 Locate the centroid of the curved slender rod shown in Fig. P5-94.

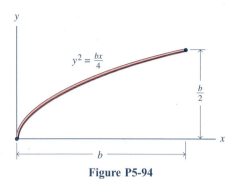

Figure P5-94

5-95* Locate the centroid of the volume of the tetrahedron shown in Fig. P5-95.

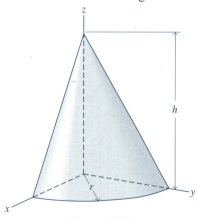

Figure P5-95

5-96* Locate the centroid of the volume of the portion of a right circular cone shown in Fig. P5-96.

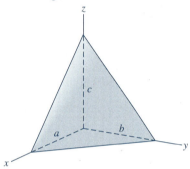

Figure P5-96

5-97 Locate the centroid of the volume obtained by revolving the shaded area shown in Fig. P5-97 about the x-axis.

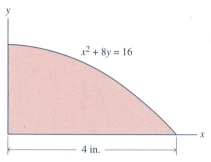

Figure P5-97

5-98 Locate the centroid of the volume obtained by revolving the shaded area shown in Fig. P5-98 about the x-axis.

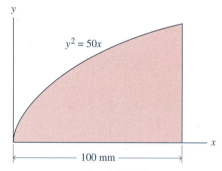

Figure P5-98

5-8 CENTROIDS OF COMPOSITE BODIES

The centroid of an area, line, or volume is a property of the shape of the area, line, or volume and is independent of the coordinate system used to compute it. For example, no matter what coordinate system is used to compute the location of the centroid of a rectangle, it will always be found to be halfway between the opposite sides, as shown in Fig. 5-42a. Similarly, the centroid of a quarter circle will always be $\dfrac{4r}{3\pi}$ from the radial lines bounding the quarter circle (Fig. 5-42b), and the centroid of a triangle will always be one third of the way (measured perpendicularly to the side) from any side of the triangle to the opposite vertex (Fig. 5-42c). Once the centroids of these common shapes are known, they need not be continually recalculated. In fact, the known location of the centroids of common simple shapes can be used to locate the centroids of more complicated shapes.

For example, the area shown in Fig. 5-43 consists of two parts—a triangle and a rectangle. Integrating to find the area and the centroid gives

$$A = \int_{x=0}^{x=a} dA + \int_{x=a}^{x=b} dA = A_{\text{tri}} + A_{\text{rect}}$$

$$M_y = x_C A = \int_{x=0}^{x=a} x \, dA + \int_{x=a}^{x=b} x \, dA$$

$$= (x_C A)_{\text{tri}} + (x_C A)_{\text{rect}}$$

The location of the centroid of the composite area, x_C, consisting of the triangle and rectangle is the sum of the first moments of the two component parts divided by the total area of the two component parts. In general, if an area is composed of n simple shapes whose centroid locations, x_{Ci}, are known (or can be looked up easily), the centroid of the larger area can be found from

$$x_C A = \sum_{i=1}^{n} x_{Ci} A_i \qquad A = \sum_{i=1}^{n} A_i \qquad (5\text{-}40)$$

Composite areas with holes are also handled easily by using Eq. (5-40). Consider the composite area of Fig. 5-44b, which consists of a circular hole cut out of a rectangle and triangle. Denoting the triangle by t, the rectangle by r, the circular hole by h, and the composite body by c, the centroid location of the solid triangle–rectangle area (Fig. 5-44a) is found from

$$(x_C A)_t + (x_C A)_r = (x_C A)_c + (x_C A)_h$$

The first, second, and last terms are easily computed using known properties of simple shapes. Therefore, the centroid of the composite body can be found from

$$(x_C A)_c = (x_C A)_t + (x_C A)_r - (x_C A)_h$$

$$A_c = A_t + A_r - A_h$$

These equations are identical to Eq. (5-40) if the area of the hole is considered to be a negative quantity.

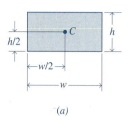

(a)

(b)

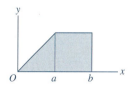

(c)

Figure 5-42

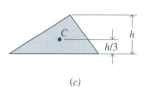

Figure 5-43

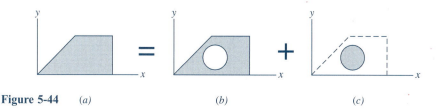

Figure 5-44   (a)   (b)   (c)

Similar equations can be developed for composite lines, volumes, masses, and weights. The final results would show the A's of Eqs. (5-40) replaced with L's, V's, m's, and W's, respectively. Tables containing a listing of centroid locations for some common shapes are presented on the inside front and back covers of this book.

The following examples illustrate the procedure for determining the locations of centroids of composite lines, areas, and volumes and centers of mass and centers of gravity for composite bodies.

Example Problem 5-18

Locate the centroid of the composite area shown in Fig. 5-45a.

SOLUTION

The composite area can be divided into four parts: a rectangle, a triangle, a quarter circle, and a circle, as shown in Fig. 5-45b. Recall that the area of the hole is negative since it must be subtracted from the area of the rectangle. The centroid locations for each of these parts can be determined from the relationships listed on the inside front cover.

For the triangle:

$$x_C = \frac{b}{3} = \frac{50}{3} = 16.67 \text{ mm}$$

$$y_C = 50 + \frac{h}{3} = 50 + \frac{50}{3} = 66.67 \text{ mm}$$

For the quarter circle:

$$x_C = -\frac{4r}{3\pi} = -\frac{4(50)}{3\pi} = -21.22 \text{ mm}$$

$$y_C = 50 + \frac{4r}{3\pi} = 50 + \frac{4(50)}{3\pi} = 71.22 \text{ mm}$$

The centroid for the composite area is determined by listing the area, centroid location, and first moments for the individual parts in a table and applying Eqs. (5-40). Thus,

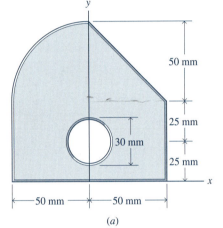

(a)

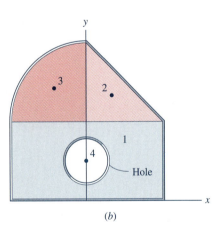

(b)

Figure 5-45

Part	A_i (mm²)	x_{Ci} (mm)	M_y (mm³)	y_{Ci} (mm)	M_x (mm³)
1	5000	0	0	25	125,000
2	1250	16.67	20,838	66.67	83,338
3	1963	−21.22	−41,655	71.22	139,805
4	−707	0	0	25	−17,675
	7506		−20,817		330,468

From Eqs. (5-40),

$$x_C = \frac{M_y}{A} = \frac{-20,817}{7506} = -2.77 \text{ mm} \qquad \textbf{Ans.}$$

$$y_C = \frac{M_x}{A} = \frac{330,468}{7506} = 44.0 \text{ mm} \ ■ \qquad \textbf{Ans.}$$

Example Problem 5-19

A slender steel rod is bent into the shape shown in Fig. 5-46a. Locate the centroid of the rod.

SOLUTION

The rod can be divided into three parts as shown in Fig. 5-46b. The centroid locations for each of these parts are known or can be determined from the relationships listed on the inside front cover. For the semicircular arc,

$$L_3 = \pi r = \pi(9.90) = 31.1 \text{ in.}$$

$$y_{C3} = 7 + \frac{r \sin \alpha}{\alpha} \cos 45° = 7 + \frac{9.90 \sin (\pi/2)}{\pi/2} \cos 45° = 11.457 \text{ in.}$$

$$z_{C3} = 7 + \frac{r \sin \alpha}{\alpha} \sin 45° = 7 + \frac{9.90 \sin (\pi/2)}{\pi/2} \sin 45° = 11.457 \text{ in.}$$

The centroid for the composite rod can be determined by listing the length, centroid location, and first moments for the individual parts in a table and applying Eqs. (5-40). Thus,

Part	L_i (in.)	x_{Ci} (in.)	M_{yz} (in.²)	y_{Ci} (in.)	M_{zx} (in.²)	z_{Ci} (in.)	M_{xy} (in.²)
1	16.0	8	128	0	0	0	0
2	14.0	0	0	7	98	0	0
3	31.1	0	0	11.457	356.3	11.457	356.3
	61.1		128		454.3		356.3

From Eqs. (5-40),

$$x_C = \frac{M_{yz}}{L} = \frac{128}{61.1} = 2.09 \text{ in.} \qquad \textbf{Ans.}$$

$$y_C = \frac{M_{zx}}{L} = \frac{454.3}{61.1} = 7.44 \text{ in.} \qquad \textbf{Ans.}$$

$$z_C = \frac{M_{xy}}{L} = \frac{356.3}{61.1} = 5.83 \text{ in.} \ \blacksquare \qquad \textbf{Ans.}$$

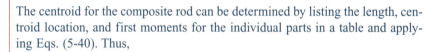

(a)

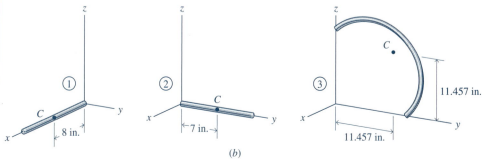

(b)

Figure 5-46

PROBLEMS

Standard Problems

5-99*   Locate the centroid of the slender rod shown in Fig. P5-99.

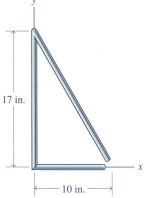

Figure P5-99

5-100*   Locate the centroid of the slender rod shown in Fig. P5-100.

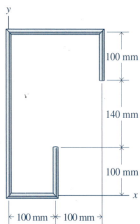

Figure P5-100

5-101   Locate the centroid of the slender rod shown in Fig. P5-101.

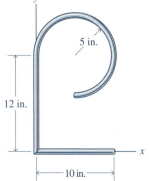

Figure P5-101

5-102   Locate the centroid of the slender rod shown in Fig. P5-102.

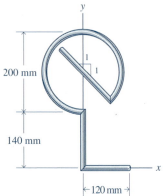

Figure P5-102

5-103*   Locate the centroid of the shaded area shown in Fig. P5-103.

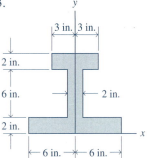

Figure P5-103

5-104*   Locate the centroid of the shaded area shown in Fig. P5-104.

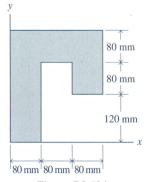

Figure P5-104

5-105   Locate the centroid of the shaded area shown in Fig. P5-105.

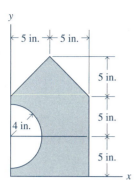

Figure P5-105

5-106 Locate the centroid of the shaded area shown in Fig. P5-106.

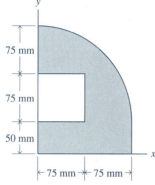

Figure P5-106

5-107* Locate the centroid of the shaded area shown in Fig. P5-107.

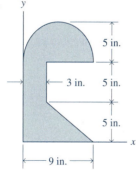

Figure P5-107

5-108* Locate the centroid of the shaded area shown in Fig. P5-108.

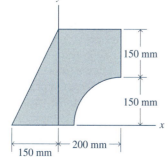

Figure P5-108

Challenging Problems

5-109* A bracket is made of brass ($\gamma = 0.316$ lb/in.^{3}) and aluminum ($\gamma = 0.100$ lb/in.^{3}) plates as shown in Fig. P5-109.
(a) Locate the centroid of the bracket.
(b) Locate the center of gravity of the bracket.

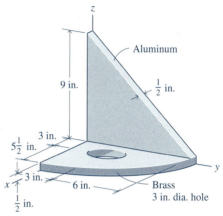

Figure P5-109

5-110* A cylinder with a hemispherical cavity and a conical cap is shown in Fig. P5-110.
(a) Locate the centroid of the composite volume if $R = 140$ mm, $L = 250$ mm, and $h = 300$ mm.
(b) Locate the center of mass of the composite volume if the cylinder is made of steel ($\rho = 7870$ kg/m^3) and the cap is made of aluminum ($\rho = 2770$ kg/m^3).

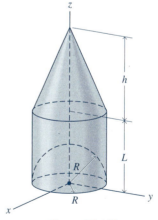

Figure P5-110

5-111 A bracket is made of steel ($\gamma = 0.284$ lb/in.^{3}) and aluminum ($\gamma = 0.100$ lb/in.^{3}) plates as shown in Fig. P5-111.
(a) Locate the centroid of the bracket.
(b) Locate the center of gravity of the bracket.

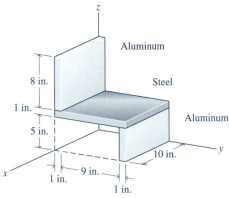

Figure P5-111

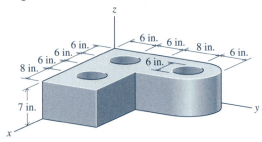

5-113* Locate the center of gravity of the bracket shown in Fig. P5-113 if the holes have 6-in. diameters.

Figure P5-113

5-112 A cylinder with a conical cavity and a hemispherical cap is shown in Fig. P5-112.
(a) Locate the centroid of the composite volume if R = 200 mm and h = 250 mm.
(b) Locate the center of mass of the composite volume if the cylinder is made of brass (ρ = 8750 kg/m^3) and the cap is made of aluminum (ρ = 2770 kg/m^3).

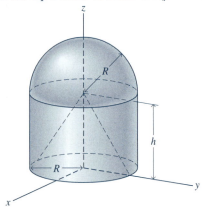

Figure P5-112

5-114* Locate the center of mass of the machine component in Fig. P5-114. The brass (ρ = 8750 kg/m^3) disk C is mounted on the steel (ρ = 7870 kg/m^3) shaft B.

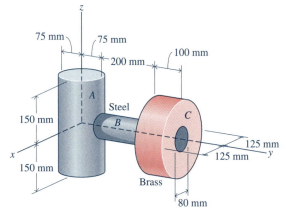

Figure P5-114

5-9 DISTRIBUTED LOADS ON STRUCTURAL MEMBERS

When a load is applied to a rigid body such as a structural member, it is often distributed along a line or over an area A. In many instances, it is convenient to replace this distributed load with a single equivalent force that is externally equivalent to the distributed load. The methods used to replace the distributed weight force with an equivalent single force work for any distributed force as long as the elements of the distributed force are all parallel to each other. The quantities to be determined are the magnitude of the equivalent force (the direction of the force is the same as that of all of the components) and the location of its line of action.

Consider the beam shown in Fig. 5-47a. A beam is a structural member whose length is large compared to its cross-sectional dimensions. Since distrib-

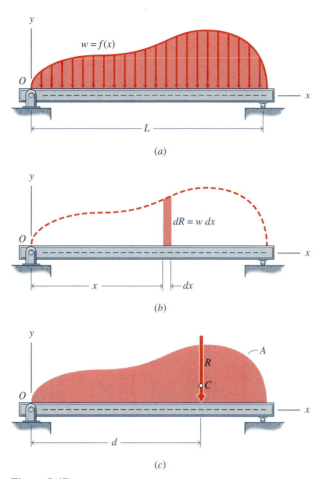

Figure 5-47

uted loads on beams do not vary across the width of the cross section of the beam, the actual load intensity on the beam can be multiplied by the width of the beam to yield a distributed line load w (N/m or lb/ft) whose magnitude varies only with position x along the beam. The distributed load w versus position x graph (see Fig. 5-47a) is known as a *load diagram*. The magnitude of the differential force $d\mathbf{R}$ exerted on the beam by the distributed load w in an increment of length dx (see Fig. 5-47b) is

$$dR = w\,dx$$

Therefore, the magnitude of the single concentrated resultant force $\mathbf{R}$ that is equivalent to the distributed load w is

$$R = \int_0^L w\,dx \qquad\qquad (5\text{-}41)$$

That is, the magnitude of the resultant force $\mathbf{R}$ is equal to the area A under the load diagram, as shown in Fig. 5-47c.

The location of the line of action of the equivalent force $\mathbf{R}$ can be determined (using the principle of moments) by equating the moment of the equivalent force $\mathbf{R}$ about an arbitrary point O to the moment of the distributed load about the same point O. The moment produced by the force $dR = w\,dx$ about point O of Fig. 5-47b is

$$dM_O = x\,dR$$

and the total moment produced by the distributed load w about point O is

$$M_O = \int dM_O = \int x\,dR = x_C R \qquad\qquad (a)$$

where x_C is the distance along the beam from point O to the centroid of the area A under the load diagram. The moment produced by the equivalent force $\mathbf{R}$ about point O is

$$M_O = Rd \qquad\qquad (b)$$

where d is the distance along the beam from point O to the line of action of the equivalent force $\mathbf{R}$. Setting Eqs. (a) and (b) equal gives

$$M_O = Rd = \int x\,dR = x_C R$$

or

$$d = x_C = \frac{M_O}{R} \qquad\qquad (5\text{-}42)$$

That is, for the purpose of calculating the sum of forces and the sum of moments, the distributed force is equivalent to a single force whose magnitude is equal to the area under the load diagram and whose line of action passes through the centroid of the load diagram.

The following examples illustrate the procedure for determining the single concentrated resultant force $\mathbf{R}$ that is equivalent to the distributed load w and the location of its line of action.

Example Problem 5-20

A beam is subjected to a system of loads that can be represented by the load diagram shown in Fig. 5-48a. Determine the resultant of this system of distributed loads and locate its line of action with respect to the left support of the beam.

SOLUTION

The magnitude of the resultant $\mathbf{R}$ of the distributed load shown in Fig. 5-48a is equal to the area under the load diagram. The load diagram can be divided into two triangles and a rectangle. Thus, from the relationships listed on the inside front cover:

$$F_1 = A_1 = \frac{1}{2}b_1h_1 = \frac{1}{2}(2)(300) = 300 \text{ N}$$

$$x_{C1} = \frac{2}{3}b_1 = \frac{2}{3}(2) = 1.333 \text{ m}$$

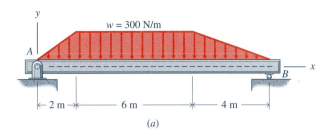

(a)

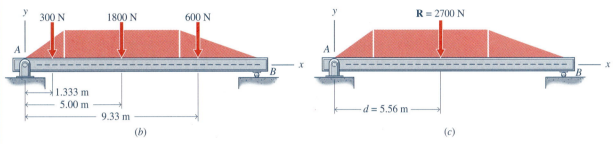

(b)   (c)

Figure 5-48

$$F_2 = A_2 = b_2 h_2 = (6)(300) = 1800 \text{ N}$$

$$x_{C2} = 2 + \frac{1}{2}b_2 = 2 + \frac{1}{2}(6) = 5.00 \text{ m}$$

$$F_3 = A_3 = \frac{1}{2}b_3 h_3 = \frac{1}{2}(4)(300) = 600 \text{ N}$$

$$x_{C3} = 8 + \frac{1}{3}b_3 = 8 + \frac{1}{3}(4) = 9.33 \text{ m}$$

The equivalent forces for the three different areas and the locations of their lines of action are shown in Fig. 5-48b. Thus,

$$R = F_1 + F_2 + F_3 = 300 + 1800 + 600 = 2700 \text{ N} \qquad \textbf{Ans.}$$

The line of action of the resultant with respect to the left support is located by summing moments about point A. Thus,

$$
\begin{aligned}
M_A = Rd &= F_1 x_{C1} + F_2 x_{C2} + F_3 x_{C3} \\
&= 300(1.333) + 1800(5.00) + 600(9.33) = 15{,}000 \text{ N} \cdot \text{m}
\end{aligned}
$$

Finally,

$$d = x_C = \frac{M_A}{R} = \frac{15{,}000}{2700} = 5.56 \text{ m} \qquad \textbf{Ans.}$$

The resultant force **R** and its line of action are shown in Fig. 5-48c. ■

Example Problem 5-21

A beam is subjected to a system of loads that can be represented by the load diagram shown in Fig. 5-49a. Determine the resultant of this system of loads and locate its line of action with respect to the left support of the beam.

SOLUTION

The magnitude of the resultant **R** of the distributed load shown in Fig. 5-49a is equal to the area under the load diagram, and the line of action of the resultant passes through the centroid of the area. Since the area under this load diagram and the location of its centroid do not normally appear in tables of areas and centroid locations, it is necessary to use integration methods to determine the magnitude of the resultant and the location of its line of action.

The area under the load diagram is determined by using the element of area shown in Fig. 5-49b. Thus,

$$A = \int_A w\ dx = \int_0^L w_{max} \sin \frac{\pi x}{2L}\ dx = \frac{2Lw_{max}}{\pi}\left[-\cos \frac{\pi x}{2L} \right]_0^L = \frac{2Lw_{max}}{\pi}$$

Thus,

$$R = A = \frac{2Lw_{max}}{\pi} = 0.637w_{max}L \qquad \textbf{Ans.}$$

The moment of the area about support A is

$$M_A = \int_A x\ (w\ dx) = \int_0^L w_{max}\ x \sin \frac{\pi x}{2L}\ dx$$

$$= w_{max}\left[\frac{4L^2}{\pi^2} \sin \frac{\pi x}{2L} - \frac{2L}{\pi}x \cos \frac{\pi x}{2L} \right]_0^L = \frac{4L^2 w_{max}}{\pi^2}$$

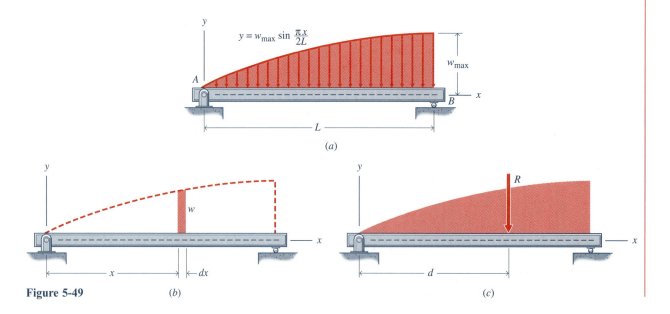

Figure 5-49

From Eq. (5-38):

$$d = x_C = \frac{M_A}{A} = \frac{4L^2 w_{max}/\pi^2}{2Lw_{max}/\pi} = \frac{2L}{\pi} = 0.637\,L \qquad \text{Ans.}$$

The results are shown in Fig. 5-49c. ■

PROBLEMS

Standard Problems

5-115* A beam is subjected to a system of loads that can be represented by the load diagram shown in Fig. P5-115. Determine the resultant **R** of this system of distributed loads and locate its line of action with respect to the left support of the beam.

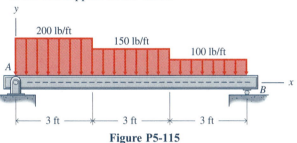

Figure P5-115

5-116* A beam is subjected to a system of loads that can be represented by the load diagram shown in Fig. P5-116. Determine the resultant **R** of this system of distributed loads and locate its line of action with respect to the left support of the beam.

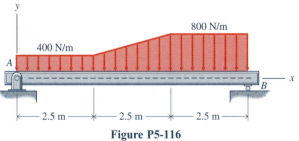

Figure P5-116

5-117* A beam is subjected to a system of loads that can be represented by the load diagram shown in Fig. P5-117. Determine the resultant **R** of this system of distributed loads and locate its line of action with respect to the left support of the beam.

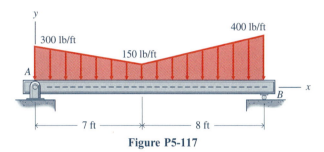

Figure P5-117

5-118* A beam is subjected to a system of loads that can be represented by the load diagram shown in Fig. P5-118. Determine the resultant **R** of this system of distributed loads and locate its line of action with respect to the left support of the beam.

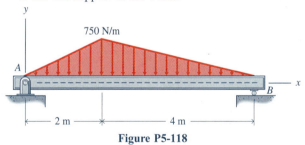

Figure P5-118

5-119 A beam is subjected to a system of loads that can be represented by the load diagram shown in Fig. P5-119. Determine the resultant **R** of this system of distributed loads and locate its line of action with respect to the left support of the beam.

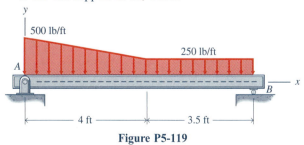

Figure P5-119

5-120 A beam is subjected to a system of loads that can be represented by the load diagram shown in Fig. P5-120. Determine the resultant **R** of this system of distributed loads and locate its line of action with respect to the left support of the beam.

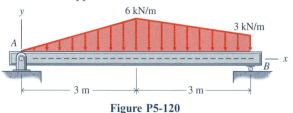

Figure P5-120

5-121* A beam is subjected to a system of loads that can be represented by the load diagram shown in Fig. P5-121. Determine the resultant **R** of this system of distributed loads and locate its line of action with respect to the left support of the beam.

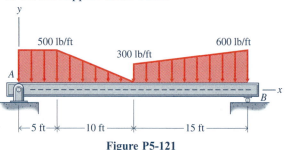

Figure P5-121

5-122* A beam is subjected to a system of loads that can be represented by the load diagram shown in Fig. P5-122. Determine the resultant **R** of this system of distributed loads and locate its line of action with respect to the left support of the beam.

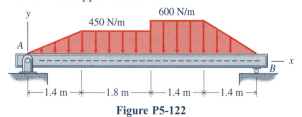

Figure P5-122

5-123 A beam is subjected to a system of loads that can be represented by the load diagram shown in Fig. P5-123. Determine the resultant **R** of this system of distributed loads and locate its line of action with respect to the left support of the beam.

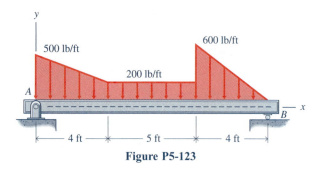

Figure P5-123

5-124 A beam is subjected to a system of loads that can be represented by the load diagram shown in Fig. P5-124. Determine the resultant **R** of this system of distributed loads and locate its line of action with respect to the left support of the beam.

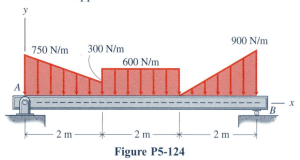

Figure P5-124

Challenging Problems

5-125 A beam is subjected to a system of loads that can be represented by the load diagram shown in Fig. P5-125. Determine the resultant **R** of this system of distributed loads and locate its line of action with respect to the left support of the beam.

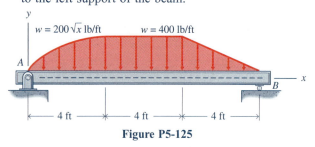

Figure P5-125

5-126 A beam is subjected to a system of loads that can be represented by the load diagram shown in Fig. P5-126. Determine the resultant **R** of this system of distributed loads and locate is line of action with respect to the left support of the beam.

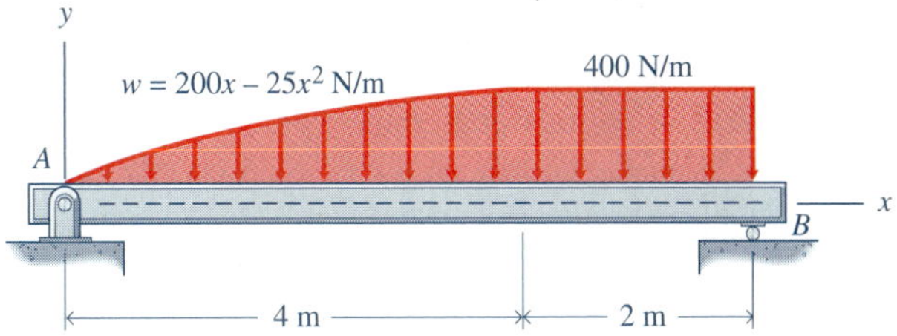

Figure P5-126

5-10 SUMMARY

For a three-dimensional body, the particle idealization discussed in Chapter 3 is not valid, in general, since the forces acting on the body are usually not concurrent. For a general force system, $\mathbf{R} = 0$ is a necessary but not a sufficient condition for equilibrium of the body. A second restriction related to the tendency of a force to produce rotation of a body must also be satisfied and gives rise to the concept of a moment. In this chapter, the moment of a force about a point and the moment of a force about a line (axis) were defined and methods were developed for finding the resultant forces and the resultant moments (couples) for any general force system that may be applied to a body.

A moment is a vector quantity since it has both a magnitude and a direction, and adds according to the parallelogram law of addition. The moment of a force $\mathbf{F}$ about a point O can be represented by the vector cross product

$$\mathbf{M}_O = \mathbf{r} \times \mathbf{F} \tag{5-3}$$

where $\mathbf{r}$ is a position vector from point O to any point on the line of action of the force $\mathbf{F}$. The cross product of the two intersecting vectors $\mathbf{r}$ and $\mathbf{F}$ is

$$\mathbf{M}_O = \mathbf{r} \times \mathbf{F} = M_x\,\mathbf{i} + M_y\,\mathbf{j} + M_z\,\mathbf{k} = M_O\,\mathbf{e} \tag{5-9}$$

where

$$M_x = r_y F_z - r_z F_y \qquad M_y = r_z F_x - r_x F_z \qquad M_z = r_x F_y - r_y F_x$$

are the three scalar components of the moment. The magnitude of the moment $\mathbf{M}_O$ is

$$M_O = |\mathbf{M}_O| = \sqrt{M_x^2 + M_y^2 + M_z^2} \tag{5-11}$$

The direction cosines associated with the unit vector $\mathbf{e}$ are

$$\cos\theta_x = \frac{M_x}{|\mathbf{M}_O|} \qquad \cos\theta_y = \frac{M_y}{|\mathbf{M}_O|} \qquad \cos\theta_z = \frac{M_z}{|\mathbf{M}_O|} \tag{5-14}$$

The moment $\mathbf{M}_{OB}$ of a force $\mathbf{F}$ about a line OB in a direction n specified by the unit vector $\mathbf{e}_n$ is

$$\mathbf{M}_{OB} = (\mathbf{M}_O \cdot \mathbf{e}_n)\,\mathbf{e}_n = [(\mathbf{r} \times \mathbf{F}) \cdot \mathbf{e}_n]\,\mathbf{e}_n = M_{OB}\,\mathbf{e}_n \tag{5-16}$$

Two equal parallel forces of opposite sense are called a couple. A couple has no tendency to translate a body in any direction but tends only to rotate the body on which it acts. Several transformations of a couple can be made without changing any of the external effects of the couple on the body. A couple (1) can be translated to a parallel position in its plane or to any parallel plane, and (2) can be rotated in its plane. Also, (3) the magnitude F of the two forces of a couple and the distance d between them can be changed provided the product Fd remains constant. Any system of couples in a plane or in space can be combined into a single resultant couple **C.**

Any force **F** can be resolved into a parallel force **F** and a couple **C.** Alternatively, a force and a couple in the same plane can be combined into a single force. The sole effect of combining a couple with a force is to move the action line of the force into a parallel position. The magnitude and sense of the force remain unchanged.

The resultant of a force system acting on a rigid body is the simplest force system that can replace the original system without altering the external effect of the system on the body. For a coplanar system of forces, the resultant is either a force **R** or a couple **C.** For a three-dimensional system of forces, the resultant may be a force **R,** a couple **C,** or both a force **R** and a couple **C.**

In many instances, surface loads on a body are not concentrated at a point but are distributed along a length or over an area. Other forces, known as body forces, are distributed over the volume of the body. A distributed force at any point is characterized by its intensity and its direction.

Previously, moments of forces about points or axes were considered. In engineering analysis, equations are also encountered that represent moments of masses, forces, volumes, areas, or lines with respect to axes or planes. Such moments are called first moments of the quantity being considered since the first power of a distance is used in the expression.

The term "center of mass" is used to denote the point in a physical body where the mass can be conceived to be concentrated so that the moment of the concentrated mass with respect to an axis or plane equals the moment of the distributed mass with respect to the same axis or plane. The term "center of gravity" is used to denote the point in the body through which the weight of the body acts, regardless of the position (or orientation) of the body. The location of the center of gravity in a body is determined by using equations of the form

$$M_{yz} = Wx_G = \int x \, dW \qquad \text{or} \qquad x_G = \frac{1}{W} \int x \, dW \qquad (5\text{-}33b)$$

If the density ρ of a body is constant, Eqs. (5-33) reduce to

$$x_C = \frac{1}{V} \int x \, dV \qquad y_C = \frac{1}{V} \int y \, dV \qquad z_C = \frac{1}{V} \int z \, dV \qquad (5\text{-}37)$$

Equations (5-37) indicate that the coordinates x_C, y_C, and z_C depend only on the geometry of the body and are independent of the physical properties. The point located by such a set of coordinates is known as the "centroid" of the volume of the body. The term centroid is usually used in connection with geometrical figures (volumes, areas, and lines), whereas the terms "center of mass" and "center of gravity" are used in connection with physical bodies. The centroid of a volume has the same position as the center of gravity of the body if the body is homogeneous. If the density is variable, the center of gravity of the body and the centroid of the volume will be at different points.

When a load, applied to a rigid body such as a structural member, is distributed along a line or over an area A, it is often convenient for purposes of static analysis to replace this distributed load with a resultant force $\mathbf{R}$ that is equivalent to the distributed load w. For a beam with a distributed load along its length, the magnitude of the resultant force is determined from the expression

$$R = \int_0^L w \, dx \qquad (5\text{-}41)$$

which indicates that the magnitude of the resultant force is equal to the area under the load diagram used to represent the distributed load. The line of action of the resultant force passes through the centroid of the area under the load diagram.

REVIEW PROBLEMS

5-127* A 2500-lb jet engine is suspended from the wing of an airplane as shown in Fig. P5-127. Determine the moment produced by the engine about point A in the wing when the plane is
(a) On the ground with the engine not operating.
(b) In flight with the engine developing a thrust $\mathbf{T}$ of 15 kip.

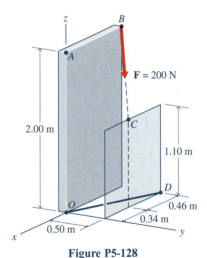

Figure P5-128

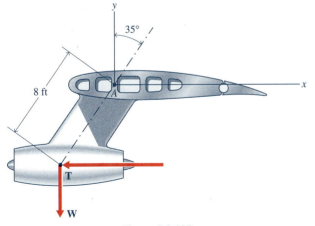

Figure P5-127

5-128* A 200-N force is applied at corner B of a rectangular plate as shown in Fig. P5-128. Determine
(a) The moment of the force about point O.
(b) The moment of the force about line OD.

5-129 Four forces are applied to a circular disk as shown in Fig. P5-129. Determine the magnitude and direction of the resultant of the four forces and the distance x_R from point O to the intercept of the line of action of the resultant with the x-axis.

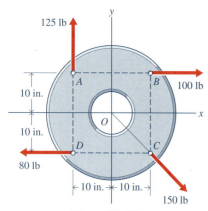

Figure P5-129

5-130 A bent rod supports two forces as shown in Fig. P5-130. Determine
(a) The moment of the resultant of two forces about point O.
(b) The moment of the resultant of two forces about line OA.

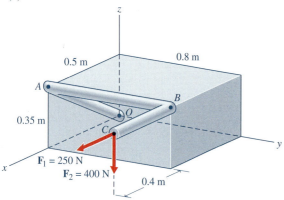

Figure P5-130

5-131* Determine the resultant of the parallel force system shown in Fig. P5-131 and locate the intersection of its line of action with the xy-plane.

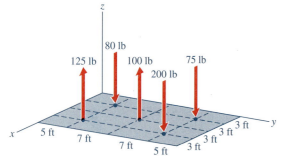

Figure P5-131

5-132* A bent rod supports a 450-N force as shown in Fig. P5-132.
(a) Replace the 450-N force with a force $\mathbf{R}$ through point O and a couple $\mathbf{C}$.

(b) Determine the twisting moments produced by force $\mathbf{F}$ in the three different segments of the rod.

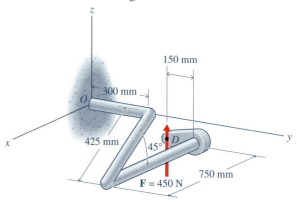

Figure P5-132

5-133 Four forces and a couple are applied to a frame as shown in Fig. P5-133. Determine
(a) The magnitude and direction of the resultant.
(b) The perpendicular distance d_R from point A to the line of action of the resultant.

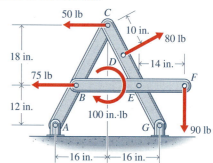

Figure P5-133

5-134 The force $\mathbf{F}$ shown in Fig. 5-134 has a magnitude of 763 N. Replace the force $\mathbf{F}$ by a force $\mathbf{F}_O$ at point O and a couple $\mathbf{C}$.
(a) Express the force $\mathbf{F}_O$ and the couple $\mathbf{C}$ in Cartesian vector form.
(b) Determine the direction angles θ_x, θ_y, and θ_z associated with the unit vector $\mathbf{e}$ that describes the aspect of the plane of the couple.

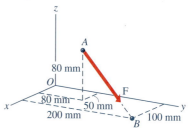

Figure P5-134

5-135* Locate the centroid of the shaded area shown in Fig. P5-135.

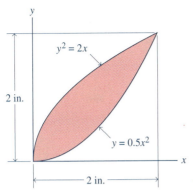

Figure P5-135

5-138 Locate the centroid of the volume shown in Fig. P5-138 if $R = 250$ mm and $h = 800$ mm.

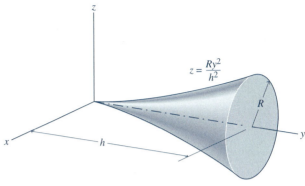

$$z = \frac{Ry^2}{h^2}$$

Figure P5-138

5-136* Locate the centroid of the shaded area shown in Fig. P5-136.

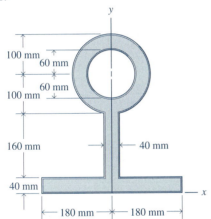

Figure P5-136

5-139* Determine the resultant **R** of the system of distributed loads on the beam of Fig. P5-139 and locate its line of action with respect to the left support of the beam.

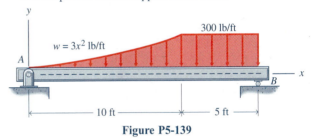

$w = 3x^2$ lb/ft, 300 lb/ft, 10 ft, 5 ft

Figure P5-139

5-140 Locate the centroid and the mass center of the volume shown in Fig. P5-140 that consists of an aluminum ($\rho = 2770$ kg/m^3) cylinder and a steel ($\rho = 7870$ kg/m^3) cylinder and sphere.

5-137 Locate the centroid of the slender rod shown in Fig. P5-137.

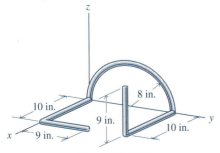

Figure P5-137

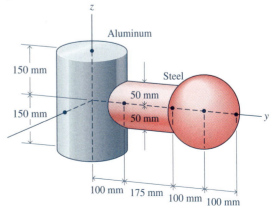

Figure P5-140

EQUILIBRIUM: RIGID AND DEFORMABLE BODIES

6-1 INTRODUCTION

The concept of equilibrium was introduced in Chapter 3 and applied to a system of forces acting on a particle. Since any system of forces acting on a particle is a concurrent force system, a particle is in equilibrium if the resultant force $\mathbf{R}$ of the force system acting on the particle is zero. For the case of a rigid body, it was shown in Chapter 5 that the most general force system can be expressed in terms of a resultant force $\mathbf{R}$ and a resultant couple $\mathbf{C}$. Therefore, for a rigid body to be in equilibrium, both the resultant force $\mathbf{R}$ and the resultant couple $\mathbf{C}$ must vanish. These two conditions are expressed by the two vector equations

$$\mathbf{R} = \Sigma F_x\mathbf{i} + \Sigma F_y\mathbf{j} + \Sigma F_z\mathbf{k} = \mathbf{0}$$
$$\mathbf{C} = \Sigma M_x\mathbf{i} + \Sigma M_y\mathbf{j} + \Sigma M_z\mathbf{k} = \mathbf{0}$$

(6-1)

Equations (6-1) can be expressed in scalar form as

$$\Sigma F_x = 0 \qquad \Sigma F_y = 0 \qquad \Sigma F_z = 0$$
$$\Sigma M_x = 0 \qquad \Sigma M_y = 0 \qquad \Sigma M_z = 0$$

(6-2)

Equations (6-2) are the necessary conditions for equilibrium of a rigid body. If all of the forces acting on a body can be determined from these equations, then they are also the sufficient conditions for equilibrium.

The forces and moments that act on a rigid body are either external or internal. Forces applied to a rigid body by another body or by the earth are external forces. Fluid pressure on the wall of a tank or the force applied by a truck wheel to a road surface are examples of external forces. The weight of a body is another example of an external force. If the body of interest is composed of several parts, the forces holding the parts together are defined as internal forces. In addition, internal forces hold the particles forming the body together. In Chapter 4, the intensity of these internal forces was called *stress*. If the intensity of the internal force is too large, the body may rupture or deform excessively.

External forces can be divided into applied forces and reaction forces. External forces that tend to cause a body to move in some direction or rotate about some axis are applied forces. External forces exerted on a body by supports or connections are reaction forces. Reaction forces oppose the tendency of applied forces to cause motion. They hold the body in equilibrium. Our concern in this chapter is with external forces and the moments these external forces produce.

Since internal forces occur as equal and opposite pairs, they have no effect on equilibrium of the overall rigid body.

The best way to identify all forces acting on a body of interest is to use the free-body diagram approach introduced in Chapter 3, where equilibrium of a particle was considered. This free-body diagram of the body of interest must show all of the applied forces and all of the reaction forces exerted on the body by the supports. In Section 6-2, free-body diagrams for a number of different rigid bodies are considered and the reaction forces exerted on them by a number of different types of supports are specified.

Once the free-body diagram has been drawn and the reaction forces have been found using the equations of equilibrium, the body of interest can be sectioned and the resultant of the internal forces can be determined. The ability of the body to withstand the intensity of the internal forces can be investigated, and the deformation characteristics of the body can be studied.

6-2 FREE-BODY DIAGRAMS

The concept of a free-body diagram was introduced in Chapter 3 and used to solve equilibrium problems involving bodies that could be idealized as a particle. In all these problems, the external forces (applied forces and reaction forces) could be represented as a concurrent force system. In the more general case of a rigid body, systems of forces other than concurrent force systems are encountered and the free-body diagrams become much more complicated. The basic procedure for drawing the diagram, however, remains the same and consists essentially of the following four steps (slightly modified from Section 3-2 to account for a rigid body):

Step 1. Decide which body or combination of bodies is to be shown on the free-body diagram.

Step 2. Prepare a drawing or sketch of the outline of this isolated or free body.

Step 3. Carefully trace around the boundary of the free body and identify all the forces and moments exerted by contacting or interacting bodies that were removed during the isolation process.

Step 4. Choose the set of coordinate axes to be used in solving the problem and indicate these directions on the free-body diagram.

Application of these four steps to a statics problem will produce a complete and correct free-body diagram, which is *an essential first step for the solution of any problem*.

Forces that are known should be added to the diagram and labeled with their proper magnitudes and directions. Letter symbols can be used to represent magnitudes of forces that are unknown. If the correct sense of an unknown force is not obvious, the sense can be arbitrarily assumed. The algebraic sign of the calculated value of the unknown force will indicate the sense of the force. A plus sign indicates that the force is in the direction assumed, and a minus sign indicates that the force is in a direction opposite from that assumed.

When connections or supports are removed from the isolated body, the actions of these connections or supports must be represented by forces and/or mo-

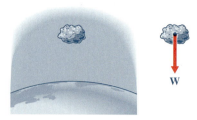

Figure 6-1

ments on the free-body diagram. The forces and moments used to represent the actions of common connections and supports used with bodies subjected to two-dimensional force systems are identified and discussed in Section 6-2-1. A similar discussion for connections and supports used with bodies subjected to three-dimensional force systems is presented in Section 6-2-2.

A common type of force, regardless of the type of support or connection, is the weight of the body, which is the gravitational attraction of the earth on the body. As shown in Fig. 6-1, the line of action of the weight, **W**, passes through the center of gravity of the body and is directed toward the center of the earth.

6-2-1 Idealization of Two-Dimensional Supports and Connections

Common types of supports and connections used with rigid bodies subjected to two-dimensional force systems, together with the forces and moments used to represent the actions of these supports and connections on a free body, are listed in Table 6-1.

TABLE 6-1 Two-Dimensional Reactions at Supports and Connections

(1) FLEXIBLE CORD, ROPE, CHAIN, OR CABLE

Figure 6-2

A flexible cord, rope, chain, or cable (Fig. 6-2) always exerts a tensile force **R** on the body. The line of action of the force **R** is known; it is tangent to the cord, rope, chain, or cable at the point of attachment.

(2) RIGID LINK

Figure 6-3

A rigid link (Fig. 6-3) can exert either a tensile or a compressive force **R** on the body. The line of action of the force **R** is known; it must be directed along the axis of the link (see Section 6-3-1 for proof).

(3) BALL, ROLLER, OR ROCKER

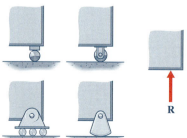

Figure 6-4

A ball, roller, or rocker (Fig. 6-4) can exert a compressive force **R** on the body. The line of action of the force **R** is perpendicular to the surface supporting the ball, roller, or rocker.

TABLE 6-1 (continued)

(4) SMOOTH SURFACE

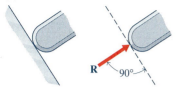

Figure 6-5

A smooth surface, either flat or curved (Fig. 6-5), can exert a compressive force **R** on the body. The line of action of the force **R** is perpendicular to the smooth surface at the point of contact between the body and the smooth surface.

(5) SMOOTH PIN

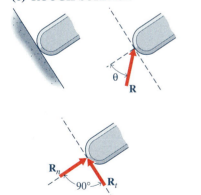

Figure 6-6

A smooth pin (Fig. 6-6) can exert a force **R** of unknown magnitude R and direction θ on the body. As a result, the force **R** is usually represented on a free-body diagram by its rectangular components $\mathbf{R}_x$ and $\mathbf{R}_y$.

(6) ROUGH SURFACE

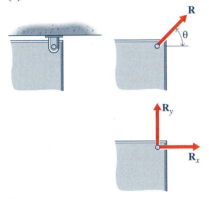

Figure 6-7

Rough surfaces (Fig. 6-7) are capable of supporting a tangential frictional force $\mathbf{R}_t$ as well as a compressive normal force $\mathbf{R}_n$. As a result, the force **R** exerted on a body by a rough surface is a compressive force **R** at an unknown angle θ. The force **R** is usually represented on a free-body diagram by its rectangular components $\mathbf{R}_n$ and $\mathbf{R}_t$.

(7) PIN IN A SMOOTH GUIDE

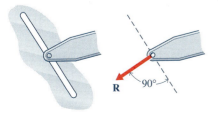

Figure 6-8

A pin in a smooth guide (Fig. 6-8) can transmit only a force **R** perpendicular to the surfaces of the guide. The sense of **R** is assumed on the figure and may be either downward to the left or upward to the right.

Table 6-1 (continues next page)

TABLE 6-1 (continued)

(8) COLLAR ON A SMOOTH SHAFT

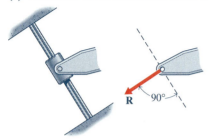

Figure 6-9

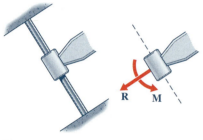

Figure 6-10

A collar on a smooth shaft (Fig. 6-9) that is pin-connected to a body can transmit only a force **R** perpendicular to the axis of the shaft. When the connection between the collar and the body is fixed (Fig. 6-10), the collar can transmit both a force **R** and a moment **M** perpendicular to the axis of the shaft. If the shaft is not smooth, a tangential frictional force $\mathbf{R}_t$ as well as a normal force $\mathbf{R}_n$ can be transmitted.

(9) FIXED SUPPORT

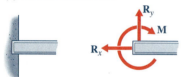

Figure 6-11

A fixed support (Fig. 6-11) can exert both a force **R** and a couple **C** on the body. The magnitude R and the direction θ of the force **R** are not known. Therefore, the force **R** is usually represented on a free-body diagram by its rectangular components $\mathbf{R}_x$ and $\mathbf{R}_y$ and the couple **C** by its moment **M**.

(10) LINEAR ELASTIC SPRING

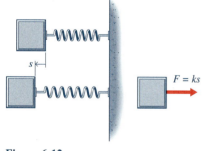

Figure 6-12

The force **R** exerted on a body by a linear elastic spring (Fig. 6-12) is proportional to the change in length of the spring. The spring will exert a tensile force if lengthened and a compressive force if shortened. The line of action of the force is along the axis of the spring.

(11) IDEAL PULLEY

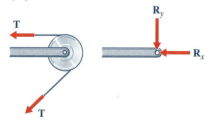

Figure 6-13

Pulleys (Fig. 6-13) are used to change the direction of a rope or cable. The pin connecting an ideal pulley to a member can exert a force **R** of unknown magnitude R and direction θ on the body. The force **R** is usually represented on a free-body diagram by its rectangular components $\mathbf{R}_x$ and $\mathbf{R}_y$. Also, since the pin is smooth (frictionless), the tension T in the cable must remain constant to satisfy moment equilibrium about the axis of the pulley.

6-2-2 Idealization of Three-Dimensional Supports and Connections

Common types of supports and connections used with rigid bodies subjected to three-dimensional force systems, together with the forces and moments used to represent the actions of these supports and connections on a free-body diagram, are listed in Table 6-2.

TABLE 6-2 Three-Dimensional Reactions at Supports and Connections

(1) BALL AND SOCKET

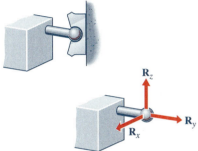

Figure 6-14

A ball and socket joint (Fig. 6-14) can transmit a force **R** but no moment. The force **R** is usually represented on a free-body diagram by its three rectangular components $\mathbf{R}_x$, $\mathbf{R}_y$, and $\mathbf{R}_z$.

(2) HINGE

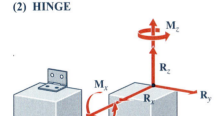

Figure 6-15

A hinge (Fig. 6-15) is normally designed to transmit a force **R** in a direction perpendicular to the axis of the hinge pin. The design may also permit a force component to be transmitted along the axis of the pin. Individual hinges also have the ability to transmit small moments about axes perpendicular to the axis of the pin. However, properly aligned pairs of hinges transmit only forces under normal conditions of use. Thus, the action of a hinge is represented on a free-body diagram by the force components $\mathbf{R}_x$, $\mathbf{R}_y$, and $\mathbf{R}_z$ and the moments $\mathbf{M}_x$, and $\mathbf{M}_z$ when the axis of the pin is in the y-direction.

(3) BALL BEARING

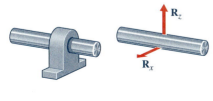

Figure 6-16

Ideal (smooth) ball bearings (Fig. 6-16) are designed to transmit a force **R** in a direction perpendicular to the axis of the bearing. The action of the bearing is represented on a free-body diagram by the force components $\mathbf{R}_x$ and $\mathbf{R}_z$ when the axis of the bearing is in the y-direction.

Table 6-2 (continues next page)

TABLE 6-2 (continued)

(4) JOURNAL BEARING

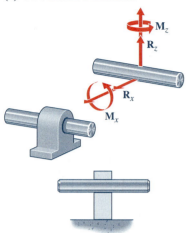

Figure 6-17

Journal bearings (Fig. 6-17) are designed to transmit a force $\mathbf{R}$ in a direction perpendicular to the axis of the bearing. Individual journal bearings also have the ability to transmit small moments about axes perpendicular to the axis of the shaft. However, properly aligned pairs of bearings transmit only forces perpendicular to the axis of the shaft under normal conditions of use. Therefore, the action of a journal bearing is represented on a free-body diagram by the force components $\mathbf{R}_x$ and $\mathbf{R}_z$ and the couple moments $\mathbf{M}_x$ and $\mathbf{M}_z$ when the axis of the bearing is in the y-direction.

(5) THRUST BEARING

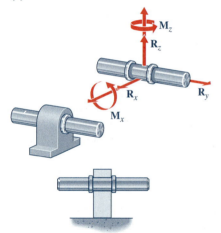

Figure 6-18

A thrust bearing (Fig. 6-18), as the name implies, is designed to transmit force components both perpendicular and parallel (thrust) to the axis of the bearing. Individual thrust bearings also have the ability to transmit small moments about axes perpendicular to the axis of the shaft. However, properly aligned pairs of bearings transmit only forces under normal conditions of use. Therefore, the action of a thrust bearing is represented on a free-body diagram by the force components $\mathbf{R}_x$, $\mathbf{R}_y$, and $\mathbf{R}_z$ and the couple moments $\mathbf{M}_x$ and $\mathbf{M}_z$ when the axis of the bearing is in the y-direction.

(6) SMOOTH PIN AND BRACKET

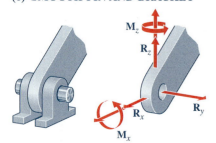

Figure 6-19

A pin and bracket (Fig. 6-19) is designed to transmit a force $\mathbf{R}$ in a direction perpendicular to the axis of the pin but may also transmit a force component along the axis of the pin. The unit also has the ability to transmit small moments about axes perpendicular to the axis of the pin. Therefore, the action of a smooth pin and bracket is represented on a free-body diagram by the force components $\mathbf{R}_x$, $\mathbf{R}_y$, and $\mathbf{R}_z$ and the couple moments $\mathbf{M}_x$ and $\mathbf{M}_z$ when the axis of the pin is in the y-direction.

TABLE 6-2 (continued)

(7) FIXED SUPPORT

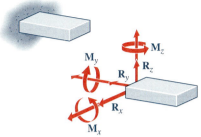

Figure 6-20

A fixed support (Fig. 6-20) can resist both a force **R** and a couple **C**. The magnitudes and directions of the force and couple are not known. Thus, the action of a fixed support is represented on a free-body diagram by the force components $\mathbf{R}_x$, $\mathbf{R}_y$, and $\mathbf{R}_z$ and the moment components $\mathbf{M}_x$, $\mathbf{M}_y$, and $\mathbf{M}_z$.

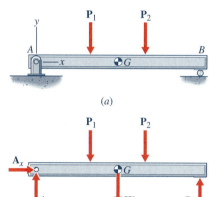

(a)

(b)

Figure 6-21

 ## Example Problem 6-1

Draw a free-body diagram for the beam shown in Fig. 6-21a.

SOLUTION

Two concentrated forces $\mathbf{P}_1$ and $\mathbf{P}_2$ are applied to the beam. The weight of the beam is represented by the force **W**, which has a line of action that passes through the center of gravity G of the beam. The beam is supported at the left end with a smooth pin and bracket and at the right end with a roller. The action of the left support is represented by the forces $\mathbf{A}_x$ and $\mathbf{A}_y$. The action of the roller is represented by the force $\mathbf{B}_y$, which acts normal to the surface of the beam. A complete free-body diagram for the beam is shown in Fig. 6-21b. ■

Example Problem 6-2

A cylinder is supported on a smooth inclined surface by a two-bar frame as shown in Fig. 6-22a. Assume that the cylinder has weight W and that the two bars have negligible weight. Draw a free-body diagram for

(a) The cylinder.   (b) The two-bar frame.   (c) The pin at C.

SOLUTION

(a) The free-body diagram for the cylinder is shown in Fig. 6-22b. The weight **W** of the cylinder acts through the center of gravity G. The forces $\mathbf{N}_1$ and $\mathbf{N}_2$ act normal to the smooth surfaces at the points of contact.

(b) The free-body diagram for the two-bar frame is shown in Fig. 6-22c. The action of the smooth pin and bracket supports at points A and C are represented by forces $\mathbf{A}_x$ and $\mathbf{A}_y$ and $\mathbf{C}_x$ and $\mathbf{C}_y$, respectively. Note that the pin forces at B are internal and do not appear on the free-body diagram shown in Fig. 6-22c.

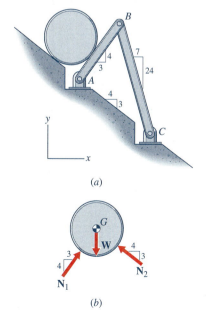

(a)

(b)

Figure 6-22

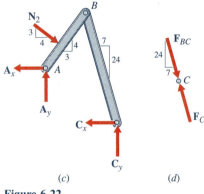

(c)   (d)

Figure 6-22

(c) Since bar BC is a link, the resultant $\mathbf{F}_C$ of forces $\mathbf{C}_x$ and $\mathbf{C}_y$ must have a line of action along the axis of the link. As a result, the free-body diagram for pin C can be drawn as shown in Fig. 6-22d. ∎

Example Problem 6-3

Draw the free-body diagram for the curved bar AC shown in Fig. 6-23a which is supported by a ball and socket joint at A, a flexible cable at B, and a pin and bracket at C. Neglect the weight of the bar.

SOLUTION

The action of the ball and socket joint at support A is represented by three rectangular force components $\mathbf{A}_x$, $\mathbf{A}_y$, and $\mathbf{A}_z$. The action of the pin and bracket at support C can be represented by force components $\mathbf{C}_x$, $\mathbf{C}_y$, and $\mathbf{C}_z$ and moment components $\mathbf{M}_x$ and $\mathbf{M}_z$. The action of the cable is represented by the cable tension $\mathbf{T}$. A complete free-body diagram for bar AC is shown in Fig. 6-23b. ∎

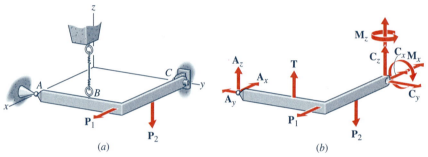

(a)   (b)

Figure 6-23

PROBLEMS

Draw complete free-body diagrams for the bodies specified in the following problems. Include the weight of the member on the diagram except where the problem statement indicates that the weight of the member is to be neglected. Assume that all surfaces are smooth unless indicated otherwise.

6-1 The cantilever beam shown in Fig. P6-1.

Figure P6-1

6-2 The bar shown in Fig. P6-2.

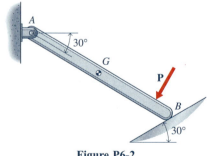

Figure P6-2

6-3 The beam shown in Fig. P6-3.

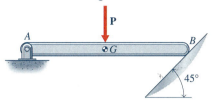

Figure P6-3

6-4 The beam shown in Fig. P6-4.

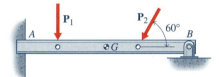

Figure P6-4

6-5 (a) The bar AB and (b) the post CD shown in Fig. P6-5.

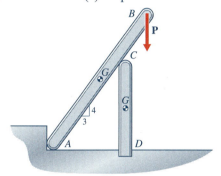

Figure P6-5

6-6 (a) The cylinder and (b) the frame shown in Fig. P6-6. Neglect the weight of the frame.

Figure P6-6

6-7 (a) The cylinder and (b) the bar shown in Fig. P6-7.

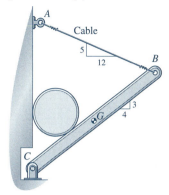

Figure P6-7

6-8 (a) The bar AB and (b) the bar CB shown in Fig. P6-8. Neglect the weights of the bars.

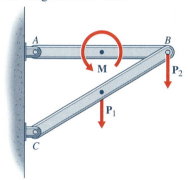

Figure P6-8

6-9 The block shown in Fig. P6-9. The support at A is a ball and socket. The support at B is a pin and bracket.

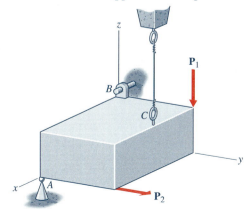

Figure P6-9

6-10 The door shown in Fig. P6-10. Assume that the hinges at supports C and D are properly aligned so that they are not required to transmit moment components.

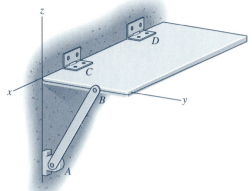

Figure P6-10

6-11 The door shown in Fig. P6-11.

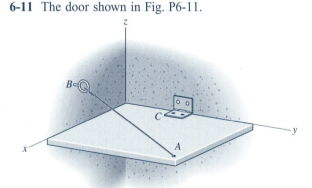

Figure P6-11

6-12 The bent bar shown in Fig. P6-12. The support at A is a journal bearing and the supports at B and C are ball bearings. Neglect the weight of the bar.

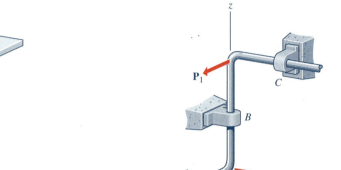

Figure P6-12

6-3 EQUILIBRIUM IN TWO DIMENSIONS

The term "two dimensional" is used to describe problems in which the forces involved are contained in a plane (say the xy-plane) and the axes of all couples are perpendicular to the plane containing the forces. For two-dimensional problems, since a force in the xy-plane has no z-component and produces no moments about the x- or y-axes, Eqs. (6-1) reduce to

$$\mathbf{R} = \Sigma F_x\,\mathbf{i} + \Sigma F_y\,\mathbf{j} = 0$$
$$\mathbf{C} = \Sigma M_z\,\mathbf{k} = 0$$

(6-3)

Thus, three of the six scalar equations of equilibrium [Eqs. (6-2)] are automatically satisfied; namely,

$$\Sigma F_z = 0 \qquad \Sigma M_x = 0 \qquad \Sigma M_y = 0$$

Therefore, there are only three independent scalar equations of equilibrium for a rigid body subjected to a two-dimensional system of forces. The three equations can be expressed as

$$\Sigma F_x = 0 \qquad \Sigma F_y = 0 \qquad \Sigma M_A = 0 \qquad (6\text{-}4)$$

The third equation represents the sum of the moments of all forces about a z-axis through any point A on or off the body. Equations (6-4) are both the necessary and sufficient conditions for equilibrium of a rigid body subjected to a two-dimensional system of forces.

There are two additional ways in which the equations of equilibrium can be expressed for a body subjected to a two-dimensional system of forces. The resultant force $\mathbf{R}$ and the resultant couple $\mathbf{C}$ at point A of a rigid body subjected to a general two-dimensional force system are shown in Fig. 6-24a. The resultant can be represented in terms of its scalar components as shown in Fig. 6-24b. If the condition $\Sigma M_A = 0$ is satisfied, $\mathbf{C} = \mathbf{0}$. If, in addition, the condition $\Sigma F_x = 0$ is satisfied, $\mathbf{R} = \Sigma F_y \, \mathbf{j}$. For any point B on or off the body that does not lie on the y-axis, the equation $\Sigma M_B = 0$ can be satisfied only if $\Sigma F_y = 0$. Thus, an alternative set of scalar equilibrium equations for two-dimensional problems is

$$\Sigma F_x = 0 \qquad \Sigma M_A = 0 \qquad \Sigma M_B = 0 \qquad (6\text{-}5)$$

where points A and B must have different x-coordinates.

The conditions of equilibrium for a two-dimensional force system can also be expressed by using three moment equations. Again, if the condition $\Sigma M_A = 0$ is satisfied, $\mathbf{C} = \mathbf{0}$. In addition, for a point B (see Fig. 6-24c) on the x-axis on or off the body (except at point A), the equation $\Sigma M_B = 0$ can be satisfied, once $\Sigma M_A = 0$, only if $\Sigma F_y = 0$. Thus, the resultant can only have an x-component, $\mathbf{R} = \Sigma F_x \, \mathbf{i}$. Finally, for any point C (see Fig. 6-24 on or off the body that does not lie on the x-axis, the equation $\Sigma M_C = 0$ can be satisfied only if $\Sigma F_x = 0$. Thus, there is a second set of alternative scalar equilibrium equations for two-dimensional problems:

$$\Sigma M_A = 0 \qquad \Sigma M_B = 0 \qquad \Sigma M_C = 0 \qquad (6\text{-}6)$$

where A, B, and C are any three points not on the same straight line.

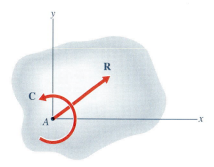

(a)

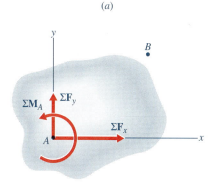

(b)

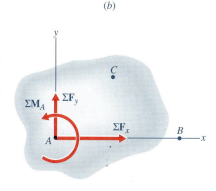

(c)

Figure 6-24

6-3-1 Two-Force Members

Equilibrium of a body under the action of two forces occurs with sufficient frequency to warrant special attention. For example, consider a link AB with negligible weight, as shown in Fig. 6-25a. Any forces exerted on the link by the frictionless pins at A and B can be resolved into components along and perpendicular to the axis of the link, as shown in Fig. 6-25b. From the equilibrium equations

$$+\nearrow \Sigma F_x = A_x - B_x = 0 \qquad A_x = B_x$$
$$+\nwarrow \Sigma F_y = A_y - B_y = 0 \qquad A_y = B_y$$

Forces A_y and B_y, however, form a couple that must be zero when the link is in equilibrium; therefore, $A_y = B_y = 0$. Thus, for two-force members, equilibrium

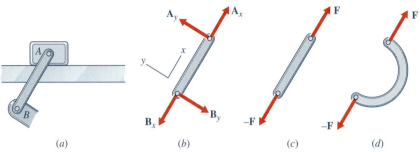

(a)   (b)   (c)   (d)

Figure 6-25

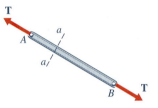

Figure 6-26

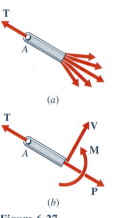

(a)

(b)

Figure 6-27

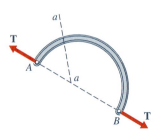

Figure 6-28

requires that the forces be equal, opposite, and collinear, as shown in Fig. 6-25c. The shape of the member, as shown in Fig. 6-25d, has no effect on this simple requirement; however, the weights of the members must be negligible.

When forces are directed away from the ends of a straight two-force member as shown in Fig. 6-26, the member is said to be in tension and the force **T** is referred to as the tension force in the member and not just the tension force on the end of the member. That this description of the force is correct is easily shown by considering equilibrium of a portion of the member to the left of section *a-a*. In general, there will be some complex force distribution acting on the cut surface as shown on the free-body diagram of Fig. 6-27a. However, this system of forces can be replaced with an equivalent force-couple as shown on the free-body diagram of Fig. 6-27b. Then equilibrium of forces in the direction perpendicular to the axis of the member requires that **V**, the shear component of the equivalent force-couple, must be zero. Similarly, equilibrium of forces in the direction along the axis of the member requires that **P**, the axial component of the equivalent force-couple, must be equal in magnitude and opposite in direction to **T**, the force applied to the end of the member. Finally, moment equilibrium requires that **M**, the couple component of the equivalent force-couple, must be zero. That is, if the forces at the ends of a straight two-force member are pulling on the member, then the resultant of the forces on any cut section of the member must also be an axial force pulling on the cut section of the member, regardless of where the member is cut. Therefore, it is proper to talk about the force **T** as the tension in the member. If the forces are directed toward the ends of a straight two-force member, the member is said to be in compression.

Axially loaded structural members such as the straight two-force member were previously discussed in Chapter 4. Once **T** (Fig. 6-26) has been found using equilibrium principles, the internal normal force P on a transverse cross section is known since **P** = **T**. The internal resistance and deformation of the member are then found using the equations of Chapter 4—that is, $\sigma = P/A$ (Eq. 4-2) and $\delta = PL/EA$ (Eq. 4-20b)—where σ is the normal stress on a transverse cross section and δ is the elongation (or contraction) of the member having cross-sectional area A, length L, and modulus of elasticity E. Thus, for straight two-force members the principles developed in Chapter 4 are sufficient to determine stress and deformation.

When a two-force member is curved, however, the forces at the ends of the member do not act along the axis of the member. Instead, the forces act along the line joining the points where the forces are applied, as shown in Fig. 6-28. If the member is cut normal to its axis at section *a-a* there will be a complex force distribution acting on the cut surface, as shown in Fig. 6-29a. This system of

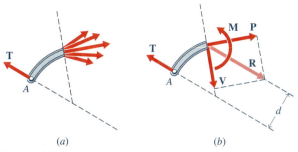

(a)   (b)

Figure 6-29

forces can be replaced with an equivalent force-couple, as shown on the free-body diagram of Fig. 6-29b. However, force equilibrium now requires that the resultant **R** of the axial and shear components **P** and **V** of the equivalent force-couple be equal in magnitude and opposite in direction to the force **T** at the end of the member, as shown in Fig. 6-29b. Since the forces **R** and **T** are not collinear, moment equilibrium now requires that $M = Td \neq 0$.

As a result of the previous discussion it is obvious that the design of straight two-force members need only consider axial forces, but curved two-force members must be designed to withstand shearing forces V and bending moments M as well as axial forces P. To further complicate the problem, the magnitudes of the shear forces, bending moments, and axial forces in curved members depend on where the member is cut. Analyzing the internal resistance of a curved two-force member requires concepts more advanced than those discussed in Chapter 4. The stress and deformation characteristics of these members will be developed throughout the remainder of this book.

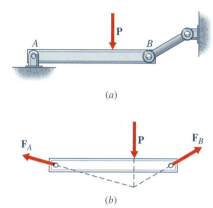

(a)

(b)

Figure 6-30

6-3-2 Three-Force Members

Equilibrium of a body under the action of three forces (see Fig. 6-30) also represents a special situation. If a body is in equilibrium under the action of three forces, then the lines of action of the three forces must be concurrent (i.e., meet at a common point); otherwise, the nonconcurrent force would exert a moment about the point of concurrency of the other two forces. A body subjected to three parallel forces represents a special case of the three-force body. For this case, the point of concurrency is assumed to be at infinity.

6-3-3 Statically Indeterminate Reactions and Partial Constraints

Consider a body subjected to a system of coplanar forces $\mathbf{F}_1$, $\mathbf{F}_2$, $\mathbf{F}_3$, $\mathbf{F}_4$, . . . , $\mathbf{F}_n$, as shown in Fig. 6-31a. This system of forces can be replaced by an equivalent force-couple system at an arbitrary point A as shown in Fig. 6-31b. The resultant force has been represented by its rectangular components $\mathbf{R}_x$ and $\mathbf{R}_y$ and the couple by $\mathbf{M}_A$. For the body to be in equilibrium, the supports must be capable of exerting an equal and opposite force-couple system on the body. As an example, consider the supports shown in Fig. 6-32a. The pin support at A, as shown in Fig. 6-32b, can exert forces in the x- and y-directions to prevent a translation of the body but no moment to prevent a rotation about an axis through A. The link at B, which exerts a force in the y-direction, produces the moment about

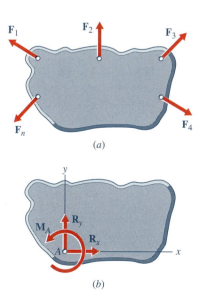

(a)

(b)

Figure 6-31

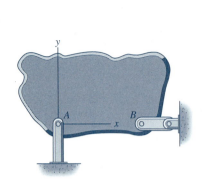

(a)

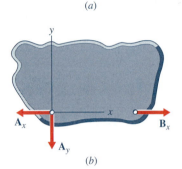

(b)

Figure 6-33

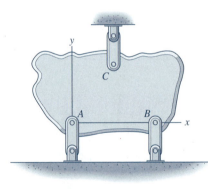

(a)

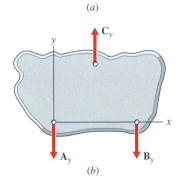

(b)

Figure 6-34

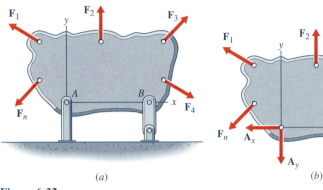

(a)

(b)

Figure 6-32

point A required to prevent rotation. Thus, all motion of the body is prevented, and the free-body diagram shown in Fig. 6-32b is in equilibrium under the action of the forces shown. The reaction forces exerted on the body by the supports are called constraint forces. When the equations of equilibrium are sufficient to determine the unknown forces at the supports, as in Fig. 6-32, the body is said to be statically determinate with adequate (proper) constraints.

Three support reactions for a body subjected to a coplanar system of forces does not always guarantee that the body is statically determinate with adequate constraints. For example, consider the body and supports shown in Fig. 6-33a. The pin support at A (see Fig. 6-33b) can exert forces in the x- and y-directions to prevent a translation of the body, but since the line of action of force $\mathbf{B}_x$ passes through point A it does not exert the moment required to prevent rotation of the body about point A. Similarly, the three links shown in Fig. 6-34a can prevent rotation of the body about any point (see Fig. 6-34b) and translation of the body in the y-direction but not translation of the body in the x-direction. The bodies in Figs. 6-33 and 6-34 are partially (improperly) constrained and the equations of equilibrium are not sufficient to determine all of the unknown reactions. A body with an adequate number of reactions is improperly constrained when the constraints are arranged in such a way that the support forces are either concurrent or parallel. Partially constrained bodies can be in equilibrium for specific systems of forces. For example, the reactions R_A and R_B for the beam shown in Fig. 6-35a can be determined by using the equilibrium equations $\Sigma F_y = 0$ and $\Sigma M_A = 0$. The beam is improperly constrained, however, since motion in the x-direction would occur if any of the applied loads had a small x-component.

Finally, if the link at B in Fig. 6-32a is replaced with a pin support, as shown in Fig. 6-36a, an additional reaction $\mathbf{B}_x$ (see Fig. 6-36b) is obtained that is not required to prevent movement of the body. Obviously, the three independent equations of equilibrium will not provide sufficient information to determine the four unknowns. Constrained bodies with extra supports are statically indeterminate since the equations of equilibrium are not sufficient to solve for all of the support reactions. Relations involving physical properties of the body, in addition to the equations of equilibrium, are required to determine some of the unknown reactions. This was done in Chapter 4 for axially loaded bodies. The supports not required to maintain equilibrium of the body are called *redundant supports*. Typical examples of redundant supports for beams are shown in Fig. 6-37. The roller at support B of the cantilever beam shown in Fig. 6-37a can

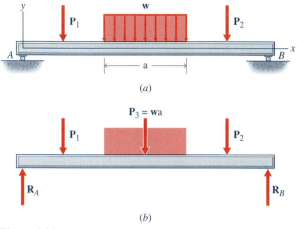

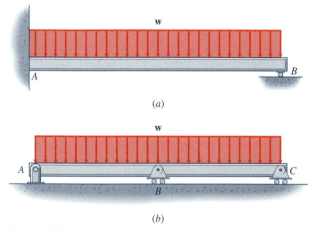

Figure 6-35

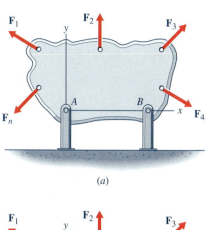

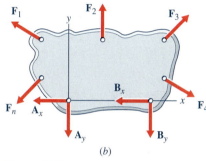

Figure 6-36

Figure 6-37

be removed and the beam will remain in equilibrium. Similarly, the roller support at either B or C (but not both) could be removed from the beam of Fig. 6-37b and the beam would remain in equilibrium.

6-3-4 Problem Solving

In Section 1-5, a procedure was outlined for solving engineering problems. The procedure consisted of three phases:

1. Problem definition and identification.
2. Model development and simplification.
3. Mathematical solution and interpretation of results.

Application of this procedure to equilibrium problems yields the following steps:

1. Read the problem carefully. Many student difficulties arise from failure to observe this preliminary step.

2. Identify the result requested.

3. Prepare a scaled sketch and tabulate the information provided.

4. Identify the equilibrium equations to be used to obtain the result.

5. Draw the appropriate free-body diagram. Carefully label all applied forces and support reactions. Establish a convenient set of coordinate axes. Use a right-handed system in case vector cross products must be employed. Compare the number of unknowns on the free-body diagram with the number of independent equations of equilibrium. Draw additional diagram(s) if needed.

6. Apply the appropriate force and moment equations of equilibrium.

7. Report the answer with the appropriate number of significant figures and the appropriate units.

8. Study the answer and determine whether it is reasonable. As a check, write some other equilibrium equations and see if they are satisfied by the solution.

Bodies subjected to coplanar force systems are not very complex, so a scalar solution is usually suitable for analysis. This results from the fact that moments can be expressed as scalars instead of vectors. For the more general case of rigid bodies subjected to three-dimensional force systems (discussed in a later section of this chapter), vector analysis methods are usually more appropriate.

In certain situations, normal stress, bearing stress, shearing stress, and deformation can be determined. For example, if a member is axially loaded (a straight two-force member), the methods developed in Sections 4-2 and 4-3 can be used to compute normal stress and shearing stress on a section of the member. If a two-force member is curved, the methods developed in Sections 4-2 and 4-3 are not sufficient to calculate the stresses; the same is true for a three-force member (members with a combination of axial forces, shear forces, and bending moments will be considered in later chapters).

In addition, the deformation of a straight two-force member can be found using the methods developed in Section 4-6. The deformation of members with a combination of axial forces, shear forces, and bending moments cannot be found using the results of Section 4-6; methods developed throughout the remainder of this book are needed to solve this class of problems.

Regardless of how a body is loaded, the methods of Sections 4-2-2 and 4-2-3 can be used to determine shearing stresses in pins and bearing stresses at pinned connections. For example, the reaction at A of Fig. 6-38 can be found using the principles of rigid-body equilibrium, and then the shearing stress in the pin at A can be found using the results of Section 4-2-2. The bearing stress between the pin at A and beam AB, or between the pin at A and the support bracket, can be found using the results of Section 4-2-3.

6-3-5 Equations of Equilibrium Applied to a Rigid Body/Deformable Body

In most engineering applications, a body is assumed to be rigid when the equations of equilibrium are used to determine support reactions even though it is a fact that the body deforms when the loads are applied. For example, consider the lever and cable system shown in Fig. 6-39a and assume that lever ABC is rigid

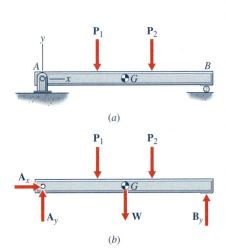

(a)

(b)

Figure 6-38

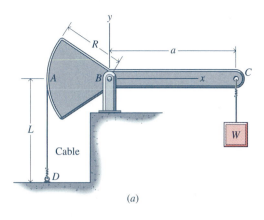

(a)

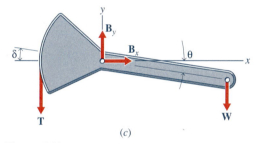

(b)

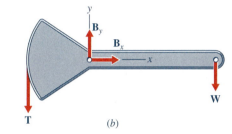

(c)

Figure 6-39

and that the weights of the lever and cable are negligible compared to the applied load. The cable is fastened to and wraps around the circular sector at the left end of the lever. The no-load ($\mathbf{W} = \mathbf{0}$) position of the system is shown in Fig. 6-39a. Equilibrium of the system under two conditions will be investigated: (1) the cable is rigid and (2) the cable deforms.

CONDITION (1): When both the cable and the lever are assumed to be rigid, the free-body diagram for the lever is as shown in Fig. 6-39b. The moment equilibrium equation for lever ABC is

$$+\!\!\int\!\Sigma M_B = TR - Wa = 0$$

Thus, the tension T in the rigid cable is

$$T = \frac{Wa}{R} \qquad\qquad\qquad\text{(a)}$$

CONDITION (2): When the cable is assumed to be deformable, the free-body diagram for the lever after the cable deforms (deformed state of equilibrium) is as shown in Fig. 6-39c. The moment equilibrium for lever ABC is then

$$+\downarrow\Sigma M_B = TR - Wa \cos \theta = 0$$

and the tension in the deformed cable is

$$T = \frac{Wa}{R} \cos \theta \qquad (b)$$

Equation (b) cannot be solved for T, since θ is unknown. Since the remaining equations of equilibrium do not provide the additional information needed to solve for T, the problem is statically indeterminate. A review of Section 4-6-1 reveals that statically indeterminate problems are solved by using the equilibrium equations [in this case Eq. (b) above] together with equations obtained from the deformation of the member. The deformation of the cable is given by Eq. (4-20b) as

$$\delta = \frac{TL}{EA} \qquad (c)$$

where A is the cross-sectional area and E is the modulus of elasticity of the cable. Combining Eqs. (b) and (c) gives

$$\frac{\delta AE}{L} = \frac{Wa}{R} \cos \theta \qquad (d)$$

which has two unknowns, δ and θ. However, the cable wraps around a circular sector on the lever, therefore

$$\delta = R\theta \qquad (e)$$

Substituting Eq. (e) into Eq. (d), and rearranging yields

$$R^2AE\theta = WaL \cos \theta \qquad (f)$$

Equation (f) can be solved for θ (if the geometric parameters R, A, a and L, the material property E, and the weight W are known) by using trial and error, by using numerical methods, or by plotting both sides of Eq. (f) versus θ and locating the intersection of the two curves. Once θ is known, Eq. (b) is used to find the tension in the cable.

Even though Eq. (a) for the rigid cable and Eq. (b) for the deformable cable are similar, the computations required to find tension T in Eq. (b) are somewhat lengthy. Computational difficulties aside, are the results obtained by using Eq. (a) significantly different from those obtained by using Eq. (b) in typical engineering situations? To help answer this question, several examples will be considered.

EXAMPLE 1: The cable is rigid.

If $W = 100$ lb, $a = 30$ in., and $R = 15$ in.,
Eq. (a) yields: $T = 200$ lb

EXAMPLE 2: The cable is a 3/32 in.–diameter steel (E = 29,000 ksi)
wire.

If $W = 100$ lb, $a = 30$ in., $R = 15$ in., and $L = 45$ in.
Eq. (f) yields: $\theta = 0.002997$ rad $= 0.1717°$
Eq. (b) yields: $T = 199.999$ lb
The percent difference in T in the two Examples is

$$\%D = \frac{200 - 199.999}{199.999}(100) = 0.0005\%$$

This error is acceptable for practical engineering problems. A check on the normal stress in the wire ($\sigma = P/A$) yields $\sigma = 29.0$ ksi. This stress level is within the linear range of the stress–strain behavior of the steels used to produce wire products. The normal stress must be in the linear range of the stress–strain diagram for Eq. (c) to be valid.

EXAMPLE 3: The cable is a 3/32 in.–diameter aluminum
(E = 10,600 ksi) wire.

If $W = 100$ lb, $a = 30$ in., $R = 15$ in., and $L = 45$ in.
Eq. (f) yields: $\theta = 0.00820$ rad $= 0.4698°$
Eq. (b) yields: $T = 199.993$ lb.
The percent difference in T for Examples 1 and 3 is
$\%D = 0.0035\%$.

A check on the normal stress in the aluminum wire yields $\sigma = 29.0$ ksi, which is again within the linear range of the stress–strain behavior of aluminum so that Eq. (c) is valid. Examples 2 and 3 indicate that the tension in the wire changes very little when the stiffness of the wire is changed by a factor of almost three and that both values are essentially the same as that obtained using the rigid wire assumption (Example 1).

In Chapter 4, a rigid body was used when calculating support reactions and internal forces; these forces were then used to determine stresses and deformations. Equilibrium requirements should be satisfied when a body is in the deformed configuration. However, the previous example illustrates that forces may be determined, within engineering accuracy, using the equilibrium equations and a free-body diagram in the undeformed configuration. These forces may then be used to determine stresses and deformations with sufficient accuracy for most engineering applications.

Example Problem 6-4

A pin-connected truss is loaded and supported as shown in Fig. 6-40a. The body W has a mass of 100 kg. Determine the components of the reactions at supports A and B. Neglect the masses of the members of the truss.

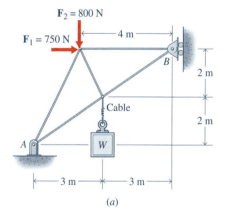

(a)

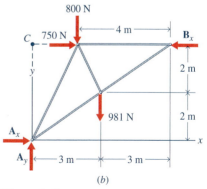

(b)

Figure 6-40

SOLUTION

The tension T in the cable between the body W and the truss is

$$T = mg = 100(9.81) = 981 \text{ N}$$

A free-body diagram of the truss is shown in Fig. 6-40b. The action of the pin support at A is represented by force components $\mathbf{A}_x$ and $\mathbf{A}_y$. The action of the roller support at B is represented by the force $\mathbf{B}_x$, which acts perpendicular to the vertical surface at B. Since the truss is subjected to a general coplanar force system, three equilibrium equations are available to solve for the unknown magnitudes of the three forces $\mathbf{A}_x$, $\mathbf{A}_y$, and $\mathbf{B}_x$.

Determination of $\mathbf{B}_x$:

$$+\!\!\curvearrowleft\Sigma M_A = B_x(4) - 800(2) - 750(4) - 981(3) = 0$$

$$B_x = +1885.8 \text{ N} \qquad \mathbf{B}_x = 1886 \text{ N} \leftarrow \qquad \textbf{Ans.}$$

Determination of $\mathbf{A}_x$:

$$+\!\!\rightarrow\Sigma F_x = A_x + 750 - B_x = A_x + 750 - 1885.8 = 0$$

$$A_x = +1135.8 \text{ N} \qquad \mathbf{A}_x = 1136 \text{ N} \rightarrow \qquad \textbf{Ans.}$$

Determination of $\mathbf{A}_y$:

$$+\!\!\uparrow\Sigma F_y = A_y - 800 - 981 = 0$$

$$A_y = +1781 \text{ N} \qquad \mathbf{A}_y = 1781 \text{ N} \uparrow \qquad \textbf{Ans.}$$

The component $\mathbf{A}_x$ can also be determined directly by summing moments about point C.

$$+\!\!\curvearrowleft\Sigma M_C = A_x(4) - 800(2) - 981(3) = 0$$

$$A_x = +1135.8 \text{ N} \qquad \mathbf{A}_x = 1136 \text{ N} \rightarrow \qquad \textbf{Ans.}$$

Solution by Vector Analysis

From Eqs. (6-3):

$$\mathbf{R} = \Sigma F_x \mathbf{i} + \Sigma F_y \mathbf{j} = \mathbf{0} \qquad \mathbf{C} = \Sigma M_z \mathbf{k} = \mathbf{0}$$

The applied loads can be written in Cartesian vector form as

$$\mathbf{F}_1 = 750\mathbf{i} \text{ N} \qquad \mathbf{F}_2 = -800\mathbf{j} \text{ N} \qquad \mathbf{F}_3 = -981\mathbf{j} \text{ N}$$

The reactions can be written as

$$\mathbf{A} = A_x\mathbf{i} + A_y\mathbf{j} \qquad \mathbf{B} = -B_x\mathbf{i}$$

Therefore, from Eqs. (6-3),

$$\mathbf{R} = \Sigma F_x \mathbf{i} + \Sigma F_y \mathbf{j}$$
$$= (A_x - B_x + 750)\mathbf{i} + (A_y - 800 - 981)\mathbf{j} = \mathbf{0}$$

Summing moments about point A to minimize the number of unknowns:

$$\mathbf{C} = \Sigma M_A \mathbf{k}$$
$$= \mathbf{r}_1 \times \mathbf{F}_1 + \mathbf{r}_2 \times \mathbf{F}_2 + \mathbf{r}_3 \times \mathbf{F}_3 + \mathbf{r}_B \times \mathbf{B}$$
$$= [(2\mathbf{i} + 4\mathbf{j}) \times (750\mathbf{i})] + [(2\mathbf{i} + 4\mathbf{j}) \times (-800\mathbf{j})]$$
$$+ [(3\mathbf{i} + 2\mathbf{j}) \times (-981\mathbf{j})] + [(6\mathbf{i} + 4\mathbf{j}) \times (-B_x \mathbf{i})]$$
$$= -3000\mathbf{k} - 1600\mathbf{k} - 2943\mathbf{k} + 4B_x\mathbf{k} = (-7543 + 4B_x)\mathbf{k} = \mathbf{0}$$

Equating the coefficients of $\mathbf{i}$, $\mathbf{j}$, and $\mathbf{k}$ to zero gives

$$A_x - B_x + 750 = 0$$
$$A_y - 800 - 981 = 0$$
$$-7543 + 4B_x = 0$$

Solving simultaneously yields:

$$A_x = 1136 \text{ N} \qquad A_y = 1781 \text{ N} \qquad B_x = 1886 \text{ N}$$

Therefore:

$$\mathbf{A} = 1136\mathbf{i} + 1781\mathbf{j} \text{ N} \qquad \mathbf{B} = -1886\mathbf{i} \text{ N} \qquad \textbf{Ans.}$$

The results are obviously identical to those obtained using a scalar analysis. For two-dimensional problems, scalar methods are usually preferred. ∎

Example Problem 6-5

A bar weighing 250 lb is supported by a post and cable as shown in Fig. 6-41a. Assume that all surfaces are smooth. Determine the tension in the cable and the forces on the bar at the contacting surfaces.

SOLUTION

A free-body diagram of the bar is shown in Fig. 6-41b. All surfaces are smooth; therefore, the reaction at A is a vertical force $\mathbf{A}$ and the reaction at C is a force $\mathbf{C}$ perpendicular to the bar. The cable exerts a tension $\mathbf{T}$ on the bar in the direction of the cable. Since the bar is subjected to a general coplanar force system, three equilibrium equations are available to solve for the unknown magnitudes of forces $\mathbf{A}$, $\mathbf{C}$, and $\mathbf{T}$.

Solution Using Eqs. (6-4):

$$+\rightarrow \Sigma F_x = T - C \sin 50° = 0$$
$$T - 0.7660C = 0 \qquad \text{(a)}$$

$$+\uparrow \Sigma F_y = A_y + C \cos 50° - 250 = 0$$
$$A_y + 0.6428C = 250 \qquad \text{(b)}$$

$$+\circlearrowleft \Sigma M_A = C(8) - T(2 \sin 50°) - 250(6 \cos 50°) = 0$$
$$8C - 1.5321T = 964.2 \qquad \text{(c)}$$

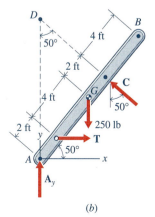

(a)

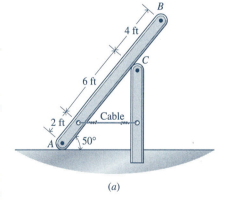

(b)

Figure 6-41

Solving Eqs. (a) and (c) simultaneously yields

$$T = +108.2 \text{ lb} \qquad \mathbf{T} = 108.2 \text{ lb} \rightarrow \qquad \textbf{Ans.}$$

$$C = +141.2 \text{ lb} \qquad \mathbf{C} = 141.2 \text{ lb} \searrow 40° \qquad \textbf{Ans.}$$

Once C is known, A_y is determined from Eq. (b) as

$$A_y = 250 - 0.6428C = 250 - 0.6428(141.2)$$

$$A_y = +159.2 \text{ lb} \qquad \mathbf{A}_y = 159.2 \text{ lb} \uparrow \qquad \textbf{Ans.}$$

Solution Using Eqs. (6-6):

Determination of **T**:

The determination of the cable tension **T** can be simplified by taking moments about the point of concurrence (off the bar) of forces **A** and **C**. Thus,

$$+\!\!\downarrow\!\Sigma M_D = T[(8/\sin 50°) - 2 \sin 50°] - 250(6 \cos 50°) = 0$$

$$T = +108.2 \text{ lb} \qquad \mathbf{T} = 108.2 \text{ lb} \rightarrow \qquad \textbf{Ans.}$$

Determination of $\mathbf{A}_y$:

$$+\!\!\downarrow\!\Sigma M_C = -A_y(8 \cos 50°) + T(6 \sin 50°) + 250(2 \cos 50°)$$

$$= -A_y(8 \cos 50°) + 108.2(6 \sin 50°) + 250(2 \cos 50°) = 0$$

$$A_y = +159.2 \text{ lb} \qquad \mathbf{A}_y = 159.2 \text{ lb} \uparrow \qquad \textbf{Ans.}$$

Determination of **C**:

$$+\!\!\downarrow\!\Sigma M_A = -T(2 \sin 50°) - 250(6 \cos 50°) + C(8)$$

$$= -108.2(2 \sin 50°) - 250(6 \cos 50°) + C(8) = 0$$

$$C = +141.2 \text{ lb} \qquad \mathbf{C} = 141.2 \text{ lb} \searrow 40° \qquad \textbf{Ans.}$$

In many problems, use of Eqs. (6-6) eliminates the need to solve simultaneous equations, as illustrated in this example problem. Once T is known, the equilibrium equations $\Sigma F_x = 0$ and $\Sigma F_y = 0$ could be used to determine C and A instead of using the equations $\Sigma M_A = 0$ and $\Sigma M_C = 0$. Thus,

$$+\!\!\rightarrow\!\Sigma F_x = T - C \sin 50° = 0$$

$$C = \frac{T}{\sin 50°} = \frac{108.2}{\sin 50°}$$

$$C = +141.2 \text{ lb} \qquad \mathbf{C} = 141.2 \text{ lb} \searrow 40° \qquad \textbf{Ans.}$$

$$+\!\!\uparrow\!\Sigma F_y = A_y + C \cos 50° - 250 = 0$$

$$A_y = 250 - C \cos 50° = 250 - 141.2 \cos 50°$$

$$A_y = +159.2 \text{ lb} \qquad \mathbf{A}_y = 159.2 \text{ lb} \uparrow \blacksquare \qquad \textbf{Ans.}$$

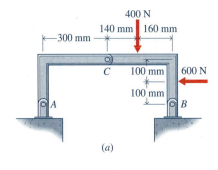

(a)

Example Problem 6-6

A pin-connected two-bar frame is loaded and supported as shown in Fig. 6-42a. Determine the reactions at supports A and B. The masses of the two bars are negligible.

SOLUTION

At first glance it may appear that the two-bar frame shown in Fig. 6-42a is statically indeterminate with support reactions A_x, A_y, B_x, and B_y. Once it is observed, however, that loads are applied to member AC only at pins A and C, member AC is identified as a two-force member with the support force at A having a line of action along the line joining the two pins.

A free-body diagram of the frame is shown in Fig. 6-42b. The action of the pin support at A is represented by the single force **A** with a slope

$$\theta_A = \tan^{-1} \frac{200}{300} = 33.69°$$

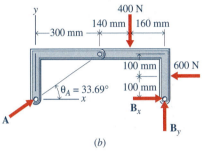

(b)

The action of the pin support at B is represented by force components B_x and B_y. Since the frame is subjected to a general coplanar force system, three equilibrium equations are available to solve for the unknown magnitudes of forces **A**, B_x, and B_y.

Determination of **A**:

$$+\circlearrowleft \Sigma M_B = -A \sin 33.69°(0.6) + 400(0.16) + 600(0.1) = 0$$

$$A = +372.6 \text{ N} \qquad \mathbf{A} = 373 \text{ N} \measuredangle 33.7° \qquad \textbf{Ans.}$$

Determination of B_x:

$$+\rightarrow \Sigma F_x = A \cos 33.69° + B_x - 600$$
$$= 372.6 \cos 33.69° + B_x - 600 = 0$$

$$B_x = +290 \text{ N} \qquad \mathbf{B}_x = 290 \text{ N} \rightarrow$$

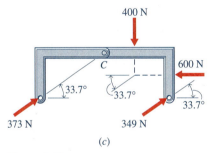

(c)

Figure 6-42

Determination of B_y:

$$+\circlearrowleft \Sigma M_A = B_y(0.6) - 400(0.44) + 600(0.1) = 0$$

$$B_y = +193.3 \text{ N} \qquad \mathbf{B}_y = 193.3 \text{ N} \uparrow$$

The reaction at support B is

$$B = \sqrt{(B_x)^2 + (B_y)^2} = \sqrt{(290)^2 + (193.3)^2} = 348.5 \text{ N} \cong 349 \text{ N}$$

$$\theta_B = \tan^{-1} \frac{B_y}{B_x} = \tan^{-1} \frac{193.3}{290} = 33.69°$$

$$\mathbf{B} = 349 \text{ N} \measuredangle 33.7° \qquad \textbf{Ans.}$$

The results are shown in Fig. 6-42c. ∎

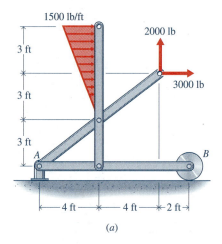

(a)

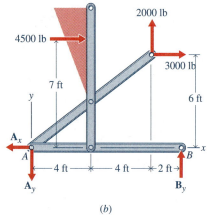

(b)

Figure 6-43

Example Problem 6-7

A pin-connected three-bar frame is loaded and supported as shown in Fig. 6-43a. Determine

(a) The reactions at supports A and B.

(b) The diameter required for the pin at B, which is in double shear, if the wheel is $\frac{3}{4}$-in. thick and the shearing stress on a cross section of the pin and the bearing stress between the pin and the wheel must be limited to 7500 psi and 8000 psi, respectively.

SOLUTION

(a) A free-body diagram of the frame is shown in Fig. 6-43b. The action of the pin support at A is represented by force components $\mathbf{A}_x$ and $\mathbf{A}_y$. The action of the roller support at B is represented by the force $\mathbf{B}_y$, which acts perpendicular to the horizontal surface at B. The distributed load can be represented temporarily on the free-body diagram by a resultant $\mathbf{R}$ (area under load diagram), with a line of action at a distance y_C (centroid of area under load diagram) above support A. Thus,

$$R = \text{Area} = \frac{1}{2}(1500)(6) = 4500 \text{ lb} \qquad y_C = 3 + \frac{2}{3}(6) = 7.00 \text{ ft}$$

Since the frame is subjected to a general coplanar force system, three equilibrium equations are available to solve for A_x, A_y, and B_y.
Determination of $\mathbf{B}_y$:

$$+\curvearrowleft \Sigma M_A = B_y(10) - 4500(7) + 2000(8) - 3000(6) = 0$$

$$B_y = +3350 \text{ lb} \qquad \mathbf{B} = \mathbf{B}_y = 3350 \text{ lb} \uparrow \qquad \textbf{Ans.}$$

Determination of $\mathbf{A}_x$:

$$+\rightarrow \Sigma F_x = -A_x + 4500 + 3000 = 0$$

$$A_x = +7500 \text{ lb} \qquad \mathbf{A}_x = 7500 \text{ lb} \leftarrow$$

Determination of $\mathbf{A}_y$:

$$+\curvearrowleft \Sigma M_B = A_y(10) - 4500(7) - 2000(2) - 3000(6) = 0$$

$$A_y = +5350 \text{ lb} \qquad \mathbf{A}_y = 5350 \text{ lb} \downarrow$$

The reaction at support A is

$$A = \sqrt{(A_x)^2 + (A_y)^2} = \sqrt{(7500)^2 + (5350)^2} = 9213 \text{ lb} \cong 9210 \text{ lb}$$

$$\theta = \tan^{-1}\frac{A_y}{A_x} = \tan^{-1}\frac{5350}{7500} = 35.5°$$

$$\mathbf{A} = 9210 \text{ lb} \measuredangle 35.5° \qquad \textbf{Ans.}$$

(b) The shearing stress on the pin at B (double shear) is given by Eq. (4-4) as

$$\tau = \frac{V}{A_s} = \frac{B}{2(\pi/4)d^2} = \frac{3350}{2(\pi/4)d^2} = 7500$$

$$d = 0.533 \text{ in.}$$

The bearing stress on the wheel at B is given by Eq. (4-6) as

$$\sigma_b = \frac{F}{A_b} = \frac{B}{dt} = \frac{3350}{d(0.750)} = 8000$$

$$d = 0.558 \text{ in.}$$

Therefore,

$$d_{min} = 0.558 \text{ in.} \ \blacksquare \qquad\qquad \textbf{Ans.}$$

Example Problem 6-8

A beam is loaded and supported as shown in Fig. 6-44a. Determine

(a) The components of the reactions at supports A and B.

(b) The shearing stress on a cross section of the 40-mm diameter pin at support A if the pin is in double shear.

(c) The bearing stress between the beam and the 40×200-mm bearing plate at support B.

SOLUTION

(a) A free-body diagram of the beam is shown in Fig. 6-44b. The action of the pin support at A is represented by force components $\mathbf{A}_x$ and $\mathbf{A}_y$. The action of the roller support at B is represented by the force $\mathbf{B}_y$, which acts per-

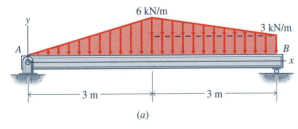

(a)

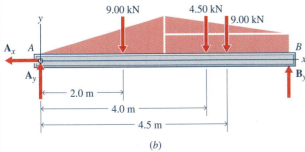

(b)

Figure 6-44

pendicular to the horizontal surface at B. The distributed loads can be represented temporarily on the free-body diagram by resultants $\mathbf{R}_1$, $\mathbf{R}_2$, and $\mathbf{R}_3$ with lines of action at distances x_{C1}, x_{C2}, and x_{C3}, respectively, from the left support. Thus,

$$R_1 = A_1 = \frac{1}{2}(6)(3) = 9.00 \text{ kN} \qquad x_{C1} = \frac{2}{3}(3) = 2.0 \text{ m}$$

$$R_2 = A_2 = \frac{1}{2}(3)(3) = 4.50 \text{ kN} \qquad x_{C2} = 3 + \frac{1}{3}(3) = 4.0 \text{ m}$$

$$R_3 = A_3 = 3(3) = 9.00 \text{ kN} \qquad x_{C3} = 3 + \frac{1}{2}(3) = 4.5 \text{ m}$$

The beam is subjected to a coplanar system of parallel applied forces in the y-direction; therefore, $\mathbf{A}_x = \mathbf{0}$. The two remaining equilibrium equations are available to solve for $\mathbf{A}_y$ and $\mathbf{B}_y$.
Determination of $\mathbf{B}_y$:

$$+\downarrow\Sigma M_A = B_y(6) - 9.00(2.0) + 4.50(4.0) - 9.00(4.5) = 0$$

$$B_y = +12.75 \text{ kN} \qquad \mathbf{B}_y = 12.75 \text{ kN} \uparrow \qquad\qquad \textbf{Ans.}$$

Determination of $\mathbf{A}_y$:

$$+\downarrow\Sigma M_B = -A_y(6) + 9.00(4.0) + 4.5(2.0) + 9.00(1.5) = 0$$

$$A_y = +9.75 \text{ kN} \qquad \mathbf{A}_y = 9.75 \text{ kN} \uparrow \qquad\qquad \textbf{Ans.}$$

Alternatively (or as a check):

$$+\uparrow\Sigma F_y = A_y - 9.00 - 4.50 - 9.00 + 12.75 = 0$$

$$A_y = +9.75 \text{ kN} \qquad \mathbf{A}_y = 9.75 \text{ kN} \uparrow$$

(b) The shearing stress on the pin at A is given by Eq. (4-4) as

$$\tau = \frac{V}{A_s} = \frac{A}{2(\pi/4)d^2} = \frac{9.75(10^3)}{2(\pi/4)(0.040)^2}$$
$$= 3.88(10^6) \text{ N/m}^2 = 3.88 \text{ MPa} \qquad\qquad \textbf{Ans.}$$

(c) The bearing stress on the beam at B is given by Eq. (4-6) as

$$\sigma_b = \frac{F}{A_b} = \frac{B}{dt} = \frac{12.75(10^3)}{0.040(0.200)}$$
$$= 1.594(10^6) \text{ N/m}^2 = 1.594 \text{ MPa C} \quad\blacksquare \quad \textbf{Ans.}$$

Example Problem 6-9

The structure shown in Fig. 6-45a is used to support a 9100-lb load. If the diameter of member AB is 1.25 in. and the diameter of the bolt at C is 1.75 in., determine

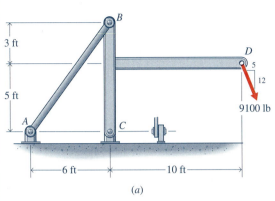

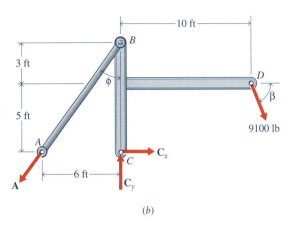

(a)   (b)

Figure 6-45

(a) The normal stress on a cross section of member AB.

(b) The shearing stress on a cross section of bolt C.

(c) The deformation of member AB if it is made of steel having a modulus of elasticity of 29,000 ksi.

SOLUTION

A free-body diagram of the complete structure is shown in Fig. 6-45b. From the geometry of the structure

$$\beta = \tan^{-1}\frac{12}{5} = 67.38° \qquad \phi = \tan^{-1}\frac{6}{8} = 36.87°$$

The action of the pin at A is represented by a force **A** with a line of action along the axis of member AB since member AB is a two-force member. The pin at C is represented by force components $\mathbf{C}_x$ and $\mathbf{C}_y$. Since the structure is subjected to a general coplanar force system, three equilibrium equations are available to solve for **A**, $\mathbf{C}_x$, and $\mathbf{C}_y$.
Determination of **A**:

$$+\!\!\downarrow\!\Sigma M_C = A(8 \sin \phi) - 9100 \sin \beta(10) - 9100 \cos \beta(5)$$
$$= A(8 \sin 36.87°) - 9100 \sin 67.38°(10) - 9100 \cos 67.38°(5) = 0$$

$$A = +21{,}150 \text{ lb} \qquad \mathbf{A} = 21{,}150 \text{ lb} \measuredangle 53.1°$$

Determination of $\mathbf{C}_x$:

$$+\!\!\downarrow\!\Sigma M_B = C_x(8) - 9100 \sin \beta(10) + 9100 \cos \beta(3)$$
$$= C_x(8) - 9100 \sin 67.38°(10) + 9100 \cos 67.38°(3) = 0$$

$$C_x = +9187 \text{ lb} \qquad \mathbf{C}_x = 9187 \text{ lb} \rightarrow$$

Determination of $\mathbf{C}_y$:

$$+\!\!\uparrow\!\Sigma F_y = C_y - 9100 \sin \beta - A \cos \phi$$
$$= C_y - 9100 \sin 67.38° - 21{,}150 \cos 36.87° = 0$$

$$C_y = +25{,}320 \text{ lb} \qquad \mathbf{C}_y = 25{,}320 \text{ lb} \uparrow$$

The reaction at support C is

$$C = \sqrt{(C_x)^2 + (C_y)^2} = \sqrt{(9187)^2 + (25,320)^2}$$
$$= 26,940 \text{ lb}$$

(a) The normal stress on a cross section of member AB is given by Eq. (4-2) as

$$\sigma_{AB} = \frac{F_{AB}}{A_{AB}} = \frac{21,150}{(\pi/4)(1.25)^2} = 17.23(10^3) \text{ psi} = 17.23 \text{ ksi T} \qquad \textbf{Ans.}$$

(b) The shearing stress on a cross section of bolt C is given by Eq. (4-4). Since the bolt is in single shear,

$$\tau = \frac{C}{A_s} = \frac{26,490}{(\pi/4)(1.75)^2} = 11.20(10^3) \text{ psi} = 11.20 \text{ ksi} \qquad \textbf{Ans.}$$

(c) The deformation of member AB is given by Eq. (4-20b) as

$$\delta = \frac{F_{AB}L_{AB}}{E_{AB}A_{AB}} = \frac{21,150(120)}{29,000,000(\pi/4)(1.25)^2} = 0.0713 \text{ in.} \quad \blacksquare \qquad \textbf{Ans.}$$

PROBLEMS

In the following problems neglect the weights of the members unless otherwise specified.

Standard Problems

6-13* A beam is loaded and supported as shown in Fig. P6-13. Determine the reactions at supports A and B.

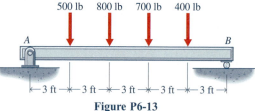

Figure P6-13

6-14* A beam is loaded and supported as shown in Fig. P6-14. Determine the reaction at support A.

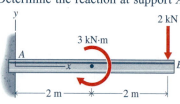

Figure P6-14

6-15* A curved slender bar is loaded and supported as shown in Fig. P6-15. Determine the reaction at support A.

Figure P6-15

6-16* A curved slender bar is loaded and supported as shown in Fig. P6-16. Determine the reactions at supports A and B.

Figure P6-16

6-17 A beam is loaded and supported as shown in Fig. P6-17. Determine the reactions at supports A and B.

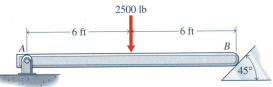

Figure P6-17

6-18 A beam is loaded and supported as shown in Fig. P6-18. Determine the reactions at supports A and B.

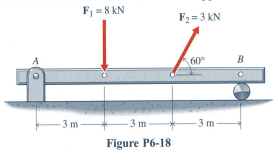

Figure P6-18

6-19* A beam is loaded and supported as shown in Fig. P6-19. Determine the reactions at supports A and B.

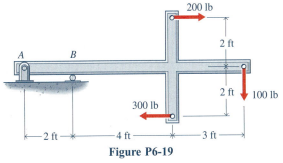

Figure P6-19

6-20* A beam is loaded and supported as shown in Fig. P6-20. Determine the reactions at supports A and B.

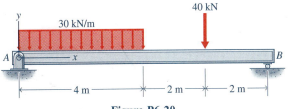

Figure P6-20

6-21 A beam is loaded and supported as shown in Fig. P6-21. Determine the reactions at supports A and B.

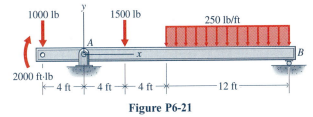

Figure P6-21

6-22 A beam is loaded and supported as shown in Fig. P6-22. Determine the reactions at supports A and B.

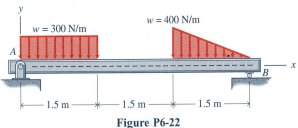

Figure P6-22

6-23* A beam is loaded and supported as shown in Fig. P6-23. Determine the reactions at supports A and B.

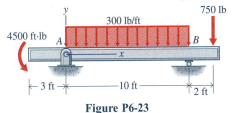

Figure P6-23

6-24* A pipe strut BC is loaded and supported as shown in Fig. P6-24. Determine
(a) The reactions at supports A and C.
(b) The shearing stress on a cross section of the pin at C, which is in double shear, if the pin has a diameter of 10 mm.
(c) The elongation of cable AB if it is made of aluminum alloy ($E = 73$ GPa) and has a diameter of 6 mm.

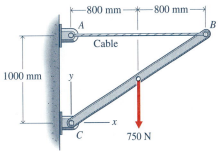

Figure P6-24

6-25 A 75-lb load is supported by an angle bracket, pulley, and cable as shown in Fig. P6-25. Determine
(a) The force exerted on the bracket by the pin at C.
(b) The reactions at supports A and B of the bracket.
(c) The shearing stress on a cross section of the pin at C, which is in double shear, if the pin has a diameter of $\frac{1}{4}$ in.

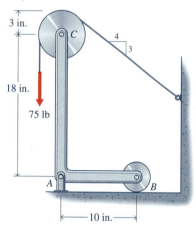

Figure P6-25

6-26 A cylinder is supported by a bracket as shown in Fig. P6-26. The mass of the cylinder is 50 kg. If all surfaces are smooth, determine
(a) The reaction at support A of the bracket.
(b) The force exerted on the cylinder at contact point B.
(c) The shearing stress on a cross section of the pin at A, which is in double shear, if the pin has a diameter of 10 mm.

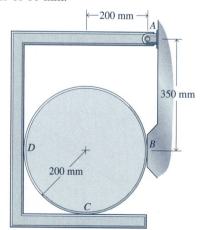

Figure P6-26

6-27* A rope and pulley system is used to support a body W as shown in Fig. P6-27. Each pulley is free to rotate and the rope is continuous over the pulleys. Determine the tension **T** in the rope required to hold body W in equilibrium if the weight of body W is 400 lb. Assume that all rope segments are vertical.

Figure P6-27

6-28* A rope and pulley system is used to support a body W as shown in Fig. P6-28. Each pulley is free to rotate. One rope is continuous over pulleys A and B; the other is continuous over pulley C. Determine the tension **T** in the rope over pulleys A and B required to hold body W in equilibrium if the mass of body W is 175 kg.

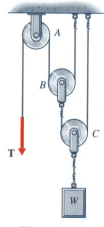

Figure P6-28

6-29 Pulleys *A* and *B* of the chain hoist shown in Fig. P6-29 are connected and rotate as a unit. The chain is continuous and each of the pulleys contain slots that prevent the chain from slipping. Determine the force **F** required to hold a 1000-lb block *W* in equilibrium if the radii of pulleys *A* and *B* are 3.5 and 4.0 in., respectively.

Figure P6-29

6-30 Pulleys 1 and 2 of the rope and pulley system shown in Fig. P6-30 are connected and rotate as a unit. The radii of pulleys 1 and 2 are 100 mm and 300 mm, respectively. Rope *A* is wrapped around pulley 1 and is

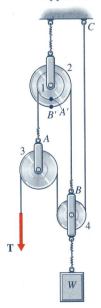

Figure P6-30

fastened to pulley 1 at point *A'*. Rope *B* is wrapped around pulley 2 and is fastened to pulley 2 at point *B'*. Rope *C* is continuous over pulleys 3 and 4. Determine the tension **T** in rope *C* required to hold body *W* in equilibrium if the mass of body *W* is 225 kg.

6-31* Three pipes are supported in a pipe rack as shown in Fig. P6-31. Each pipe weighs 100 lb. Determine
(a) The reactions at supports *A* and *B*.
(b) The shearing stress on a cross section of the pin at *A*, which is in single shear, if the pin has a diameter of $\frac{3}{8}$ in.

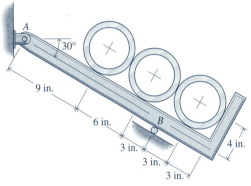

Figure P6-31

6-32* A beam is loaded and supported as shown in Fig. P6-32. The beam has a uniform cross section and a mass of 120 kg. Determine
(a) The reactions at supports *A* and *B*.
(b) The shearing stress on a cross section of the pin at *A*, which is in double shear, if the pin has a diameter of 10 mm.

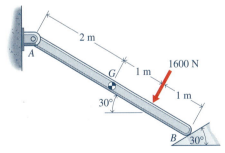

Figure P6-32

6-33 A homogeneous cylinder is supported by a bar of uniform cross section as shown in Fig. P6-33. The weight of the cylinder is 100 lb and the weight of the bar is 20 lb. If all surfaces are smooth, determine the reactions at supports *A* and *B* of the bar.

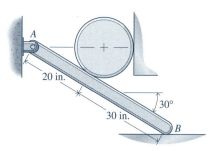

Figure P6-33

6-34 A curved bar is loaded and supported as shown in Fig. P6-34. The bar has a uniform cross section and a mass of 75 kg. Determine
(a) The reactions at supports A and B.
(b) The shearing stress on a cross section of the pin at A, which is in double shear, if the pin has a diameter of 8 mm.
(c) The elongation of link B if it is made of aluminum alloy ($E = 73$ GPa) and has a diameter of 10 mm.

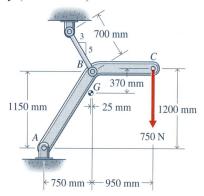

Figure P6-34

Challenging Problems

6-35* A homogeneous cylinder is supported by a bar of uniform cross section and a cable as shown in Fig. P6-35. The weight of the cylinder is 150 lb and the

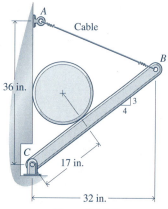

Figure P6-35

weight of the bar is 20 lb. If all surfaces are smooth, determine
(a) The reaction at support C of the bar and the tension **T** in the cable.
(b) The elongation of cable AB if it is made of steel ($E = 30,000$ ksi) and has a diameter of $\frac{1}{8}$ in.

6-36* A homogeneous cylinder with a mass of 50 kg is supported on an inclined surface by a pin-connected two-bar frame as shown in Fig. P6-36. Assume that all surfaces are smooth. Determine
(a) The forces exerted on the cylinder by the contacting surfaces.
(b) The reactions at supports A and C of the two-bar frame.
(c) The shearing stress on a cross section of the pin at B, which is in single shear, if the pin has a diameter of 6 mm.

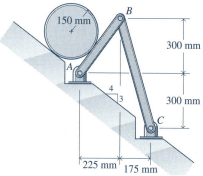

Figure P6-36

6-37 Two beams are loaded and supported as shown in Fig. P6-37. Determine
(a) The reactions at supports A, B, and C.
(b) The shearing stress on a cross section of the pin at B, which is in double shear, if the pin has a diameter of $\frac{1}{4}$ in.
(c) The bearing stress between the pin at A and the beam if the beam is $\frac{1}{2}$-in. thick and the pin has a diameter of $\frac{3}{8}$ in.

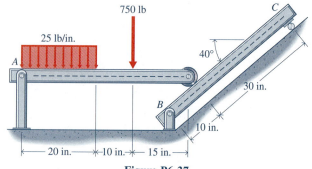

Figure P6-37

6-38 A bracket is loaded and supported as shown in Fig. P6-38. Determine
(a) The reactions at supports A and B.
(b) The shearing stress on a cross section of the pin at A, which is in double shear, if the pin has a diameter of 10 mm.
(c) The bearing stress between the beam and the $25 \times 50 \times 10$ mm–thick bearing plate at support B.

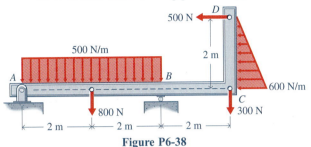

Figure P6-38

6-39* Three bars are connected with smooth pins to form the frame shown in Fig. P6-39. The weights of the bars are negligible. Determine
(a) The reactions at supports A and D.
(b) The shearing stress on a cross section of the pin at D, which is in double shear, if the pin has a diameter of $\frac{1}{2}$ in.
(c) The bearing stress between the pin at A and bar ABC if the bar is $\frac{1}{2}$-in. thick and the pin has a diameter of $\frac{1}{2}$ in.

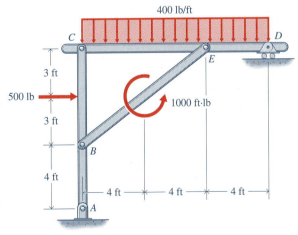

Figure P6-39

6-40* Three bars are connected with smooth pins to form the frame shown in Fig. P6-40. The weights of the bars are negligible. Determine
(a) The reactions at supports A and E.
(b) The shearing stress on a cross section of the pin at A, which is in single shear, if the pin has a diameter of 6 mm.

(c) The bearing stress between the pin at E and the wheel if the wheel is 6-mm thick and the pin has a diameter of 8 mm.

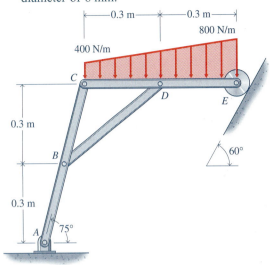

Figure P6-40

6-41 A hoist is made from a 4×4-in. wood post BC and a 1 in.-diameter steel ($E = 30,000$ ksi) eyebar AB as shown in Fig. P6-41. The hoist supports a load W of 7000 lb. Determine
(a) The axial stresses in members AB and BC.
(b) The shearing stress in the 1 in.-diameter bolt at A, which is in double shear.
(c) The deformation of member AB.

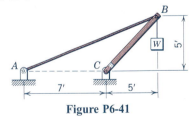

Figure P6-41

6-42 A lever is attached to the shaft of a steel gate valve with a square key as shown in Fig. P6-42. If the shearing stress in the key must not exceed 125 MPa, determine the minimum dimension "a" that must be used if the key is 20 mm long.

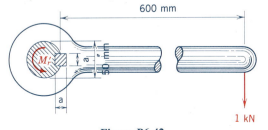

Figure P6-42

6-43* A frame is loaded and supported as shown in Fig. P6-43. Rod CD has a $\frac{1}{2}$-in. diameter and is made of aluminum alloy ($E = 10{,}600$ ksi). Determine

(a) The reactions at supports A and E.
(b) The axial stress in rod CD.
(c) The elongation of rod CD.
(d) The shearing stress on a cross section of the $\frac{1}{2}$ in.-diameter bolt at A, which is in single shear.

6-44 The axial stresses are 12 MPa C in the wood post B and 150 MPa T in the steel bar A of Fig. P6-44. Member CDE is rigid. Determine

(a) The load P.
(b) The minimum diameter for pin C if it is in single shear and the shearing stress is limited to 70 MPa.
(c) The minimum diameter for pin D if it is in double shear and the shearing stress is limited to 70 MPa.

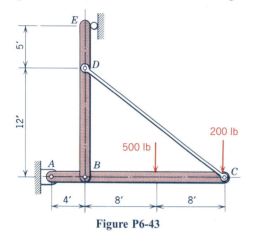

Figure P6-43

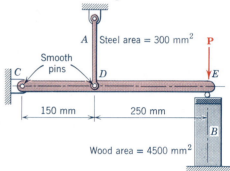

Figure P6-44

6-4 PLANE TRUSSES

A truss is a structure composed of straight members joined together at their end points and loaded only at the joints (see Fig. 6-46). The airy structure of a truss provides greater strength over large spans than would more solid types of structures. Trusses are commonly seen supporting the roofs of buildings as well as television towers, antennas, aircraft frames, and highway bridges. Although not commonly seen, trusses also form the skeletal structure of many large buildings.

Planar trusses lie in a single plane and all applied loads must lie in the same plane. Planar trusses are often used in pairs to support bridges, as shown in Fig. 6-47. All members of the truss $ABCDEF$ lie in the same vertical plane. Loads on the floor of the bridge are carried by means of the floor construction to the joints A, B, C, and D. The loads thus transmitted to the joints lie in the same vertical plane as the truss.

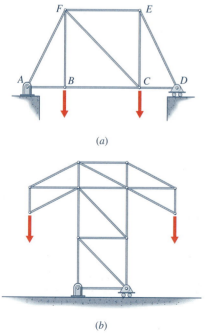

(a)

(b)

Figure 6-46

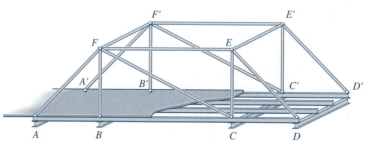

Figure 6-47

Four main assumptions are made in the analysis of trusses. One result of the assumptions is that all members of the idealized truss are two-force members. Although the assumptions are idealizations of actual structures, real trusses behave according to these idealizations to a high degree of approximation. The resulting error is usually small enough to justify the assumptions.

Truss members are connected at their ends only.

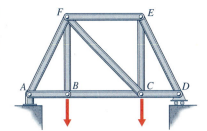

Figure 6-48

The first assumption means that the truss of Figs. 6-46a and 6-47 should be drawn as shown in Fig. 6-48. In actual practice, the main top and bottom chords frequently consist of members that span several joints, such as member *ABCD* of Fig. 6-47, rather than a series of shorter members between joints. The members of a truss are usually long and slender, however, and can support little lateral load or bending moment. Hence, the noncontinuous member assumption is usually acceptable.

Truss members are connected by frictionless pins.

In real trusses, the members are usually bolted, welded, or riveted to a gusset plate, as shown in Fig. 6-49a, rather than connected by an idealized frictionless pin, as shown in Fig. 6-49b. However, experience has shown the frictionless pin to be an acceptable idealization as long as the axes of the members all intersect at a single point.

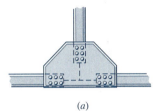

(a)

Trusses are loaded only at the joints.

As stated earlier, the members that make up a truss are usually long and slender. Like cables, such members can withstand large tensile (axial) load but cannot withstand moments or large lateral loads. Loads must either be applied directly to the joints as indicated in the diagrams of Fig. 6-46 or must be carried to the joints by a floor structure as shown in Fig. 6-47.

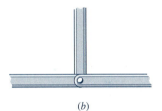

(b)

Figure 6-49

The weights of the members may be neglected.

Frequently in the analysis of trusses the weights of the members are neglected. While this may be acceptable for small trusses, it may not be acceptable for a large bridge truss. Again because the members can withstand little bending moment or lateral load, experience has shown that little error results from assuming the load acts at the joints of the truss. Common practice is to assume that half of the weight of each member acts at each joint.

The result of these four assumptions is that forces act only at the ends of the members. Also, because the pins are assumed to be frictionless, there is no moment applied to the ends of the members. Therefore, by the analysis of Section 6-3, each member is a two-force member supporting only an axial force, as shown in Fig. 6-50. In its simplest form, a truss (such as that shown in Fig. 6-51a) consists of a collection of two-force members held together by frictionless pins as shown in Fig. 6-51b. The forces acting on the pins of the truss shown in Fig. 6-51a are also shown in Fig. 6-51b.

For general two-force members, the forces act along the line joining the points where the forces are applied. Since truss members are usually straight, however, the forces will act along the axis of the member as shown in Figs.

Figure 6-50

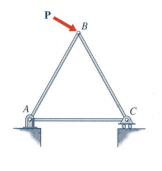

(a)

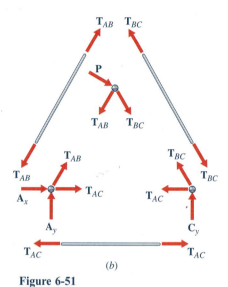

(b)

Figure 6-51

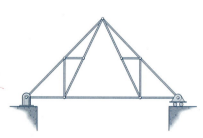

Figure 6-52

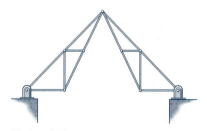

Figure 6-53

6-50 and 6-51.[1] Forces that pull on the ends of a member (as in Figs. 6-50 and 6-51) are called *tensile* and tend to elongate the member. Such forces are called *tensile forces*, and the member is said to be in *tension*. Forces that push on the ends of the member tend to shorten the member. Such forces are called *compressive forces* and the member is said to be in *compression*. It may be noted from Fig. 6-51 that when a joint exerts a force that pulls on the end of a member, the member exerts a force that pulls on the joint.

It is important to distinguish between truss members that are in tension and those that are in compression. The long slender members that make up a truss are very strong in tension but tend to bend or buckle under large compressive loads. Truss members in compression either must be made thicker than the other truss members or must be braced to prevent buckling. Buckling will be discussed in Chapter 11.

One end of a bridge truss is usually allowed to "float" on a rocker or roller support, as shown in Fig. 6-46. Aside from the mathematical requirement (in a planar equilibrium problem, only three support reactions can be determined), such a support is needed to allow for the expansion or contraction of the structure due to temperature variations.

To retain their shape and support the large loads applied to them, trusses must be rigid structures. The simplest structure that is rigid (independent of how it is supported) is a triangle. Of course, the word "rigid" does not mean that a truss will not deform under loading. It will undergo very small deformations, but will very nearly retain its original shape.

"Rigid" is often interpreted also to mean that the truss will retain its shape when removed from its supports or when one of the supports is free to slide. In this sense, the truss of Fig. 6-52 is rigid, whereas the truss of Fig. 6-53 is not. The truss of Fig. 6-53 is called a *compound truss* and the lack of internal rigidity is made up for by an extra external support reaction.

The basic building block of all trusses is a triangle. Large trusses are constructed by attaching several triangles together. One method of construction starts with a basic triangular element, such as triangle *ABC* of Fig. 6-54. Additional triangular elements are added one at a time by attaching one new joint (for example, *D*) to the truss, using two new members (for example, *BD* and *CD*). A truss that can be constructed in this fashion is called a *simple truss*. While it might appear that all trusses composed of triangles are simple trusses, such is not the case. For example, neither of the trusses of Figs. 6-52 or 6-53 is a simple truss.

The importance of a simple truss is that it allows a simple way to check the rigidity and solvability of a truss. Clearly, since a simple truss is constructed solely of triangular elements, it is always rigid. Also, since each new joint brings two new members with it, a simple relationship exists between the number of joints j and the number of members m in a simple plane truss:

$$m = 2j - 3 \qquad (6\text{-}7)$$

In the discussion of the method of joints that follows, this will be seen to be exactly the condition necessary to guarantee that the number of equations to be

---

[1]For curved two-force members, however, the line joining the ends is not the axis of the member. All of the trusses considered in this chapter will contain only straight two-force members.

solved ($2j$) is the same as the number of unknowns to be solved for (m member forces and 3 support reactions).

Although Eq. (6-7) ensures that a simple plane truss is rigid and solvable, it is neither sufficient nor necessary to ensure that a nonsimple plane truss is rigid and solvable. For example, the nonsimple plane trusses of Figs. 6-52 and 6-53 are both rigid (at least while attached to their supports) and solvable, although one (Fig. 6-52) satisfies Eq. (6-7) and the other (Fig. 6-53) does not. A tempting generalization of Eq. (6-7) is

$$m = 2j - r \qquad (6-8)$$

where r is the number of support reactions. Both trusses of Figs. 6-52 and 6-53 satisfy Eq. (6-8), as do all simple trusses that have the customary three support reactions. However, constructions can be envisioned for which even Eq. (6-8) is not a proper test of the solvability of a truss.

6-4-1 Method of Joints

Consider the truss of Fig. 6-55a, whose free-body diagram is shown in Fig. 6-55b. Since the entire truss is a rigid body in equilibrium, each part must also be in equilibrium. The method of joints consists of taking the truss apart, drawing separate free-body diagrams of each part—each member and each pin, as in Fig. 6-56—and applying the equations of equilibrium to each part of the truss in turn.

The free-body diagrams of the members in Fig. 6-56 have only axial forces applied to their ends because of assumptions about how a truss is constructed and loaded. The symbol T_{BC} is used to represent the unknown force in member BC. (No significance is attached to the order of the subscripts; that is, $T_{BC} = T_{CB}$). Since the line of action of the member forces are all known, the force in each member is completely specified by giving its magnitude and sense; that is, whether the force points away from the member as in Fig. 6-56 or toward the member. Thus the force (a vector) is represented by the scalar symbol T_{BC}. The sense of the force will be taken from the sign of T_{BC}; positive indicates the direction drawn on the free-body diagram, negative indicates the opposite direction.

Forces that point away from a member, as in Fig. 6-56, tend to stretch the member and are called *tensile*. Forces that point toward a member tend to compress the member and are called *compressive*. Whether a member is in compression or tension is usually not known ahead of time. Although some people try to guess and draw some of the forces in tension and others in compression, it is not necessary to do so. In this book, all free-body diagrams will be drawn as though all members are in tension. A negative value for a force in the solution will indicate that the member was really in compression. This can be reported either by saying that $T_{BC} = -2500$ lb or by saying that $T_{BC} = 2500$ lb (C). The latter is preferable since it does not depend on whether member BC was assumed to be in tension or in compression in the free-body diagram.

According to Newton's third law (of action and reaction), the force exerted on a member by a pin and the force exerted on a pin by a member are equal and opposite. Therefore, the same symbol T_{AB} is used for the force exerted by the member AB on pin B and for the force exerted by pin B on member AB. Having drawn the free-body diagrams of the members as two-force members ensures that the members are in equilibrium. No further information is obtained from the

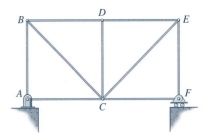

Figure 6-54

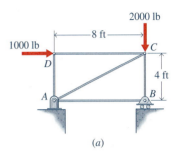

(a)

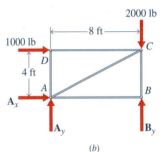

(b)

Figure 6-55

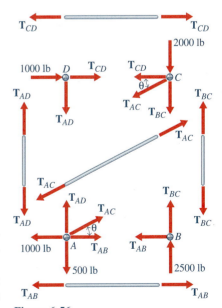

Figure 6-56

free-body diagrams of the two-force members; therefore, they may be discarded for the remainder of the analysis. The analysis of the truss reduces to considering the equilibrium of the joints that make up the truss—hence the name "Method of Joints."

Equilibrium of the joints that make up the truss is expressed by drawing a separate free-body diagram for each joint and writing the equilibrium equation

$$\Sigma \mathbf{F} = \mathbf{0} \qquad (6\text{-}9)$$

for each joint. Since each joint consists of concurrent forces in a plane, moment equilibrium gives no useful information and Eq. (6-9) has only two independent components. Therefore, for a plane truss containing j pins, there will be a total of $2j$ independent scalar equations available. But according to Eq. (6-7), this is precisely the number of independent equations needed to solve for the m member forces and 3 support reactions of a simple truss.

Solution of the $2j$ equations is significantly simplified if a joint can be found on which only two unknown forces and one or more known forces act (for example, joint D of Fig. 6-56). In this case the two equations for this joint can be solved independently of the rest of the equations. If such a joint is not readily available, one can usually be created by solving the equilibrium equations for the entire truss first, that is, solve for the support reactions. Once two of the unknown forces have been determined they can be treated as known forces on the free-body diagrams of the other joints. The joints are solved sequentially in this fashion until all forces are known.

As mentioned earlier, a negative value for a force indicates that the member is in compression rather than in tension. It is unnecessary to go back to the free-body diagram and change the direction of the arrow. In fact, doing so is likely only to cause confusion. The free-body diagrams should all be drawn consistently. A negative value for a symbol on one free-body diagram translates to the same negative value for the same symbol on another free-body diagram.

Once all of the forces have been determined, a summary should be made listing the magnitude of the force in each member and whether the member is in tension or compression (see the Example Problems).

Finally, it must be noted that the equations of equilibrium for the entire truss are contained in the equations of equilibrium of the joints (see Problem 6-62). That is, if all the joints are in equilibrium and all the members are in equilibrium, then the entire truss is also in equilibrium. A consequence of this is that the three support reactions can be determined along with the m member forces from the $2j$ equations of equilibrium of the joints. Overall equilibrium in this case may be used as a check of the solution. However, if overall equilibrium is used first to determine the support reactions and help start the method of joints, then three of the $2j$ joint equations of equilibrium will be redundant and may be used as a check of the solution.

6-4-2 Zero-Force Members

Frequently, certain members of a given truss carry no load. Zero-force members in a truss usually arise in one of two general ways. The first is

> When only two members form a noncollinear truss joint and no external load or support reaction is applied to the joint, then both members must be zero-force members.

The truss of Fig. 6-57a is an example of this condition. The free-body diagram of pin C is drawn in Fig. 6-57b. The equations of equilibrium for this joint,

$$+\rightarrow \Sigma F_x = -T_{BC} - T_{CD} \cos 30° = 0$$

$$+\uparrow \Sigma F_y = -T_{CD} \sin 30° = 0$$

are trivially solved to get

$$T_{CD} = 0 \quad \text{and} \quad T_{BC} = 0$$

That is, for this particular truss and for this particular loading, the two members BC and CD could be removed without affecting the solution or even (in this particular case) the stability of the truss.

The second way in which zero-force members normally arise in a truss is as follows:

> When three members form a truss joint for which two of the members are collinear and the third forms an angle with the first two, then the non-collinear member is a zero-force member, provided no external force or support reaction is applied to that joint. The two collinear members carry equal loads (either both tension or both compression).

Such a condition arises, for example, when the load of Fig. 6-57a is moved from pin B to pin C, as in Fig. 6-58a. The free-body diagram of pin B is drawn in Fig. 6-58b. The equations of equilibrium for this joint are

$$+\rightarrow \Sigma F_x = -T_{AB} + T_{BC} = 0$$

$$+\uparrow \Sigma F_y = -T_{BD} = 0$$

Thus, since joint B is now unloaded, the force in member BD vanishes and the forces in members AB and BC are equal in magnitude—either both tension (both positive) or both compression (both negative).

Once it is known that BD is a zero-force member, the same reasoning can then be used to show that member AD carries no load. The free-body diagram of pin D is drawn in Fig. 6-58c. To simplify the calculations, coordinate axes are chosen along and normal to the collinear members CD and DE. The equations of equilibrium are then

$$+\nearrow \Sigma F_{x'} = -T_{DE} - T_{AD} \cos 60° + T_{BD} \cos 60° + T_{CD} = 0$$

$$+\nwarrow \Sigma F_{y'} = T_{AD} \sin 60° + T_{BD} \sin 60° = 0$$

But since $T_{BD} = 0$ (BD is already known to be a zero-force member), then

$$T_{AD} = 0 \quad \text{and} \quad T_{DE} = T_{CD}$$

Thus, for the loading of Fig. 6-58a, both members AD and BD are zero-force members.

These zero-force members cannot simply be removed from the truss and discarded; they are needed to guarantee the stability of the truss. If members AD and BD were removed, there would be nothing to prevent some small disturbance

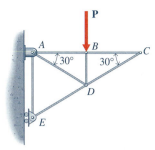

(a)

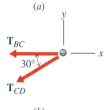

(b)

Figure 6-57

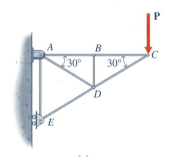

(a)

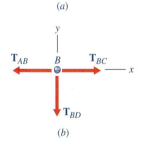

(b)

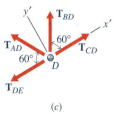

(c)

Figure 6-58

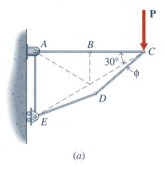

(a)

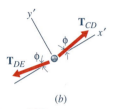

(b)

Figure 6-59

from moving pin D slightly out of alignment, as in Fig. 6-59a. Then the free-body diagram of pin D would look like Fig. 6-59b. Again choosing axes along and normal to the line CE gives the equilibrium equations

$$+\nearrow\Sigma F_{x'} = -T_{DE} \cos \phi + T_{CD} \cos \phi = 0$$

$$+\nwarrow\Sigma F_{y'} = T_{DE} \sin \phi + T_{CD} \sin \phi = 0$$

The first of these equations requires that $T_{CD} = T_{DE}$, while the second requires that $T_{CD} = -T_{DE}$. The only way both of these equations can be satisfied is if both forces equal zero. But equilibrium of pin C requires that T_{CD} not be zero. What has happened, of course, is that the truss is no longer in static equilibrium. Pin D will continue to buckle outward and the truss will collapse.

A seemingly trivial solution to the stability problem would be to replace the two members CD and DE with a single member CE and to replace the two members AB and BC with a single member AC. While this solution would satisfy the statics part of the problem, it would not take care of the tendency for long slender members to buckle when subjected to large compressive loads. Therefore, long members such as member CE of Fig. 6-58 are usually replaced by a pair of shorter members and the mid-joint braced if analysis of the truss indicates the member is likely to be in compression for some expected loading. Long members such as member AC of Fig. 6-57 must also be replaced by a pair of shorter members and the mid-joint braced if it is ever desired to load the truss at some point along the long member.

Thus one must not be too quick to discard truss members just because they carry no load for a given configuration. These members are often needed to carry part of the load when the applied loading changes and they are almost always needed to guarantee the stability of the truss.

While recognizing these and other special joint-loading conditions can simplify the analysis of a truss, such recognition is not required to solve the truss. If one does not recognize that a member is a zero-force member, drawing the free-body diagram and writing the equilibrium equations will immediately show that it is a zero-force member. Also, these shortcuts should be applied with care. If there is any doubt about whether or not a member is a zero-force member, the prudent choice is to draw the free-body diagram and solve for the member force.

Example Problem 6-10

Identify all zero-force members in the truss shown in Fig. 6-60a.

SOLUTION

Joint B connects three members, two of which (AB and BC) are collinear. The joint is unloaded; therefore, the force in the noncollinear member (BJ) must be zero. This is easily checked by drawing the free-body diagram for pin B (Fig. 6-60b) and writing the equation of equilibrium in the direction perpendicular to the collinear members AB and BC. Thus,

$$+\searrow\Sigma F = T_{BJ} \cos \theta = 0 \qquad T_{BJ} = 0 \qquad \textbf{Ans.}$$

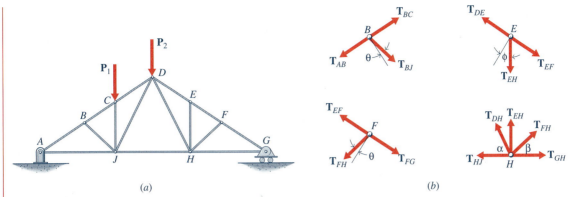

Figure 6-60

Similarly at joints E and F, which are unloaded, two members of these joints are collinear and the third member is noncollinear. Free-body diagrams for pins E and F (Fig. 6-60b) and equations of equilibrium in a direction perpendicular to the collinear members give

From a free-body diagram for joint E:

$$+ \swarrow \Sigma F = T_{EH} \cos \phi = 0 \qquad T_{EH} = 0 \qquad \textbf{Ans.}$$

From a free-body diagram for joint F:

$$+ \swarrow \Sigma F = T_{FH} \cos \theta = 0 \qquad T_{FH} = 0 \qquad \textbf{Ans.}$$

Finally, from a free-body diagram for joint H:

$$+ \uparrow \Sigma F = T_{DH} \sin \alpha + T_{EH} + T_{FH} \sin \beta = 0$$

Since $T_{EH} = T_{FH} = 0$,

$$T_{DH} = 0 \qquad \textbf{Ans.}$$

Therefore, the zero-force members in the truss for the given loading are BJ, EH, FH, and DH. ∎

Example Problem 6-11

Use the method of joints to find the force in each member of the truss of Fig. 6-61a.

SOLUTION

Truss $ABCD$ is a simple truss with $m = 5$ members and $j = 4$ joints. Therefore, the eight equations obtained from equilibrium of the four joints can be solved for the three support reactions as well as the forces in all five members.

The first step is to draw a free-body diagram of the entire truss (Fig. 6-61b) and write the equilibrium equations

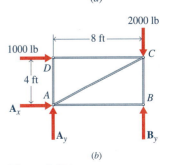

Figure 6-61

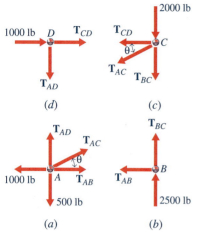

(d)

(c)

(a)

(b)

Figure 6-62

$$+\rightarrow\Sigma F_x = 1000 + A_x = 0$$

$$+\uparrow\Sigma F_y = -2000 + A_y + B_y = 0$$

$$+\downarrow\Sigma M_A = -4(1000) - 8(2000) + 8B_y = 0$$

These equations can be solved to get

$$A_x = -1000 \text{ lb} \qquad B_y = 2500 \text{ lb} \qquad A_y = -500 \text{ lb}$$

Next, draw a free-body diagram of pin D (Fig. 6-62d) and solve the equilibrium equations

$$+\rightarrow\Sigma F_x = 1000 + T_{CD} = 0$$

$$+\uparrow\Sigma F_y = -T_{AD} = 0$$

to get

$$T_{CD} = -1000 \text{ lb} \qquad T_{AD} = 0 \text{ lb}$$

Next draw a free-body diagram of pin C (Fig. 6-62c) and solve the equilibrium equations

$$+\rightarrow\Sigma F_x = -T_{CD} - T_{AC}\cos\theta = 0$$

$$+\uparrow\Sigma F_y = -2000 - T_{BC} - T_{AC}\sin\theta = 0$$

where

$$\sin\theta = \frac{AD}{AC} = \frac{4}{\sqrt{4^2 + 8^2}} = 0.4472$$

and

$$\cos\theta = \frac{CD}{AC} = \frac{8}{\sqrt{4^2 + 8^2}} = 0.8944$$

But $T_{CD} = -1000$ lb, so

$$T_{AC} = -\frac{T_{CD}}{\cos\theta} = -\frac{-1000}{0.8944} = 1118 \text{ lb}$$

and

$$T_{BC} = -2000 - 1118(0.4472) = -2500 \text{ lb}$$

Next, draw a free-body diagram of pin B (Fig. 6-62b) and write the equilibrium equations

$$+\rightarrow\Sigma F_x = -T_{AB} = 0$$

$$+\uparrow\Sigma F_y = T_{BC} + 2500 = 0$$

The first of these equations can be solved to get

$$T_{AB} = 0 \text{ lb}$$

The second equation contains no unknowns since the value of T_{BC} has already been found. The second equation can be used to check the consistency of the answers

$$-2500 + 2500 = 0 \text{ (check)}$$

Finally, draw a free-body diagram of pin A (Fig. 6-62a) and write the equilibrium equations

$$+\rightarrow \Sigma F_x = T_{AB} + T_{AC} \cos \theta - 1000 = 0$$

$$+\uparrow \Sigma F_y = T_{AD} + T_{AC} \sin \theta - 500 = 0$$

Again, there are no unknowns in these equations since the values of T_{AB}, T_{AC}, and T_{AD} have already been found. These two equations again reduce to a check of the consistency of the solution:

$$0 + 1118(0.8944) - 1000 = -0.0608$$

$$0 + 1118(0.4472) - 500 = -0.0304$$

The difference is less than the rounding performed on T_{AB}, T_{AC}, and T_{AD} and so the solution checks. The desired answers then are

AB, AD:	0 lb	**Ans.**
AC:	1118 lb (T)	**Ans.**
BC:	2500 lb (C)	**Ans.**
CD:	1000 lb (C)	**Ans.**

The fact that T_{AB} and T_{AD} both came out zero is a peculiarity of the loading and does not mean that members AB and AD should be eliminated from the truss. For a slightly different loading situation, the forces in these members will not be zero. Even for the given loading condition, members AB and AD are necessary to ensure the rigidity of the truss. Without member AB, for example, the truss would collapse if the roller support at B is disturbed slightly to the right or left. ■

Example Problem 6-12

The truss shown in Fig. 6-63a supports one side of a bridge; an identical truss supports the other side. Floor beams carry vehicle loads to the truss joints. A 2000-kg car is stopped on the bridge. Calculate the force in each member of the truss using the method of joints.

SOLUTION

The truss of Fig. 6-63a is a simple truss with $m = 7$ members and $j = 5$ joints. Therefore, the ten equations obtained from equilibrium of the five joints can be solved for the three support reactions as well as the forces in all seven members.

The first step is to divide the weight of the car between the joints of the truss. Half of the car's weight—$\frac{1}{2}(2000)(9.81) = 9810$ N—is carried by the truss shown and the other half is carried by the truss on the other side of the bridge.

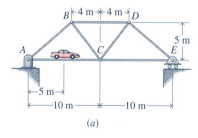

(a)

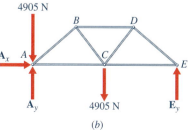

(b)

Figure 6-63

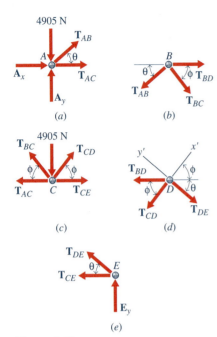

(a)

(b)

(c)

(d)

(e)

Figure 6-64

Since the car is mid-way between joints A and C, $\frac{1}{2}(9810) = 4905$ N will be applied to joint A and 4905 N will be applied to joint C.

The next step is to draw a free-body diagram of the entire truss (Fig. 6-63b) and write the equilibrium equations

$$+\rightarrow\Sigma F_x = A_x = 0$$

$$+\uparrow\Sigma F_y = A_y - 4905 - 4905 + E_y = 0$$

$$+\downarrow\Sigma M_A = 20E_y - 10(4905) = 0$$

These equations can be solved immediately to get

$$A_x = 0 \text{ N} \qquad E_y = 2453 \text{ N} \qquad A_y = 7357 \text{ N}$$

Now the free-body diagram (Fig. 6-64a) of pin A is drawn and the equilibrium equations are written

$$+\rightarrow\Sigma F_x = A_x + T_{AC} + T_{AB} \cos \theta = 0$$

$$+\uparrow\Sigma F_y = A_y - 4905 + T_{AB} \sin \theta = 0$$

where $A_x = 0$ N, $A_y = 7357$ N, and $\theta = \tan^{-1}\left(\frac{5}{6}\right) = 39.81°$. This gives

$$T_{AB} = -3{,}830 \text{ N} \qquad T_{AC} = 2{,}942 \text{ N}$$

Since one of the three forces applied at pin B is now known, the free-body diagram of pin B (Fig. 6-64b) is drawn next. The equilibrium equations for this pin are

$$+\rightarrow\Sigma F_x = -T_{AB} \cos \theta + T_{BD} + T_{BC} \cos \phi = 0$$

$$+\uparrow\Sigma F_y = -T_{AB} \sin \theta - T_{BC} \sin \phi = 0$$

where $T_{AB} = -3{,}830$ N and $\phi = \tan^{-1}\left(\frac{5}{4}\right) = 51.34°$. The equations are solved to get

$$T_{BC} = 3140 \text{ N} \qquad T_{BD} = -4904 \text{ N}$$

At this point, either pin C, pin D, or pin E could be used since each of these pins has only two forces whose values have not been determined. For this example the free-body diagram of pin D (Fig. 6-64d) will be considered next. Writing the standard horizontal and vertical components of the equilibrium equations gives

$$+\rightarrow\Sigma F_x = -(-4904) - T_{CD} \cos \phi + T_{DE} \cos \theta = 0$$

$$+\uparrow\Sigma F_y = -T_{CD} \sin \phi - T_{DE} \sin \theta = 0$$

Both of these equations contain the unknown forces T_{CD} and T_{DE}. While the solution of this pair of equations is not particularly difficult, the calculations can be simplified if the equilibrium equations are written in terms of components that are along and perpendicular to member CD. This gives

$$+\nearrow \Sigma F_{x'} = -(-4904)\cos\phi - T_{CD} + T_{DE}\cos(\theta + \phi) = 0$$

$$+\nwarrow \Sigma F_{y'} = -4904\sin\phi - T_{DE}\sin(\theta + \phi) = 0$$

The second of these equations can be solved immediately to get

$$T_{DE} = -3830 \text{ N}$$

Then

$$T_{CD} = 3140 \text{ N}$$

Moving to pin C, the free-body diagram (Fig. 6-64c) is drawn and the equilibrium equations are written

$$+\rightarrow\Sigma F_x = -2942 - 3140\cos\phi + 3140\cos\phi + T_{CE} = 0$$

$$+\uparrow\Sigma F_y = 3140\sin\phi - 4905 + 3140\sin\phi = 0$$

The first of these equations gives

$$T_{CE} = 2942 \text{ N}$$

Since the values of all of the forces in the second equation have already been found, this equation reduces to a check of the consistency of the results:

$$3140\sin 51.34° - 4905 + 3140\sin 51.34° = -1.1570$$

The small number -1.1570 is due to rounding all of the intermediate answers to four significant figures. Keeping more accuracy in the intermediate values would reduce the residual and so the solution checks.

Finally, draw the free-body diagram for pin E (Fig. 6-64e) and write the equilibrium equations

$$+\rightarrow\Sigma F_x = -T_{CE} - T_{DE}\cos\theta = 0$$

$$+\uparrow\Sigma F_y = T_{DE}\sin\theta + E_y = 0$$

Again, there are no unknowns left to be solved for. These equations are used simply as a check:

$$-2942 - (-3830)\cos 39.81° = 0.09796$$

$$-3830\sin 39.81° + 2543 = 0.8663$$

and again the solution checks. The required answers (to three significant figures) are

AB, DE:	3830 N (C)	**Ans.**
AC, CE:	2940 N (T)	**Ans.**
BC, CD:	3140 N (T)	**Ans.**
BD:	4900 N (C) ■	**Ans.**

PROBLEMS

For Problems 6-45 through 6-50, use the method of joints to determine the force in each member of the truss. State whether each member is in tension or compression.

Standard Problems

6-45* The truss shown in Fig. P6-45 if $a = 20$ ft and $P = 2000$ lb.

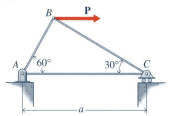

Figure P6-45

6-46* The truss shown in Fig. P6-46 if $a = 6.25$ m, $P = 1$ kN, and $\theta = 75°$.

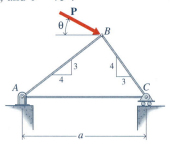

Figure P6-46

6-47* The truss shown in Fig. P6-47 if $a = 5$ ft and $P = 600$ lb.

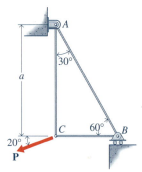

Figure P6-47

6-48* The truss shown in Fig. P6-48 if $a = 2$ m and $P = 5$ kN.

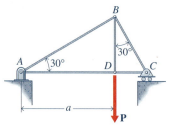

Figure P6-48

6-49 The truss shown in Fig. P6-49 if $a = 8$ ft, $P_1 = 800$ lb, and $P_2 = 600$ lb.

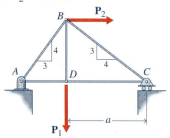

Figure P6-49

6-50 The truss shown in Fig. P6-50 if $a = 4$ m, $P_1 = 3$ kN, and $P_2 = 4$ kN.

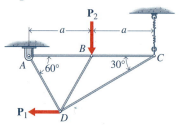

Figure P6-50

6-51* Determine the force in each member of the truss shown in Fig. P6-51.

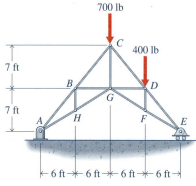

Figure P6-51

6-52* Determine the force in each member of the truss shown in Fig. P6-52. All members are 3 m long.

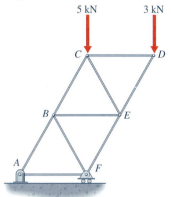

Figure P6-52

6-53 A 4000-lb crate is attached by light, inextensible cables to the truss of Fig. P6-53. Determine the force in each member of the truss.

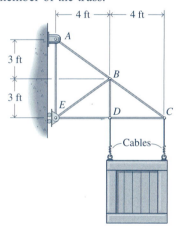

Figure P6-53

6-54 A crate with a mass of 1800 kg is attached by light, inextensible cables to the truss of Fig. P6-54. Determine the force in each member of the truss.

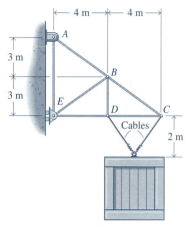

Figure P6-54

6-55* Determine the forces in members CD, CF, and FG of the inverted Mansard truss of Fig. P6-55.

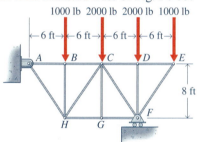

Figure P6-55

6-56* Determine the forces in members BC, CG, and FG of the Warren bridge truss in Fig. P6-56.

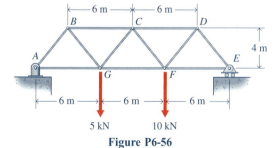

Figure P6-56

Challenging Problems

6-57 The Gambrel truss shown in Fig. P6-57 supports one side of a bridge; an identical truss supports the other side. Floor beams carry vehicle loads to the truss joints. Calculate the forces in members BC, BG, and CG when a truck weighing 7500 lb is stopped in the middle of the bridge as shown. The center of gravity of the truck is midway between the front and rear wheels.

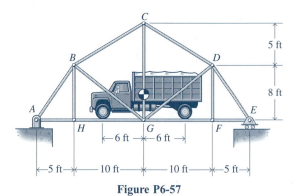

Figure P6-57

6-58 The Gambrel truss shown in Fig. P6-58 supports one side of a bridge; an identical truss supports the other side. Floor beams carry vehicle loads to the truss joints. Calculate the forces in members *BC*, *BG*, and *CG* when a truck having a mass of 3500 kg is stopped in the middle of the bridge as shown. The center of gravity of the truck is 1 m in front of the rear wheels.

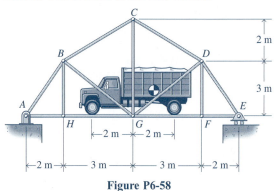

Figure P6-58

6-59* Determine the forces in members *DE*, *DF*, and *EF* of the scissors truss shown in Fig. P6-59.

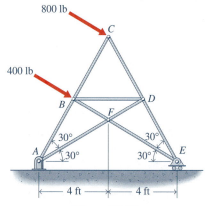

Figure P6-59

6-60* Determine the forces in members *CD*, *CI*, and *CJ* of the truss shown in Fig. P6-60.

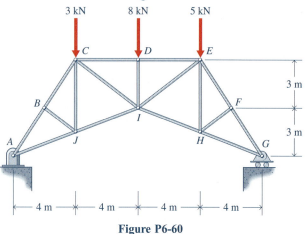

Figure P6-60

6-61 Determine the force in each member of the truss shown in Fig. P6-61.

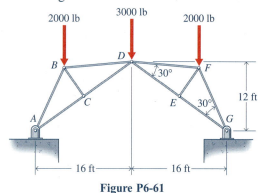

Figure P6-61

6-62 For the simple truss of Fig. P6-62 show that overall equilibrium of the truss is a consequence of equilibrium of all of the pins; hence the equations of overall equilibrium give no new information. (*Hint:* Write equations of equilibrium for each of the pins and eliminate the unknown member forces from these equations.)

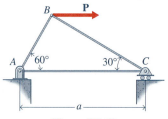

Figure P6-62

6-4-3 Method of Sections

As stated in the Section on the Method of Joints, if an entire truss is in equilibrium, then each and every part of the truss is also in equilibrium. That does not mean, however, that the truss must be broken up into its most elemental parts—individual members and pins. In the method of sections, the truss will be divided up into just two pieces. Each of these pieces is also a rigid body in equilibrium.

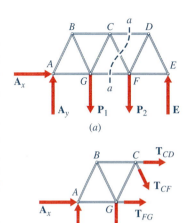

For example, the truss of Fig. 6-65a can be divided into two parts by passing an imaginary section a-a through some of its members. Of course, the section must pass entirely through the truss so that complete free-body diagrams can be drawn for each of the two pieces. Since the whole truss is in equilibrium, the part of the truss to the left of section a-a and the part of the truss to the right of section a-a are both in equilibrium also.

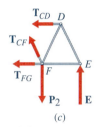

Figure 6-65

Free-body diagrams of the two parts are drawn in Figs. 6-65b and 6-65c, and include the forces on each cut member that was exerted by the other part of the member, which was cut away. Since the members are all straight two-force members, the forces in these members must act along the members as shown. Forces in members that have not been cut are internal to the rigid bodies and are not shown on the free-body diagrams. Thus, in order to determine the force in a member, the section must cut through that member.

As is the case with any rigid body in plane equilibrium, three independent equations of equilibrium can be written for each rigid body. The resulting six equations are sufficient to solve for the six unknowns—the forces in the three cut members and the three support reactions. As with the method of joints, the solution of the equations can be simplified if the support reactions are determined from overall equilibrium before the truss is sectioned. Then the equilibrium equations for either free-body diagram will yield the remaining three unknown forces. In this case, the equilibrium equations for the remaining portion of the truss give no new information; they merely repeat the other equilibrium equations (see Problem 6-82).

If a section cuts through four or more members whose forces are unknown, then the method of sections will not generate enough equations of equilibrium to solve for all of the unknown forces. While it still might be possible to obtain values for one or two of the forces (see Example Problem 6-15), it is usually best to use a section that cuts through no more than three members whose forces are unknown.

It will often happen that a section that cuts no more than three members and that passes through a given member of interest cannot be found. In such a case it may be necessary to draw a section through a nearby member and solve for the forces in it first. Then the method of joints can be used to find the forces in the members next to the cut section, or the truss can be further sectioned to find the force in the member of interest (see Example Problem 6-16).

One of the principal advantages of using the method of sections is that the force in a member near the center of a large truss usually can be determined without first obtaining the forces in the rest of the truss. As a result, the calculation of the force is independent of any errors in other internal forces previously calculated. To find the same force using the method of joints, however, would require that the force in a large number of other members be determined first. Any errors made in the determination of one member force will cause all subsequent forces to be in error as well.

Finally, the method of sections may be used as a spot check when the method of joints or a computer program is used to solve a truss problem with a large number of members. Although it is unlikely that a computer will make an error in its computation, it is quite possible that the input data may be in error. Most often these errors occur when an operator incorrectly enters the coordinates of a joint, incorrectly specifies how the joints are connected, or incorrectly applies a load to the truss. In such cases, the method of sections can be used to check independently the forces in one or two interior members.

Since the members of a truss are straight two-force members, the methods of Sections 4-2 and 4-6 may be used to determine stresses (in the pins and members) and deformations (elongations or contractions) of the members, respectively. The procedure will be illustrated in Example Problem 6-17.

Example Problem 6-13

The roof truss of Fig. 6-66a is composed of 30°–60°–90° right triangles and is loaded as shown. Determine the forces in members CD, CE, and EF.

SOLUTION

A free-body diagram of the entire truss (Fig. 6-66b) is used to solve for the support reactions at A and B. Dimensions a and b needed for these calculations are

$$b = 10/\cos 30° = 11.547 \text{ ft}$$

$$a = 11.547 \tan 30° = 6.667 \text{ ft}$$

Summing moments about A gives

$$+\!\downarrow\!\Sigma M_A = B_y(40) - 600 \sin 30° \, (40) + 600 \cos 30° \, (11.547)$$
$$- 600 \sin 30° \, (33.333) - 800(33.333) - 800(20) - 800(6.667) = 0$$

which can be solved for B_y to get

$$B_y = 1600.0 \text{ lb} = 1600 \text{ lb} \uparrow$$

Summing forces in the x- and y-directions gives

$$+\!\rightarrow\!\Sigma F_x = A_x - 2(600 \cos 30°) = 0$$

$$A_x = 1039.2 \text{ lb} \cong 1039 \text{ lb} \rightarrow$$

$$+\!\uparrow\!\Sigma F_y = A_y + 1600 - 3(800) - 2(600 \sin 30°) = 0$$

$$A_y = 1400.0 \text{ lb} = 1400 \text{ lb} \uparrow$$

Section a-a of Fig. 6-66b passes through members CD, CE, and EF. A free-body diagram for a part of the truss to the right of this section is shown in Fig. 6-66c. The force in member CE can be found by summing forces in the y-direction. Thus,

$$+\!\uparrow\!\Sigma F_y = T_{CE} \sin 30° + 1600 - 800 - 2(600) \sin 30° = 0$$
$$T_{CE} = -400.0 \text{ lb}$$

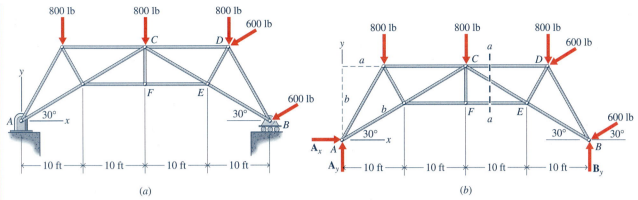

(a)

Figure 6-66

The force in member *EF* can be found by summing moments about point *C* in Fig. 6-66*c*. Thus,

$$+\!\!\!\!\downarrow \Sigma M_C = -T_{EF}(10 \tan 30°) - 800(13.333) - 600 \sin 30° \, (13.333)$$
$$- \, 600 \cos 30° \, (20 \tan 30°) - 600 \sin 30° \, (20) + 1600(20) = 0$$

$$T_{EF} = 923.8 \text{ lb}$$

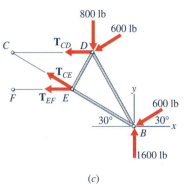

(c)

Finally, the force in member *CD* can be found by summing forces in the *x*-direction or by summing moments about *E*. Summing forces in the *x*-direction and using Fig. 6-66*c* gives

$$+\!\!\rightarrow\!\Sigma F_x = -T_{CD} - T_{EF} - T_{CE} \cos 30° - 2(600 \cos 30°)$$
$$= -T_{CD} - 923.8 + 400 \cos 30° - 2(600 \cos 30°) = 0$$

$$T_{CD} = -1616.6 \text{ lb}$$

Alternatively, summing moments about *E* gives

$$+\!\!\!\!\downarrow \Sigma M_E = T_{CD}(5.774) - 800(3.333) - 600 \sin 30° \, (3.333)$$
$$+ \, 600 \cos 30° \, (5.774) - 600 \sin 30° \, (10) - 600 \cos 30° \, (5.774)$$
$$+ \, 1600(10) = 0$$

$$T_{CD} = -1616.5 \text{ lb}$$

The desired answers are

CD:	1617 lb (C)	**Ans.**
CE:	400 lb (C)	**Ans.**
EF:	924 lb (T)	**Ans.**

Since overall equilibrium was first used to find the support reactions, equilibrium of either the part of the truss to the left of section *a-a* or the part of the truss to the right of section *a-a* can be used to solve for the member forces. Usually the part with the fewest forces acting on it will result in the simplest equations of equilibrium. ■

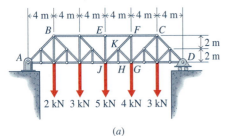

(a)

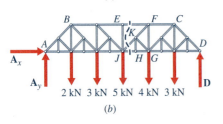

(b)

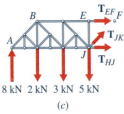

(c)

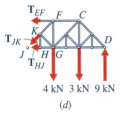

(d)

Figure 6-67

Example Problem 6-14

Use the method of sections to find the forces in members EF, JK, and HJ of the Baltimore truss shown in Fig. 6-67a.

SOLUTION

A free-body diagram of the entire truss (Fig. 6-67b) will be used to solve for the support reactions at A and D. Summing moments about A gives

$$+\!\!\zeta\Sigma M_A = 24D - 4(2) - 8(3) - 12(5) - 16(4) - 20(3) = 0$$

which can be solved for D to get

$$D = 9 \text{ kN} = 9 \text{ kN} \uparrow$$

Then force equilibrium gives

$$+\!\!\rightarrow\Sigma F_x = A_x = 0$$
$$A_x = 0 \text{ kN}$$

$$+\!\!\uparrow\Sigma F_y = A_y - 2 - 3 - 5 - 4 - 3 + D = 0$$
$$A_y = 8 \text{ kN} = 8 \text{ kN} \uparrow$$

A section through members EF, JK, and HJ is shown on Fig. 6-67b. Free-body diagrams of the parts of the truss to the left and right of the section are shown in Figs. 6-67c and 6-67d, respectively. The force in member JK can be found by summing forces in the vertical direction for either free-body diagram. For example, from Fig. 6-67c,

$$+\!\!\uparrow\Sigma F_y = 8 - 2 - 3 - 5 + T_{JK} \sin 45 = 0$$

which gives

$$T_{JK} = 2.828 \text{ kN}$$

Alternatively, from Fig. 6-67d,

$$+\!\!\uparrow\Sigma F_y = 9 - 3 - 4 - T_{JK} \sin 45 = 0$$

which also gives

$$T_{JK} = 2.828 \text{ kN}$$

The force in member HJ can be found by summing moments about point F in Fig. 6-67d:

$$+\!\!\zeta\Sigma M_F = 8(9) - 4(3) - 4T_{HJ} = 0$$

which gives

$$T_{HJ} = 15.00 \text{ kN}$$

Note that this is the same result as that obtained by summing moments about point F in Fig. 6-67c:

$$+\,\curvearrowright\Sigma M_F = 12(2) + 8(3) + 4(5) - 16(8) + 4T_{HJ} = 0$$

which again gives

$$T_{HJ} = 15.00 \text{ kN}$$

Finally, the force in member EF can be found by summing forces in the horizontal direction or by summing moments about J. Choosing the latter method and using Fig. 6-67d gives

$$+\,\curvearrowright\Sigma M_J = 12(9) + 4T_{EF} - 8(3) - 4(4) = 0$$

or

$$T_{EF} = -17.00 \text{ kN}$$

The desired answers are

EF:	17.00 kN (C)	**Ans.**
HJ:	15.00 kN (T)	**Ans.**
JK:	2.83 kN (T) ■	**Ans.**

Example Problem 6-15

Use the method of sections to find the forces in members CD and FG of the truss in Fig. 6-68a.

SOLUTION

Cut a section through members CD, DE, EF, and FG as shown in Fig. 6-68a and draw a free-body diagram for the upper part of the truss (Fig. 6-68b). Summing moments about D

$$+\,\curvearrowright\Sigma M_D = 4(500 \cos 30°) - 6(500 \sin 30°) - 8T_{FG}$$
$$- 12(500 \cos 30°) - 6(500 \sin 30°) = 0$$

gives

$$T_{FG} = -808.0 \text{ lb}$$

Then summing moments about F

$$+\,\curvearrowright\Sigma M_F = 12(500 \cos 30°) - 6(500 \sin 30°) + 8T_{CD}$$
$$- 4(500 \cos 30°) - 6(500 \sin 30°) = 0$$

gives

$$T_{CD} = -58.01 \text{ lb}$$

The consistency of these answers can be checked by summing forces in the y-direction. Thus,

$$+\,\uparrow\Sigma F_y = -2(500 \cos 30°) - (-808.0) - (-58.01) = -0.01540$$

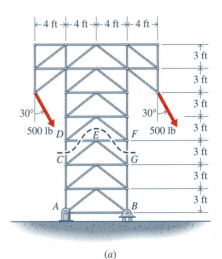

(a)

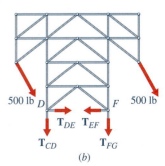

(b)

Figure 6-68

which is within the accuracy of the answers above. The desired answers are

CD:	58.0 lb (C)	**Ans.**
FG:	808 lb (C)	**Ans.**

Note that it was not necessary in this problem to first find the support reactions using overall equilibrium. Also note that neither T_{DE} nor T_{EF} can be found from this section. Either additional sections or the method of joints would be needed to find these forces if they needed to be found. ■

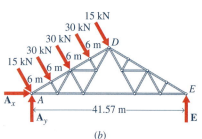

(a)

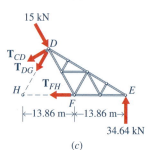

(b)

Example Problem 6-16

Find the forces in members *BC* and *BG* in the Fink truss of Fig. 6-69a. All triangles are either equilateral or 30°–60°–90° right triangles and the loads are all perpendicular to side *ABCD*.

SOLUTION

First find the support reactions by drawing a free-body diagram of the entire truss (Fig. 6-69b) and writing the equilibrium equations:

$$+\!\!\downarrow\!\!\Sigma M_A = 41.57E - 6(30) - 12(30) - 18(30) - 24(15) = 0$$
$$E = 34.64 \text{ kN} = 34.64 \text{ kN} \uparrow$$

$$+\!\!\rightarrow\!\!\Sigma F_x = A_x + (15 + 30 + 30 + 30 + 15)(\sin 30°) = 0$$
$$A_x = -60.00 \text{ kN} = 60.00 \text{ kN} \leftarrow$$

$$+\!\!\uparrow\!\!\Sigma F_y = A_y - (15 + 30 + 30 + 30 + 15)(\cos 30°) + E = 0$$
$$A_y = 69.28 \text{ kN} = 69.28 \text{ kN} \uparrow$$

A section through members *BC*, *BG*, *GH*, and *HF* (section b-b of Fig. 6-69a) would expose four unknown internal forces, two of which are wanted. The equilibrium equations cannot be completely solved until one or more of these forces are determined by some other means. A combination of the method of sections and the method of joints will be used to find the desired forces.

First, a section will be cut through the middle of the truss near the members in which the forces are desired. Section a-a of Fig. 6-69a cuts through members *CD*, *DG*, and *FH*. A free-body diagram of the part of the truss to the right of this section is shown in Fig. 6-69c. Summing moments about point *H*

$$+\!\!\downarrow\!\!\Sigma M_H = 27.72(34.64) - 13.86 \cos 30°(15)$$
$$+ 13.86(T_{CD} \sin 30°) = 0$$

gives

$$T_{CD} = -112.58 \text{ kN}$$

Next, draw a free-body diagram of pin *C* (Fig. 6-69d) and write the equilibrium equations for *x'*- and *y'*-axes along and perpendicular to members *BC* and *CD*:

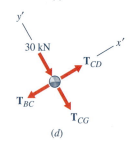

(c)

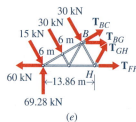

(d)

(e)

Figure 6-69

$$+\nearrow\Sigma F_{x'} = T_{CD} - T_{BC} = 0$$

$$+\nwarrow\Sigma F_{y'} = -30 - T_{CG} = 0$$

which are immediately solved to get

$$T_{BC} = T_{CD} = -112.58 \text{ kN} \quad \text{and} \quad T_{CG} = -30.00 \text{ kN}$$

Finally, cut a section *b-b* (Fig. 6-69*a*) through members *BC*, *BG*, *GH*, and *FH* and draw the free-body diagram of the part of the truss (Fig. 6-69*d*) to the left of the section. Again summing moments about point *H*

$$+\,\!\mathrel{\raisebox{-0.3ex}{$\downarrow$}}\Sigma M_H = -13.86(69.28) + 12(15) + 6(30)$$
$$- 13.86 \sin 30° T_{BC} - 6T_{BG} = 0$$

gives

$$T_{BG} = 29.99 \text{ kN}$$

Thus, the desired answers are

BC:	112.6 kN (C)	**Ans.**
BG:	30.0 kN (T) ◼	**Ans.**

Example Problem 6-17

All members of the inverted Mansard truss of Fig. 6-70*a* are made of structural steel. Determine

(a) The axial stress in member *CH* if the cross-sectional area of this member is 2.50 in.^{2}.

(b) The deformation of member *BH* if the cross-sectional area of this member is 2.50 in.^{2}.

SOLUTION

First find the support reactions by drawing a free-body diagram of the entire truss (Fig. 6-70*b*) and writing the equilibrium equations:

$$+\,\!\mathrel{\raisebox{-0.3ex}{$\downarrow$}}\Sigma M_A = F_y(18) - 10(6) - 8(12) - 4(18) - 2(24) = 0$$

$$F_y = 15.333 \text{ kip} = 15.333 \text{ kip} \uparrow$$

$$+\rightarrow\Sigma F_x = A_x = 0$$
$$A_x = 0 \text{ kip}$$

$$+\uparrow\Sigma F_y = A_y - 10 - 8 - 4 - 2 + F_y = 0$$

$$A_y = 8.667 \text{ kip} = 8.667 \text{ kip} \uparrow$$

(a) Section *a-a* of Fig. 6-70*b* passes through members *BC*, *CH*, and *GH*. A free-body diagram for the part of the truss to the left of this section is shown

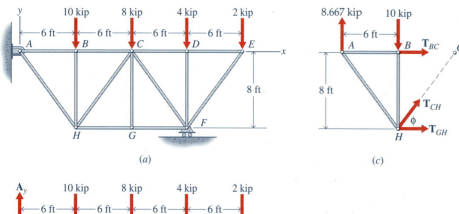

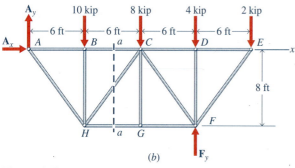

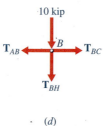

Figure 6-70

in Fig. 6-70c. The force in member *CH* can be found by summing forces in the *y*-direction. Thus,

$$\phi = \tan^{-1}\frac{8}{6} = 53.13°$$

$$+\uparrow\Sigma F_y = T_{CH}\sin 53.13° + 8.667 - 10 = 0$$

$$T_{CH} = 1.6663 \text{ kip} = 1666.3 \text{ lb (T)}$$

The axial stress in member *CH* is given by Eq. (4-2) as

$$\sigma = \frac{T_{CH}}{A_{CH}} = \frac{1666.3}{2.5} = 666.5 \text{ psi} \cong 667 \text{ psi (T)} \qquad \textbf{Ans.}$$

(b) The force in member *BH* can be determined from a free-body diagram of the pin at *B* (Fig. 6-70d) and the equilibrium equation $\Sigma F_y = 0$. Thus,

$$+\uparrow\Sigma F_y = -T_{BH} - 10 = 0$$

$$T_{BH} = -10.00 \text{ kip} = 10,000 \text{ lb (C)}$$

The deformation of member *BH* is given by Eq. (4-20b) as

$$\delta = \frac{T_{BH}L_{BH}}{E_{BH}A_{BH}} = \frac{-10,000(8)(12)}{29,000,000(2.5)} = -0.01324 \text{ in.} \quad \blacksquare \qquad \textbf{Ans.}$$

PROBLEMS

Solve the following problems by the method of sections. Unless directed otherwise, neglect the weight of the members compared with the forces they support. Be sure to indicate whether the members are in tension or compression.

Standard Problems

6-63* Each truss member in Fig. P6-63 is 5 ft long. Find the forces in members CD and EF.

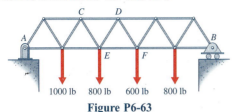

Figure P6-63

6-64* Determine the forces in members CD and EF of the truss (Fig. P6-64) that serves to support the deck of a bridge.

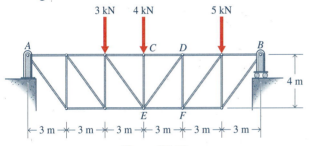

Figure P6-64

6-65 The Howe roof truss of Fig. P6-65 supports the vertical loading shown. Determine the forces in members CD and CK.

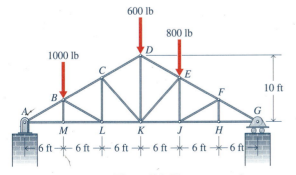

Figure P6-65

6-66 Determine the forces in members BC, BG, and GH of the bridge truss shown in Fig. P6-66.

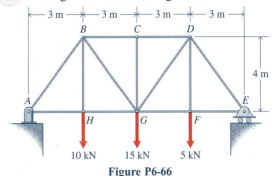

Figure P6-66

6-67* Determine the forces in members CD, DF, and EF of the bridge truss shown in Fig. P6-67.

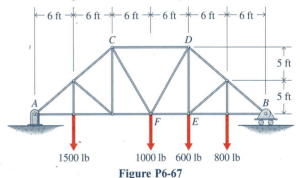

Figure P6-67

6-68* Find the forces in members EJ and HJ of the roof truss shown in Fig. P6-68.

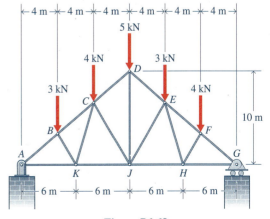

Figure P6-68

6-69 Find the forces in members *BC* and *EF* of the stairs truss of Fig. P6-69.

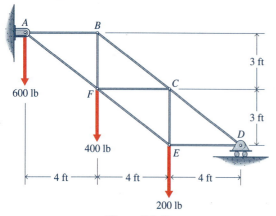

Figure P6-69

6-70 Find the forces in members *CD*, *CE*, and *FG* of the Fink truss of Fig. P6-70.

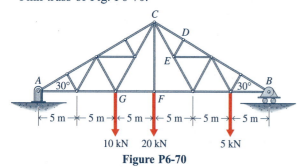

Figure P6-70

6-71* A pin-connected truss is loaded and supported as shown in Fig. P6-71. Determine

(a) The axial stress in member *GF* if it has a cross-sectional area of 2.50 in.²

(b) The minimum cross-sectional area for member *CD* if the axial stress is limited to 7500 psi.

(c) The elongation (or contraction) of member *CF* if it is made of structural steel and has a cross-sectional area of 2.75 in.²

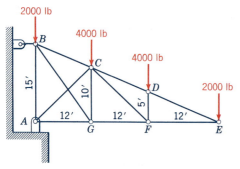

Figure P6-71

6-72* The truss shown in Fig. P6-72 supports an automobile whose mass is 1800 kg. Determine

(a) The axial stress in member *AB* if it has a cross-sectional area of 200 mm².

(b) The minimum cross-sectional area for member *EF* if the axial stress is limited to 50 MPa.

(c) The minimum cross-sectional area for member *AG* if the axial stress is limited to 150 MPa.

(d) The elongation (or contraction) of member *DE* if it is made of structural steel and has a cross-sectional area of 200 mm².

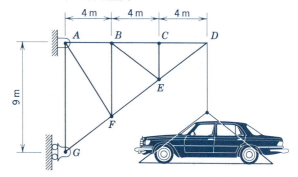

Figure P6-72

6-73 A pin-connected truss is loaded and supported as shown in Fig. P6-73. Determine

(a) The axial stress in member *FG* if it has a cross-sectional area of 7.50 in.²

(b) The minimum cross-sectional area for member *BC* if the axial stress is limited to 10 ksi.

(c) The minimum cross-sectional area for member *BG* if the axial stress is limited to 5 ksi.

(d) The elongation (or contraction) of member *FG* if it is made of structural steel.

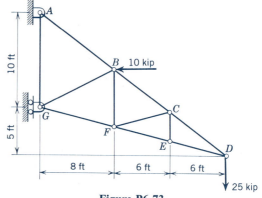

Figure P6-73

6-74 Member *AD* of the timber truss shown in Fig. P6-74 is framed into the 100 × 150-mm bottom chord *ABC* as shown in the insert. Determine the dimension "*a*" that must be used if the average shearing stress parallel to the grain at the ends of chord *ABC* is not to exceed 2.25 MPa.

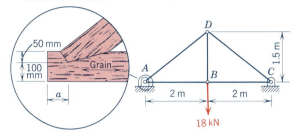

Figure P6-74

Challenging Problems

6-75* Find the forces in members *CD* and *FG* of the K-truss shown in Fig. P6-75. (*Hint:* Use section *a-a*.)

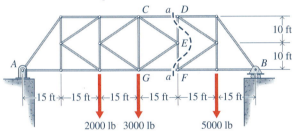

Figure P6-75

6-76* Find the forces in members *AB* and *FG* of the truss shown in Fig. P6-76. (*Hint:* Use section *a-a*.)

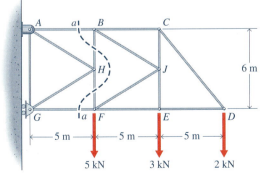

Figure P6-76

6-77 Find the maximum load *P* that can be supported by the truss of Fig. P6-77 without producing a force of more than 2500 lb in member *CD*.

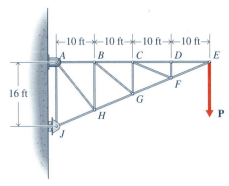

Figure P6-77

6-78 Determine the forces in members *CD*, *DE*, and *DF* of the roof truss shown in Fig. P6-78. Triangle *CDF* is an equilateral triangle and joints *E* and *G* are at the midpoints of their respective sides.

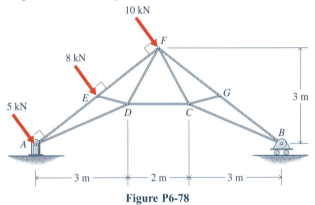

Figure P6-78

6-79* Find the forces in members *DE*, *DJ*, and *JK* of the truss of Fig. P6-79.

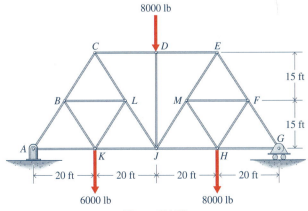

Figure P6-79

6-80* The transmission line truss of Fig. P6-80 supports the load shown. Determine
(a) The forces in members *CD*, *DF*, and *EF*.
(b) The axial stress in member *DG* if it has a cross-sectional area of 300 mm².
(c) The elongation (or contraction) of member *FG* if it is made of structural steel and has a cross-sectional area of 300 mm².

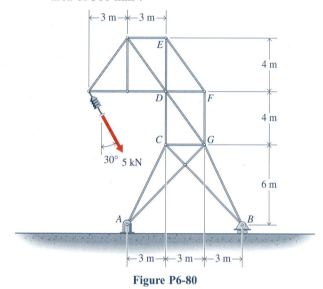

Figure P6-80

6-81 All members of the scissors truss (Fig. P6-81) are made of structural steel and have cross-sectional areas of 2.00 in.². Determine
(a) The forces in members *BC* and *BF*.
(b) The axial stress and elongation (or contraction) of member *CF*.

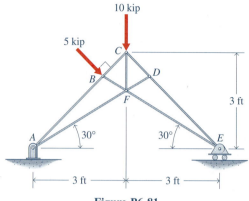

Figure P6-81

6-82 Show that the overall equilibrium of a truss is a consequence of the equilibrium of the two separate parts generated by the method of sections. That is, section the bridge truss of Fig. P6-82 as indicated, and write the equilibrium equations for each piece. Eliminate the member forces from the resulting six equations and show that the result is equivalent to the equilibrium of the whole truss.

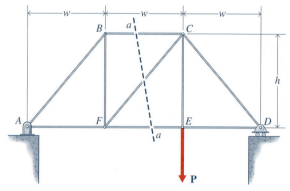

Figure P6-82

6-5 FRAMES AND MACHINES

Structures that contain members other than two-force members are called *frames* or *machines*. While frames and machines may also contain one or more two-force members, they always contain at least one member that is acted upon by forces at more than two points or is acted upon by both forces and moments.

The main distinction between frames and machines is that frames are rigid structures while machines are not. For example, the structure shown in Fig. 6-71*a* is a frame. Since it is a rigid body, three support reactions (Fig. 6-71*b*) are sufficient to fix it in place and overall equilibrium is sufficient to determine the three support reactions.

The structure of Fig. 6-72*a* is a machine although it is also referred to as a nonrigid frame or a linkage. It is nonrigid in the sense that it depends on its

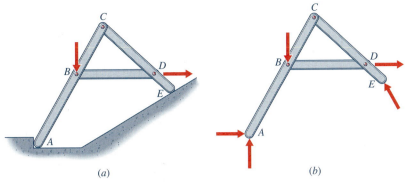

(a)

(b)

Figure 6-71

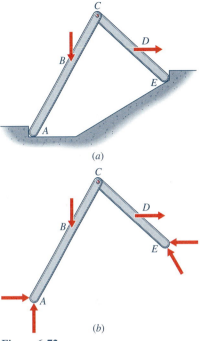

(a)

(b)

Figure 6-72

supports to maintain its shape. The lack of internal rigidity is compensated for by an extra support reaction (Fig. 6-72b). In this case, overall equilibrium is not sufficient to determine all four support reactions. The structure must be taken apart and analyzed even if the only information desired is the support reactions.

In a more specific sense, the term "machine" is usually used to describe devices such as pliers, clamps, nutcrackers, and so on that are used to magnify the effect of forces. In each case, a force (input) is applied to the handle of the device and a much larger force (output) is applied by the device somewhere else. Like nonrigid frames, these machines must be taken apart and analyzed even if the only information desired is the relationship between the input and output forces.

As with the analysis of trusses, the method of solution for frames and machines consists of taking the structures apart, drawing free-body diagrams of each of the components, and writing the equations of equilibrium for each of the free-body diagrams. In the case of trusses, the direction of the force in all members was known and the method of joints reduced to solving a series of particle equilibrium problems. Since some of the members of frames and machines are not two-force members, the directions of the forces on these members are not known. The analysis of frames and machines will consist of solving for the equilibrium of a system of rigid bodies rather than a system of particles.

6-5-1 Frames

The method of analysis for frames can be demonstrated by using the table shown in Fig. 6-73a. None of the members that make up the table are two-force members, so the structure is definitely not a truss. Although the table can be folded up by unhooking the top from the leg, in normal use the table is a stable, rigid structure. Therefore, the table is a frame.

The analysis will be started by first drawing the free-body diagram of the entire table (Fig. 6-73b) for which the equations of equilibrium

$$+\rightarrow \Sigma F_x = A_x = 0$$

$$+\uparrow \Sigma F_y = A_y + D_y - W = 0$$

$$+\downarrow \Sigma M_A = 24D_y - 12W = 0$$

yield the support reactions

$$A_x = 0 \qquad A_y = W/2 \qquad \text{and} \qquad D_y = W/2$$

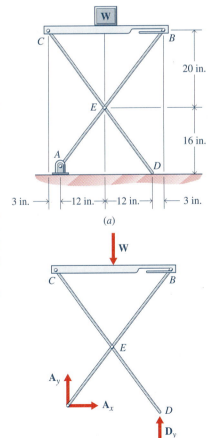

(a)

(b)

Figure 6-73

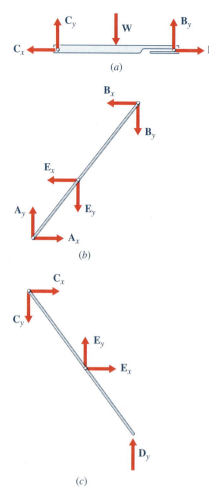

Figure 6-74

Next, the table is taken apart and free-body diagrams of each of its parts are drawn (Fig. 6-74). Since none of the members are two-force members, the directions of the forces at joints B, C, and E are not known—they are not directed along the members! Although the forces may be represented in terms of any convenient components, the free-body diagrams must take into account Newton's third law of action and reaction. That is, when drawing free-body diagrams, the forces exerted by one member on a second must be equal in magnitude and opposite in direction to the forces exerted by the second member on the first. For Fig. 6-74 this is effected by showing the components of the force exerted by member AB on member CD at joint E (Fig. 6-74c) to have equal magnitude and opposite direction to the components of the force exerted by member CD on member AB at joint E (Fig. 6-74b) and similarly for the other joints.

The smooth floor at D can only exert an upward force on the leg, and the force should be shown as such on the free-body diagrams. Similarly, the horizontal component of force exerted by the slot on the leg AB can only act to the left and should be shown as such. If the values of these forces turn out to be negative, either the solution is in error or the table is not in equilibrium.

The proper directions of the other force components are not as clear. While it is easy to guess that the vertical force components B_y and C_y act upward on the table top BC and downward on the legs, it is not as easy to decide whether to draw E_y acting upward or downward on leg AB. At this point it doesn't matter since the frictionless pin connections can support a force in either direction. As in the case of trusses, the direction that a force is shown on one member is unimportant as long as the force is represented consistent with Newton's third law on each part of the structure. If a force component is assumed in the wrong direction, its symbol will simply end up having a negative value. This can be accounted for in the report of final answers as shown in the Example Problems.

Although not all of the members of a frame can be two-force members, it is possible and quite likely that one or more of the members will be two-force members. Take advantage of any such members and show that force as acting in its known direction. But, be sure that all forces are not directed along the members. Perhaps one of the most common mistakes that students make is to treat frames like trusses; that is, to draw all forces as acting along the members and trying to apply the method of joints.

Unlike the analysis of trusses, the free-body diagrams of the pins of a frame are usually not drawn and analyzed separately. In the case of a truss, equilibrium of the members was assured by the two-force member assumption, the direction of the force exerted by each member on a pin was known, and equilibrium of the pins contained all of the useful information of the problem. None of these statements apply to the analysis of frames, however, and it is seldom useful to analyze the equilibrium of the pins separately.

In most cases, it doesn't matter to which member a pin is attached when the structure is taken apart. There are, however, a few special situations in which it does matter:

When a pin connects a support and two or more members, the pin must be assigned to (left attached to) one of the members. The support reactions are applied to the pin on this member.

When a pin connects two or more members and a load is applied to the pin, the pin must be assigned to one of the members. The load is applied to the pin on this member.

Following these simple rules will avoid confusion as to where the loads and support reactions should be applied.

Special care is also warranted when one or more of the members that meet at a joint is a two-force member:

Pins should never be assigned to two-force members.

When all of the members meeting at a joint are two-force members, the pin should be removed and analyzed separately as in the method of joints for a truss.

While these last two "rules" are not strictly necessary, following them will prevent confusion when dealing with two-force members in frames.

Finally, the equations of equilibrium are written for each part of the frame and are solved for the joint forces. There are three independent equations of equilibrium (two force and one moment) for each part; hence, for the table parts of Fig. 6-74 there will be nine equations to solve for the six remaining unknown forces (B_x, B_y, C_x, C_y, E_x, and E_y). Prior solution of the overall equilibrium of the frame for the support reactions will have reduced three of these equations to a check of the consistency of the answers.

6-5-2 Machines

The method described for frames is also used to analyze machines and other nonrigid structures. In each case, the structure is taken apart, free-body diagrams are drawn for each part, and the equations of equilibrium are applied to each free-body diagram. For machines and nonrigid structures, however, the structure must be taken apart and analyzed even if the only information desired is the support reactions or the relationship between the external forces acting upon it.

The method of analysis for machines can be demonstrated by using the simple garlic press shown in Fig. 6-75a. Forces H_1 and H_2 applied to the handles (the input forces) are converted into forces G_1 and G_2 applied to the garlic clove (the output forces). Equilibrium of the entire press only gives that $H_1 = H_2$; it gives no information about the relationship between the input forces and the output forces.

To determine the relationship between the input forces and the output forces, the machine must be taken apart and free-body diagrams drawn for each of its parts, as shown in Fig. 6-75b. Then the sum of moments about B of the top handle gives

$$(a + b)H_1 = b\, G_1$$

or

$$G_1 = \frac{a + b}{b}H_1$$

The ratio of the output and input forces is called the mechanical advantage (M.A.) of the machine:

$$\text{mechanical advantage} = \frac{\text{output force}}{\text{input force}}$$

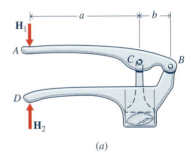

(a)

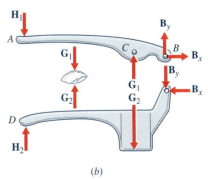

(b)

Figure 6-75

For the garlic press, the mechanical advantage is just

$$\text{M.A.} = \frac{a + b}{b}$$

6-5-3 Stress and Deformation: Frames and Machines

As stated previously, frames and machines contain members other than two-force members. However, frames and machines may also contain one or more two-force members. In Chapter 4, methods were developed to determine stress and deformation for two-force members. For members more complex than two-force members—for example, any of the members of the frame shown in Fig. 6-73, or members AB or DB of the machine of Fig. 6-75—the calculations of stress and deformation are more complicated than the methods presented in Chapter 4. Methods to solve these problems will be developed later in this book. However, certain stress calculations for frames and machines can be made. For example, shearing stress can be found for a pin, and the bearing stress between a pin and a member can be determined. Example Problem 6-19 will illustrate stress and deformation calculations for a frame.

Example Problem 6-18

A bag of potatoes is sitting on the chair of Fig. 6-76a. The force exerted by the potatoes on the frame at one side of the chair is equivalent to horizontal and vertical forces of 24 N and 84 N, respectively, at E and a force of 28 N perpendicular to member BH at G (as shown in the free-body diagram of Fig. 6-76b). Find the forces acting on member BH.

SOLUTION

The equations of equilibrium for the entire chair are

$$+\rightarrow \Sigma F_x = 24 - 28 \cos \theta = 0$$

$$+\uparrow \Sigma F_y = A + B - 84 - 28 \sin \theta = 0$$

$$+\circlearrowleft \Sigma M_B = 0.2(84) - 0.5(24) - 0.4\,A + \left(0.3 + \frac{0.5}{\cos \theta}\right)(28) = 0$$

where $\theta = \tan^{-1}(\frac{3}{5}) = 30.96°$. The first equation is satisfied identically. The remaining two equations give

$$A = 73.82 \text{ N} \qquad B = 24.58 \text{ N}$$

Next the chair is disassembled and free-body diagrams are drawn for each part (Fig. 6-77). For member DF, the equilibrium equations can be written

$$+\rightarrow \Sigma F_x = D_x - F_x + 24 = 0$$

$$+\uparrow \Sigma F_y = F_y + D_y - 84 = 0$$

$$+\circlearrowleft \Sigma M_D = 0.4(84) - 0.5\,F_y = 0$$

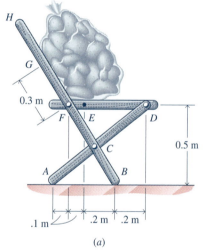

H

G

0.3 m

F *E* *D*

C

0.5 m

A *B*

.1 m .2 m .2 m

(a)

28 N

θ

84 N

24 N

θ

A **B**

(b)

Figure 6-76

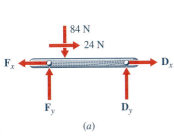

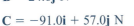

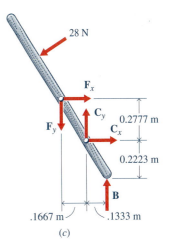

Figure 6-77

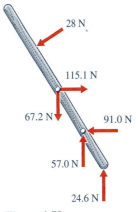

which gives

$$F_y = 67.2 \text{ N} \quad D_y = 16.80 \text{ N} \quad D_x = F_x - 24 \text{ N}$$

Now the equations of equilibrium for member BH are

$$+\rightarrow\Sigma F_x = F_x + C_x - 28 \cos\theta = 0$$

$$+\uparrow\Sigma F_y = 24.58 + C_y - 67.2 - 28 \sin\theta = 0$$

$$+\downharpoonleft\Sigma M_C = \left(0.3 + \frac{0.1667}{\sin\theta}\right)(28) + 0.1333(24.58)$$

$$+ 0.1667(67.2) - 0.2777F_x = 0$$

which have only three unknowns remaining and can be solved to get

$$F_x = 115.1 \text{ N} \quad C_x = -91.0 \text{ N} \quad C_y = 57.0 \text{ N}$$

Then the forces acting on member BH are

$$\mathbf{B} = 24.6\mathbf{j} \text{ N} \qquad \textbf{Ans.}$$

$$\mathbf{C} = -91.0\mathbf{i} + 57.0\mathbf{j} \text{ N} \qquad \textbf{Ans.}$$

$$\mathbf{F} = 115.1\mathbf{i} - 67.2\mathbf{j} \text{ N} \qquad \textbf{Ans.}$$

plus the applied force of 28 N perpendicular to the bar at G. These forces are shown on the "report diagram" of Fig. 6-78. ∎

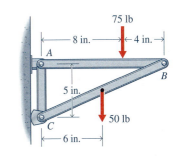

Figure 6-78

Example Problem 6-19

The weight of books on a shelf bracket is equivalent to a vertical force of 75 lb as shown on Fig. 6-79a. In addition, a vertical load of 50 lb is suspended from the middle of the lower brace BC. All members are made of 195-T6 cast aluminum and all pins have $\frac{1}{4}$-in. diameters. Determine

(a) All forces acting on all three members of this frame.

(b) The shearing stress on a cross section of pin B, which is in single shear.

(c) The change in length of member AC as a result of the loads if the member has a $\frac{1}{8} \times \frac{1}{2}$-in. rectangular cross section.

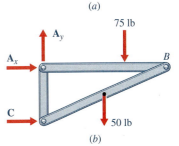

Figure 6-79

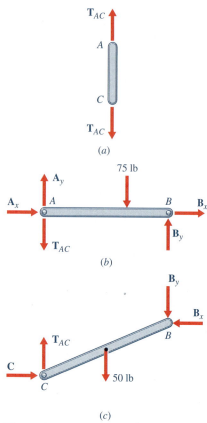

Figure 6-80

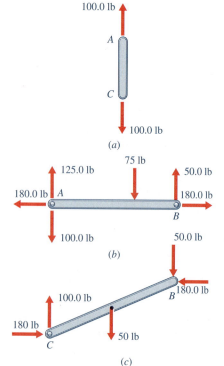

Figure 6-81

SOLUTION

(a) First draw the free-body diagram of the entire shelf bracket as in Fig. 6-79b. The equations of overall equilibrium are

$$+\circlearrowleft\Sigma M_A = 5\,C - 8(75) - 6(50) = 0$$

$$+\rightarrow\Sigma F_x = A_x + C = 0$$

$$+\uparrow\Sigma F_y = A_y - 75 - 50 = 0$$

which are solved to get the support reactions

$$A_x = -180.0 \text{ lb} \quad A_y = 125.0 \text{ lb} \quad C = 180.0 \text{ lb} \qquad \textbf{Ans.}$$

Next, dismember the bracket and draw separate free-body diagrams of each member (Fig. 6-80). Pin A connects a support and two members and member AC is a two-force member; therefore, pin A will be assigned to (left attached to) member AB and likewise pin C will be assigned to member BC. It doesn't matter which member pin B is assigned to since no two-force members are attached to joint B and joint B is neither loaded nor attached to a support. Just to be definite, pin B will be assigned to member AB, as shown in Fig. 6-80b. Then the equations of equilibrium can be written for member AB

$$+\circlearrowleft\Sigma M_B = 4(75) + 12\,T_{AC} - 12(125.0) = 0$$

$$+\circlearrowleft\Sigma M_A = 12\,B_y - 8(75) = 0$$

$$+\rightarrow\Sigma F_x = -180.0 + B_x = 0$$

from which

$$T_{AC} = 100.0 \text{ lb} \quad B_x = 180.0 \text{ lb} \quad B_y = 50.0 \text{ lb} \qquad \textbf{Ans.}$$

It is easily verified that these values also satisfy the equations of equilibrium for other free-body diagrams. These forces are all shown on the "report diagram" of Fig. 6-81.

(b) The force transmitted by a cross section of pin B is

$$B = \sqrt{(B_x)^2 + (B_y)^2} = \sqrt{(180.0)^2 + (50.0)^2} = 186.82 \text{ lb}$$

The shearing stress on a cross section of pin B is determined by using Eq. (4-4):

$$\tau = \frac{B}{A_s} = \frac{186.82}{(\pi/4)(1/4)^2} = 3806 \text{ psi} \cong 3810 \text{ psi} \qquad \textbf{Ans.}$$

(c) The change in length of member AC is determined by using Eq. (4-20b):

$$\delta = \frac{T_{AC}L_{AC}}{E_{AC}A_{AC}} = \frac{100.0(5)}{10{,}300{,}000(1/8)(1/2)} = 0.000777 \text{ in.} \qquad \textbf{Ans.}$$

Alternative Solution

(a) Pins A and C can be assigned to the two-force member AC as shown on the free-body diagram of Fig. 6-82 if care is taken with the representation of the internal forces. The tension T_{AC} shown on Fig. 6-80a represents the force exerted on member AC by the pin C. Since the pin is now part of the member, the force T_{AC} is an internal force and is not shown on the free-body diagram of Fig. 6-82a. The symbols A_x and A_y have already been used to represent the components of the support reaction, so F_{Ax} and F_{Ay} will be used for the components of the forces of action and reaction. Similarly, F_{Cx} and F_{Cy} will be used for the components of the forces of action and reaction at pin C so as not to confuse these forces with the support reaction C.

As stated before, since no supports or two-force members are attached to pin B and no loads are applied at pin B, it doesn't matter to which member pin B is assigned. To illustrate this point, this time pin B will be assigned to member BC. Now the equations of equilibrium for member AB are

$$+\!\downarrow\!\Sigma M_A = 12\,B_y - 8(75) = 0$$

$$+\!\downarrow\!\Sigma M_B = 4(75) - 12\,F_{Ay} = 0$$

$$+\!\rightarrow\!\Sigma F_x = F_{Ax} + B_x = 0$$

from which

$$B_y = 50.0\ \text{lb} \qquad F_{Ay} = 25.0\ \text{lb} \qquad B_x = -F_{Ax}$$

Next, equilibrium of member AC gives the equations

$$+\!\downarrow\!\Sigma M_C = 5\,F_{Ax} - 5(-180.0) = 0$$

$$+\!\downarrow\!\Sigma M_A = 5(180.0) - 5\,F_{Cx} = 0$$

$$+\!\uparrow\!\Sigma F_y = 125.0 - 25.0 - F_{Cy} = 0$$

which gives

$$F_{Ax} = -180.0\ \text{lb} \qquad F_{Cx} = 180.0\ \text{lb} \qquad F_{Cy} = 100.0\ \text{lb}$$

Finally, returning to the free-body diagram of AB, the horizontal component of force equilibrium

$$+\!\rightarrow\!\Sigma F_x = F_{Ax} + B_x = 0$$

gives

$$B_x = 180.0\ \text{lb}$$

These forces are shown on the "report diagram" of Fig. 6-83. Although the forces at B are clearly the same on Fig. 6-81 and Fig. 6-83, the forces at A and C appear to be different on the two diagrams. However, the resultant forces are actually the same on both diagrams; they are only expressed using different components. ∎

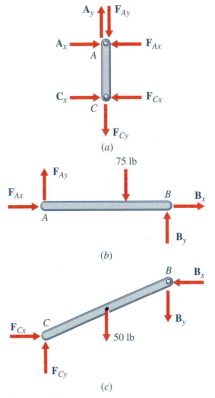

Figure 6-82

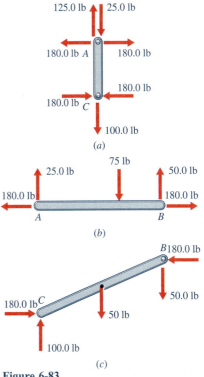

Figure 6-83

■ PROBLEMS

Standard Problems

6-83* In the linkage of Fig. P6-83, $a = 2.0$ ft, $b = 1.5$ ft, $\theta = 30°$, and $P = 40$ lb. Determine all forces acting on member BCD.

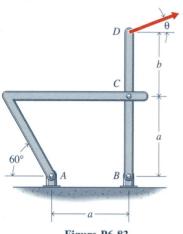

Figure P6-83

6-84* In the linkage of Fig. P6-84, $a = 50$ mm, $P_1 = 500$ N, and $P_2 = 250$ N. Determine all forces acting on member ABC.

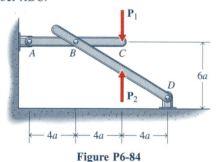

Figure P6-84

6-85 Determine all forces acting on member $ABCD$ of the frame of Fig. P6-85.

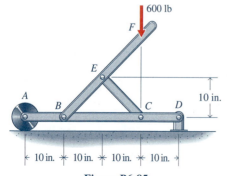

Figure P6-85

6-86 Determine all forces acting on member ABE of the frame of Fig. P6-86.

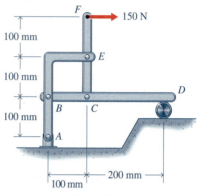

Figure P6-86

6-87* The spring clamp of Fig. P6-87 is used to hold the block H against the floor. The force in the spring is $F = k(\ell - \ell_O)$, where ℓ is the present length of the spring, $\ell_O = 3$ in. is the unstretched length of the spring, and $k = 240$ lb/ft is the spring constant. Determine all forces acting on member ABC of the spring clamp and the force exerted by the spring clamp on the block H.

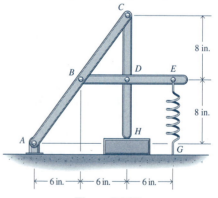

Figure P6-87

6-88* The spring clamp of Fig. P6-88 is used to hold block E in the corner. The force in the spring is $H = k(\ell - \ell_O)$, where ℓ is the present length of the spring, $\ell_O = 15$ mm is the unstretched length of the spring, and $k = 5000$ N/m is the spring constant. Determine all forces acting on member ABC of the spring clamp and the force exerted by the spring clamp on the block E.

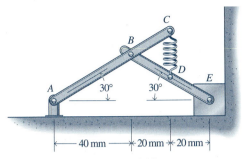

Figure P6-88

6-89 Determine all forces acting on member *ABCD* of the frame of Fig. P6-89.

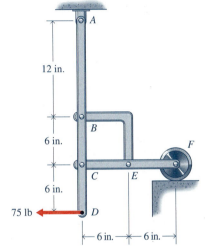

Figure P6-89

6-90 The sand on the tray of Fig. P6-90 can be treated as a triangular distributed load with a maximum intensity of 800 N/m. The wheel at *C* is frictionless. Determine all forces acting on member *ABC*.

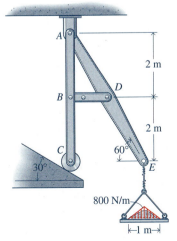

Figure P6-90

6-91* The hoist pulley structure of Fig. P6-91 is rigidly attached to the wall at *C*. A load of sand hangs from the cable that passes around the 1 ft-diameter, frictionless pulley at *D*. The weight of the sand can be treated as a triangular distributed load with a maximum intensity of 70 lb/ft. Determine all forces acting on member *ABC*.

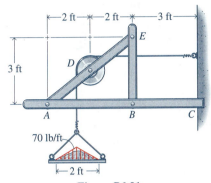

Figure P6-91

6-92* A simple frame supports a 450-kN load as shown in Fig. P6-92. Members *AC* and *BC* have cross-sectional areas of 2000 mm^2. The wheel at *C* is frictionless. Determine

(a) The axial stress in member *AC*.
(b) The change in length of member *BC* if it is made of structural steel.
(c) The shearing stress on a cross section of pin *B* if it has a diameter of 50 mm and is loaded in double shear.

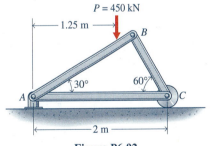

Figure P6-92

6-93 In Fig. P6-93, a cable is attached to the structure at *D*, passes around a 1-ft diameter, frictionless pulley, and is then attached to a 250-lb weight *W*. Determine

(a) All forces acting on member *ABCDE*.
(b) The axial stress in member *AG* if it has a cross-sectional area of 2 in.^{2}.
(c) The shearing stress on a cross section of pin *C* if it has a diameter of $\frac{1}{4}$ in. and is loaded in double shear.

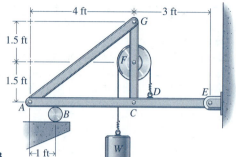

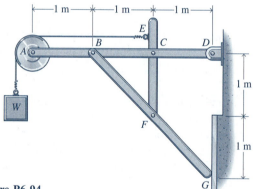

Figure P6-93

6-94 In Fig. P6-94, a cable is attached to the structure at *E*, passes around the 0.8 m–diameter, frictionless pulley at *A*, and then is attached to a 1000-N weight *W*. Determine
(a) All forces acting on member *ABCD*.
(b) The shearing stress on a cross section of pin *F* if it has a diameter of 10 mm and is loaded in single shear.

Figure P6-94

6-95* The fold-down chair of Fig. P6-96 weighs 25 lb and has its center of gravity at *G*. Determine all forces acting on member *ABC*.

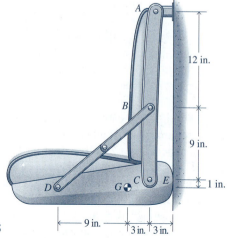

Figure P6-95

6-96* Forces of 5 N are applied to the handles of the paper punch of Fig. P6-95. Determine the force exerted on the paper at *D* and the force exerted on the pin at *B* by handle *ABC*.

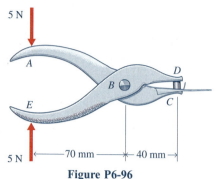

Figure P6-96

Challenging Problems

6-97 Forces of 50 lb are applied to the handles of the bolt cutter of Fig. P6-97. Determine
(a) All forces acting on the handle *ABC*.
(b) The force exerted on the bolt at *E*.
(c) The axial stress in the links at *D* (one on each side) if each has a $\frac{1}{8} \times \frac{3}{4}$-in. rectangular cross section.
(d) The change in length of the links at *D* if they are 4 in. long and made of SAE 4340 heat-treated steel.

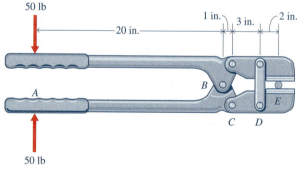

Figure P6-97

6-98* The tower crane of Fig. P6-98 is rigidly attached to the building at *F*. A cable is attached at *D* and passes over small frictionless pulleys at *A* and *E*. The object suspended from *C* weighs 15 kN. Determine
(a) All forces acting on member *ABCD*.
(b) The shearing stress on a cross section of pin *A* if it has a diameter of 15 mm and is loaded in double shear.

(c) The shearing stress on a cross section of pin E if it has a diameter of 25 mm and is loaded in double shear.

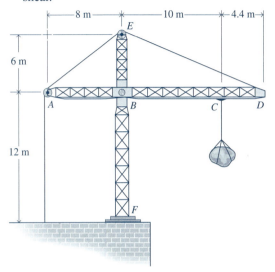

Figure P6-98

6-99* The mechanism of Fig. P6-99 is designed to keep its load level while raising it. A pin on the rim of the 4 ft-diameter pulley fits in a slot on arm ABC. Arms ABC and DE are each 4 feet long and the package being lifted weighs 80 lb. The mechanism is raised by pulling on the rope that is wrapped around the pulley. Determine the force P applied to the rope and all forces acting on the arm ABC when the package has been lifted 4 feet, as shown.

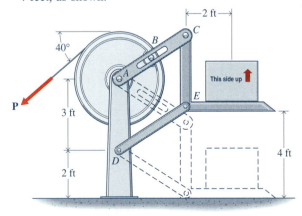

Figure P6-99

6-100 Fig. P6-100 is a simplified sketch of the mechanism used to raise the bucket of a bulldozer. The bucket and its contents weigh 10 kN and have a center of gravity at H. Arm $ABCD$ has a weight of 2 kN and a cen-

ter of gravity at B; arm $DEFG$ has a weight of 1 kN and a center of gravity at E. The weight of the hydraulic cylinders can be ignored. Calculate the force in the horizontal cylinders CJ and EI and all forces acting on arm $DEFG$ for the position shown.

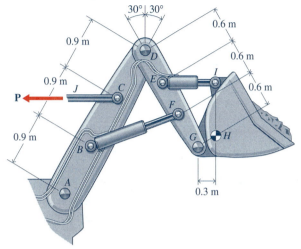

Figure P6-100

6-101* All members of the frame shown in Fig. P6-101 are made of structural steel and have rectangular cross sections 2 in. wide by $\frac{1}{2}$ in. thick. All pins have $\frac{1}{2}$-in. diameters. Determine
(a) The axial stress in member CD.
(b) The shearing stress on a cross section of pin F if it is loaded in double shear.
(c) The change in length of member BE.

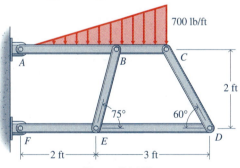

Figure P6-101

6-102* Member BD of the frame shown in Fig. P6-102 is made of structural steel and has a rectangular cross section 50 mm wide by 15 mm thick. All pins have 15-mm diameters. Determine
(a) The axial stress in member BD.
(b) The shearing stress on a cross section of pin C if it is loaded in double shear.
(c) The change in length of member BD.

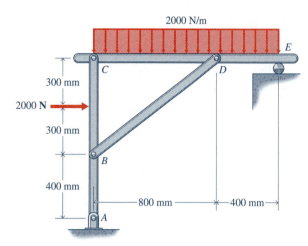

Figure P6-102

6-103 A triangular plate is supported in a vertical plane by a bar and a $\frac{1}{16}$ in.–diameter stainless steel cable as shown in Fig. P6-103. The plate weighs 175 lb. Determine

(a) The axial stress in the cable between pins B and D.
(b) The shearing stress on a cross section of pin A if it has a $\frac{1}{2}$-in. diameter and is loaded in double shear.
(c) The change in length of the cable.

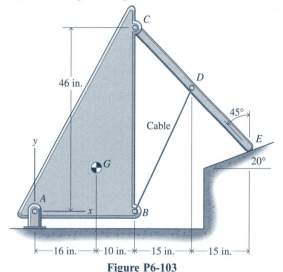

Figure P6-103

6-104 Member BD of the frame shown in Fig. P6-104 is made of 195-T6 cast aluminum and has a rectangular cross section 50 mm wide by 15 mm thick. All pins have 12-mm diameters. Determine

(a) The axial stress in member BD.
(b) The shearing stress on a cross section of pin A if it is loaded in double shear.
(c) The change in length of member BD.

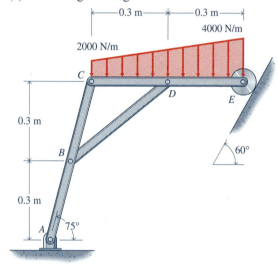

Figure P6-104

6-6 EQUILIBRIUM IN THREE DIMENSIONS

Any three-dimensional system of forces $\mathbf{F}_1$, $\mathbf{F}_2$, ..., $\mathbf{F}_n$ and couples $\mathbf{C}_1$, $\mathbf{C}_2$, ..., $\mathbf{C}_n$ can be replaced by an equivalent system that consists of three mutually perpendicular concurrent forces and three mutually perpendicular couples. The resultant $\mathbf{R}$ of the concurrent force system can be expressed as

$$\mathbf{R} = \Sigma F_x \mathbf{i} + \Sigma F_y \mathbf{j} + \Sigma F_z \mathbf{k} \qquad (6\text{-}10)$$

The resultant **C** of the system of couples can be expressed as

$$\mathbf{C} = \Sigma M_x \mathbf{i} + \Sigma M_y \mathbf{j} + \Sigma M_z \mathbf{k} \qquad (6\text{-}11)$$

The resultant force **R** and the resultant couple **C**, together, constitute the resultant of the general three-dimensional force system. Equations (6-10) and (6-11) indicate that the resultant of the force system may be a force **R**, a couple **C**, or both a force **R** and a couple **C**. Thus, a rigid body subjected to a general three-dimensional system of forces will be in equilibrium if **R** = **C** = **0**, which requires that

$$\Sigma F_x = 0 \qquad \Sigma F_y = 0 \qquad \Sigma F_z = 0$$

and $\qquad\qquad\qquad\qquad\qquad\qquad\qquad\qquad\qquad$ (6-2)

$$\Sigma M_x = 0 \qquad \Sigma M_y = 0 \qquad \Sigma M_z = 0$$

Thus, there are six independent scalar equations of equilibrium for a rigid body subjected to a general three-dimensional system of forces. The first three equations express the requirement that the x-, y-, and z-components of the resultant force **R** must be zero for a body to be in equilibrium. The second three equations express the further equilibrium requirement that there be no couple components acting on the body about any of the coordinate axes or about axes parallel to the coordinate axes. These six equations are both the necessary and sufficient conditions for equilibrium of the body. The six equations are independent since each can be satisfied independently of the others.

The following examples illustrate the solution of three-dimensional problems. A vector analysis is used for Example Problem 6-20, while a scalar analysis is used for Example Problem 6-21. Simple stress and deformation calculations are illustrated in Example Problem 6-20.

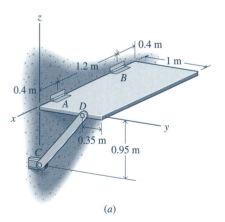

(a)

Example Problem 6-20

The homogeneous door shown in Fig. 6-84a has a mass of 25 kg and is supported in a horizontal position by two hinges and a bar. The hinges have been properly aligned; therefore, they exert only force reactions on the door. Assume that the hinge at B resists any force along the axis of the hinge pins. Determine

(a) The reactions at supports A, B, and D.
(b) The axial stress and deformation in bar CD if it is made of structural steel and has a 25 × 8-mm rectangular cross section.
(c) The shearing stress on a cross section of the pin at D if the diameter of the pin is 6 mm.

SOLUTION

(a) A free-body diagram of the door is shown in Fig. 6-84b. The weight, the hinge reactions at A and B, and the bar reaction at D can be written as

$$\mathbf{W} = -mg\mathbf{k} = -25(9.81)\ \mathbf{k} = -245.3\ \mathbf{k}\ \text{N}$$

$$\mathbf{A} = A_y\,\mathbf{j} + A_z\,\mathbf{k}\ \text{N}$$

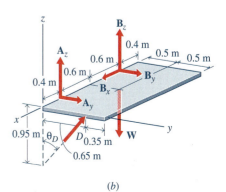

(b)

Figure 6-84

$$\mathbf{B} = B_x\,\mathbf{i} + B_y\,\mathbf{j} + B_z\,\mathbf{k}\ \text{N}$$

$$\mathbf{D} = 0.5647D\,\mathbf{j} + 0.8253D\,\mathbf{k}\ \text{N}$$

For the door to be in equilibrium, Eq. (6-10) must be satisfied. Thus,

$$
\begin{aligned}
\mathbf{R} &= \Sigma F_x\,\mathbf{i} + \Sigma F_y\,\mathbf{j} + \Sigma F_z\,\mathbf{k} \\
&= (B_x)\,\mathbf{i} + (A_y + B_y + 0.5647D)\,\mathbf{j} \\
&\quad + (A_z + B_z + 0.8253D - 245.3)\,\mathbf{k} = \mathbf{0}
\end{aligned}
\tag{a}
$$

Summing moments about B to eliminate the maximum number of unknowns

$$
\begin{aligned}
\mathbf{C} &= \Sigma C_x\mathbf{i} + \Sigma C_y\mathbf{j} + \Sigma C_z\mathbf{k} \\
&= (\mathbf{r}_{A/B} \times \mathbf{A}) + (\mathbf{r}_{W/B} \times \mathbf{W}) + (\mathbf{r}_{D/B} \times \mathbf{D})
\end{aligned}
$$

$$
=
\begin{vmatrix}
\mathbf{i} & \mathbf{j} & \mathbf{k} \\
1.2 & 0 & 0 \\
0 & A_y & A_z
\end{vmatrix}
+
\begin{vmatrix}
\mathbf{i} & \mathbf{j} & \mathbf{k} \\
0.6 & 0.5 & 0 \\
0 & 0 & -245.3
\end{vmatrix}
+
\begin{vmatrix}
\mathbf{i} & \mathbf{j} & \mathbf{k} \\
1.6 & 0.65 & 0 \\
0 & 0.5647D & 0.8253D
\end{vmatrix}
$$

$$
\begin{aligned}
&= (-122.65 + 0.5364D)\,\mathbf{i} + (-1.2A_z + 147.18 - 1.3205D)\,\mathbf{j} \\
&\quad + (1.2A_y + 0.9035D)\,\mathbf{k} = \mathbf{0}
\end{aligned}
\tag{b}
$$

Equating the coefficients of $\mathbf{i}$, $\mathbf{j}$, and $\mathbf{k}$ to zero in Eqs. (a) and (b) and solving yields

$$
\begin{aligned}
B_x &= 0 & B_x &= 0 \\
A_y + B_y + 0.5647D &= 0 & B_y &= 43.03\ \text{N} \\
A_z + B_z + 0.8253D - 245.3 &= 0 & B_z &= 185.56\ \text{N} \\
-122.65 + 0.5364D &= 0 & D &= 228.65\ \text{N} \\
-1.2A_z + 147.18 - 1.3205D &= 0 & A_z &= -128.96\ \text{N} \\
1.2A_y + 0.9035D &= 0 & A_y &= -172.15\ \text{N}
\end{aligned}
$$

The reactions at hinges A and B and the force exerted by the bar at D are

$$\mathbf{A} = -172.2\,\mathbf{j} - 129.0\,\mathbf{k}\ \text{N} \qquad |\mathbf{A}| = 215\ \text{N} \qquad \textbf{Ans.}$$

$$\mathbf{B} = 43.0\,\mathbf{j} + 185.6\,\mathbf{k}\ \text{N} \qquad |\mathbf{B}| = 190.5\ \text{N} \qquad \textbf{Ans.}$$

$$\mathbf{D} = 129.1\,\mathbf{j} + 188.7\,\mathbf{k}\ \text{N} \qquad |\mathbf{D}| = 229\ \text{N} \qquad \textbf{Ans.}$$

A scalar analysis can frequently be used in three-dimensional problems when a single unknown is the only quantity required. For example, the force exerted by the bar at support D can be determined by summing moments about the axis of the hinges. Thus,

$$
\begin{aligned}
+ \curvearrowright \Sigma M_x &= D(0.95 \sin \theta_D) - W(0.50) \\
&= D(0.95 \sin 34.38°) - 245.3(0.50) = 0
\end{aligned}
$$

$$D = 228.6 \cong 229\ \text{N}$$

(b) The axial stress and deformation in bar CD is given by Eqs. (4-2) and (4-20b) as

$$\sigma = \frac{F_{CD}}{A_{CD}} = \frac{228.6}{0.025(0.008)} = 1.143(10^6) \text{ N/m}^2 = 1.143 \text{ MPa} \qquad \textbf{Ans.}$$

$$L_{CD} = \sqrt{(0.65)^2 + (0.95)^2} = 1.151 \text{ m}$$

$$E_{CD} = 200 \text{ GPa (see Table B-18)}$$

$$\delta = \frac{F_{CD}L_{CD}}{E_{CD}A_{CD}} = \frac{228.6(1.151)}{200(10^9)(0.025)(0.008)}$$

$$= 0.00658(10^{-3}) \text{ m} = 0.00658 \text{ mm} \qquad \textbf{Ans.}$$

(c) The shearing stress on a cross section of the pin at D, which is in single shear, is given by Eq. (4-4) as

$$\tau = \frac{F_{CD}}{A_s} = \frac{228.6}{(\pi/4)(0.006)^2} = 8.085(10^6) \text{ N/m}^2 \cong 8.09 \text{ MPa} \quad \blacksquare \qquad \textbf{Ans.}$$

Example Problem 6-21

A homogeneous flat plate which weighs 500 lb is supported by a shaft AB and a cable C as shown in Fig. 6-85a. The bearing at A is a ball bearing and the bearing at B is a thrust bearing. The bearings are properly aligned; therefore, they transmit only force components. When the three forces shown in Fig. 6-85a are applied to the plate, determine the reactions at bearings A and B and the tension in cable C. Use a scalar analysis for the solution but express the final results in Cartesian vector form.

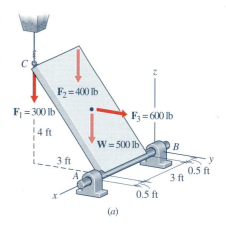

(a)

SOLUTION

A free-body diagram for the plate is shown in Fig. 6-85b. The bearing at A is a ball bearing; therefore, this reaction is represented by two force components $\mathbf{A}_y$ and $\mathbf{A}_z$. The bearing at B is a thrust bearing; therefore, this reaction is represented by three force components $\mathbf{B}_x$, $\mathbf{B}_y$, and $\mathbf{B}_z$. The force in the cable is represented by tension $\mathbf{T}_C$. The plate is subjected to a general, three-dimensional system of forces; therefore, six equilibrium equations [Eqs. (6-2)] are available for determining the six unknown force components indicated on the free-body diagram. Summing forces in the positive x-, y-, and z-directions and summing moments about the x-, y-, and z-axes in accordance with the right-hand rule yields

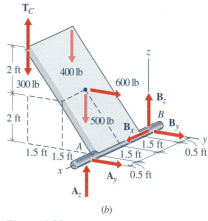

(b)

$$\Sigma F_x = B_x = 0 \qquad\qquad\qquad B_x = 0 \qquad \text{(a)}$$

$$\Sigma F_y = A_y + B_y + 600 = 0 \qquad\qquad A_y + B_y = -600 \text{ lb} \quad \text{(b)}$$

$$\Sigma F_z = A_z + B_z + T_C - 300 - 400 - 500 = 0 \qquad A_z + B_z + T_C = 1200 \text{ lb} \quad \text{(c)}$$

$$\Sigma M_x = 300(3) + 400(3) - 600(2) + 500(1.5) - T_C(3) = 0 \qquad T_C = 550 \text{ lb} \quad \text{(d)}$$

Figure 6-85

$$\Sigma M_y = 300(3.5) + 400(0.5) + 500(2) - T_C(3.5) - A_z(4) = 0$$
$$4A_z + 3.5T_C = 2250 \text{ ft} \cdot \text{lb} \quad (e)$$

$$\Sigma M_z = A_y(4) + 600(2) = 0 \qquad\qquad A_y = -300 \text{ lb} \quad (f)$$

From Eqs. (b) and (f):

$$A_y + B_y = -300 + B_y = -600 \text{ lb}$$
$$B_y = -300 \text{ lb}$$

From Eqs. (d) and (e):

$$4A_z + 3.5T_C = 4A_z + 3.5(550) = 2250 \text{ ft} \cdot \text{lb}$$
$$A_z = 81.3 \text{ lb}$$

From Eq. (d):

$$A_z + B_z + T_C = 81.3 + B_z + 550 = 1200 \text{ lb}$$
$$B_z = 568.7 \text{ lb}$$

$$\mathbf{A} = -300 \mathbf{j} + 81.3 \mathbf{k} \text{ lb} \qquad\qquad \textbf{Ans.}$$

$$\mathbf{B} = -300 \mathbf{j} + 569 \mathbf{k} \text{ lb} \qquad\qquad \textbf{Ans.}$$

$$\mathbf{T}_C = 550 \mathbf{k} \text{ lb} \ \blacksquare \qquad\qquad \textbf{Ans.}$$

PROBLEMS

Standard Problems

6-105* Determine the reaction at support A of the pipe system shown in Fig. P6-105 when the force applied to the pipe wrench is 50 lb.

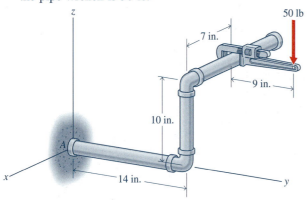

Figure P6-105

6-106* The bent bar shown in Fig. P6-106 is supported with two brackets that exert only force reactions on the bar. The end of the bar at C rests against smooth horizontal and vertical surfaces. Determine the reactions at supports A, B, and C.

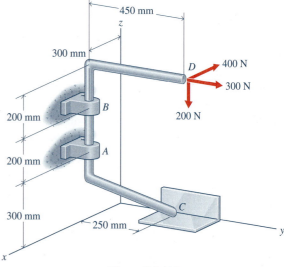

Figure P6-106

6-107 A shaft is loaded through a pulley and a lever (Fig. P6-107) that are fixed to the shaft. Friction between the belt and pulley prevents slipping of the belt. Determine the force **P** required for equilibrium and the reactions at supports A and B. The support at A is a ball bearing and the support at B is a thrust bearing. The bearings exert only force reactions on the shaft.

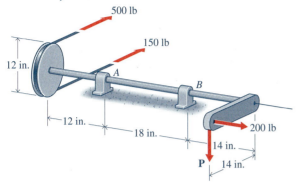

Figure P6-107

6-108 A bar is supported by a ball-and-socket joint and two cables as shown in Fig. P6-108. Determine
(a) The tensions in the two cables.
(b) The reaction at support A (the ball-and-socket joint).
(c) The axial stress and deformation in each of the cables if they are made of steel ($E = 210$ GPa) and have 8-mm diameters.

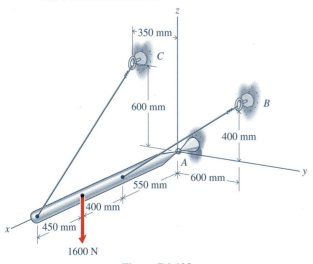

Figure P6-108

6-109* A beam is supported by a ball-and-socket joint and two cables as shown in Fig. P6-109. Determine
(a) The tensions in the two cables.
(b) The reaction at support A (the ball-and-socket joint).

(c) The axial stress and deformation in each of the cables if they are made of steel ($E = 30,000$ ksi) and have $\frac{1}{4}$-in. diameters.

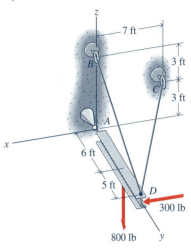

Figure P6-109

6-110* The rectangular plate of uniform thickness shown in Fig. P6-110 has a mass of 500 kg. Determine the tensions in the three cables supporting the plate.

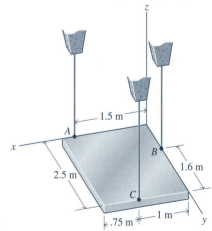

Figure P6-110

Challenging Problems

6-111 The plate shown in Fig. P6-111 weighs 200 lb and is supported in a horizontal position by a hinge and a cable. Determine
(a) The reactions at the hinge and the tension in the cable.
(b) The axial stress and deformation in cable BC if it is made of steel ($E = 30,000$ ksi) and has a $\frac{1}{8}$-in. diameter.

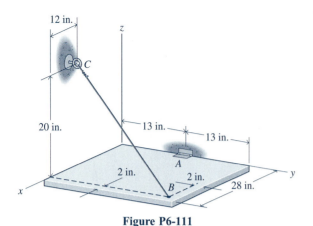

Figure P6-111

(b) The axial stress and deformation in cable CD if it is made of steel ($E = 210$ GPa) and has an 8-mm diameter.

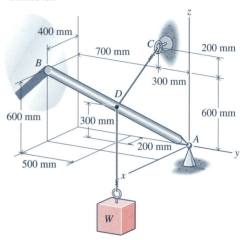

Figure P6-112

6-112   The block W shown in Fig. P6-112 has a mass of 250 kg. Bar AB rests against a smooth vertical wall at end B and is supported at end A with a ball-and-socket joint. The two cables are attached to a point on the bar midway between the ends. Determine

(a) The reactions at supports A and B and the tension in cable CD.

6-7 FRICTION

Thus far in this book, two surfaces in contact have been either perfectly smooth or perfectly rough. A perfectly smooth or frictionless surface that exerts only normal forces on bodies is a useful model for a large number of problems. However, frictional forces that act tangent to the surface are present in the contact between all real surfaces. Whether the friction forces are large or small depends on a number of things, including the types of materials in contact.

Friction forces act to oppose the tendency of contacting surfaces to slip relative to one another and can be either good or bad. Without friction it would be impossible to walk, ride a bicycle, drive a car, or pick up objects. In some machine applications, such as brakes and belt drives, a design consideration is to maximize the friction forces. In many other machine applications, however, friction is undesirable. Friction causes energy loss and wears down sliding surfaces in contact. In these cases, a primary design consideration is to minimize the friction forces.

Two main types of friction are encountered in engineering practice—dry friction and fluid friction. As its name suggests, dry friction (or Coulomb friction) describes the tangential component of the contact force that exists when two dry surfaces slide or tend to slide relative to one another. Coulomb friction is the primary concern of this section and will be studied in considerable detail.

Fluid friction describes the tangential component of the contact force that exists between adjacent layers in a fluid that are moving at different velocities relative to each other, as in the thin layer of oil between bearing surfaces. The tangential forces developed between the adjacent fluid layers oppose the relative motion and are dependent primarily on the relative velocity between the two layers. Fluid friction is one of the primary concerns in the study of fluid mechanics and is more properly treated in a course in fluid mechanics.

6-7-1 Characteristics of Coulomb Friction

To investigate the behavior of frictional forces, consider a simple experiment consisting of a solid block of mass m resting on a rough horizontal surface and acted on by a horizontal force **P** (Fig. 6-86). Equilibrium of the block requires a force having both a normal component ($N = mg$) and a horizontal or friction component ($F = P$) acting on the contact surface. When the horizontal force **P** is zero, no horizontal component of force F will be required for equilibrium, and friction will not exert a force on the surface. As the force **P** increases, the friction force F also increases, as shown in the graph of Fig. 6-87. The friction force cannot increase indefinitely, however, and it eventually reaches its maximum value F_{max}. The maximum value is also called the *limiting value of static friction*. In other words, F_{max} is the maximum value of the friction force for which static equilibrium exists.

The condition in which the friction force is at its maximum value is called the *condition of impending motion*. That is, if **P** increases beyond the point $P = F_{max}$ then friction can no longer supply the amount of force necessary for equilibrium. Therefore the block will no longer be in equilibrium, but will start moving in the direction of the force **P**. When the block starts moving, the friction force F normally decreases in magnitude by about 20 to 25 percent. From this point on, the block will slide with increasing speed while the friction force (the kinetic friction force) remains approximately constant (see Fig. 6-87).

Repeating the experiment with a second block of mass $m_2 = 2m$ would produce similar results, but the limiting force at which the block starts to move would be observed to be twice as great. Repeating the experiment with two blocks of different sizes but the same mass and material would yield the same limiting force for both blocks. That is, the value of the limiting friction force is proportional to the normal force at the contact surface:

$$F_{max} = \mu_s N \qquad (6\text{-}12)$$

The constant of proportionality μ_s is called the *coefficient of static friction* and it depends on the types of material in contact. However, μ_s is observed to be relatively independent of both the normal force and the area of contact.

To understand how μ_s can be independent of the area of contact, one must consider where the friction forces come from. It is generally believed that dry friction results primarily from the roughness between two surfaces and to a lesser extent from attraction between the molecules of the two surfaces[2]. Even two surfaces that are considered to be smooth have small irregularities, as the (idealized) enlargement of the contact surfaces of Fig. 6-88 shows. Therefore, contact

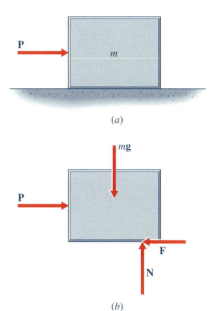

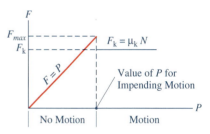

Figure 6-86

Figure 6-87

---

[2]D. Halliday, R. Resnick, and J. Walker, *Fundamentals of Physics,* 4th ed., John Wiley and Sons, Inc., New York, 1993, pp. 133–134.

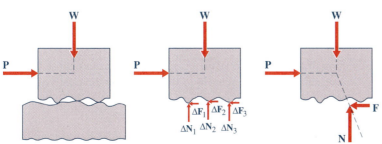

Figure 6-88

between the block and the surface takes place only over a few very small areas of the common surface. The friction force F is then the resultant of the tangential components of the forces acting at each of these tiny contact points, just as the normal force N is the resultant of the normal components of the forces acting at each of the contact points. (Normal and tangential here are relative to the overall contact plane and not the individual tiny contact points.) Increasing the number of contact points just means that the normal and frictional components at each point are proportionately smaller, but their sums F and N do not change. Therefore, μ_s will not change either.

Before going on it must be noted that the normal force N is the resultant of a distributed force. If the force distribution is uniform, N will act at the center of the surface. In general, however, N will not act at the center of the surface or through the center of the body. Since the actual distribution of forces is generally not known, the location of N must be determined using moment equilibrium.

In many friction problems, it is easily recognized that a body is in no danger of tipping over. Since only force equilibrium is considered, it doesn't matter where the normal force is drawn on the free-body diagram. Even so, the student should not get into the habit of showing the force as always acting through the center of the body.

It must also be noted that friction is a resistive force. That is, friction always acts to oppose motion; it never acts to create motion. Equation (6-12) only tells how much friction is available $F_{\text{avail}} = F_{\text{max}} = \mu_s N$ to prevent motion. No matter how much frictional force is available on a surface, however, the frictional force actually exerted is never greater than that required to satisfy the equations of equilibrium

$$F \le \mu_s N \qquad (6\text{-}13)$$

where the equality holds only at the point of impending motion.

Once the block starts to slip relative to the surface, the friction force will decrease to

$$F = \mu_k N \qquad (6\text{-}14)$$

where μ_k is called the *coefficient of kinetic friction*. This coefficient is again independent of the normal force and is also independent of the speed of the relative motion—at least for low speeds. Of course the presence of any oil or moisture on the surface can change the problem from one of dry friction, in which the friction force is independent of the speed of the body, to one of fluid friction, in which the friction force is a function of the speed. At higher speeds, the effect of lubrication by an intervening fluid film (such as oil, surface moisture, or even air) can become appreciable.

These results are summarized as follows.

Coulomb's Laws of Dry Friction

The direction of the friction force on a surface is such as to oppose the tendency of one surface to slide relative to the other. It is the relative motion or the impending relative motion of one body relative to another that is important.

The friction force is never greater than just sufficient to prevent motion.

For the static equilibrium case in which the two surfaces are stationary with respect to one another, the normal and tangential components of the contact force satisfy

$$F \le \mu_s N$$

where the equality holds for the case of impending motion in which the contacting surfaces are on the verge of sliding relative to each other.

For the case where two contacting surfaces are sliding over each other, the normal and tangential components of the contact force satisfy

$$F = \mu_k N$$

where $\mu_k < \mu_s$.

TABLE 6-3 Coefficients of Friction for Common Surfaces

Materials	μ_s
Metal on metal	0.5
Wood on metal	0.5
Wood on wood	0.4
Leather on wood	0.4
Rubber on metal	0.5
Rubber on wood	0.5
Rubber on pavement	0.7

Of course, Coulomb's laws apply only when N is positive, that is, when the surfaces are being pressed together.

The values of μ_s and μ_k must be determined experimentally for each pair of contacting surfaces. Average values of μ_s for various types of materials are given in Table 6-3. Reported values for μ_s vary widely, however, depending on the exact nature of the contacting surfaces. Values for μ_k are generally 20 to 25 percent less than those reported for μ_s. Values for μ_k are not listed in Table 6-3 since the uncertainty in μ_k is much larger (by as much as 100 percent in some cases) than the difference between it and μ_s.

Because of the uncertainty in the values of μ_s and μ_k, Table 6-3 should only be used to get a rough estimate of the magnitude of the friction forces. If more accurate values are needed, experiments should be performed using the actual surfaces being studied.

Since the coefficients of friction are the ratio of two forces, they are dimensionless quantities and can be used with either the SI or U.S. customary system of units.

In many simple friction problems it is convenient to use the resultant of the friction and normal forces rather than their separate components. In the case of the block in Fig. 6-89, this leaves only three forces acting on the block. Moment equilibrium is established simply by making the three forces concurrent, and only force equilibrium need be considered. Since N and F are rectangular components of the resultant **R**, the magnitude and direction of the resultant are given by

$$R = \sqrt{N^2 + F^2} \quad \text{and} \quad \tan \theta = F/N \quad (6\text{-}15)$$

At the point of impending motion, Eq. (6-15) becomes

$$R = \sqrt{N^2 + F^2_{\max}} = \sqrt{N^2 + (\mu_s N)^2} = N\sqrt{1 + \mu_s^2} \quad (6\text{-}16a)$$

$$\tan \phi_s = \frac{F}{N} = \frac{\mu_s N}{N} = \mu_s \quad (6\text{-}16b)$$

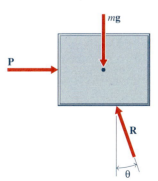

Figure 6-89

where ϕ_s, the angle between the resultant and the normal to the surface, is called the *angle of static friction*. For a given normal force N, if the friction force is less than the maximum ($F < \mu_s N$), then the angle of the resultant will be less than the angle of static friction: $\theta < \phi_s$. In no case can the angle of the resultant

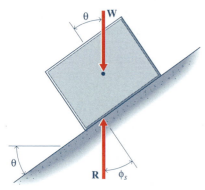

Figure 6-90

θ be greater than ϕ_s for a body in equilibrium. A similar relation is obtained in the case of kinetic friction:

$$\tan \phi_k = \mu_k$$

where ϕ_k is called the *angle of kinetic friction*.

When a block sits on an inclined surface and is acted upon only by gravity, the resultant of the normal and friction force must be collinear with the weight, as shown in Fig. 6-90. The angle between the resultant and the normal to the surface can never be greater than the angle of static friction ϕ_s. Thus, the steepest inclination θ for which the block will be in equilibrium is equal to the angle of static friction. This angle is called the *angle of repose*.

There are three typical types of friction problems encountered in engineering analysis:

1. **Impending motion is not assumed.** This first case is just the type of equilibrium problem solved in previous chapters of this book. The required friction force F_{req} ($\neq \mu_s N$) and the normal force N are drawn on the free-body diagram and are determined using force equilibrium. The normal force should not be drawn as acting through the center of the body. Instead, its location is determined using moment equilibrium. The amount of friction required for equilibrium F_{req} and the location of the normal force are then checked against their maximum values. Three possibilities exist.

   a. If the amount of friction required for equilibrium F_{req} is smaller than or equal to the maximum amount of friction available $F_{avail} = F_{max} = \mu_s N$ and the location of the normal force is on the body, then the body is in equilibrium. In this case, the actual friction force supplied by the surface is

   $$F_{actual} = F_{req} < F_{avail} = \mu_s N$$

   That is, the surface supplies just enough friction force (resistance) to keep the body from moving.

   b. If the amount of friction required for equilibrium F_{req} is smaller than or equal to the maximum amount of friction available $F_{avail} = F_{max} = \mu_s N$ but the location of the normal force is not somewhere on the body, then the body is not in equilibrium and will tip over.

   c. If the amount of friction required for equilibrium F_{req} is greater than the maximum amount of friction available $F_{avail} = F_{max} = \mu_s N$, then the body is not in equilibrium and will slide. In this case, the actual friction force supplied by the surface is the kinetic friction force $F_{actual} = \mu_k N$.

2. **Impending slipping is known to occur at all surfaces of contact.** Since impending slipping is known to occur at all surfaces of contact, the magnitude of the friction forces can be shown as $\mu_s N$ on the free-body diagrams. The equations of equilibrium are then written and

   a. If all of the applied forces are given but μ_s is unknown, the equations of equilibrium can be solved for N and μ_s. This μ_s is the smallest coefficient of static friction for which the body will be in equilibrium.

   b. If the coefficient of static friction is given but one of the applied forces is unknown, the equations of equilibrium can be solved for N and the unknown applied force.

3. **Impending motion is known to exist but the type of motion or surface of slip is not known.** Since it is not known whether the body tips or slips, the free-body diagrams must be drawn as in Case 1 above. That is, the friction forces must not be shown as $\mu_s N$ on the free-body diagrams. At this point the three equations of equilibrium will contain more than three unknowns. Assumptions must be made about the type of motion that is about to occur until the number of equations equals the number of unknowns. The equations are then solved for the remaining unknowns and checked against the assumptions made about slipping or tipping. If F_{req} comes out greater than $F_{avail} = \mu_s N$ at some surface or if the location of the normal force is not on the body, then the assumptions must be changed and the problem solved again.

In both of the last two cases (2b and 3), the friction force is treated as if it is a known force. Care must be taken to be sure that it opposes the tendency of the other forces to cause motion. The result will not include a negative sign to indicate that the friction is in the wrong direction. If the direction of the friction force is drawn incorrectly, incorrect answers will result. The direction is determined by pretending for a moment that friction does not exist. Then apply the friction in such a direction as to oppose the motion that would occur in the absence of friction.

 ## Example Problem 6-22

The pickup truck of Fig. 6-91 is traveling at a constant speed of 50 mi/h and is carrying a 60-lb box in the back. The box projects 1 ft above the cab of the pickup. The wind resistance on the box can be approximated as a uniformly distributed force of 25 lb/ft on the exposed edge of the box. Calculate the minimum coefficient of friction required to keep the box from sliding on the bed of the pickup. Also determine whether or not the box will tip over.

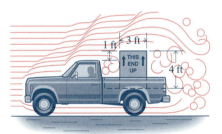

Figure 6-91

SOLUTION

The free-body diagram of the box is drawn in Fig. 6-92. The equilibrium equations for the box

$$+\rightarrow \Sigma F_x = 25(1) - C_f = 0$$

$$+\uparrow \Sigma F_y = C_n - 60 = 0$$

$$+\!\!\downarrow\!\Sigma M_A = 60(1.5) - 25(1)(3.5) - C_n x_C = 0$$

are solved to get

$$C_f = 25.0 \text{ lb} \qquad C_n = 60.0 \text{ lb}$$

$$x_C = 0.042 \text{ ft} = 0.500 \text{ in.}$$

Thus the required coefficient of friction is

$$\mu_s = \frac{C_f}{C_n} = \frac{25.0}{60.0} = 0.417$$ **Ans.**

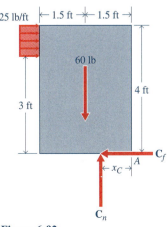

and since x_C is positive, the box will not tip.   **Ans.**   **Figure 6-92**

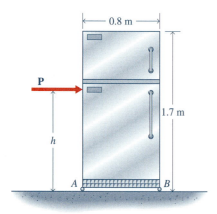

Figure 6-93

Note: The friction and normal force must act on the box; thus, x_C must be a number between 0 and 3 ft. If the solution had given x_C to be negative, then no normal force on the bottom of the box could satisfy moment equilibrium and the box would tip over. ∎

Example Problem 6-23

The wheels of the refrigerator of Fig. 6-93 are stuck and will not turn. The refrigerator weighs 600 N. Assume a coefficient of friction between the wheels and the floor of 0.6 and determine the force necessary to cause the refrigerator to just begin to move (impending motion). Also determine the maximum height h at which the force can be applied without causing the refrigerator to tip over.

SOLUTION

The free-body diagram of the refrigerator is drawn in Fig. 6-94. The equilibrium equations

$$+\rightarrow \Sigma F_x = P - A_f - B_f = 0$$

$$+\uparrow \Sigma F_y = A_n + B_n - 600 = 0$$

$$+\downarrow \Sigma M_B = 600(0.4) - P(h) - A_n(0.8) = 0$$

give

$$A_n + B_n = 600$$

$$P = A_f + B_f = 0.6(A_n + B_n) = 360 \text{ N} \qquad \textbf{Ans.}$$

and

$$h = \frac{240 - 0.8 A_n}{360}$$

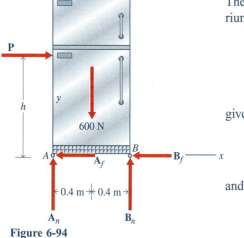

Figure 6-94

When $h = 0$, $A_n = B_n = 300$ N, and the wheels share the load of the weight equally. As h increases, A_n gets smaller. However, the force at A cannot be negative, so

$$h < \frac{240}{360} = 0.667 \text{ m} \qquad \textbf{Ans.}$$

Note that at the point of impending tipping, none of the weight is carried by the wheel at A; it has all been shifted to the wheel at B. ∎

Example Problem 6-24

A 20-lb homogeneous box has tipped and is resting against a 40-lb homogeneous box (Fig. 6-95). The coefficient of friction between box A and the floor is 0.7; between box B and the floor, 0.4. Treat the contact surface between the two boxes as smooth and determine whether the boxes are in equilibrium.

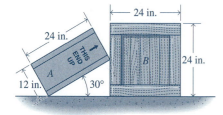

Figure 6-95

SOLUTION

The free-body diagram of box A is drawn in Fig. 6-96a. The equilibrium equations

$$+\rightarrow \Sigma F_x = A_f - C = 0$$

$$+\uparrow \Sigma F_y = A_n - 20 = 0$$

$$+\!\!\downarrow \Sigma M_D = C(12) - 20(7.392) = 0$$

are solved to get

$$C = 12.32 \text{ lb} \qquad A_n = 20.00 \text{ lb} \qquad A_f = 12.32 \text{ lb}$$

The friction force available at this surface is

$$F_{max} = \mu_s A_n = 0.7(20.00) = 14.00 \text{ lb}$$

Since the friction force required (12.32 lb) is less than the friction force available (14.00 lb), box A is in equilibrium.

The free-body diagram of box B is drawn in Fig. 6-96b. The equilibrium equations for this box

$$+\rightarrow \Sigma F_x = C - B_f = 12.32 - B_f = 0$$

$$+\uparrow \Sigma F_y = B_n - 40 = 0$$

$$+\!\!\downarrow \Sigma M_E = B_n(x_B) - 12.23(12) - 40(12) = 0$$

give

$$B_f = 12.32 \text{ lb} \qquad B_n = 40.0 \text{ lb} \qquad x_B = 15.669 \text{ in.}$$

The friction force available at this surface is

$$F_{max} = \mu_s B_n = 0.4(40.0) = 16.00 \text{ lb}$$

Again the friction force available (16.00 lb) is greater the friction force required (12.32 lb) and box B is also in equilibrium.

Thus, both boxes are in equilibrium.   **Ans.**

Note that while the normal force B_n does not act at the center of the crate, it does act on the bottom of the crate. ■

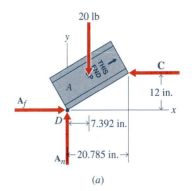

(a)

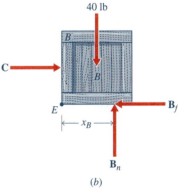

(b)

Figure 6-96

Example Problem 6-25

A 500-N weight (Fig. 6-97a) is supported by a lightweight rope wrapped around the inner cylinder of a 1000-N homogeneous drum. The coefficient of friction between the weightless brake arm and the outer cylinder of the drum is 0.40. The force P just prevents motion of the weight. The pins at A and D have diameters of 8 mm. Determine the shearing stress on a cross section of each pin if both pins are in double shear.

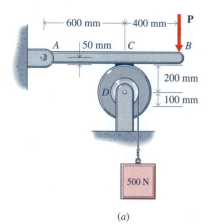

(a)

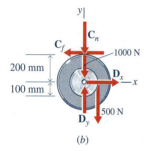

(b)

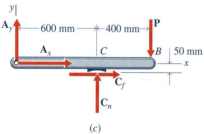

(c)

Figure 6-97

SOLUTION

A free-body diagram of the drum is shown in Fig. 6-97b. The friction force C_f prevents rotation of the drum; therefore, from the equilibrium equation

$$+\!\downarrow\!\Sigma M_D = C_f(200) - 500(100) = 0 \qquad\qquad C_f = 250 \text{ N}$$

And since motion is impending,

$$C_f = \mu_s C_n = 0.40\,C_n = 250\ N \qquad\qquad C_n = 625\ N$$

From the remaining equilibrium equations:

$$+\!\rightarrow\!\Sigma F_x = D_x - C_f = D_x - 250 = 0 \qquad\qquad D_x = 250 \text{ N}$$

$$\begin{aligned}+\!\uparrow\!\Sigma F_y &= D_y - C_n - 500 - 1000 \\ &= D_y - 625 - 500 - 1000 = 0 \qquad D_y = 2125 \text{ N}\end{aligned}$$

$$F_D = \sqrt{(D_x)^2 + (D_y)^2} = \sqrt{(250)^2 + (2125)^2} = 2140 \text{ N}$$

The shearing stress on a cross section of the pin at D, which is in double shear, is then given by Eq. (4-4) as

$$\tau = \frac{F_D}{A_s} = \frac{2140}{2(\pi/4)(0.008)^2} = 21.29(10^6) \text{ N/m}^2 \cong 21.3 \text{ MPa} \qquad \textbf{Ans.}$$

A free-body diagram of the brake arm is shown in Fig. 6-97c. From the equilibrium equation

$$\begin{aligned}+\!\downarrow\!\Sigma M_A &= C_f(50) + C_n(600) - P(1000) \\ &= 250(50) + 625(600) - P(1000) = 0 \qquad P = 387.5 \text{ N}\end{aligned}$$

From the remaining equilibrium equations

$$+\!\rightarrow\!\Sigma F_x = A_x + C_f = A_x + 250 = 0 \qquad A_x = -250 \text{ N} = 250 \text{ N} \leftarrow$$

$$\begin{aligned}+\!\uparrow\!\Sigma F_y &= A_y + C_n - P \\ &= A_y + 625 - 387.5 = 0 \qquad A_y = -237.5 \text{ N} = 237.5 \text{ N} \downarrow\end{aligned}$$

$$F_A = \sqrt{(A_x)^2 + (A_y)^2} = \sqrt{(250)^2 + (237.5)^2} = 344.8 \text{ N}$$

The shearing stress on a cross section of the pin at A, which is in double shear, is then given by Eq. (4-4) as

$$\tau = \frac{F_A}{A_s} = \frac{344.8}{2(\pi/4)(0.008)^2} = 3.430(10^6) \text{ N/m}^2 \cong 3.43 \text{ MPa} \quad\blacksquare \qquad \textbf{Ans.}$$

PROBLEMS

Standard Problems

6-113* A light, inextensible cord passes over a friction-less pulley (Fig. P6-113). One end of the cord is attached to a block; a force P is applied to the other end. Block A weighs 600 lb and block B weighs 1000 lb. The coefficient of friction between the blocks is 0.33, while the coefficient of friction between B and the floor is 0.25. Determine
(a) Whether the system would be in equilibrium for $P = 400$ lb.
(b) The maximum force P for which the system is in equilibrium.

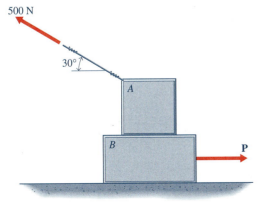

Figure P6-113

6-114* A force of 500 N is applied to the end of the cord that is attached to block A of Fig. P6-114. Block A weighs 2000 N and block B weighs 4000 N. The coefficient of friction between the blocks is 0.3, while the coefficient of friction between B and the floor is 0.2.

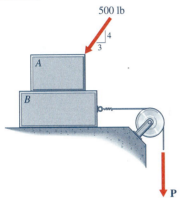

Figure P6-114

Determine
(a) Whether the system would be in equilibrium for $P = 2500$ N.
(b) The maximum force P for which the system is in equilibrium.

6-115 The block in Fig. P6-115 weighs 500 lb and the coefficient of friction between the block and the inclined plane is 0.2. Determine
(a) Whether the system would be in equilibrium for $P = 400$ lb.
(b) The minimum force P to prevent motion.
(c) The maximum force P for which the system is in equilibrium.

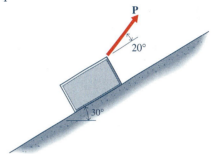

Figure P6-115

6-116 A 20-kg triangular block sits on top of a 10-kg rectangular block (Fig. P6-116). The coefficient of friction is 0.40 at all surfaces. Determine the maximum horizontal force P for which motion will not occur.

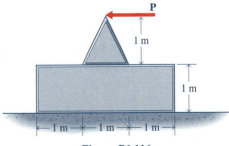

Figure P6-116

6-117* Workers are pulling a 400-lb crate up an incline as shown in Fig. P6-117. The coefficient of friction between the crate and the surface is 0.2 and the rope on which the workers are pulling is horizontal.
(a) Determine the force P that the workers must exert to start sliding the crate up the incline.

(b) If one of the workers lets go of the rope for a moment, determine the minimum force **P** the other workers must exert to keep the crate from sliding back down the incline.

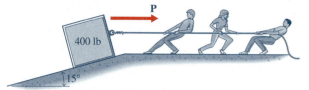

Figure P6-117

6-118* A 75-kg man starts climbing a 5-m long ladder leaning against a wall (Fig. P6-118). The coefficient of friction is 0.25 at both surfaces. Neglect the weight of the ladder and determine how far up the ladder the man can climb before the ladder starts to slip.

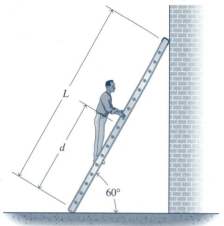

Figure P6-118

6-119 The 200-lb crate of Fig. P6-119 is being moved by a rope that passes over a smooth pulley. The coefficient of friction between the crate and the floor is 0.30.

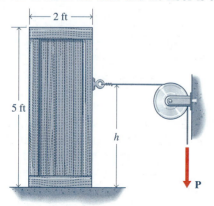

Figure P6-119

(a) Assume that $h = 4$ ft and determine the force **P** necessary to produce impending motion.
(b) Determine the value of h for which impending motion by slipping and by tipping would occur simultaneously if the pulley is adjusted so that the rope attached to the crate stays horizontal.

6-120 A 50-kg uniform plank rests on rough supports at A and B (Fig. P6-120). The coefficient of friction is 0.60 at both surfaces. If a man weighing 800 N pulls on the rope with a force of $P = 400$ N, determine
(a) The minimum and maximum angles θ_{min} and θ_{max} for which the system will be in equilibrium.
(b) The minimum coefficient of friction that must exist between the man's shoes and the ground for each of the cases in part (a).

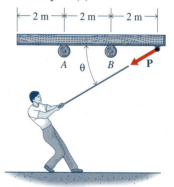

Figure P6-120

6-121* A 120-lb girl is walking up a 48-lb uniform beam (Fig. P6-121). The coefficient of friction is 0.2 at all surfaces. Determine how far up the beam the girl can walk before the beam starts to slip.

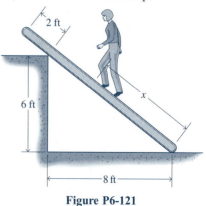

Figure P6-121

6-122* A worker is lifting a 35-kg uniform beam with a force that is perpendicular to the beam (Fig. P6-122).

The beam is 5 m long and the coefficient of friction between the beam and the ground is 0.2. Determine the height h at which the beam will begin to slip.

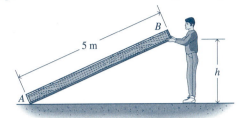

Figure P6-122

6-123 A uniform plank 8 ft long is balanced on a corner (Fig. P6-123) using a horizontal force of 75 lb. If the plank weighs 45 lb, determine
(a) The angle θ for equilibrium.
(b) The minimum coefficient of friction for which the plank is in equilibrium.

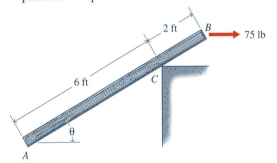

Figure P6-123

6-124 A block of mass M rests on a 50-kg block, which in turn rests on an inclined plane (Fig. P6-124). The coefficient of friction between the blocks is 0.40; between the 50-kg block and the incline, 0.30. The pulley is frictionless and the weight of the cord can be neglected. Determine the minimum mass M_{min} necessary to prevent slipping.

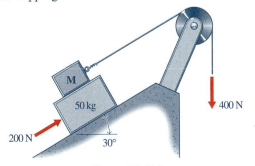

Figure P6-124

6-125* In Fig. P6-125 box A weighs 10 lb and rests on an inclined surface, while box B weighs 20 lb and rests on a level surface. The coefficient of friction between box A and the surface is 0.45; between box B and the surface, 0.5. The pulleys are frictionless. Determine the maximum weight of box C such that no motion will occur. Motion of which box would occur first?

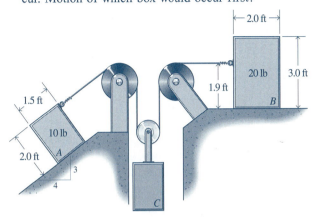

Figure P6-125

6-126* The three blocks of Fig. P6-126 are connected by light, inextensible cords. The left block weighs 160 N and the center block weighs 300 N. The coefficient of friction between the 300-N block and the floor is 0.20; the two pulleys are frictionless.
(a) Assume that $h = 1.8$ m and determine the minimum and maximum weight of the right block, W_{min} and W_{max}, such that motion does not occur.
(b) Determine the value of h for which impending motion by slipping and by tipping would occur simultaneously.

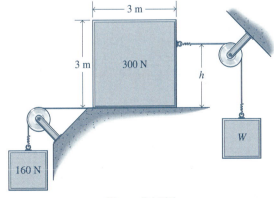

Figure P6-126

6-127 The block A in Fig. P6-127 is pressed against the incline by a 65-lb force on the handle. The pin is frictionless and the weight of the handle can be neglected.

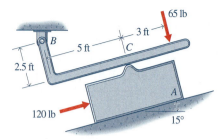

Figure P6-127

The coefficient of friction is 0.2 at all surfaces. Determine

(a) The minimum weight, W_A, necessary to prevent slipping.
(b) The shearing stress on a cross section of the pin at B when motion is impending if the pin has a $\frac{1}{8}$-in. diameter and is in double shear.

6-128 The 25-kg block in Fig. P6-128 is held against the wall by the brake arm. The coefficient of friction between the wall and the block is 0.20; between the block and the brake arm, 0.50. Neglect the weight of the brake arm. Determine

(a) Whether the system would be in equilibrium for $P = 230$ N.
(b) The minimum force P for which the system would be in equilibrium.
(c) The maximum force P for which the system would be in equilibrium.
(d) The shearing stress on a cross section of the pin at A when impending motion of the block is downward if the pin has a 6-mm diameter and is in double shear.

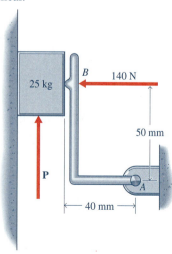

Figure P6-128

Challenging Problems

6-129* A 250-lb weight is suspended from a lightweight rope wrapped around the inner cylinder of a drum (Fig. P6-129). A brake arm is pressed against the outer cylinder of the drum by a hydraulic cylinder. The coefficient of friction between the brake arm and the drum is 0.40. Determine

(a) The smallest force in the hydraulic cylinder necessary to prevent motion.
(b) The shearing stress on a cross section of the pin at A when motion is impending if the drum weighs 75 lb and the pin has a $\frac{1}{2}$-in. diameter and is in double shear.

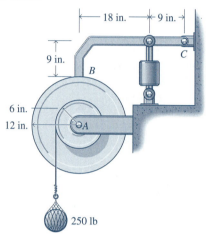

Figure P6-129

6-130* The rods of Fig. P6-130 are lightweight and all pins are frictionless. The coefficient of friction between the 40-kg slider block and the floor is 0.40.

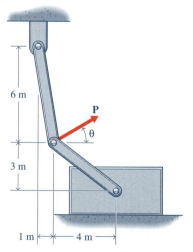

Figure P6-130

(a) Assume that the force **P** is horizontal ($\theta = 0$) and determine the maximum force **P** for which motion does not occur.
(b) Determine the angle θ that gives the absolute greatest force **P** for which motion does not occur.

6-131* A lightweight rope is wrapped around a 100-lb drum, passes over a frictionless pulley, and is attached to a weight W (Fig. P6-131). The coefficient of friction between the drum and each surface is 0.50. Determine
(a) The maximum amount of weight that can be supported by this arrangement.
(b) The shearing stress on a cross section of the pin supporting the pulley when the maximum weight is applied if the pin has a $\frac{1}{4}$-in. diameter and is in double shear.

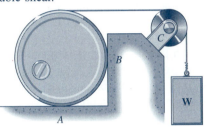

Figure P6-131

6-132 A 100-kg cylinder has a 300-mm diameter and rests against a wall and a plate (Fig. P6-132). The coefficient of friction is 0.30 at both surfaces. The plate rests on frictionless rollers. Neglect the weight of the plate and determine the minimum force **P** necessary to move the plate.

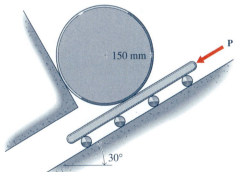

Figure P6-132

6-133 When a drawer is pulled by only one of the handles, it tends to twist and rub as shown (highly exaggerated) in Fig. P6-133, which is a top view of the drawer. The weight of the drawer and its contents is 2 lb and is uniformly distributed. The coefficient of friction between the sides of the drawer and the sides of the dresser is 0.6; between the bottom of the drawer and the side rails the drawer rides on, $\mu_s = 0.1$. Determine the minimum amount of force necessary to pull the drawer out.

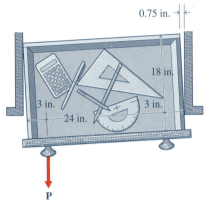

Figure P6-133

6-134 Three identical cylinders are stacked as shown in Fig. P6-134. The cylinders each weigh 100 N and are 200 mm in diameter and the coefficient of friction is $\mu_s = 0.40$ at all surfaces. Determine the maximum force **P** that the cylinders can support without moving.

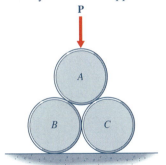

Figure P6-134

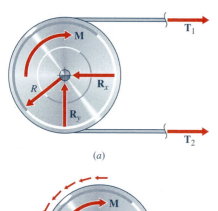

(a)

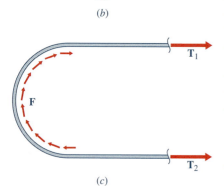

(b)

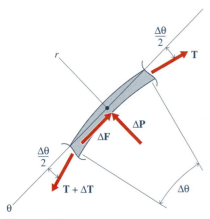

(c)

Figure 6-98

6-7-2 Flat Belts and V-Belts

Many types of power machinery rely on belt drives to transfer power from one piece of equipment to another. Without friction, the belts would slip on their pulleys and no power transfer would be possible. Maximum torque is applied to the pulley when the belt is at the point of impending slip.

Although the analysis presented is for flat belts, it also applies to any shape belt as well as circular ropes as long as the only contact between the belt and the pulley is on the bottom surface of the belt. This section ends with a brief discussion of V-belts which indicates the kind of modifications required when friction acts on the sides of the belt instead of the bottom.

Figure 6-98a shows a flat belt passing over a circular drum. The tensions in the belt on either side of the drum are T_1 and T_2 and the bearing reaction is **R**. Friction in the bearing is neglected for this analysis, but a torque **M** is applied to the drum to keep it from rotating. If there is no friction between the belt and the drum, $T_1 = T_2$ and no torque is required for moment equilibrium and **M** = **0**. If there is friction between the belt and the drum, however, then the two tensions need not be equal and a torque $M = R(T_2 - T_1)$ is needed to satisfy moment equilibrium. Assuming that $T_2 > T_1$, this means that friction forces must exert a counterclockwise moment on the drum (Fig. 6-98b) and the drum will exert an opposite frictional resistance on the belt (Fig. 6-98c). Because the friction force depends on the normal force and the normal force varies around the drum, care must be taken in adding up the total frictional resistance.

The free-body diagram (Fig. 6-99) of a small segment of the belt includes the friction force Δ**F** and the normal force Δ**P**. The tension in the belt increases from **T** on one side of the segment to **T** + Δ**T** on the other side. Equilibrium in the radial direction gives

$$+\nwarrow \Sigma F_r = \Delta P - T \sin(\Delta\theta/2) - (T + \Delta T)\sin(\Delta\theta/2) = 0$$

or

$$\Delta P = 2T \sin(\Delta\theta/2) + \Delta T \sin(\Delta\theta/2) \qquad (a)$$

while equilibrium in the circumferential (θ-) direction gives

$$+\swarrow \Sigma F_\theta = (T + \Delta T)\cos(\Delta\theta/2) - T \cos(\Delta\theta/2) - \Delta F = 0$$

or

$$\Delta T \cos(\Delta\theta/2) = \Delta F \qquad (b)$$

In the limit as $\Delta\theta \to 0$ the normal force ΔP on the small segment of the belt must vanish according to Eq. (a). But when the normal force vanishes ($\Delta P \to 0$), there can be no friction on the belt either ($\Delta F \to 0$). Therefore, the change in tension across the small segment of the belt must also vanish ($\Delta T \to 0$) in the limit as $\Delta\theta \to 0$ according to Eq. (b).

Assuming that slip is impending gives $\Delta F = \mu_s \Delta P$ and Eqs. (a) and (b) can be combined to give

$$\Delta T \cos(\Delta\theta/2) = \mu_s 2T \sin(\Delta\theta/2) + \mu_s \Delta T \sin(\Delta\theta/2) \qquad (c)$$

which after dividing through by $\Delta\theta$ is

Figure 6-99

$$\frac{\Delta T}{\Delta \theta} \cos (\Delta\theta/2) = \mu_s T \frac{\sin (\Delta\theta/2)}{(\Delta\theta/2)} + \frac{\mu_s \Delta T \sin (\Delta\theta/2)}{2(\Delta\theta/2)} \qquad (d)$$

Finally, taking the limit as $\Delta\theta \to 0$ and recalling that $\lim\limits_{\Delta\theta\to 0} \dfrac{\Delta T}{\Delta\theta} = \dfrac{dT}{d\theta}$;

$$\lim_{x\to 0} \cos x = 1 \qquad \lim_{x\to 0} \sin x = x \qquad \lim_{x\to 0} \frac{\sin x}{x} = 1$$

gives

$$\frac{dT}{d\theta} = \mu_s T \qquad (e)$$

Equation (e) can be rearranged in the form

$$\frac{dT}{T} = \mu_s d\theta \qquad (6\text{-}17)$$

which, since the coefficient of friction is a constant, can be immediately integrated from θ_1 where the tension is T_1 to θ_2 where the tension is T_2 to get

$$\ln \left(\frac{T_2}{T_1}\right) = \mu_s(\theta_2 - \theta_1) = \mu_s\beta \qquad (6\text{-}18)$$

or

$$T_2 = T_1 e^{\mu_s\beta} \qquad (6\text{-}19)$$

where $\beta = \theta_2 - \theta_1$ is the central angle of the drum for which the belt is in contact with the drum. The angle of wrap β must be measured in radians and must obviously be positive. Angles greater than 2π radians are possible and simply mean that the belt is wrapped more than one complete revolution around the drum.

It must be emphasized that Eq. (6-19) assumes impeding slip at all points along the belt surface and therefore gives the maximum change in tension that the belt can have. Since the exponential function of a positive value is always greater than 1, Eq. (6-19) gives that T_2 (the tension in the belt on the side toward which slip tends to occur) will always be greater than T_1 (the tension in the belt on the side away from which slip tends to occur). Of course, if slip is not known to be impending, then Eq. (6-19) does not apply and T_2 may be larger or smaller than T_1.

V-belts, as shown in Fig. 6-100a, are handled similarly to the above. A view of the belt cross section (Fig. 6-100b), however, shows that there are now two normal forces and there will also be two frictional forces (acting along the edges of the belt and pointing into the plane of the figure). Equilibrium in the circumferential (θ-) direction now gives

$$\Delta T \cos (\Delta\theta/2) = 2\,\Delta F$$

while equilibrium in the radial direction gives

$$2\,\Delta P \sin (\alpha/2) = 2T \sin (\Delta\theta/2) + \Delta T \sin (\Delta\theta/2)$$

(a)

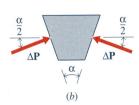

(b)

Figure 6-100

Continuing as above results finally in

$$T_2 = T_1 \, e^{(\mu_s)_{enh}\beta} \qquad (6\text{-}20)$$

in which $(\mu_s)_{enh} = \left[\dfrac{\mu_s}{\sin(\alpha/2)}\right] > \mu_s$ is an enhanced coefficient of friction. That is, V-belts always give a larger T_2 than flat belts for a given coefficient of friction μ_s and a given angle of wrap β.

Equations (6-19) and (6-20) can also be used when slipping is actually occurring by replacing the static coefficient of friction μ_s with the kinetic coefficient of friction μ_k.

Example Problem 6-26

A sport utility vehicle is prevented from moving by pulling on a rope that is wrapped $n + \frac{1}{4}$ times around the stump of a tree (Fig. 6-101). The coefficient of friction between the rope and the tree is 0.35 and the force exerted by the vehicle is 750 lb. If it is desired that the force exerted on the rope be no more than 25 lb, determine n, the number of times the rope must be wrapped around the tree stump.

Figure 6-101

SOLUTION

The angle of wrap of the rope around the stump to hold the vehicle is found from Eq. (6-19). Thus,

$$T_2 = T_1 \, e^{\mu_s \beta}$$
$$750 = 25e^{0.35\,\beta}$$

or

$$\beta = \frac{\ln \dfrac{750}{25}}{0.35} = 9.718 \text{ radians}$$

which is 1.547 times around the tree stump. Any angle less than this will require a resisting force greater than 25 lb, while any angle greater than this will require less force. Thus

$$n = 2 \qquad \qquad \textbf{Ans.}$$

will be sufficient.

Example Problem 6-27

The band brake shown in Fig. 6-102a is used to control the rotation of a drum. The coefficient of friction between the drum and the flat belt is 0.25. The brake arm AB may be considered weightless but the drum weighs 50 N. The couple **M** applied to the drum causes impending motion of the drum. If the diameters of the pins at C and D are 8 mm, determine the shearing stress on a cross section of each of the pins. Pin C is in single shear and pin D is in double shear.

SOLUTION

A free-body diagram of the brake arm is shown in Fig. 6-102b. From the equilibrium equation

$$+\left(\Sigma M_C = T_B(225) - T_A(75) - 100(400) = 0\right.$$

The second relationship between T_A and T_B is provided by the belt friction equation [Eq. (6-19)]

$$T_B = T_A e^{\mu_s \beta} = T_A e^{0.25(\pi)} = 2.193\ T_A$$

Solving yields

$$T_A = 95.60\ \text{N} \qquad T_B = 209.6\ \text{N}$$

From the remaining equilibrium equations

$$+\rightarrow \Sigma F_x = C_x = 0 \qquad\qquad C_x = 0\ \text{N}$$

$$+\uparrow \Sigma F_y = C_y + T_A + T_B - 100$$
$$= C_y + 95.60 + 209.6 - 100 = 0 \qquad C_y = -205\ \text{N} = 205\ \text{N} \downarrow$$

The shearing stress on a cross section of the pin at C, which is in single shear, is then given by Eq. (4-4) as

$$\tau = \frac{F_C}{A_s} = \frac{205}{(\pi/4)(0.008)^2} = 4.082(10^6)\ \text{N/m}^2 \cong 4.08\ \text{MPa} \qquad \textbf{Ans.}$$

A free-body diagram of the drum is shown in Fig. 6-102c. From the equilibrium equations

$$+\rightarrow \Sigma F_x = D_x = 0 \qquad\qquad D_x = 0\ \text{N}$$

$$+\uparrow \Sigma F_y = D_y - T_A - T_B - 50$$
$$= D_y - 95.60 - 209.6 - 50 = 0 \qquad D_y = 355.2\ \text{N}$$

The shearing stress on a cross section of the pin at D, which is in double shear, is then given by Eq. (4-4) as

$$\tau = \frac{F_D}{A_s} = \frac{355.2}{2(\pi/4)(0.008)^2} = 3.533(10^6)\ \text{N/m}^2 \cong 3.53\ \text{MPa} \qquad \blacksquare \qquad \textbf{Ans.}$$

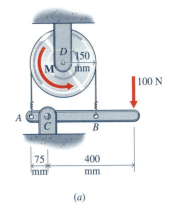

(a)

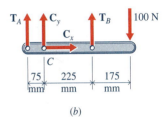

(b)

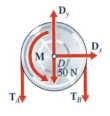

(c)

Figure 6-102

PROBLEMS

Standard Problems

6-135* An 80-lb child is sitting on a swing suspended by a rope that passes over a tree branch (Fig. P6-135). The coefficient of friction between the rope and the branch (which can be modeled as a flat belt over a drum) is 0.5 and the weight of the rope can be ignored. Determine the minimum force that must be applied to the other end of the rope to keep the child suspended.

Figure P6-135

6-136* A rope attached to a 220-kg block passes over a fixed drum (Fig. P6-136). If the coefficient of friction between the rope and the drum is 0.30, determine the minimum force **P** that must be used to
(a) Keep the block from falling.
(b) Begin to raise the block.

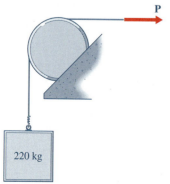

Figure P6-136

6-137 A rope attached to a 35-lb crate passes around two fixed pegs (Fig. P6-137). The 45-lb crate is attached to a wall by a second cord. The coefficient of friction between the two crates is 0.25; between the crate and the floor, 0.25; and between the rope and the pegs, 0.20. Determine the minimum force P_{min} that must be used to cause motion.

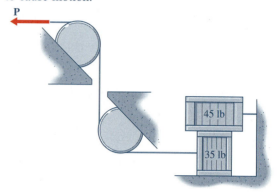

Figure P6-137

6-138 The scaffolding of Fig. P6-138 is raised using an electric motor that sits on the scaffolding. Frictionless wheels at the ends of the scaffold restrict horizontal motion. The 250 mm–diameter pulley at the top is jammed and will not rotate. The coefficient of friction between the rope and the pulley is 0.25 and the weight of the motor, scaffold, and supplies is 2500 N. Determine the minimum torque that must be supplied by the motor:
(a) To raise the scaffold at a constant rate.
(b) To lower the scaffold at a constant rate.

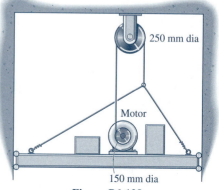

Figure P6-138

6-139* A rope connecting two blocks is wrapped $\frac{3}{4}$-turn around a fixed peg (Fig. P6-139). The coefficient of friction between the peg and the rope is 0.15; between the 65-lb block and the floor, 0.40. Determine the minimum and maximum weight of block *B* for which no motion occurs.

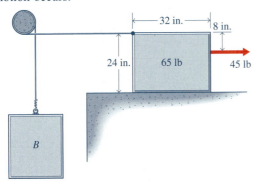

Figure P6-139

Challenging Problems

6-140* A rope connecting two blocks passes around a frictionless pulley and a fixed drum (Fig. P6-140). The coefficient of friction between the rope and the drum and between the 10-kg block and the wall is 0.30; the pin supporting the 30-kg block is frictionless. Determine the maximum force **P** for which no motion occurs.

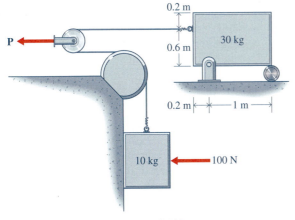

Figure P6-140

6-141 The band wrench of Fig. P6-141 is used to unscrew an oil filter from a car. (The filter acts as if it were a wheel with a resisting torque of 40 lb · ft.) The weight of the handle can be neglected. Determine

(a) The minimum coefficient of friction between the band and the filter that will prevent slippage.

(b) The shearing stress on a cross section of the pin at *B* when slipping is impending if the pin has a $\frac{1}{4}$-in. diameter and is loaded in double shear.

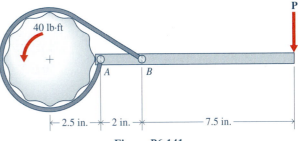

Figure P6-141

6-142 The band brake of Fig. P6-142 is used to control rotation of a drum. The coefficient of friction between the belt and the drum is 0.35. The brake arm may be considered weightless but the drum weighs 100 N. If a force of 200 N is applied to the end of the handle, determine

(a) The maximum torque for which no motion occurs if the torque is applied counterclockwise.

(b) The shearing stress on a cross section of the pin at *B* when slipping is impending if the pin has a 25-mm diameter and is loaded in single shear.

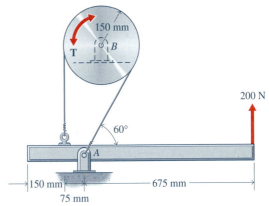

Figure P6-142

6-143* The band brake of Fig. P6-143 is used to control rotation of a drum. The coefficient of friction between the belt and the drum is 0.25. The brake arm may be considered weightless but the drum weighs 25 lb. If a force of 50 lb is applied to the end of the handle, determine

(a) The maximum torque for which no motion occurs if the torque is applied clockwise.

(b) The shearing stress on a cross section of the pin at B when slipping is impending if the pin has a $\frac{3}{4}$-in. diameter and is loaded in double shear.

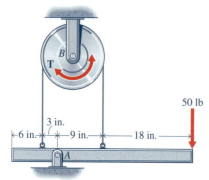

Figure P6-143

6-144 A uniform belt 2 m long and weighing 20 N hangs over a small fixed peg (Fig. P6-144). If the coefficient of friction between the belt and the peg is 0.40, determine the maximum distance d between the two ends for which the belt will not slip off the peg.

Figure P6-144

6-8 DESIGN PROBLEMS

Design was previously discussed in Section 4-8. As before, design will be limited to selecting a material and/or properly proportioning a member to perform a specified function without failure. Within the context of this book failure will refer to elastic failure, slip failure, or failure by fracture. Furthermore, at this point in the text, design will be limited to pins and two-force members. Design will be extended to more complicated members and loading situations in later chapters of this book. The following example illustrates the use of design principles.

Example Problem 6-28

All members, including pins, of the structure shown in Fig. 6-103a are made of structural steel. All pins are in single shear. For rod CD the allowable normal stress is 18 ksi, and for the pins the allowable shearing stress is 10 ksi. Determine the diameter of rod CD and the diameters of the pins at A, B, C, and D required to support the forces shown. Assume that all other members of the structure are of adequate size to support the applied forces.

SOLUTION

A free-body diagram of the complete structure is shown in Fig. 6-103b. From the equilibrium equations

$$+\!\!\downarrow \Sigma M_A = E_x(17) - 800(12) - 450(20) = 0 \qquad E_x = 1094.1 \text{ lb}$$

$$+\!\!\rightarrow \Sigma F_x = A_x - E_x = A_x - 1094.1 = 0 \qquad A_x = 1094.1 \text{ lb}$$

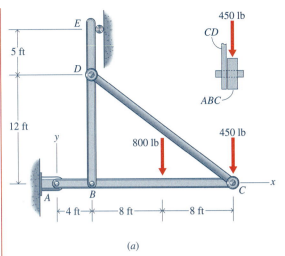

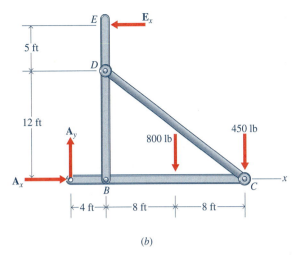

(a)   (b)

Figure 6-103

$$+\uparrow\Sigma F_y = A_y - 800 - 450 = 0 \qquad A_y = 1250.0 \text{ lb}$$

$$F_A = \sqrt{(A_x)^2 + (A_y)^2} = \sqrt{(1094.1)^2 + (1250.0)^2} = 1661.2 \text{ lb}$$

The shearing stress on a cross section of the pin at A, which is in single shear, is then given by Eq. (4-4) as

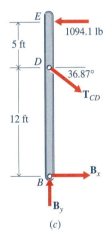

$$\tau_A = \frac{F_A}{A_s} = \frac{1661.2}{(\pi/4)(d_A)^2} = 10{,}000 \text{ psi} \qquad d_A \geq 0.460 \text{ in.} \qquad \textbf{Ans.}$$

A free-body diagram of member BDE is shown in Fig. 6-103c. From the equilibrium equations

$$+\downarrow\Sigma M_B = 1094.1(17) - T_{CD}\cos 36.87° (12) = 0 \qquad T_{CD} = 1937.5 \text{ lb}$$

$$+\rightarrow\Sigma F_x = B_x + T_{CD}\cos 36.87° - 1094.1$$
$$= B_x + 1937.5 \cos 36.87° - 1094.1 = 0 \qquad B_x = -455.9 \text{ lb}$$

$$+\uparrow\Sigma F_y = B_y - T_{CD}\sin 36.87°$$
$$= B_y - 1937.5 \sin 36.87° = 0 \qquad B_y = 1162.5 \text{ lb}$$

$$F_B = \sqrt{(B_x)^2 + (B_y)^2} = \sqrt{(-455.9)^2 + (1162.5)^2} = 1248.7 \text{ lb}$$

The shearing stress on a cross section of the pin at B, which is in single shear, is then given by Eq. (4-4) as

$$\tau_B = \frac{F_B}{A_s} = \frac{1248.7}{(\pi/4)(d_B)^2} = 10{,}000 \text{ psi} \qquad d_B \geq 0.399 \text{ in.} \qquad \textbf{Ans.}$$

The shearing stresses on cross sections of the pins at C and D, which are in single shear, are then given by Eq. (4-4) as

$$\tau_C = \tau_D = \frac{T_{CD}}{A_s} = \frac{1937.5}{(\pi/4)(d_C)^2} = 10{,}000 \text{ psi} \qquad d_C = d_D \geq 0.497 \text{ in.} \qquad \textbf{Ans.}$$

The axial stress in rod CD is given by Eq. (4-2) as

$$\sigma_{CD} = \frac{T_{CD}}{A_{CD}} = \frac{1937.5}{(\pi/4)(d_{CD})^2} = 18{,}000 \text{ psi} \qquad d_{CD} \geq 0.370 \text{ in.} \qquad \textbf{Ans.}$$

The previous results indicate that $\frac{1}{2}$ in.–diameter pins (a standard size) could be used for all of the connections and either a $\frac{3}{8}$ or a $\frac{1}{2}$ in.–diameter rod could be used for member CD. ∎

PROBLEMS

Standard Problems

6-145 A rigid handle is used to twist a 2.5 in.–diameter valve, as shown in Fig. P6-145. A pair of oppositely directed parallel forces $P = 100$ lb is needed to twist the valve. Force is transmitted from the handle to the valve shaft by means of a 1.0-in. long (into the page) key with a square cross section. The key is to be made of 2024-T4 wrought aluminum. If failure is by fracture, and the factor of safety is to be 3.0, determine the cross-sectional dimensions of the square key to the nearest $\frac{1}{64}$ in.

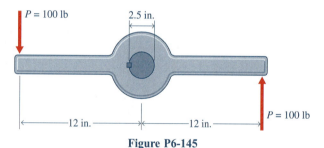

Figure P6-145

6-146 The structure shown in Fig. P6-146 consists of a circular tie rod AB and a rigid member BC. If the structure is to support a load $P = 40$ kN, determine the required diameters of the pins at A, B, and C, and the required diameter of the tie rod. The tie rod is made of a structural steel with an allowable normal stress of 140 MPa and the pins are made of 0.2% C hardened steel with an allowable shearing stress of 105 MPa. All pins are in double shear. The tie rod is adequately reinforced around the pins so that failure does not occur at the pins.

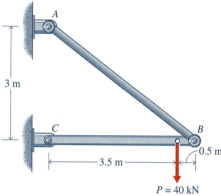

Figure P6-146

6-147 The assembly shown in Fig. P6-147 consists of a rigid member AB and a standard steel pipe CD. The assembly is to support a load $P = 9000$ lb. Load is transferred to the standard-weight steel pipe through a roller resting on a rigid plate attached to the pipe. The pin at A is made of 0.4% C hot-rolled steel (allowable shearing stress is 18 ksi), and is in double shear. The allowable normal stress for the steel pipe is 12 ksi. Determine

(a) The smallest standard size pin that can be used if pins are available with diameters in increments of $\frac{1}{32}$ in.

(b) The smallest nominal diameter available for the steel pipe.

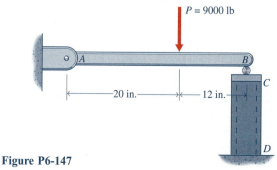

Figure P6-147

6-148 The rigid beam *AB* shown in Fig. P6-148 is subjected to a uniformly distributed load of magnitude 15 kN/m. The load is transmitted through a roller and rigid steel plate to an air-dried Douglas fir member *CD*. Load is transferred from the timber member to the ground by means of a concrete footing. Determine the smallest size standard structural timber that can be used if failure is by fracture and a factor of safety of 1.75 is specified.

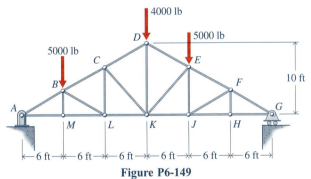

Figure P6-148

6-149 A Howe roof truss is loaded and supported as shown in Fig. P6-149. Each member of the truss is an American standard beam (S-shape) made of structural steel with an allowable stress of 9 ksi. Determine the lightest S-shapes that can be used for members *DE*, *EK*, and *JK*.

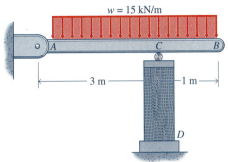

Figure P6-149

6-150 Each member of the truss of Fig. P6-150 is an American standard beam (S-shape) made of structural steel. The pin at *A* is made of 0.2% C hardened steel with an allowable shearing stress of 100 MPa. The allowable axial stress for the beams is 100 MPa. Determine

(a) The lightest S-shape that can be used for member *EJ*.

(b) The smallest standard-diameter pin that can be used at joint *A* if pins are available with diameters in increments of 1 mm.

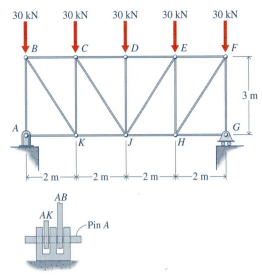

Figure P6-150

6-151 In Fig. P6-151, a cable is attached to rigid member *DEBFG* at *E*, passes around the 2 ft–diameter frictionless pulley at *C*, and is attached to a weight $W = 6000$ lb. A solid circular rod made of structural steel is attached to the vertical member at *F* and to rigid member *ABC* at *A* by pins. The pins at *B*, *F*, and *C* are in single shear, whereas the pin at *G* is in double shear. All pins are made of 2024-T4 wrought aluminum. If the mode of failure is by slip (pins and rod), and a factor of safety of 1.75 is specified, determine

(a) The required diameter for rod *AF*, to the nearest $\frac{1}{32}$ in.

(b) The required diameters for pins *B*, *C*, and *G*, to the nearest $\frac{1}{64}$ in.

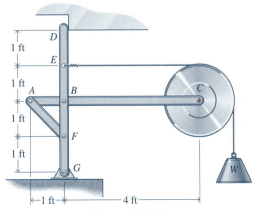

Figure P6-151

6-152 A pair of vise grip pliers is shown in Fig. P6-152. All members of the pliers, except for the pins and spring, may be treated as rigid. It is anticipated that the largest applied force on the handles will be $P = 200$ N. If each of the three pins is in double shear and made of 0.4% C hardened steel with an allowable shearing stress of 100 MPa, determine the required diameters for the pins to the nearest mm. In the analysis, neglect the spring force.

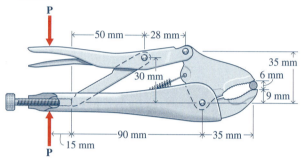

Figure P6-152

6-153 A pin-connected system of levers and bars is used as a toggle for a press to crush cans, as shown in Fig. P6-153. The system is designed to provide a crushing force of 550 lb. The handle may be considered rigid. The pin at D is to be made of 2024-T4 wrought aluminum. If failure is by slip, and the factor of safety is to be 1.5, determine the required diameter of pin D to the nearest $\frac{1}{32}$ in.

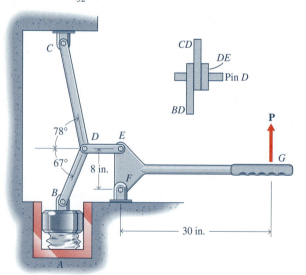

Figure P6-153

6-154 The brake shown in Fig. P6-154 is used to control the motion of block B. The mass of the block is 1000 kg, the coefficient of static friction between the brake arm and drum is 0.35, and the coefficient of kinetic friction

is 0.30. A force P is applied to the rigid brake arm such that the block moves downward at a constant velocity. Determine the required diameter for the pin at the end of the brake arm (to the nearest mm) if it is in double shear and is made of 0.2% C hardened steel with an allowable shearing stress of 80 MPa.

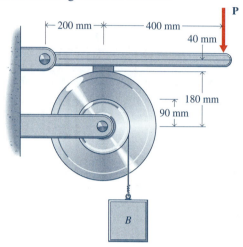

Figure P6-154

Challenging Problems

6-155 The structure shown in Fig. P6-155 consists of three members pinned at A, B, C, and D, and subjected to a load $P = 20$ kip. All members are made of structural steel with an allowable axial stress of 9 ksi. The tension members are solid circular rods and the compression

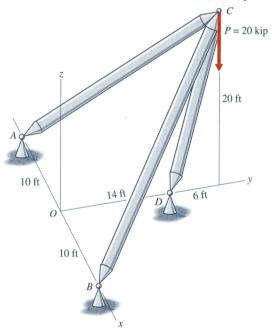

Figure P6-155

member is to be a standard steel pipe. Determine the required diameters for the solid rods, and the smallest nominal-size steel pipe required to support the applied load. Assume that *CD* does not buckle.

6-156 The flat roof of a building is supported by a series of parallel plane trusses spaced 2 m apart (only one such truss is shown in Fig. P6-156). Water of density 1000 kg/m^3 may collect on the roof to a depth of 0.2 m. All members of the truss are to be the same size. If the truss members are made of structural steel, failure is by slip, and the factor of safety is 3.0, determine the smallest diameter solid circular rod, to the nearest mm, that can be used. The members are braced so that buckling does not occur.

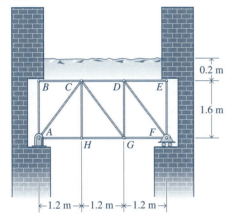

Figure P6-156

6-157 The truss shown in Fig. P6-157 is subjected to four applied forces. Each member of the truss is made of structural steel with an allowable stress of 9 ksi, and each member is to have the same S-shape. Determine the lightest S-shape that can be used.

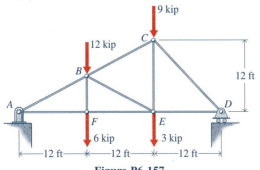

Figure P6-157

6-158 An overhead crane consists of an I-beam supported by a truss as shown in Fig. P6-158. The uniform I-beam weighs 1800 N. The 35-kN load applied to the I-beam may move along the beam such that $0 \le d \le 3$ m, and thus the forces in the members of the truss change as d changes. Determine the lightest structural steel S-shape that may be used for member *CD*. Failure is by slip, and the factor of safety is 4.0.

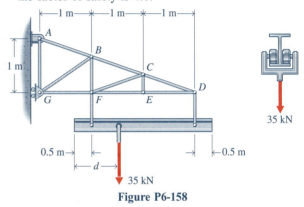

Figure P6-158

6-159 Solid circular bars *AB* and *BC* of Fig. P6-159 are pinned at the ends, and the structure is subjected to a load $P = 6000$ lb. The angle θ may vary, but pin *A* is always directly above pin *C*. Both bars are made of 6061-T6 wrought aluminum alloy with an allowable stress of 10,000 psi. In the force analysis assume that the weights of the bars are negligible with respect to the applied loads. Determine

(a) The angle θ for a minimum weight structure.
(b) The diameters of the bars.
(c) The weights of the bars.

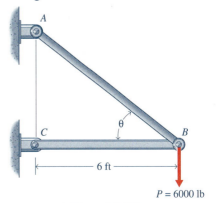

Figure P6-159

6-9 SUMMARY

Any system of forces acting on a rigid body can be expressed in terms of a resultant force $\mathbf{R}$ and a resultant couple $\mathbf{C}$. Therefore, for a rigid body to be in equilibrium, both the resultant force $\mathbf{R}$ and the resultant couple $\mathbf{C}$ must vanish. These two conditions can be expressed by the two vector equations

$$\mathbf{R} = \Sigma F_x \mathbf{i} + \Sigma F_y \mathbf{j} + \Sigma F_z \mathbf{k} = \mathbf{0}$$
$$\mathbf{C} = \Sigma M_x \mathbf{i} + \Sigma M_y \mathbf{j} + \Sigma M_z \mathbf{k} = \mathbf{0}$$

(6-1)

Equations (6-1) can be expressed in scalar form as

$$\Sigma F_x = 0 \qquad \Sigma F_y = 0 \qquad \Sigma F_z = 0$$
$$\Sigma M_x = 0 \qquad \Sigma M_y = 0 \qquad \Sigma M_z = 0$$

(6-2)

Equations (6-2) are the necessary conditions for equilibrium of a rigid body. They are also the sufficient conditions for equilibrium if all of the forces acting on the body can be determined from these equations.

To study the force system acting on a body, it is necessary to identify all forces, both known and unknown, that act on the body. The best way to identify all forces acting on a body is to use the free-body diagram approach. Special care must be exercised when representing the actions of connections and supports on the free-body diagram.

The term "two dimensional" is used to describe problems in which the forces involved are contained in a plane (say the xy-plane) and the axes of all couples are perpendicular to the plane containing the forces. For two-dimensional problems, the equations of equilibrium reduce to

$$\mathbf{R} = \Sigma F_x \mathbf{i} + \Sigma F_y \mathbf{j} = \mathbf{0}$$
$$\mathbf{C} = \Sigma M_A \mathbf{k} = \mathbf{0}$$

(6-3)

or in scalar form to

$$\Sigma F_x = 0 \qquad \Sigma F_y = 0 \qquad \Sigma M_A = 0$$

(6-4)

The third equation represents the sum of the moments of all forces about a z-axis through any point A on or off the body. Equations (6-4) are both the necessary and sufficient conditions for equilibrium of a body subjected to a two-dimensional system of forces. Alternative forms of Eqs. (6-4) are

$$\Sigma F_x = 0 \qquad \Sigma M_A = 0 \qquad \Sigma M_B = 0$$

(6-5)

where points A and B must have different x-coordinates, and

$$\Sigma M_A = 0 \qquad \Sigma M_B = 0 \qquad \Sigma M_C = 0$$

(6-6)

where A, B, and C are any three points not on the same straight line.

Two broad categories of engineering structures were considered in this chapter; namely, trusses and frames. Four main assumptions are made in the analysis of trusses:

1. Truss members are connected only at their ends; no member is continuous through a joint.
2. Members are connected by frictionless pins.
3. The truss structure is loaded only at the joints.
4. The weight of the members may be neglected.

Because of these assumptions, truss members are modeled as two-force members with the forces acting at the ends of the member and directed along the axis of the member.

One method of analysis (the method of joints) for trusses is performed by drawing a free-body diagram for each pin (joint). Application of the vector equilibrium equation, $\Sigma \mathbf{F} = \mathbf{0}$, at each joint yields two algebraic equations that can be solved for two unknowns. The pin forces are solved sequentially starting from a pin on which only two unknown forces and one or more known forces act. Once these forces are determined, their values can be applied to adjacent joints and treated as known quantities. This process is repeated until all unknown forces have been determined. The method of joints is most often used when the forces in all of the members of a truss are to be determined.

A second method of analysis for trusses is the method of sections. When the method of sections is used, the truss is divided into two parts by passing an imaginary plane or curved section through the members of interest. Free-body diagrams may then be drawn for either or both parts of the truss. Since each part is a rigid body, three independent equations of equilibrium can be written for either part. Therefore, a section that cuts through no more than three members should be used.

It will often happen that a section cutting no more than three members and passing through a given member of interest cannot be found. In such a case it may be necessary to draw a section through a nearby member and solve for the force in that member first. Then the method of joints can be used to find the force in the member of interest. One of the principal advantages of using the method of sections is that the force in a member near the center of a large truss usually can be determined without first obtaining the forces in the rest of the truss.

Structures that are not constructed entirely of two-force members are called *frames* or *machines*. While frames and machines may also contain one or more two-force members, they always contain at least one member that is acted upon by forces at more than two points or is acted upon by both forces and moments. The main distinction between frames and machines is that frames are rigid structures while machines are not.

As with the analysis of trusses, the method of solution for frames and machines consists of taking the structures apart, drawing free-body diagrams of each of the components, and writing the equations of equilibrium for each of the free-body diagrams. Since some of the members of frames and machines are not two-force members, however, the directions of the forces in these members is not known. The analysis of frames and machines will consist of solving the equilibrium equations of a system of rigid bodies.

For machines and nonrigid structures, the structure must be taken apart and analyzed even if the only information desired is the support reactions or the relationship between the external forces (input and output forces) acting on it.

Tangential forces due to friction are always present at the interface between two contacting bodies and these forces always act in a direction to oppose the tendency of the contacting surfaces to slip relative to one another. In some types of machine elements, such as bearings and skids, it is desired to minimize friction, while in other types, such as brakes and belt drives, it is desired to maximize friction. Two main types of friction are commonly encountered in engineering practice—dry friction and fluid friction. Dry friction is encountered when the unlubricated surfaces of two solids are in contact and a condition of sliding or tendency to slide exists. Fluid friction develops when adjacent layers in a fluid move at different velocities.

Since friction forces cannot increase without limit, they eventually reach a maximum value $F_{\max}$. The condition when a friction force is at its maximum value is called a *condition of impending motion.* Beyond that point, friction can no longer supply the amount of force required for equilibrium. The value of limiting friction force is proportional to the normal force N at the contact surface. Thus,

$$F_{\max} = \mu_s N \qquad (6\text{-}12)$$

The constant of proportionality μ_s, the coefficient of static friction, depends on the types of material in contact but is independent of both the normal force and the area of contact. The frictional force actually exerted is never greater than that required to satisfy the equations of equilibrium.

$$F \leq \mu_s N \qquad (6\text{-}13)$$

The equality in Eq. (6-13) holds only at the point of impending motion. Thus, friction belongs to a class of forces known as resistances; it only operates to oppose motion, never to produce it. Once a body starts to slip relative to its supporting surface, the friction force will decrease approximately 25 percent to

$$F = \mu_k N \qquad (6\text{-}14)$$

where μ_k is the coefficient of kinetic friction. This coefficient is again independent of the normal force and is also independent of the speed of the relative motion—at least for low speeds. The presence of any moisture or oil on the surface, however, can change the problem from one of dry friction, in which the friction force is independent of the speed of the body, to one of fluid friction, in which the friction force is directly proportional to the speed.

For some problems, the methods developed in Chapter 4 can be used to determine stress and/or deformation. For example, both normal stress and deformation can be calculated for straight two-force members. For members loaded in a fashion more complex than axially loaded members, the calculation of stress and deformation is more complicated than indicated in Chapter 4. The methods to determine stress and deformation for these members will be developed throughout the remainder of this book. However, shearing stresses in pins and bearing stresses between pins and members may be found using the results of Chapter 4, regardless of the type of loading on the member.

REVIEW PROBLEMS

6-160* A beam is loaded and supported as shown in Fig. P6-160. Determine the reactions at supports A and B when $m_1 = 75$ kg and $m_2 = 225$ kg.

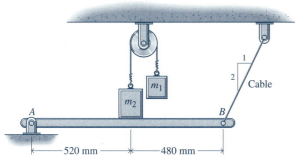

Figure P6-160

6-161* A bar is loaded and supported as shown in Fig. P6-161. The bar has a uniform cross section and weighs 100 lb. Determine the reactions at supports A, B, and C.

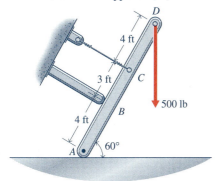

Figure P6-161

6-162 Bar AB of Fig. P6-162 has a uniform cross section, a mass of 25 kg, and a length of 1 m. Determine the angle θ for equilibrium.

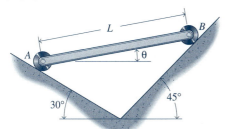

Figure P6-162

6-163 Determine the force **P** required to pull the 250-lb roller over the step shown in Fig. P6-163.

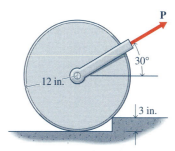

Figure P6-163

6-164* Determine the reactions at supports A and B of the truss shown in Fig. P6-164.

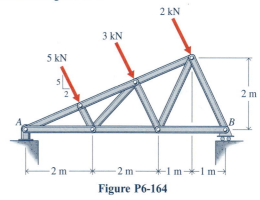

Figure P6-164

6-165* The crane and boom, shown in Fig. P6-165, weigh 12,000 lb and 600 lb, respectively. When the boom is in the position shown, determine

(a) The maximum load that can be lifted by the crane.
(b) The tension in the cable used to raise and lower the boom when the load being lifted is 3600 lb.
(c) The pin reaction at boom support A when the load being lifted is 3600 lb.

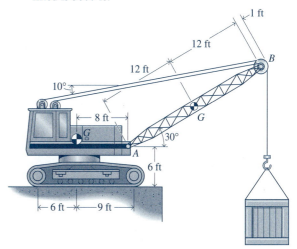

Figure P6-165

6-166 The masses of cartons 1, 2, and 3, which rest on the platform shown in Fig. P6-166, are 300 kg, 100 kg, and 200 kg, respectively. The mass of the platform is 500 kg. Determine the tensions in the three cables *A, B,* and *C* that support the platform.

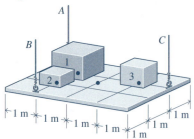

Figure P6-166

6-167 A 500-lb homogeneous circular plate is supported by three cables as shown in Fig. P6-167. Determine the tensions in the three cables.

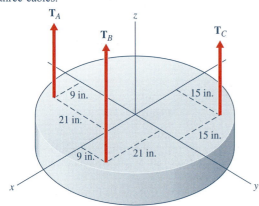

Figure P6-167

6-168* The plate shown in Fig. P6-168 has a mass of 75 kg. The brackets at supports *A* and *B* exert only force reactions on the plate. Each of the brackets can resist a force along the axis of pins in one direction only. Determine the reactions at supports *A* and *B* and the tension in the cable.

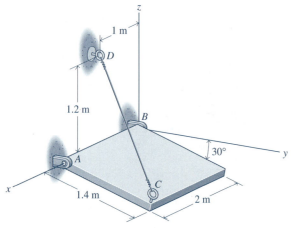

Figure P6-168

6-169* The 570-lb block, shown in Fig. P6-169, is supported by a ball-and-socket joint at *A*, by a smooth pin at *B*, and by a cable at *C*. Determine the components of the reactions at supports *A* and *B* and the force in the cable at *C*.

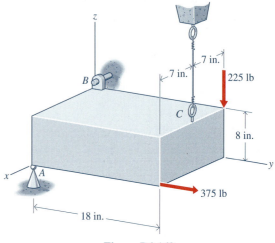

Figure P6-169

6-170 Determine the forces in members *BC, BG, CG,* and *CF* of the truss shown in Fig. P6-170.

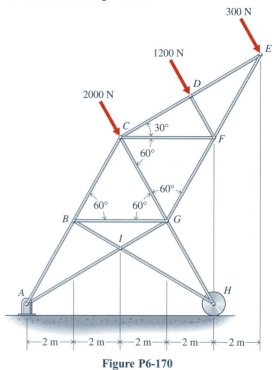

Figure P6-170

6-171 Determine the force in each member of the truss shown in Fig. P6-171.

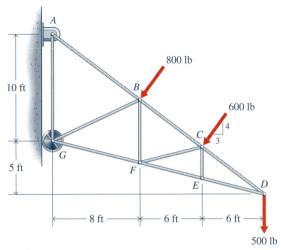

Figure P6-171

6-172* Find the forces in members *CD* and *CG* of the stairs truss of Fig. P6-172.

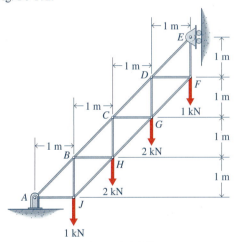

Figure P6-172

6-173* Determine the forces in members *BC, CF, FG,* and *GE* of the truss shown in Fig. P6-173.

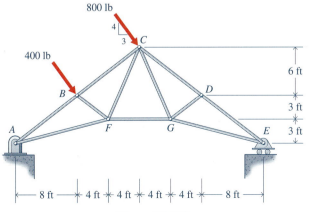

Figure P6-173

6-174* Three bars are connected with smooth pins to form the frame shown in Fig. P6-174. The weights of the bars are negligible. Determine
(a) The force exerted by the pin at *D* on member *CDE*.
(b) The reactions at supports *A* and *E*.

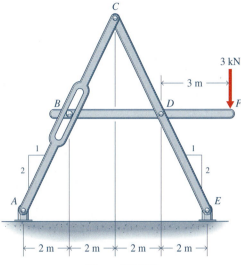

Figure P6-174

6-175* A device for lifting rectangular objects such as bricks and concrete blocks is shown in Fig. P6-175. Determine the minimum coefficient of static friction between the contacting surfaces required to make the device work.

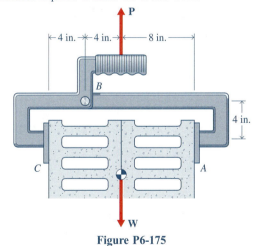

Figure P6-175

6-176 The static coefficient of friction between the brake pad and the brake drum of Fig. P6-176 is 0.40. When a force of 350 N is being applied to the brake arm, determine the couple required to initiate rotation of the drum if the direction of rotation is (a) clockwise and (b) counterclockwise.

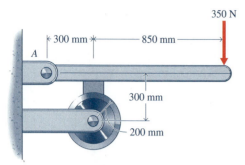

350 N

300 mm — 850 mm

A

300 mm

200 mm

Figure P6-176

6-177 The device shown in Fig. P6-177 is used to raise boxes and crates between floors of a factory. The frame, which slides on the 4 in.–diameter vertical post, weighs 50 lb. If the coefficient of friction between the post and the frame is 0.10, determine the force **P** required to raise a 150-lb box.

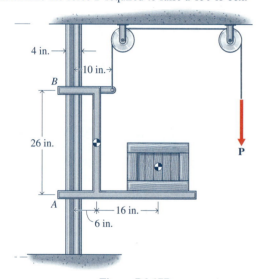

4 in.

10 in.

B

26 in.

P

A

16 in.

6 in.

Figure P6-177

6-178* A conveyor belt is driven with the 200 mm–diameter multiple-pulley drive shown in Fig. P6-178. Couples C_A and C_B are applied to the system at pulleys A and B, respectively. The angle of contact between the belt and a pulley is 225° for each pulley, and the coefficient of friction is 0.30. Determine the maximum force **T** that can be developed by the drive and the magnitudes of input couples C_A and C_B.

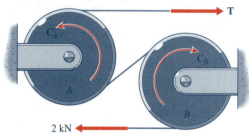

T

C_A

C_B

A

B

2 kN

Figure P6-178

6-179* The electric motor shown in Fig. P6-179 weighs 30 lb and delivers 50 in. · lb of torque to pulley A of a furnace blower by means of a V-belt. The effective diameters of the 36° pulleys are 5 in. The coefficient of friction is 0.30. Determine the minimum distance a to prevent slipping of the belt if the rotation of the motor is clockwise.

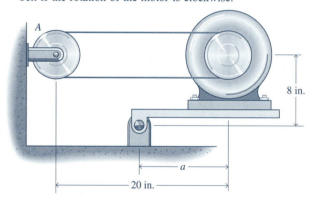

A

8 in.

a

20 in.

Figure P6-179

COMPUTER PROBLEMS

C6-180 The lever shown in Fig. P6-180 is a quarter circular arc of radius 450 mm. The 25 mm–diameter pin at A is smooth and frictionless and is in single shear.
(a) Plot A, the magnitude of the pin force at A, and B, the force on the smooth support B, as functions of θ ($0° \le \theta \le 80°$), the angle at which the support is located.
(b) Plot τ, the average shear stress in the pin at A, as a function of θ ($0° \le \theta \le 80°$).

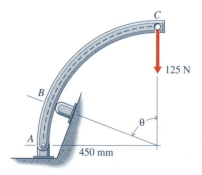

C

125 N

B

θ

A

450 mm

Figure P6-180

C6-181 The crane and boom shown in Fig. P6-181 weigh 12,000 lb and 600 lb, respectively. The pulleys at D and E are small and the cables attached to them are essentially parallel. The 3 in.–diameter pin at A is smooth and frictionless and is in double shear. The distributed force exerted on the treads by the ground is equivalent to a single resultant force N acting at some distance d behind point C.
(a) Plot d, the location of the equivalent force N relative to point C, as a function of the boom angle θ ($0° \le \theta \le 80°$), when the crane is lifting a 3600-lb load.
(b) Plot τ, the average shear stress in the pin at A, as a function of θ ($0° \le \theta \le 80°$) when the crane is lifting a 3600-lb load.
(c) It is desired that the resultant force on the tread always be at least 1 ft behind C to ensure that the crane is never in danger of tipping over. Plot W_{max}, the maximum load that may be lifted, as a function of θ ($0° \le \theta \le 80°$).

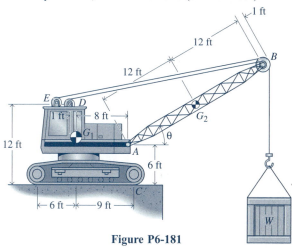

Figure P6-181

C6-182 An extension ladder arrangement is being raised into position by a hydraulic cylinder as shown in Fig. P6-182. The center of gravity of the 500-kg ladder is at G. The 25

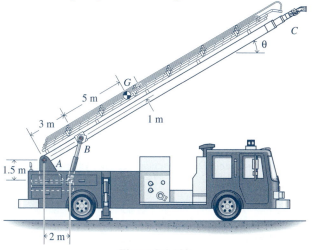

Figure P6-182

mm–diameter pin at A is smooth, frictionless, and in double shear.
(a) Plot σ_n, the normal stress in the 40 mm–diameter piston rod of the hydraulic cylinder, as a function of the angle θ ($10° \le \theta \le 90°$).
(b) Plot τ, the shear stress in the pin at A, as a function of θ ($10° \le \theta \le 90°$).

C6-183 The wrecker truck of Fig. P6-183 has a weight of 15,000 lb and a center of gravity at G. The force exerted on the rear (drive) wheels by the ground consists of both a normal component B_y and a tangential component B_x, while the force exerted on the front wheel consists of a normal force A_y only. Plot P, the maximum amount of pull the wrecker can exert as a function of θ ($0° \le \theta \le 90°$), if B_x cannot exceed 0.8 B_y (because of friction considerations) and the wrecker does not tip over backwards (the front wheels remain in contact with the ground).

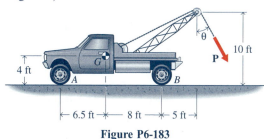

Figure P6-183

C6-184 Three bars are connected by smooth frictionless pins as shown in Fig. P6-184. Bar $BCDE$ is rigid, bar AB is aluminum ($E = 73$ GPa; $L = 1.5$ m; $d = 30$ mm; $\sigma_{max} = 280$ MPa), and bar EF is steel ($E = 200$ GPa; $L = 1.2$ m; $d = 20$ mm; $\sigma_{max} = 250$ MPa). The 25 mm–diameter pivot pin D is aluminum ($\tau_{max} = 160$ MPa) and is in double shear.
(a) Plot σ_{AB}, σ_{EF}, and τ_D as functions of P ($0 < P < 300$ kN).
(b) Plot δ_{AB} and δ_{EF} as functions of P ($0 < P < 300$ kN).
(c) What is the maximum P that can be applied to the bar?

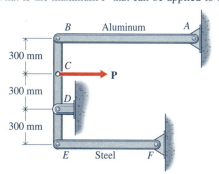

Figure P6-184

C6-185 Three bars are connected by smooth frictionless pins as shown in Fig. P6-185. Bar BCD is rigid, bar AC is aluminum [$E = 12,000$ ksi; $L = 36$ in.; $d = 1$ in.; $\sigma_{max} = 20$ ksi; $\alpha = 12.5(10^{-6})/°F$], and bar DE is steel [$E = 30,000$ ksi; $L = 30$ in.; $d = \frac{1}{2}$ in.; $\sigma_{max} = 10$ ksi; $\alpha = 6.6(10^{-6})/°F$]. The $\frac{5}{8}$ in.–diameter pivot pin B is steel ($\tau_{max} = 5$ ksi) and is in single shear. If the temperature drops after the unit is assembled,
(a) Plot σ_{AC}, σ_{DE}, and τ_B as functions of the temperature drop ΔT ($0 > \Delta T > -60°F$).
(b) Plot δ_{AC} and δ_{DE} as functions of ΔT ($0 > \Delta T > -60°$ F).
(c) What is the maximum temperature drop the system can withstand?

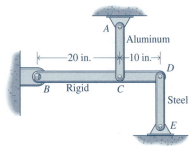

Figure P6-185

C6-186 Three bars are connected by smooth frictionless pins as shown in Fig. P6-186. Bar BCD is rigid, bar AB is aluminum ($E = 73$ GPa; $L = 750$ mm; $d = 40$ mm; $\sigma_{max} = 240$ MPa), and bar DE is steel ($E = 200$ GPa; $L = 500$ mm; $d = 30$ mm; $\sigma_{max} = 400$ MPa). The 50 mm–diameter pivot pin C is aluminum ($\tau_{max} = 180$ MPa) and is in double shear. After the unit is assembled, the nut D is slowly tightened.
(a) Plot σ_{AB}, σ_{DE}, and τ_C as functions of the distance that the nut D advances δ_{nut} ($0 < \delta_{nut} < 2$ mm).
(b) Plot δ_{AB} and δ_{DE} as functions of δ_{nut} ($0 < \delta_{nut} < 2$ mm).
(c) What is the maximum amount δ_{nut} that the nut can be tightened?

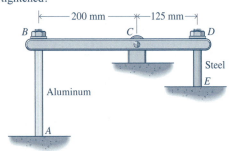

Figure P6-186

C6-187 Three bars are connected by smooth frictionless pins as shown in Fig. P6-187. Bar $ABCD$ is rigid, bar AE is aluminum ($E = 10,600$ ksi; $L = 24$ in.; $d = \frac{1}{2}$ in.; $\sigma_{max} = 40$ ksi), and bar CF is steel ($E = 30,000$ ksi; $L = 18$ in.; $d = \frac{3}{4}$ in.; $\sigma_{max} = 50$ ksi). The $\frac{3}{4}$ in.–diameter pivot pin B is steel ($\tau_{max} = 25$ ksi)

and is in single shear. The holes in bar CF are slightly overdrilled so that pin C moves down 0.06 in. before contact is made with bar CF.
(a) Plot σ_{AE}, σ_{CF}, and τ_B as functions of P ($0 < P < 10$ kip).
(b) Plot δ_{AE} and δ_{CF} as functions of P ($0 < P < 10$ kip).
(c) What is the maximum force P that the system can withstand?

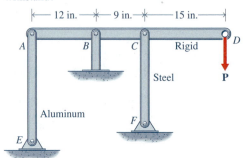

Figure P6-187

C6-188 Three bars are connected by smooth frictionless pins as shown in Fig. P6-188. Bar $ABCD$ is rigid, bar BE is aluminum ($E = 73$ GPa; $L = 0.500$ m; $d = 25$ mm; $\sigma_{max} = 90$ MPa), and bar CF is steel ($E = 200$ GPa; $L = 1.00$ m; $d = 40$ mm; $\sigma_{max} = 150$ MPa). The pivot pin A is aluminum ($\tau_{max} = 60$ MPa) and is in double shear.
(a) Plot σ_{BE} and σ_{CF} as functions of P ($0 < P < 140$ kN).
(b) Plot δ_{BE} and δ_{CF} as functions of P ($0 < P < 140$ kN).
(c) What is the minimum diameter pin A that must be used?

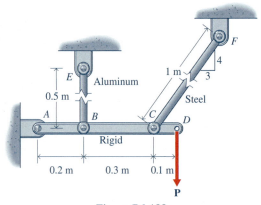

Figure P6-188

C6-189 The garage door $ABCD$ shown in Fig. P6-189 is being raised by a cable DE. The one-piece door is a homogeneous rectangular slab weighing 225 lb. Frictionless rollers (B and C) run in tracks at each side of the door as shown.
(a) Plot T, the tension in the cable, as a function of d ($0 \le d \le 100$ in.).
(b) Plot B and C, the forces on the frictionless rollers, as a function of d ($0 \le d \le 100$ in.).

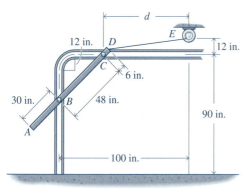

Figure P6-189

C6-190 A hand winch is used to raise a 75-kg load as shown in Fig. P6-190. If the force P is always perpendicular to both the handle DE and the arm CD, plot A and B, the magnitudes of the bearing forces, as a function of the angle θ ($0 \leq \theta \leq 360°$).

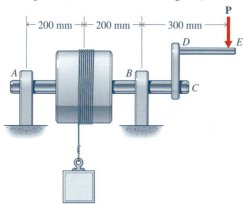

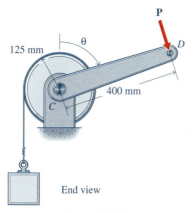

End view

Figure P6-190

C6-191 An overhead crane consists of an I-beam supported by a simple truss as shown in Fig. P6-191. If the uniform I-beam weighs 400 lb, plot the force in members AB, BC, EF, and FG as a function of the position d ($0 \leq d \leq 8$ ft).

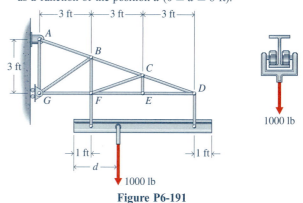

Figure P6-191

C6-192 The Gambrel truss shown in Fig. P6-192 supports one side of a bridge; an identical truss supports the other side. A 3400-kg truck is stopped on the bridge at the location shown and floor beams carry the vehicle load to the truss joints. If the center of gravity of the truck is located 1.5 m in front of the rear wheels, plot the force in members BC, BG, and GH as a function of the truck's location d ($0 \leq d \leq 20$ m).

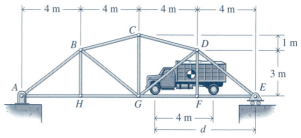

Figure P6-192

C6-193 The mechanism shown in Fig. P6-193 is designed to keep its load level while raising it. A pin on the rim of the 4 ft–diameter pulley fits in a slot on arm ABC. Arms ABC and DE are each 4 ft long and the package being lifted weighs 80 lb. The mechanism is raised by pulling on the rope that is wrapped around the pulley.

(a) Plot P, the force required to hold the platform as a function of the platform height h ($0 \leq h \leq 5.75$ ft).

(b) Plot A, C, and E, the magnitudes of the pin reaction forces at A, C, and E as a function of h ($0 \leq h \leq 5.75$ ft).

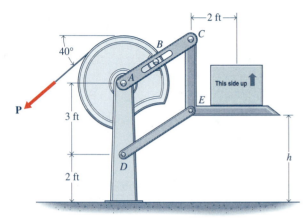

Figure P6-193

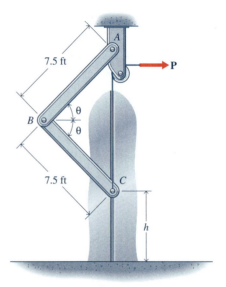

Figure P6-195

C6-194 Forces of $P = 100$ N are being applied to the handles of the vise grip pliers shown in Fig. P6-194. Plot the force applied on the bolt by the jaws as a function of the distance d ($20 \le d \le 30$ mm).

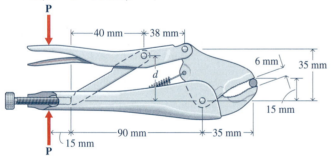

Figure P6-194

C6-195 The door to an airplane hangar consists of two uniform sections that are hinged at the middle as shown in Fig. P6-195. The door is raised by means of a cable attached to a bar along the bottom edge of the door. Smooth rollers at the ends of the bar C run in a smooth vertical channel. If the door is 30 ft wide, 15 ft tall, and weighs 1620 lb,
(a) Plot P, the force required to hold the door open, as a function of the door opening height h ($0.5 \le h \le 13.5$ ft).
(b) Plot A and B, the hinge forces, as a function of the height h ($0.5 \le h \le 13.5$ ft).
(c) What is the maximum height h that the door can be opened if the force in the hinges is not to exceed 5000 lb?

C6-196 A 30-kg block is suspended from a lightweight rope that is securely wrapped around a drum (Fig. P6-196). Rotation of the drum is controlled by the lightweight brake arm ABC. If the coefficient of friction between the brake arm and the drum is $\mu_s = 0.3$,
(a) Plot P, the minimum force that must be applied to the brake arm to hold the weight, as a function of a, the location of the drum ($10 \le a \le 250$ mm).
(b) What happens to the system at $a = 75$ mm?
(c) What does the solution for $a < 75$ mm mean?
(d) Repeat the problem for the case in which the block is suspended from the other side of the drum.

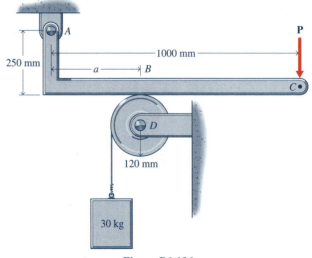

Figure P6-196

C6-197 The simple mechanism of Fig. P6-197 is often used to hold the handles of brooms, mops, shovels, and other such tools. The weight of the tool causes the two otherwise free cylinders to become wedged into the corner between the handle and the rails. Although no amount of downward force will cause the handle to slip, the tool can be removed easily by lifting it upward and pulling it forward. The coefficient of friction is the same between the cylinder and the broom handle and between the cylinder and the side rail. If the broom shown weighs 1.5 lb and the weight of the two small cylinders may be neglected,
(a) Plot μ_s, the minimum coefficient of friction for which the system is in equilibrium, as a function of the rail angle θ ($5° \leq \theta \leq 50°$).
(b) Plot A_n, the normal component of the force exerted on the broom handle by one of the cylinders, as a function of the angle θ ($5° \leq \theta \leq 50°$).

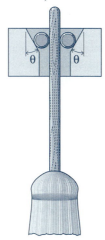

Figure P6-197

C6-198 A 10-kg cylinder rests on a thin lightweight piece of cardboard as shown in Fig. P6-198. The coefficient of the friction is the same at all surfaces.
(a) Plot P, the maximum force that may be applied to the cardboard without moving it, as a function of the coefficient of friction μ_s ($0.05 \leq \mu_s \leq 0.8$).
(b) On the same graph, plot $(A_f)_{actual}$ and $(B_f)_{actual}$, the actual amounts of friction force that act at points A and B, and $(A_f)_{avail}$ and $(B_f)_{avail}$, the maximum amounts of friction force available for equilibrium at points A and B.
(c) What happens to the system at $\mu_s \cong 0.364$?
(d) What does the solution for $\mu_s > 0.364$ mean?

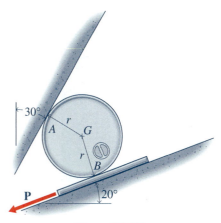

Figure P6-198

C6-199 The hand brake of Fig. P6-199 is used to control the rotation of a drum. The coefficient of friction between the belt and the drum is $\mu_s = 0.15$, and the weight of the handle may be neglected. If a counterclockwise torque of 5 lb·ft is applied to the drum,
(a) Plot P, the minimum force that must be applied to the handle to prevent motion, as a function of a, the location of the pivot point ($0.5 \leq a \leq 5.8$ in.).
(b) Repeat the problem for a clockwise torque of 5 lb·ft. What happens to the system at $a \cong 4.6$ in.? What does the solution mean for $a > 4.6$ in.?

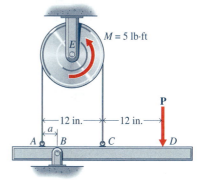

Figure P6-199

TORSIONAL LOADING: SHAFTS

7-1 INTRODUCTION

The problem of transmitting a torque (a couple) from one plane to a parallel plane is frequently encountered in the design of machinery. The simplest device for accomplishing this function is a circular shaft such as that connecting an electric motor with a pump, compressor, or other machine. A modified free-body diagram (the weight and bearing reactions are not shown because they contribute no useful information to the torsion problem) of a shaft used to transmit a torque from a driving motor A to a coupling B is shown in Fig. 7-1. The resultant of the electromagnetic forces applied to armature A of the motor is a couple resisted by the resultant of the bolt forces (another couple) acting on the flange coupling B. The circular shaft transmits the torque from the armature to the coupling. Typical torsion problems involve determinations of significant stresses in and deformations of shafts.

A segment of the shaft between transverse planes a-a and b-b of Fig. 7-1 will be studied. The complicated stress distributions at the locations of the torque-applying devices are beyond the scope of this elementary treatment of the torsion problem. A free-body diagram of the segment of the shaft between sections a-a and b-b is shown in Fig. 7-2 with the torque applied by the armature indicated on the left end as T. The resisting torque T_r at the right end of the segment is the resultant of the differential forces dF acting on the transverse plane b-b. The force dF is equal to $\tau \, dA$ where τ is the shearing stress on the transverse plane at a distance ρ from the center of the shaft and dA is a differential area. For circular sections, the shearing stress at a point on any transverse plane is always perpendicular to the radius to the point. If the shaft is in equilibrium, a summation of moments about the axis of the shaft indicates that

$$T = T_r = \int_{\text{area}} \rho \, dF = \int_{\text{area}} \rho \, \tau \, dA \qquad (7\text{-}1)$$

The law of variation of the shearing stress on the transverse plane (τ as a function of radial position ρ) must be known before the integral of Eq. (7-1) can be evaluated. Thus, the problem of determining the relationship between torque and shearing stress is statically indeterminate. Recalling the procedures developed in Chapter 4, the solution of a statically indeterminate problem requires the use of

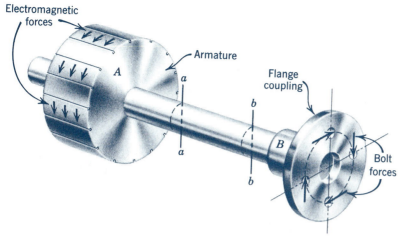

Figure 7-1

the equations of equilibrium [Eq. (7-1) for torsion], an analysis of deformation, and the relationship between stress and strain.

In 1784 C. A. Coulomb, a French engineer, developed (experimentally) a relationship between applied torque and angle of twist for circular bars.[1] In a paper published in 1820[1] A. Duleau, another French engineer, derived the same relationship analytically by making the assumption that a plane section before twisting remains plane after twisting and a diameter remains straight. Visual examination of twisted models indicates that these assumptions are correct for circular sections either solid or hollow (provided the hollow section is circular and symmetrical with respect to the axis of the shaft), but incorrect for any other shape. Compare, for example, the distortions of rubber models with circular and rectangular cross sections shown in Fig. 7-3. Figure 7-3*b* shows the circular shaft

---

[1]From *History of Strength of Materials,* S. P. Timoshenko, McGraw-Hill, New York, 1953.

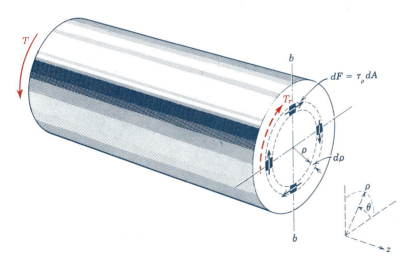

Figure 7-2

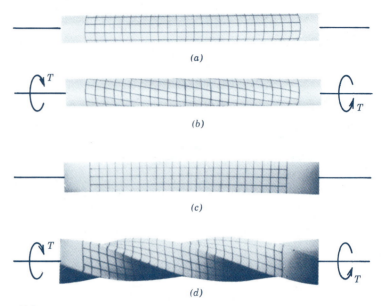

Figure 7-3

after loading, and illustrates that plane sections remain plane. For the rectangular shaft, plane sections before loading (Fig. 7-3*c*) become warped after loading (Fig. 7-3*d*).

7-2 TORSIONAL SHEARING STRAIN

If a plane transverse section before twisting remains plane after twisting and a diameter of the section remains straight, the distortion of the shaft of Fig. 7-2 will be as indicated in Fig. 7-4*a*, where points *B* and *D* on a common radius in a plane move to points *B'* and *D'* in the same plane and still on the same radius. The angle ϕ is called the *angle of twist*. The surface *ABB'* of Fig. 7-4*a* is shown in plan view in Fig. 7-4*b*, in which a differential element of the material at *B* (Fig. 7-4*c*) is distorted due to shearing stress at *B'* (Fig. 7-4*d*). Clearly, the angle ϕ of Fig. 7-4*b* is the same as the shearing strain γ_c of Fig. 7-4*d*. Similar figures could be drawn for the surface *EDD'*. It is recommended that the reader review the concept of shearing strain in Section 4-4.

At this point the assumption is made that all longitudinal elements (*AB*, *ED*, etc.) have the same length *L* (which limits the results to straight shafts of constant diameter). From Fig. 7-4, the shearing strain γ_ρ at a distance ρ from the center of the shaft and γ_c at the surface of the shaft ($\rho = c$) are related to the angle of twist θ by

$$\tan \gamma_c = \frac{BB'}{AB} = \frac{c\theta}{L}$$

and

$$\tan \gamma_\rho = \frac{DD'}{ED} = \frac{\rho\theta}{L}$$

or, if the strain is small (tan $\gamma \cong \sin \gamma \cong \gamma$, γ in radians),

$$\gamma_c = \frac{c\theta}{L} \qquad (7\text{-}2a)$$

and

$$\gamma_\rho = \frac{\rho\theta}{L} \qquad (7\text{-}2b)$$

Combining Eqs. (7-2a) and (7-2b) gives

$$\theta = \frac{\gamma_c L}{c} = \frac{\gamma_\rho L}{\rho}$$

which indicates that the shearing strain

$$\gamma_\rho = \frac{\gamma_c}{c}\rho \qquad (7\text{-}3)$$

is zero at the center of the shaft and increases linearly with respect to the distance ρ from the axis of the shaft. Equation (7-3) is the result of the deformation analysis of a circular shaft subjected to torsional loading. This equation can be combined with Eq. (7-1) once the relationship between shearing stress τ and shearing strain γ is known.

Up to this point, no assumptions have been made about the relationship between stress and strain or about the type of material of which the shaft is made. Therefore, Eq. (7-3) is valid for elastic or inelastic action and for homogeneous or heterogeneous materials, provided the strains are not too large (tan $\gamma \cong \gamma$). Problems in this book will be assumed to satisfy this requirement.

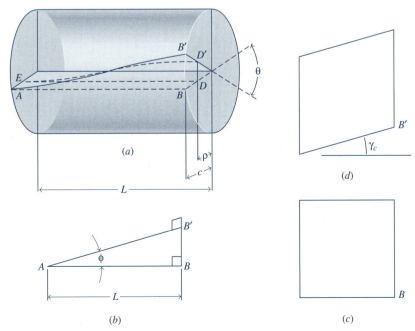

Figure 7-4

7-3 TORSIONAL SHEARING STRESS— THE ELASTIC TORSION FORMULA

If the assumption is now made that Hooke's law applies (the accompanying limitation is that the stresses must be below the proportional limit of the material), the shearing stress τ is related to the shearing strain γ by the expression $\tau = G\gamma$ [Eq. (4-15c)]. Then, multiplying Eq. (7-3) by the shear modulus (modulus of rigidity) G gives

$$\tau_\rho = \frac{\tau_c}{c}\rho \tag{7-4}$$

When Eq. (7-4) is substituted into Eq. (7-1), the result is

$$T = T_r = \frac{\tau_c}{c}\int \rho^2 \, dA = \frac{\tau_\rho}{\rho}\int \rho^2 \, dA \tag{a}$$

The integral in Eq. (a) is called the *polar second moment of area*[2] and is given the symbol J. For a solid circular shaft such as is shown in Fig. 7-5a, the polar second moment J is

$$J = \int \rho^2 \, dA = \int_0^c \rho^2 \, (2\pi\rho \, d\rho) = \frac{\pi c^4}{2} \tag{7-5a}$$

For a circular annulus such as shown in Fig. 7-5b, the polar second moment J is

$$J = \int \rho^2 \, dA = \int_b^c \rho^2 \, (2\pi\rho \, d\rho)$$
$$= \frac{\pi c^4}{2} - \frac{\pi b^4}{2} = \frac{\pi}{2}(r_o^4 - r_i^4) \tag{7-5b}$$

where r_o and r_i are the outer and inner radii, respectively, of the circular annulus. In terms of the polar second moment J, Eq. (a) can be written as

$$T = T_r = \frac{\tau_c J}{c} = \frac{\tau_\rho J}{\rho} \tag{b}$$

or solving for the unknown shearing stress τ

$$\tau_\rho = \frac{T\rho}{J} \qquad \text{and} \qquad \tau_c = \frac{Tc}{J} \tag{7-6}$$

Like the shearing strain γ_ρ, the shearing stress τ_ρ is zero at the center of the shaft and increases linearly with respect to the distance ρ from the axis of the shaft.

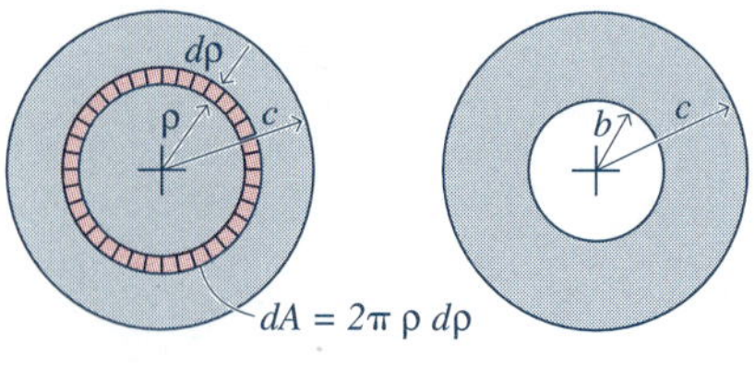

(a)   (b)

Figure 7-5

dA = 2π ρ dρ

---

[2]Integrals of the type $\int x^2 \, dA$ arise often in mechanics and are given the general name "second moments of area." Second moments of area are discussed further in Chapter 8 where they are used to relate stresses to internal forces and moments in beams. Second moments of area are sometimes (improperly) called "moments of inertia" since they are closely related to the moment of inertia integral $\int \rho^2 \, dm$ which arises in dynamics.

Both the shearing strain γ and the shearing stress τ are maximum when $\rho = c$. Equation (7-6) is known as the "elastic torsion formula," in which τ_ρ is the shearing stress on a transverse plane at a distance ρ from the axis of the shaft, and T is the resisting torque (the torque produced on the transverse plane by the shearing stresses). Even though the resisting torque is T_r, the equivalent symbol T is used in Eq. (7-6); see Eq. (7-1). Equation (7-6) is valid for both solid and hollow circular shafts. The resisting torque T_r is generally different than the external torques applied to various points along the shaft and must be obtained from a free-body diagram and an equilibrium equation. The procedure for calculating the resisting torque is illustrated in Example Problem 7-1. Note that Eq. (7-6) applies only for linearly elastic action in homogeneous and isotropic materials since Hooke's law $\tau = G\gamma$ was used in its development.

7-4 TORSIONAL DISPLACEMENTS

Frequently the amount of twist in a shaft is of paramount importance. Therefore, determination of the angle of twist is a common problem for the machine designer. The fundamental approach to such problems is provided by the following equations

$$\gamma_\rho = \rho \frac{\theta}{L} \quad \text{or} \quad \gamma_\rho = \rho \frac{d\theta}{dL} \qquad (7\text{-}2)$$

$$\tau_\rho = \frac{T\rho}{J} \quad \text{or} \quad \tau_c = \frac{Tc}{J} \qquad (7\text{-}6)$$

$$G = \frac{\tau}{\gamma} \qquad (4\text{-}15)$$

The second form of Eq. (7-2) is used when the torque or the cross section varies as a function of position along the length of the shaft. Equation (7-2) is valid for both elastic and inelastic action. Equation (7-6) is the elastic torsion formula that provides the shearing stress τ_ρ on a transverse plane at a distance ρ from the axis of the shaft. Equation (4-15) is Hooke's law for shearing stresses. The last two expressions are limited to stresses below the proportional limit of the material (linearly elastic action). The three equations can be combined to give several different relationships; for example,

$$\theta = \frac{\gamma_\rho L}{\rho} = \frac{\tau_\rho L}{G\rho} \qquad (7\text{-}7a)$$

or

$$\theta = \frac{TL}{GJ} \qquad (7\text{-}7b)$$

The angle of twist θ determined from the above expressions is for a length L of shaft, of constant diameter ($J = $ constant), constant material properties ($G = $ constant), and carrying a torque T. The resisting torque $T = T_r$ is the torque produced on the transverse plane by the shear stresses and is generally different than the external torques applied to the shaft at various sections by gears, pulleys, or couplings. Ideally, the length of shaft should not include sections too near to (within about one-half shaft diameter of) places where mechanical devices are attached. For practical purposes, however, it is customary to neglect local distortions at all connections and to compute angles as if there were no discontinuities.

If T, G, or J is not constant along the length of the shaft, Eq. (7-7b) takes the form

$$\theta = \sum_{i=1}^{n} \frac{T_i L_i}{G_i J_i} \qquad (7\text{-}7c)$$

where each term in the summation is for a length L where T, G, and J are constant. If T, G, or J is a function of x (the distance along the length of the shaft), the angle of twist is found using

$$\theta = \int_{0}^{L} \frac{T \, dx}{GJ} \qquad (7\text{-}7d)$$

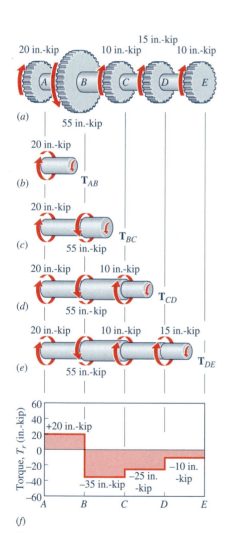

(a)

(b)

(c)

(d)

(e)

(f)

Figure 7-6

Example Problem 7-1

A steel shaft is used to transmit torque from a motor to operating units in a factory. The torque is input at gear B (see Fig. 7-6a) and is removed at gears A, C, D, and E.

(a) Determine the torques transmitted by cross sections (resisting torques) in intervals AB, BC, CD, and DE of the shaft.

(b) Draw a torque diagram for the shaft.

SOLUTION

(a) The torques transmitted by cross sections, or resisting torques, in intervals AB, BC, CD, and DE of the shaft shown in Fig. 7-6a are obtained by using the four free-body diagrams shown in Figs. 7-6b, c, d, and e. The moment equilibrium equation $\Sigma M = 0$ about the axis of the shaft yields

$$+\downarrow\!\Sigma M = T_{AB} - 20 = 0 \qquad\qquad T_{AB} = +20.0 \text{ in.} \cdot \text{kip} \quad \textbf{Ans.}$$

$$+\downarrow\!\Sigma M = T_{BC} - 20 + 55 = 0 \qquad\qquad T_{BC} = -35.0 \text{ in.} \cdot \text{kip} \quad \textbf{Ans.}$$

$$+\downarrow\!\Sigma M = T_{CD} - 20 + 55 - 10 = 0 \qquad T_{CD} = -25.0 \text{ in.} \cdot \text{kip} \quad \textbf{Ans.}$$

$$+\downarrow\!\Sigma M = T_{DE} - 20 + 55 - 10 - 15 = 0 \qquad T_{DE} = -10.00 \text{ in.} \cdot \text{kip} \quad \textbf{Ans.}$$

For all of the above calculations, a free-body diagram of the part of the bar to the left of the imaginary cut has been used. A free-body diagram of the part of the bar to the right of the cut would have yielded identical results. In fact, for the determination of T_{CD} and T_{DE}, the free-body diagram to the right of the cut would have been more efficient since fewer torques would have appeared on the diagram.

(b) A torque diagram is a graph in which abscissas represent distances along the shaft and ordinates represent the internal or resisting torques at the corresponding cross sections. Positive torques point outward from the cross section when represented as a vector according to the right-hand rule. A torque diagram for the shaft of Fig. 7-6a, constructed by using the results from part (a), is shown in Fig. 7-6f. Note in the diagram that the abrupt changes in torque are equal to the applied torques at gears A, B, C, D, and E. Thus, the torque diagram could have been drawn directly below the sketch of the shaft of Fig. 7-6a, without the aid of the free-body diagrams shown in Figs. 7-6b, c, d, and e, by using the applied torques at gears A, B, C, D, and E. However, care must be exercised with the signs used for torque, since Fig. 7-6f represents resisting torques. ∎

Example Problem 7-2

A hollow steel shaft with an outside diameter of 400 mm and an inside diameter of 300 mm is subjected to a torque of 300 kN · m as shown in Fig. 7-7. The modulus of rigidity G (shear modulus) for the steel is 80 GPa. Determine

(a) The maximum shearing stress in the shaft.

(b) The shearing stress on a transverse cross section at the inside surface of the shaft.

(c) The magnitude of the angle of twist in a 2-m length.

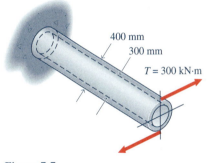

400 mm
300 mm
$T = 300$ kN·m

Figure 7-7

SOLUTION

Equations for shearing stress and angle of twist in a circular shaft subjected to a torque contain the polar second moment J of the cross section, which is given by Eq. (7-5b) as

$$J = \frac{\pi}{2}(r_o^4 - r_i^4) = \frac{\pi}{2}(200^4 - 150^4) = 1718.1(10^6) \text{ mm}^4 = 1718.1(10^{-6}) \text{ m}^4$$

(a) The resisting torque on all cross sections of the shaft is $T_r = T = 300$ kN · m. The maximum shearing stress occurs on a transverse cross section at the outer surface of the shaft and is given by Eq. (7-6) as

$$\tau_{\max} = \frac{Tc}{J} = \frac{300(10^3)(200)(10^{-3})}{1718.1(10^{-6})}$$

$$= 34.9(10^6) \text{ N/m}^2 = 34.9 \text{ MPa} \qquad \textbf{Ans.}$$

(b) The shearing stress on a transverse cross section at the inner surface of the shaft is given by Eq. (7-6) as

$$\tau_\rho = \frac{T\rho}{J} = \frac{300(10^3)(150)(10^{-3})}{1718.1(10^{-6})} = 26.2(10^6) \text{ N/m}^2 = 26.2 \text{ MPa} \qquad \textbf{Ans.}$$

(c) The angle of twist in a 2-m length is given by Eq. (7-7b) as

$$\theta = \frac{TL}{GJ} = \frac{300(10^3)(2)}{80(10^9)1718.1(10^{-6})} = 0.00437 \text{ rad} \quad \blacksquare \qquad \textbf{Ans.}$$

Example Problem 7-3

A solid steel shaft 14 ft long has a diameter of 6 in. for 9 ft of its length and a diameter of 4 in. for the remaining 5 ft. The shaft is in equilibrium when subjected to the three torques shown in Fig. 7-8a. The modulus of rigidity (shear modulus) of the steel is 12,000 ksi. Determine

(a) The maximum shearing stress in the shaft.
(b) The rotation of end B of the 6-in. segment with respect to end A.
(c) The rotation of end C of the 4-in. segment with respect to end B.
(d) The rotation of end C with respect to end A.

SOLUTION

In general, free-body diagrams should be drawn to evaluate the resisting torque in each section of the shaft. Such diagrams are shown in Figs. 7-8b and c. From the free-body diagram of Fig. 7-8b,

$$+ \langle \Sigma M_x = T_6 - 20 + 5 = 0 \qquad T_6 = 15 \text{ ft} \cdot \text{kip} \; -\langle -$$

From the free-body diagram of Fig. 7-8c,

$$+ \langle \Sigma M_x = -T_4 + 5 = 0 \qquad T_4 = 5 \text{ ft} \cdot \text{kip} \; -\langle -$$

A torque diagram, such as the one shown in Fig. 7-8d, provides a pictorial representation of the levels of resisting torque being transmitted by each of the sections and serves as an aid for stress and deformation calculations.

Equations for the shearing stress and angle of twist in a circular shaft subjected to a torque contain the polar second moment J of the cross section, which is given by Eq. (7-5a) as

$$J_4 = \frac{\pi}{2} r^4 = \frac{\pi}{2} (2^4) = 25.13 \text{ in.}^4$$

$$J_6 = \frac{\pi}{2} r^4 = \frac{\pi}{2} (3^4) = 127.23 \text{ in.}^4$$

(a) The location of the maximum shearing stress is not apparent; hence, the stress must be checked at both sections. The maximum shearing stress on a transverse cross section occurs at the outer surface of the shaft and is given by Eq. (7-6) as

$$\tau_{AB} = \frac{T_{AB} c_{AB}}{J_{AB}} = \frac{15(12)(3)}{127.23} = 4.244 \text{ ksi}$$

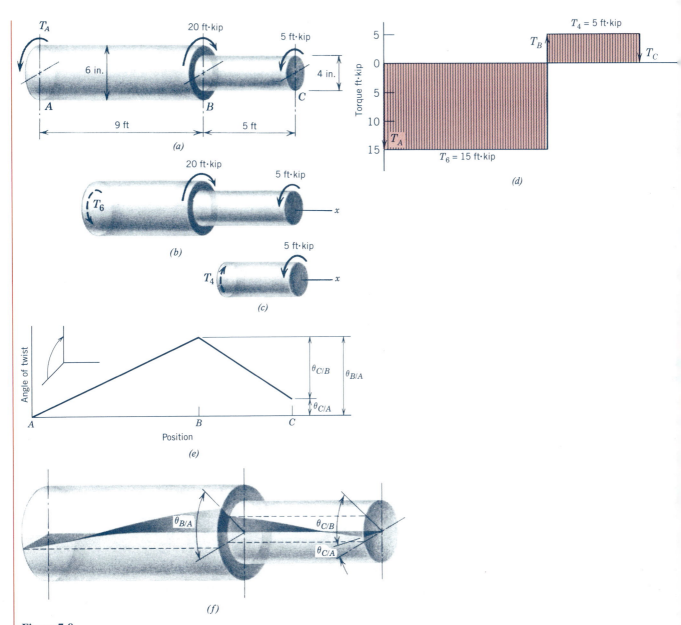

Figure 7-8

$$\tau_{BC} = \frac{T_{BC}c_{BC}}{J_{BC}} = \frac{5(12)(2)}{25.13} = 4.775 \text{ ksi}$$

Therefore,

$$\tau_{max} = \tau_{BC} = 4.775 \text{ ksi} \cong 4.78 \text{ ksi} \qquad \textbf{Ans.}$$

(b) As the resisting torque of 15 ft · kip is transmitted from section to section in segment AB of the shaft, the section at B twists relative to the section at A by an amount $\theta_{B/A}$ as shown on the angle of twist diagram of Fig. 7-8e.

The slope of the angle of twist diagram is constant since the term T/GJ (in Eq. 7-7b) is constant. The rotation of the section at B (angle of twist in the 6-in. section) is given by Eq. (7-7b) as

$$\theta_{B/A} = \frac{T_{AB}L_{AB}}{G_{AB}J_{AB}} = \frac{15(12)(9)(12)}{12,000(127.23)} = 0.012733 \text{ rad} \,—\!\!\curvearrowright\!\!— \qquad \textbf{Ans.}$$

(c) Similarly for segment BC:

$$\theta_{C/B} = \frac{T_{BC}L_{BC}}{G_{BC}J_{BC}} = \frac{5(12)(5)(12)}{12,000(25.13)} = 0.011938 \text{ rad} \,—\!\!\curvearrowleft\!\!— \qquad \textbf{Ans.}$$

(d) If there was no resisting torque being transmitted by segment BC it would rotate as a rigid body through angle $\theta_{B/A}$. However, the resisting torque of 5 ft · kip causes the section at C to rotate relative to the section at B by an amount $\theta_{C/B}$ as segment BC deforms, as shown in Fig. 7-8e. The resultant of the deformations in the two segments of the shaft is

$$\theta_{C/A} = \theta_{B/A} - \theta_{C/B}$$
$$= 0.012733 - 0.011938 = 0.000795 \text{ rad} \,—\!\!\curvearrowright\!\!— \qquad \textbf{Ans.}$$

The distortion for the entire shaft is pictorially shown in Fig. 7-8f. ∎

Example Problem 7-4

A solid steel shaft has a 100-mm diameter for 2 m of its length and a 50-mm diameter for the remaining 1 m of its length, as shown in Fig. 7-9a. A 100-mm long pointer CD is attached to the end of the shaft. The shaft is attached to a rigid support at the left end and is subjected to a 16-kN · m torque at the right end of the 100-mm section and a 4-kN · m torque at the right end of the 50-mm section. The modulus of rigidity G of the steel is 80 GPa. Determine

(a) The maximum shearing stress in the 50-mm section of the shaft.
(b) The maximum shearing stress in the 100-mm section of the shaft.
(c) The rotation of a cross section at B with respect to its no-load position.
(d) The movement of point D with respect to its no-load position.

SOLUTION

Equations for the shearing stress and angle of twist in a circular shaft subjected to a torque contain the polar second moment J of the cross section, which is given by Eq. (7-5a) as

$$J_{50} = \frac{\pi}{2}r^4 = \frac{\pi}{2}(25^4) = 0.6136(10^6) \text{ mm}^4 = 0.6136(10^{-6}) \text{ m}^4$$

$$J_{100} = \frac{\pi}{2}r^4 = \frac{\pi}{2}(50^4) = 9.817(10^6) \text{ mm}^4 = 9.817(10^{-6}) \text{ m}^4$$

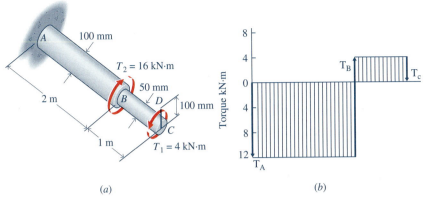

Figure 7-9

(a) The magnitude of the resisting torque on all cross sections of the shaft in the 50-mm section is $T_r = T_1 = 4$ kN · m (see Fig. 7-9b). The torque diagram of Fig. 7-9b was drawn directly by using the fact that the abrupt changes in the diagram are equal to the applied torques. The maximum shearing stress occurs at the outer surface of the shaft and is given by Eq. (7-6) as

$$\tau_{max} = \frac{T_{BC}c_{BC}}{J_{BC}} = \frac{4(10^3)(25)(10^{-3})}{0.6136(10^{-6})}$$
$$= 163.0(10^6) \text{ N/m}^2 = 163.0 \text{ MPa}$$ **Ans.**

(b) The magnitude of the resisting torque on all cross sections of the shaft in the 100-mm section is $T_r = T_2 - T_1 = 16 - 4 = 12$ kN · m (see Fig. 7-9b). The maximum shearing stress occurs at the outer surface of the shaft and is given by Eq. (7-6) as

$$\tau_{max} = \frac{T_{AB}c_{AB}}{J_{AB}} = \frac{12(10^3)(50)(10^{-3})}{9.817(10^{-6})}$$
$$= 61.1(10^6) \text{ N/m}^2 = 61.1 \text{ MPa}$$ **Ans.**

(c) The rotation of the section at B (angle of twist in the 100-mm section) is given by Eq. (7-7b) as

$$\theta_B = \frac{T_{AB}L_{AB}}{G_{AB}J_{AB}} = \frac{12(10^3)(2)}{80(10^9)(9.817)(10^{-6})}$$
$$= 0.03056 \text{ rad} \cong 0.0306 \text{ rad} \;-\!\!\curvearrowright\!\!-$$ **Ans.**

(d) The rotation of the section at C (angle of twist in the 100-mm section minus the angle of twist in the 50-mm section) is given by Eq. (7-7c) as

$$\theta_C = \sum_{i=1}^{n} \frac{T_i L_i}{G_i J_i} = \frac{T_{AB}L_{AB}}{G_{AB}J_{AB}} - \frac{T_{BC}L_{BC}}{G_{BC}J_{BC}}$$
$$= \frac{12(10^3)(2)}{80(10^9)(9.817)(10^{-6})} - \frac{4(10^3)(1)}{80(10^9)(0.6136)(10^{-6})}$$
$$= 0.03056 - 0.08149 = -0.05093 \cong 0.0509 \text{ rad} \;-\!\!\curvearrowleft\!\!-$$

The movement of point D with respect to its no-load position is

$$s_D = r\theta_C = 100(0.05093) = 5.09 \text{ mm} —\!\!\smile\!\!— \blacksquare$$ **Ans.**

Example Problem 7-5

Two 1.50 in.–diameter steel ($G = 12{,}000$ ksi) shafts are connected with gears as shown in Fig. 7-10a. End D of shaft CD is fastened to a rigid support that prevents rotation. The diameters of gears B and C are 10 in. and 6 in., respectively. If an input torque of $T_A = 750$ ft · lb is applied at section A of shaft AB, determine

(a) The maximum shearing stress on a cross section of shaft CD.
(b) The rotation of section A of shaft AB with respect to its no-load position.

SOLUTION

(a) The torque at section D of shaft CD required for equilibrium of the system can be determined from equilibrium considerations for the two shafts. As an aid for these considerations, free-body diagrams for gears B and C are shown in Figs. 7-10b and c, respectively. The input torque T_A in shaft AB is transferred to shaft CD by means of the gear tooth force F shown in the two diagrams. Thus, from a summation of moments about the axis of each of the shafts:

For shaft AB,

$$T_A - r_B F = 0 \qquad\qquad (a)$$

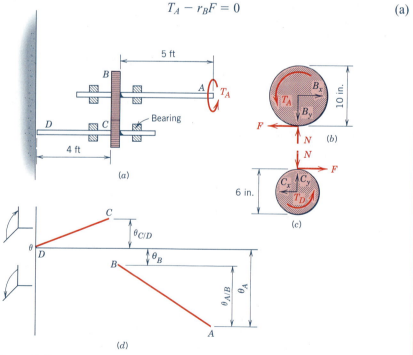

Figure 7-10

For shaft CD,

$$T_D - r_C F = 0 \qquad \text{(b)}$$

Since the force F in Eqs. (a) and (b) must be equal,

$$T_D = (r_C/r_B)T_A = (3/5)(750) = 450 \text{ ft} \cdot \text{lb}$$

The magnitude of the resisting torque on all cross sections of the shaft CD is $T_{CD} = 450$ ft · lb. The maximum shearing stress on a transverse cross section of shaft CD occurs at the outside surface of the shaft and is given by Eq. (7-6) as

$$\tau_{max} = \frac{T_{CD}c_{CD}}{J_{CD}} = \frac{450(12)(0.75)}{(\pi/2)(0.75)^4} \qquad \text{Ans.}$$
$$= 8149 \text{ psi} \cong 8150 \text{ psi}$$

(b) The quantities required for the determination of the rotation of section A of shaft AB with respect to its no-load position are illustrated on the angle of twist diagram shown in Fig. 7-10d. The rotation of the section at C relative to the section at D in shaft CD is given by Eq. (7-7b) as

$$\theta_{C/D} = \frac{T_{CD}L_{CD}}{G_{CD}J_{CD}} = \frac{450(12)(4)(12)}{12,000,000(\pi/2)(0.75)^4} = 0.04346 \text{ rad} —\curvearrowright—$$

The teeth on gears B and C must move through the same arc length. Therefore,

$$s = r_B\theta_B = r_C\theta_{C/D}$$

from which

$$\theta_B = (r_C/r_B)(\theta_{C/D}) = (3/5)(0.04346) = 0.02608 \text{ rad} —\curvearrowleft—$$

The magnitude of the resisting torque on all cross sections of the shaft AB is $T_{AB} = 750$ ft · lb, and the rotation of the section at A relative to the section at B in shaft AB is given by Eq. (7-7b) as

$$\theta_{A/B} = \frac{T_{AB}L_{AB}}{G_{AB}J_{AB}} = \frac{750(12)(5)(12)}{12,000,000(\pi/2)(0.75)^4} = 0.09054 \text{ rad} —\curvearrowleft—$$

Finally,

$$\theta_A = \theta_B + \theta_{A/B}$$
$$= 0.02608 + 0.09054 = 0.11662 \text{ rad} = 6.68° —\curvearrowleft— \blacksquare \qquad \text{Ans.}$$

PROBLEMS

Standard Problems

7-1* For the steel shaft shown in Fig. P7-1,
(a) Determine the torques transmitted by cross sections in intervals *AB, BC, CD,* and *DE* of the shaft.
(b) Draw a torque diagram for the shaft.

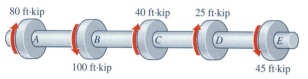

80 ft·kip   40 ft·kip   25 ft·kip

100 ft·kip   45 ft·kip

Figure P7-1

7-2* The motor shown in Fig. P7-2 supplies a torque of 500 N · m to shaft *BCDE.* The torques removed at gears *C, D,* and *E* are 100 N · m, 150 N · m, and 250 N · m, respectively.
(a) Determine the torques transmitted by cross sections in intervals *BC, CD,* and *DE* of the shaft.
(b) Draw a torque diagram for the shaft.

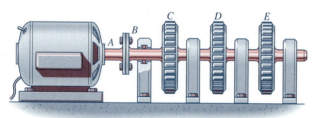

Figure P7-2

7-3 For the steel shaft shown in Fig. P7-3,
(a) Determine the maximum torque transmitted by any transverse cross section of the shaft.
(b) Draw a torque diagram for the shaft.

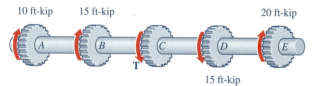

10 ft-kip   15 ft-kip   20 ft-kip

T

15 ft-kip

Figure P7-3

7-4 For the steel shaft shown in Fig. P7-4,
(a) Determine the maximum torque transmitted by any transverse cross section of the shaft.
(b) Draw a torque diagram for the shaft.

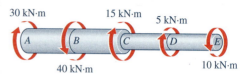

30 kN·m   15 kN·m

5 kN·m

40 kN·m   10 kN·m

Figure P7-4

7-5* A solid circular steel shaft 2 in. in diameter is subjected to a torque of 18,000 in. · lb. The modulus of rigidity (shear modulus) for the steel is 12,000 ksi. Determine
(a) The maximum shearing stress in the shaft.
(b) The magnitude of the angle of twist in a 6-ft length.

7-6* A hollow steel shaft has an outside diameter of 150 mm and an inside diameter of 100 mm. The shaft is subjected to a torque of 35 kN · m. The modulus of rigidity (shear modulus) for the steel is 80 GPa. Determine
(a) The shearing stress on a transverse cross section at the outside surface of the shaft.
(b) The shearing stress on a transverse cross section at the inside surface of the shaft.
(c) The magnitude of the angle of twist in a 2.5-m length.

7-7 Determine the maximum allowable torque to which a solid circular steel ($G = 12,000$ ksi) bar 9 ft long and 3 in. in diameter can be subjected when it is specified that the shearing stress must not exceed 12,000 psi and the angle of twist must not exceed 0.075 rad.

7-8 Specifications for a solid circular aluminum alloy ($G = 28$ GPa) rod 1.5 m long require that it shall be adequate to resist a torque of 3.0 kN · m without twisting more than 15° or exceeding a shearing stress of 100 MPa. What minimum diameter is required?

7-9* A torque of 50,000 in. · lb is supplied to the 3 in.–diameter factory drive shaft of Fig. P7-9 by a belt that drives pulley *A*. A torque of 30,000 in. · lb is taken off by pulley *B* and the remainder by pulley *C*. Shafts *AB* and *BC* are 4 ft and 3 ft long, respectively. Both shafts are made of steel ($G = 12,000$ ksi). Determine
(a) The maximum shearing stress in each of the shafts.
(b) The magnitudes of the angles of twist of pulleys *B* and *C* with respect to pulley *A*.

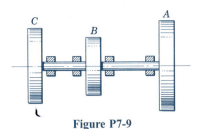

Figure P7-9

7-10* A torque of 10 kN · m is supplied to the steel ($G = 80$ GPa) factory drive shaft of Fig. P7-10 by a belt that drives pulley A. A torque of 6.5 kN · m is taken off by pulley B and the remainder by pulley C. Shafts AB and AC are 1.75 m and 1.25 m long, respectively. For an allowable shearing stress of 75 MPa, determine

(a) The minimum permissible diameters of the two shafts.
(b) The rotation of pulley B with respect to pulley A if shaft AB has a diameter of 75 mm.
(c) The rotation of pulley C with respect to pulley B if both shafts have diameters of 80 mm.

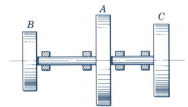

Figure P7-10

7-11 A torque of 90,000 in. · lb is supplied to the driving gear B of Fig. P7-11 by a motor. Gear A takes off 30,000 in. · lb of torque and the remainder is taken off by gear C. Both shafts are made of steel ($G = 12,000$ ksi). Determine

(a) The maximum shearing stress in each of the shafts.
(b) The angle of twist of gear C with respect to gear A.

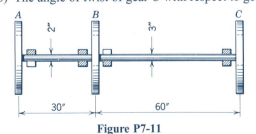

Figure P7-11

7-12 A torque of 12.0 kN · m is supplied to the driving gear B of Fig. P7-12 by a motor. Gear A takes off 4.0 kN · m of torque, and the remainder is taken off by gear C. For an allowable shearing stress of 80 MPa, determine

(a) The minimum permissible diameters for the two shafts.
(b) The angle of twist of gear A with respect to gear C if both shafts are made of steel ($G = 80$ GPa) and have diameters of 75 mm.

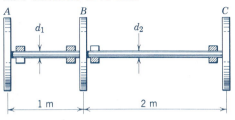

Figure P7-12

7-13* The hollow circular steel ($G = 12,000$ ksi) shaft of Fig. P7-13 is in equilibrium under the torques indicated. Determine

(a) The maximum shearing stress in the shaft.
(b) The rotation of a section at D with respect to a section at B.

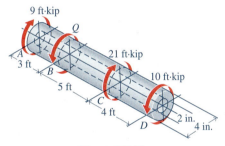

Figure P7-13

7-14* The hollow circular steel ($G = 80$ GPa) shaft of Fig. P7-14 is in equilibrium under the torques indicated. Determine

(a) The minimum permissible outside diameter d if the maximum shearing stress in the shaft is not to exceed 100 MPa.
(b) The rotation of a section at D with respect to a section at A for a shaft with an outside diameter of 120 mm.

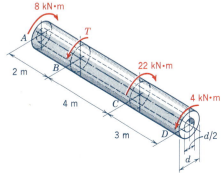

Figure P7-14

7-15 A solid circular steel ($G = 12,000$ ksi) shaft with diameters as shown in Fig. P7-15 is subjected to a torque T. The allowable shearing stress is 10,000 psi and the maximum allowable angle of twist in the 7-ft length is 0.05 rad. Determine the maximum allowable value of T.

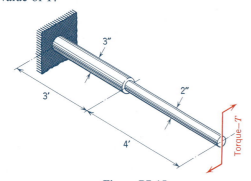

Figure P7-15

7-16 A motor supplies a torque of 5.5 kN · m to the constant diameter steel ($G = 80$ GPa) line shaft shown in Fig. P7-16. Three machines are driven by gears B, C, and D on the shaft and they require torques of 3 kN · m, 1.5 kN · m, and 1 kN · m, respectively. Determine
(a) The minimum diameter required if the maximum shearing stress in the shaft is limited to 100 MPa.
(b) The rotation of gear D with respect to the coupling at A if the coupling and gears are spaced at 2-m intervals and the shaft diameter is 75 mm.

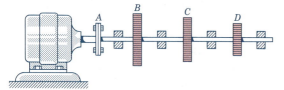

Figure P7-16

7-17* The shaft shown in Fig. P7-17 consists of a brass ($G = 5600$ ksi) tube AB that is securely connected to a solid stainless steel ($G = 12,500$ ksi) bar BC. The tube AB has an outside diameter of 5 in. and an inside diameter of 2.5 in. Bar BC has an outside diameter of 3.5 in. Torques T_1 and T_2 are 100 in. · kip and 40 in. · kip, respectively, in the directions shown. Determine
(a) The maximum shearing stress in the shaft.
(b) The rotation of a section at C with respect to its no-load position.

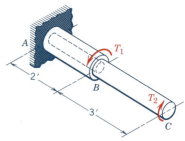

Figure P7-17

7-18* The solid circular shaft and the hollow tube shown in Fig. P7-18 are both attached to a rigid circular plate at the left end. A torque $T_A = 2$ kN · m applied to the right end of the shaft is resisted by a torque T_B at the right end of the tube. The shaft is made of steel ($G = 80$ GPa) and the tube is made of an aluminum alloy ($G = 28$ GPa). If the shaft has a diameter of 50 mm and the tube has an outside diameter of 80 mm, determine
(a) The maximum inside diameter that can be used for the tube if the maximum shearing stress in the tube must be limited to 50 MPa.
(b) The maximum inside diameter that can be used for the tube if the rotation of the right end of the shaft with respect to the right end of the tube must be limited to 0.25 rad.

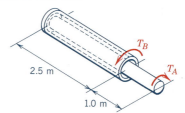

Figure P7-18

7-19 The 4 in.–diameter shaft shown in Fig. P7-19 is composed of brass ($G = 5000$ ksi) and steel ($G = 12,000$ ksi) sections rigidly connected. Determine the maximum allowable torque, applied as shown, if the maximum shearing stresses in the brass and the steel

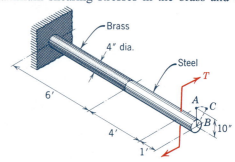

Figure P7-19

are not to exceed 7500 psi and 10,000 psi, respectively, and the arc length AC, through which the end of the 10-in. pointer AB moves, is not to exceed 0.60 in.

7-20 A stepped steel ($G = 80$ GPa) shaft has the dimensions and is subjected to the torques shown in Fig. P7-20. Determine
(a) The maximum shearing stress on a section 3 m from the left end of the shaft.
(b) The rotation of a section 2 m from the left end of the shaft with respect to its no-load position.
(c) The rotation of the section at the right end of the shaft with respect to its no-load position.

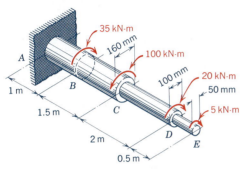

Figure P7-20

Challenging Problems

7-21* The maximum shearing stress in the stepped steel ($G = 12,000$ ksi) shaft shown in Fig. P7-21 is 20,000 psi. Determine
(a) The magnitude of torque T_3.
(b) The rotation of a section at B with respect to its no-load position.
(c) The rotation of a section at D with respect to its no-load position.

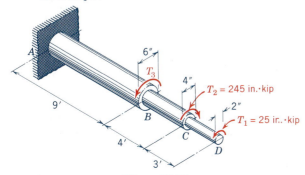

Figure P7-21

7-22* The steel ($G = 80$ GPa) shaft of Fig. P7-22 is in equilibrium under the torques applied as shown. The maximum shearing stress in the 160-mm section is 120 MPa, and the rotation of end C with respect to the section at A is 0.018 rad counterclockwise looking left. Determine the torques T_1 and T_2.

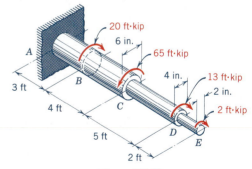

Figure P7-22

7-23 A steel ($G = 12,000$ ksi) shaft is loaded and supported as shown in Fig. P7-23. Determine
(a) The maximum shearing stress in the shaft.
(b) The rotation of the right end of the shaft with respect to its no-load position.
(c) The rotation of a section 5 ft from the left end of the shaft with respect to its no-load position.

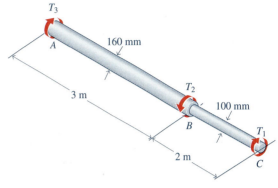

Figure P7-23

7-24 The 100 mm-diameter shaft shown in Fig. P7-24 is composed of aluminum alloy ($G = 28$ GPa) and steel ($G = 80$ GPa) sections rigidly connected. Determine
(a) The maximum shearing stress in the shaft.
(b) The rotation of a section 3 m from the left end of the shaft with respect to its no-load position.
(c) The rotation of a section at the right end of the shaft with respect to its no-load position.

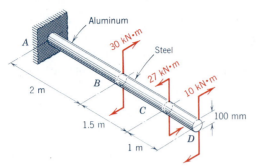

Figure P7-24

7-25* A torque of 40 ft · lb is applied through gear A to the left end of the gear train shown in Fig. P7-25. The diameters of gears B and C are 5 in. and 2 in., respectively. If the maximum shearing stresses in the aluminum alloy ($G = 3800$ ksi) shafts AB and CD are limited to 15 ksi, determine
(a) The minimum permissible diameter for shaft AB.
(b) The minimum permissible diameter for shaft CD.
(c) The maximum length for shaft CD if the rotation of a section at D with respect to a section at C must not exceed 0.5 rad.

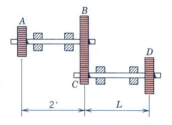

Figure P7-25

7-26* Torque is applied to the steel ($G = 80$ GPa) shaft shown in Fig. P7-26 through gear C and is removed through gears A and B. If the torque applied to gear C by the motor is 20 kN · m and the torque removed through gear B is 12 kN · m, determine

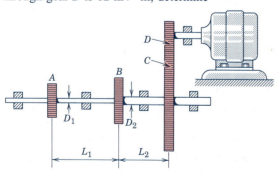

Figure P7-26

(a) The minimum permissible diameter for each section of the shaft if the maximum shearing stresses must not exceed 125 MPa.
(b) The minimum permissible uniform diameter for a shaft with $L_1 = 1.5$ m and $L_2 = 1.25$ m if the rotation of gear A relative to gear C must be less than 0.15 rad.

7-27 A motor supplies a torque of 31.4 in. · kip to the steel ($G = 12,000$ ksi) shaft AB shown in Fig. P7-27. Two machines are powered by gears D and E with the torque removed at gear E being 5.7 in. · kip. The diameters of gears B and C are 10 in. and 5 in., respectively. If the maximum shearing stresses in shafts AB and DCE must be limited to 20 ksi, determine
(a) The minimum satisfactory diameter for shaft AB.
(b) The minimum satisfactory diameter for shaft DCE.
(c) The rotation of gear B with respect to the coupling at A if a shaft with a 2.5-in. diameter is used.

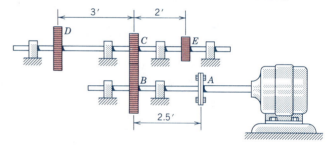

Figure P7-27

7-28 The motor shown in Fig. P7-28 supplies a torque of 45 kN · m to shaft AB. Two machines are powered by gears D and E. The torque delivered by gear E to the machine is 8 kN · m. Shafts AB and DCE are made of steel ($G = 80$ GPa) and have 150-mm and 80-mm diameters, respectively. If the diameters of gears B and C are 450 mm and 150 mm, respectively, determine
(a) The maximum shearing stress in shaft AB.
(b) The maximum shearing stress in shaft DCE.
(c) The rotation of gear E relative to gear D.

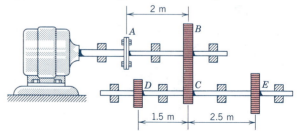

Figure P7-28

7-5 STRESSES ON OBLIQUE PLANES

At this point it is necessary to ascertain whether the transverse plane is a plane of maximum shearing stress and whether there are other significant stresses induced by torsion. For this study, the stresses at point A in the shaft of Fig. 7-11a will be analyzed. Figure 7-11b shows a differential element taken from the shaft at A and the stresses acting on transverse and longitudinal planes. The shearing stress τ_{xy} can be determined by means of the elastic torsion formula.[3] The equality of shearing stresses on orthogonal planes can be demonstrated by applying the equations of equilibrium to a free-body diagram of the differential element of Fig. 7-11b, which has length dx, height dy, and thickness dz. If a shearing force $V_x = \tau_{yx} \, dx \, dz$ is applied to the top surface of the element, the equation $\Sigma F_x = 0$ will require application of an oppositely directed force V_x to the bottom of the element, thus leaving the element subjected to a clockwise couple. This clockwise couple must be balanced by a counterclockwise couple composed of the oppositely directed forces $V_y = \tau_{xy} \, dy \, dz$ applied to the vertical faces of the element. Finally, application of the equation $\Sigma M_z = 0$ yields

$$\tau_{yx}(dx \, dz) \, dy = \tau_{xy}(dy \, dz) \, dx$$

from which

$$\tau_{yx} = \tau_{xy} \tag{7-8}$$

Therefore, if a shearing stress exists at a point on any plane, a shearing stress of the same magnitude must also exist at this point on an orthogonal plane. This statement is also valid when normal stresses are acting on the planes, since the normal stresses occur in collinear but oppositely directed pairs, and thus have zero moment with respect to any axis.

If the equations of equilibrium are applied to the free-body diagram of Fig. 7-11c (which is a wedge-shaped part of the differential element of Fig. 7-11b with dA being the area of the inclined face), the following results are obtained:

$$+\nwarrow\Sigma F_t = 0$$
$$\tau_{nt} \, dA - \tau_{xy}(dA \cos \alpha) \cos \alpha + \tau_{yx}(dA \sin \alpha) \sin \alpha = 0$$

from which

$$\tau_{nt} = \tau_{xy}(\cos^2 \alpha - \sin^2\alpha) = \tau_{xy} \cos 2\alpha \tag{7-9}$$
$$+\nearrow\Sigma F_n = 0$$
$$\sigma_n dA - \tau_{xy}(dA \cos \alpha) \sin \alpha - \tau_{yx}(dA \sin \alpha) \cos \alpha = 0$$

from which

$$\sigma_n = 2\tau_{xy} \sin \alpha \cos \alpha = \tau_{xy} \sin 2\alpha \tag{7-10}$$

In the development of Eqs. (7-9) and (7-10) stresses were multiplied by areas to produce forces on the free-body diagram of Fig. 7-11c; forces must always be used when applying the equations of equilibrium. In Eqs. (7-9) and (7-10)

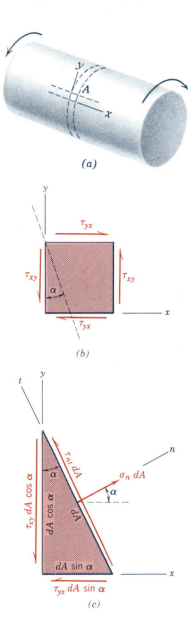

(a)

(b)

(c)

Figure 7-11

---

[3]The double subscript on the shearing stress is used to designate both the plane on which the stress acts and the direction of the stress. The first subscript indicates the plane (or rather the normal to the plane), and the second subscript indicates the direction of the stress.

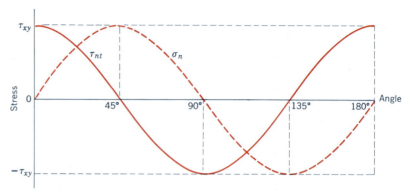

Figure 7-12

σ_n is the normal stress on the inclined plane and τ_{nt} is the shearing stress on the same plane. The shearing stress τ_{xy} is found using the elastic torsion formula, Eq. (7-6). At a given point of the circular shaft τ_{xy} is constant, and thus Eqs. (7-9) and (7-10) show that the stresses σ_n and τ_{nt} are functions of the angle of the inclined plane, α. The directions of all stresses and the angle α in Eqs. (7-9) and (7-10) and Fig. 7-11c are considered to be positive. The results obtained from Eqs. (7-9) and (7-10) are shown in the graph of Fig. 7-12, from which it is apparent that the maximum shearing stress occurs on both transverse ($\alpha = 0$) and longitudinal ($\alpha = 90°$) (diametral) planes. The graph also shows that maximum normal stresses occur on planes oriented at 45° with the axis of the bar and perpendicular to the surface of the bar. On one of these planes ($\alpha = 45°$ on Fig. 7-12b), the stress is tension, and on the other ($\alpha = 135°$), the stress is compression. Furthermore, all of these maximum stresses have the same magnitude; hence, the elastic torsion formula gives the magnitude of both the maximum normal stress and the maximum shearing stress at a point in a circular shaft subjected to pure torsion (the only loading is a torque).

Any of the stresses discussed previously may be significant in a given problem. Compare, for example, the failures shown in Fig. 7-13. In Fig. 7-13a, the

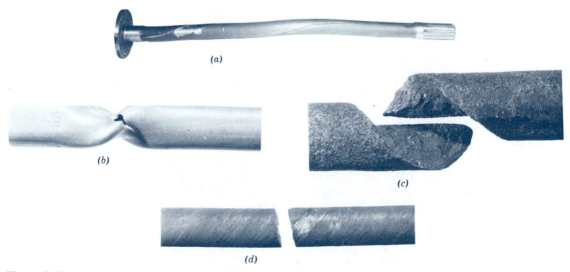

(a)

(b)

(c)

(d)

Figure 7-13

steel rear axle of a truck split longitudinally. One would also expect this type of failure to occur in a shaft of wood with the grain running longitudinally. In Fig. 7-13b, the compressive stress caused the thin-walled aluminum alloy tube to buckle along one 45° plane, while the tensile stress caused tearing on the other 45° plane. Buckling of thin-walled tubes subjected to torsional loading is a matter of paramount concern to the designer. In Fig. 7-13c, the tensile stresses caused the gray cast iron to fail in tension—typical of any brittle material subjected to torsion. In Fig. 7-13d, the low-carbon steel failed in shear on a plane that is almost transverse—a typical failure for a ductile material. The reason the fracture in Fig. 7-13d did not occur on a transverse plane is that under the large plastic twisting deformation before rupture (note the spiral lines indicating elements originally parallel to the axis of the bar), longitudinal elements were subjected to both torsion and axial tensile loading because the grips of the testing machine would not permit the bar to shorten as the elements were twisted into spirals. This axial tensile stress (not shown in Fig. 7-11) changes the plane of maximum shearing stress from a transverse to an oblique plane (resulting in a warped surface of rupture).[4]

Example Problem 7-6

A cylindrical tube is fabricated by butt-welding a 6 mm–thick steel plate along a spiral seam as shown in Fig. 7-14. If the maximum compressive stress in the tube must be limited to 80 MPa, determine

(a) The maximum torque T that can be applied to the tube.
(b) The factor of safety with respect to failure by fracture for the weld, when a torque of 12 kN · m is applied, if the ultimate strengths of the weld metal are 205 MPa in shear and 345 MPa in tension.

SOLUTION

(a) For the cylindrical tube:

$$J = \frac{\pi}{2}(75^4 - 69^4) = 14.096(10^6) \text{ mm}^4 = 14.096(10^{-6}) \text{ m}^4$$

The maximum compressive stress in the tube is given by Eq. (7-6) as

$$\sigma_{max} = \tau_{max} = \frac{T_{max}c}{J} = 80 \text{ MPa} = 80(10^6) \text{ N/m}^2$$

Thus,

$$T_{max} = \frac{\sigma_{max}J}{c} = \frac{80(10^6)(14.096)(10^{-6})}{75(10^{-3})}$$
$$= 15.036(10^3) \text{ N} \cdot \text{m} \cong 15.04 \text{ kN} \cdot \text{m} \qquad \textbf{Ans.}$$

---

[4]The tensile stress is not entirely due to the grips because the plastic deformation of the outer elements of the bar is considerably greater than that of the inner elements. This results in a spiral tensile stress in the outer elements and a similar compressive stress in the inner elements.

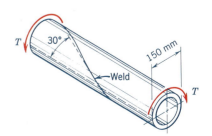

Figure 7-14

(b) The normal stress σ_n and the shear stress τ_{nt} on the weld surface are given by Eqs. (7-9) and (7-10) as

$$\sigma_n = \tau_{xy} \sin 2\alpha = \frac{Tc}{J} \sin 2\alpha = \frac{12(10^3)(75)(10^{-3})}{14.096(10^{-6})} \sin 2(60°)$$
$$= 55.29(10^6) \text{ N/m}^2 = 5.29 \text{ MPa (T)}$$

$$\tau_{nt} = \tau_{xy} \cos 2\alpha = \frac{Tc}{J} \cos 2\alpha = \frac{12(10^3)(75)(10^{-3})}{14.096(10^{-6})} \cos 2(60°)$$
$$= -31.92(10^6) \text{ N/m}^2 = -31.92 \text{ MPa}$$

The minus sign indicates that the direction of τ_{nt} is opposite to that shown on Fig. 7-11c. The factors of safety with respect to failure by fracture for the weld are

$$FS_\sigma = \frac{\sigma_{ult}}{\sigma_n} = \frac{345}{55.29} = 6.24 \qquad \textbf{Ans.}$$

$$FS_\tau = \frac{\tau_{ult}}{\tau_{nt}} = \frac{205}{31.92} = 6.42 \ \blacksquare \qquad \textbf{Ans.}$$

PROBLEMS

Standard Problems

7-29* An aluminum alloy ($G = 3800$ ksi) tube is to be used to transmit a torque in a control mechanism. The tube has an outside diameter of 1.25 in. and a wall thickness of 0.065 in. Because of the tendency of thin sections to buckle, the maximum compressive stress must be limited to 8000 psi. Determine
(a) The maximum torque that can be applied.
(b) The angle of twist in a 3-ft length when a torque of 1000 in. · lb is applied.

7-30* Determine the maximum torque that can be resisted by a hollow circular shaft having an inside diameter of 60 mm and an outside diameter of 100 mm without exceeding a normal stress of 70 MPa T or a shearing stress of 75 MPa.

7-31 A solid circular steel ($G = 12,000$ ksi) shaft with diameters as shown in Fig. P7-31 is subjected to a torque $T = 15$ in. · kip. Determine
(a) The maximum tensile stress in section AB of the shaft.
(b) The maximum compressive stress in section BC of the shaft.

(c) The rotation of a section at C with respect to its no-load position.

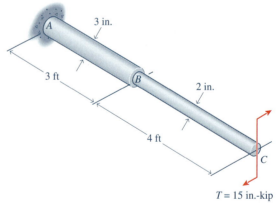

Figure P7-31

7-32 The hollow circular steel ($G = 80$ GPa) shaft shown in Fig. P7-32 has an outside diameter of 120 mm and an inside diameter of 60 mm. Determine
(a) The maximum compressive stress in the shaft.
(b) The maximum compressive stress in the shaft after the inside diameter is increased to 100 mm.

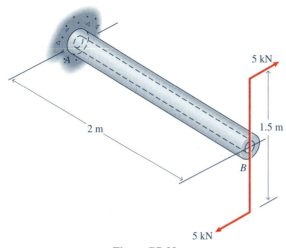

Figure P7-32

Challenging Problems

7-33* When the two torques are applied to the steel ($G = 12{,}000$ ksi) shaft of Fig. P7-33, point A moves 0.172 in. in the direction indicated by torque T_1. Determine

(a) The torque T_1.
(b) The maximum tensile stress in section BC of the shaft.
(c) The maximum compressive stress in section CD of the shaft.

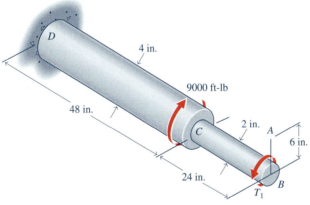

Figure P7-33

7-34* A solid circular stepped steel ($G = 80$ GPa) shaft has the dimensions and is subjected to the torques shown in Fig. P7-34. Determine

(a) The maximum tensile stress in section AB of the shaft.
(b) The maximum compressive stress in section BC of the shaft.
(c) The rotation of a section at C with respect to its no-load position.

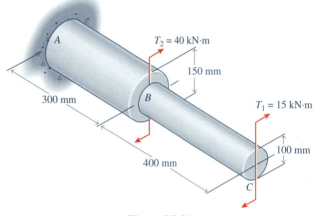

Figure P7-34

7-6 POWER TRANSMISSION

Almost anyone will recognize that one of the most common uses for the circular shaft is the transmission of power; therefore, no discussion of torsion would be complete without including this topic. Power is defined as the time rate of doing work. In everyday life, the word "work" is applied to any form of activity that requires the exertion of muscular or mental effort. In mechanics, however, the term is used in a much more restricted sense.

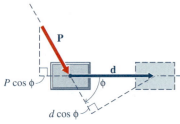

Figure 7-15

7-6-1 Work of a Force

In mechanics, a force does work only when the particle to which the force is applied moves. For example, when a constant force **P** is applied to a particle that moves a distance d in a straight line, as shown in Fig. 7-15, the work, U, done on the particle by the force **P** is defined by the scalar product

$$U = \mathbf{P} \cdot \mathbf{d} = Pd \cos \phi$$
$$= P_x d_x + P_y d_y + P_z d_z \tag{7-11}$$

where ϕ is the angle between the vectors **P** and **d**. Equation (7-11) is usually interpreted as

> The work done by a force **P** is the product of the magnitude of the force (P) and the magnitude of the rectangular component of the displacement in the direction of the force ($d \cos \phi$) (see Fig. 7-15).

However, $\cos \phi$ can also be associated with the force **P** instead of with the displacement **d**. Then Eq. (7-11) would be interpreted as

> The work done by a force **P** is the product of the magnitude of the displacement (d) and the magnitude of the rectangular component of the force in the direction of the displacement ($P \cos \phi$) (see Fig. 7-15).

Since work is defined as the scalar product of two vectors, work is a scalar quantity with only magnitude and algebraic sign. When the sense of the displacement and the sense of the force component in the direction of the displacement are the same ($0 \le \phi < 90°$), the work done by the force is positive. When the sense of the displacement and the sense of the force component in the direction of the displacement are opposite ($90° < \phi \le 180°$), the work done by the force is negative. When the direction of the force is perpendicular to the direction of the displacement ($\phi = 90°$), the component of the force in the direction of the displacement is zero and the work done by the force is zero. Of course, the work done by the force is also zero if the displacement is zero, $d = 0$.

Work has the dimensions of force times length. In the SI system of units, this combination of dimensions is called a joule (1 J = 1 N · m).[5] In the U.S. customary system of units, there is no special unit for work. It is expressed simply as ft · lb.

If the force is not constant or if the displacement is not in a straight line, Eq. (7-11) gives the work done by the force only during an infinitesimal part of the displacement, $d\mathbf{r}$ (see Fig. 7-16):

$$dU = \mathbf{P} \cdot d\mathbf{r} = P \, ds \cos \phi = P_t \, ds$$
$$= P_x \, dx + P_y \, dy + P_z \, dz \tag{7-12}$$

where $d\mathbf{r} = ds \, \mathbf{e}_t = dx \, \mathbf{i} + dy \, \mathbf{j} + dz \, \mathbf{k}$. The total work done by the force as

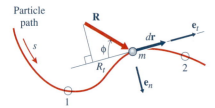

Figure 7-16

---

[5]It may be noted that work and moment of a force have the same dimensions: they are both force times length. However, work and moment are two toally different concepts and the special unit joule should be used only to describe work. The moment of a force must always be expressed as N · m in the SI system of units.

the particle moves from position 1 to position 2 is obtained by integrating Eq. (7-12) along the path of the particle

$$U_{1 \to 2} = \int_1^2 dU = \int_{s_1}^{s_2} P_t \, ds$$
$$= \int_{x_1}^{x_2} P_x \, dx + \int_{y_1}^{y_2} P_y \, dy + \int_{z_1}^{z_2} P_z \, dz \qquad (7\text{-}13)$$

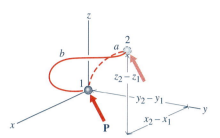

Figure 7-17

For the special case in which the force **P** is constant (both in magnitude and direction), the force components can be taken outside the integral signs in Eq. (7-13). Then Eq. (7-13) gives

$$U_{1 \to 2} = P_x \int_{x_1}^{x_2} dx + P_y \int_{y_1}^{y_2} dy + P_z \int_{z_1}^{z_2} dz$$
$$= P_x (x_2 - x_1) + P_y (y_2 - y_1) + P_z (z_2 - z_1) \qquad (7\text{-}14)$$

Note that the evaluation of the work done by a constant force depends on the co-ordinates at the end points of the particle's path but not on the actual path traveled by the particle. For the constant force **P** shown in Fig. 7-17, it doesn't matter if the particle moves along path a from position 1 to position 2 or along path b or along some other path. The work done by the force is always the same. Forces for which the work done is independent of the path are called *conservative forces*.

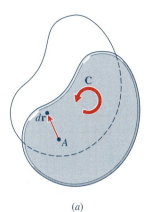

(a)

The weight W of a particle is a particular example of a constant force. When bodies move near the surface of the earth, the force of the earth's gravity is essentially constant ($P_x = 0$, $P_y = 0$, and $P_z = -W$). Therefore, the work done on a particle by its weight is $-W(z_2 - z_1)$. When $z_2 > z_1$, the particle moves upward (opposite the gravitational force), and the work done by gravity is negative. When $z_2 < z_1$, the particle moves downward (in the direction of the gravitational force), and the work done by gravity is positive.

Examples of forces that do work when a body moves from one position to another include the weight of the body, friction between the body and other surfaces, and externally applied loads. Examples of forces that do no work include forces at fixed points ($ds = 0$) and forces acting in a direction perpendicular to the displacement ($\cos \phi = 0$).

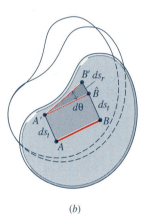

(b)

7-6-2 Work of a Couple

The work done by a couple is obtained by calculating the work done by each force of the couple separately and adding their works together. For example, consider a couple **C** acting on a rigid body, as shown in Fig. 7-18a. During some small time dt the body translates and rotates. If the displacement of point A is $d\mathbf{r} = ds \, \mathbf{e}_t$, choose a second point B such that the line AB is perpendicular to $d\mathbf{r}$. Then the motion that takes A to A' will take B to B'. This motion may be considered in two parts: first a translation that takes the line AB to $A'\hat{B}$, followed by a rotation $d\theta$ about A' that takes $\hat{B}$ to B' (see Fig. 7-18b).

Now represent the couple by a pair of forces of magnitude $P = C/b$ in the direction perpendicular to the line AB (see Fig. 7-18c). During the translational part of the motion, one force will do positive work $P \, ds_t$, and the other will do

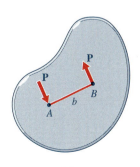

(c)

Figure 7-18

negative work $-P\, ds_t$: therefore, the sum of the work done on the body by the pair of forces during the translational part of the motion is zero. During the rotational part of the motion, A' is a fixed point and the force applied at A' does no work. The work done by the force at B is $dU = P\, ds_r \cong Pb\, d\theta$, where $d\theta$ is in radians and $C = Pb$ is the magnitude of the moment of the couple. Therefore, when a body is simultaneously translated and rotated, the couple does work only as a result of the rotation.

The total work done by the couple during the differential motion is

$$dU = P\, ds_t - P\, ds_t + Pb\, d\theta = C\, d\theta \tag{7-15}$$

The work is positive if the angular displacement $d\theta$ is in the same direction as the sense of rotation of the couple and negative if the displacement is in the opposite direction. No work is done if the couple is translated or rotated about an axis parallel to the plane of the couple.

The work done on the body by the couple as the body rotates through a finite angle $\Delta\theta = \theta_2 - \theta_1$ is obtained by integrating Eq. (7-15):

$$U_{1\to2} = \int_1^2 dU = \int_{\theta_1}^{\theta_2} C\, d\theta \tag{7-16}$$

If the couple is constant, then C can be taken outside the integral sign and Eq. (7-16) becomes

$$U_{1\to2} = C\int_{\theta_1}^{\theta_2} d\theta = C(\theta_2 - \theta_1) = C\,\Delta\theta \tag{7-17}$$

If the body rotates in space, the component of the infinitesimal angular displacement $d\boldsymbol{\theta}$ in the direction of the couple $\mathbf{C}$ is required. For this case, the work done is determined by using the dot product relationship,

$$dU = \mathbf{C} \cdot d\boldsymbol{\theta} = M\, d\theta \cos\phi \tag{7-18}$$

where M is the magnitude of the moment of the couple, $d\theta$ is the magnitude of the infinitesimal angular displacement, and ϕ is the angle between $\mathbf{C}$ and $d\boldsymbol{\theta}$. For planar motion (in the xy-plane), $\mathbf{C} = C\mathbf{k}$, $d\boldsymbol{\theta} = d\theta\, \mathbf{k}$, $\mathbf{C} \cdot d\boldsymbol{\theta} = C\, d\theta$, and Eq. (7-18) gives the same result as Eq. (7-17)

Since work is a scalar quantity, the work done on a rigid body by a system of external forces and couples is the algebraic sum of work done by the individual forces and couples.

Example Problem 7-7

A 500-lb block A is held in equilibrium on an inclined surface with a cable and weight system and a force P as shown in Fig. 7-19a. When the force P is removed and the system is disturbed, block A slides slowly down the incline at a constant speed for a distance of 10 ft. The coefficient of friction between block A and the inclined surface is 0.20. Determine

(a) The work done by the cable on block A.

(b) The work done by gravity on block A.

(c) The work done by the surface of the incline on block A.

(d) The work done by all forces on block A.

(e) The work done by the cable on block B.

(f) The work done by gravity on block B.

(g) The work done by all forces on block B.

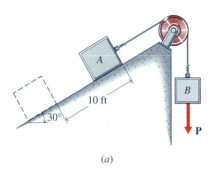

(a)

SOLUTION

Free-body diagrams for blocks A and B are shown in Figs. 7-19b and c. Four forces act on block A: the cable tension T, the weight W_A, and the normal and frictional forces N and F at the surface of contact between the block and the inclined surface. Two forces act on block B: the cable tension T and the weight W_B. Since the blocks move at a constant velocity they are in equilibrium and the equilibrium equations applied to block A yield

$$N = W_A \cos 30° = 500 \cos 30° = 433 \text{ lb}$$
$$F = \mu N = 0.20(433) = 86.6 \text{ lb}$$
$$T = W_A \sin 30° - F = 500 \sin 30° - 86.6 = 163.4 \text{ lb}$$

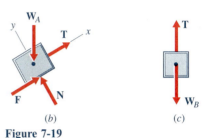

(b)   (c)

Figure 7-19

The equilibrium equations applied to block B yield

$$W_B = T = 163.4 \text{ lb}$$

Since all of the forces are constant in both magnitude and direction during the movements of the blocks, the work can be computed by using Eq. (7-14) (in two dimensions so that $z_2 - z_1 = 0$):

$$U_{1 \to 2} = P_x(x_2 - x_1) + P_y(y_2 - y_1)$$

Therefore, for block A, $x_2 - x_1 = -10$ ft, $y_2 - y_1 = 0$ ft, and

(a) $U_T = 163.4(-10) = -1634 \text{ ft} \cdot \text{lb}$ **Ans.**

(b) $U_W = (-500 \cos 60°)(-10) = 2500 \text{ ft} \cdot \text{lb}$ **Ans.**

Alternatively, the weight force acts vertically downward and block A drops a vertical distance $h = 10 \sin 30°$ so that

$$U_W = W_A h = 500(10 \sin 30°) = 2500 \text{ ft} \cdot \text{lb}$$

(c) $U_F = 86.6(-10) = -866 \text{ ft} \cdot \text{lb}$ **Ans.**

   $U_N = 0(-10) + 433(0) = 0 \text{ ft} \cdot \text{lb}$ **Ans.**

(d) $U_{\text{total}} = U_T + U_W + U_F + U_N = -1634 + 2500 - 866 + 0 = 0$ **Ans.**

and for block B, $x_2 - x_1 = 0$ ft, $y_2 - y_1 = 10$ ft, and

(e) $U_T = 163.4(10) = 1634 \text{ ft} \cdot \text{lb}$ **Ans.**

(f) $U_W = 163.4(-10) = -1634 \text{ ft} \cdot \text{lb}$ **Ans.**

(g) $U_{\text{total}} = U_T + U_W = -1634 + 1634 = 0$ ∎   **Ans.**

Example Problem 7-8

A constant couple

$$\mathbf{C} = 25\mathbf{i} + 35\mathbf{j} - 50\mathbf{k} \text{ N} \cdot \text{m}$$

acts on a rigid body. The unit vector associated with the fixed axis of rotation of the body for an infinitesimal angular displacement $d\boldsymbol{\theta}$ is

$$\mathbf{e}_\theta = 0.667\mathbf{i} + 0.333\mathbf{j} + 0.667\mathbf{k}$$

Determine the work done on the body by the couple during an angular displacement of 2.5 radians.

SOLUTION

The magnitude (moment M) of the couple is

$$M = \sqrt{(25)^2 + (35)^2 + (-50)^2} = 65.95 \text{ N} \cdot \text{m}$$

The unit vector associated with the couple is

$$\mathbf{e}_C = \frac{+25}{65.95}\mathbf{i} + \frac{+35}{65.95}\mathbf{j} + \frac{-50}{65.95}\mathbf{k} = 0.379\mathbf{i} + 0.531\mathbf{j} - 0.758\mathbf{k}$$

The cosine of the angle between the axis of the couple and the axis of rotation of the body is

$$\begin{aligned}
\cos \alpha &= \mathbf{e}_C \cdot \mathbf{e}_\theta \\
&= (0.379\mathbf{i} + 0.531\mathbf{j} - 0.758\mathbf{k}) \cdot (0.667\mathbf{i} + 0.333\mathbf{j} + 0.667\mathbf{k}) \\
&= -0.0760
\end{aligned}$$

Therefore,

$$\alpha = 94.36°$$

The work done by the couple during the finite rotation of 2.5 radians can be obtained by using Eq. (7-18). Thus,

$$dU = M \cos \alpha \, d\theta = 65.95(-0.0760)d\theta = -5.0122 \, d\theta$$

Therefore,

$$U = \int_0^{2.5} dU = \int_0^{2.5} (M \cos \alpha) \, d\theta$$

$$= -5.0122 \int_0^{2.5} d\theta = -12.53 \text{ N} \cdot \text{m} \qquad \textbf{Ans.}$$

Alternatively,

$$U = \int_0^{2.5} \mathbf{C} \cdot d\boldsymbol{\theta}$$

$$= \int_0^{2.5} (25\mathbf{i} + 35\mathbf{j} - 50\mathbf{k}) \cdot (0.667\mathbf{i} + 0.333\mathbf{j} + 0.667\mathbf{k})d\theta$$

$$= \int_0^{2.5} -5.02 \, d\theta = -12.55 \text{ N} \cdot \text{m} \qquad \textbf{Ans.}$$

The slight difference between this answer and the previous answer is due to the way the numbers were rounded off before being multiplied together. ■

PROBLEMS

Standard Problems

7-35* A 175-lb man climbs a flight of stairs 12 ft high. Determine
(a) The work done by the man.
(b) The work done on the man by gravity.

7-36* A box with a mass of 600 kg is dragged up an incline 12 m long and 4 m high by using a cable that is parallel to the incline. The force in the cable is 2500 N. Determine
(a) The work done on the box by the cable.
(b) The work done on the box by gravity.

7-37 A 100-lb block is pushed at constant speed for a distance of 20 ft along a level floor by a force that makes an angle of 35° with the horizontal, as shown in Fig. P7-37. The coefficient of friction between the block and the floor is 0.35. Determine
(a) The work done on the block by the force.
(b) The work done on the block by gravity.
(c) The work done on the block by the floor.

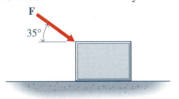

Figure P7-37

7-38 A block with a mass of 100 kg slides down an inclined surface that is 5 m long and makes an angle of 25° with the horizontal. A man pushes horizontally on the block so that it slides down the incline at a constant speed. The coefficient of friction between the block and the inclined surface is 0.15. Determine
(a) The work done on the block by the man.
(b) The work done on the block by gravity.
(c) The work done on the block by the inclined surface.

7-39* A crate weighing 300 lb is supported by a rope that is 40 ft long. A man pushes the crate 8 ft horizontally and holds it there. Determine
(a) The work done on the crate by the man.
(b) The work done on the crate by the rope.
(c) The work done on the crate by gravity.

7-40 A steel bar of uniform cross section is 2 m long and has a mass of 25 kg. The bar is supported in a vertical position by a horizontal pin at the top end of the bar. Determine the work done on the bar by gravity as the bar rotates 60° about the pin in a vertical plane.

Challenging Problems

7-41* A constant force acting on a particle can be expressed in Cartesian vector form as $\mathbf{F} = 8\mathbf{i} - 6\mathbf{j} + 2\mathbf{k}$ lb. Determine the work done by the force on the particle if the displacement of the particle can be expressed in Cartesian vector form as $\mathbf{s} = 5\mathbf{i} + 4\mathbf{j} + 6\mathbf{k}$ ft.

7-42* A constant force acting on a particle can be expressed in Cartesian vector form as $\mathbf{F} = 3\mathbf{i} + 5\mathbf{j} - 4\mathbf{k}$ N. Determine the work done by the force on the particle if the displacement of the particle can be expressed in Cartesian vector form as $\mathbf{s} = 4\mathbf{i} - 2\mathbf{j} + 3\mathbf{k}$ m.

7-43 A constant couple $\mathbf{C} = 120\mathbf{i} + 75\mathbf{j} - 150\mathbf{k}$ ft · lb acts on a rigid body. The unit vector associated with the fixed axis of rotation of the body for an infinitesimal angular displacement $d\boldsymbol{\theta}$ is $\mathbf{e}_\theta = 0.600\mathbf{i} + 0.300\mathbf{j} - 0.742\mathbf{k}$. Determine the work done on the body by the couple during an angular displacement of 0.75 radians.

7-44 A constant couple $\mathbf{C} = 200\mathbf{i} + 300\mathbf{j} + 350\mathbf{k}$ N · m acts on a rigid body. The unit vector associated with the fixed axis of rotation of the body for an infinitesimal angular displacement $d\boldsymbol{\theta}$ is $\mathbf{e}_\theta = 0.250\mathbf{i} + 0.350\mathbf{j} - 0.903\mathbf{k}$. Determine the work done on the body by the couple during an angular displacement of 1.5 radians.

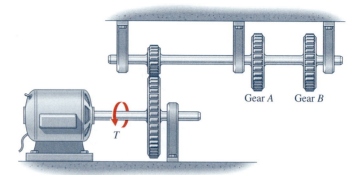

Figure 7-20

7-6-3 Power Transmission by Torsional Shafts

As previously mentioned, power is defined as the time rate of doing work. If a constant torque T acts on a circular shaft (see Fig. 7-20), the work done on the shaft by the torque is given by Eq. (7-17) as $U = T\theta$, where θ is the angular displacement of the shaft in radians. The time derivative of Eq. (7-17) gives

$$\text{power} = \frac{dU}{dt} = T\frac{d\theta}{dt} = T\omega \qquad (7-19)$$

where dU/dt is the power in ft $\cdot$ lb per min (or similar units), T is a constant torque in ft $\cdot$ lb, and ω is the angular velocity (of the shaft) in radians per minute. All units, of course, may be changed to any other consistent set of units. Since the angular velocity is usually given in revolutions per minute (rpm), the conversion of revolutions to radians will often be necessary. Also, in the U.S. customary system, power is usually given in units of horsepower, and the relation 1 hp = 33,000 ft $\cdot$ lb per min will be found useful. In the SI system, the basic unit of power is the watt (1 watt = 1 N $\cdot$ m/s).

Solution of a power transmission problem is illustrated in the following example problem.

Example Problem 7-9

A diesel engine for a small commercial boat is to operate at 200 rpm and deliver 800 hp through a gearbox with a ratio of 4 to 1 to the propeller shaft as shown in Fig. 7-21. Both the shaft from the engine to the gearbox and the propeller shaft are to be solid and made of heat-treated alloy steel. Determine the minimum permissible diameters for the two shafts if the allowable stress is 20 ksi and the angle of twist in a 10-ft length of the propeller shaft is not to exceed 4°. Neglect power loss in the gearbox and assume (incorrectly because of thrust stresses) that the propeller shaft is subjected to pure torsion.

SOLUTION

The first step is the determination of the torques to which the shafts are to be subjected. The torques are obtained by using Eq. (7-19):

$$\text{power} = T\omega = T(2\pi N)$$

$$T_1 = \frac{\text{power}}{2\pi N} = \frac{33{,}000(\text{hp})}{2\pi N} = \frac{33{,}000(800)}{2\pi(200)} = 21{,}010 \text{ ft} \cdot \text{lb}$$

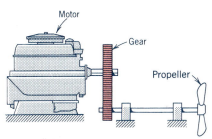

Figure 7-21

which is the torque at the crankshaft of the engine. Because the propeller shaft speed is four times that of the crankshaft and power loss in the gearbox is to be neglected, the torque on the propeller shaft is one fourth that on the crankshaft. Thus

$$T_2 = \tfrac{1}{4} T_1 = \tfrac{1}{4}(21{,}010) = 5252 \text{ ft} \cdot \text{lb}$$

Equation (7-6) will be used to determine the shaft sizes needed to satisfy the stress specification.
For the main shaft,

$$\frac{J}{c} = \frac{T}{\tau}$$

$$\frac{(\pi/2)c_1^4}{c_1} = \frac{21{,}010(12)}{20(10^3)}$$

which yields

$$c_1^3 = 8.024 \qquad \text{and} \qquad c_1 = 2.002 \text{ in.}$$

Thus, the minimum diameter of the shaft from the engine to the gearbox should be

$$d_m = 2c_1 = 2(2.002) = 4.004 \cong 4.00 \text{ in.} \qquad \textbf{Ans.}$$

The torque on the propeller shaft is one fourth that on the main shaft, and this is the only change in the expression for c_1^3; therefore,

$$c_2^3 = 8.024/4 \qquad \text{and} \qquad c_2 = 1.261 \text{ in.}$$

The propeller shaft size needed to satisfy the distortion specification will be determined from Eq. (7-7b)

$$\frac{\theta}{L} = \frac{T}{GJ}$$

which gives

$$\frac{4(\pi/180)}{10(12)} = \frac{5252(12)}{12(10^6)(\pi c_3^4/2)}$$

$$c_3^4 = 5.747 \qquad \text{and} \qquad c_3 = 1.548 > 1.261$$

Therefore, the minimum diameter of the propeller shaft must be

$$d_p = 2c_3 = 2(1.548) = 3.096 \cong 3.10 \text{ in.} \quad \blacksquare \qquad \textbf{Ans.}$$

PROBLEMS

Standard Problems

7-45* Determine the horsepower that a 10 in.–diameter shaft can transmit at 200 rpm if the maximum shearing stress in the shaft must be limited to 15,000 psi.

7-46* The shaft of a diesel engine is being designed to transmit 250 kW at 240 rpm. Determine the minimum diameter required if the maximum shearing stress in the shaft is not to exceed 80 MPa.

7-47 A steel (G = 12,000 ksi) shaft with a 4-in. diameter must not twist more than 0.06 rad in a 20-ft length. Determine the maximum power that the shaft can transmit at 270 rpm.

7-48 A solid circular steel (G = 80 GPa) shaft transmits 225 kW at 180 rpm. Determine the minimum diameter required if the angle of twist in a 3-m length must not exceed 0.035 rad.

7-49* A solid circular steel (G = 12,000 ksi) shaft is 3 in. in diameter and 4 ft long. If the maximum shearing stress must be limited to 10 ksi and the angle of twist over the 4-ft length must be limited to 0.024 rad, determine the maximum horsepower that this shaft can deliver when rotating at 225 rpm.

7-50* A 3 m–long hollow steel (G = 80 GPa) shaft has an outside diameter of 100 mm and an inside diameter of 60 mm. The maximum shearing stress in the shaft is 80 MPa and the angular velocity is 200 rpm. Determine
(a) The power being transmitted by the shaft.
(b) The magnitude of the angle of twist in the shaft.

7-51 The hydraulic turbines in a water-power plant rotate at 60 rpm and are rated at 20,000 hp with an overload capacity of 25,000 hp. The 30 in.–diameter shaft between the turbine and the generator is made of steel (G = 12,000 ksi) and is 20 ft long. Determine
(a) The maximum shearing stress in the shaft at rated load.
(b) The maximum shearing stress in the shaft at maximum overload.
(c) The angle of twist in the 20-ft length at maximum overload.

7-52 A solid circular steel (G = 80 GPa) shaft 1.5 m long transmits 200 kW at a speed of 400 rpm. If the allowable shearing stress is 70 MPa and the allowable angle of twist is 0.045 rad, determine
(a) The minimum permissible diameter for the shaft.

(b) The speed at which this power can be delivered if the stress is not to exceed 50 MPa in a shaft with a diameter of 75 mm.

7-53* The engine of an automobile supplies 162 hp at 3800 rpm to the drive shaft. If the maximum shearing stress in the drive shaft must be limited to 5 ksi, determine
(a) The minimum diameter required for a solid drive shaft.
(b) The maximum inside diameter permitted for a hollow drive shaft if the outside diameter is 3 in.
(c) The percent savings in weight realized if the hollow shaft is used instead of the solid shaft.

7-54* A hollow shaft of aluminum alloy (G = 28 GPa) is to transmit 1200 kW at 1800 rpm. The shearing stress is not to exceed 100 MPa and the angle of twist is not to exceed 0.20 rad in a 3-m length. Determine the minimum permissible outside diameter if the inside diameter is to be three fourths of the outside diameter.

7-55 A motor delivers 350 hp at 1800 rpm to a gearbox that reduces the speed to 200 rpm to drive a ball mill. If the maximum shearing stress in the shafts (G = 12,000 ksi) is not to exceed 15 ksi and the angle of twist in a 10-ft length is not to exceed 0.10 rad, determine the minimum permissible diameter for each of the shafts.

7-56 A motor delivers 120 kW at 1800 rpm to a gearbox that reduces the speed to 200 rpm to drive a crusher. If the maximum shearing stress in the shafts (G = 80 GPa) is not to exceed 70 MPa and the angle of twist in a 3-m length is not to exceed 0.075 rad, determine the minimum permissible diameter for each of the shafts.

Challenging Problems

7-57* A motor delivers 200 hp at 250 rpm to gear B of the factory drive shaft shown in Fig. P7-57. Gears A and

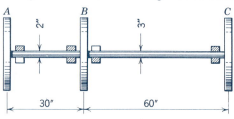

Figure P7-57

C transfer 120 hp and 80 hp, respectively, to operating machinery in the factory. Determine
(a) The maximum shearing stress in shaft AB.
(b) The maximum shearing stress in shaft BC.
(c) The rotation of gear C with respect to gear A if the shafts are both made of steel ($G = 12,000$ ksi).

7-58* A motor supplies 200 kW at 250 rpm to gear A of the factory drive shaft shown in Fig. P7-58. Gears B and C transfer 125 kW and 75 kW, respectively, to operating machinery in the factory. For an allowable shearing stress of 75 MPa, determine
(a) The minimum permissible diameter d_1 for shaft AB.
(b) The minimum permissible diameter d_2 for shaft BC.
(c) The rotation of gear C with respect to gear A if both shafts are made of steel ($G = 80$ GPa) and have diameters of 75 mm.

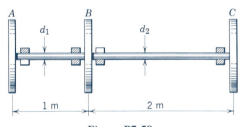

Figure P7-58

7-59 The motor shown in Fig. P7-59 develops 100 hp at a speed of 360 rpm. Gears A and B deliver 40 hp and 60 hp, respectively, to operating units in a factory. If the maximum shearing stress in the shafts must be limited to 12 ksi, determine
(a) The minimum satisfactory diameter for the motor shaft.

(b) The minimum satisfactory diameter for the power shaft.

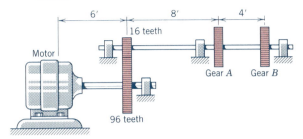

Figure P7-59

7-60 A motor provides 180 kW of power at 400 rpm to the drive shafts shown in Fig. P7-60. The maximum shearing stress in the three solid steel ($G = 80$ GPa) shafts must not exceed 70 MPa. Gears A, B, and C supply 40 kW, 60 kW, and 80 kW, respectively, to operating units in the plant. Determine
(a) The minimum satisfactory diameter for shaft A.
(b) The minimum satisfactory diameter for shaft B.
(c) The minimum satisfactory diameter for shaft C.

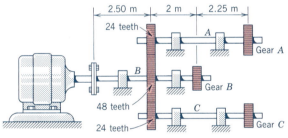

Figure P7-60

7-7 STATICALLY INDETERMINATE MEMBERS

All problems discussed in the preceding sections of this chapter were statically determinate; therefore, only the equations of equilibrium were required to determine the resisting torque at any section. Occasionally, torsionally loaded members are constructed and loaded such that the member is statically indeterminate (the number of independent equilibrium equations is less than the number of unknowns). When this occurs, distortion equations, which involve angles of twist, must be written until the total number of equations agrees with the number of unknowns to be determined. A simplified angle of twist diagram will often be of assistance in obtaining the correct equations. The following examples illustrate the procedures to be followed in solving statically indeterminate torsion problems.

Example Problem 7-10

The circular shaft AC of Fig. 7-22a is fixed to rigid walls at A and C. The solid section AB is made of annealed bronze ($G_{AB} = 45$ GPa), and the hollow section BC is made of aluminum alloy ($G_{BC} = 28$ GPa). There is no stress in the shaft before the 30-kN · m torque T is applied. Determine the maximum shearing stresses in both the bronze and aluminum portions of the shaft after the torque T is applied.

SOLUTION

A free-body diagram of the shaft is shown in Fig. 7-22b. The torques T_A and T_C at the supports are unknown. A summation of moments about the axis of the shaft, as shown in the torque diagram of Fig. 7-22c, gives

$$T_A + T_C = 30(10^3) \tag{a}$$

This is the only independent equation of equilibrium relating the two unknown torques T_A and T_C; therefore, the problem is statically indeterminate. A second equation can be obtained from the deformation of the shaft since the left and right portions of the shaft undergo the same angle of twist, as shown in Fig. 7-22d. Thus,

$$\theta_{B/A} = \theta_{B/C}$$

Since Eq. (a) is expressed in terms of T_A and T_C, the convenient form of the angle of twist equation for use in this example is Eq. (7-7b). Thus,

$$\frac{T_A L_{AB}}{G_{AB} J_{AB}} = \frac{T_C L_{BC}}{G_{BC} J_{BC}}$$

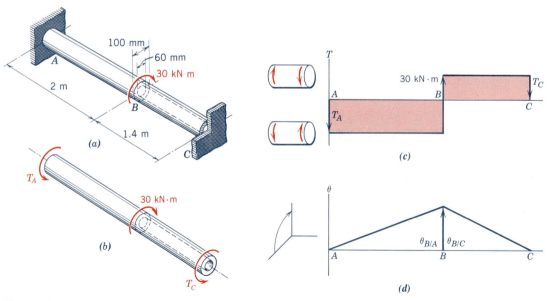

Figure 7-22

For the two segments of the shaft the polar second moments of area are

$$J_{AB} = (\pi/2)(50^4) = 9.817(10^6) \text{ mm}^4 = 9.817(10^{-6}) \text{ m}^4$$

$$J_{BC} = (\pi/2)(50^4 - 30^4) = 8.545(10^6) \text{ mm}^4 = 8.545(10^{-6}) \text{ m}^4$$

Therefore,

$$\frac{T_A(2)}{45(10^9)(9.817)(10^{-6})} = \frac{T_C(1.4)}{28(10^9)(8.545)(10^{-6})}$$

from which

$$T_C = 0.7737\, T_A \qquad \qquad \text{(b)}$$

Solving Eqs. (a) and (b) simultaneously yields

$$T_A = 16{,}914 \text{ N} \cdot \text{m} = 16.914 \text{ kN} \cdot \text{m}$$

$$T_C = 13{,}086 \text{ N} \cdot \text{m} = 13.086 \text{ kN} \cdot \text{m}$$

The stresses in the two portions of the shaft can then be obtained by using Eq. (7-6). Thus,

$$\tau_{AB} = \frac{T_A c_{AB}}{J_{AB}} = \frac{16.914(10^3)(50)(10^{-3})}{9.817(10^{-6})}$$

$$= 86.1(10^6) \text{ N/m}^2 = 86.1 \text{ MPa} \qquad \textbf{Ans.}$$

$$\tau_{BC} = \frac{T_C c_{BC}}{J_{BC}} = \frac{13.086(10^3)(50)(10^{-3})}{8.545(10^{-6})}$$

$$= 76.7(10^6) \text{ N/m}^2 = 76.7 \text{ MPa} \; \blacksquare \qquad \textbf{Ans.}$$

Example Problem 7-11

A hollow circular aluminum alloy ($G_a = 4000$ ksi) cylinder has a steel ($G_s = 11{,}600$ ksi) core as shown in Fig. 7-23a. The steel and aluminum parts are securely connected at the ends. If the allowable stresses in the steel and aluminum must be limited to 14 ksi and 10 ksi, respectively, determine

(a) The maximum torque T that can be applied to the right end of the composite shaft.

(b) The rotation of the right end of the composite shaft when the torque of part (a) is applied.

SOLUTION

A free-body diagram of the shaft is shown in Fig. 7-23b. Since torques and stresses will be related by the elastic torsion formula, which is limited to cross sections of homogeneous material, two unknown torques—the torque in the aluminum T_a and the torque in the steel T_s—have been placed on the left end of the shaft. Summing moments with respect to the axis of the shaft yields

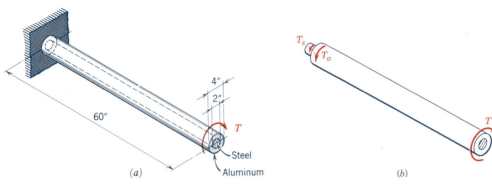

Figure 7-23

$$T_a + T_s = T \qquad\qquad (a)$$

Since Eq. (a) is the only independent equation of equilibrium, the problem is statically indeterminate. A second equation can be obtained from the deformation of the shaft. The fact that the shaft is nonhomogeneous does not invalidate the assumptions of plane cross sections remaining plane and diameters remaining straight. As a result, strains remain proportional to the distance from the axis of the shaft; however, stresses are not proportional to the radii throughout the entire cross section since G is not single-valued. The steel and aluminum parts of the shaft experience the same angle of twist because of the secure connections at the ends. Thus,

$$\theta_s = \theta_a$$

Since maximum shearing stresses are specified, the convenient form of the angle of twist equation for use in this example is Eq. (7-7a). Thus,

$$\frac{\tau_s L_s}{G_s c_s} = \frac{\tau_a L_a}{G_a c_a}$$

$$\frac{\tau_s(60)}{11.6(10^6)(1)} = \frac{\tau_a(60)}{4.0(10^6)(2)}$$

from which

$$\tau_s = 1.45\,\tau_a \qquad\qquad (b)$$

It is obvious from Eq. (b) that the shearing stress in the steel controls; therefore,

$$\tau_s = 14 \text{ ksi} \qquad \text{and} \qquad \tau_a = 14/1.45 = 9.655 \text{ ksi} < 10 \text{ ksi}$$

(a) Once the maximum shearing stresses in the steel and aluminum portions of the shaft are known, Eq. (7-6) can be used to determine the torques transmitted by the two parts of the shaft. Thus,

$$T_s = \frac{\tau_s J_s}{c_s} = \frac{14{,}000(\pi/2)(1^4)}{1} = 21{,}990 \text{ in.} \cdot \text{lb}$$

$$T_a = \frac{\tau_a J_a}{c_a} = \frac{9655(\pi/2)(2^4 - 1^4)}{2} = 113{,}750 \text{ in.} \cdot \text{lb}$$

From Eq. (a)

$$T = T_a + T_s = 21,990 + 113,750$$
$$= 135,740 \text{ in.} \cdot \text{lb} = 135.7 \text{ in.} \cdot \text{kip} \qquad \text{Ans.}$$

(b) The rotation of the right end of the shaft with respect to its no-load position can be determined by using either Eq. (7-7a) or Eq. (7-7b). If Eq. (7-7a) is used,

$$\theta = \theta_a = \theta_s = \frac{T_s L_s}{G_s c_s} = \frac{14,000(60)}{11.6(10^6)(1)} = 0.0724 \text{ rad} \quad \blacksquare \qquad \text{Ans.}$$

▮ Example Problem 7-12

The torsional assembly shown in Fig. 7-24a consists of a solid bronze ($G_B = 45$ GPa) shaft CD and a hollow aluminum alloy ($G_A = 28$ GPa) shaft EF that has a steel ($G_S = 80$ GPa) core. The ends C and F are fixed to rigid walls and the steel core of shaft EF is connected to the flange at E so that the aluminum and steel parts act as a unit. The two flanges D and E are bolted together, and the bolt clearance permits flange D to rotate through 0.03 rad before EF carries any of the load. Determine the maximum shearing stress in each of the shaft materials when the torque $T = 54$ kN · m is applied to flange D.

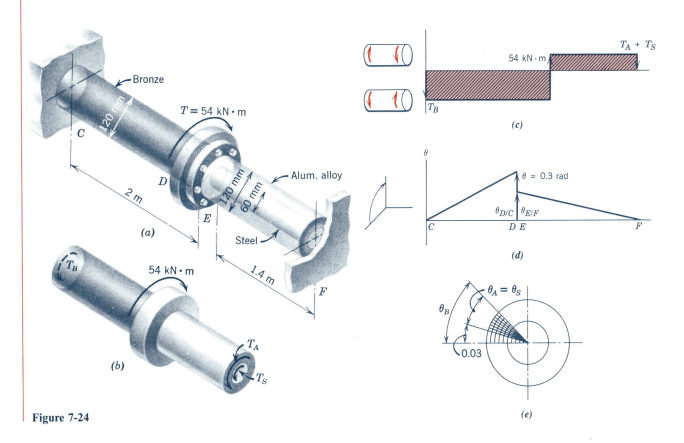

Figure 7-24

SOLUTION

A free-body diagram for the assembly is shown in Fig. 7-24b. An unknown torque T_B is shown at the left support and two unknown torques T_A and T_S are shown at the right support. Summing moments with respect to the axis of the shaft, as shown in the torque diagram of Fig. 7-24c, gives

$$T_B + T_A + T_S = 54(10^3) \tag{a}$$

Equation (a) is the only independent equilibrium equation that can be written relating the three unknown torques T_A, T_B, and T_S. Since there are three unknown torques and only one equilibrium equation, two distortion equations are needed to solve the problem. Two different types of angle of twist diagrams are shown in Figs. 7-24d and e. Angles of rotation for all cross sections of the shafts are shown in Fig. 7-24d. Relationships between angles $\theta_{D/C}$, $\theta_{E/F}$, and the rotation $\theta = 0.03$ rad permitted by the bolt clearance are required for solution of this example. The same quantities are shown in a polar form of representation in Fig. 7-24e. The angle of twist in the bronze shaft θ_B is the same as the rotation of coupling D with respect to the support at C. The angles of twist in the aluminum and steel shafts θ_A and θ_S, respectively, are equal and the same as the rotation of coupling E with respect to the support at F. Thus, the two distortion equations required for solution of the problem are

$$\theta_B = \theta_A + 0.03 \tag{b}$$

and

$$\theta_A = \theta_S \tag{c}$$

Equations (a), (b), and (c) can be written in terms of the same three unknowns (torque, angle, or stress) and solved simultaneously. Since maximum stresses are required, Eqs. (a), (b), and (c) will be written in terms of the maximum stress in each material by using Eqs. (7-6) and (7-7b), in which the polar second moments of area are

$$J_A = (\pi/2)(60^4 - 30^4) = 19.085(10^6)\ \text{mm}^4 = 19.085(10^{-6})\ \text{m}^4$$

$$J_B = (\pi/2)(60^4) = 20.36(10^6)\ \text{mm}^4 = 20.36(10^{-6})\ \text{m}^4$$

$$J_S = (\pi/2)(30^4) = 1.2723(10^6)\ \text{mm}^4 = 1.2723(10^{-6})\ \text{m}^4$$

Thus,

$$\frac{\tau_B J_B}{c_B} + \frac{\tau_A J_A}{c_A} + \frac{\tau_S J_S}{c_S} = 54(10^3)$$

$$\frac{\tau_B (20.36)(10^{-6})}{60(10^{-3})} + \frac{\tau_A(19.085)(10^{-6})}{60(10^{-3})} + \frac{\tau_S(1.2723)\ (10^{-6})}{30(10^{-3})} = 54(10^3)$$

from which

$$16\tau_B + 15\tau_A + 2\tau_S = 2.546(10^9) \tag{d}$$

Similarly,

$$\frac{\tau_B L_B}{G_B c_B} = \frac{\tau_A L_A}{G_A c_A} + 0.03$$

$$\frac{\tau_B(2)}{45(10^9)(60)(10^{-3})} = \frac{\tau_A(1.4)}{28(10^9)(60)(10^{-3})} + 0.03$$

from which

$$8\tau_B = 9\tau_A + 324(10^6) \qquad\qquad \text{(e)}$$

Also

$$\frac{\tau_A L_A}{G_A c_A} = \frac{\tau_S L_S}{G_S c_S}$$

$$\frac{\tau_A(1.4)}{28(10^9)(60)(10^{-3})} = \frac{\tau_S(1.4)}{80(10^9)(30)(10^{-3})}$$

from which

$$10\tau_A = 7\tau_S \qquad\qquad \text{(f)}$$

Solving Eqs. (d), (e), and (f) simultaneously yields

$$\tau_A = 52.93(10^6) \text{ N/m}^2 = 52.9 \text{ MPa} \qquad\qquad \textbf{Ans.}$$

$$\tau_B = 100.0(10^6) \text{ N/m}^2 = 100.0 \text{ MPa} \qquad\qquad \textbf{Ans.}$$

$$\tau_S = 75.62(10^6) \text{ N/m}^2 = 75.6 \text{MPa} \ \blacksquare \qquad\qquad \textbf{Ans.}$$

▌ PROBLEMS

Standard Problems

7-61* The 2 in.–diameter steel ($G = 12,000$ ksi) shaft shown in Fig. P7-61 is fixed to rigid walls at both ends. When a torque of 3500 ft · lb is applied as shown, determine

(a) The maximum shearing stress in the shaft.
(b) The angle of rotation of the section where the torque is applied with respect to its no-load position.

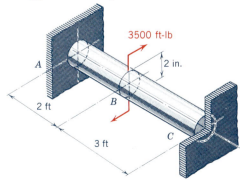

Figure P7-61

7-62* Two 80 mm–diameter solid circular steel ($G = 80$ GPa) and bronze ($G = 45$ GPa) shafts are rigidly

connected and supported as shown in Fig. P7-62. A torque T is applied at the junction of the two shafts as indicated. The allowable shearing stresses are 125 MPa for the steel and 40 MPa for the bronze. Determine

(a) The maximum torque T that can be applied.
(b) The angle of rotation of the section where the torque is applied with respect to its no-load position.

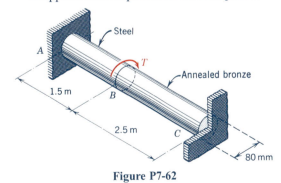

Figure P7-62

7-63 A hollow circular aluminum alloy ($G = 4000$ ksi) cylinder has a steel ($G = 12,000$ ksi) core as shown in Fig. P7-63. The steel and aluminum parts are securely connected at the ends. If the allowable stresses in the steel and aluminum must be limited to 12 ksi and 10 ksi, respectively, determine

(a) The maximum torque that can be applied to the right end of the composite shaft.

(b) The rotation of the right end of the composite shaft when the torque of part (a) is applied.

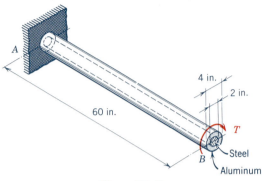

Figure P7-63

7-64 A composite shaft consists of a hollow bronze ($G = 40$ GPa) cylinder with a 150-mm outside diameter and a 100-mm inside diameter over a 50 mm–diameter solid steel ($G = 80$ GPa) core. The two members are rigidly connected to a bar at the right end and to the wall at the left end as shown in Fig. P7-64. When the couple shown is applied to the shaft, determine

(a) The rotation of bar AB.

(b) The maximum shearing stress in each of the members.

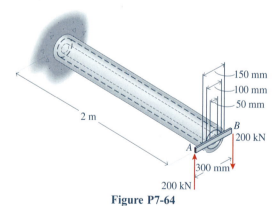

Figure P7-64

7-65* A hollow circular brass ($G = 5600$ ksi) tube with an outside diameter of 4 in. and an inside diameter of 2 in. is attached at the ends to a solid 2 in.–diameter steel ($G = 12,000$ ksi) core as shown in Fig. P7-65. Determine the maximum shearing stress in the tube when the composite shaft is transmitting a torque of 10 ft · kip.

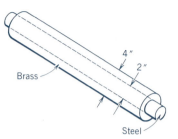

Figure P7-65

7-66* A steel ($G = 80$ GPa) tube with an inside diameter of 100 mm and an outside diameter of 125 mm is encased in a Monel ($G = 65$ GPa) tube with an inside diameter of 125 mm and an outside diameter of 175 mm as shown in Fig. P7-66. The tubes are connected at the ends to form a composite shaft. The shaft is subjected to a torque of 15 kN · m. Determine

(a) The maximum shearing stress in each material.

(b) The angle of twist in a 5-m length.

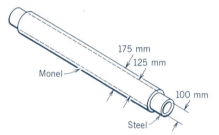

Figure P7-66

7-67 A 3 in.–diameter cold-rolled steel ($G = 11,600$ ksi) shaft, for which the maximum allowable shearing stress is 15 ksi, exhibited severe corrosion in a certain installation. It is proposed to replace the shaft with one in which an aluminum alloy ($G = 4000$ ksi) tube $\frac{1}{4}$-in. thick is bonded to the outer surface of the cold-rolled steel shaft to produce a composite shaft. If the maximum allowable shearing stress in the aluminum alloy tube is 12 ksi, determine

(a) The maximum torque that the original shaft can transmit.

(b) The maximum torque that the replacement shaft can transmit.

7-68 A composite shaft consists of a bronze ($G = 45$ GPa) sleeve with an outside diameter of 80 mm and an inside diameter of 60 mm over a solid aluminum ($G = 28$ GPa) rod with an outside diameter of 60 mm. If the allowable shearing stress in the bronze is 150 MPa, determine

(a) The maximum torque T that can be transmitted by the composite shaft.
(b) The maximum shearing stress in the aluminum rod when the maximum torque is being transmitted.

7-69* The composite shaft shown in Fig. P7-69 consists of a solid brass ($G = 5600$ ksi) core with an outside diameter of 2 in. covered by a steel ($G = 11,600$ ksi) tube with an inside diameter of 2 in. and a wall thickness of $\frac{1}{2}$ in. that is in turn covered by an aluminum alloy ($G = 4000$ ksi) sleeve with an inside diameter of 3 in. and a wall thickness of $\frac{1}{4}$ in. The three materials are bonded so that they act as a unit. Determine
(a) The maximum shearing stress in each material when the assembly is transmitting a torque of 10 ft · kip.
(b) The angle of twist in a 10-ft length when the assembly is transmitting a torque of 8 ft · kip.

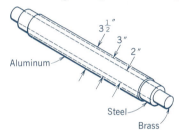

Figure P7-69

7-70* A hollow steel ($G = 80$ GPa) tube with an outside diameter of 100 mm and an inside diameter of 50 mm is covered with a Monel ($G = 65$ GPa) tube that has an outside diameter of 125 mm and an inside diameter of 102 mm. The tubes are connected at the ends to form a composite shaft. If the maximum allowable shearing stress in the steel is 70 MPa and the maximum allowable shearing stress in the Monel is 85 MPa, determine
(a) The maximum torque T that the composite shaft can transmit.
(b) The angle of twist in a 2.5-m length when the composite shaft is transmitting the maximum torque.

7-71 A solid aluminum alloy ($G = 4000$ ksi) rod with an outside diameter of 2 in. is used as a shaft. A hollow steel ($G = 12,000$ ksi) tube with an inside diameter of 2 in. is placed over the rod to increase the torque-transmitting capacity of the shaft. The tube and the rod are attached at the ends to form a composite shaft. Determine the tube thickness required to permit the torque-transmitting capacity of the shaft to be increased by 50 percent.

7-72 A solid Monel ($G = 65$ GPa) rod with an outside diameter of 60 mm is used as a shaft. A hollow stainless

steel ($G = 86$ GPa) tube with an inside diameter of 60 mm is placed over the rod to increase the torsional stiffness of the shaft. The tube and rod are attached at the ends to form a composite shaft. Determine the tube thickness required if the angle of twist for a given torque must be decreased by 50 percent.

7-73* The steel ($G = 12,000$ ksi) shaft shown in Fig. P7-73 is attached to rigid walls at both ends. The right 10 ft of the shaft is hollow, having an inside diameter of 2 in. Determine
(a) The maximum shearing stress in the shaft.
(b) The angle of rotation of the section where the torque is applied with respect to its no-load position.

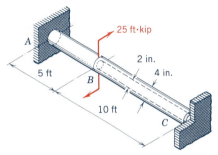

Figure P7-73

7-74* A composite shaft consists of a bronze ($G = 45$ GPa) shell that has an outside diameter of 100 mm bonded to a solid steel ($G = 80$ GPa) core. Determine the diameter of the steel core when the torque resisted by the steel core is equal to the torque resisted by the bronze shell.

7-75 The aluminum alloy ($G = 4000$ ksi) shaft of Fig. P7-75 is rigidly fixed to the wall at C, but the flange at A allows the left end of the shaft to rotate 0.012 rad before the bolts provide rigid support. Determine the maximum torque that can be applied at section B if the shearing stress is not to exceed 12 ksi in either section of the shaft.

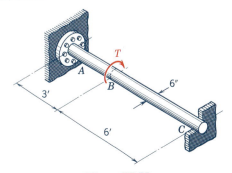

Figure P7-75

7-76 The torsional assembly of Fig. P7-76 consists of an aluminum alloy ($G = 28$ GPa) segment AB securely connected to a steel ($G = 80$ GPa) segment BCD by means of a flange coupling with four bolts. The diameters of both segments are 80 mm, the cross-sectional area of each bolt is 130 mm², and the bolts are located 65 mm from the center of the shaft. If the shearing stress in the bolts is limited to 55 MPa, determine

(a) The maximum torque T that can be applied at section C.

(b) The maximum shearing stress in the steel.

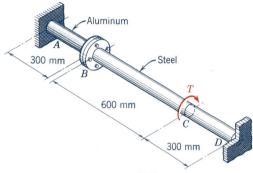

Figure P7-76

Challenging Problems

7-77* The steel ($G = 12,000$ ksi) shaft shown in Fig. P7-77 will be used to transmit a torque of 1000 in. · lb. The hollow portion AE of the shaft is connected to the solid portion BF with two pins at C and D as shown. If the average shearing stress in the pins must be limited to 25 ksi, determine the minimum satisfactory diameters for the pins.

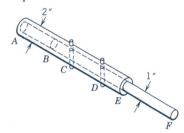

Figure P7-77

7-78* A torque T of 8 kN · m is applied to the steel ($G = 80$ GPa) shaft of Fig. P7-78 without the brass ($G = 40$ GPa) shell. The brass shell is then slipped into place and attached to the steel. After the original torque is released, determine the maximum shearing stress in the brass shell and the maximum shearing stress in the steel shaft.

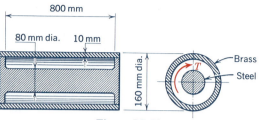

Figure P7-78

7-79 The 4 in.–diameter shaft of Fig. P7-79 is composed of brass ($G = 6000$ ksi) and steel ($G = 12,000$ ksi) segments. Determine the maximum permissible magnitude for the torque T, applied at C, if the allowable shearing stresses are 5 ksi for the brass and 12 ksi for the steel.

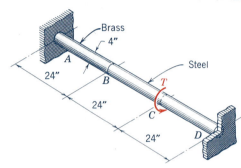

Figure P7-79

7-80 The circular shaft of Fig. P7-80 consists of a steel ($G = 80$ GPa) segment ABC securely connected to a bronze ($G = 40$ GPa) segment CD. Ends A and D of the shaft are fastened securely to rigid supports. Determine the maximum shearing stresses in the bronze and in the steel when T has a magnitude of 22 kN · m.

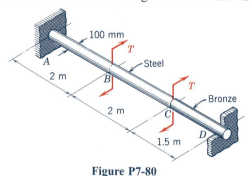

Figure P7-80

7-81* The shaft of Fig. P7-81 consists of a 6-ft hollow steel ($G = 12,000$ ksi) section AB and a 4-ft solid aluminum alloy ($G = 4000$ ksi) section CD. The torque T of 34 ft · kip is applied initially only to the steel section AB. Section CD is then connected and the torque T is released. When the torque is released, the connection slips 0.012 rad before the aluminum section takes any load. Determine the maximum shearing stress in the aluminum alloy.

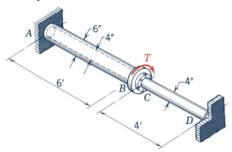

Figure P7-81

7-82* A hollow steel ($G = 80$ GPa) tube with an inside diameter of 60 mm, an outside diameter of 75 mm, and a length of 500 mm is encased with a brass ($G = 40$ GPa) tube that has an inside diameter of 75 mm and an outside diameter of 100 mm. The brass and steel tubes, while unstressed, are brazed together at one end. A couple is then applied to the other end of the brass tube, which twists one degree with respect to the steel tube. The tubes are then brazed together in that position. Determine the maximum shearing stresses in the steel tube and in the brass tube after the couple is removed from the brass tube.

7-83 The inner surface of the aluminum alloy ($G = 4000$ ksi) sleeve A and the outer surface of the steel ($G = 12,000$ ksi) shaft B of Fig. P7-83 are smooth. Both the sleeve and the shaft are rigidly fixed to the wall at D. The 0.500 in.–diameter pin C fills a hole drilled completely through a diameter of the sleeve and shaft. If the average shearing stress on the cross-sectional area of the pin at the interface between the shaft and the sleeve must not exceed 5000 psi, determine

(a) The maximum torque T that can be applied to the right end of the steel shaft B.
(b) The maximum shearing stress in the aluminum sleeve A when the torque of part (a) is applied.
(c) The rotation of the right end of the shaft when the maximum torque T is applied.

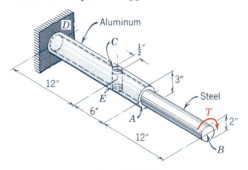

Figure P7-83

7-84 The 160 mm–diameter steel ($G = 80$ GPa) shaft shown in Fig. P7-84 has a 100 mm–diameter bronze ($G = 40$ GPa) core inserted in 3 m of the right end and securely bonded to the steel. Determine
(a) The maximum shearing stress in each of the materials.
(b) The rotation of the free end of the shaft.

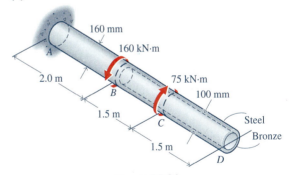

Figure P7-84

7-8 DESIGN PROBLEMS

Design, within a limited context, has been discussed previously in Chapters 4 and 6. In those chapters, design was limited to axially loaded members and to pins. This chapter will extend design to solid or hollow circular bars subjected to static torsional loading. Design of these bars will be limited to proportioning a torsionally loaded member to perform a specified function without failure. As before, failure will refer to elastic failure, slip failure, or failure by fracture.

Furthermore, design will be limited to circular bars made of ductile materials for which the significant failure stress is the shearing stress at yield or fracture. This book will not address the important issue of fatigue loading. That topic is covered in later courses.

Example Problem 7-13

A solid circular shaft 4 ft long made of 2014-T4 wrought aluminum is subjected to a torsional load of 10,000 in. · lb. If failure is by slip and a factor of safety (FS) of 2.0 is specified, select a suitable diameter for the shaft, if 2014-T4 wrought aluminum bars are available with diameters in increments of $\frac{1}{8}$ in.

SOLUTION

Since failure is by slip, the significant failure stress is the shearing yield stress. The shearing yield stress for 2014-T4 wrought aluminum is listed in Table B-17 as 24 ksi. Thus, the allowable shearing stress in the shaft is

$$\tau_{all} = \frac{\tau_{yield}}{FS} = \frac{24}{2.0} = 12 \text{ ksi}$$

since the factor of safety for the shaft is 2.0. For torsional loading the shearing stress in the shaft is given by Eq. (7-6) as

$$\tau_{all} = \frac{Tc}{J} \qquad \text{or} \qquad \frac{J}{c} = \frac{T}{\tau_{all}}$$

The polar second moment J of the cross section is given by Eq. (7-5a) as

$$J = \frac{\pi}{2} c^4$$

Thus,

$$\frac{\pi c^3}{2} = \frac{T}{\tau_{all}} = \frac{10,000}{12,000}$$

from which

$$c = 0.8095 \text{ in.}$$

Therefore,

$$d = 2c = 2(0.8095) = 1.6190 \text{ in.}$$

To the next largest $\frac{1}{8}$ in.:

$$d = 1\frac{5}{8} \text{ in.} \quad \blacksquare \qquad\qquad \textbf{Ans.}$$

Example Problem 7-14

A solid circular shaft 2 m long is to transmit 1000 kW at 600 rpm. The shaft is made of structural steel with an allowable shearing stress of 55 MPa. If a maximum angle of twist of 3.0° in the 2-m length is specified, select a suitable diameter for the shaft, if structural steel bars are available with diameters in increments of 10 mm.

SOLUTION

The shearing stress in the shaft must be limited to the allowable shearing stress for structural steel, which is 55 MPa. The torque to be transmitted by the shaft is given by Eq. (7-19) as

$$T = \frac{\text{Power}}{\omega} = \frac{1000(10^3)}{2\pi(600)/60} = 15{,}915 \text{ N} \cdot \text{m}$$

For torsional loading, the shearing stress in the shaft is given by Eq. (7-6)

$$\tau_{all} = \frac{Tc}{J} \quad \text{or} \quad \frac{J}{c} = \frac{T}{\tau_{all}}$$

The polar second moment J of the cross section is given by Eq. (7-5a) as

$$J = \frac{\pi}{2} c^4$$

Thus,

$$\frac{J}{c} = \frac{\pi c^3}{2} = \frac{T}{\tau_{all}} = \frac{15{,}915}{55(10^6)}$$

from which

$$c = 56.90(10^{-3}) \text{ m} = 56.90 \text{ mm}$$

For torsionally loaded shafts, the angle of twist is given by Eq. (7-7b) as

$$\theta = \frac{TL}{GJ} = \frac{15{,}915(2)}{76(10^9)(\pi/2)(0.05690)^4} = 0.02544 \text{ rad} = 1.457° < 3.0°$$

Therefore, $c = 56.90$ mm satisfies both the stress and angle of twist requirements. The required diameter is

$$d = 2c = 2(56.90) = 113.80 \text{ mm}$$

To the next largest 10 mm:

$$d = 120.0 \text{ mm} \quad \blacksquare \qquad\qquad \textbf{Ans.}$$

Example Problem 7-15

A stepped shaft is subjected to the torques shown in Fig. 7-25. Both segments of the shaft are made of 6061-T6 wrought aluminum. For a factor of safety of 2.0 against failure by slip, determine the required diameters for the two segments of the shaft.

SOLUTION

For failure by slip, the significant stress in the shaft is the shearing yield stress. From Table B-17, the shearing yield stress for 6061-T6 wrought aluminum is

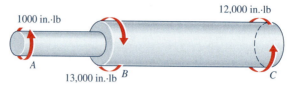

Figure 7-25

26 ksi. Since the specified factor of safety is $FS = 2.0$, the allowable shearing stress is

$$\tau_{\text{all}} = \frac{\tau_{\text{yield}}}{FS} = \frac{26}{2.0} = 13 \text{ ksi}$$

The torques transmitted by cross sections in intervals AB and BC of the shaft are

$$T_{AB} = 1000 \text{ in.} \cdot \text{lb} \qquad T_{BC} = 12,000 \text{ in.} \cdot \text{lb}$$

The shearing stress in a circular shaft is related to the resisting torque on a cross section by the elastic torsion formula, Eq. (7-6). Thus,

$$\tau_{\text{all}} = \frac{Tc}{J}$$

The polar second moment J of a circular cross section [see Eq. (7-5a)] is

$$J = \frac{\pi c^4}{2}$$

Thus,

$$\frac{J}{c} = \frac{\pi c^3}{2} = \frac{T}{\tau_{\text{all}}}$$

For segment AB of the shaft:

$$\frac{J}{c} = \frac{\pi c^3}{2} = \frac{1000}{13,000}$$

$$c_{AB} = 0.3659 \text{ in.}$$

$$d_{AB} = 2(0.3659) = 0.7318 \cong 0.732 \text{ in.} \qquad \textbf{Ans.}$$

For segment CD of the shaft:

$$\frac{J}{c} = \frac{\pi c^3}{2} = \frac{12,000}{13,000}$$

$$c_{CD} = 0.8376 \text{ in.}$$

$$d_{CD} = 2(0.8378) = 1.6752 \cong 1.675 \text{ in.} \qquad \blacksquare \qquad \textbf{Ans.}$$

Example Problem 7-16

A steel pipe will be used as a shaft to transmit 100 kW at 120 rpm. For an allowable shearing stress of 50 MPa, determine the lightest weight standard steel pipe that can be used for the shaft.

SOLUTION

The torque to be transmitted by the shaft is given by Eq. (7-19) as

$$T = \frac{\text{Power}}{\omega} = \frac{100(10^3)}{2\pi(120)/60} = 7958 \text{ N} \cdot \text{m}$$

For torsional loading, the shearing stress in the shaft is given by Eq. (7-6) as

$$\tau_{\text{all}} = \frac{Tc}{J} \quad \text{or} \quad \frac{J}{c} = \frac{T}{\tau_{\text{all}}}$$

The polar second moment J of the cross section is given by Eq. (7-5b) as

$$J = \frac{\pi}{2}(r_o^4 - r_i^4)$$

This equation can be satisfied for an infinite number of hollow pipes having different radius ratios. For this problem

$$\frac{J}{c} = \frac{T}{\tau_{\text{all}}} = \frac{7958}{50(10^6)} = 159.16(10^{-6}) \text{ m}^3 = 159.16(10^3) \text{ mm}^3$$

Table B-14 can be used to select a pipe that satisfies the requirement that $J/c \geq 159.16(10^3)$ mm^3. However, Table B-14 lists the properties I and S, where I is the rectangular second moment of area with respect to a diameter of the pipe, and S is the section modulus, $S = I/c$. Since $J = 2I$ for circular sections (either solid or hollow), the required section modulus for the pipe is

$$S = \frac{I}{c} = \frac{1}{2}\frac{J}{c} = \frac{1}{2}(159.16)(10^3) = 79.58(10^3) \text{ mm}^3$$

The lightest weight pipe in Table B-14, with $S \geq 79.58(10^3)$ mm^3, is a 127-mm nominal diameter pipe. Any pipe in Table B-14 that has a section modulus greater than $S = 79.58(10^3)$ mm^3 would satisfy the stress requirement, but the 127-mm pipe is the lightest since it has the smallest cross-sectional area. ∎

Ans.

PROBLEMS

Standard Problems

7-85 A 5 ft–long standard steel ($E = 11,000$ ksi) pipe is subjected to a torque of 1600 ft · lb at each end. The allowable shearing stress is 8000 psi and the allowable angle of twist is 3.0°. Determine the nominal diameter of the lightest standard weight steel pipe that can be used for the shaft.

7-86 The motor shown in Fig. P7-86 supplies a torque of 1000 N · m to shaft ABCDE. The torques removed at C, D, and E are 500 N · m, 300 N · m, and 200 N · m, respectively. The shaft is the same diameter throughout, and is made of steel alloy with an allowable shearing stress of 60 MPa. Select a suitable diameter for the shaft, if shafts are available with diameters from 20 mm to 120 mm in increments of 10 mm.

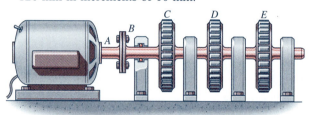

Figure P7-86

7-87 A torque of 30,000 in. · lb is supplied to the factory drive shaft of Fig. P7-87 by a belt that drives pulley A. A torque of 10,000 in. · lb is removed by pulley B and 20,000 in. · lb by pulley C. The shaft is made of structural steel and has a constant diameter over its length. Segment AB of the shaft is 3 ft long, and segment BC is 4 ft long. The allowable shearing stress for the structural steel is 8000 psi and the allowable rotation of end A with respect to end C is 2.0°. Select a suitable diameter for the shaft, if shafts are available with diameters from 2 in. to 6 in. in increments of $\frac{1}{8}$ in.

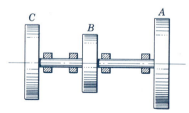

Figure P7-87

7-88 A standard steel pipe must transmit 150 kW at 60 rpm without exceeding an allowable shearing stress of 55 MPa.

(a) Select the lightest standard weight steel pipe that can be used.
(b) If solid structural steel shafts are available with diameters in increments of 10 mm, determine the minimum diameter that can be used.
(c) Compare the weights of the two shafts.

7-89 A shaft is to transmit 100 hp at 200 rpm. The designer has a variety of solid bars and standard steel pipes to select from. Both the bars and the pipes are made of structural steel with an allowable shearing stress of 8000 psi.

(a) Select the lightest standard weight steel pipe that can be used.
(b) Select a suitable solid shaft, if they are available with diameters in increments of $\frac{1}{8}$ in.
(c) If weight is critical, which shaft should be used?

7-90 A motor is to transmit 150 kW to a piece of mechanical equipment. The power is transmitted through a solid steel shaft with an allowable shearing stress of 50 MPa. The designer has the freedom to operate the motor at 60 rpm or 6000 rpm. For each case, determine the minimum shaft diameter needed to satisfy the stress requirement. Shafts are available with diameters in increments of 5 mm. If weight is important, which speed would be used?

7-91 Solid steel shafts are available with diameters in increments of $\frac{1}{8}$ in. A designer must select a shaft to transmit 100 hp without exceeding a shearing stress of 8000 psi. The designer can operate the system at 60 rpm or 6000 rpm. For each case, select the minimum diameter shaft needed to satisfy the stress conditions. If weight is important, which speed would be used?

Challenging Problems

7-92 The motor shown in Fig. P7-92 delivers 150 kW of power at 300 rpm to a piece of equipment at B. Shafts made from the materials listed below are available with diameters in increments of 5 mm. The length of the shaft between A and B is 1 m. Determine the shaft diameter required and the material to be used if the maximum shearing stress in the shaft must not exceed the allowable shearing stress in the material, and the angle of twist between A and B must not exceed 1.5°.

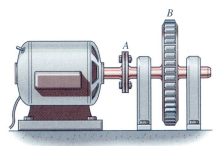

Figure P7-92

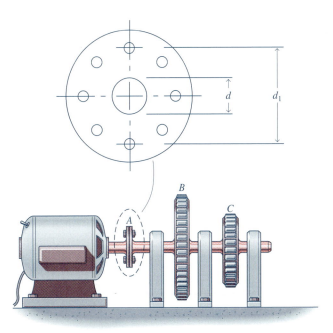

Figure P7-94

Material	Allowable stress in shear, MPa (ksi)	Modulus of rigidity GPa (ksi)
Structural steel	175 (10.8)	76 (11,000)
0.4% C hot-rolled steel	108 (15.9)	80 (11,600)
SAE 4340 heat-treated steel	273 (39.6)	76 (11,000)
Monel	102 (15)	65 (9,500)
Annealed titanium	279 (40.5)	36 (5,300)

7-93 A motor delivers 200 hp at a speed of 300 rpm to a piece of industrial equipment at B as shown in Fig. P7-92. Shafts made from the materials listed in Problems 7-92 are available with diameters in increments of $\frac{1}{8}$ in. The length of the shaft between A and B is 36 in. Determine the shaft diameter required and the material to be used if the maximum shearing stress in the shaft must not exceed the allowable shearing stress in the material, and the angle of twist between A and B must not exceed 3.5°.

7-94 A motor supplies a torque of 5 kN · m to the constant-diameter structural steel line shaft shown in Fig. P7-94. Two machines are driven by gears B and C on the shaft and they require torques of 3 kN · m and 2 kN · m, respectively. The allowable shearing stress for the shaft is 80 MPa and the allowable angle of twist between the coupling at A and gear C is 2.0°. The length of segment BC is 1.5 m and the length of segment AB is 1.0 m. Determine
(a) The minimum permissible diameter for the shaft, if shafts are available with diameters in increments of 5 mm.
(b) The diameter of the bolts in the coupling, if eight bolts are used, and the allowable shearing stress for the bolts is 42 MPa. The diameter of the bolt circle is $d_1 = 120$ mm, and bolts are available with diameters in increments of 5 mm.

7-95 The motor shown in Fig. P7-95 supplies a torque of 380 lb · ft to shaft BCD. The torques removed at gears C and D are 220 lb · ft and 160 lb · ft, respectively. The shaft BCD has a constant diameter and is made of 0.4% C hot-rolled steel with an allowable shearing stress of 15 ksi. The angle of twist between B and D cannot exceed 3.0°. Each of the segments BC and CD of the shaft has a length of 4 ft. Determine

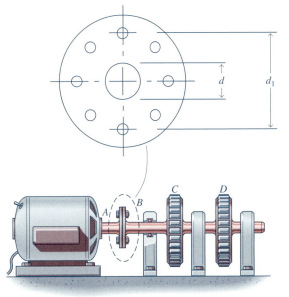

Figure P7-95

(a) The minimum permissible diameter of the shaft, if shafts are available with diameters in increments of $\frac{1}{8}$ in.

(b) The diameter of the bolts used in the coupling, if eight bolts are used, and the allowable shearing stress for the bolts is 7000 psi. The diameter of the bolt circle is $d_1 = 3.5$ in. and bolts are available with diameters in increments of $\frac{1}{16}$ in.

7-96 A shaft used to transmit power is constructed by joining two solid segments of shaft with a collar, as shown in Fig. P7-96. The collar has an inside diameter equal to the diameter of the shaft, and both the collar and the shaft are made of the same material. The collar is se-

curely bonded to the shaft segments. Determine the ratio of the diameters of the collar and shaft such that the splice can transmit the same power as the shaft and at the same maximum shearing stress level. Is the solution dependent upon the material selected?

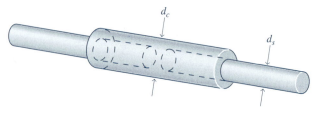

Figure P7-96

7-9 SUMMARY

The problem of transmitting a torque (a couple) from one plane to a parallel plane is frequently encountered in the design of machinery. The simplest device for accomplishing this function is a circular shaft. If the shaft is in equilibrium, a summation of moments about the axis of the shaft indicates that

$$T = T_r = \int_{\text{area}} \rho\, \tau\, dA \tag{7-1}$$

The law of variation of the shearing stress on the transverse plane (τ as a function of radial position ρ) must be known before the integral of Eq. (7-1) can be evaluated. If the assumption is made that a plane transverse cross section before twisting remains plane after twisting and a diameter of the section remains straight, the distortion of the shaft can be expressed as

$$\gamma_c = \frac{c\theta}{L} \quad \text{and} \quad \gamma_\rho = \frac{\rho\theta}{L} \tag{7-2}$$

or

$$\gamma_\rho = \frac{\gamma_c}{c}\, \rho \tag{7-3}$$

The angle θ is called the angle of twist. Equation (7-3) indicates that the shearing strain is zero at the center of the shaft and increases linearly with respect to the distance ρ from the axis of the shaft. This equation can be combined with Eq. (7-1) once the relationship between shearing stress τ and shearing strain γ is known. Since no assumptions have been made about the relationship between stress and strain or about the type of material of which the shaft is made, Eq. (7-3) is valid for elastic or inelastic action and for homogeneous or heterogeneous materials, provided the strains are not too large ($\tan \gamma \cong \gamma$). If the assumption is made that Hooke's law ($\tau = G\gamma$) applies (stresses must be below the proportional limit of the material), Eq. (7-3) can be written

$$\tau_\rho = \frac{\tau_c}{c}\, \rho \tag{7-4}$$

When Eq. (7-4) is substituted into Eq. (7-1), the result is

$$\tau_\rho = \frac{T\rho}{J} \quad \text{and} \quad \tau_c = \frac{Tc}{J} \tag{7-6}$$

where J is the polar second moment of the cross-sectional area of the shaft. Equation (7-6) indicates that the shearing stress τ_ρ, like the shearing strain γ_ρ, is zero at the center of the shaft and increases linearly with respect to the distance ρ from the axis of the shaft. Both the shearing strain γ and the shearing stress τ are maximum when $\rho = c$. Equation (7-6) is known as the elastic torsion formula and is valid for both solid and hollow circular shafts.

Frequently, the amount of twist in a shaft is important. Equations (7-2), (7-6), and Hooke's law ($\tau = G\gamma$) can be combined to give

$$\theta = \frac{\gamma_\rho L}{\rho} = \frac{\tau_\rho L}{\rho G} \quad \text{or} \quad \theta = \frac{TL}{GJ} \tag{7-7a, b}$$

The angle of twist determined from the above expressions is for a length of shaft, of constant diameter ($J = $ constant), constant material properties ($G = $ constant), and carrying a torque T. Ideally, the length of shaft should not include sections too near to (within about one-half shaft diameter of) places where mechanical devices (gears, pulleys, or couplings) are attached. For practical purposes, however, it is customary to neglect distortions at connections and to compute angles as if there were no discontinuities.

If T, G, or J is not constant along the length of the shaft, Eq. (7-7b) takes the form

$$\theta = \sum_{i=1}^{n} \frac{T_i L_i}{G_i J_i} \tag{7-7c}$$

where each term in the summation is for a length L where T, G, and J are constant. If T, G, or J is a function of x (the distance along the length of the shaft), the angle of twist is found using

$$\theta = \int_0^L \frac{T \, dx}{GJ} \tag{7-7d}$$

One of the most common uses for the circular shaft is the transmission of power. Power is defined as the time rate of doing work, and the basic relationship for work done by a constant torque T is $U = T\theta$, where U is work and θ is the angular displacement of the shaft in radians. The derivative of U with respect to time t gives

$$\frac{dU}{dt} = T\frac{d\theta}{dt} = T\omega \tag{7-19}$$

where dU/dt is power, T is a constant torque, and ω is the angular velocity of the shaft. In the U.S. customary system of units, power is usually given in units of horsepower (1 hp $= 33,000$ ft $\cdot$ lb per min). In the SI system of units, power is given in watts (N $\cdot$ m/s).

REVIEW PROBLEMS

7-97* A hollow steel ($G = 12{,}000$ ksi) shaft with an outside diameter of 3.20 in. and an inside diameter of 2.40 in. is subjected to a torque of 30 in. · kip. Determine
(a) The maximum shearing stress in the shaft.
(b) The shearing stress on a cross section at the inside surface of the shaft.
(c) The magnitude of the angle of twist in an 8-ft length.

7-98* A 2 m–long steel ($E = 76$ GPa) pipe will be used for a shaft that is to carry a torque of 5 kN · m. If the maximum shearing stress in the shaft must not exceed 50 MPa and the angle of twist must not exceed 2.0°, determine the lightest standard weight steel pipe that can be used for the shaft.

7-99 A solid circular stepped steel ($G = 12{,}000$ ksi) shaft is loaded and supported as shown in Fig. P7-99. Determine
(a) The maximum shearing stress in the shaft.
(b) The rotation of a section at B with respect to its no-load position.
(c) The rotation of end C with respect to its no-load position.

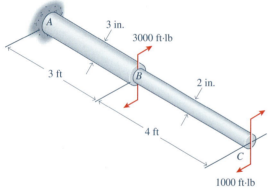

Figure P7-99

7-100 A 150 mm–diameter solid circular aluminum alloy ($G = 28$ GPa) shaft is loaded and supported as shown in Fig. P7-100. Determine
(a) The maximum shearing stress in the shaft.
(b) The rotation of a section at B with respect to its no-load position.
(c) The rotation of end C with respect to its no-load position.

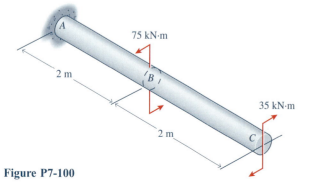

Figure P7-100

7-101* A stepped steel ($G = 12{,}000$ ksi) shaft has the dimensions and is subjected to the torques shown in Fig. P7-101. If the maximum shearing stress in the shaft must be limited to 7500 psi and the rotation of a section at C with respect to its no-load position must be limited to 0.007 rad, determine the maximum torque T_2 that can be applied.

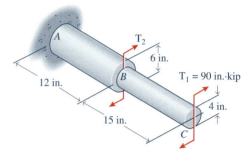

Figure P7-101

7-102* A solid circular steel ($G = 80$ GPa) shaft is fastened securely to a solid circular bronze ($G = 40$ GPa) shaft and loaded as indicated in Fig. P7-102. Determine
(a) The maximum shearing stress in the shaft.
(b) The rotation of a section at B with respect to its no-load position.
(c) The rotation of end C with respect to its no-load position.

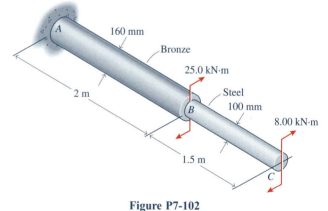

Figure P7-102

7-103 A steel ($G = 12{,}000$ ksi) shaft is loaded and supported as shown in Fig. P7-103. Determine
(a) The maximum shearing stress in the shaft.
(b) The rotation of a section at the right end with respect to its no-load position.
(c) The rotation of a section 7 ft from the left end with respect to its no-load position.

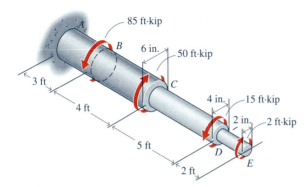

Figure P7-103

7-104 The 100 mm–diameter shaft shown in Fig. P7-104 is composed of aluminum alloy ($G = 28$ GPa) and steel ($G = 80$ GPa) sections rigidly connected. Determine
(a) The maximum shearing stress in the shaft.
(b) The rotation of a section 3 m from the left end with respect to its no-load position.
(c) The rotation of a section at the right end with respect to its no-load position.

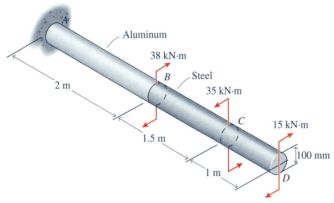

Figure P7-104

7-105* A solid circular aluminum alloy ($G = 4000$ ksi) shaft is 2.5 in. in diameter and 3 ft long. The maximum permissible angle of twist is 0.052 rad and the allowable shearing stress is 10,000 psi. Determine the maximum horsepower that this shaft can deliver when rotating at 500 rpm.

7-106* A 3 m–long hollow steel ($G = 80$ GPa) shaft has an outside diameter of 200 mm and an inside diameter of 120 mm. The maximum shearing stress in the shaft is 75 MPa when it is rotating at an angular velocity of 450 rpm. Determine
(a) The power being transmitted by the shaft.
(b) The magnitude of the angle of twist in the shaft.

7-107 For the steel ($G = 12,000$ ksi) and bronze ($G = 6500$ ksi) shaft of Fig. P7-107, which is attached to rigid supports at both ends, determine
(a) The maximum shearing stress in the shaft.
(b) The rotation of the section where the torque is applied with respect to its no-load position.

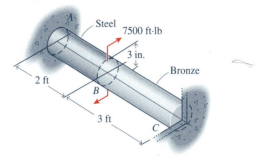

Figure P7-107

7-108 The aluminum alloy ($G = 28$ GPa) shaft of Fig. P7-108 is attached to rigid supports at both ends. The right 2 m of the shaft, which is hollow, has an inside diameter of 80 mm. If the maximum shearing stress in the shaft must not exceed 80 MPa, determine
(a) The maximum torque T that can be applied to the shaft.
(b) The rotation of the section where the torque is applied with respect to its no-load position when $T = 60$ kN $\cdot$ m.

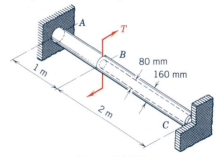

Figure P7-108

7-109 The band brake shown in Fig. P7-109 is part of a hoisting machine. The coefficient of friction between the 20 in.–diameter drum and the flat belt is 0.20. The maximum actuating force P that can be applied to the brake arm is 110 lb. Rotation of the drum is clockwise. What size shaft should be used to transmit the resisting torque developed by the brake to the machine if the shaft is to be made of a steel with an allowable shearing stress of 8,000 psi? Circular steel bars are available with diameters in increments of $\frac{1}{8}$ in.

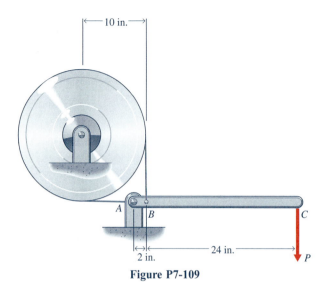

←—— 10 in. ——→

A

B

2 in.

←—— 24 in. ——→

C

P

Figure P7-109

7-110 A disk is supported by two circular shafts attached to rigid walls as shown in Fig. P7-110. Shaft A is made of brass ($G = 39$ GPa) and has a diameter of 100 mm and a length of 500 mm. Shaft B is made of Monel ($G = 65$ GPa) and has a diameter of 75 mm and a length of 675 mm. If torque $T = 20$ kN · m, determine

(a) The maximum shearing stress in each of the shafts.
(b) The rotation of the disk with respect to its no-load position.

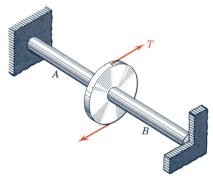

Figure P7-110

COMPUTER PROBLEMS

C7-111 A hollow circular steel ($G = 11,000$ ksi) shaft 3 ft long is being designed to transmit a torque T of 3000 ft · lb. The outer radius r_o of the shaft can vary (1 in. $\leq r_o \leq 4$ in.), but the cross-sectional area A of the shaft must remain constant ($A = 3$ in.²). Compute and plot:

(a) The angle of twist θ for the 3-ft length as a function of the outer radius r_o (1 in. $\leq r_o \leq 4$ in.).
(b) The maximum shearing stress τ_{max} in the shaft as a function of the outer radius r_o (1 in. $\leq r_o \leq 4$ in.).

C7-112 A hollow circular brass ($G = 40$ GPa) shaft 2 m long is being designed to transmit a torque T of 7500 N · m. The outer radius r_o of the shaft must be fixed ($r_o = 50$ mm); however, the inner radius r_i of the shaft can vary (0 mm $\leq r_i \leq 45$ mm). Compute and plot:

(a) The angle of twist θ for the 2-m length as a function of the radius ratio r_i/r_o ($0 \leq r_i/r_o \leq 0.9$).
(b) The maximum shearing stress τ_{max} in the shaft as a function of the radius ratio r_i/r_o ($0 \leq r_i/r_o \leq 0.9$).

C7-113 The solid circular steel ($G = 12,000$ ksi) shaft shown in Fig. P7-113 is subjected to a torque T of 4000 ft · lb. The diameter d_c of the 2-ft center section of the shaft can vary (2 in. $\leq d_c \leq 4$ in.), while the diameters d_o of the 2-ft end sections of the shaft remain fixed at 3 in. Compute and plot the angle of twist θ as a function of position x along the shaft

(0 ft $\leq x \leq 6$ ft) when the center section diameter d_c equals 2 in., 2.5 in., 3 in., 3.5 in., and 4 in.

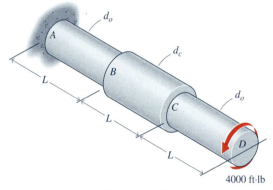

Figure P7-113

C7-114 The shaft of Fig. P7-114 is turned out of aluminum ($G = 28$ GPa). Section AC is 1.5 m long and 100 mm in diameter (AB is hollow, BC is solid); section CD is 0.50 m long and 75 mm in diameter.

(a) If the diameter of the hole from A to B is 75 mm, calculate and plot the angle of twist $\theta_{D/A}$ as a function of the distance L_{AB} ($0 \leq L_{AB} \leq 1.4$ m).
(b) Repeat for a 90 mm–diameter hole.

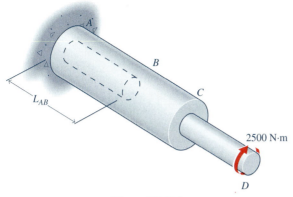

L_{AB}

2500 N·m

D

Figure P7-114

C7-115 A 2 in.–diameter solid steel ($G = 12,000$ ksi, $\tau_{max} = 30$ ksi) shaft and a 3 in.–diameter hollow aluminum ($G = 4000$ ksi, $\tau_{max} = 24$ ksi) shaft are fastened together by a $\frac{1}{2}$ in.–diameter brass pin ($\tau_{max} = 36$ ksi) as shown in Fig. P7-115. Calculate and plot:
(a) The maximum shear stresses τ_a in the aluminum and τ_s in the steel segments of the shaft as a function of the torque applied to the end of the shaft T ($0 \le T \le 20$ in. · kip).
(b) The average shear stress τ_b on the cross-sectional area of the pin at the interface between the shafts as a function of T ($0 \le T \le 20$ in. · kip).
(c) The angle of twist $\theta_{B/D}$ of end B as a function of T ($0 \le T \le 20$ in. · kip).
(d) What is the maximum torque that can be applied to the shaft without exceeding the maximum shear stress in either shaft or the pin?

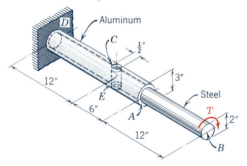

Figure P7-115

C7-116 The hollow circular brass ($G = 40$ GPa) shaft shown in Fig. P7-116 is 3 m long and is attached to rigid supports at both ends. An 8-kN·m torque T is applied at section C, which is located 1 m from the left end. The outer radius r_o of the shaft can vary (25 mm $\le r_o \le$ 100 mm); but the cross-sectional area A of the shaft must remain constant at 1500 mm^2. Compute and plot:
(a) The rotation θ of a section at C with respect to its no-load position as a function of the outer radius r_o (25 mm $\le r_o \le$ 100 mm).
(b) The maximum shearing stress τ_{max} in the shaft as a function of the outer radius r_o (25 mm $\le r_o \le$ 100 mm).

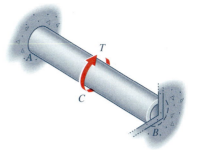

Figure P7-116

C7-117 The hollow circular aluminum alloy ($G = 4000$ ksi) shaft shown in Fig. P7-116 is 5 ft long and is attached to rigid supports at both ends. A 7500 ft · lb torque T is applied at section C, which is located 2 ft from the right end. The outer radius r_o of the shaft must be 2 in.; however, the inner radius r_i can vary ($0 \le r_i \le 1.90$ in.). Compute and plot:
(a) The rotation θ of a section at C with respect to its no-load position as a function of the radius ratio r_i/r_o ($0 \le r_i/r_o \le 0.95$).
(b) The maximum shearing stress τ_{max} in the shaft as a function of the radius ratio r_i/r_o ($0 \le r_i/r_o \le 0.95$).

C7-118 A composite shaft consists of a 2 m–long solid circular steel ($G = 80$ GPa) section securely fastened to a 2 m–long solid circular bronze ($G = 40$ GPa) section as shown in Fig. P7-118. Both ends of the composite shaft are attached to rigid supports. The maximum shearing stress τ_{max} in the shaft must not exceed 60 MPa and the rotation θ of any cross section in the shaft must not exceed 0.04 rad. The ratio of the diameters d_b/d_s of the two sections can vary ($\frac{1}{2} \le d_b/d_s \le 2$), but the average diameter ($d_b + d_s$)/2 must be 100 mm. Compute and plot:
(a) The maximum allowable torque T as a function of the diameter ratio d_b/d_s ($\frac{1}{2} \le d_b/d_s \le 2$).
(b) The rotation θ of a section at C as a function of the diameter ratio d_b/d_s ($\frac{1}{2} \le d_b/d_s \le 2$).
(c) The maximum shearing stress τ_b in the bronze shaft as a function of the diameter ratio d_b/d_s ($\frac{1}{2} \le d_b/d_s \le 2$).
(d) The maximum shearing stress τ_s in the steel shaft as a function of the diameter ratio d_b/d_s ($\frac{1}{2} \le d_b/d_s \le 2$).

Figure P7-118

C7-119 A hollow steel ($G = 12,000$ ksi) shaft is stiffened by filling its center with aluminum ($G = 4000$ ksi) as shown in Fig. P7-119. If the steel and aluminum rotate as a single unit, calculate and plot:
(a) The angle of twist $\theta_{B/A}$ of end B as a function of the diameter of the aluminum shaft d_a ($0 \le d_a \le 3.75$ in.).
(b) The shear stress at the interface between the aluminum and the steel τ as a function of d_a ($0 \le d_a \le 3.75$ in.).

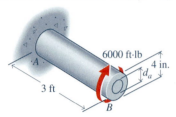

Figure P7-119

C7-120 The torsional assembly shown in Fig. P7-120 consists of a solid steel ($G = 210$ GPa) shaft CD and a hollow aluminum alloy ($G = 28$ GPa) shaft EF that has a bronze ($G = 45$ GPa) core. The ends C and F are fixed to rigid walls and the bronze core of shaft EF is connected to the flange at E so that the aluminum and bronze parts act as a unit. Pins on the flange D fit through holes in the flange E. The holes are slightly overdrilled so that flange D rotates 3° before EF carries any

of the load. If a torque T_D is applied to the flange D, calculate and plot:
(a) The maximum shearing stress in each of the materials as a function of the torque T_D ($0 \le T_D \le 30$ kN · m).
(b) The angle of twist $\theta_{D/C}$ of a section at D as a function of T_D ($0 \le T_D \le 30$ kN · m).

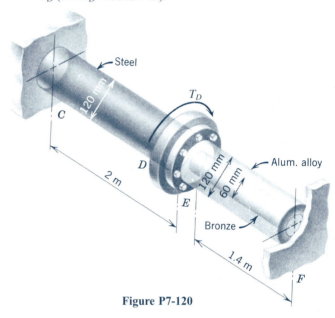

Figure P7-120

FLEXURAL LOADING: STRESSES IN BEAMS

8

8-1 INTRODUCTION

A member subjected to loads applied transverse to the long dimension of the member and which causes the member to bend is known as a "beam." The beam, or flexural member, is frequently encountered in structures and machines, and its elementary stress analysis constitutes one of the more interesting facets of mechanics of materials. Figure 8-1 is a photograph of an I-beam, *AB,* simply supported in a testing machine and loaded at the one-third points. Figure 8-2 depicts the shape (exaggerated) of the beam when loaded.

Before proceeding with a discussion of stress analysis for flexural members, it may be well to classify some of the various types of beams and loadings encountered in practice. Beams are frequently classified on the basis of their supports or reactions. A beam supported by a pin, roller, or smooth surface at the ends and having one span is called a *simple beam* (Fig. 8-3*a*). A simple support (a pin or roller) will develop a reaction normal to the beam but will not produce a couple. If either or both ends of the beam project beyond the supports, it is called a *simple beam with overhang* (Fig. 8-3*b*). A beam with more than two simple supports is a *continuous beam* (Fig. 8-3*c*). A *cantilever beam* is one in which

Figure 8-1

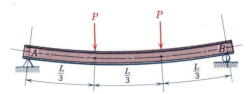

Figure 8-2

one end is built into a wall or other support so that the built-in end cannot move transversely or rotate (Fig. 8-3d). The built-in end is said to be *fixed* if no rotation occurs and *restrained* if a limited amount of rotation occurs. The supports shown in Figs. 8-3d, e, and f represent fixed ends unless otherwise stated. The beams in Figs. 8-3d, e, and f are, in order, a cantilever beam, a beam fixed (or restrained) at the left end and simply supported near the other end (which has an overhang), and a beam fixed (or restrained) at both ends.

Cantilever beams and simple beams have only two reactions (two forces or one force and a couple), and these reactions can be obtained from a free-body diagram of the beam by applying the equations of equilibrium. Such beams are said to be statically determinate since the reactions can be obtained from the equations of equilibrium. Beams with more than two reaction components are called statically indeterminate since there are not enough equations of equilibrium to determine the reactions.

All of the beams shown in Fig. 8-3 are subjected to both uniformly distributed loads and concentrated loads, and, although shown as horizontal, they

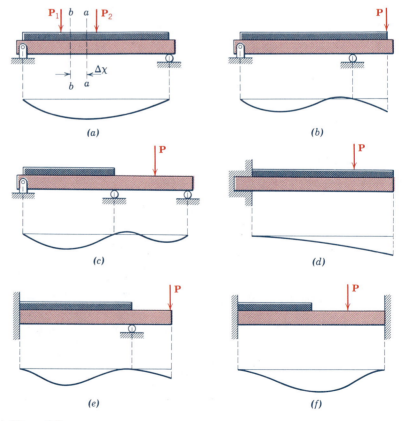

Figure 8-3

may have any orientation. Distributed loads will be shown on the side of the beam on which they are acting; that is, if drawn on the bottom of the beam, the load is pushing upward and if drawn on the right side of a vertical beam, the load is pushing to the left. Deflection curves (greatly exaggerated) are shown beneath the beams to assist in visualizing the shapes of the loaded beams.

A free-body diagram of the portion of the beam of Fig. 8-3a between the left end and plane a-a is shown in Fig. 8-4a. A study of this diagram shows that a transverse force V_r and a couple M_r at the cut section and a force R (a reaction) at the left support are needed to maintain equilibrium. The force V_r is the resultant of the shearing stresses acting on the cut section (on plane a-a) and is called the *resisting shear*. The couple M_r is the resultant of the normal stresses acting on the cut section (on plane a-a) and is called the *resisting moment*. The magnitudes and senses of V_r and M_r are obtained from the equations of equilibrium $\Sigma F_y = 0$ and $\Sigma M_O = 0$ where O is any axis perpendicular to the xy-plane. The reaction R must be evaluated from a free body of the entire beam.

The normal and shearing stresses σ and τ on plane a-a are related to the resisting moment M_r and the shear V_r by the equations

$$V_r = -\int_{\text{Area}} \tau \, dA \tag{8-1a}$$

$$M_r = -\int_{\text{Area}} y \, \sigma \, dA \tag{8-1b}$$

The resisting shear and moment (V_r and M_r), as shown on Fig. 8-4a, will be defined as positive quantities later in Section 8-4. The normal and shear stresses

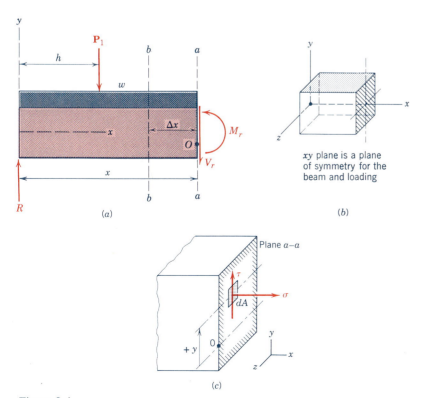

Figure 8-4

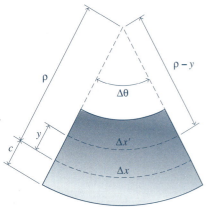

Figure 8-5

(σ and τ), as shown on Fig. 8-4c, are defined as positive stresses. The minus signs in Eqs. (8-1a) and (8-1b) are required to bring these two definitions into agreement. It is obvious from Eqs. (8-1) that the laws of variation of the normal and shearing stresses must be known before the integrals can be evaluated. Thus, the problem is statically indeterminate. For the present, the shearing stresses will be ignored while the normal stresses are studied.

8-2 FLEXURAL STRAINS

A segment of the beam of Fig. 8-4a, between planes a-a and b-b, is shown in Fig. 8-5 with the distortion greatly exaggerated. When Fig. 8-5 was drawn, the assumption was made that a plane section before bending remains a plane after bending. For this to be strictly true, it is necessary that the beam be bent only with couples (no shear on transverse planes). Also, the beam must be so proportioned that it will not buckle and the loads applied so that no twisting occurs (this last limitation will be satisfied if the loads are applied in a plane of symmetry—a sufficient though not a necessary condition). When a beam is bent only with couples, the deformed shape of all longitudinal elements (also referred to as "fibers") is an arc of a circle.

Precise experimental measurements indicate that at some distance c above the bottom of the beam, longitudinal elements undergo no change in length. The curved surface formed by these elements (at radius ρ in Fig. 8-5) is referred to as the neutral surface of the beam, and the intersection of this surface with any cross section is called the *neutral axis of the section*. All elements (fibers) on one side of the neutral surface are compressed, and those on the opposite side are elongated. As shown in Fig. 8-5, the fibers above the neutral surface of the beam of Fig. 8-4a (on the same side as the center of curvature) are compressed and the fibers below the neutral surface (on the side opposite the center of curvature) are elongated.

Finally, the assumption is made that all longitudinal elements have the same initial length. This assumption imposes the restriction that the beam be initially straight and of constant cross section; however, in practice, considerable deviation from these last restrictions is often tolerated.

The longitudinal strain ϵ_x experienced by a longitudinal element that is located a distance y from the neutral surface of the beam is determined by using the definition of normal strain as expressed by Eq. (4-11). Thus,

$$\epsilon_x = \frac{\Delta L}{L} = \frac{L_f - L_i}{L_i}$$

where L_f is the final length of the fiber after the beam is loaded and L_i is the initial length of the fiber before the beam is loaded. From the geometry of the beam segment shown in Fig. 8-5,

$$\epsilon_x = \frac{\Delta x' - \Delta x}{\Delta x} = \frac{(\rho - y)(\Delta \theta) - \rho(\Delta \theta)}{\rho(\Delta \theta)} = -\frac{1}{\rho}y \qquad (8\text{-}2)$$

Equation (8-2) indicates that the strain developed in a fiber is directly proportional to the distance of the fiber from the neutral surface of the beam. This

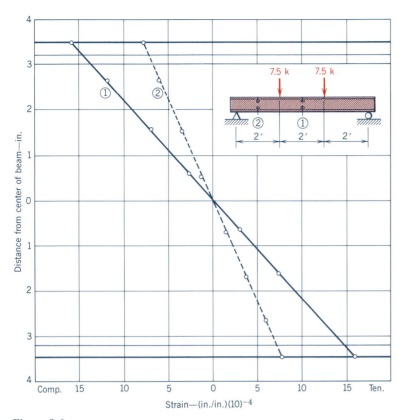

Figure 8-6

variation can be demonstrated experimentally by means of strain gages attached to a beam, as shown in Fig. 8-1. The strains, as measured by gages on two different sections, are plotted against the vertical position of the gages on the beam in Fig. 8-6. Curve 1 represents strains on a section at the center of the beam where pure bending occurs (no transverse shear), and curve 2 shows strains at a section near one end of the beam where both flexural (normal) stresses and transverse shearing stresses exist. These curves are both straight lines within the limits of the accuracy of the measuring equipment.[1] Note that Eq. (8-2) is valid for elastic or inelastic action so long as the beam does not twist or buckle and the transverse shearing stresses are small. Problems in this book will be assumed to satisfy these restrictions.

8-3 FLEXURAL STRESSES

With the acceptance of the premise that the longitudinal strain ϵ_x is proportional to the distance of the fiber from the neutral surface of the beam, the law of variation of the normal stress $\sigma = \sigma_x$ on the transverse plane can be determined by

---

[1]A more exact analysis using principles developed in the theory of elasticity indicates that curve 2 should be curved slightly. *Note:* Other experiments indicate that a plane section of an initially curved beam will also remain plane after bending and that the deformations will still be proportional to the distance of the fiber from the neutral surface. The strain, however, will not be proportional to this distance, since each deformation must be divided by a different original length.

using a tensile–compressive stress–strain diagram for the material used in fabricating the beam. For most real materials, the tension and compression stress–strain
diagrams are identical in the linearly elastic range. Although the diagrams may
differ somewhat in the inelastic range, the differences can be neglected for most
real problems. For beam problems in this book, the compressive stress–strain diagram will be assumed to be identical to the tensile diagram.

For the special case of linearly elastic action, the relationship between stress
σ_x and strain ϵ_x is given by Hooke's law, Eq. (4-15a) (since the state of stress is
uniaxial) as

$$\sigma_x = E\epsilon_x \tag{a}$$

Substituting Eq. (8-2) into Eq. (a) yields

$$\sigma_x = E\epsilon_x = -\frac{E}{\rho}\,y \tag{8-3}$$

Equation (8-3) shows that the normal stress σ_x on the transverse cross section of
the beam varies linearly with distance y from the neutral surface. Also, since
plane cross sections remain plane, the normal stress σ_x is uniformly distributed
in the z-direction (see Fig. 8-4c).

With the law of variation of flexural stress known, Fig. 8-4 can now be redrawn as shown in Fig. 8-7. The forces F_C and F_T are the resultants of the compressive and tensile flexural stresses, respectively. Since the sum of the forces in
the x-direction must be zero, F_C is equal to F_T; hence, they form a couple of
magnitude M_r.

The resisting moment M_r developed by the normal stresses in a typical
beam with loading in a plane of symmetry but of arbitrary cross section, such as
the one shown in Fig. 8-8, is given by Eq. (8-1) as

$$M_r = -\int_A y\,\sigma_x\,dA \tag{b}$$

Since y is measured from the neutral surface, it is first necessary to locate this
surface by means of the equilibrium equation $\Sigma F_x = 0$, which gives

$$\int_A \sigma_x\,dA = 0 \tag{c}$$

Substituting Eq. (8-3) into Eq. (c) yields

$$\int_A \sigma_x\,dA = \int_A -\frac{E}{\rho}\,y\,dA$$

$$= -\frac{E}{\rho}\int_A y\,dA = -\frac{E}{\rho}\,y_C\,A = 0 \tag{8-4}$$

where y_C is the distance from the neutral axis to the centroidal axis (c-c) of the
cross section (Fig. 8-8) that is perpendicular to the plane of bending. Since neither (E/ρ) nor A are zero, y_C must equal zero. Thus, for flexural loading and linearly elastic action, the neutral axis passes through the centroid of the cross section.

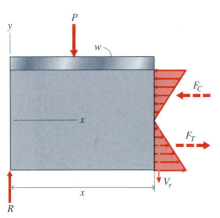

Figure 8-7

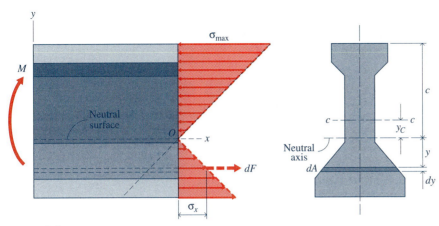

Figure 8-8

Since the normal stress σ_x varies linearly with distance y from the neutral surface, the maximum normal stress σ_{max} on the cross section can be written as

$$\sigma_{max} = -\frac{E}{\rho}c \qquad (8\text{-}5)$$

where c is the distance to the surface of the beam (top or bottom) farthest from the neutral surface. If the quantity (E/ρ) is eliminated from Eqs. (8-3) and (8-5), a useful relationship between the maximum stress σ_{max} on a transverse cross section and the stress σ_x at an arbitrary distance y from the neutral surface is obtained. Thus,

$$\sigma_x = \frac{y}{c}\,\sigma_{max} = \frac{y}{c}\,\sigma_c \qquad (8\text{-}6)$$

Substitution of Eq. (8-6) into Eq. (8-1) yields

$$M_r = -\int_A y\,\sigma_x\,dA = -\frac{\sigma_c}{c}\int_A y^2\,dA \qquad (8\text{-}7)$$

in which $\sigma_{max} = \sigma_c$ (equals σ_x evaluated at $y = c$). The integral $\int y^2\,dA$ is called the *second moment of area*.[2] The second moments of area of several common shapes are given on the back inside cover of this book. Second moments of more complex areas can usually be derived (without integration) from combinations of these simple shapes as shown in the next section. After a discussion of second

---

[2]In Chapter 5, the centroid for an area was located by evaluating an integral of the form $\int_A x\,dA$, which is called the *first moment of the area with respect to the y-axis*. By analogy, the integral $\int_A x^2\,dA$ is called the *second moment of the area with respect to the y-axis*. In the analysis of the angular motion of rigid bodies, an integral of the form $\int_m r^2\,dm$ occurs in which dm represents an element of mass and r represents the distance from the element to some axis. Euler gave the name "moment of inertia" to these integrals. Because of the similarity between the two types of integrals, both have become widely known as moments of inertia. In this book, integrals involving areas will be referred to as second moments of area and less frequently as area moments of inertia. Integrals involving masses will be referred to as mass moments of inertia or simply moments of inertia.

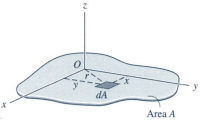

Figure 8-9

moments of area, attention will be focused on the application of Eq. (8-7) to calculate the normal stress acting on a transverse section of a beam.

8-4 SECOND MOMENTS OF AREAS

The second moment of an area with respect to an axis in the plane of the area will be denoted by the symbol I. The particular axis about which the second moment is taken will be denoted by subscripts. Thus, the second moments of the area A shown in Fig. 8-9 with respect to x- and y-axes in the plane of the area are

$$I_x = \int_A y^2 \, dA \quad \text{and} \quad I_y = \int_A x^2 \, dA \tag{8-8}$$

where y is the distance of the area dA from the x-axis and x is the distance of the area dA from the y-axis. The quantities I_x and I_y are sometimes referred to as rectangular second moments[3] of the area A.

The second moment of an area can be visualized as the sum of a number of terms each consisting of an area multiplied by a distance squared. Thus, the dimensions of a second moment are a length raised to the fourth power (L^4). Common units are mm[4] and in.[4]. Also, the sign of each term summed to obtain the second moment is positive since either a positive or negative distance squared is positive. Therefore, the second moment of an area is always positive.

8-4-1 Radius of Gyration

Since the second moment of an area has the dimensions of length to the fourth power, it can be expressed as the area A multiplied by a length k squared. That is,

$$I_x = \int_A y^2 \, dA = A \, k_x^2 \quad \text{or} \quad k_x = \sqrt{\frac{I_x}{A}} \tag{8-9a}$$

$$I_y = \int_A x^2 \, dA = A \, k_y^2 \quad \text{or} \quad k_y = \sqrt{\frac{I_y}{A}} \tag{8-9b}$$

The distance k is called the *radius of gyration*. The subscript denotes the axis about which the second moment of area is taken. The radius of gyration does not identify a physical point on the area A. Instead, it can be visualized as the distance from the axis to a point where a concentrated area of the same size could be placed and would have the same second moment of area with respect to the given axis.

---

[3]The second moment of an area with respect to an axis perpendicular to the plane of the area is denoted by the symbol J. For example, the second moment of the area A shown in Fig. 8-9 with respect to a z-axis that is perpendicular to the plane of the area at the origin O of the xy-coordinate system is

$$J_z = \int_A r^2 \, dA = \int_A (x^2 + y^2) \, dA = \int_A x^2 \, dA + \int_A y^2 \, dA = I_y + I_x$$

The quantity J_z is known as the polar second moment of the area A and was used in Chapter 7 in the calculation of the stress in circular shafts transmitting torques.

8-4-2 Parallel-Axis Theorem for Second Moments of Area

When the second moment of an area has been determined with respect to a given axis, the second moment with respect to a parallel axis can be obtained by means of the parallel axis theorem (also known as the transfer formula). If one of the axes (say the x-axis) passes through the centroid of the area as shown in Fig. 8-10, the second moment of the area about a parallel x'-axis is

$$I_{x'} = \int_A (y + y_C)^2 \, dA$$

$$= \int_A y^2 \, dA + 2 y_C \int_A y \, dA + y_C^2 \int_A dA \qquad \text{(a)}$$

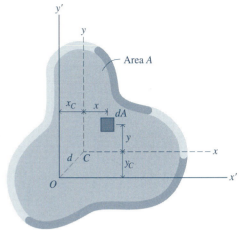

Area A

Figure 8-10

where y_C has been taken outside the integral signs since it is the same for every element of area dA. The integral in the first term is the second moment of the area with respect to the (centroidal) x-axis $\int_A y^2 \, dA = I_{xC}$, the integral in the last term is the total area $\int_A dA = A$, and the integral $\int_A y \, dA$ is the first moment of the area with respect to the x-axis. Since the x-axis passes through the centroid C of the area, the first moment is zero (the middle term vanishes) and Eq. (a) becomes

$$I_{x'} = I_{xC} + y_C^2 \, A \qquad \text{(8-10)}$$

The parallel axis theorem, Eq. (8-10), states that the second moment of an area with respect to any axis in the plane of the area is equal to the second moment of the area with respect to a parallel axis through the centroid of the area added to the product of the area and the square of the perpendicular distance between the two axes. The theorem also indicates that the second moment of an area with respect to an axis through the centroid of the area is less than that for any parallel axis, since

$$I_{xC} = I_{x'} - y_C^2 \, A$$

and I_{xC}, $y_C^2 \, A$, and $I_{x'}$ are all positive. As a point of caution, note that the transfer formula is valid only for transfers to or from a centroidal axis. That is, if x'' is a second axis parallel to the x'-axis and y_{2C} is the distance between x'' and the centroidal axis (Fig. 8-11), then

$$I_{x''} = I_{xC} + y_{2C}^2 \, A = (I_{x'} - y_C^2 \, A) + y_{2C}^2 \, A$$
$$= I_{x'} + (y_{2C}^2 - y_C^2) \, A \neq I_{x'} + (y_{2C} - y_C)^2 \, A$$

8-4-3 Second Moments of Composite Areas

Frequently in engineering practice, an irregular area A will be encountered that can be broken up into a series of simple areas $A_1, A_2, A_3, \ldots, A_n$ for which the second-moment integrals have been evaluated and tabulated. The second moment of the irregular area (the composite area) with respect to any axis is equal to the sum of the second moments of the separate parts of the area with respect to the specified axis

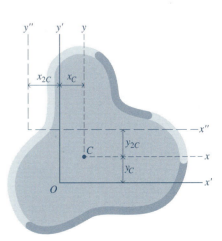

Figure 8-11

Figure 8-12

$$I_x = \int_A y^2 \, dA$$

$$= \int_{A_1} y^2 \, dA + \int_{A_2} y^2 \, dA + \int_{A_3} y^2 \, dA + \cdots + \int_{A_n} y^2 \, dA$$

$$= I_{x1} + I_{x2} + I_{x3} + \cdots + I_{xn}$$

The inside back cover contains a listing of the values of the integrals for frequently encountered shapes such as rectangles, triangles, circles, and semicircles. Tables listing second moments of area and other properties for common structural shapes can be found in engineering handbooks. Properties of selected structural shapes are listed in Tables B1 through B16 of Appendix B for use in solving problems.

When an area such as a hole is removed from a larger area, its second moment must be subtracted from the second moment of the larger area to obtain the resulting second moment. For example, for the case of a square plate with a circular hole (Fig. 8-12)

$$I_{\blacksquare} = I_{\square} + I_{\odot}$$

Rearranging gives

$$I_{\square} = I_{\blacksquare} - I_{\odot}$$

Of course, all second moments must be evaluated with respect to the same axis before they are added or subtracted. If necessary, the parallel-axis theorem can be used to relate the second moments tabulated on the inside back cover to the axes required in particular problems.

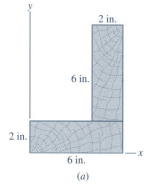

(a)

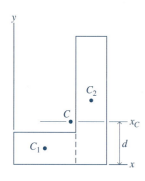

(b)

Figure 8-13

Example Problem 8-1

A beam is constructed by gluing a 2 × 6-in. wooden plank 10 ft long to a second wooden plank 2 × 6-in. also 10 ft long forming the cross-section shown in Fig. 8-13a. Determine the second moment of the cross-sectional area with respect to

(a) The x-axis.

(b) The y-axis.

(c) The x_C-axis, which passes through the centroid of the area and is parallel to the x-axis.

SOLUTION

(a) As shown in Fig. 8-13b, the cross-sectional area can be divided into two simple rectangles. Since both rectangles have an edge along the x-axis their second moments of area are just $bh^3/3$ where b is the base and h the height of the rectangle. Therefore, the second moment of area of the entire area with respect to the x-axis is

$$I_x = I_{x1} + I_{x2} = \frac{1}{3}(4)(2)^3 + \frac{1}{3}(2)(8)^3 = 352 \text{ in.}^4 \qquad \textbf{Ans.}$$

(b) Using the same division of areas as in part (a), the first area has an edge along the y-axis and its second moment of area is given by $bh^3/3$. However, the parallel-axis theorem is needed for the second rectangle since neither the centroid of the rectangle nor either edge is along the y-axis. Therefore, the second moment of area of the entire area with respect to the y-axis is

$$I_y = I_{y1} + I_{y2} = \frac{1}{3}(2)(4)^3 + \left[\frac{1}{12}(8)(2)^3 + 16(5)^2\right]$$

$$= 448 \text{ in.}^4 \qquad \textbf{Ans.}$$

(c) The centroid will be located using the principle of moments as applied to areas and the same division of areas as above

$$d(8 + 16) = 1(8) + 4(16) \qquad d = 3 \text{ in.}$$

Then, using the parallel axis theorem for both rectangles, the second moment of area of the entire area with respect to the x_C-axis is

$$I_{xC} = I_{xC1} + I_{xC2}$$

$$= \left[\frac{1}{12}(4)(2)^3 + 8(3-1)^2\right]$$

$$+ \left[\frac{1}{12}(2)(8)^3 + 16(4-3)^2\right] = 136.0 \text{ in.}^4 \qquad \textbf{Ans.}$$

Note that, since the x_C-axis passes through the centroid of the entire area, the second moments of area I_x and I_{xC} are related by the parallel axis theorem

$$I_x = I_{xC} + Ad^2$$
$$= 136.0 + 24(3)^2 = 352 \text{ in.}^4 \quad\blacksquare$$

Example Problem 8-2

Determine the second moment of the shaded area shown in Fig. 8-14a with respect to

(a) The x-axis.
(b) The y-axis.

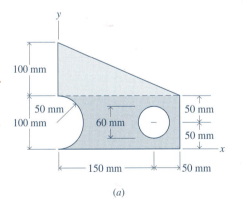

(a)

SOLUTION

As shown in Fig. 8-14b, the shaded area can be divided into a 100 × 200-mm rectangle (A) with a 60 mm–diameter circle (B) and a 100 mm–diameter half circle (C) removed and a 100 × 200-mm triangle (D). The second moments for these four areas, with respect to the x- and y-axes, can be obtained by using information from the inside back cover, as follows.

(a) For the rectangle (shape A):

$$I_{x1} = \frac{bh^3}{3} = \frac{200(100^3)}{3} = 66.667(10^6) \text{ mm}^4$$

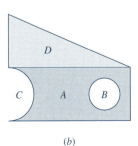

(b)

Figure 8-14

For the circle (shape B):

$$I_{x2} = I_{xC} + y_C^2 A = \frac{\pi R^4}{4} + y_C^2(\pi R^2)$$

$$= \frac{\pi(30^4)}{4} + (50^2)(\pi)(30^2) = 7.705(10^6) \text{ mm}^4$$

For the half circle (shape C):

$$I_{x3} = I_{xC} + y_C^2 A = \frac{\pi R^4}{8} + y_C^2 \left[\frac{\pi R^2}{2} \right]$$

$$= \frac{\pi(50^4)}{8} + (50)^2 \left[\frac{\pi(50)^2}{2} \right] = 12.272(10^6) \text{ mm}^4$$

For the triangle (shape D):

$$I_{x4} = I_{xC} + y_C^2 A = \frac{bh^3}{36} + y_C^2 \left[\frac{bh}{2} \right]$$

$$= \frac{200(100^3)}{36} + \left[100 + \frac{100}{3} \right]^2 \left[\frac{200(100)}{2} \right] = 183.333(10^6) \text{ mm}^4$$

For the composite area:

$$I_x = I_{x1} - I_{x2} - I_{x3} + I_{x4}$$
$$= 66.667(10^6) - 7.705(10^6) - 12.272(10^6) + 183.333(10^6)$$
$$= 230.023(10^6) \cong 230(10^6) \text{ mm}^4 \qquad \textbf{Ans.}$$

(b) For the rectangle (shape A):

$$I_{y1} = \frac{b^3 h}{3} = \frac{200^3(100)}{3} = 266.667(10^6) \text{ mm}^4$$

For the circle (shape B):

$$I_{y2} = I_{yC} + x_C^2 A = \frac{\pi R^4}{4} + x_C^2(\pi R^2)$$

$$= \frac{\pi(30^4)}{4} + (150^2)(\pi)(30^2) = 64.253(10^6) \text{ mm}^4$$

For the half circle (shape C):

$$I_{y3} = \frac{\pi R^4}{8} = \frac{\pi(50^4)}{8} = 2.454(10^6) \text{ mm}^4$$

For the triangle (shape D):

$$I_{y4} = \frac{bh^3}{12} = \frac{100(200^3)}{12} = 66.667(10^6) \text{ mm}^4$$

For the composite area:

$$I_y = I_{y1} - I_{y2} - I_{y3} + I_{y4}$$
$$= 266.667(10^6) - 64.253(10^6) - 2.454(10^6) + 66.667(10^6)$$
$$= 266.627(10^6) \cong 267(10^6) \text{ mm}^4 \quad \blacksquare \qquad \text{Ans.}$$

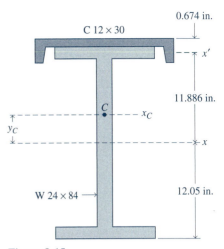

Figure 8-15

Example Problem 8-3

A column with the cross section shown in Fig. 8-15 is constructed from a W24 × 84 wide-flange section and a C12 × 30 channel. Determine the second moments and radii of gyration of the cross-sectional area with respect to horizontal and vertical axes through the centroid of the cross section.

SOLUTION

Properties and dimensions for the structural shapes can be obtained from Tables B-1 and B-5 of Appendix B. In Fig. 8-15, the x-axis passes through the centroid of the wide-flange section and the x'-axis passes through the centroid of the channel. The centroidal x_C-axis for the composite section can be located by using the principle of moments as applied to areas.
The total area A_T for the composite section is

$$A_T = A_{WF} + A_{CH} = 24.7 + 8.82 = 33.52 \text{ in.}^2$$

The moment of the composite area about the x-axis is

$$M_x = A_{WF}(y_C)_{WF} + A_{CH}(y_C)_{CH} = 24.7(0) + 8.82(11.886) = 104.835 \text{ in.}^3$$

The distance y_C from the x-axis to the centroid of the composite section is

$$y_C = \frac{M_x}{A_T} = \frac{104.835}{33.52} = 3.128 \text{ in.}$$

The second moment I_{xCWF} for the wide-flange section about the centroidal x_C-axis of the composite section is

$$I_{xCWF} = I_{xWF} + (y_C)_{WF}^2 A_{WF} = 2370 + (3.128)^2(24.7) = 2611.7 \text{ in.}^4$$

The second moment I_{xCCH} for the channel about the centroidal x_C-axis of the composite section is

$$I_{xCCH} = I_{x'CH} + [(y_C)_{CH} - (y_C)]^2 A_{CH}$$
$$= 5.14 + (11.886 - 3.128)^2(8.82) = 681.7 \text{ in.}^4$$

For the composite area:

$$I_{xC} = I_{xCWF} + I_{xCCH}$$
$$= 2611.7 + 681.7 = 3293.4 \cong 3290 \text{ in.}^4 \qquad \text{Ans.}$$

The y-axis passes through the centroid of each of the areas; therefore, the second moment I_{yC} for the composite section is

$$I_{yC} = I_{yCWF} + I_{yCCH}$$
$$= 94.4 + 162 = 256.4 \cong 256 \text{ in.}^4 \qquad \textbf{Ans.}$$

The radius of gyration about the x_C-axis for the composite section is

$$k_{xC} = \left[\frac{I_{xC}}{A_T}\right]^{1/2} = \left[\frac{3293.4}{33.52}\right]^{1/2} = 9.912 \cong 9.91 \text{ in.} \qquad \textbf{Ans.}$$

The radius of gyration about the y_C-axis for the composite section is

$$k_{yC} = \left[\frac{I_{yC}}{A_T}\right]^{1/2} = \left[\frac{256.4}{33.52}\right]^{1/2} = 2.766 \cong 2.77 \text{ in.} \; \blacksquare \qquad \textbf{Ans.}$$

PROBLEMS

Standard Problems

8-1* Determine the second moments of the shaded area shown in Fig. P8-1 with respect to x- (horizontal) and y- (vertical) axes through the centroid of the area.

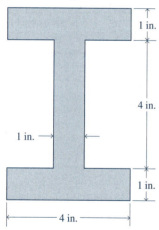

1 in.

4 in.

1 in.

1 in.

4 in.

Figure P8-1

8-2* Determine the second moments of the shaded area shown in Fig. P8-2 with respect to x- (horizontal) and y- (vertical) axes through the centroid of the area.

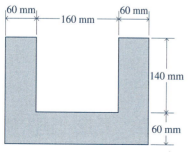

60 mm · 160 mm · 60 mm

140 mm

60 mm

Figure P8-2

8-3 Determine the second moment of the shaded area shown in Fig. P8-3 with respect to
(a) The x-axis. (b) The y-axis.

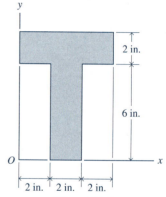

y

2 in.

6 in.

O

2 in. 2 in. 2 in.

x

Figure P8-3

8-4 Determine the second moment of the shaded area shown in Fig. P8-4 with respect to
(a) The x-axis. (b) The y-axis.

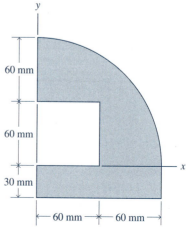

y

60 mm

60 mm

30 mm

60 mm · 60 mm

x

Figure P8-4

8-5* Determine the second moments of the shaded area shown in Fig. P8-5 with respect to x- (horizontal) and y- (vertical) axes through the centroid of the area.

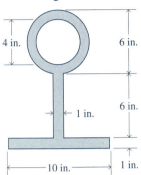

4 in.

6 in.

6 in.

1 in.

10 in.

1 in.

Figure P8-5

8-6* Four C305 × 45 channels are welded together to form the cross section shown in Fig. P8-6. Determine the second moments of the area with respect to x- (horizontal) and y- (vertical) axes through the centroid of the area.

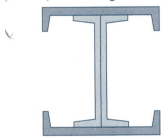

Figure P8-6

8-7 Two 10 × 1-in. steel plates are welded to the flanges of an S18 × 70 I-beam as shown in Fig. P8-7. Determine the second moments of the area with respect to x- (horizontal) and y- (vertical) axes through the centroid of the area.

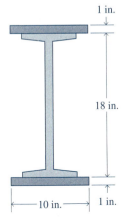

1 in.

18 in.

10 in. 1 in.

Figure P8-7

8-8 Two 250 × 25-mm steel plates and two C254 × 45 channels are welded together to form the cross section shown in Fig. P8-8. Determine the second moments of the area with respect to x- (horizontal) and y- (vertical) axes through the centroid of the area.

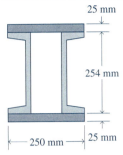

25 mm

254 mm

250 mm 25 mm

Figure P8-8

Challenging Problems

8-9* Determine the second moments of the shaded area shown in Fig. P8-9 with respect to the x- and y-axes.

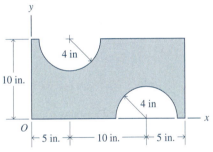

y

4 in

10 in.

4 in

O

5 in. 10 in. 5 in.

Figure P8-9

8-10* Determine the second moments of the shaded area shown in Fig. P8-10 with respect to the x- and y-axes.

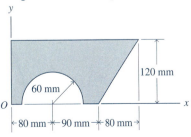

y

120 mm

60 mm

O

80 mm 90 mm 80 mm

x

Figure P8-10

8-11 Determine the second moments of the shaded area shown in Fig. P8-11 with respect to
(a) The x- and y-axes shown on the figure.
(b) The x- and y-axes through the centroid of the area.

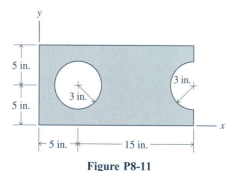

Figure P8-11

8-12 Determine the second moments of the shaded area shown in Fig. P8-12 with respect to
(a) The x- and y-axes shown on the figure.
(b) The x- and y-axes through the centroid of the area.

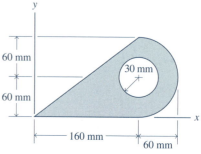

Figure P8-12

8-5 THE ELASTIC FLEXURE FORMULA

When the integral $\int_A y^2 \, dA$ in Eq. (8-7) is replaced by the symbol I, the elastic flexure formula is obtained by combining Eqs. (8-6) and (8-7) as

$$\sigma_x = -\frac{M_r y}{I} \qquad (8\text{-}11)$$

where σ_x is the normal stress on a transverse plane, at a distance y from the neutral surface, M_r is the resisting moment of the section, and I is the second moment of area of the transverse section with respect to the neutral axis. Recall that the neutral axis passes through the centroid of the area.

At any section of the beam, the flexural stress, or normal stress, will be maximum (have the greatest magnitude) at the surface farthest from the neutral axis ($y = c$), and Eq. (8-11) becomes

$$\sigma_{\max} = -\frac{M_r c}{I} = -\frac{M_r}{S} \qquad (8\text{-}12)$$

where $S = I/c$ is called the *section modulus of the beam*. Although the section modulus can be readily calculated for a given section, values are often included in tables to simplify calculations. Observe that for a given area, S becomes larger as the shape is altered to concentrate more of the area as far as possible from the neutral axis. Commercial rolled shapes such as I- and WF-beams and the various built-up sections are intended to optimize the area–section modulus relation.

Thus far, the discussion of flexural behavior has been limited to structural members with symmetric cross sections that are loaded in a plane of symmetry. Many other shapes are subjected to flexural loadings, and methods are needed to determine stress distributions in these nonsymmetric shapes. The flexure formula [Eq. (8-11)] provides a means for relating the resisting moment M_r at a section of a beam to the normal stress at a point on the transverse cross section. Further insight into the applicability of the flexure formula to nonsymmetric sections can be gained by considering the requirements for equilibrium when the applied moment M does not have a component about the y-axis of the cross section. Thus, from Eq. (8-6) and the equilibrium equation $\Sigma M_y = 0$,

$$\int_A z \, \sigma_x \, dA = \int_A z \, \frac{\sigma_c}{c} y \, dA = \frac{\sigma_c}{c} \int_A zy \, dA = \frac{\sigma_c}{c} I_{yz} = 0 \qquad (a)$$

The quantity I_{yz} is commonly known as the *mixed second moment* of the cross-sectional area with respect to the centroidal y- and z-axes. Obviously, Eq. (a) can be satisfied only if $I_{yz} = 0$. For symmetric cross sections, $I_{yz} = 0$ when the y- and z-axes coincide with the axes of symmetry. For nonsymmetric cross sections, $I_{yz} = 0$ when the y- and z-axes are centroidal principal axes for the cross section. Thus, the flexure formula is valid for any cross section, provided y is measured along a principal direction, I is a principal second moment of area, and M is a moment about a principal axis. Problems of this type will not be discussed in this book.

Example Problem 8-4

A timber beam consists of four 2×8-in. planks fastened together to form a box section 8 in. wide by 12 in. deep as shown in Fig. 8-16. If the flexural (normal) stress at point A of the cross section is 1250 psi T, determine:

(a) The flexural stress at point B of the cross section.

(b) The flexural stress at point C of the cross section.

(c) The flexural stress at point D of the cross section.

SOLUTION

The flexural stress σ_x at a distance y from the neutral surface on a transverse cross section of a beam is given by Eq. (8-11) as

$$\sigma_x = -\frac{M_r y}{I} \qquad \text{or} \qquad M_r = -\frac{\sigma_x I}{y}$$

For the cross section the second moment of area with respect to the neutral axis is:

$$I = \frac{8(12)^3}{12} - \frac{4(8)^3}{12} = 981.3 \text{ in.}^4$$

The resisting moment is:

$$M_r = -\frac{\sigma_x I}{y_A} = \frac{1250(981.3)}{6} = -204.4(10^3) \text{ in.} \cdot \text{lb}$$

(a) For point B:

$$\sigma_x = -\frac{M_r y_B}{I} = -\frac{-204.4(10^3)(-4)}{981.3} = -833 \text{ psi} = 833 \text{ psi C} \qquad \textbf{Ans.}$$

(b) For point C:

$$\sigma_x = -\frac{M_r y_C}{I} = -\frac{-204.4(10^3)(2)}{981.3} = +417 \text{ psi} = 417 \text{ psi T} \qquad \textbf{Ans.}$$

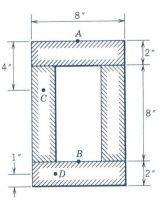

Figure 8-16

(c) For point D:

$$\sigma_x = -\frac{M_r y_D}{I} = -\frac{-204.4(10^3)(-5)}{981.3} = -1042 \text{ psi} = 1042 \text{ psi C} \quad \textbf{Ans.}$$

Alternatively from Eq. (8-11), note that on a given cross section

$$\sigma_x = -\frac{M_r y}{I} \qquad \text{or} \qquad \frac{\sigma_x}{y} = -\frac{M_r}{I} = \text{constant}$$

Therefore, if the stress is known at a point on the cross section it can be determined at any other point without knowing either the resisting moment M_r or the second moment I for the cross section. Thus,

$$\frac{\sigma_A}{y_A} = \frac{\sigma_B}{y_B} = \frac{\sigma_C}{y_C} = \frac{\sigma_D}{y_D}$$

(a) $\sigma_B = \dfrac{y_B}{y_A} \sigma_A = \dfrac{-4}{6}(1250) = -833 \text{ psi} = 833 \text{ psi C}$ **Ans.**

(b) $\sigma_C = \dfrac{y_C}{y_A} \sigma_A = \dfrac{2}{6}(1250) = +417 \text{ psi} = 417 \text{ psi T}$ **Ans.**

(c) $\sigma_D = \dfrac{y_D}{y_A} \sigma_A = \dfrac{-5}{6}(1250) = -1042 \text{ psi} = 1042 \text{ psi C}$ ■   **Ans.**

Example Problem 8-5

The maximum flexural stress at a certain section (Fig. 8-17) in a rectangular beam 100 mm wide by 200 mm deep is 15 MPa. Determine

(a) The resisting moment M_r developed at the section.
(b) The percentage decrease in M_r if the dotted central portion of the cross section shown in Fig. 8-17 is removed.

SOLUTION

The normal stress σ_x at a distance y from the neutral surface on a transverse cross section of a beam is given by Eq. (8-11) as

$$\sigma_x = -\frac{M_r y}{I} \qquad \text{or} \qquad M_r = -\frac{\sigma_x I}{y}$$

(a) For the original cross section:

$$I = \frac{100(200)^3}{12} = 66.67(10^6) \text{ mm}^4 = 66.67(10^{-6}) \text{ m}^4$$

$$|M_r| = \frac{\sigma_x I}{y_A} = \frac{15(10^6)(66.67)(10^{-6})}{100(10^{-3})}$$
$$= 10.00(10^3) \text{ N} \cdot \text{m} = 10.00 \text{ kN} \cdot \text{m} \quad \textbf{Ans.}$$

where $|M_r|$ is the magnitude of the resisting moment.

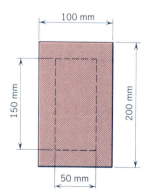

Figure 8-17

(b) For the modified cross section:

$$I = \frac{100(200)^3}{12} - \frac{50(150)^3}{12} = 52.60(10^6) \text{ mm}^4 = 52.60(10^{-6}) \text{ m}^4$$

$$|M_r| = \frac{\sigma_x I}{y_A} = \frac{15(10^6)(52.60)(10^{-6})}{100(10^{-3})}$$
$$= 7.89(10^3) \text{ N} \cdot \text{m} = 7.89 \text{ kN} \cdot \text{m}$$

$$\text{percent decrease} = D = \frac{10.00 - 7.89}{10.00}(100) = 21.1\% \quad \blacksquare \qquad \textbf{Ans.}$$

Example Problem 8-6

Determine the maximum flexural stress produced by a resisting moment of $-30 \text{ ft} \cdot \text{kip}$ if the beam has the cross section shown in Fig. 8-18.

SOLUTION

The neutral axis is horizontal and passes through the centroid of the cross section. The centroid is located by using the principle of moments as applied to areas. The total area A of the cross section is

$$A = 4(1) + 8(1) + 8(1) = 20 \text{ in.}^2$$

The moment of the area M_A about the bottom edge of the cross section is

$$M_A = 8(1)(0.5) + 8(1)(5) + 4(1)(9.5) = 82 \text{ in.}^3$$

The distance y_C from the bottom edge of the cross section to the centroid is

$$y_C = \frac{M_A}{A} = \frac{82}{20} = 4.1 \text{ in.}$$

The second moment I of the cross section about the neutral axis is

$$I = \frac{1}{3}(8)(4.1)^3 - \frac{1}{3}(7)(3.1)^3 + \frac{1}{3}(4)(5.9)^3 - \frac{1}{3}(3)(4.9)^3 = 270.5 \text{ in.}^4$$

The distance c from the neutral axis to the fiber farthest from the neutral axis is 5.9 in. The flexural stress σ_{max} is given by Eq. (8-11) as

$$\sigma_{max} = -\frac{M_r c}{I} = -\frac{-30(10^3)(12)(5.9)}{270.5}$$
$$= +7852 \text{ psi} \cong 7850 \text{ psi T} \qquad \textbf{Ans.}$$

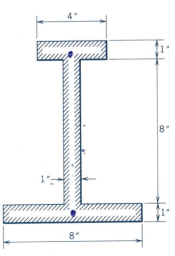

Figure 8-18

The maximum flexural, or normal, stress occurs at the top edge of the transverse cross section, and is constant across the 4-in. segment of the section. $\blacksquare$

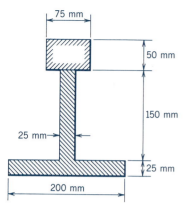

Figure 8-19

Example Problem 8-7

A beam has the cross section shown in Fig. 8-19. On a section where the resisting moment is $-75 \text{ kN} \cdot \text{m}$, determine

(a) The maximum tensile flexural stress.
(b) The maximum compressive flexural stress.

SOLUTION

The neutral axis is horizontal and passes through the centroid of the cross section. The centroid is located by using the principle of moments as applied to areas. The total area A of the cross section is

$$A = 200(25) + 150(25) + 50(75) = 12{,}500 \text{ mm}^2$$

The moment of the area M_A about the bottom edge of the cross section is

$$M_A = 200(25)(12.5) + 25(150)(100) + 75(50)(200) = 1{,}187{,}500 \text{ mm}^3$$

The distance y_C from the bottom edge of the cross section to the centroid is

$$y_C = \frac{M_A}{A} = \frac{1{,}187{,}500}{12{,}500} = 95 \text{ mm}$$

The second moment I of the cross section about the neutral axis is

$$I = \frac{1}{3}(200)(95)^3 - \frac{1}{3}(175)(70)^3 + \frac{1}{3}(75)(130)^3 - \frac{1}{3}(50)(80)^3$$
$$= 83.54(10^6) \text{ mm}^4 = 83.54(10^{-6}) \text{ m}^4$$

(a) The maximum tensile flexural stress occurs at the top of the beam and is

$$\sigma_{\text{max}} = -\frac{M_r c}{I} = -\frac{-75(10^3)(130)(10^{-3})}{83.54(10^{-6})}$$
$$= +116.7(10^6) \text{ N/m}^2 = 116.7 \text{ MPa T} \qquad \text{Ans.}$$

(b) The maximum compressive flexural stress occurs at the bottom of the beam and is

$$\sigma_{\text{max}} = -\frac{M_r c}{I} = -\frac{-75(10^3)(-95)(10^{-3})}{83.54(10^{-6})}$$
$$= -85.3(10^6) \text{ N/m}^2 = 85.3 \text{ MPa C} \blacksquare \qquad \text{Ans.}$$

PROBLEMS

Note: In problems involving rolled shapes (see Appendix B), unless otherwise stated, the beam is oriented so that the maximum section modulus applies.

Standard Problems

8-13* The maximum flexural stress on a transverse plane in a beam with a rectangular cross section 4 in. wide by 8 in. deep is 1500 psi. Determine the resisting moment M_r transmitted by the plane.

8-14* The maximum flexural stress on a transverse plane in a beam with a rectangular cross section 150 mm wide by 300 mm deep is 15 MPa. Determine the resisting moment M_r transmitted by the plane.

8-15 Determine the maximum flexural stress produced by a resisting moment M_r of $+5000$ ft · lb if the beam has the cross section shown in Fig. P8-15.

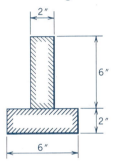

Figure P8-15

8-16 Determine the maximum flexural stress produced by a resisting moment M_r of -15 kN · m if the beam has the cross section shown in Fig. P8-16.

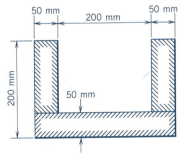

Figure P8-16

8-17* A timber beam is made of three 2×6-in. planks fastened together to form an I-beam 6 in. wide by 10 in. deep. If the maximum flexural stress must not ex-

ceed 1200 psi determine the maximum resisting moment M_r that the beam can support.

8-18* A timber beam consists of three 50×200-mm planks fastened together to form an I-beam 200 mm wide by 300 mm deep as shown in Fig. P8-18. If the flexural stress at point A of the cross section is 7.5 MPa C, determine
(a) The flexural stress at point B of the cross section.
(b) The flexural stress at point C of the cross section.
(c) The flexural stress at point D of the cross section.

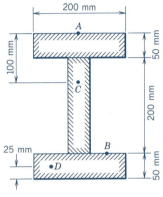

Figure P8-18

8-19 The maximum flexural stress on a transverse cross section of a beam with a 2×4-in. rectangular cross section must not exceed 10 ksi. Determine the maximum resisting moment M_r that the beam can support if the neutral axis is
(a) Parallel to the 4-in. side.
(b) Parallel to the 2-in. side.

8-20 A hardened steel ($E = 210$ GPa) bar with a 50 mm–square cross section is subjected to a flexural form of loading that produces a flexural strain of $+1200$ μm/m at a point on the top surface of the beam. Determine
(a) The maximum flexural stress at the point.
(b) The resisting moment M_r developed in the beam on a transverse cross section through the point.

8-21* A steel ($E = 29,000$ ksi) bar with a rectangular cross section is bent over a rigid mandrel ($R = 10$ in.) as shown in Fig. P8-21. If the maximum flexural stress in the bar is not to exceed the yield strength ($\sigma_y = 36$ ksi) of the steel, determine the maximum allowable thickness h for the bar.

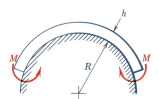

Figure P8-21

8-22* An aluminum alloy ($E = 73$ GPa) bar with a rectangular cross section is bent over a rigid mandrel as shown in Fig. P8-21. The thickness h of the bar is 25 mm. If the maximum flexural stress in the bar must be limited to 200 MPa, determine the minimum allowable radius R for the mandrel.

8-23 A wide-flange beam will be used to resist a bending moment M_r of 55,000 ft · lb. If the maximum flexural stress must not exceed 18,000 psi, select the most economical wide-flange section listed in Appendix B.

8-24 A pair of channels fastened back-to-back will be used as a beam to resist a bending moment M_r of 60 kN · m. If the maximum flexural stress must not exceed 120 MPa, select the most economical channel section listed in Appendix B.

8-25* A beam has the cross section shown in Fig. P8-25. On a section where the resisting moment is +12.5 ft · kip, determine
(a) The maximum tensile flexural stress.
(b) The maximum compressive flexural stress.

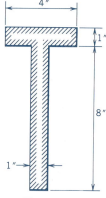

Figure P8-25

8-26* Two L102 × 102 × 12.7-mm structural steel angles (see Appendix B for dimensions) are attached back to back to form a T-section. Determine the maximum resisting moment M_r that can be supported by the beam if the maximum flexural stress must be limited to 125 MPa.

8-27 The load carrying capacity of an S24 × 80 American Standard beam (see Appendix B for dimensions) is to be increased by fastening two $8 × \frac{3}{4}$-in. plates to the flanges of the beam. The maximum flexural stress in both the original and modified beams must be limited to 15 ksi. Determine
(a) The maximum resisting moment that the original beam can support.
(b) The maximum resisting moment that the modified beam can support.

8-28 The beam of Fig. P8-28 is made of a material that has a yield strength of 250 MPa. Determine the maximum resisting moment that the beam can support if yielding must be avoided.

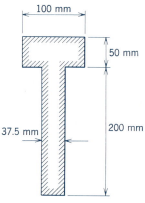

Figure P8-28

Challenging Problems

8-29* Determine the percentage of the resisting moment carried by the flanges of a W36 × 160 wide-flange beam (see Appendix B for dimensions).

8-30* Determine the percentage of the resisting moment carried by the flanges of a W838 × 226 wide-flange beam (see Appendix B for dimensions).

8-31 An I-beam is fabricated by welding two 16 × 2-in. flange plates to a 24 × 1-in. web plate. The beam is loaded in the plane of symmetry parallel to the web. Determine the percentage of the resisting moment carried by the flanges.

8-32 Determine the maximum flexural stress produced by a resisting moment of +100 kN · m if the beam has the cross section shown in Fig. P8-32.

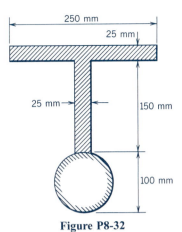

250 mm

25 mm

25 mm

150 mm

100 mm

Figure P8-32

8-33* A beam has the cross section shown in Fig. P8-33. On a section where the resisting moment is −30 ft · kip, determine

(a) The maximum tensile flexural stress.
(b) The maximum compressive flexural stress.

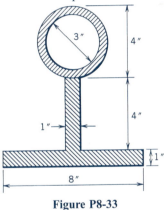

3″

4″

4″

1″

8″

1″

Figure P8-33

8-34* A beam with a solid rectangular cross section (width b and depth h) is being designed to support a maximum resisting moment of 4 kN · m. The maximum flexural stress in the beam must not exceed 50 MPa. Prepare a design curve showing the acceptable combinations of width b and depth h. The width b of the beam cannot exceed 75 mm.

8-35 A beam with a hollow circular cross section is being designed to support a maximum resisting moment of 100 in. · kip. The maximum flexural stress in the beam must not exceed 40 ksi. Prepare a design curve showing the acceptable combinations of inside diameter d_i and outside diameter d_o. The maximum diameter of the beam cannot exceed 5 in.

8-36 A beam with a hollow square cross section is being designed to support a maximum resisting moment of 10 kN · m. The maximum flexural stress in the beam cannot exceed 100 MPa. Prepare a design curve showing the acceptable combinations of outside dimension a_o and inside dimension a_i. The outside dimension a_o cannot exceed 120 mm.

8-6 SHEAR FORCES AND BENDING MOMENTS IN BEAMS

The method for determining flexural stresses outlined in Section 8-5 is adequate if one wishes to determine the flexural stresses on any specified transverse cross section of the beam. However, if the maximum flexural stress is required in a beam subjected to a loading that produces a resisting moment that varies with position along the beam, it is desirable to have a method for determining the maximum resisting moment. Similarly, the maximum transverse shearing stress will occur at a section where the resisting shear (V_r of Fig. 8-4a) is maximum, and a method for determining such sections is likewise desirable.

In Section 8-6-1 equations for the resisting shear V_r and the resisting moment M_r will be obtained using equilibrium equations. In Section 8-6-2 relationships will be developed between loads applied to the beam and the resisting

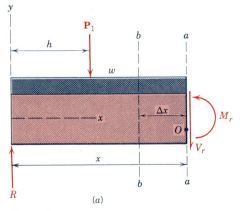

Figure 8-4

shear V_r at a section and between the resisting shear V_r at a section and the resisting moment M_r.

8-6-1 Shear Force and Bending Moment: An Equilibrium Approach

When the equilibrium equation $\Sigma F_y = 0$ is applied to the free-body diagram of Fig. 8-4a, the result can be written as

$$R - wx - P = V_r \qquad \text{or} \qquad V = V_r$$

where $V = R - wx - P$ is the resultant of the external transverse forces acting on the part of the beam to either side of a transverse section and is called the *transverse shear* or just the shear at the section.

As seen from the definitions of V and V_r, the shear V is equal in magnitude and opposite in sense to V_r. Since these shear forces are always equal in magnitude, they are frequently treated as though they were identical. For simplicity the symbol V will be used henceforth to represent both the transverse shear and the resisting shear. The sign convention to be selected will have sufficient generality to apply to both quantities.

The resultant of the flexural stresses on any transverse section has been shown to be a couple (if only transverse loads are considered) and has been designated as M_r. When the equilibrium equation $\Sigma M_O = 0$ (where O is any axis parallel to the neutral axis of the section) is applied to the free-body diagram of Fig. 8-4a, the result can be written as

$$Rx - \frac{wx^2}{2} - P(x - h) = M_r \qquad \text{or} \qquad M = M_r$$

where $M = Rx - \dfrac{wx^2}{2} - P(x - h)$ is the algebraic sum of the moments of the external forces acting on the part of the beam to either side of the transverse section, with respect to an axis in the section and is called the *bending moment* or just the moment at the section. The axis O is usually taken to be the neutral axis of the cross section.

As seen from the definitions of M and M_r, the bending moment M is equal in magnitude and opposite in sense to M_r. Since these moments are always equal in magnitude, they are frequently treated as though they were identical. For simplicity the symbol M will be used henceforth to represent both the bending moment and the resisting moment. The sign convention for moments will have sufficient generality to apply to both quantities.

The bending moment M and the transverse shear V are normally shown on a free-body diagram. It is also customary procedure to show each external force individually as indicated in Fig. 8-4. The variation of V and M along the beam can be shown conveniently by means of equations or by means of shear and bending moment diagrams (graphs of V and M as functions of x).

A sign convention is necessary for the correct interpretation of results obtained from equations or diagrams for shear and moment. The following convention will give consistent results regardless of whether one proceeds from left to right or from right to left along the beam. By definition, the shear at a section is positive when the portion of the beam to the left of the section (for a hori-

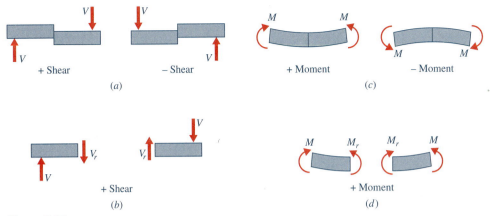

Figure 8-20

zontal beam) tends to move upward with respect to the portion to the right of the section, as shown in Fig. 8-20a. Then for a positive shear force, the transverse shear pushes up on the portion of the beam to the left of the section while the resisting shear pushes downward on the section, as shown in Fig. 8-20b. The converse applies when the portion of the beam is taken to the right of the section. Also by definition, the bending moment in a horizontal beam is positive at sections for which the top of the beam is in compression and the bottom is in tension, as shown in Fig. 8-20c. Then for a positive bending moment, the bending moment acting on the portion of the beam to the left of the section must act clockwise while the resisting moment on the section must act counterclockwise, as shown in Fig. 8-20d. Observe that the signs of the terms in the preceding equations for V and M agree with these definitions (sign conventions).

Since M and V vary with x, they are functions of x, and equations for M and V can be obtained from free-body diagrams of portions of the beam. As illustrated in the Example Problems that follow,

- The beam is cut at an arbitrary location x.
- A free-body diagram is drawn for the portion of the beam to the left (or for the portion to the right) of the transverse cross section. The shear force and bending moment must be drawn according to the sign convention established earlier.
- Equilibrium equations $\Sigma F_y = 0$ and $\Sigma M_{\text{cut}} = 0$ are written from the free-body diagrams and solved to get the shear force and bending moment at the location x. The equations obtained for $V(x)$ and $M(x)$ will be valid for a range of x for which the nature of the loading does not change. That is, if the beam is cut in a region in which there are no concentrated or distributed loads, the equations will be valid for the entire region $x_L < x < x_R$, where x_L is the location of the nearest load to the left of the cut and x_R is the location of the nearest load to the right of the cut. No matter where in the region $x_L < x < x_R$ the beam is cut, the free-body diagram will look exactly the same and the equations will be exactly the same.
- The process must be repeated for each different segment of the beam.

Example Problem 8-8 will illustrate the procedure to calculate the shear force and bending moment for any section of a beam.

Example Problem 8-8

A beam is loaded and supported as shown in Fig. 8-21a. Write equations for the shear V and bending moment M for any section of the beam in the interval BC.

SOLUTION

A free-body diagram, or load diagram, for the beam is shown in Fig. 8-21b. The reactions shown on the load diagram are determined from the equations of equilibrium.

$$+\curvearrowright \Sigma M_D = R_1(8) - 400(6)(3) - 2000(2) = 0$$

$$+\curvearrowleft \Sigma M_A = R_2(8) - 400(6)(5) - 2000(6) = 0$$

$$R_1 = 1400 \text{ lb} \quad \text{and} \quad R_2 = 3000 \text{ lb}$$

A free-body diagram of a portion of the beam from the left end to any section between B and C is shown in Fig. 8-21c. Note that the origin of the xy-coordinate system has been placed at the left support and the resisting shear V and the resisting moment M are shown as positive values. As discussed in Section 8-6-1, the subscript r has been dropped from the resisting shear and resisting moment. From the definition of V, or from the equilibrium equation $\Sigma F_y = 0$,

$$V = 1400 - 400(x - 2) = 2200 - 400x \qquad 2 < x < 6 \text{ ft} \qquad \textbf{Ans.}$$

From the definition of M, or from the equilibrium equation $\Sigma M_O = 0$,

$$M = 1400x - 400(x - 2)\left[\frac{x - 2}{2}\right]$$

$$= -200x^2 + 2200x - 800 \qquad 2 < x < 6 \text{ ft} \qquad \textbf{Ans.}$$

The equations for V and M in the other intervals can be determined in a similar manner. ■

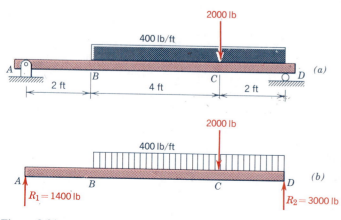

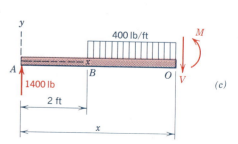

Figure 8-21

Example Problem 8-9

An S152 $\times$ 19 steel beam is loaded and supported as shown in Fig. 8-22. On a section 2 m to the right of B, determine

(a) The flexural stress at a point 25 mm below the top of the beam.
(b) The maximum flexural stress on the section.

SOLUTION

On a section 2 m to the right of B, the resisting moment M_r is

$$M_r = -8 + 3(2)(3) + 3(2) - 4(2)(1) = 8 \text{ kN} \cdot \text{m}$$

For an S152 $\times$ 19 section (see Table B-4 of Appendix B),

$$I = 9.20(10^6) \text{ mm}^4 = 9.20(10^{-6}) \text{ m}^4$$
$$S = 121(10^3) \text{ mm}^3 = 121(10^{-6}) \text{ m}^3,$$

and the depth of the beam is 152.4 mm = 0.1524 m.

(a) At a point 25 mm below the top of the beam, $y = 51.2$ mm = 0.0512 m; therefore, with all quantities expressed in newtons and meters, the stress is

$$\sigma = -\frac{My}{I} = -\frac{8(10^3)(0.0512)}{9.20(10^{-6})} = -44.5(10^6) \text{ N/m}^2 = 44.5 \text{ MPa C} \quad \textbf{Ans.}$$

(b) The maximum flexural stress is

$$\sigma_{max} = \frac{M}{S} = \frac{8(10^3)}{121(10^{-6})} = 66.1(10^6) \text{ N/m}^2$$
$$= 66.1 \text{ MPa T on the bottom} \quad \textbf{Ans.}$$
$$= 66.1 \text{ MPa C on the top} \quad \blacksquare \quad \textbf{Ans.}$$

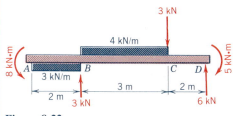

Figure 8-22

PROBLEMS

Note: In problems involving rolled shapes (see Appendix B), the beam is so oriented that the maximum section modulus applies.

Standard Problems

8-37* A beam is loaded and supported as shown in Fig. P8-37. Using the coordinate axes shown, write equations for the shear V and bending moment M for any section of the beam in the interval $4 < x < 8$ ft.

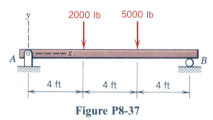

Figure P8-37

8-38* A beam is loaded and supported as shown in Fig. P8-38. Using the coordinate axes shown, write equations for the shear V and bending moment M for any section of the beam in the interval $0 < x < 4$ m.

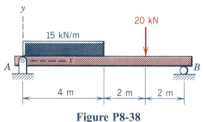

Figure P8-38

8-39 A beam is loaded as shown in Fig. P8-39. Using the coordinate axes shown, write equations for the shear V and bending moment M for any section of the beam in the interval $0 < x < 10$ ft.

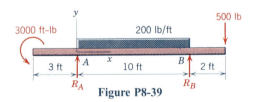

Figure P8-39

8-40 A beam is loaded and supported as shown in Fig. P8-40. Using the coordinate axes shown, write equations for the shear V and bending moment M for any section of the beam in the interval $0 < x < 4$ m.

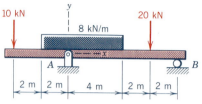

Figure P8-40

8-41* A beam is loaded and supported as shown in Fig. P8-41. Using the coordinate axes shown, write equations for the shear V and bending moment M for any section of the beam in the interval $2 < x < 8$ ft.

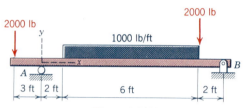

Figure P8-41

8-42* A beam is loaded and supported as shown in Fig. P8-42. Using the coordinate axes shown, write equations for the shear and bending moment for any section of the beam
(a) In the interval $0 < x < L$.
(b) In the interval $L < x < 2L$.

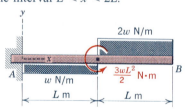

Figure P8-42

8-43 A beam is loaded and supported as shown in Fig. P8-43. Using the coordinate axes shown, write equations for the shear V and bending moment M for any section of the beam
(a) In the interval $5 < x < 10$ ft.
(b) In the interval $10 < x < 20$ ft.

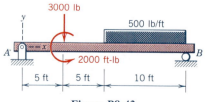

Figure P8-43

8-44 A beam is loaded and supported as shown in Fig. P8-44. Using the coordinate axes shown, write equations for the shear V and bending moment M for any section of the beam
(a) In the interval $0 < x < 4$ m.
(b) In the interval $4 < x < 6$ m.

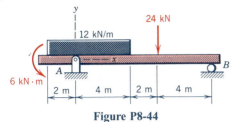

Figure P8-44

8-45* A beam is loaded as shown in Fig. P8-45. Using the coordinate axes shown, write equations for the shear and bending moment for any section of the beam between the supports at A and B.

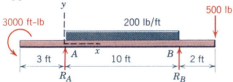

Figure P8-45

8-46* A beam is loaded as shown in Fig. P8-46. Using the coordinate axes shown, write equations for the shear and bending moment for any section of the beam in the interval from B to C.

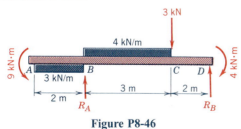

Figure P8-46

8-47 A beam is loaded and supported as shown in Fig. P8-47. Using the coordinate axes shown,
(a) Write equations for the shear V and bending moment M for any section of the beam.
(b) Determine the magnitudes and locations of the maximum shear and the maximum bending moment in the beam.

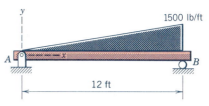

Figure P8-47

8-48 A beam is loaded and supported as shown in Fig. P8-48. Using the coordinate axes shown,
(a) Write equations for the shear V and bending moment M for any section of the beam.
(b) Determine the magnitudes and locations of the maximum shear and the maximum bending moment in the beam.

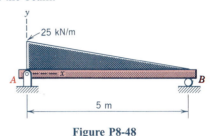

Figure P8-48

8-49* A beam is loaded and supported as shown in Fig. P8-49. The allowable flexural stresses on the section at the load P are 4000 psi T and 6000 psi C. Determine the maximum allowable load P.

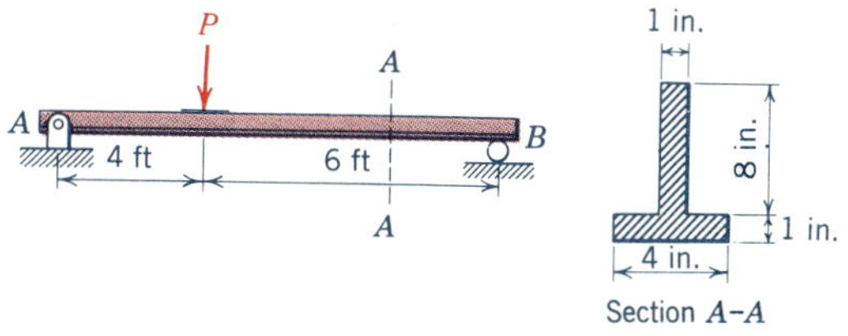

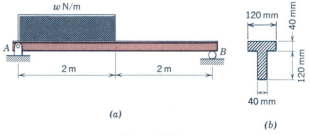

Figure P8-49

8-50* A beam is loaded and supported as shown in Fig. P8-50. The allowable flexural stresses on a section at the middle of the span are 15 MPa T and 10 MPa C. Determine the maximum permissible value for the load w.

Figure P8-50

8-51 A timber beam is loaded and supported as shown in Fig. P8-51a. The beam has the cross section shown in Fig. P8-51b. On a transverse cross section 4 ft from the right end, determine
(a) The flexural stress at point A of the cross section.
(b) The flexural stress at point B of the cross section.

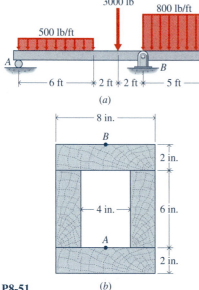

(a)

(b)

Figure P8-51

8-52 A timber beam is loaded and supported as shown in Fig. P8-52a. The beam has the cross section shown in Fig. P8-52b. On a transverse cross section 1.2 m from the left end, determine
(a) The flexural stress at point A of the cross section.
(b) The flexural stress at point B of the cross section.

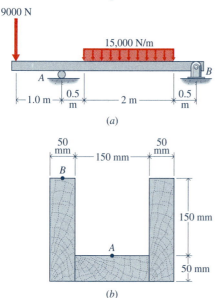

(a)

(b)

Figure P8-52

8-53[*] A timber beam is loaded as shown in Fig. P8-53a. The beam has the cross section shown in Fig. P8-53b. On a transverse cross section 1 ft from the left end, determine
(a) The flexural stress at point A of the cross section.
(b) The flexural stress at point B of the cross section.

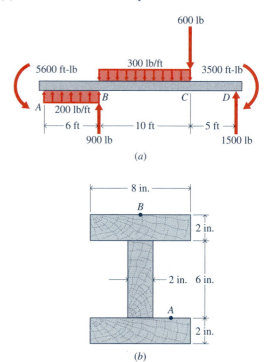

(a)

(b)

Figure P8-53

8-54[*] A timber beam is loaded and supported as shown in Fig. P8-54a. The beam has the cross section shown in Fig. P8-54b. On a transverse cross section 0.6 m from the left end, determine
(a) The flexural stress at point A of the cross section.
(b) The flexural stress at point B of the cross section.

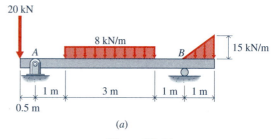

(a)

Figure P8-54

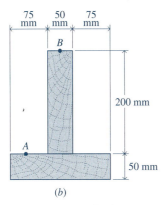

75 mm   50 mm   75 mm

200 mm

50 mm

(b)

Figure P8-54

8-55 A simply supported timber beam with a rectangular cross section and a span of 20 ft must support a concentrated load of 2500 lb at the center of the span. If the depth of the cross section is 12 in., determine the minimum acceptable width for the cross section if the maximum flexural stress must be limited to 1250 psi.

8-56 A simply supported timber beam with a circular cross section and a span of 6 m must support a uniformly distributed load of 7.5 kN/m. Determine the minimum acceptable diameter for the timber beam if the maximum flexural stress must be limited to 15 MPa.

8-57* A steel pipe with an outside diameter of 6 in. and an inside diameter of 5 in. rests horizontally on simple supports near the ends of the pipe. Steel weighs approximately 490 lb/ft³. If the maximum flexural stress in the pipe due to its own weight must be limited to 7500 psi, determine the maximum permissible distance between supports.

8-58 A simply supported timber beam with a rectangular cross section and a span of 6 m must support a concentrated load of 10 kN at the center of the span. If the width of the cross section is 200 mm, determine the minimum acceptable depth for the cross section if the maximum flexural stress must be limited to 10 MPa.

Challenging Problems

8-59* A beam is loaded and supported as shown in Fig. P8-59. Using the coordinate axes shown,
(a) Write equations for the shear V and bending moment M for any section of the beam.
(b) Determine the magnitudes and locations of the maximum shear and the maximum bending moment in the beam.

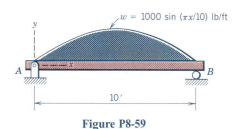

$w = 1000 \sin (\pi x/10)$ lb/ft

10′

Figure P8-59

8-60* A beam is loaded and supported as shown in Fig. P8-60. Using the coordinate axes shown,
(a) Write equations for the shear V and bending moment M for any section of the beam.
(b) Determine the magnitudes and locations of the maximum shear and the maximum bending moment in the beam.

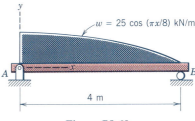

$w = 25 \cos (\pi x/8)$ kN/m

4 m

Figure P8-60

8-61 A beam is loaded and supported as shown in Fig. P8-61. Using the coordinate axes shown,
(a) Write equations for the shear V and bending moment M for any section of the beam.
(b) Determine the magnitudes and locations of the maximum shear and the maximum bending moment in the beam.

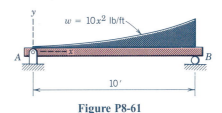

$w = 10x^2$ lb/ft

10′

Figure P8-61

8-62 A beam is loaded and supported as shown in Fig. P8-62. Using the coordinate axes shown,
(a) Write equations for the shear V and bending moment M for any section of the beam.
(b) Determine the magnitudes and locations of the maximum shear and the maximum bending moment in the beam.

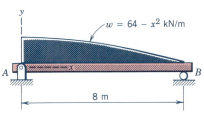

Figure P8-62

8-63* Two C10 × 15.3 steel channels (see Appendix B) are placed back to back to form a 10 in.–deep beam. The beam is 10 ft long and is simply supported at its ends. The beam carries a uniformly distributed load of 1000 lb/ft over its entire length and a concentrated load P at the center of the span. If the allowable flexural

stress on the section at the center of the span is 16,000 psi, determine the maximum permissible value of the concentrated load.

8-64 Two 50 × 200-mm structural timbers are used to fabricate a beam with an inverted-T cross section (flange on the bottom). The beam is simply supported at the ends and is 4 m long. If the maximum flexural stress must be limited to 30 MPa, determine
(a) The maximum moment that can be resisted by the beam.
(b) The largest concentrated load P that can be supported at the center of the span.
(c) The largest uniformly distributed load (over the entire span) that can be supported by the beam.

8-6-2 Load, Shear Force, and Bending Moment Relationships

The equilibrium approach is a fairly simple and straightforward method of getting equations for the shear force and bending moment in a beam. However, if the loading on the beam is complex, the equilibrium approach can require several cuts and several free-body diagrams. An alternative approach is to derive mathematical relationships between the loads acting on the beam and the shear forces in the beam and relationships between the shear forces and bending moments in the beam.

Consider the beam loaded and supported as shown in Fig. 8-23a. At some location x, the beam is acted on by a distributed load $w(x)$, a concentrated load P, and a concentrated couple C. A free-body diagram of a segment of the beam centered at the location x is shown in Fig. 8-23b. The upward direction is considered positive for the applied loads $w(x)$ and P; the shears and moments shown are positive according to the sign convention established earlier; and Δx, ΔV, and ΔM may be large or small. The element must be in equilibrium, and force equilibrium $(+\uparrow\Sigma F_y = 0)$ gives

$$V_L + w_{avg}\Delta x + P - (V_L + \Delta V) = 0$$

from which

$$\Delta V = P + w_{avg}\Delta x \qquad (8\text{-}13a)$$

 Four important relationships are obtained from Eq. (8-13a). First, if the concentrated force P and the distributed force w are both zero in some region of the beam (Δx large or small), then

$$\Delta V = 0 \qquad \text{or} \qquad V_L = V_R \qquad (8\text{-}13b)$$

That is, in any segment of a beam where there are no loads, the shear force is constant.

Second, if the concentrated load P is not zero, then in the limit as $\Delta x \rightarrow 0$,

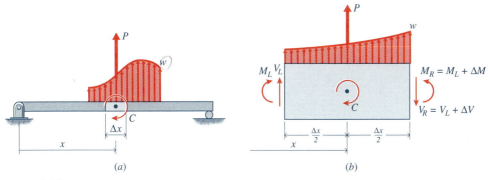

Figure 8-23

$$\Delta V = P \quad \text{or} \quad V_R = V_L + P \tag{8-13c}$$

That is, across any concentrated load P, the shear force graph (shear force versus x) jumps by the amount of the concentrated load. Furthermore, moving from left to right along the beam, the shear force graph jumps in the direction of the concentrated load.

Third, if the concentrated load P is zero, then in the limit as $\Delta x \to 0$,

$$\Delta V = w_{\text{avg}} \Delta x \to 0$$

and the shear force is a continuous function at x. Dividing through by Δx then gives

$$\lim_{\Delta x \to 0} \frac{\Delta V}{\Delta x} = \frac{dV}{dx} = w \tag{8-13d}$$

That is, the slope of the shear force graph at any location (section) x in the beam is equal to the intensity of loading at that section of the beam. Moving from left to right along the beam, if the distributed force is upward, then the slope of the shear force graph ($dV/dx = w$) is positive and the shear force graph is increasing (moving upward). If the distributed force is zero, then the slope of the shear force graph ($dV/dx = 0$) is zero and the shear force is constant.

Finally, in any region of the beam in which Eq. (8-13d) is valid (any region in which there are no concentrated loads), the equation can be integrated between definite limits to obtain

$$V_2 - V_1 = \int_{V_1}^{V_2} dV = \int_{x_1}^{x_2} w \, dx \tag{8-13e}$$

That is, for any section of the beam acted on by a distributed load w and no concentrated force ($P = 0$), the change in shear between sections at x_1 and x_2 is equal to the area under the load diagram between the two sections.

Similarly, applying moment equilibrium ($+\langle\Sigma M_{\text{center}} = 0$) to the free-body diagram of Fig. 8-23b gives

$$(M_L + \Delta M) - M_L - C - V_L \frac{\Delta x}{2} - (V_L + \Delta V)\frac{\Delta x}{2} + a(w_{\text{avg}}\Delta x) = 0$$

or

$$\Delta M = C + V_L \Delta x + \Delta V \frac{\Delta x}{2} - a(w_{avg}\Delta x) \qquad (8\text{-}14a)$$

in which $\dfrac{-\Delta x}{2} < a < \dfrac{\Delta x}{2}$ and in the limit as $\Delta x \to 0$, $a \to 0$ and $w_{avg} \to w$.

Three important relationships are obtained from Eq. (8-14a). First, if the concentrated couple C is not zero, then in the limit as $\Delta x \to 0$,

$$\Delta M = C \quad \text{or} \quad M_R = M_L + C \qquad (8\text{-}14b)$$

That is, across any concentrated couple C, the bending moment graph (bending moment versus x) jumps by the amount of the concentrated couple. Furthermore, moving from left to right along the beam, the bending moment graph jumps upward for a clockwise concentrated couple and jumps downward for a counterclockwise concentrated couple.

Second, if the concentrated couple C and concentrated force P are both zero,[4] then in the limit as $\Delta x \to 0$,

$$\Delta M = V_L \Delta x + \Delta V \frac{\Delta x}{2} - a(w_{avg}\Delta x) \to 0$$

and the bending moment is a continuous function at x. Dividing through by Δx then gives

$$\lim_{\Delta x \to 0} \frac{\Delta M}{\Delta x} = \frac{dM}{dx} = V \qquad (8\text{-}14c)$$

That is, the slope of the bending moment graph at any location x in the beam is equal to the value of the shear force at that section of the beam. Moving from left to right along the beam, if the shear force is positive, then $dM/dx = V$ is positive and the bending moment graph is increasing.

Finally, in any region of the beam in which Eq. (8-14c) is valid (any region in which there are no concentrated loads or couples), the equation can be integrated between definite limits to obtain

$$M_2 - M_1 = \int_{M_1}^{M_2} dM = \int_{x_1}^{x_2} V\, dx \qquad (8\text{-}14d)$$

That is, for any section of the beam in which the shear force is continuous ($C = P = 0$), the change in bending moment between sections at x_1 and x_2 is equal to the area under the shear force graph between the two sections.

Note that Eqs. (8-13) through (8-14) were derived with the x-axis positive to the right, the applied loads positive upward, and the shear and moment with signs as indicated in Fig. 8-20. If one or more of these assumptions are changed, the algebraic signs in the equations may need to be altered.

---

[4]If the concentrated force P is not zero, the bending moment will still be continuous at x but it will not be continuously differentiable at x. For a point slightly to the left of P, $dM/dx = V_L$, while for a point slightly to the right of P, $dM/dx = V_R$.

Equations (8-13) through (8-14) are used to draw shear and moment diagrams and to compute values of shear and moment at various sections along a beam.

8-6-3 Shear Force and Bending Moment Diagrams

Shear and moment diagrams provide a method for obtaining maximum values of shear and moment. A shear diagram is a graph in which abscissas represent distances along the beam and ordinates represent the transverse shear at the corresponding sections. A moment diagram is a graph in which abscissas represent distances along the beam and ordinates represent the bending moment at the corresponding sections.

Shear and moment diagrams can be drawn by calculating values of shear and moment at various sections along the beam and plotting enough points to obtain a smooth curve. Such a procedure is rather time-consuming; therefore, other more rapid methods will be developed using the load, shear force, and bending moment relationships developed in Section 8-6-2.

A convenient arrangement for constructing shear and moment diagrams is to draw a free-body diagram of the entire beam and construct shear and moment diagrams directly below. Two methods of procedure are used.

The first method consists of writing algebraic equations for the shear V and the moment M and constructing curves from the equations. This method has the disadvantage that unless the load is uniformly distributed or varies according to a known equation along the entire beam, no single elementary expression can be written for the shear V or the moment M, which applies to the entire length of the beam. Instead, it is necessary to divide the beam into intervals bounded by the abrupt changes in the loading. An origin should be selected, positive directions should be shown for the coordinate axes, and the limits of the abscissa (usually x) should be indicated for each interval.

Complete shear and moment diagrams should indicate values of shear and moment at each section where the load changes abruptly and at sections where they are maximum or minimum (negative maximum values). Sections where the shear and moment are zero should also be located.

The second method consists of drawing the shear diagram from the load diagram and the moment diagram from the shear diagram by means of Eqs. (8-13) and (8-14). This latter method, though it may not produce a precise curve, is less time-consuming than the first and it does provide the information usually required.

When all loads and reactions are known, the shear and moment at the ends of the beam can be determined by inspection. Both shear and moment are zero at the free end of a beam unless a force or a couple is applied there; in this case, the shear is the same as the force and the moment the same as the couple. At a simply supported or pinned end, the shear must equal the end reaction and the moment must be zero. At a built-in or fixed-end, the reactions are the shear and moment values.

Once a starting point for the shear diagram is established, the diagram can be sketched by using the definition of shear and the fact that the slope of the shear diagram can be obtained from the load diagram. When positive directions are chosen as upward and to the right, a positive distributed load (acting upward) produces a positive slope on the shear diagram. Similarly, a negative load (acting downward) produces a negative slope on the shear diagram. A concentrated force produces an abrupt change in shear. The change in shear between any two sections is given by the area under the load diagram between the same two sections. The change in shear at a concentrated force is equal to the concentrated force.

A moment diagram is drawn from the shear diagram in the same manner. The slope at any point on the moment diagram is given by the shear at the corresponding point on the shear diagram: a positive shear produces a positive slope and a negative shear produces a negative slope, when upward and to the right are positive. The change in moment between any two sections is given by the area under the shear diagram between the two sections. A couple applied to a beam at any section will cause the moment at the section to change abruptly by an amount equal to the moment of the couple.

The choice of which method to use depends on the type of information needed. If only the maximum values of shear force or bending moment are needed, then the second method usually gives these values more easily than the first. If equations of the bending moment are needed (they will be needed in Chapter 9 for finding the deflected shape of the beam), then the equilibrium approach must be used.

Example Problems 8-10 and 8-11 illustrate the two methods for drawing shear and moment diagrams.

Example Problem 8-10

A beam is loaded and supported as shown in Fig. 8-24a.

(a) Write equations for the shear and bending moment for any section of the beam in the interval AB.

(b) Write equations for the shear and bending moment for any section of the beam in the interval BC.

(c) Draw complete shear and moment diagrams for the beam.

SOLUTION

A free-body diagram for the beam is shown in Fig. 8-24b. It is not necessary to compute the reactions on the cantilever beam to write shear and bending moment equations; however, the reactions provide a convenient check. Thus,

$$+\uparrow\Sigma F_y = -500 - 200(6) - V_C = 0$$

$$+\curvearrowleft\Sigma M_C = 500(10) + 200(6)(3) + M_C = 0$$

from which

$$V_C = -1700 \text{ lb} \quad \text{and} \quad M_C = -8600 \text{ ft} \cdot \text{lb}$$

(a) A free-body diagram of a portion of the beam from the left end to any section between A and B is shown in Fig. 8-24c. The resisting shear V and the resisting moment M are shown as positive values.

From the equilibrium equation $\Sigma F_y = 0$,

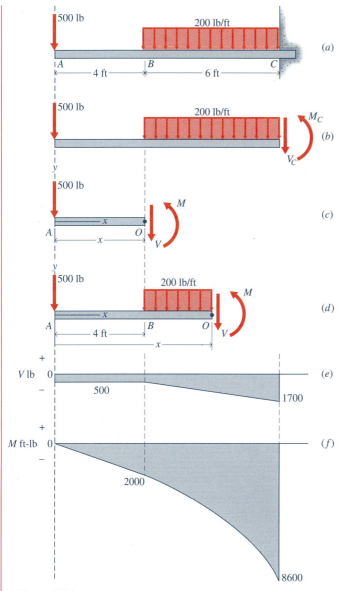

Figure 8-24

$$+\uparrow\Sigma F_y = -500 - V = 0$$

$$V = -500 \text{ lb} \qquad 0 < x < 4 \text{ ft} \qquad \textbf{Ans.}$$

From the equilibrium equation $\Sigma M_O = 0$,

$$+\smallint\Sigma M_O = 500(x) + M = 0$$

$$M = -500x \text{ ft} \cdot \text{lb} \qquad 0 < x < 4 \text{ ft} \qquad \textbf{Ans.}$$

(b) A free-body diagram of a portion of the beam from the left end to any section between B and C is shown in Fig. 8-24d.

From the equilibrium equation $\Sigma F_y = 0$,

$$+\uparrow\Sigma F_y = -500 - 200(x - 4) - V = 0$$

$$V = 300 - 200x \text{ lb} \qquad 4 < x < 10 \text{ ft} \qquad \textbf{Ans.}$$

From the equilibrium equation $\Sigma M_O = 0$,

$$+\downarrow\Sigma M_O = 500(x) + 200(x - 4)(x - 4)/2 + M = 0$$

$$M = -100x^2 + 300x - 1600 \text{ ft} \cdot \text{lb} \qquad 4 < x < 10 \text{ ft} \qquad \textbf{Ans.}$$

(c) The equations for V can be plotted in the appropriate intervals to give the shear diagram of Fig. 8-24e. Likewise, the equations for M can be plotted to give the moment diagram of Fig. 8-24f.

The shear and moment diagrams can also be drawn from the load diagram of Fig. 8-24b without writing the shear and moment equations. The shear just to the right of the 500-lb load is −500 lb. The slope of the shear diagram is equal to the load, and since no load is applied between A and B, the slope of the diagram is zero. From B to C the load is uniform; therefore, the slope of the shear diagram is constant. The change of shear from B to C is equal to the area under the load diagram between B and C [200(6) = 1200 lb]; therefore, the shear at C is −1700 lb. This shear is the same as the reaction at C, which provides a check.

The moment is zero at the free end A. From A to B the shear is constant (−500); therefore, the slope of the moment diagram is constant, and the moment diagram is a straight line. The shear increases from −500 lb at B to −1700 lb at C; therefore, the slope of the moment diagram is negative from B to C and increases uniformly in magnitude. The change in moment from A to B is equal to the area under the shear diagram and is 500(4) = 2000 ft · lb. The area under the shear diagram from B to C is $(\frac{1}{2})(500 + 1700)(6) = 6600$ ft · lb. Thus, the moment at C is −2000 + (−6600) = −8600 ft · lb. This moment is the same as the reaction at C. ■

Example Problem 8-11

A beam is loaded and supported as shown in Fig. 8-25a.

(a) Write equations for the shear and bending moment for any section of the beam in the interval CD.

(b) Draw complete shear and moment diagrams for the beam.

SOLUTION

The reactions are determined by using the free-body diagram of the beam shown in Fig. 8-25b. From the equilibrium equations,

$$+\curvearrowright\Sigma M_D = R_1(5) + 4 - 4(7)(1.5) - 8(1.5) = 0$$

$$+\downarrow\Sigma M_A = R_2(5) - 4 - 4(7)(3.5) - 8(3.5) = 0$$

$$R_1 = 10.00 \text{ kN} \qquad \text{and} \qquad R_2 = 26.0 \text{ kN}$$

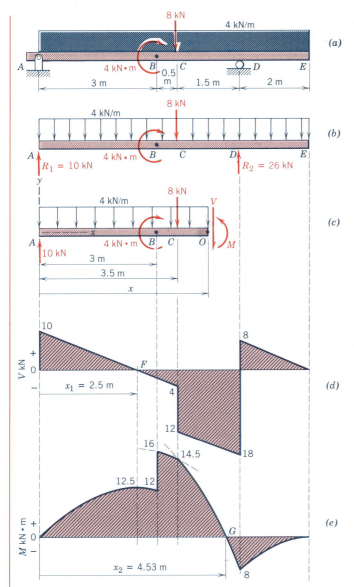

Figure 8-25

(a) A free-body diagram of a portion of the beam from the left end to any section between C and D is shown in Fig. 8-25c.

From the equilibrium equation $\Sigma F_y = 0$,

$$+\uparrow\Sigma F_y = 10 - 4(x) - 8 - V = 0$$

$$V = -4x + 2 \text{ kN} \qquad 3.5 < x < 5 \text{ m} \qquad \textbf{Ans.}$$

From the equilibrium equation $\Sigma M_O = 0$,

$$+\textstyle\int\Sigma M_O = -10(x) + 4(x)(x/2) - 4 + 8(x - 3.5) + M = 0$$

$$M = -2x^2 + 2x + 32 \text{ kN} \cdot \text{m} \qquad 3.5 < x < 5 \text{ m} \qquad \textbf{Ans.}$$

(b) The equations for V and M in the other intervals can be written and the shear and moment diagrams obtained by plotting these equations. In this example, the shear diagram will be drawn directly from the load diagram (Fig. 8-25b). The shear just to the right of A is 10 kN. From A to C the shear decreases at a rate of 4 kN/m. Thus, the shear just to the left of C is

$$V_C = V_A + \Delta V = 10 - 4(3.5) = -4 \text{ kN}$$

The concentrated downward load at C causes the shear to change suddenly from -4 kN just to the left of C to -12 kN just to the right of C. From C to D the shear continues to decrease at a rate of 4 kN/m. Thus, the shear just to the left of D is

$$V_D = V_C + \Delta V = -12 - 4(1.5) = -18 \text{ kN}$$

The reaction at D causes the shear to change suddenly from -18 kN just to the left of D to $+8$ kN just to the right of D. From D to E the shear continues to decrease at a rate of 4 kN/m. Thus, the shear at E is

$$V_E = V_D + \Delta V = +8 - 4(2) = 0$$

Since the distributed load is uniform over the entire beam, the slope of the shear diagram is constant. Points of zero shear, such as point F in Fig. 8-25d, are located from the geometry of the shear diagram. For example, the slope of the shear diagram is 4 kN/m. Therefore,

$$x_1 = 10/4 = 2.50 \text{ m}$$

The moment is zero at A, and the slope of the moment diagram (equal to the shear) is 10 kN · m/m. From A to C the shear and, hence, the slope of the moment diagram decreases uniformly to zero at F and to -4 at C. The abrupt change of shear at C indicates a sudden change of slope of the moment diagram; thus, the two parts of the moment diagram at C are not tangent. The slope of the moment diagram changes from -12 at C to -18 just to the left of D. From D to E the slope changes from $+8$ to zero. The change of moment from A to F is equal to the area under the shear diagram from A to F; therefore,

$$M_F = M_A + \Delta M = 0 + 10(2.5)/2 = 12.5 \text{ kN} \cdot \text{m}$$

Similarly,

$$V_B = V_A + \Delta V = 10 - 4(3) = -2 \text{ kN}$$
$$M_B = M_F + \Delta M = 12.5 - 2(0.5)/2 = 12.0 \text{ kN} \cdot \text{m}$$

The 4-kN · m couple (point moment) is applied at B. Since the moment just to the left of B is positive and the couple contributes an additional positive moment to sections to the right of B, the moment changes abruptly from $+12$ kN · m to $+16$ kN · m. In a similar manner moments at C, D, and E are determined from the shear diagram areas as

$$M_C = M_B + \Delta M = 16.0 - (4 + 2)(0.5)/2 = 14.5 \text{ kN} \cdot \text{m}$$

$$M_D = M_C + \Delta M = 14.5 - (12 + 18)(1.5)/2 = -8.0 \text{ kN} \cdot \text{m}$$

$$M_E = M_D + \Delta M = -8.0 + 8(2)/2 = 0$$

There is no moment applied at end E of the beam, and if the moment at E is not zero, it indicates that an error has occurred.

The point G, where the moment is zero, can be determined by setting the expression for the moment from part (a) equal to zero and solving for x. The result is

$$x_2 = 0.5 \pm \sqrt{16.25} = 4.531 \cong 4.53 \text{ m} \qquad \textbf{Ans.}$$

Note in this example that maximum and minimum moments may occur at sections where the shear curve passes through zero. In general, the shear curve may pass through zero at a number of points along the beam, and each such crossing indicates a point of possible maximum moment (in engineering, the moment with the largest absolute value is the maximum moment). It should be emphasized that the shear curve does not indicate the presence of abrupt discontinuities in the moment curve; hence, the maximum moment may occur where a couple is applied to the beam, rather than where the shear passes through zero. All possibilities should be examined to determine the maximum moment.

Since the flexural stress is zero at sections where the bending moment is zero, if a beam must be spliced, the splice should be located at or near such a section. ■

PROBLEMS

Note: In problems involving rolled shapes (see Appendix B), the beam is so oriented that the maximum section modulus applies.

Standard Problems

8-65* Draw complete shear and bending moment diagrams for the beam shown in Fig. P8-65.

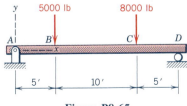

Figure P8-65

8-66* Draw complete shear and bending moment diagrams for the beam shown in Fig. P8-66.

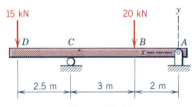

Figure P8-66

8-67 Draw complete shear and bending moment diagrams for the beam shown in Fig. P8-67.

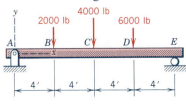

Figure P8-67

8-68 Draw complete shear and bending moment diagrams for the beam shown in Fig. P8-68.

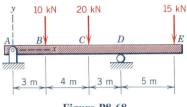

Figure P8-68

8-69* Draw complete shear and bending moment diagrams for the beam shown in Fig. P8-69.

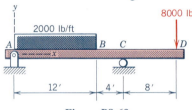

Figure P8-69

8-70* Draw complete shear and bending moment diagrams for the beam shown in Fig. P8-70.

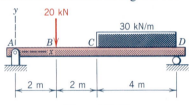

Figure P8-70

8-71 Draw complete shear and bending moment diagrams for the beam shown in Fig. P8-71.

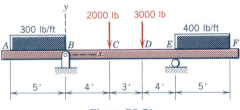

Figure P8-71

8-72 Draw complete shear and bending moment diagrams for the beam shown in Fig. P8-72.

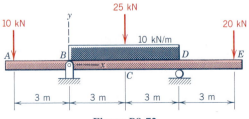

Figure P8-72

8-73* Draw complete shear and bending moment diagrams for the beam shown in Fig. P8-73.

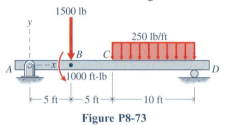

Figure P8-73

8-74* Draw complete shear and bending moment diagrams for the beam shown in Fig. P8-74.

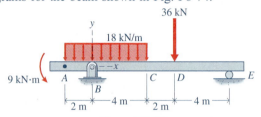

Figure P8-74

8-75 Draw complete shear and bending moment diagrams for the beam shown in Fig. P8-75.

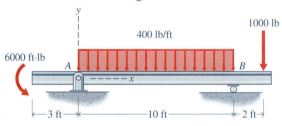

Figure P8-75

8-76 Draw complete shear and bending moment diagrams for the beam shown in Fig. P8-76.

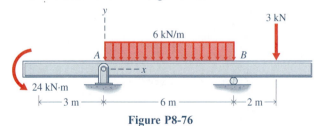

Figure P8-76

8-77* Draw complete shear and bending moment diagrams for the beam shown in Fig. P8-77.

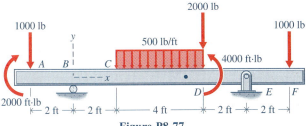

Figure P8-77

8-78* Draw complete shear and bending moment diagrams for the beam shown in Fig. P8-78.

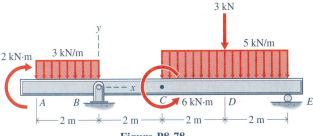

Figure P8-78

8-79 Draw complete shear and bending moment diagrams for the beam shown in Fig. P8-79.

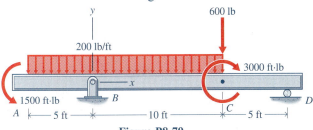

Figure P8-79

8-80 Draw complete shear and bending moment diagrams for the beam shown in Fig. P8-80.

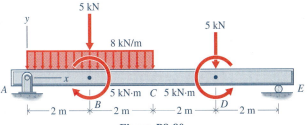

Figure P8-80

8-81* Select the lightest steel wide-flange or American standard beam (see Appendix B) that may be used for the beam of Fig. P8-81 if the flexural stress must be limited to 10 ksi.

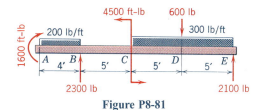

Figure P8-81

8-82* Select the lightest steel wide-flange or American standard beam (see Appendix B) that may be used for the beam of Fig. P8-82 if the flexural stress must be limited to 75 MPa.

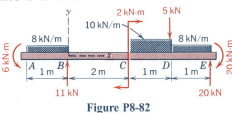

Figure P8-82

8-83 A timber beam with a rectangular cross section (width approximately equal to one half the depth) is to be loaded and supported as shown in Fig. P8-83. If the allowable flexural stress in the wood is 1500 psi, determine the lightest standard size timber (see Appendix B) required for the beam.

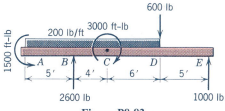

Figure P8-83

8-84 Select the lightest pair of structural-steel angles (see Appendix B) that may be used for the beam of Fig. P8-84 if the flexural stress must be limited to 60 MPa. The angles will be fastened back-to-back to form a T-section.

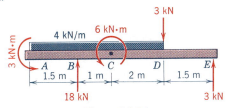

Figure P8-84

8-85[*] The beam shown in Fig. P8-85a has the cross section shown in Fig. P8-85b. Determine
(a) The maximum tensile flexural stress in the beam.
(b) The maximum compressive flexural stress in the beam.

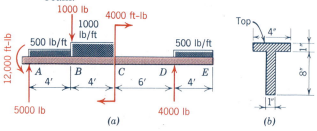

Figure P8-85

8-86 The beam shown in Fig. P8-86a has the cross section shown in Fig. P8-86b. Determine
(a) The maximum tensile flexural stress in the beam.
(b) The maximum compressive flexural stress in the beam.

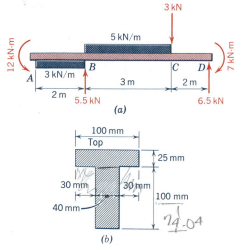

Figure P8-86

Challenging Problems

8-87[*] Draw complete shear and bending moment diagrams for the beam shown in Fig. P8-87.

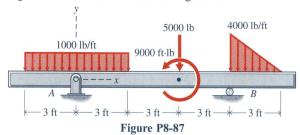

Figure P8-87

8-88[*] Draw complete shear and bending moment diagrams for the beam shown in Fig. P8-88.

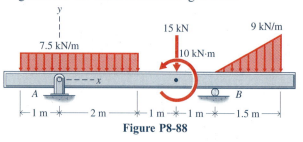

Figure P8-88

8-89 Draw complete shear and bending moment diagrams for the beam shown in Fig. P8-89.

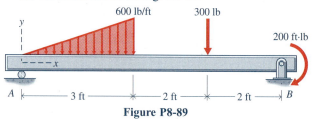

Figure P8-89

8-90 Draw complete shear and bending moment diagrams for the beam shown in Fig. P8-90.

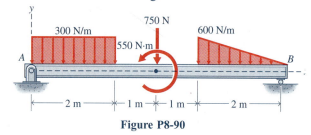

Figure P8-90

8-91[*] Draw complete shear and bending moment diagrams for the beam shown in Fig. P8-91.

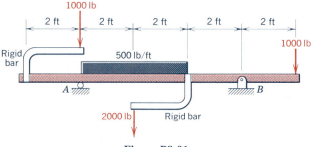

Figure P8-91

8-92 Draw complete shear and bending moment diagrams for the beam shown in Fig. P8-92.

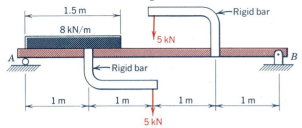

Figure P8-92

8-93* A W8 × 40 steel beam (see Appendix B) is loaded and supported as shown in Fig. P8-93. The total length of the beam is 10 ft. If the allowable flexural stress is 12 ksi, determine the maximum permissible value for the distributed load w.

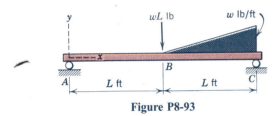

Figure P8-93

8-94* An S203 × 34 steel beam (see Appendix B) is loaded and supported as shown in Fig. P8-94. The total length of the beam is 4 m. If the allowable flexural stress is 125 MPa, determine the maximum permissible value for the distributed load w.

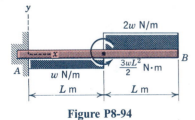

Figure P8-94

8-95 An S15 × 50 steel beam (see Appendix B) is loaded and supported as shown in Fig. P8-95. The total length of the beam is 15 ft. If the allowable flexural stress is 15,000 psi, determine the maximum permissible value for the distributed load w.

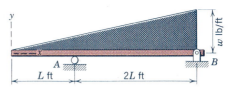

Figure P8-95

8-96 The beam shown in Fig. P8-96 has a pin-connected joint at D. Select the lightest steel wide-flange or American standard beam (see Appendix B) that may be used for the beam if the allowable flexural stress is 75 MPa.

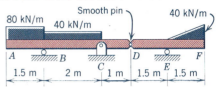

Figure P8-96

8-97 A timber beam with a rectangular cross section (width approximately equal to one half the depth) is to be loaded and supported as shown in Fig. P8-97. If the allowable flexural stress in the wood is 1500 psi, determine the lightest standard size timber (see Appendix B) required for the beam.

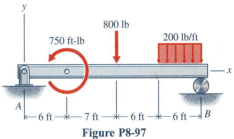

Figure P8-97

8-98 Select the lightest steel wide-flange or American standard beam (see Appendix B) that may be used for the beam of Fig. P8-98 if the allowable flexural stress is 120 MPa.

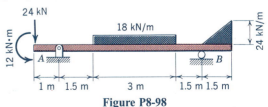

Figure P8-98

8-7 SHEARING STRESSES IN BEAMS

The discussion of shearing stresses in beams was delayed while flexural stresses were studied in Section 8-3. This procedure seems to be in keeping with the historical record on the study of beam stresses. From the time of Coulomb's paper, which contained the correct theory of the distribution of flexural stresses, approximately seventy years elapsed before the Russian engineer D. J. Jourawski (1821–1891), while designing timber railroad bridges in 1844–1850, developed the elementary shear stress theory used today. In 1856, Saint-Venant developed a rigorous solution for shearing stresses in beams; however, the elementary solution of Jourawski is the one in general use today by engineers and architects since it yields adequate results and is much easier to apply. The method requires use of the elastic flexure formula in its development; therefore, the formula developed is limited to linearly elastic action. For practical engineering problems, the shearing stress evaluation discussed in this section is important only (1) for the evaluation of principal elastic stresses at interior points in certain metallic beams (see Section 10-7), and (2) for the design of timber beams, because of the longitudinal plane of low shear resistance.

If one constructs a beam by stacking flat slabs one on top of another without fastening them together, and then loads this beam in a direction normal to the surface of the slabs, the resulting deformation will appear somewhat like that in Fig. 8-26a. This same type of deformation can be observed by taking a pack of cards and bending them, and noting the relative motion of the ends of the cards with respect to each other. The fact that a solid beam does not exhibit this rela-

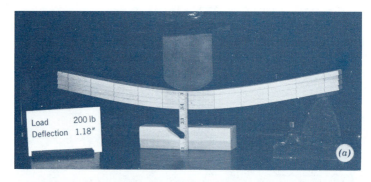

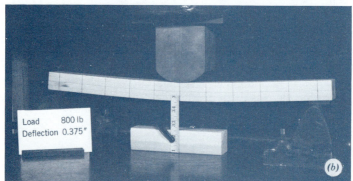

Figure 8-26

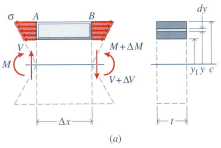
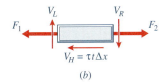

(a)   (b)

Figure 8-27

tive movement of longitudinal elements (see Fig. 8-26b, in which the beam is identical to that of Fig. 8-26a except that the layers are glued together) indicates the presence of shearing stresses on longitudinal planes. Evaluation of these shearing stresses will be determined by means of equilibrium and the free-body diagram of the short portion of a beam with a rectangular cross section shown in Fig. 8-27a.

The normal force dF acting on a differential area $dA = t\,dy$ on a cross section of the beam is equal to $\sigma\,dA$. The resultant of these differential forces is $F = \int \sigma\,dA$ integrated over the area of the cross section, where σ is the flexural stress at a distance y from the neutral surface and is given by the expression $\sigma = -My/I$. Therefore, the resultant normal force F_1 on the left end of the segment from y_1 to the top of the beam is[5]

$$F_1 = -\frac{M}{I}\int y\,dA = -\frac{M}{I}\int_{y_1}^{c} y\,(t\,dy)$$

Similarly, the resultant force F_2 on the right side of the element is

$$F_2 = -\frac{(M + \Delta M)}{I}\int_{y_1}^{c} y\,(t\,dy)$$

These forces are shown on the free-body diagram of Fig. 8-27b. Also shown on the free-body diagram of Fig. 8-27b are the resultants of the vertical shear stresses on the left V_L and right V_R sides of the element and the resultant of the horizontal shear stress V_H on the bottom of the element. A summation of forces in the horizontal direction yields

$$V_H = F_2 - F_1 = -\frac{\Delta M}{I}\int_{y_1}^{c} y\,(t\,dy)$$

The average shearing stress τ_{avg} is the horizontal shear force V_H divided by the horizontal shear area $A_s = t\,\Delta x$ between sections A and B. Thus,

$$\tau_{\text{avg}} = \frac{V_H}{A_s} = -\frac{\Delta M}{It\,\Delta x}\int_{y_1}^{c} y\,(t\,dy)$$

---

[5]If M is positive, then the normal stress above the neutral axis (where y is positive) will be negative (compression) and the force F_1 will be a compressive force. If M is negative, then the normal stress above the neutral axis will be positive (tension) and the force F_1 will be a tensile force as drawn.

In the limit as $\Delta x \rightarrow 0$

$$\tau = \lim_{\Delta x \to 0} \frac{\Delta M}{\Delta x} \left[-\frac{1}{It} \right] \int_{y_1}^{c} ty \, dy = \frac{dM}{dx} \left[-\frac{1}{It} \right] \int_{y_1}^{c} ty \, dy \qquad \text{(a)}$$

The shear V at the beam section where the stress is to be evaluated is given by Eq. (8-14c) as $V = dM/dx$. The integral of Eq. (a) is the first moment of the portion of the cross-sectional area between the transverse line where the stress is to be evaluated and the extreme fiber of the beam. This integral is designated Q, and when values of V and Q are substituted into Eq. (a), the formula for the horizontal (or longitudinal) shearing stress becomes

$$\tau_H = -\frac{VQ}{It} \qquad \text{(b)}$$

The minus sign in Eq. (b) is needed to satisfy Eq. (8-1a) and is consistent with the sign convention for shearing stresses (see Fig. 8-4c). At each point in the beam, the horizontal (longitudinal) and vertical (transverse) shearing stresses have the same magnitude [Eq. (7-8)]; hence, Eq. (b) also gives the vertical shearing stress at a point in a beam (averaged across the width).[6] For the balance of this chapter, magnitudes of V and Q will be used to determine the magnitude of the shearing stress τ, and Eq. (b) will be written as

$$\tau = \frac{VQ}{It} \qquad \text{(8-15)}$$

The sense of the stress τ will be determined from the sense of the shear V on transverse planes and from Eq. (7-8) on longitudinal planes.

Because the flexure formula was used in the derivation of Eq. (8-15) it is subject to the same assumptions and limitations as the flexure formula. Although the stress given by Eq. (8-15) is associated with a particular point in a beam, it is averaged across the thickness t and hence is accurate only if t is not too great. For a rectangular section having a depth twice the width, the maximum stress as computed by Saint-Venant's more rigorous method is about 3 percent greater than that given by Eq. (8-15). If the beam is square, the error is about 12 percent. If the width is four times the depth, the error is almost 100 percent, from which one must conclude that, if Eq. (8-15) were applied to a point in the flange of an I-beam or T-section, the result would be worthless. Furthermore, if Eq. (8-15) is applied to sections where the sides of the beam are not parallel, such as a triangular section, the average transverse shearing stress is subject to additional error because the variation of transverse shearing stress is greater when the sides are not parallel.

The variation of shearing stress on a transverse cross section of a beam will be demonstrated by using the inverted T-shaped beam shown in Fig. 8-28a. For this section

---

[6]If the shear force V is positive (downward on section B), then the horizontal shear stress will be negative (to the right on the bottom of the element) and the vertical shear stress will also be negative (downward on section B—in the same direction as the shear force).

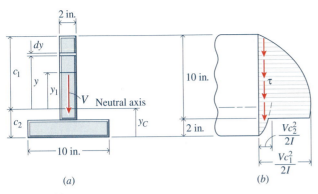

Figure 8-28

$$y_C = \frac{2(10)(7) + 10(2)(1)}{2(10) + 10(2)} = 4 \text{ in.}$$

$$c_1 = 8 \text{ in.} \quad \text{and} \quad c_2 = 4 \text{ in.}$$

Note that in Eq. (8-15), V and I are constant for any section, and only Q and t vary for different points in the section. The transverse shearing stress at any point in the stem of the section a distance y_1 from the neutral axis is from Fig. 8-28a and Eq. (8-15),

$$\tau = \frac{V}{It} \int_{y_1}^{c_1} y\, t\, dy$$
$$= \frac{V}{2I}(c_1^2 - y_1^2) = \frac{V}{2I}(8^2 - y_1^2) \quad (-2 \le y_1 \le 8)$$

An expression for the average shearing stress in the flange can be written in a similar manner and is

$$\tau = \frac{V}{2I}(c_2^2 - y_1^2) = \frac{V}{2I}(4^2 - y_1^2) \quad (-4 \le y_1 \le -2)$$

These are parabolic equations for the theoretical stress distribution, and the results are shown in Fig. 8-28b. The diagram has a discontinuity at the junction of the flange and stem because the thickness of the section changes abruptly. The distribution in the flange is fictitious since the stress at the top of the flange must be zero (a free surface). From Fig. 8-28b and Eq. (8-15), one may conclude that, in general, the maximum[7] longitudinal and transverse shearing stress occurs at the neutral surface at a section where the transverse shear V is maximum. There may be exceptions such as a beam with a cross section in the form of a Greek cross with Q/t at the neutral surface less than the value some distance from the neutral surface. The following example illustrates the use of the shearing stress formula, Eq. (8-15).

---

[7] In this book the term "maximum," as applied to a longitudinal and transverse shearing stress, will mean the average stress across the thickness t at a point where such average has the maximum value.

Example Problem 8-12

If the part of the beam of Fig. 8-29a has the T cross section shown in Fig. 8-29b, determine

(a) The average shearing stress on a horizontal plane 4 in. above the bottom of the beam.

(b) The average shearing stress on the horizontal plane in the stem at the junction between the flange and the stem.

(c) The maximum vertical shearing stress on the cross section.

SOLUTION

The second moment of the cross-sectional area about the neutral axis is

$$I = \frac{1}{3}(2)(8)^3 + \frac{1}{3}(10)(4)^3 - \frac{1}{3}(8)(2)^3 = 533.3 \text{ in.}^4$$

(a) The first moment Q for the bottom 4 in. of the stem is

$$Q_4 = y_C A = 6[(2)(4)] = 48 \text{ in.}^3$$

The average shearing stress on the plane is then given by Eq. (8-15) as

$$\tau = \frac{VQ}{It} = \frac{900(48)}{533.3(2)} = 40.5 \text{ psi} \qquad \textbf{Ans.}$$

(b) The first moment Q for the flange of the beam is

$$Q_F = y_C A = 3[(10)(2)] = 60 \text{ in.}^3$$

The average shearing stress on the plane is then given by Eq. (8-15) as

$$\tau = \frac{VQ}{It} = \frac{900(60)}{533.3(2)} = 50.6 \text{ psi} \qquad \textbf{Ans.}$$

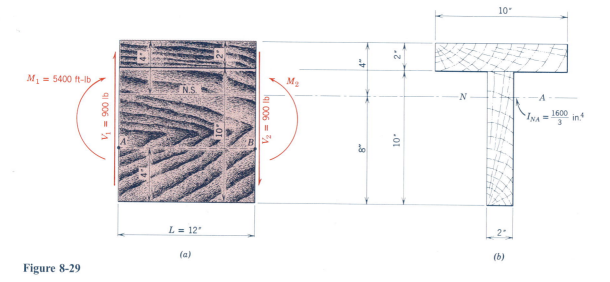

$M_1 = 5400$ ft-lb
$V_1 = 900$ lb
N.S.
A B
$V_2 = 900$ lb
M_2
$L = 12''$

(a)

$10''$
$4''$ $2''$
N A
$I_{NA} = \frac{1600}{3}$ in.^{4}
$8''$ $10''$
$2''$

(b)

Figure 8-29

(c) The maximum stress will occur at the neutral surface. The first moment Q for the part of the stem below the neutral axis is

$$Q_S = y_C A = 4[(2)(8)] = 64 \text{ in.}^3$$

$$\tau_{max} = \frac{900(64)}{533.3(2)} = 54.0 \text{ psi} \qquad \textbf{Ans.}$$

In the computation of Q, it is immaterial whether one takes the area above or below a transverse line. For example, Q for the area above the neutral axis is $Q_{flange} + Q_{stem} = 3[(10)(2)] + 1[(2)(2)] = 64$, which is the same as that for the part of the stem below the neutral axis. ∎

PROBLEMS

Standard Problems

8-99* The transverse shear V at a certain section of a timber beam is 6000 lb. If the beam has the cross section shown in Fig. P8-99, determine
  (a) The vertical shearing stress 3 in. below the top of the beam.
  (b) The maximum vertical shearing stress on the cross section.

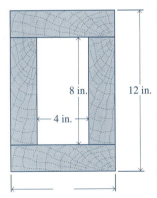

Figure P8-99

8-100* A timber beam 4 m long is simply supported at its ends and carries a uniformly distributed load w of 8 kN/m over its entire length. If the beam has the cross section shown in Fig. P8-100, determine
  (a) The maximum horizontal shearing stress in the glued joints between the web and the flanges of the beam.
  (b) The maximum horizontal shearing stress in the beam.

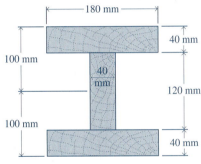

Figure P8-100

8-101 A timber beam is simply supported and carries a uniformly distributed load w of 360 lb/ft over its entire 18-ft span. If the beam has the cross section shown in Fig. P8-101, determine
  (a) The vertical shearing stress at a point 2 ft from the right end and 3 in. below the top surface of the beam.
  (b) The maximum horizontal shearing stress in the beam.

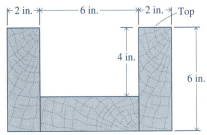

Figure P8-101

8-102 A timber beam is fabricated from two 40 × 120-mm and one 40 × 80-mm pieces of lumber to form the cross section shown in Fig. P8-102. In a region of constant shear, the two ends of a 400-mm segment of the beam are subjected to negative bending moments of 8 and 6 kN · m, respectively. Determine the average shearing stress in each of the two glued joints between the flanges and the web.

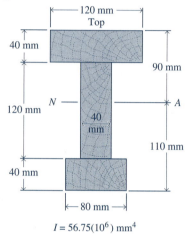

$$I = 56.75(10^6) \text{ mm}^4$$

Figure P8-102

8-103* The cross section of a timber beam is a rectangle 6-in. wide by 12-in. deep. The beam is simply supported and carries a concentrated load of 18 kip at the center of a 12-ft span. Determine

(a) The horizontal shearing stress at a point 3 in. from the top of the beam and 3 ft from the left support.

(b) The maximum vertical shearing stress in the beam.

(c) The maximum tensile flexural stress in the beam.

8-104* The cross section of a timber beam is a rectangle 200-mm wide by 500-mm deep. The beam is simply supported and carries a uniformly distributed load of 40 kN/m over its entire 6-m span. Determine

(a) The maximum vertical shearing stress in the beam.

(b) The vertical shearing stress at a point 80 mm above the bottom of the beam and 1 m from the left support.

(c) The maximum compressive flexural stress in the beam.

8-105 A structural steel cantilever beam 6 ft long supports a concentrated load of 5 kip at the free end. If the beam has the cross section shown in Fig. P8-105, determine

(a) The horizontal shearing stress at a point 1 in. from the top of the beam and 1 ft from the fixed support.

(b) The maximum horizontal shearing stress in the beam.

(c) The maximum tensile flexural stress in the beam.

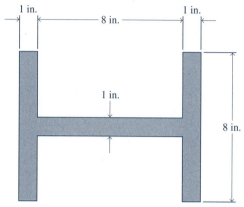

Figure P8-105

8-106 A timber beam is simply supported and carries a concentrated load of 50 kN at the center of a 4-m span. If the beam has the cross section shown in Fig. P8-106, determine

(a) The horizontal shearing stress in the glued joint 50 mm below the top of the beam and 1 m from the left support.

(b) The maximum horizontal shearing stress in the beam.

(c) The maximum tensile flexural stress in the beam.

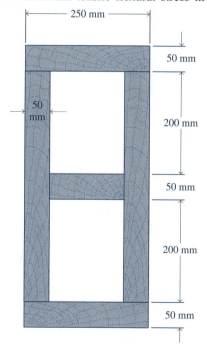

Figure P8-106

8-107* A laminated wood beam consists of six 2 × 6-in. planks glued together to form a section 6 in. wide by 12 in. deep as shown in Fig. P8-107. If the strength of the glue in shear is 120 psi, determine
(a) The maximum uniformly distributed load w that can be applied over the full length of the beam if the beam is simply supported and has a span of 15 ft.
(b) The horizontal shearing stress in the glued joint 2 in. below the top of the beam and 2 ft from the left support when the load of part (a) is applied.
(c) The maximum tensile flexural stress in the beam when the load of part (a) is applied.

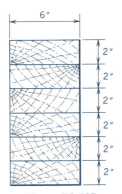

6″

2″
2″
2″
2″
2″
2″

Figure P8-107

8-108* A timber beam is simply supported and carries a uniformly distributed load of 10 kN/m over the full length of the beam. If the beam has the cross section shown in Fig. P8-108 and a span of 6 m, determine
(a) The horizontal shearing stress in the glued joint 50 mm below the top of the beam and 1 m from the left support.
(b) The horizontal shearing stress in the glued joint 50 mm above the bottom of the beam and $\frac{1}{2}$ m from the left support.
(c) The maximum horizontal shearing stress in the beam.
(d) The maximum tensile flexural stress in the beam.

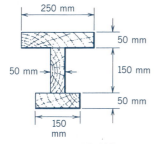

250 mm

50 mm

50 mm→ ←   150 mm

50 mm

150 mm

Figure P8-108

8-109 The segment of a timber beam shown in Fig. P8-109a has the cross section shown in Fig. P8-109b. At section b-b, the shear is −2800 lb, and the bending moment is +6400 ft · lb. Determine
(a) The flexural stress and the vertical shearing stress at point A of the cross section at section a-a.
(b) The magnitude and location of the maximum horizontal shearing stress in the beam segment.
(c) The magnitude and location of the maximum compressive stress in the beam segment.

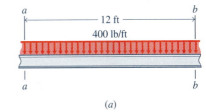

a   b

12 ft

400 lb/ft

a   b

(a)

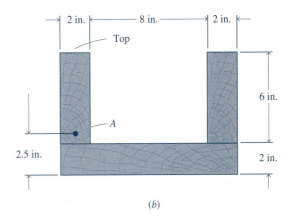

2 in.   8 in.   2 in.

Top

6 in.

A

2.5 in.

2 in.

(b)

Figure P8-109

8-110 The segment of a timber beam shown in Fig. P8-110a has the cross section shown in Fig. P8-110b. At section a-a, the shear is +8 kN, and the bending moment is −6 kN · m. Determine
(a) The flexural stress and the vertical shearing stress 40 mm below the top of the beam at section b-b.
(b) The magnitude and location of the maximum horizontal shearing stress in the glued joint between the flange and the stem in the beam segment.
(c) The magnitude and location of the maximum tensile flexural stress in the beam segment.

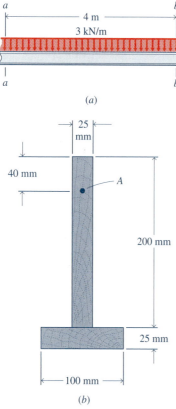

(b)

Figure P8-110

8-111[*] A beam shown in Fig. P8-111a is composed of two 1 × 6-in. and two 1 × 3-in. hard maple boards that are glued together as shown in Fig. P8-111b. Determine the magnitude and location of
(a) The maximum tensile flexural stress in the beam.
(b) The maximum horizontal shearing stress in the beam.

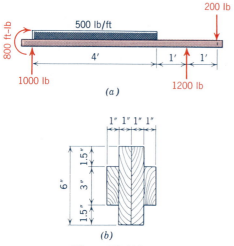

Figure P8-111

8-112[*] The beam shown in Fig. P8-112a is composed of three pieces of timber that are glued together as shown in Fig. P8-112b. If the maximum horizontal shearing stresses are not to exceed 700 kPa in the glued joints or 800 kPa in the wood and if the maximum tensile and compressive flexural stresses in the beam are not to exceed 12 MPa, determine the maximum permissible value for P.

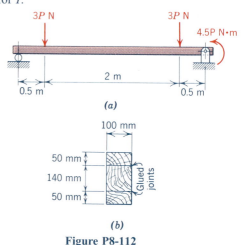

Figure P8-112

8-113 The timber beam shown in Fig. P8-113a is fabricated by gluing two 1 × 5-in. and two 1 × 4-in. boards together as shown in Fig. P8-113b. If the maximum horizontal shearing stresses are not to exceed 200 psi in the glued joints or 240 psi in the wood, and if the maximum tensile and compressive flexural stresses in the beam are not to exceed 2400 psi, determine the maximum permissible magnitude for the distributed load w.

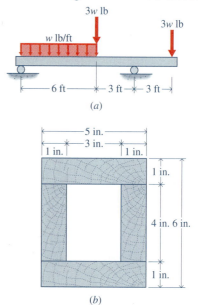

Figure P8-113

8-114 The cross section of the simple beam shown in Fig. P8-114 is a T-section formed by gluing two 40×120-mm boards together. If the maximum horizontal shearing stresses are not to exceed 800 kPa in the glued joint or 900 kPa in the wood, and if the maximum tensile and compressive flexural stresses in the beam are not to exceed 13.5 MPa, determine the maximum permissible value for the distributed load w.

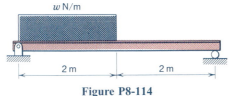

Figure P8-114

Challenging Problems

8-115* A timber beam is fabricated from one 2×8-in. and two 2×6-in. pieces of lumber to form the cross section shown in Fig. P8-115. The flanges of the beam are fastened to the web with nails that can safely transmit a shear force of 100 lb. If the beam is simply supported and carries a 1000-lb load at the center of a 12-ft span, determine
(a) The shear force transferred by the nails from the flange to the web in a 12-in. length of the beam.
(b) The spacing required for the nails.

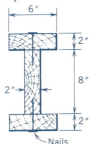

Figure P8-115

8-116* A cantilever beam is used to support a concentrated load of 20 kN. The beam is fabricated by bolting two C457 $\times$ 86 steel channels (see Appendix B) back-to-back to form the H-section shown in Fig. P8-116. If the pairs of bolts are spaced at 300-mm intervals along the beam, determine
(a) The shear force carried by each of the bolts.
(b) The bolt diameter required if the shear and bearing stresses for the bolts must be limited to 60 MPa and 125 MPa, respectively.

Figure P8-116

8-117 A W18 $\times$ 97 steel beam (see Appendix B) will have $\frac{1}{2} \times 10$-in. cover plates welded to the top and bottom flanges as shown in Fig. P8-117. A 20-in. length of the beam for which the shear is constant will be subjected to bending moments of $+4600$ in. $\cdot$ kip and $+2300$ in. $\cdot$ kip at the ends. The fillet weld has an allowable load of 2400 lb/in. Determine the number of fillet welds, each 2-in. long, required on each side of the cover plate.

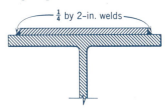

Figure P8-117

8-118 A box beam will be fabricated by bolting two 15×260-mm steel plates to two C305 $\times$ 45 steel channels (see Appendix B) as shown in Fig. P8-118. The beam will be simply supported at the ends and will carry a concentrated load of 125 kN at the center of a 5-m span. Determine the bolt spacing required if the bolts have a diameter of 20 mm and an allowable shearing stress of 150 MPa.

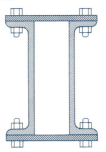

Figure P8-118

8-119* A W21 $\times$ 101 steel beam (see Appendix B) is simply supported at the ends and carries a concentrated load at the center of a 20-ft span. The concentrated load must be increased to 125 kip, which requires that the beam be strengthened. It has been decided that two $\frac{3}{4} \times 16$-

in. steel plates will be bolted to the flanges, as shown in Fig. P8-119. Determine the bolt spacing required if the bolts have a diameter of $\frac{3}{4}$ in. and an allowable shearing stress of 17.5 ksi.

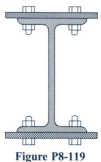

Figure P8-119

8-120   A W356 × 122 steel beam (see Appendix B) has a C381 × 74 channel bolted to the top flange, as shown in Fig. P8-120. The beam is simply supported at the ends and carries a concentrated load of 96 kN at the center of an 8-m span. If the pairs of bolts are spaced at 500-mm intervals along the beam, determine

(a) The shear force carried by each of the bolts.

(b) The bolt diameter required if the shear and bearing stresses for the bolts must be limited to 60 MPa and 125 MPa, respectively.

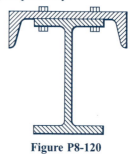

Figure P8-120

8-8 DESIGN PROBLEMS

Again in this chapter, design will be limited to proportioning a member (in this case, a beam) to perform a specified function without exceeding specified levels of deformation or stress. Failure by excessive deformation (excessive elastic deflection) will be discussed in Chapter 9. Slip failure or failure by fracture that results from excessive normal (flexural) stresses or shearing stresses must be considered when designing beams. The elastic flexure formula and the shearing stress formula developed in Sections 8-5 and 8-7 and used to calculate flexural stresses and transverse (or longitudinal) shearing stresses in beams are

$$\sigma_{\max} = \frac{Mc}{I} = \frac{M}{S} \qquad \text{and} \qquad \tau = \frac{VQ}{It}$$

To determine maximum normal and shearing stresses in beams, sections must be located where M and V are maximum (critical sections). A shear diagram is used to locate the critical section of the beam where V is maximum. The critical section for flexure, the section where M is maximum, is found from a moment diagram. In general, the absolute maximum value of M is used for design purposes. However, care must be exercised for beams with cross sections that are nonsymmetric with respect to the neutral axis (such as T-beams) and beams made of materials with different properties in tension and compression. In these cases, both the largest positive and the largest negative values of M must be considered.

Experience indicates that beam design is usually governed by flexural stresses. Thus, a beam is usually designed for flexure using Eq. (8-12), and then checked for shearing stress using Eq. (8-15). If the shearing stress is less than the allowable shearing stress, this procedure is adequate. If the allowable shearing stress has been exceeded, the beam is redesigned and the process is repeated.

However, the shearing stress (longitudinal) may be the controlling factor for beams made of timber. The examples that follow illustrate the procedures for designing beams.

Example Problem 8-13

An air-dried Douglas fir beam of rectangular cross section is to support the load shown in Fig. 8-30a. If the allowable flexural stresses in tension and compression are 1200 psi and the allowable shearing stress is 100 psi, determine the lightest weight standard structural timber that can be used.

SOLUTION

First, load (free-body), shear force, and bending moment diagrams are drawn, as shown in Figs. 8-30b, c, and d, respectively. From these diagrams it is determined that the maximum shear force is 900 lb and the maximum bending moment is 3375 ft · lb. Using the maximum bending moment in Eq. (8-12) yields the required minimum section modulus for the beam as

$$S = \frac{M_{max}}{\sigma_{all}} = \frac{3375(12)}{1200} = 33.75 \text{ in.}^3$$

Table B-15 in Appendix B contains a listing of standard structural timbers together with values for their section modulus S. Note that the properties listed are for dressed (actual size) timbers. Since the lightest weight beam is wanted, select a beam with $S \geq 33.75$ in.^{3} and the smallest weight per unit length. Thus, select a 2 × 12-in. nominal size timber. For this beam, the actual values of S and I are 35.8 in.^{3} and 206 in.^{4}, respectively. Equation (8-15) can now be used to see whether or not the allowable shearing stress requirement is met. The maximum shearing stress for a beam with a rectangular cross section occurs at the neutral axis and is given by

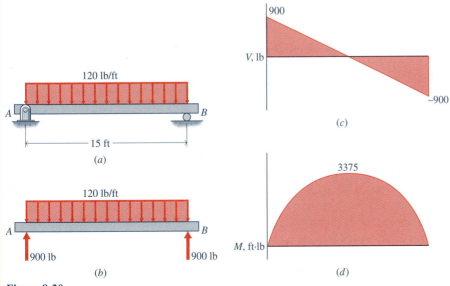

(a)

(b)

(c)

(d)

Figure 8-30

$$\tau_{max} = \frac{VQ}{It} = \frac{V[b(h/2)(h/4)]}{(bh^3/12)(b)} = 1.5\frac{V}{bh} = 1.5\frac{V}{A}$$

where V is the absolute value of the maximum shear force and A is the cross-sectional area of the beam. Thus,

$$\tau_{max} = 1.5\frac{V}{A} = 1.5\frac{900}{(1.625)(11.5)} = 72.24 \text{ psi} < 100 \text{ psi}$$

The maximum shearing stress is within the allowable limit; therefore, the beam selected is satisfactory. However, the analysis assumed that the beam was weightless, whereas the beam selected weighs 5.19 lb/ft. The bending moment diagram for the uniformly distributed weight of the beam is similar to Fig. 8-30d and M_{max} for the weight is M_{max} = 146 ft · lb. Adding the maximum bending moments for the applied loading and the beam weight gives M_{max} = 3375 + 146 = 3521 ft · lb. The required section modulus then becomes

$$S = \frac{M_{max}}{\sigma_{all}} = \frac{3521(12)}{1200} = 35.21 \text{ in.}^3$$

Similarly, V_{max} = 900 + $\frac{1}{2}$(5.19)(15) = 939 lb. Thus,

$$\tau_{max} = 1.5\frac{V}{A} = 1.5\frac{939}{(1.625)(11.5)} = 75.37 \text{ psi} < 100 \text{ psi}$$

The original beam selected had a section modulus of S = 35.8 in.^{3}. Thus, the 2 × 12-in. nominal beam is satisfactory. If the value of S (35.21 in.^{3} for this example) had been greater than S for the original beam selected (35.8 in.^{3} for this example), the entire procedure would be repeated based on the section modulus for the beam with the applied loading plus the weight. Thus, the procedure is a trial-and-error process. ■

Example Problem 8-14

Select the lightest wide-flange beam that can be used to support the load shown in Fig. 8-31a. The allowable flexural stresses in tension and compression are 160 MPa and the allowable shearing stress is 82 MPa.

SOLUTION

Load (free-body), shear force, and bending moment diagrams for the beam are shown in Figs. 8-31b, c, and d, respectively. From these diagrams it is determined that the maximum shear force is 37.5 kN and the maximum bending moment is 22.50 kN · m. Using the maximum bending moment in Eq. (8-12) yields the required minimum section modulus for the beam as

$$S = \frac{M_{max}}{\sigma_{all}} = \frac{22.5(10^3)}{160(10^6)} = 140.63(10^{-6}) \text{ m}^3 = 140.63(10^3) \text{ mm}^3$$

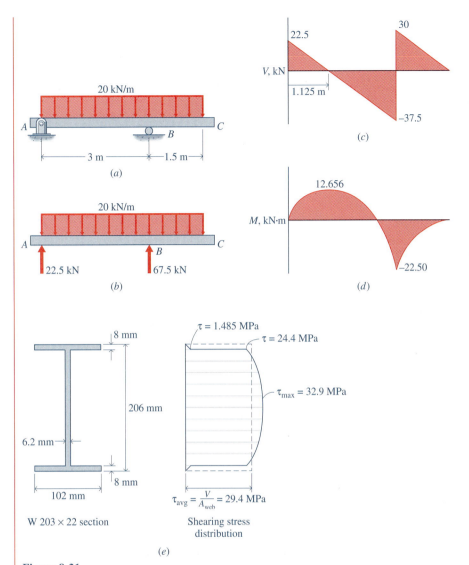

Figure 8-31

Table B-2 in Appendix B contains a listing of the properties of wide-flange sections and from this listing it is determined that the lightest section with $S \geq$ $140.63(10^3)$ mm^3 (with respect to the x-x axis) is a W203 × 22 section.

When designing wide-flange beams or American Standard beams, it is often assumed that the entire shear load is carried by the web of the beam and that it is uniformly distributed. The validity of this assumption for the W203 × 22 section can be verified by using Eq. (8-15) and calculating τ at various distances y from the neutral axis of the beam. For example, at the neutral axis on the section where V is maximum:

$$Q_{NA} = 102(8)(99) + 95(6.2)(47.5) = 108.76(10^3) \text{ mm}^3 = 108.76(10^{-6}) \text{ m}^3$$

$$\tau_{NA} = \frac{VQ}{It} = \frac{37.5(10^3)(108.76)(10^{-6})}{20(10^{-6})(6.2)(10^{-3})} = 32.9(10^6) \text{ N/m}^2 = 32.9 \text{ MPa}$$

Similarly, at the junction between the flange and the web:

$$Q_J = 102(8)(99) = 80.78(10^3) \text{ mm}^3 = 80.78(10^{-6}) \text{ m}^3$$

$$\tau_J = \frac{VQ}{It} = \frac{37.5(10^3)(80.78)(10^{-6})}{20(10^{-6})(6.2)(10^{-3})} = 24.4(10^6) \text{ N/m}^2 = 24.4 \text{ MPa}$$

The shear stress distribution for the complete cross section of the beam is shown in Fig. 8-31e together with the average shearing stress on the cross section obtained by dividing the shear force V by the area of the web (that is, by the thickness of the web times the total depth of the beam):

$$\tau_{avg} = \frac{V}{A_{web}} = \frac{37.5(10^3)}{6.2(10^{-3})(206)(10^{-3})} = 29.4(10^6) \text{ N/m}^2 = 29.4 \text{ MPa}$$

The maximum stress and the average stress differ by approximately 11 percent. Since both τ_{max} and τ_{avg} are less than $\tau_{allowable} = 82$ MPa, the W203 × 22 section is satisfactory.

As in the previous example, the weight of the beam should be considered. The mass per unit length for this beam is 22 kg/m, and thus the weight per unit length is $W = mg = 22(9.81) = 215.8$ N/m, or 0.2158 kN/m. The weight per unit length of the beam is about 1.1 percent of the applied force per unit length. The actual section modulus $S = 193(10^3)$ mm^3 is about 27 percent higher than the required minimum value of $S_{min} = 140.63(10^3)$ mm^3. Thus, the maximum flexural stress is less than the allowable value even when the weight of the beam is considered and the W203 × 22 wide-flange section is acceptable. It will be shown in Chapter 10 (combined static loading), that for W- or S-shapes, the maximum normal stress may occur at the junction of the web and flange. Thus, Eq. (8-12) does not give the maximum value of normal stress. Instead, the maximum normal stress at the junction of the web and flange is a combination of stresses given by Eqs. (8-12) and (8-15). However, in design codes, the allowable stresses are set such that Eqs. (8-12) and (8-15) are adequate for design purposes. ■

PROBLEMS

Standard Problems

8-121 A simply supported timber beam 16 ft long is loaded with 1500 lb concentrated loads applied 6 ft from each support. If the allowable flexural stress is 1300 psi and the allowable shearing stress is 85 psi, select the lightest standard structural timber that can be used to support the loading.

8-122 A 6-m long simply supported timber beam is loaded with 8 kN concentrated loads applied 2 m from each support. The allowable flexural stress is 13.1 MPa and the allowable shearing stress is 0.6 MPa. Select the lightest standard structural timber that can be used to support the loading.

8-123 A 16-ft simply supported beam is loaded with a uniform load of 4000 lb/ft over its entire length. If the allowable flexural stress is 22 ksi and the allowable shearing stress is 14.5 ksi, select the lightest structural steel wide-flange beam that can be used to support the loading.

8-124 A structural steel beam is subjected to the loading shown in Fig. P8-124. The allowable flexural stress is 152 MPa and the allowable shearing stress is 100 MPa. Select the lightest American Standard beam that can be used to support the loading.

22 kN   40 kN   22 kN

A   B

1.25 m   1.5 m   1.5 m   1.25 m

Figure P8-124

8-125 A structural steel beam is to carry the loading shown in Fig. P8-125. Select the lightest wide-flange beam that can be used to support the loading if the allowable flexural stress is 22 ksi and the allowable shearing stress is 14 ksi.

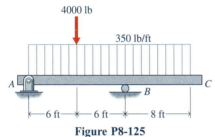

4000 lb

350 lb/ft

A   B   C

6 ft   6 ft   8 ft

Figure P8-125

8-126 Select the lightest wide-flange beam that can be used to support the loading shown in Fig. P8-126. The allowable flexural stress is 152 MPa and the allowable shearing stress is 100 MPa.

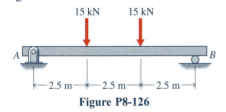

15 kN   15 kN

A   B

2.5 m   2.5 m   2.5 m

Figure P8-126

8-127 The 8-ft pole shown in Fig. P8-127 is used to support a clothesline. The pole is embedded in a concrete base. The estimated maximum force to be applied to the

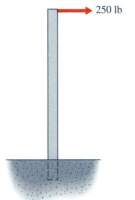

250 lb

Figure P8-127

pole is 250 lb. Select an appropriate standard steel pipe if the allowable flexural stress is 20 ksi. Neglect the effects of shear.

8-128 The lever shown in Fig. P8-128 is used to lift the 275-kg rock. Select a standard steel pipe to perform the task. The allowable flexural stress is 135 MPa. Neglect the effects of shear.

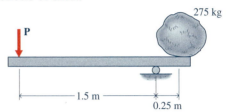

275 kg

P

1.5 m

0.25 m

Figure P8-128

Challenging Problems

8-129 The floor-framing plan for a residential dwelling is shown in Fig. P8-129. The floor decking is to be supported by 2-in. nominal width joists spaced 16 in. apart. Each joist is to span 12 ft, and is simply supported at the ends. The floor decking is subjected to a uniform loading of 60 lb/ft^2, which includes the live load plus an allowance for the dead load of the flooring system. The joists are made of construction grade Douglas fir with an allowable flexural stress of 1200 psi and an allowable shearing stress of 120 psi. Determine the required nominal depth of the joists.

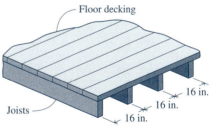

Floor decking

16 in.

16 in.

Joists   16 in.

Figure P8-129

8-130 A floor-framing plan for a building is shown in Fig. P8-130. Four columns support the floor, along with seven I-beams. Design specifications call for a uniform live load of 3600 N/m^2. In addition, it is estimated that the flooring system produces a uniform load of 3600 N/m^2. Select the lightest American Standard beam to support the loading if the allowable flexural stress is 165 MPa and the allowable shearing stress is 96 MPa. Assume that the beams are simply supported at the ends.

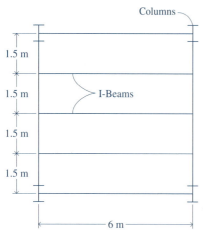

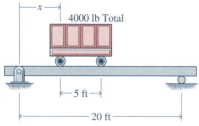

Figure P8-130

8-131 A carriage moves slowly along a simply supported I-beam, as shown in Fig. P8-131. Select the lightest permissible wide-flange beam to support the loading, if the

Figure P8-131

allowable flexural stress is 24 ksi and the allowable shearing stress is 14.5 ksi. Neglect the weight of the beam. Note that the shear force and bending moment are functions of x, the position of the left-hand wheel.

8-132 A 15-kN load is supported by a roller on an I-beam, as shown in Fig. P8-132. The roller moves slowly along the beam, thereby causing the shear force and bending moment to be functions of x. Select the lightest permissible American Standard beam to support the loading. The allowable flexural stress is 152 MPa and the allowable shearing stress is 100 MPa. Neglect the weight of the beam.

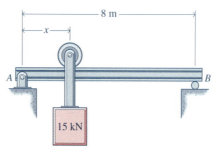

Figure P8-132

8-9 SUMMARY

A member subjected to loads applied transverse to the long dimension of the member and which cause the member to bend is known as a *beam*. A beam supported by pins, rollers, or smooth surfaces at the ends is called a *simple beam*. A simple support will develop a reaction normal to the beam but will not produce a couple. A cantilever beam has one end built into a wall or other support. The built-in end is said to be fixed if no rotation occurs and restrained if a limited amount of rotation occurs.

Cantilever beams and simple beams have only two reactions (two forces or one force and a couple), and these reactions can be obtained from a free-body diagram of the beam by applying the equations of equilibrium. Such beams are said to be statically determinate. Beams with more than two reaction components are called statically indeterminate since there are not enough equations of equilibrium to determine the reactions.

A free-body diagram of a portion of a beam with a cross section exposed by an imaginary cut shows that a transverse force V_r and a couple M_r at the cut section are needed to maintain equilibrium. The force V_r is the resultant of the shearing stresses. The couple M_r is the resultant of the normal stresses. The magnitudes and senses of V_r and M_r are obtained from the equations of equilibrium $\Sigma F_y = 0$ and $\Sigma M_O = 0$, where O is any axis perpendicular to the xy-plane.

The normal and shearing stresses σ and τ on a transverse plane of a beam are related to the resisting moment M_r and the shear V_r by the equations

$$V_r = -\int_{\text{Area}} \tau \, dA \tag{8-1a}$$

$$M_r = -\int_{\text{Area}} y \, \sigma \, dA \tag{8-1b}$$

It is obvious from Eqs. (8-1) that the laws of variation of the normal and shearing stresses must be known before the integrals can be evaluated.

The variation of normal stress on a plane is obtained by assuming that a plane section before bending remains a plane after bending. For this to be strictly true, it is necessary that the beam be bent only with couples. When a beam is bent with couples, the deformed shape of all longitudinal elements (also referred to as fibers) is an arc of a circle. Precise experimental measurements indicate that at some distance c above the bottom of the beam, longitudinal elements undergo no change in length. The curved surface formed by these elements is referred to as the *neutral surface of the beam*, and the intersection of this surface with any cross section is called the *neutral axis of the section*. All elements (fibers) on one side of the neutral surface are compressed, while those on the opposite side are elongated. As a result, the normal strain at any point on the plane can be expressed as

$$\epsilon_x = -\frac{1}{\rho} y \tag{8-2}$$

Equation (8-2) indicates that the strain in a fiber is proportional to the distance of the fiber from the neutral surface of the beam. Equation (8-2) is valid for elastic or inelastic action so long as the transverse shearing stresses are small.

Since the longitudinal strain ϵ_x is proportional to the distance of the fiber from the neutral surface of the beam, the normal stress σ_x on the plane (for linearly elastic action) is given by Hooke's law as

$$\sigma_x = E\epsilon_x = -\frac{E}{\rho} y \tag{8-3}$$

Substituting Eq. (8-3) into Eq. (8-1) yields

$$M_r = -\int_A y \, \sigma_x \, dA = -\frac{E}{\rho} \int_A y^2 \, dA$$

The integral $\int y^2 \, dA$ is called the second moment of area. When the integral $\int_A y^2 \, dA$ is replaced by the symbol I, the elastic flexure formula is obtained as

$$\sigma_x = -\frac{M_r y}{I} \tag{8-11}$$

where σ_x is the flexural stress at a distance y from the neutral surface and on a transverse plane, M_r is the resisting moment of the section, and I is the second moment of area of the transverse section with respect to the neutral axis.

At any section of the beam, the flexural stress will be maximum (have the greatest magnitude) at the surface farthest from the neutral axis ($y = c$), and Eq. (8-11) becomes

$$\sigma_{max} = \frac{M_r c}{I} = \frac{M_r}{S} \tag{8-12}$$

where $S = I/c$ is called the *section modulus of the beam.*

If the maximum flexural stress is required in a beam subjected to a loading that produces a bending moment that varies with position along the beam, it is desirable to have a method for determining the maximum moment. Similarly, the maximum transverse shearing stress will occur at a section where the resisting shear is maximum. Shear and bending moment diagrams provide a method for obtaining maximum values of shear and moment. A shear diagram is a graph in which abscissas represent distances along the beam and ordinates represent the transverse shear at the corresponding sections. A moment diagram is a graph in which abscissas represent distances along the beam and ordinates represent the bending moment at the corresponding sections. Four simple relationships developed by using equilibrium considerations that are used to construct shear and moment diagrams are:

1:
$$\frac{dV}{dx} = w \tag{8-13d}$$

That is, the slope of the shear diagram at any location x in the beam is equal to the intensity of loading at that section of the beam.

2:
$$V_2 - V_1 = \int_{x_1}^{x_2} w \, dx \tag{8-13e}$$

That is, for any section of the beam acted on by a distributed load w and no concentrated force ($P = 0$), the change in shear between sections at x_1 and x_2 is equal to the area under the load diagram between the two sections.

3:
$$\frac{dM}{dx} = V \tag{8-14c}$$

That is, the slope of the bending moment diagram at any location x in the beam is equal to the value of the shear force at that section of the beam.

4:
$$M_2 - M_1 = \int_{x_1}^{x_2} V \, dx \tag{8-14d}$$

That is, for any section of the beam in which the shear force is continuous ($C = P = 0$), the change in bending moment between sections at x_1 and x_2 is equal to the area under the shear diagram between the two sections.

At each point in a beam, the horizontal (longitudinal) and vertical (transverse) shearing stresses have the same magnitude and are given by the expression

$$\tau = \frac{VQ}{It} \qquad (8\text{-}15)$$

where Q is the first moment of the portion of the area of the cross section between the transverse line where the stress is to be evaluated and the extreme fiber of the beam. The sense of stress τ is the same as the sense of shear V on the transverse plane. Because the flexure formula was used in the derivation of Eq. (8-15) it is subject to the same limitations as the flexure formula.

REVIEW PROBLEMS

8-133* A beam has the cross section shown in Fig. P8-133. If the flexural stress at point A is 2000 psi T, determine
(a) The maximum flexural stress on the section.
(b) The resisting moment M_r at the section.

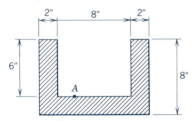

Figure P8-133

8-134* A T-beam has the cross section shown in Fig. P8-134. Determine the maximum tensile and compressive stresses on a cross section of the beam where the resisting moment being transmitted is 100 kN·m.

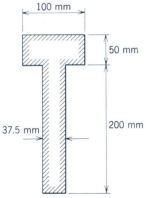

Figure P8-134

8-135 The maximum flexural stress on the cross section of the beam shown in Fig. P8-135 is 8000 psi C. Determine
(a) The resisting moment being transmitted by the section.
(b) The magnitude of the flexural force carried by the flange.

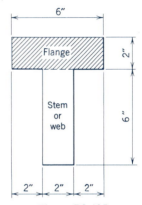

Figure P8-135

8-136 A beam is loaded and supported as shown in Fig. P8-136.
(a) Draw complete shear and bending moment diagrams for the beam.
(b) Using the coordinate axes shown, write equations for the shear and bending moment for any section of the beam in the interval $1\ \text{m} < x < 4\ \text{m}$.

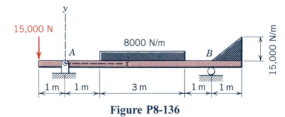

Figure P8-136

8-137* A beam is loaded as shown in Fig. P8-137.
(a) Draw complete shear and bending moment diagrams for the beam.
(b) Using the coordinate axes shown, write equations for the shear and bending moment for any section of the beam in the interval $0 < x < 10\ \text{ft}$.

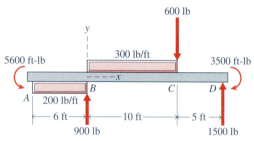

Figure P8-137

8-138* A free-body diagram for a beam is shown in Fig. P8-138.
(a) Draw complete shear and bending moment diagrams for the beam.
(b) Using the coordinate axes shown, write equations for the shear and bending moment for any section of the beam in the interval $0 < x < 2$ m.

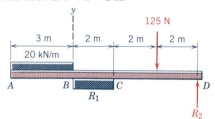

Figure P8-138

8-139 A steel pipe with an outside diameter of 4 in. and an inside diameter of 3 in. is simply supported at the ends and carries two concentrated loads as shown in Fig. P8-139. On a section 5 ft from the right support, determine
(a) The flexural stress at point A on the cross section.
(b) The flexural stress at point B on the cross section.

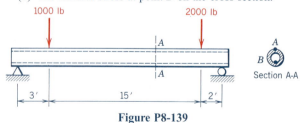

Figure P8-139

8-140 A beam is loaded and supported as shown in Fig. P8-140a. The beam has the cross section shown in Fig. P8-140b. On a section 1.5 m from the left support, determine
(a) The flexural stress at point A on the cross section.
(b) The flexural stress at point B on the cross section.

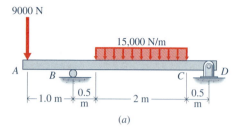

(a)

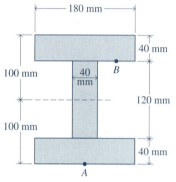

Figure P8-140

8-141* A free-body diagram for a beam is shown in Fig. P8-141a. The beam has the cross section shown in Fig. P8-141b. Determine
(a) The maximum tensile and compressive flexural stresses in the beam.
(b) The maximum vertical shearing stress in the beam.

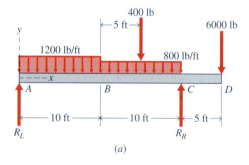

(a)

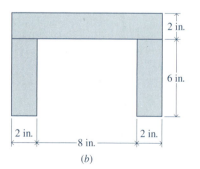

(b)

Figure P8-141

8-142* A beam is loaded and supported as shown in Fig. P8-142a. The beam has the cross section shown in Fig. P8-142b. Determine

(a) The maximum tensile and compressive flexural stresses in the beam.
(b) The maximum vertical shearing stress in the beam.

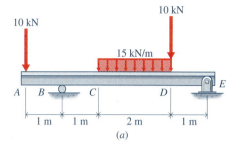

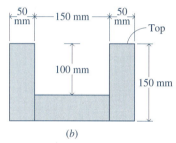

Figure P8-142

8-143 A WT12 × 52 structural tee (see Appendix B) is loaded and supported as a beam (with the flange on the bottom) as shown in Fig. P8-143. Determine

(a) The maximum tensile flexural stress in the beam.
(b) The maximum compressive flexural stress in the beam.
(c) The maximum vertical shearing stress in the beam.
(d) The vertical shearing stress at a point in the stem just above the flange on the cross section where the maximum vertical shearing stress occurs.

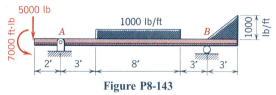

Figure P8-143

8-144 A WT305 × 77 structural tee (see Appendix B) is loaded and supported as a beam (with the flange on top) as shown in Fig. P8-144. Determine

(a) The maximum tensile flexural stress in the beam.
(b) The maximum compressive flexural stress in the beam.
(c) The maximum vertical shearing stress in the beam.
(d) The vertical shearing stress at a point in the stem just below the flange on the cross section where the maximum vertical shearing stress occurs.

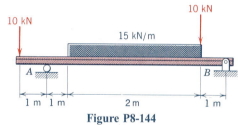

Figure P8-144

COMPUTER PROBLEMS

C8-145 The supports for the beam shown in Fig. P8-145 are symmetrically located. If the distance d from the supports to the ends of the beam is adjustable, compute and plot

(a) The bending moment $M(x)$ in the beam for $d = 2$ ft, 3 ft, and 4 ft as a function of x ($0 \le x \le 15$ ft).
(b) The maximum bending moments $(M_{max})_{AB}$ in segment AB and $(M_{max})_{BC}$ in segment BC as functions of d ($0 \le d \le 6$ ft).
(c) What value of d gives the smallest M_{max} in the beam?

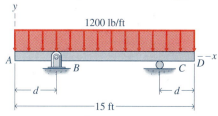

Figure P8-145

C8-146 The left support of the beam shown in Fig. P8-146 is fixed at 1 m from the end while the location of the right support is adjustable. Compute and plot

(a) The bending moment $M(x)$ in the beam for $d = 1.0$ m, 1.5 m, and 2.0 m as a function of x (0 m $\le x \le 6$ m).
(b) The maximum bending moments $(M_{max})_{AB}$ in segment AB, $(M_{max})_{BC}$ in segment BC, and $(M_{max})_{CD}$ in segment CD as functions of d (0 ft $\le d \le 6$ ft).
(c) What value of d gives the smallest M_{max} in the beam?

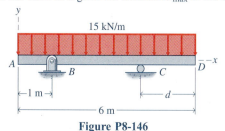

Figure P8-146

C8-147 An overhead crane consists of a carriage that moves along a beam as shown in Fig. P8-147. If the carriage moves slowly along the beam, compute and plot

(a) The bending moment $M(x)$ in the beam for $b = 50$ in., 90 in., and 120 in. as a function of x (0 in. $\leq x \leq$ 200 in.).

(b) The bending moments M_B under the left wheel and M_C under the right wheel as functions of b (10 in. $\leq b \leq$ 190 in.).

(c) What is the largest bending moment in the beam? When and where does it occur?

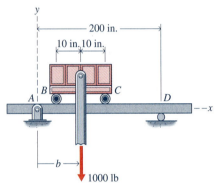

1000 lb

Figure P8-147

C8-148 A beam with a solid rectangular cross section (Fig. P8-148) is being designed to support a maximum moment of 4 kN · m. If the maximum flexural stress in the beam must not exceed 50 MPa and the width of the beam must not exceed 75 mm, prepare a design curve that shows the acceptable values of depth h as a function of the width b (0 mm $\leq b \leq$ 75 mm).

Figure P8-148

C8-149 A beam with a hollow circular cross section (Fig. P8-149) is being designed to support a maximum moment of 100 in. · kip. If the maximum flexural stress in the beam must not exceed 40 ksi and the maximum diameter of the beam must not exceed 5 in., prepare a design curve that shows the acceptable values of the inside diameter d_i as a function of the outside diameter d_o (0 in. $\leq d_o \leq$ 5 in.).

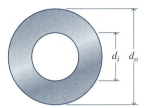

Figure P8-149

C8-150 A beam with a solid rectangular cross section (Fig. P8-150a) is being redesigned as an I-beam (Fig. P8-150b) having the same weight per meter (the same cross-sectional area). If the maximum flexural stress in the beam must not exceed 150 MPa and the thickness of the web must not be less than 6 mm, compute and plot the percent increase in load carrying capability (M_{max}) of the beam as a function of t.

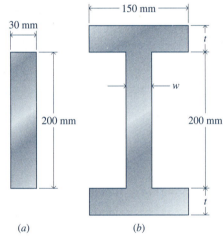

(a)   (b)

Figure P8-150

C8-151 An S4 × 9.5 structural steel beam is carrying a moment M that produces a maximum flexural stress of 20 ksi. The moment M must be increased by 75 percent, but the maximum fiber stress must not be increased. In order to strengthen the

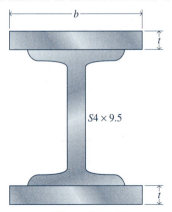

Figure P8-151

beam, rectangular steel plates are to be attached to the top and bottom flanges as shown in Fig. P8-151. If the thickness t of the plates must not exceed $\frac{1}{2}$ in., prepare a design curve that shows the acceptable values of plate width b as a function of the plate thickness t (0 in. $\leq t \leq \frac{1}{2}$ in.)

C8-152 The transverse shear V at a certain section of a timber beam is 18 kN. If the beam has the cross section shown in Fig. P8-152, compute and plot the vertical shearing stress τ as a function of distance y (-150 mm $\leq y \leq 150$ mm) from the neutral axis.

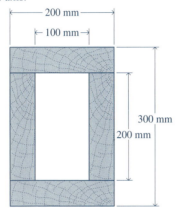

Figure P8-152

C8-153 A timber beam is simply supported and carries a uniformly distributed load w of 360 lb/ft over its entire 18-ft span (Fig. P8-153a). If the beam has the cross section shown in Fig. P8-153b, compute and plot the vertical shearing stress τ as a function of distance y from the neutral axis for a cross section 2 ft from the left end of the beam.

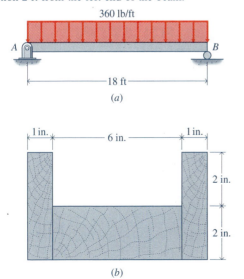

Figure P8-153

C8-154 The beam shown in Fig. P8-154a is fabricated by gluing two pieces of timber together to form the cross section shown in Fig. P8-154b. Compute and plot the vertical shearing stress τ as a function of distance y from the neutral axis for a cross section 0.5 m from the left end of the beam.

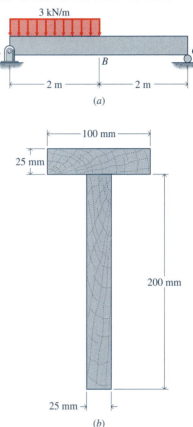

Figure P8-154

FLEXURAL LOADING: BEAM DEFLECTIONS

9-1 INTRODUCTION

Important relations between applied load and stress (flexural and shear) in a beam were presented in Chapter 8. A beam design, however, is frequently not complete until the amount of deflection has been determined for the specified load. Failure to control beam deflections within proper limits in building construction is frequently reflected by the development of cracks in plastered walls and ceilings. Beams in many machines must deflect just the right amount for gears or other parts to make proper contact. In innumerable instances the requirements for a beam involve a given load-carrying capacity with a specified maximum deflection.

The deflection of a beam depends on the stiffness of the material and the dimensions of the beam as well as on the applied loads and supports. Three common methods for calculating beam deflections owing to flexural stresses are presented here: (1) the double integration method, (2) the singularity function method, and (3) the superposition method.

9-2 THE DIFFERENTIAL EQUATION OF THE ELASTIC CURVE

When a straight beam is loaded and the action is linearly elastic, the centroidal axis of the beam is a curve defined as the elastic curve. In regions of constant bending moment, the elastic curve is an arc of a circle of radius ρ as indicated in Fig. 9-1. Since the portion AB of the beam is bent only with couples, plane sections A and B remain plane and the deformation of the fibers (elongation and compression) is proportional to the distance from the neutral surface, which is unchanged in length. From Fig. 9-1,

$$\theta = \frac{L}{\rho} = \frac{L + \delta}{\rho + c}$$

from which

$$\frac{c}{\rho} = \frac{\delta}{L} = \epsilon = \frac{\sigma}{E} = \frac{Mc}{EI} \qquad (9\text{-}1)$$

Therefore,

$$\frac{1}{\rho} = \frac{M}{EI} \qquad (9\text{-}2)$$

which relates the radius of curvature of the neutral surface of the beam to the bending moment M, the stiffness of the material E, and the second moment of the cross-sectional area I.

Equation (9-2) for the curvature of the elastic curve is useful only when the bending moment is constant for the interval of the beam involved. For most beams, the bending moment is a function of position along the beam and a more general expression is required.

The curvature from the calculus (see any standard calculus textbook) is

$$\frac{1}{\rho} = \frac{d^2y/dx^2}{[1 + (dy/dx)^2]^{3/2}}$$

For most beams the slope dy/dx is very small, and its square can be neglected in comparison to unity. With this approximation,

$$\frac{1}{\rho} = \frac{d^2y}{dx^2}$$

and Eq. (9-2) becomes

$$EI\frac{d^2y}{dx^2} = M(x) \qquad (9\text{-}3)$$

which is the differential equation for the elastic curve of a beam where the moment M is a function of x, $M(x)$.

The differential equation of the elastic curve, Eq. (9-3), can also be obtained from the geometry of the bent beam as shown in Fig. 9-2, where it is evident that $dy/dx = \tan\theta \cong \theta$ for small angles and that $d^2y/dx^2 = d\theta/dx$. Again from Fig. 9-2,

$$d\theta = \frac{dL}{\rho} \cong \frac{dx}{\rho}$$

for small angles. Therefore,

$$\frac{d^2y}{dx^2} = \frac{d\theta}{dx} = \frac{1}{\rho} = \frac{M}{EI}$$

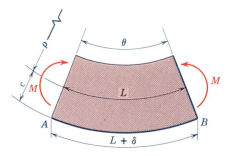

Figure 9-1

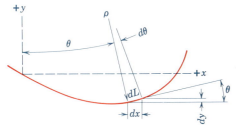

Figure 9-2

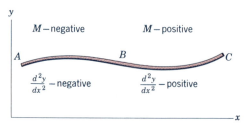

Figure 9-3

and thus

$$EI \frac{d^2y}{dx^2} = M \tag{9-3}$$

The sign convention for bending moments established in Section 8-6 will be used for Eq. (9-3). Both E and I are always positive; therefore, the signs of the bending moment and the second derivative must be consistent. With the coordinate axes shown in Fig. 9-3, the slope changes from positive to negative in the interval from A to B; therefore, the second derivative is negative, which agrees with the sign convention for the moment established in Section 8-6. For the interval BC, both d^2y/dx^2 and M are positive.

Figure 9-3 also reveals that the signs of the bending moment and the second derivative are also consistent when the origin of the coordinate system is selected at the right end of the beam with x positive to the left and y positive upward.

Equations (8-13d), (8-14c), and (9-3) provide a means for correlating the successive derivatives of the deflection y of the elastic curve with the physical quantities that they represent in beam action. They are

$$\text{deflection} = y$$

$$\text{slope} = \frac{dy}{dx}$$

$$\text{moment} = EI \frac{d^2y}{dx^2} \text{ [from Eq. (9-3)]}$$

$$\text{shear} = \frac{dM}{dx} \text{ [from Eq. (8-14c)]} = EI \frac{d^3y}{dx^3} \text{ (for } EI \text{ constant)}$$

$$\text{load} = \frac{dV}{dx} \text{ [from Eq. (8-13d)]} = EI \frac{d^4y}{dx^4} \text{ (for } EI \text{ constant)}$$

where the signs are as given in Section 8-6.

In Section 8-6-3 a method based on these differential relations was presented for starting from the load diagram and drawing first the shear diagram and then the moment diagram. This method can readily be extended to the construction of the slope diagram from the relation

$$M = EI \frac{d\theta}{dx}$$

from which

$$\int_{\theta_A}^{\theta_B} d\theta = \int_{x_A}^{x_B} \frac{M}{EI} \, dx$$

This relation shows that except for a factor EI, the area under the moment diagram between any two points along the beam gives the change in slope between the same two points. Likewise, the area under the slope diagram between two points along a beam gives the change in deflection between these points. These relations have been used to construct the complete series of diagrams shown in Fig. 9-4 for a simply supported beam with a concentrated load at the center of the span. The geometry of the beam was used to locate the points of zero slope and deflection, required as starting points for the construction. More commonly used methods for calculating beam deflections will be developed in succeeding sections.

Before developing specific methods for calculating beam deflections, it is advisable to consider the assumptions used in the development of the basic relation, Eq. (9-3). All of the limitations that apply to the flexure formula also apply to the calculation of deflections because the flexure formula was used in the derivation of Eq. (9-3). It is further assumed that

1. The square of the slope of the beam is negligible compared to unity.
2. The beam deflection due to shearing stresses is negligible (a plane section is assumed to remain plane).
3. The values of E and I remain constant for any interval along the beam. In case either of them varies and can be expressed as a function of the distance x along the beam, a solution of Eq. (9-3) that takes this variation into account may be possible.

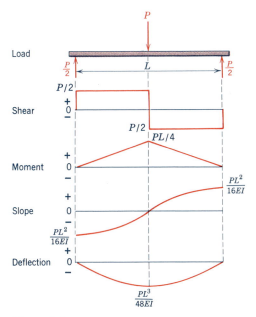

Figure 9-4

9-3 DEFLECTION BY INTEGRATION

Whenever the assumptions of the previous section are essentially correct and the bending moment can be readily expressed as an integrable function of x, Eq. (9-3) can be solved for the deflection y of the elastic curve of a beam at any point x along the beam. The constants of integration can be evaluated from the applicable boundary or matching conditions.

A boundary condition is defined as a known value for y or dy/dx at a specific location along the beam. One boundary condition can be used to determine one and only one constant of integration.

Many beams are subjected to abrupt changes in loading along the beam, such as concentrated loads, reactions, or even distinct changes in the amount of uniformly distributed load. Since the expressions for the bending moment on the left and right of any abrupt change in load are different functions of x, it is impossible to write a single equation for the bending moment in terms of ordinary algebraic functions that is valid for the entire length of the beam. This can be resolved by writing separate bending moment equations for each interval of the beam. Although the intervals are bounded by abrupt changes in load, the beam is continuous at such locations; therefore, the slope and the deflection at the junction of adjacent intervals must match. A matching condition is defined as the equality of slope or deflection, as determined at the junction of two intervals from the elastic curve equations for both intervals. One matching condition (for example, at x equals $L/3$, y from the left equation equals y from the right equation) can be used to determine one and only one constant of integration.

The procedure for obtaining beam deflections when matching conditions are required is lengthy and tedious. A method is presented in the next section in which singularity functions are used to write a single equation for the bending moment that is valid for the entire length of the beam; this eliminates the need for matching conditions and, accordingly, reduces the labor involved.

Calculating the deflection of a beam by the double integration method, that is, integrating Eq. (9-3) twice, involves four definite steps, and the following sequence for these steps is strongly recommended.

1. Select the interval or intervals of the beam to be used; next, place a set of coordinate axes on the beam with the origin at one end of an interval and then indicate the range of values of x in each interval. For example, two adjacent intervals might be

$$0 \le x \le L/3 \quad \text{and} \quad L/3 \le x \le L$$

2. List the available boundary and matching conditions (where two or more adjacent intervals are used) for each interval selected. Remember that two conditions are required to evaluate the two constants of integration for each interval used.
3. Express the bending moment as a function of x for each interval selected and equate it to $EI(d^2y/dx^2)$.
4. Solve the differential equation or equations from No. 3 and evaluate all constants of integration. Check the resulting equations for dimensional homogeneity. Calculate the deflection at specific points when required.

A roller or pin at any point in a beam (Figs. 9-5a and b) represents a simple support at which the beam cannot deflect (unless otherwise stated in the problem) but can rotate. At a fixed end, as represented by Figs. 9-5c and d, the beam can neither deflect nor rotate unless otherwise stated.

The following examples illustrate the use of the double integration method for calculating beam deflections.

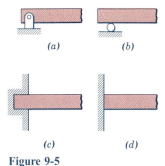

Figure 9-5

Example Problem 9-1

For the beam loaded and supported as shown in Fig 9-6a, determine

(a) The deflection midway between the supports.
(b) The point of maximum deflection between the supports.
(c) The maximum deflection in the interval between the supports.

SOLUTION

From a free-body diagram of the beam and the equation $\Sigma M_B = 0$,

$$+ \curvearrowleft \Sigma M_B = R_A(L) - wL^2/12 - wL(L/2) = 0 \qquad R_A = 7wL/12 \uparrow$$

As indicated in Fig. 9-6b, the origin of coordinates is selected at the left support, and the interval to be used is $0 \le x \le L$.

The two required boundary conditions are $y = 0$ when $x = 0$ and $y = 0$ when $x = L$. From the free-body diagram of the portion of the beam shown in Fig. 9-6b, Eq. (9-3) yields

$$EI \frac{d^2y}{dx^2} = M(x) = \frac{7wL}{12}x - \frac{wL^2}{12} - wx\left(\frac{x}{2}\right)$$

Successive integration gives

$$EI \frac{dy}{dx} = \frac{7wL}{24}x^2 - \frac{wL^2}{12}x - \frac{w}{6}x^3 + C_1$$

and

$$EIy = \frac{7wL}{72}x^3 - \frac{wL^2}{24}x^2 - \frac{w}{24}x^4 + C_1x + C_2$$

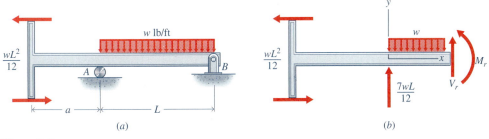

Figure 9-6

where C_1 and C_2 are constants of integration to be determined using the boundary conditions. Substitution of the boundary condition $y = 0$ when $x = 0$ yields

$$C_2 = 0$$

Substitution of the remaining boundary condition $y = 0$ when $x = L$ gives

$$C_1 = -wL^3/72$$

Therefore, the elastic curve equation is

$$EIy = \frac{7wL}{72}x^3 - \frac{wL^2}{24}x^2 - \frac{w}{24}x^4 - \frac{wL^3}{72}x$$

(a) The deflection midway between the supports is obtained by substituting $x = L/2$ into the elastic curve equation. Thus,

$$EIy = \frac{7wL}{72}(L/2)^3 - \frac{wL^2}{24}(L/2)^2 - \frac{w}{24}(L/2)^4 - \frac{wL^3}{72}(L/2)$$

from which

$$y = -\frac{3wL^4}{384EI} = -\frac{7.81(10^{-3})wL^4}{EI} = \frac{7.81(10^{-3})wL^4}{EI} \downarrow \qquad \text{Ans.}$$

(b) The maximum deflection occurs where the slope dy/dx is zero or

$$0 = \frac{7wL}{24}x^2 - \frac{wL^2}{12}x - \frac{w}{6}x^3 - \frac{wL^3}{72}$$

from which

$$12x^3 - 21Lx^2 + 6L^2x + L^3 = 0$$

The solution of this cubic equation in x gives the point of maximum deflection at

$$x = 0.541L \text{ to the right of the left support} \qquad \text{Ans.}$$

(c) The maximum deflection can readily be obtained by substituting $0.541L$ for x in the elastic curve equation. The result is

$$y = -\frac{7.88(10^{-3})wL^4}{EI} = \frac{7.88(10^{-3})wL^4}{EI} \downarrow \blacksquare \qquad \text{Ans.}$$

Example Problem 9-2

For the beam loaded and supported as shown in Fig 9-7a, determine the deflection of the right end.

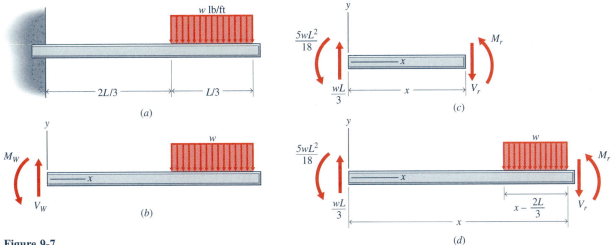

Figure 9-7

SOLUTION

From the free-body diagram of Fig. 9-7b the equations of the equilibrium give

$+\uparrow\Sigma F_y = V_W - w(L/3) = 0$ $\qquad\qquad$ $V_W = wL/3 \uparrow$

$+\mathord{\downarrow}\Sigma M_A = M_W - w(L/3)(5L/6) = 0$ $\qquad$ $M_W = 5wL^2/18 \mathord{\downarrow}$

With the sign conventions established in Section 8-6, these results are

$$V_W = +\frac{wL}{3} \qquad \text{and} \qquad M_W = -\frac{5wL^2}{18}$$

In Fig. 9-7b there are two intervals to be considered, namely, the loaded and the unloaded portions of the beam. The loaded portion of the beam must be used because it contains the point where the deflection is required. However, a quick check reveals the absence of boundary conditions in this interval. It therefore becomes necessary to use both intervals as well as matching and boundary conditions; hence, the origin of coordinates is selected at the left end of the beam, as shown. The intervals are

$$0 \le x \le 2L/3 \qquad \text{and} \qquad 2L/3 \le x \le L$$

The available boundary conditions are

$$\frac{dy}{dx} = 0 \qquad \text{when} \qquad x = 0$$

$$y = 0 \qquad \text{when} \qquad x = 0$$

The available matching conditions are when $x = 2L/3$

$$\frac{dy}{dx} \text{ from the left equation} = \frac{dy}{dx} \text{ from the right equation}$$

$$y \text{ from the left equation} = y \text{ from the right equation}$$

Four conditions (two boundary and two matching) are sufficient for the evaluation of the four constants of integration (two in each of the two elastic curve differential equations); therefore, the problem can be solved in this manner.

From the free-body diagram of Fig. 9-7c where the beam is cut in the unloaded interval, Eq. (9-3) yields for $0 \le x \le 2L/3$

$$EI\frac{d^2y}{dx^2} = M(x) = \frac{wL}{3}x - \frac{5wL^2}{18} \tag{a}$$

From the free-body diagram of Fig. 9-7d where the beam is cut in the loaded interval. Eq. (9-3) yields for $2L/3 \le x \le L$

$$EI\frac{d^2y}{dx^2} = M(x) = \frac{wL}{3}x - \frac{5wL^2}{18} - w\left(x - \frac{2L}{3}\right)\left(\frac{x - 2L/3}{2}\right) \tag{b}$$

Integration of equations (a) and (b) gives

$$EI\frac{dy}{dx} = \frac{wL}{6}x^2 - \frac{5wL^2}{18}x + C_1 \qquad\qquad 0 \le x \le 2L/3 \quad \text{(c)}$$

$$EI\frac{dy}{dx} = \frac{wL}{6}x^2 - \frac{5wL^2}{18}x - \frac{w}{6}\left(x - \frac{2L}{3}\right)^3 + C_3 \qquad 2L/3 \le x \le L \quad \text{(d)}$$

Substitution of the boundary condition $\frac{dy}{dx} = 0$ when $x = 0$ into Eq. (c) gives

$$C_1 = 0$$

Since the beam has a continuous slope at $x = 2L/3$,

$$\frac{dy}{dx}[\text{from Eq. (c) at } x = 2L/3] = \frac{dy}{dx}[\text{from Eq. (d) at } x = 2L/3]$$

which gives

$$C_1 = C_3 = 0$$

Integration of the resulting differential equations gives

$$EIy = \frac{wL}{18}x^3 - \frac{5wL^2}{36}x^2 + C_2 \qquad\qquad 0 \le x \le 2L/3$$

$$EIy = \frac{wL}{18}x^3 - \frac{5wL^2}{36}x^2 - \frac{w}{24}\left(x - \frac{2L}{3}\right)^4 + C_4 \qquad 2L/3 \le x \le L$$

Use of the remaining boundary condition yields $C_2 = 0$. Use of the final matching conditions yields $C_4 = C_2 = 0$.

The deflection of the right end of the beam can now be obtained from the elastic curve equation for the right interval by replacing x by its value L. The result is

$$EIy = \frac{wL}{18}(L^3) - \frac{5wL^2}{36}(L^2) - \frac{w}{24}\left(L - \frac{2L}{3}\right)^4$$

from which

$$y = -\frac{163}{1944}\frac{wL^4}{EI} = \frac{163}{1944}\frac{wL^4}{EI} \downarrow \blacksquare \qquad\qquad \textbf{Ans.}$$

PROBLEMS

Standard Problems

9-1* A beam is loaded and supported as shown in Fig. P9-1.
(a) Derive the equation of the elastic curve in terms of P, L, x, E, and I. Use the designated axes.
(b) Determine the slope at the left end of the beam.
(c) Determine the deflection at the left end of the beam.

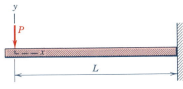

Figure P9-1

9-2* A beam is loaded and supported as shown in Fig. P9-2.
(a) Derive the equation of the elastic curve in terms of w, L, x, E, and I. Use the designated axes.
(b) Determine the slope at the left end of the beam.
(c) Determine the deflection at the left end of the beam.

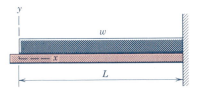

Figure P9-2

9-3 A beam is loaded and supported as shown in Fig. P9-3.
(a) Derive the equation of the elastic curve in terms of P, L, x, E, and I. Use the designated axes.
(b) Determine the slope at the left end of the beam.
(c) Determine the deflection midway between the supports.

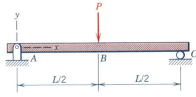

Figure P9-3

9-4 A beam is loaded and supported as shown in Fig. P9-4.
(a) Derive the equation of the elastic curve in terms of w, L, x, E, and I. Use the designated axes.
(b) Determine the slope at the left end of the beam.
(c) Determine the deflection midway between the supports.

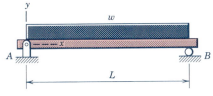

Figure P9-4

9-5* For the steel beam ($E = 30,000$ ksi and $I = 32.1$ in.^{4}) shown in Fig. P9-5, determine
(a) The radius of curvature at any point between the supports.
(b) The deflection at a section midway between the supports.

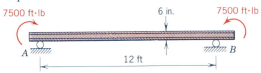

Figure P9-5

9-6* For the steel beam ($E = 200$ GPa and $I = 32.0(10^6)$ mm^4) shown in Fig. P9-6, determine
(a) The radius of curvature at any point between the supports.
(b) The deflection at a section midway between the supports.

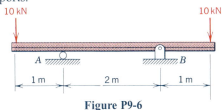

Figure P9-6

9-7 The cantilever beam shown in Fig. P9-7a is fabricated from two 1×3-in. steel ($E = 30,000$ ksi) bars as shown in Fig. P9-7b. Determine
(a) The radius of curvature of the beam.
(b) The deflection at the right end of the beam.

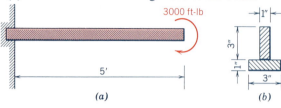

3000 ft-lb

5'

3"

1"

1"

3"

(a)

(b)

Figure P9-7

9-8 The cantilever beam shown in Fig. P9-8a is fabricated from three 30×120-mm aluminum ($E = 70$ GPa) bars as shown in Fig. P9-8b. Determine
(a) The radius of curvature of the beam.
(b) The deflection at the left end of the beam.

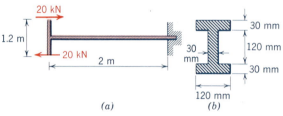

20 kN

1.2 m

20 kN

2 m

30 mm

30 mm

120 mm

30 mm

120 mm

(a)

(b)

Figure P9-8

9-9* A beam is loaded and supported as shown in Fig. P9-9.
(a) Derive the equation of the elastic curve in terms of E, I, w, and L for the interval of the beam between the supports. Use the designated axes.
(b) Determine the deflection midway between the supports.

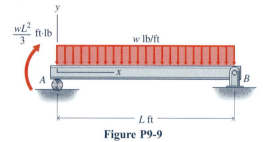

$\frac{wL^2}{3}$ ft·lb

w lb/ft

x

A

B

L ft

Figure P9-9

9-10* A beam is loaded and supported as shown in Fig. P9-10.
(a) Derive the equation of the elastic curve in terms of E, I, w, and L for the interval BC of the beam. Use the designated axes.

(b) Determine the deflection at the midpoint of interval BC.

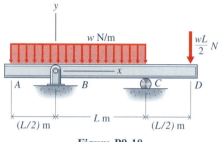

y

w N/m

$\frac{wL}{2}$ N

x

A

B

C

D

$(L/2)$ m

L m

$(L/2)$ m

Figure P9-10

9-11 A beam is loaded and supported as shown in Fig. P9-11.
(a) Derive the equation of the elastic curve in terms of E, I, w, and L for the interval of the beam between the supports. Use the designated axes.
(b) Determine the deflection midway between the supports.

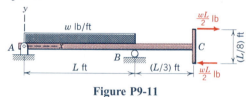

y

w lb/ft

$\frac{wL}{2}$ lb

A

x

B

C

L ft

$(L/3)$ ft

$\frac{wL}{2}$ lb

$(L/8)$ ft

Figure P9-11

9-12 A beam is loaded and supported as shown in Fig. P9-12.
(a) Derive the equation of the elastic curve in terms of E, I, w, and L for the interval of the beam between the supports. Use the designated axes.
(b) Determine the deflection midway between the supports.

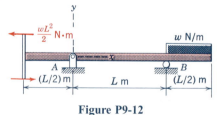

y

$\frac{wL^2}{2}$ N·m

w N/m

x

A

B

$(L/2)$ m

L m

$(L/2)$ m

Figure P9-12

9-13* A beam is loaded and supported as shown in Fig. P9-13.
(a) Derive the equation of the elastic curve in terms of w, L, x, E, and I. Use the designated axes.
(b) Determine the slope at the left end of the beam.
(c) Determine the deflection at the left end of the beam.

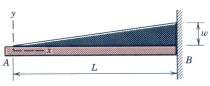

Figure P9-13

9-14* A beam is loaded and supported as shown in Fig. P9-14.
(a) Derive the equation of the elastic curve in terms of w, L, x, E, and I. Use the designated axes.
(b) Determine the slope at the right end of the beam.
(c) Determine the deflection at the right end of the beam.

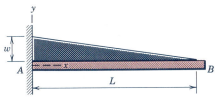

Figure P9-14

9-15 A beam is loaded and supported as shown in Fig. P9-15.
(a) Derive the equation of the elastic curve in terms of w, L, x, E, and I. Use the designated axes.
(b) Determine the slope at the right end of the beam.
(c) Determine the deflection midway between the supports.

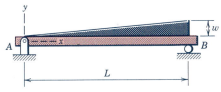

Figure P9-15

9-16 A beam is loaded and supported as shown in Fig. P9-16.
(a) Derive the equation of the elastic curve in terms of w, L, x, E, and I. Use the designated axes.
(b) Determine the slope at the right end of the beam.
(c) Determine the deflection midway between the supports.

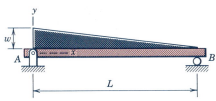

Figure P9-16

Challenging Problems

9-17* A beam is loaded and supported as shown in Fig. P9-17. Determine in terms of w, L, E, and I
(a) The slope at the left end of the beam.
(b) The maximum deflection between the supports.

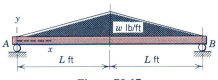

Figure P9-17

9-18* A beam is loaded and supported as shown in Fig. P9-18. Determine in terms of M, L, E, and I
(a) The slope at the left end of the beam.
(b) The maximum deflection in the left half of the beam.

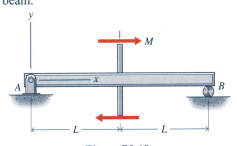

Figure P9-18

9-19 A beam is loaded and supported as shown in Fig. P9-19. Determine in terms of P, L, E, and I
(a) The deflection at the left load P.
(b) The maximum deflection between the supports.

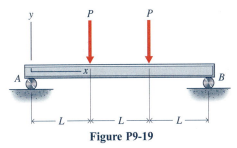

Figure P9-19

9-20 A beam is loaded and supported as shown in Fig. P9-20. Determine in terms of P, L, E, and I
(a) The maximum deflection between the supports.
(b) The deflection at the right end of the beam.

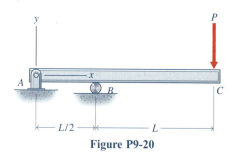

Figure P9-20

9-21* A beam is loaded and supported as shown in Fig. P9-21. Determine the deflection midway between the supports in terms of E, I, w, and a.

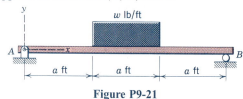

Figure P9-21

9-22* A beam is loaded and supported as shown in Fig. P9-22. Determine the deflection midway between the supports in terms of w, L, E, and I.

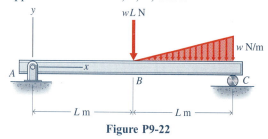

Figure P9-22

9-23 A beam is loaded and supported as shown in Fig. P9-23. Determine in terms of w, L, E, and I
(a) The deflection midway between the supports.
(b) The deflection at the left end of the beam.

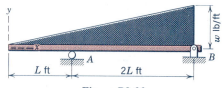

Figure P9-23

9-24 A beam is loaded and supported as shown in Fig. P9-24. Determine the deflection midway between the supports in terms of w, L, E, and I.

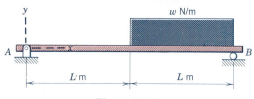

Figure P9-24

9-25* The beam AB shown in Fig. P9-25 is the flexural member of a scale that is used to obtain the weight W of food in a microwave oven. Determine
(a) The equation of the elastic curve for beam AB in terms of W, L, x, E, and I.
(b) The deflection at point C when $W = 20$ lb, $L = 2$ in., and $EI = 100$ lb $\cdot$ in.^{2}.

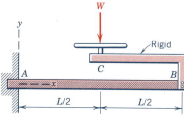

Figure P9-25

9-26 A beam AB is loaded and supported as shown in Fig. P9-26. The load P is applied through a collar that can be positioned on the load bar at any location in the interval $L/4 < a < 3L/4$. Determine
(a) The equation of the elastic curve for beam AB in terms of P, L, x, a, E, and I.
(b) The location of load P for maximum deflection at end B.
(c) The location of load P for zero deflection at end B.

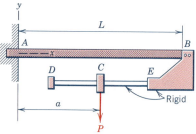

Figure P9-26

9-4 SINGULARITY FUNCTIONS

The double integration method of Section 9-3 becomes tedious and time-consuming when several intervals and several sets of matching conditions are required. The labor involved in solving problems of this type, however, can be diminished by making use of singularity functions following the method developed in 1862 by the German mathematician A. Clebsch (1833–1872).[1]

Singularity functions are closely related to the unit step function used by the British physicist O. Heaviside (1850–1925) to analyze the transient response of electrical circuits. Singularity functions will be used here for writing one bending moment equation that applies in all intervals along a beam, thus eliminating the need for matching conditions.

A singularity function of x is written as $\langle x - x_0 \rangle^n$, where n is any integer (positive or negative) including zero, and x_0 is a constant equal to the value of x at the initial boundary of a specific interval along a beam. Selected properties of singularity functions required for beam-deflection problems are listed here for emphasis and ready reference.

$$\langle x - x_0 \rangle^n = \begin{cases} (x - x_0)^n & \text{when } n > 0 \text{ and } x \geq x_0 \\ 0 & \text{when } n > 0 \text{ and } x < x_0 \end{cases}$$

$$\langle x - x_0 \rangle^0 = \begin{cases} 1 & \text{when } x \geq x_0 \\ 0 & \text{when } x < x_0 \end{cases}$$

$$\int \langle x - x_0 \rangle^n dx = \frac{1}{n+1} \langle x - x_0 \rangle^{n+1} + C \qquad \text{when } n \geq 0$$

$$\frac{d}{dx} \langle x - x_0 \rangle^n = n \langle x - x_0 \rangle^{n-1} \qquad \text{when } n \geq 1$$

Several examples of singularity functions are shown in Fig. 9-8.

---

[1] For a rather complete history of the Clebsch method and the numerous extensions thereof, see "Clebsch's Method for Beam Deflections," Walter D. Pilkey, *Journal of Engineering Education,* January 1964, p. 170.

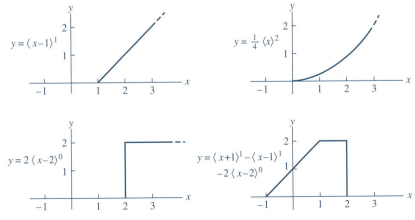

$y = \langle x-1 \rangle^1$

$y = \frac{1}{4} \langle x \rangle^2$

$y = 2 \langle x-2 \rangle^0$

$y = \langle x+1 \rangle^1 - \langle x-1 \rangle^1 - 2 \langle x-2 \rangle^0$

Figure 9-8

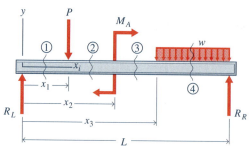

Figure 9-9

By making use of these properties of singularity functions, one is able to write a single equation for the bending moment for a beam and obtain the correct value of the moment in any interval along the beam. To amplify this statement, consider the beam of Fig. 9-9. The moment equations at the four designated sections are

$$M_1 = R_L x \qquad\qquad 0 < x < x_1$$
$$M_2 = R_L x - P(x - x_1) \qquad\qquad x_1 < x < x_2$$
$$M_3 = R_L x - P(x - x_1) + M_A \qquad\qquad x_2 < x < x_3$$
$$M_4 = R_L x - P(x - x_1) + M_A - (w/2)(x - x_3)^2 \qquad x_3 < x < L$$

These four moment equations can be combined into a single equation by means of singularity functions to give

$$M(x) = R_L x - P\langle x - x_1 \rangle^1 + M_A\langle x - x_2 \rangle^0 - \frac{w}{2}\langle x - x_3 \rangle^2 \qquad 0 \le x \le L$$

where $M(x)$ indicates that the moment is a function of x.

Distributed loadings that are sectionally continuous (the distributed load cannot be represented by a single function of x for all values of x) are readily obtained by superposition, as illustrated in the following examples.

Example Problem 9-3

For the beam loaded and supported as shown in Fig 9-10a, determine the deflection of the right end.

SOLUTION

The free-body diagram of Fig. 9-10b and the equations of equilibrium $\Sigma F_y = 0$ and $\Sigma M_O = 0$ yield

$$+\uparrow\Sigma F_y = V_W - w(L/3) = 0 \qquad V_W = wL/3 \uparrow$$
$$+\downarrow\Sigma M_O = M_W - w(L/3)(5L/6) = 0 \qquad M_W = 5wL^2/18 \downarrow$$

The boundary conditions, in the interval $0 \le x \le L$, are $dy/dx = 0$ when $x = 0$ and $y = 0$ when $x = 0$ with the axes as shown in Fig. 9-10b. When the expression for the bending moment obtained from the free-body diagram of Fig. 9-10c is substituted in Eq. (9-3), the result is

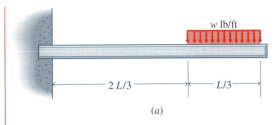

(a)

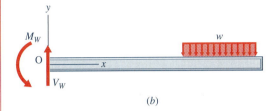

(b)

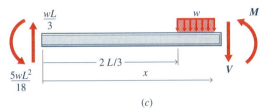

(c)

Figure 9-10

$$EI\frac{d^2y}{dx^2} = -\frac{5wL^2}{18} + \frac{wLx}{3} - \frac{w}{2}\left\langle x - \frac{2L}{3}\right\rangle^2$$

The first integration gives

$$EI\frac{dy}{dx} = -\frac{5wL^2x}{18} + \frac{wLx^2}{6} - \frac{w}{6}\left\langle x - \frac{2L}{3}\right\rangle^3 + C_1$$

The boundary condition $dy/dx = 0$ when $x = 0$ gives $C_1 = 0$ since the term in the brackets is zero when $x \le 2L/3$. Integrating again gives

$$EIy = -\frac{5wL^2x^2}{36} + \frac{wLx^3}{18} - \frac{w}{24}\left\langle x - \frac{2L}{3}\right\rangle^4 + C_2$$

The boundary condition $y = 0$ when $x = 0$ gives $C_2 = 0$ since the term in the brackets is zero when $x \le 2L/3$.

The deflection at the right end is obtained by substituting $x = L$ in the elastic curve equation. The result is

$$y = -\frac{163}{1944}\frac{wL^4}{EI} = \frac{163}{1944}\frac{wL^4}{EI}\downarrow \qquad \textbf{Ans.}$$

This result agrees with that of Example 9-2. ■

▮ Example Problem 9-4

Determine the deflection at the left end of the beam of Fig. 9-11a.

SOLUTION

The free-body diagram of Fig. 9-11b and the equilibrium equation $\Sigma M_B = 0$ yield

$$+\curvearrowleft\Sigma M_B = R_L(L) - w(L/2)(7L/4) + wL^2/2 = 0 \qquad\qquad R_L = 3wL/8 \uparrow$$

To express the moment of the distributed load at the left end of the beam in terms of singularity functions that are valid for the full length of the beam, the distributed load must be represented on the free-body diagram by equivalent distributed loads on the top and bottom of the beam as shown in Fig. 9-11c. When the expression for the bending moment obtained from the free-body diagram of Fig. 9-11c is substituted in Eq. (9-3), the result is

$$EI\frac{d^2y}{dx^2} = -\frac{w}{2}(x + L)^2 + \frac{w}{2}\left\langle x + \frac{L}{2}\right\rangle^2 + \frac{3wL}{8}\langle x - 0\rangle^1 + \frac{wL^2}{2}\left\langle x - \frac{L}{2}\right\rangle^0$$

where the first term after the equal sign represents the distributed load on the top of the beam and the second term represents the distributed load on the bottom of the beam. The effect of the two terms is to terminate the distributed load at $x = -L/2$. The boundary conditions are $y = 0$ when $x = 0$, and $y = 0$ when $x = L$. Two integrations of the moment equation give

$$EI\frac{dy}{dx} = -\frac{w}{6}(x + L)^3 + \frac{w}{6}\left\langle x + \frac{L}{2}\right\rangle^3 + \frac{3wL}{16}\langle x - 0\rangle^2 + \frac{wL^2}{2}\left\langle x - \frac{L}{2}\right\rangle^1 + C_1$$

and

$$EIy = -\frac{w}{24}(x + L)^4 + \frac{w}{24}\left\langle x + \frac{L}{2}\right\rangle^4 + \frac{wL}{16}\langle x\rangle^3 + \frac{wL^2}{4}\left\langle x - \frac{L}{2}\right\rangle^2 + C_1x + C_2$$

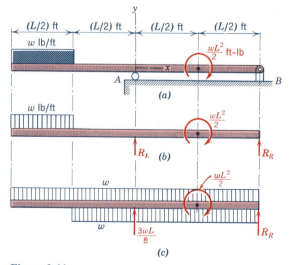

Figure 9-11

The first boundary condition, $y = 0$ when $x = 0$, gives

$$0 = -\frac{wL^4}{24} + \frac{wL^4}{384} + 0 + 0 + 0 + C_2 \qquad C_2 = +\frac{5}{128}wL^4$$

The second boundary condition, $y = 0$ when $x = L$, gives

$$0 = -\frac{16wL^4}{24} + \frac{81wL^4}{384} + \frac{wL^4}{16} + \frac{wL^4}{16} + C_1L + \frac{5wL^4}{128} \qquad C_1 = +\frac{7}{24}wL^3$$

The deflection at the left end is obtained by substituting $x = -L$ in the elastic curve equation. The result is

$$EIy = 0 + 0 + 0 + 0 + \frac{7}{24}wL^3(-L) + \frac{5}{128}wL^4 = -\frac{97}{384}wL^4$$

Thus,

$$y = -\frac{97}{384}\frac{wL^4}{EI} = \frac{97}{384}\frac{wL^4}{EI}\downarrow \qquad \blacksquare \qquad \text{Ans.}$$

Example Problem 9-5

Use singularity functions to write a single equation for the bending moment at any section of the beam shown in Fig. 9-12a.

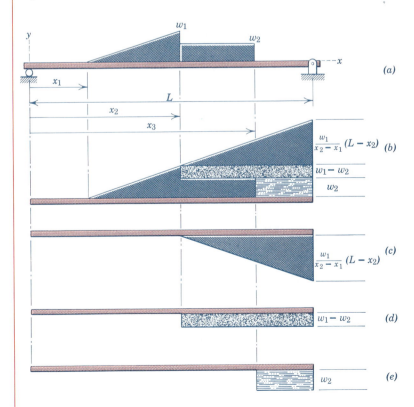

Figure 9-12

SOLUTION

The loading on the beam of Fig. 9-12a can be considered a combination of the loadings shown in Figs. 9-12b, c, d, and e, where downward-acting loads are shown on top of the beam and upward-acting loads on the bottom. The magnitude of the linearly varying load at any point $x \geq x_1$ is

$$w = \frac{w_1(x - x_1)}{x_2 - x_1}$$

The moment of the linearly varying load at any point $x \geq x_1$ is

$$M = -\frac{1}{2}\left[\frac{w_1}{x_2 - x_1}(x - x_1)\right](x - x_1)(x - x_1)/3 = -\frac{w_1}{6(x_2 - x_1)}(x - x_1)^3$$

Once the moment term for the linearly varying load is introduced in the moment equation at $x = x_1$, its effect continues to the end of the beam. As a result, terms must be introduced at the appropriate locations to terminate the effects. The effect is reduced to that of a constant distributed load of magnitude w_1 by introducing the linearly distributed load of Fig. 9-12c at $x = x_2$. The magnitude of this constant distributed load is reduced from w_1 to w_2 by introducing the constant distributed load of magnitude $w_1 - w_2$ at $x = x_2$, as shown in Fig. 9-12d. Finally, the constant distributed load of magnitude w_2 is terminated at $x = x_3$ by introducing the constant distributed load w_2, as shown in Fig. 9-12e. The moment equation for the beam is then written in terms of singularity functions as

$$M(x) = R_L x - \frac{w_1}{6(x_2 - x_1)}\langle x - x_1\rangle^3 + \frac{w_1}{6(x_2 - x_1)}\langle x - x_2\rangle^3$$
$$+ \frac{w_1 - w_2}{2}\langle x - x_2\rangle^2 + \frac{w_2}{2}\langle x - x_3\rangle^2 \quad \blacksquare \quad \textbf{Ans.}$$

PROBLEMS

Standard Problems

9-27* A cantilever beam is loaded and supported as shown in Fig. P9-27. Use singularity functions to determine the deflection, in terms of P, L, E, and I,
(a) At a distance $x = L$ from the support.
(b) At the right end of the beam.

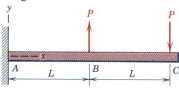

Figure P9-27

9-28* A cantilever beam is loaded and supported as shown in Fig. P9-28. Use singularity functions to determine the deflection, in terms of P, L, E, and I,
(a) At a distance $x = L$ from the support.
(b) At the right end of the beam.

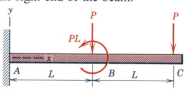

Figure P9-28

9-29 A beam is loaded and supported as shown in Fig. P9-29. Use singularity functions to determine, in terms of P, L, E, and I,
(a) The deflection under the load P.
(b) The deflection at the middle of the span.
(c) The maximum deflection of the beam.

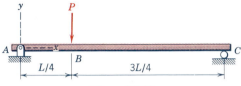

Figure P9-29

9-30 A beam is loaded and supported as shown in Fig. P9-30. Use singularity functions to determine the deflection in terms of P, L, E, and I,
(a) At a distance $x = L$ from the left support.
(b) At the middle of the span.

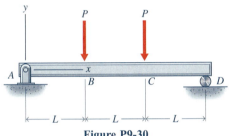

Figure P9-30

9-31* Use singularity functions to determine the deflection, in terms of w, L, E, and I, at the right end of the cantilever beam shown in Fig. P9-31.

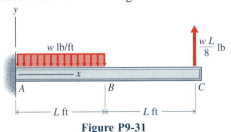

Figure P9-31

9-32* Use singularity functions to determine the deflection, in terms of w, L, E, and I, at the left end of the cantilever beam shown in Fig. P9-32.

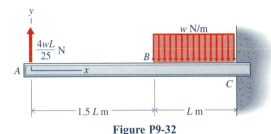

Figure P9-32

9-33 A beam is loaded and supported as shown in Fig. P9-33. Use singularity functions to determine the deflection in terms of P, L, E, and I,
(a) At the right end of the beam.
(b) At a section midway between the supports.

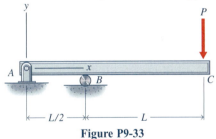

Figure P9-33

9-34 A beam is loaded and supported as shown in Fig. P9-34. Use singularity functions to determine, in terms of M, L, E, and I,
(a) The deflection at the middle of the span.
(b) The maximum deflection of the beam.

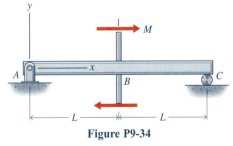

Figure P9-34

9-35* A cantilever beam is loaded and supported as shown in Fig. P9-35. Use singularity functions to determine the deflection in terms of w, L, E, and I,
(a) At a distance $x = L$ from the support.
(b) At the right end of the beam.

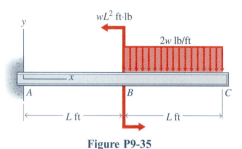

Figure P9-35

9-36* A cantilever beam is loaded and supported as shown in Fig. P9-36. Use singularity functions to determine the deflection in terms of w, L, E, and I,
(a) At a distance $x = L$ from the support.
(b) At the right end of the beam.

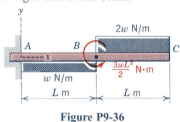

Figure P9-36

Challenging Problems

9-37 A beam is loaded and supported as shown in Fig. P9-37. Use singularity functions to determine the deflection, in terms of w, L, E, and I,
(a) At the left end of the distributed load.
(b) At a section midway between the supports.

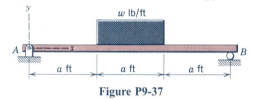

Figure P9-37

9-38 A beam is loaded and supported as shown in Fig. P9-38. Use singularity functions to determine the deflection, in terms of w, L, E, and I,
(a) At the left end of the distributed load.
(b) At a section midway between the supports.

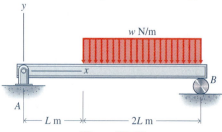

Figure P9-38

9-39* A beam is loaded and supported as shown in Fig. P9-39. Use singularity functions to determine, in terms of w, L, E, and I,
(a) The deflection midway between the supports.
(b) The maximum deflection of the beam.

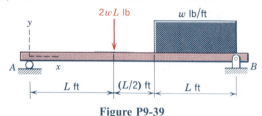

Figure P9-39

9-40* A beam is loaded and supported as shown in Fig. P9-40. Use singularity functions to determine, in terms of w, L, E, and I,
(a) The deflection midway between the supports.
(b) The maximum deflection of the beam.

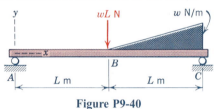

Figure P9-40

9-41 A beam is loaded and supported as shown in Fig. P9-41. Use singularity functions to determine, in terms of w, L, E, and I,
(a) The deflection at the middle of the span.
(b) The maximum deflection of the beam.

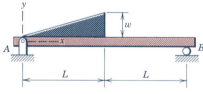

Figure P9-41

9-42 A beam is loaded and supported as shown in Fig. P9-42. Use singularity functions to determine, in terms of w, L, E, and I,
(a) The deflection at the middle of the span.
(b) The maximum deflection of the beam.

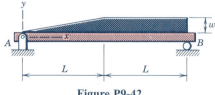

Figure P9-42

9-5 DEFLECTIONS BY INTEGRATION OF SHEAR-FORCE OR LOAD EQUATIONS

In Section 9-3, the equation of the elastic curve was obtained by integrating Eq. (9-3) and applying the appropriate boundary conditions to evaluate the two constants of integration. In a similar manner, the equation of the elastic curve can be obtained from load and shear-force equations. The differential equations that relate deflection y to load $w(x)$ or deflection y to shear-force $V(x)$ are obtained by substituting Eq. (8-14c) or Eq. (8-13d), respectively, into Eq. (9-3). Thus,

$$EI\frac{d^2y}{dx^2} = M(x) \qquad\qquad (9\text{-}3)$$

$$EI\frac{d^3y}{dx^3} = V(x) \qquad\qquad (9\text{-}4)$$

$$EI\frac{d^4y}{dx^4} = w(x) \qquad\qquad (9\text{-}5)$$

When Eqs. (9-4) or (9-5) are used to obtain the equation of the elastic curve, either three or four integrations will be required instead of the two integrations required with Eq. (9-3). These additional integrations will introduce additional constants of integration. The boundary conditions, however, now include conditions on the shear forces and bending moments, in addition to the conditions on slopes and deflections. Use of a particular differential equation is usually made on the basis of mathematical convenience or personal preference. In those instances when the expression for the load is easier to write than the expression for the moment, Eq. (9-5) would be preferred over Eq. (9-3). The following example illustrates the use of Eq. (9-5) for calculating beam deflections.

Example Problem 9-6

A beam is loaded and supported as shown in Fig. 9-13. Determine
(a) The equation of the elastic curve in terms of w, L, x, E, and I.
(b) The deflection at the right end of the beam.
(c) The support reactions V_A and M_A at the left end of the beam.

SOLUTION

Since the equation for the load distribution is given and the moment equation is not easy to write, Eq. (9-5) will be used to determine the deflections.

(a) In Section 8-6 (see Fig. 8-23), the upward direction was considered positive for a distributed load w; therefore, Eq. (9-5) is written as

$$EI\frac{d^4y}{dx^4} = w(x) = -w\cos\left(\frac{\pi x}{2L}\right)$$

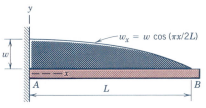

Figure 9-13

Successive integration gives

$$EI\frac{d^3y}{dx^3} = V(x) = -\frac{2wL}{\pi}\sin\left(\frac{\pi x}{2L}\right) + C_1$$

$$EI\frac{d^2y}{dx^2} = M(x) = \frac{4wL^2}{\pi^2}\cos\left(\frac{\pi x}{2L}\right) + C_1x + C_2$$

$$EI\frac{dy}{dx} = EI\theta = \frac{8wL^3}{\pi^3}\sin\left(\frac{\pi x}{2L}\right) + C_1\left(\frac{x^2}{2}\right) + C_2x + C_3$$

$$EIy = EI\delta = -\frac{16wL^4}{\pi^4}\cos\left(\frac{\pi x}{2L}\right) + C_1\left(\frac{x^3}{6}\right) + C_2\left(\frac{x^2}{2}\right) + C_3x + C_4$$

The four constants of integration are determined by applying the boundary conditions. Thus,

$$\text{At } x = 0, y = 0; \qquad \text{therefore, } C_4 = \frac{16wL^4}{\pi^4}$$

$$\text{At } x = 0, \frac{dy}{dx} = 0; \qquad \text{therefore, } C_3 = 0$$

$$\text{At } x = L, V = 0; \qquad \text{therefore, } C_1 = \frac{2wL}{\pi}$$

$$\text{At } x = L, M = 0; \qquad \text{therefore, } C_2 = -\frac{2wL^2}{\pi}$$

Thus,

$$y = -\frac{w}{3\pi^4 EI}\left[48L^4\cos\left(\frac{\pi x}{2L}\right) - \pi^3Lx^3 + 3\pi^3L^2x^2 - 48L^4\right] \qquad \textbf{Ans.}$$

(b) The deflection at the right end of the beam is

$$\delta_B = y_{x=L} = -\frac{w}{3\pi^4 EI}(-\pi^3L^4 + 3\pi^3L^4 - 48L^4)$$

$$= -\frac{(2\pi^3 - 48)wL^4}{3\pi^4 EI} = -0.0480wL^4/EI \qquad \textbf{Ans.}$$

(c) The shear force $V(x)$ and bending moment $M(x)$ at any distance x from the support are

$$V(x) = \frac{2wL}{\pi}\left[1 - \sin\left(\frac{\pi x}{2L}\right)\right]$$

$$M(x) = \frac{2wL}{\pi^2}\left[2L\cos\left(\frac{\pi x}{2L}\right) + \pi x - \pi L\right]$$

Thus, the support reactions at the left end of the beam are

$$V_A = V_{x=0} = \frac{2wL}{\pi} \qquad \textbf{Ans.}$$

$$M_A = M_{x=0} = -\frac{2(\pi - 2)wL^2}{\pi^2} \qquad ■ \qquad \textbf{Ans.}$$

PROBLEMS

Standard Problems

9-43* A beam is loaded and supported as shown in Fig. P9-43. Determine
(a) The equation of the elastic curve in terms of w, L, x, E, and I.
(b) The deflection at the left end of the beam.
(c) The support reactions V_B and M_B.

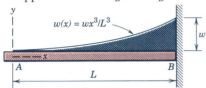

$w(x) = wx^3/L^3$

Figure P9-43

9-44* A beam is loaded and supported as shown in Fig. P9-44. Determine
(a) The equation of the elastic curve in terms of w, L, x, E, and I.
(b) The deflection midway between the supports.
(c) The maximum deflection of the beam.
(d) The support reactions R_A and R_B.

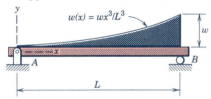

$w(x) = wx^3/L^3$

Figure P9-44

9-45 A beam is loaded and supported as shown in Fig. P9-45. Determine
(a) The equation of the elastic curve in terms of w, L, x, E, and I.
(b) The deflection at the left end of the beam.
(c) The support reactions V_B and M_B.

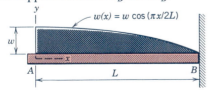

$w(x) = w \cos(\pi x/2L)$

Figure P9-45

9-46 A beam is loaded and supported as shown in Fig. P9-46. Determine
(a) The equation of the elastic curve in terms of w, L, x, E, and I.

(b) The deflection at the left end of the beam.
(c) The slope at the left end of the beam.
(d) The support reactions V_B and M_B.

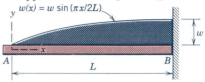

$w(x) = w \sin(\pi x/2L)$

Figure P9-46

9-47* A beam is loaded and supported as shown in Fig. P9-47. Determine
(a) The equation of the elastic curve in terms of w, L, x, E, and I.
(b) The deflection midway between the supports.
(c) The slope at the left end of the beam.
(d) The support reactions R_A and R_B.

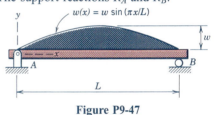

$w(x) = w \sin(\pi x/L)$

Figure P9-47

9-48 A beam is loaded and supported as shown in Fig. P9-48. Determine
(a) The equation of the elastic curve in terms of w, L, x, E, and I.
(b) The deflection midway between the supports.
(c) The maximum deflection of the beam.
(d) The slope at the left end of the beam.
(e) The support reactions R_A and R_B.

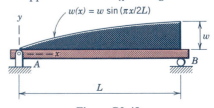

$w(x) = w \sin(\pi x/2L)$

Figure P9-48

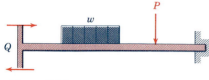

Figure 9-14

9-6 DEFLECTIONS BY SUPERPOSITION

The method of superposition is based on the fact that the resultant effect of several loads acting on a member simultaneously is the sum of the contributions from each of the loads applied individually. The results for the separate loads are frequently available from previous work or easily determined by previous methods. In such instances, the superposition method becomes a powerful concept or tool for finding stresses, deflections, and the like. The method is applicable in all cases in which a linear relation exists between the stresses or deflections and the applied loads.

To show that beam deflections can be accurately determined by the method of superposition, consider the cantilever beam of Fig. 9-14 with loads Q, w, and P. To determine the deflection at any point of this beam by the double integration method, it is necessary to express the bending moment in terms of the applied loads. For each interval along the beam, the value of M is the algebraic sum of the moments due to the separate loads. After two successive integrations, the solution for the deflection at any point will still be the algebraic sum of the contributions from each applied load. Furthermore, for any given value of x, the relation between applied load and resulting deflection will be linear. It is evident, therefore, that the deflection of a beam is the sum of the deflections produced by the individual loads. Once the deflections produced by a few typical individual loads have been determined by one of the methods already presented, the superposition method provides a means of rapidly solving a wide variety of more complicated problems by various combinations of known results. As more data become available, a wider range of problems can be solved by superposition.

The data in Appendix B (Table B-19) are provided for use in mastering the superposition method. No attempt is made to give a large number of results because such data are readily available in various handbooks. The data given and the illustrative examples are for the purpose of making the concept and methods clear.

Example Problem 9-7

A 16 ft–long, simply supported beam carries a uniformly distributed load of 500 lb/ft and a concentrated load of 1000 lb as shown in Fig. 9-15a. The beam is rough sawn (4 in. wide by 8 in. tall, $I = 170.67$ in.^{4}) out of air-dried Douglas fir ($E = 1900$ ksi). Determine the deflection y_c at the center of the beam.

SOLUTION

The deflection at the center of the beam consists of two parts, y_1 due to the distributed load and y_2 due to the concentrated load. As shown in Fig. 9-15b, the original beam with two loads can be replaced by two beams, each carrying one of the two loads.

From case 7 of Table B-19 (Appendix B), the deflection at the center of the beam due to the distributed load, with $L = 16$ ft $= 192$ in., is given by

$$y_1 = -\frac{5wL^4}{384EI} = -\frac{5(500/12)(192)^4}{384(1.9)(10^6)(170.67)} = -2.2737 \text{ in.}$$

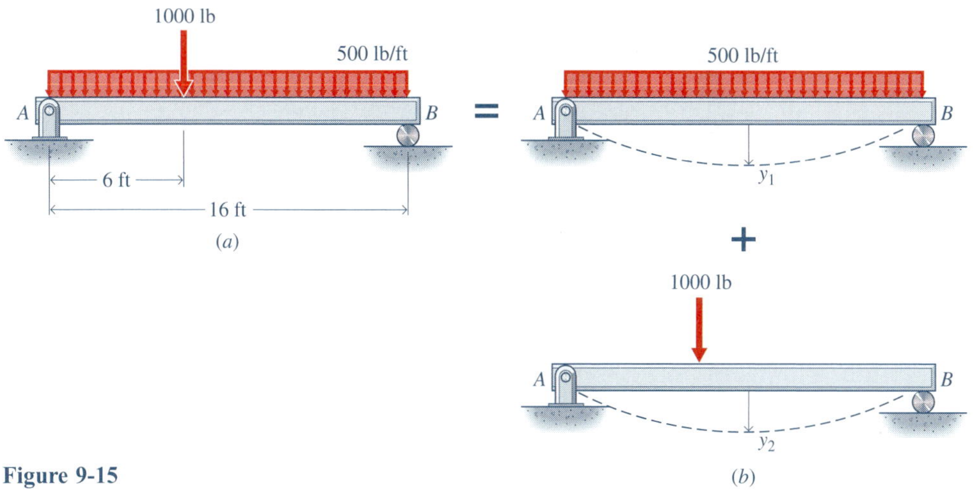

Figure 9-15

(a)   (b)

Similarly, from case 5 of Table B-19 (in which $b = 6$ ft $= 72$ in. is the shorter distance between the concentrated load and the end of the beam), the deflection at the center of the beam due to the concentrated load is given by

$$y_2 = -\frac{Pb(3L^2 - 4b^2)}{48EI} = -\frac{1000(72)[3(192)^2 - 4(72)^2]}{48(1.9)(10^6)(170.67)} = -0.4157 \text{ in.}$$

Consequently, the total deflection of the center of the beam is

$$y_c = y_1 + y_2 = (-2.2737) + (-0.4157)$$
$$= -2.689 \text{ in.} \cong 2.69 \text{ in.} \downarrow \blacksquare \qquad \textbf{Ans.}$$

Example Problem 9-8

A 5 m–long cantilever beam carries a uniformly distributed load of 7.5 kN/m and a concentrated load of 25 kN as shown in Fig. 9-16a. The steel ($E = 28$ GPa) beam is a wide-flange section [$I = 500(10^{-6})$ m^4]. Determine the deflection at the right end of the beam.

SOLUTION
As shown in Fig. 9-16b, the cantilever beam with two loads can be replaced by two beams, each carrying one of the two loads. The elastic curve for the concentrated load is shown greatly exaggerated in Fig. 9-16b. The deflection at the right end is given as $y_1 + y_2$, where y_1 is the deflection at the location of the concentrated load and y_2 is the additional deflection of the unloaded 2 m. From case 1 of Table B-19 (Appendix B),

$$y_1 = -\frac{PL^3}{3EI} = -\frac{25,000(3)^3}{3(28)(10^9)(500)(10^{-6})} = -0.016071 \text{ m}$$

From the concentrated load to the end of the beam, the slope is constant. Again from case 1 of Table B-19, the constant slope of the beam is

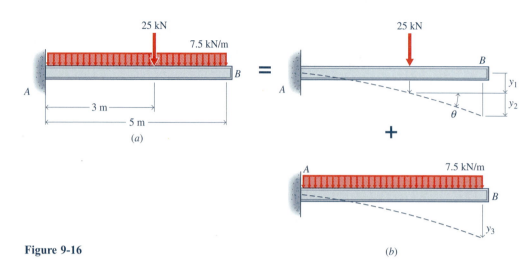

Figure 9-16

$$\theta = -\frac{PL^2}{2EI} = -\frac{25{,}000(3)^2}{2(28)(10^9)(500)(10^{-6})} = -0.008036 \text{ rad}$$

and the deflection y_2 is given by

$$y_2 = \theta L = -0.008036(2) = -0.016071 \text{ m}$$

Consequently, the total deflection at the right end of the beam due to the concentrated load is

$$y_1 + y_2 = (-0.016071) + (-0.016071) = -0.03214 \text{ m} = 32.14 \text{ mm} \downarrow$$

The elastic curve for the distributed load is also shown (greatly exaggerated) in Fig. 9-16b. From case 2 of Table B-19 (Appendix B), the deflection of the right end of the beam due to the distributed load is

$$y_3 = -\frac{wL^4}{8EI} = -\frac{7500(5)^4}{8(28)(10^9)(500)(10^{-6})} = -0.04185 \text{ m} = 41.85 \text{ mm} \downarrow$$

Finally, the total deflection of the right end of the beam due to both loads is

$$y = (y_1 + y_2) + y_3 = (-32.14) + (-41.85)$$
$$= -73.99 \text{ mm} \cong 74.0 \text{ mm} \downarrow \blacksquare \qquad \textbf{Ans.}$$

Example Problem 9-9

For the beam in Fig. 9-17a, determine the maximum deflection when E is $12(10^6)$ psi and I is 81 in.[4].

SOLUTION

From symmetry and the equilibrium equation $\Sigma F_y = 0$, the reactions are equal and 4500 lb upward. Because of the symmetrical loading, the slope of the beam is zero at the center of the span, and the right (or left) half of the beam can be

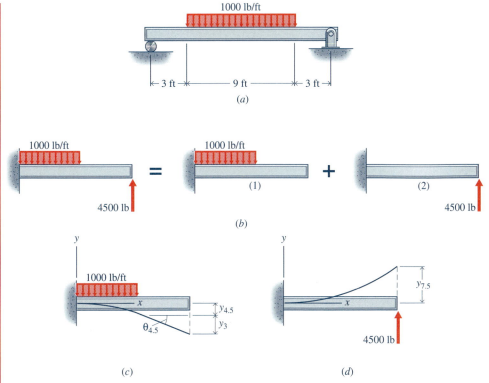

Figure 9-17

considered a cantilever beam with two loads. As shown in Fig. 9-17b, for the
right half, the cantilever with two loads can be replaced by two beams (designated 1 and 2), each carrying one of the two loads.

The elastic curve (exaggerated) for part 1, as shown in Fig. 9-17c, gives
the deflection at the right end as $y_{4.5} + y_3$, where $y_{4.5}$ is the deflection at the
end of the uniformly distributed load and y_3 is the additional deflection of the
unloaded 3 ft. From case 2 of Table B-19 of Appendix B,

$$y_{4.5} = -\frac{wL^4}{8EI} = -\frac{1000(4.5)^4(12)^3}{8(12)(10^6)(81)} = -0.09113 \text{ in.}$$

Note that in determining the deflection y, w (lb/ft) times L (ft) gives the applied
load in pounds. The remaining L^3 must be expressed in inches if E and I are in
inches and the result is to be in inches. Similarly,

$$\theta_{4.5} = -\frac{wL^3}{6EI} = -\frac{1000(4.5)^3(12)^2}{6(12)(10^6)(81)} = -0.002250 \text{ rad}$$

from which

$$y_3 = 3(12)(-0.002250) = -0.0810 \text{ in.}$$

Consequently, the total deflection of the right end is

$$y_{7.5} = y_{4.5} + y_3 = (-0.09113) + (-0.0810)$$
$$= -0.17213 \text{ in.} = 0.17213 \text{ in.} \downarrow$$

The elastic curve (exaggerated) for part 2 is shown in Fig. 9-17d. From case 1 of Table B-19 of Appendix B,

$$y_{7.5} = +\frac{PL^3}{3EI} = +\frac{4500(7.5)^3(12)^3}{3(12)(10^6)(81)} = 1.1250 \text{ in. } \uparrow$$

The algebraic sum of the deflections for parts 1 and 2 is

$$y_R = y_1 + y_2 = (-0.17213) + (1.1250)$$
$$= +0.9529 \text{ in.} \cong 0.953 \text{ in. } \uparrow$$

which means that the right end of the beam is 0.953 in. above the center. Obviously, the right end does not move, and the maximum deflection is at the center and is

$$y_{max} = 0.953 \text{ in. } \downarrow \blacksquare$$

Ans.

Example Problem 9-10

A beam is loaded and supported as shown in Fig. 9-18a. Use the method of superposition to determine the deflection, in terms of w, L, E, and I,

(a) At a point midway between the supports.

(b) At the right end of the beam.

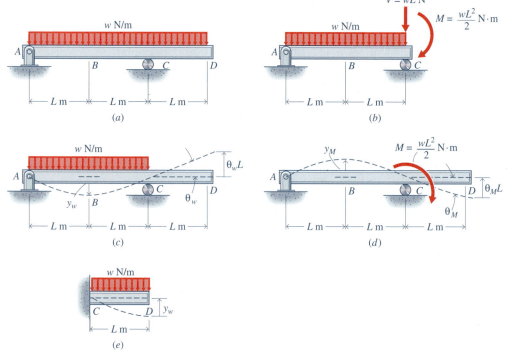

Figure 9-18

SOLUTION

(a) The deflection at a point midway between the supports is determined by using the beam shown in Fig. 9-18b. The effects of the loaded overhang on span AC of the beam can be represented by a shear force $V = wL$ and a moment $M = wL^2/2$. Since the shear force V does not contribute to the deflection at any point in span AC of the beam, the deflection at the middle of the span, as shown in Figs. 9-18c and d, can be expressed as

$$y_B = y_w + y_M$$

The deflections y_w and y_M are listed as cases 7 and 8 of Table B-19 of Appendix B, respectively. Thus,

$$y_B = -\frac{5w(2L)^4}{384EI} + \frac{(wL^2/2)(2L)^2}{16EI} = -\frac{wL^4}{12EI} = \frac{wL^4}{12EI} \downarrow \qquad \textbf{Ans.}$$

(b) The deflection at the right end of the beam is produced by the combined effects of the distributed load on the overhang and the rotation of the cross section of the beam at support C as shown in Figs. 9-18c, d, and e. Thus,

$$y_D = \theta_w L + \theta_M L + y_w$$

The angles θ_w and θ_M and the deflection y_w are listed as cases 7, 8, and 2 of Table B-19 of Appendix B, respectively. Thus,

$$y_D = \frac{w(2L)^3(L)}{24EI} - \frac{(wL^2/2)(2L)(L)}{3EI} - \frac{wL^4}{8EI} = -\frac{wL^4}{8EI} = \frac{wL^4}{8EI} \downarrow \blacksquare \quad \textbf{Ans.}$$

■ PROBLEMS

Use the method of superposition to determine deflections in the following problems.

Standard Problems

9-49* Determine the deflection at the free end of the cantilever beam of Fig. P9-49 when $w = 600$ lb/ft, $L = 10$ ft, $E = 12(10^6)$ psi, and $I = 14.4$ in.^{4}.

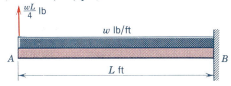

Figure P9-49

9-50* Determine the deflection at the free end of the cantilever beam shown in Fig. P9-50 in terms of P, L, E, and I.

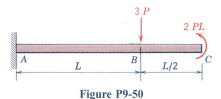

Figure P9-50

9-51 Determine the deflection at the free end of the cantilever beam shown in Fig. P9-51 when $w = 800$ lb/ft, $L = 6$ ft, $E = 30(10^6)$ psi, and $I = 100$ in.^{4}.

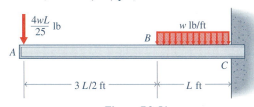

Figure P9-51

9-52 Determine the deflection at the free end of the cantilever beam shown in Fig. P9-52 in terms of w, L, E, and I.

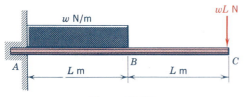

Figure P9-52

9-53* Determine the deflection midway between the supports of the beam shown in Fig. P9-53 when $P = 13.5$ kN, $L = 3$ m, $I = 80(10^6)$ mm^4, and $E = 200$ GPa.

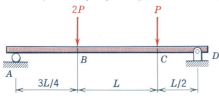

Figure P9-53

9-54* Determine the deflection at a point midway between the supports for the beam shown in Fig. P9-54 when $w = 9$ kN/m, $L = 4$ m, $E = 200$ GPa, and $I = 35(10^6)$ mm^4.

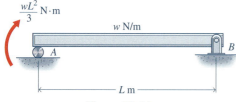

Figure P9-54

9-55 Determine the deflection at the right end of the beam shown in Fig. P9-55 in terms of w, L, E, and I.

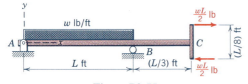

Figure P9-55

9-56 Determine the deflection at the right end of the beam shown in Fig. P9-56 in terms of w, L, E, and I.

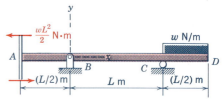

Figure P9-56

9-57* Determine the deflection at a point midway between the supports of the beam shown in Fig. P9-55 in terms of w, L, E, and I.

9-58* Determine the deflection at the left end of the beam shown in Fig. P9-56 in terms of w, L, E, and I.

9-59 Determine the deflection at a point midway between the supports of the beam shown in Fig. P9-59 in terms of w, L, E, and I.

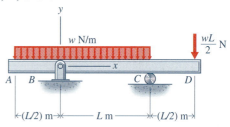

Figure P9-59

9-60 Determine the deflection at a point midway between the supports of the beam shown in Fig. P9-60 in terms of w, L, E, and I.

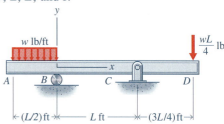

Figure P9-60

9-61* Determine the deflection at the free end of the cantilever beam shown in Fig. P9-61 when $w = 500$ lb/ft, $L = 6$ ft, $I = 305$ in.^{4}, and $E = 29,000$ ksi.

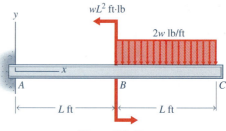

Figure P9-61

9-62 Determine the deflection at the free end of the cantilever beam shown in Fig. P9-62 when $w = 7.5$ kN/m, $L = 3$ m, $I = 180(10^6)$ mm^4, and $E = 200$ GPa.

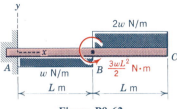

Figure P9-62

Challenging Problems

9-63* Determine the deflection at the free end of the cantilever beam shown in Fig. P9-63 in terms of w, L, E, and I.

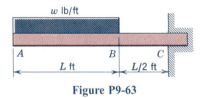

Figure P9-63

9-64* Determine the deflection at the free end of the cantilever beam shown in Fig. P9-64 in terms of w, L, E, and I.

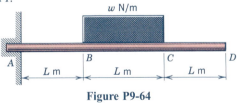

Figure P9-64

9-65 Determine the deflection at the right end of the beam shown in Fig. P9-65 in terms of P, L, E, and I.

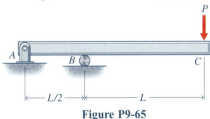

Figure P9-65

9-66 Determine the deflection at the right end of the beam shown in Fig. P9-66 in terms of w, L, E, and I.

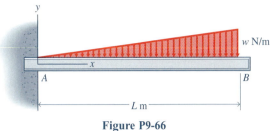

Figure P9-66

9-67* Determine the deflection at the right end of the beam shown in Fig. P9-67 in terms of w, L, E, and I.

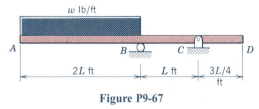

Figure P9-67

9-68* Determine the deflection at the free end of the cantilever beam shown in Fig. P9-68 in terms of w, L, E, and I.

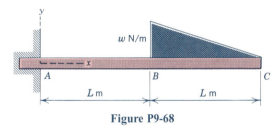

Figure P9-68

9-69 Determine the deflection at the left end of the beam shown in Fig. P9-67 in terms of w, L, E, and I.

9-70 Determine the deflection at the right end of the cantilever beam shown in Fig. P9-70 in terms of w, L, E, and I.

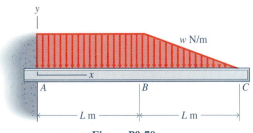

Figure P9-70

9-7 STATICALLY INDETERMINATE BEAMS

A beam, subjected only to transverse loads, with more than two reaction components, is statically indeterminate because the equations of equilibrium are not sufficient to determine all the reactions. In such cases the geometry of the deformation of the loaded beam is used to obtain the additional relations needed for an evaluation of the reactions (or other unknown forces). For problems involving elastic action, each additional constraint on a beam provides additional information concerning slopes or deflections. Such information, when used with the appropriate slope or deflection equations, yields expressions that supplement the independent equations of equilibrium.

9-7-1 The Integration Method

For statically determinate beams, known slopes and deflections were used to obtain boundary and matching conditions, from which the constants of integration in the elastic curve equation could be evaluated. For statically indeterminate beams, the procedures are identical. However, the moment equations will contain reactions or loads that cannot be evaluated from the available equations of equilibrium; one additional boundary condition is needed for the evaluation of each such unknown. For example, if a beam is subjected to a force system for which there are two independent equilibrium equations and if there are four unknown reactions or loads on the beam, two boundary or matching conditions are needed in addition to those necessary for the determination of the constants of integration. These extra boundary conditions, when substituted in the appropriate elastic curve equations (slope or deflection), will yield the necessary additional equations. The following example illustrates the method.

 Example Problem 9-11

A beam is loaded and supported as shown in Fig. 9-19a. Determine the reactions in terms of w and L.

SOLUTION

From the free-body diagram of Fig. 9-19b it is seen that there are three unknown reaction components (M, V, and R) and that only two independent equations of equilibrium are available. The additional unknown requires the use of the elastic curve equation, for which one extra boundary condition is required in addition to the two required for the constants of integration. Because three boundary conditions are available in the interval between the supports, only one elastic curve equation need be written. The origin of coordinates is arbitrarily placed at the wall and for the interval $0 \leq x \leq L$, the boundary conditions are as follows: when $x = 0$, $dy/dx = 0$; when $x = 0$, $y = 0$; and when $x = L$, $y = 0$. From Fig. 9-19c and Eq. (9-3),

$$EI\frac{d^2y}{dx^2} = M(x) = Vx + M - \frac{wx^2}{2} \qquad 0 \leq x \leq L$$

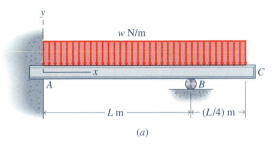

(a)

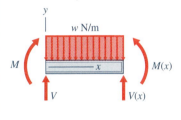

(b)

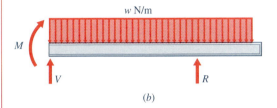

(c)

Figure 9-19

Integration gives

$$EI\frac{dy}{dx} = \frac{Vx^2}{2} + Mx - \frac{wx^3}{6} + C_1$$

The first boundary condition $dy/dx = 0$ when $x = 0$ gives $C_1 = 0$. A second integration yields

$$EIy = \frac{Vx^3}{6} + \frac{Mx^2}{2} - \frac{wx^4}{24} + C_2$$

The second boundary condition $y = 0$ when $x = 0$ gives $C_2 = 0$, and the last boundary condition $y = 0$ when $x = L$ gives

$$0 = \frac{VL^3}{6} + \frac{ML^2}{2} - \frac{wL^4}{24}$$

which reduces to

$$4VL + 12M = wL^2 \qquad\qquad (a)$$

The equation of equilibrium $\Sigma M_B = 0$ for the free-body diagram of Fig. 9-19b yields

$$+\uparrow\Sigma M_B = VL + M - w\left(\frac{5L}{4}\right)\left(\frac{3L}{8}\right) = 0$$

which reduces to

$$32VL + 32M = 15wL^2 \qquad\qquad (b)$$

Simultaneous solution of Eqs. (a) and (b) gives

$$M = -\frac{7wL^2}{64} = \frac{7wL^2}{64} \text{ N} \cdot \text{m} \downharpoonleft \qquad\qquad \textbf{Ans.}$$

and

$$V = +\frac{37wL^2}{64} = \frac{37wL^2}{64} \text{ N} \uparrow \qquad\qquad \textbf{Ans.}$$

Finally, the equation $\Sigma F_y = 0$ for Fig. 9-19b gives

$$+\uparrow\Sigma F_y = R + V - w(5L/4) = 0$$

from which

$$R = +\frac{43wL}{64} = \frac{43wL}{64} \text{ N} \uparrow \qquad\qquad \textbf{Ans.}$$

An alternate solution would be to place the origin of coordinates at the right support and write the moment equation for the interval $0 \le x \le L$. This equation would involve only one unknown, the reaction R. Upon integration and evaluation of the constants, the third boundary condition would directly yield the value of R. The two independent equilibrium equations could then be used to evaluate M and V. ■

PROBLEMS

Use the integration method (with singularity functions if needed) to solve the following problems.

Standard Problems

9-71* When the moment M is applied to the left end of the cantilever beam shown in Fig. P9-71, the slope at the left end of the beam is zero. Determine
(a) The magnitude of the moment M in terms of P and L.
(b) The deflection at the left end of the beam in terms of P, L, E, and I.

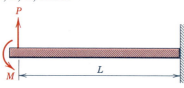

Figure P9-71

9-72* A beam is loaded and supported as shown in Fig. P9-72. Determine the magnitude of the moment M, in terms of w and L, required to
(a) Make the slope at the left end of the beam zero.
(b) Make the deflection at the left end of the beam zero.

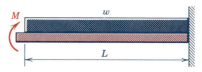

Figure P9-72

9-73 When the load P is applied to the left end of the cantilever beam shown in Fig. P9-73, the deflection at the left end of the beam is zero. Determine
(a) The magnitude of the load P in terms of w and L.
(b) The slope at the left end of the beam in terms of w, L, E, and I.

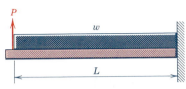

Figure P9-73

9-74 A beam is loaded and supported as shown in Fig. P9-74. Determine the magnitude of the load P, in terms of w and L, required to
(a) Make the slope at the left end of the beam zero.
(b) Make the deflection at the left end of the beam zero.

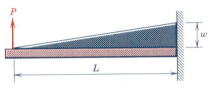

Figure P9-74

9-75* A beam is loaded and supported as shown in Fig. P9-75. Determine
(a) The reactions at supports A and B in terms of M and L.
(b) The deflection at the middle of the span in terms of M, L, E, and I.

Figure P9-75

9-76* A beam is loaded and supported as shown in Fig. P9-76. Determine
(a) The reactions at supports A and B in terms of w and L.
(b) The deflection at the middle of the span in terms of w, L, E, and I.

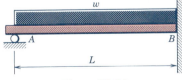

Figure P9-76

9-77 A beam is loaded and supported as shown in Fig. P9-77. Determine
(a) The reactions at supports A and B in terms of w and L.

(b) The maximum deflection in terms of w, L, E, and I.

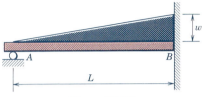

Figure P9-77

9-78 A beam is loaded and supported as shown in Fig. P9-78. Determine
(a) The reactions at supports A and B in terms of P and L.
(b) The deflection at the middle of the span in terms of P, L, E, and I.

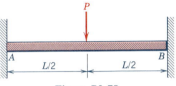

Figure P9-78

9-79* A beam is loaded and supported as shown in Fig. P9-79. Determine
(a) The reactions at supports A and B in terms of w and L.
(b) The maximum deflection in terms of w, L, E, and I.

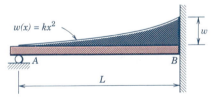

Figure P9-79

9-80* A beam is loaded and supported as shown in Fig. P9-80. Determine
(a) The reactions at supports A and B in terms of w and L.
(b) The deflection at the middle of the span in terms of w, L, E, and I.
(c) The moment at the middle of the span in terms of w and L.

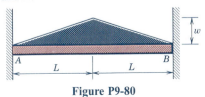

Figure P9-80

9-81 A beam is loaded and supported as shown in Fig. P9-81. Determine
(a) The reactions at supports A and B in terms of M and L.
(b) The deflection at the middle of the span in terms of M, L, E, and I.

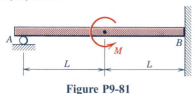

Figure P9-81

9-82 A beam is loaded and supported as shown in Fig. P9-82. Determine
(a) The reactions at supports A and B in terms of P and L.
(b) The deflection at the middle of the span in terms of P, L, E, and I.

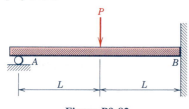

Figure P9-82

Challenging Problems

9-83* A beam is loaded and supported as shown in Fig. P9-83. Determine
(a) The reactions at supports A, B, and C in terms of w and L.
(b) The moment over the middle support in terms of w and L.
(c) The deflection at the middle of span BC in terms of w, L, E, and I.

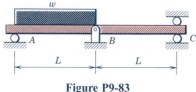

Figure P9-83

9-84* A beam is loaded and supported as shown in Fig. P9-84. Determine
(a) The reactions at supports A, B, and C in terms of w and L.

(b) The moment over the middle support in terms of w and L.
(c) The deflection at the middle of span AB in terms of w, L, E, and I.

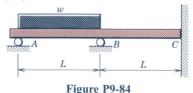

Figure P9-84

9-85 A beam is loaded and supported as shown in Fig. P9-85. Determine
(a) The reactions at supports A, B, and C in terms of w and L.
(b) The bending moment over the center support in terms of w and L.

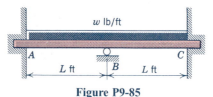

Figure P9-85

9-86 A beam is loaded and supported as shown in Fig. P9-86. Determine
(a) The reactions at supports A, B, and C in terms of w and L.
(b) The bending moment over the center support in terms of w and L.
(c) The maximum bending moment in the beam in terms of w and L.

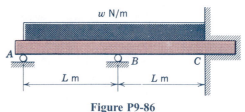

Figure P9-86

9-87* The right end of the beam shown in Fig. P9-87 is only partially fixed and under action of the load the slope becomes $wL^3/(36EI)$ upward to the right. Determine
(a) The reactions at supports A and B in terms of w and L.
(b) The maximum deflection in terms of w, L, E, and I.

(c) The maximum bending moment in the beam in terms of w and L.

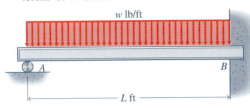

Figure P9-87

9-88* When the load P is applied to the right end of the beam shown in Fig. P9-88, the slope becomes zero over the right support. Determine
(a) The magnitude of the force P in terms of w and L.
(b) The maximum deflection between the supports in terms of w, L, E, and I.

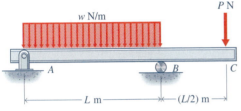

Figure P9-88

9-89 When the load P is applied to the right end of the beam shown in Fig. P9-89, the slope of the beam over the right support becomes $wL^3/(72EI)$ downward to the right. Determine
(a) The magnitude of the force P in terms of w and L.
(b) The reactions at supports A and B in terms of w and L.
(c) The deflection midway between the supports in terms of w, L, E, and I.

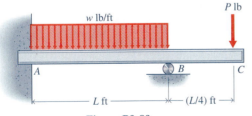

Figure P9-89

9-90 A beam is loaded and supported as shown in Fig. P9-90. Determine
(a) The reactions at supports A and B.
(b) The deflection at C if $E = 200$ GPa and $I = 150(10^6)$ mm^4.

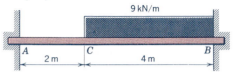

Figure P9-90

9-91* A beam is loaded and supported as shown in Fig. P9-91. Determine
(a) The reactions at supports A, B, and D.
(b) The deflection at C if $E = 30,000$ ksi and $I = 26$ in.^{4}.

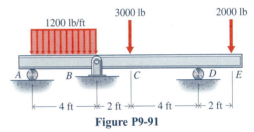

Figure P9-91

9-92 A beam is loaded and supported as shown in Fig. P9-92. Determine
(a) The reactions at supports A and D.
(b) The deflection at B if $E = 200$ GPa and $I = 350(10^6)$ mm^4.

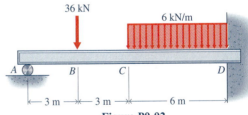

Figure P9-92

9-7-2 The Superposition Method

The concept (discussed in Section 9-6) that a slope or deflection due to several loads is the algebraic sum of the slopes or deflections due to each of the loads acting individually is frequently used to provide the deformation equations needed to supplement the equilibrium equations in the solution of statically indeterminate beam problems. To provide the necessary deformation equations, selected restraints are removed and replaced by unknown loads (forces and couples); the deformation diagrams corresponding to individual loads (both known and unknown) are sketched; and the component deflections or slopes are summed to produce the known configuration. The following examples illustrate the use of superposition for this purpose.

Example Problem 9-12

A steel (E = 30,000 ksi) beam 20 ft long is simply supported at the ends and at the midpoint. Under a load of 400 lb/ft, uniformly distributed over the entire beam, the center support is 0.12 in. above the end supports. Determine the reactions. The moment of inertia of the cross section with respect to the neutral axis is 100 in.^{4}.

SOLUTION

With three unknown reactions and only two equations of equilibrium available, the beam is statically indeterminate. Replace the center support with an unknown applied load. The resulting simply supported beam is equivalent to two beams with individual loads as shown in Fig. 9-20. The resulting deflection at the midpoint of the beam is the deflection y_R due to R_C, plus the deflection y_w due to the uniform load; that is,

$$y = y_R + y_w = 0.12 \text{ in.} \tag{a}$$

The deflection y_w can be obtained from case 7 of Table B-19 in Appendix B, and is

$$y_w = -\frac{5wL^4}{384EI} = -\frac{5[400(20)][20(12)]^3}{384(30)(10^6)(100)} = 0.4800 \text{ in.} \downarrow$$

The deflection y_R can be obtained in terms of R_C from case 6 of Table B-19 in Appendix B, and is

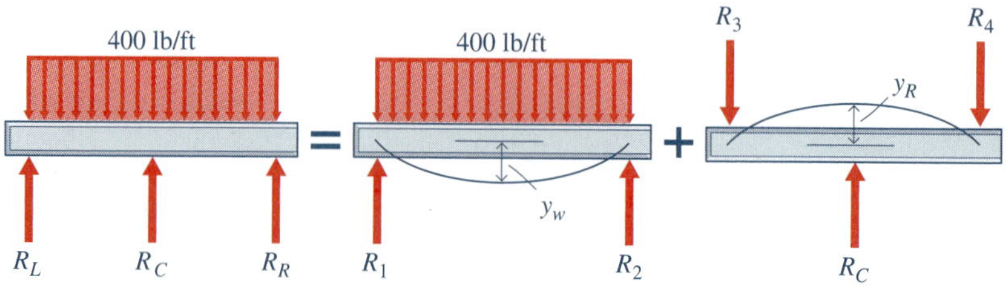

Figure 9-20

$$y_R = \frac{R_C L^3}{48EI} = \frac{R_C[20(12)]^3}{48(30)(10^6)(100)} = 96.00(10^{-6})R_C \text{ in. } \uparrow$$

When these values are substituted in Eq. (a), the result is

$$y = 96.00(10^{-6})R_C - 0.4800 = 0.12$$

from which

$$R_C = 6250 \text{ lb } \uparrow \qquad\qquad \textbf{Ans.}$$

The equilibrium equation $\Sigma F_y = 0$ and symmetry give

$$R_L = R_R = \frac{1}{2}[400(20) - 6250] = 875 \text{ lb } \uparrow \qquad\qquad \textbf{Ans.}$$

Note that in determining y_w, w (lb/ft) times L (ft) gives the applied load in pounds. The remaining L^3 must be expressed in inches if E and I are in inches and the result is to be in inches. The arithmetic will frequently be simplified if the expressions for the deflections are substituted in the deflection equation in symbol form. In this example Eq. (a) becomes

$$\frac{R_C L^3}{48EI} - \frac{5wL^4}{384EI} = 0.12$$

which reduces (when multiplied by $48EI/L^3$) to

$$R_C - \frac{5wL}{8} = \frac{0.12(48EI)}{L^3}$$

or

$$R_C = \frac{5(400)(20)}{8} + \frac{0.12(48)(30)(10^6)(100)}{[20(12)]^3} = 6250 \text{ lb } \uparrow \quad \blacksquare$$

Example Problem 9-13

A beam is loaded and supported as shown in Fig. 9-21a. Determine the reactions at supports A and B in terms of P, L, and a.

SOLUTION

There are four unknown reactions (a shear and moment at each end), and only two equations of equilibrium are available; therefore, the beam is statically indeterminate, and two deformation equations are necessary. The constraint at the right end can be replaced with an unknown force and couple. The resulting cantilever beam is equivalent to three beams with individual loads, as shown in Fig. 9-21b. Note that the unknown shear and moment at the right end are both shown as positive values so that the algebraic sign of the result will be correct. From

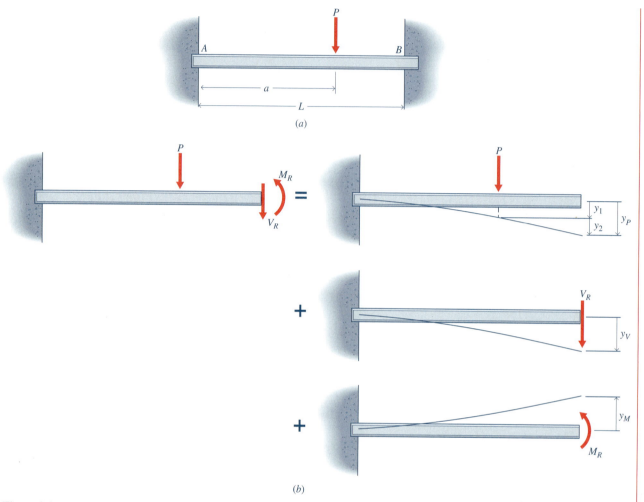

Figure 9-21

the geometry of the constrained beam, the resultant slope and the resultant deflection at the right end are both zero. The slope and deflection at the end of each of the three replacement beams can be obtained from the expressions in Table B-19 of Appendix B. Thus, the first beam with load P (see case 1 of Table B-19) has a constant slope from P to the end of the beam, which is

$$\theta_P = -\frac{Pa^2}{2EI}$$

The deflection y_P at the end is made up of two parts: y_1 for a beam of length a, and y_2 the added deflection of the tangent segment (straight line) from P to the end of the beam. This deflection is

$$y_P = y_1 + y_2 = -\frac{Pa^3}{3EI} + (L - a)\theta_P$$

$$= -\frac{Pa^3}{3EI} + (L - a)\left(-\frac{Pa^2}{2EI}\right) = \frac{Pa^3}{6EI} - \frac{Pa^2L}{2EI}$$

The slope and deflection at the end of the beam due to the shear V_R (also from case 1 of Table B-19) are

$$\theta_V = -\frac{V_R L^2}{2EI} \quad \text{and} \quad y_V = -\frac{V_R L^3}{3EI}$$

Finally, the slope and deflection at the right end of the beam due to M_R (see case 4 of Table B-19) are

$$\theta_M = \frac{M_R L}{EI} \quad \text{and} \quad y_M = \frac{M_R L^2}{2EI}$$

Since the resultant slope is zero,

$$\theta_P + \theta_V + \theta_M = -\frac{Pa^2}{2EI} - \frac{V_R L^2}{2EI} + \frac{M_R L}{EI} = 0$$

Similarly

$$y_P + y_V + y_M = \frac{Pa^3}{6EI} - \frac{Pa^2 L}{2EI} - \frac{V_R L^3}{3EI} + \frac{M_R L^2}{2EI} = 0$$

The simultaneous solution of these two equations yields

$$M_R = -\frac{Pa^2(L-a)}{L^2} \quad \text{and} \quad V_R = -\frac{Pa^2(3L-2a)}{L^3} \quad \textbf{Ans.}$$

When the equations of equilibrium are applied to a free-body diagram of the entire beam, the shear and moment at the left end are found to be

$$M_L = -\frac{Pa(L-a)^2}{L^2} \quad \text{and} \quad V_L = +\frac{P(L^3 - 3a^2 L + 2a^3)}{L^3} \quad \blacksquare \quad \textbf{Ans.}$$

PROBLEMS

Use the method of superposition to solve the following problems

Standard Problems

9-93* A beam is loaded and supported as shown in Fig. P9-93. Determine the magnitude of the moment M, in terms of w and L, required to make
(a) The slope at the left end of the beam zero.
(b) The deflection at the left end of the beam zero.

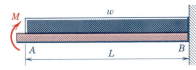

Figure P9-93

9-94* A beam is loaded and supported as shown in Fig. P9-94. Determine the magnitude of the load P, in terms of w and L, required to make
(a) The slope at the left end of the beam zero.
(b) The deflection at the left end of the beam zero.

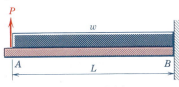

Figure P9-94

9-95 A beam is loaded and supported as shown in Fig. P9-95. Determine
(a) The reactions at supports A, B, and C in terms of w and L.
(b) The deflection at the middle of span AB in terms of w, L, E, and I.

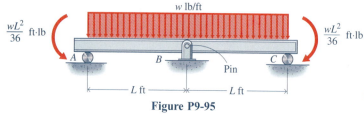

Figure P9-95

9-96 A beam is loaded and supported as shown in Fig. P9-96. Determine
(a) The reactions at supports A and C in terms of w and L.
(b) The deflection at the left end of the distributed load in terms of w, L, E, and I.

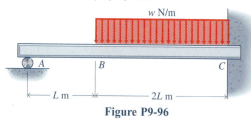

Figure P9-96

9-97* A beam is loaded and supported as shown in Fig. P9-97. Determine
(a) The reactions at supports A and D in terms of w and L.
(b) The deflection at the left end of the distributed load in terms of w, L, E, and I.

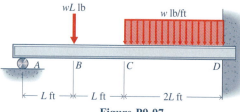

Figure P9-97

9-98* A beam is loaded and supported as shown in Fig. P9-98. Determine
(a) The reactions at supports A and B in terms of w and L.
(b) The deflection at the left end of the distributed load in terms of w, L, E, and I.

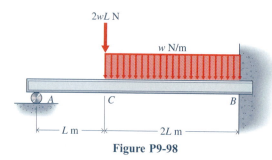

Figure P9-98

9-99 A beam is loaded and supported as shown in Fig. P9-99. When the load P is applied, the slope at the right end of the beam is zero. Determine
(a) The value of P in terms of w and L.
(b) The reactions at supports A and B in terms of w and L.

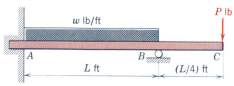

Figure P9-99

9-100 A beam is loaded and supported as shown in Fig. P9-100. Determine the magnitude of the moment M, in terms of P and L, required to make
(a) The slope at the right end of the beam zero.
(b) The deflection at the right end of the beam zero.

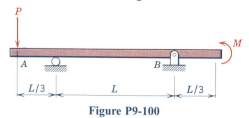

Figure P9-100

9-101* A beam is loaded and supported as shown in Fig. P9-101. Determine the reactions at supports A and B in terms of P and L.

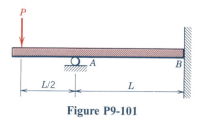

Figure P9-101

9-102* A beam is loaded and supported as shown in Fig. P9-102. Determine the reactions at supports A and B in terms of w and L.

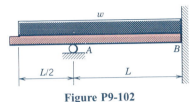

Figure P9-102

9-103 A beam is loaded and supported as shown in Fig. P9-103. Determine
(a) The reactions at supports A and B in terms of P and L.
(b) The deflection at the middle of the span in terms of P, L, E, and I.

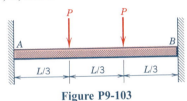

Figure P9-103

9-104 A beam is loaded and supported as shown in Fig. P9-104. Determine
(a) The reactions at supports A and B in terms of w and L.
(b) The deflection at the middle of the span in terms of w, L, E, and I.

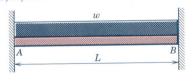

Figure P9-104

Challenging Problems

9-105* A beam is loaded and supported as shown in Fig. P9-105. Determine the reactions at supports A and C in terms of P and L.

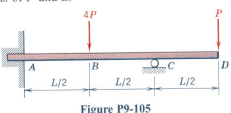

Figure P9-105

9-106* A beam is loaded and supported as shown in Fig. P9-106. Determine the reactions at supports A, B, and C in terms of P and L.

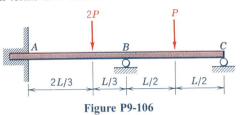

Figure P9-106

9-107 A beam is loaded and supported as shown in Fig. P9-107. Determine
(a) The reactions at supports A and C in terms of w and L.
(b) The deflection at the right end of the distributed load in terms of w, L, E, and I.

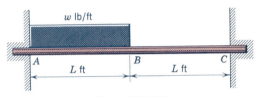

Figure P9-107

9-108 A beam is loaded and supported as shown in Fig. P9-108. Determine
(a) The reactions at supports A, B, and C in terms of w and L.
(b) The slope of the beam over support C in terms of w, L, E, and I.

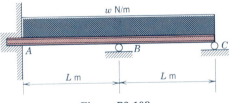

Figure P9-108

9-109* A timber ($E = 1800$ ksi) beam is loaded and supported as shown in Fig. P9-109. The tension in the $\frac{1}{2}$ in.–diameter rod BD is zero before the load is applied to the beam. Determine the reaction at B if
(a) The tie rod BD is made of steel ($E = 30{,}000$ ksi).
(b) The tie rod BD is made of aluminum alloy ($E = 10{,}000$ ksi).

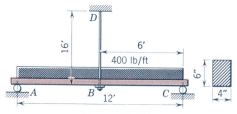

Figure P9-109

9-110* A steel [$E = 200$ GPa and $I = 50(10^6)$ mm⁴] beam is loaded and supported as shown in Fig. P9-110. The post BD is a 150×150-mm timber ($E = 10$ GPa) that is braced to prevent buckling. Determine the load carried by the post if it is unstressed before the 8-kN/m distributed load is applied.

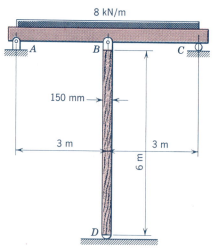

Figure P9-110

9-111 The 4-in. wide by 6-in. deep timber ($E = 1200$ ksi) beam shown in Fig. P9-111 is fixed at the left end and supported at the right end with a tie rod that has a cross-sectional area of 0.125 in.². Determine the tension in

the tie rod if it is unstressed before the load is applied to the timber beam and

(a) The tie rod is made of steel ($E = 30,000$ ksi).
(b) The tie rod is made of aluminum alloy ($E = 10,000$ ksi).

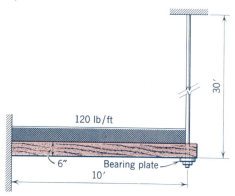

Figure P9-111

9-112 The steel ($E = 200$ GPa) beam AB of Fig. P9-112 is fixed at ends A and B and supported at the center by the pin-connected timber ($E = 10$ GPa) struts CD and CE. The cross-sectional area of each strut is 6400 mm² and the moment of inertia for the beam is $25(10^6)$ mm⁴. Determine the force in each strut after the 6-kN/m distributed load is applied to the beam.

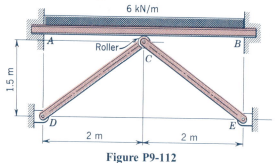

Figure P9-112

9-8 DESIGN PROBLEMS

In Chapter 8, when the design of beams was discussed, either flexural strength or shear strength was the controlling parameter. In this chapter an additional parameter, deflection, will be introduced. Thus, the design of a beam may be based on flexural stress, shearing stress, or deflection. The design procedure is similar to that presented in Chapter 8. The beam is first designed based on flexural stress, and then checked for shearing stress and deflection. If the shearing stress and deflection are within allowable limits the design is satisfactory. If the shearing stress or deflection is greater than the allowable value, the beam must be redesigned until all of the allowable limits are satisfied. Clearly, this is a trial-and-error process. The following examples illustrate the procedures for designing beams where allowable limits are given for flexural stress, shearing stress, and deflection.

Example Problem 9-14

The boards for a concrete form are to be bent to a circular curve of 15-ft radius. What maximum thickness of board can be used if the stress in the board is not to exceed 2000 psi? The modulus of elasticity for the wood is 1400 ksi.

SOLUTION

If the board is subjected to a pure moment form of loading (couples applied at the ends), the elastic curve will be an arc of a circle of radius ρ. Under these conditions Eq. (9-1) provides a relationship between the stress σ in the board and the radius of curvature ρ, the thickness t, and the modulus of elasticity E of the board as

$$\frac{c}{\rho} = \frac{t/2}{\rho} = \frac{\sigma}{E}$$

Thus,

$$t = \frac{2\rho\sigma}{E} = \frac{2(15)(12)(2000)}{1,400,000} = 0.514 \text{ in.}$$

A $\frac{1}{2}$ in.–thick board could be used for the form without exceeding a stress of 2000 psi. ■

Example Problem 9-15

An air-dried Douglas fir timber beam is loaded as shown in Fig. 9-22. If the allowable flexural stress is 8 MPa, the allowable shearing stress is 0.7 MPa, and the allowable deflection is 14 mm, determine the lightest weight standard structural timber that can be used for the beam.

SOLUTION

Load (free-body), shear force, and bending moment diagrams for the beam of Fig. 9-22 are shown in Figs. 9-23a, b, and c, respectively. Since the load is uniformly distributed over the full length of the beam,

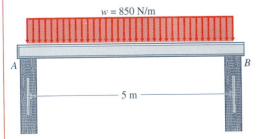

Figure 9-22

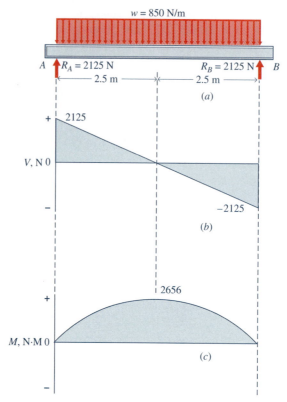

Figure 9-23

$$R_A = R_B = \frac{1}{2}(850)(5) = 2125 \text{ N}$$

$$V_{max} = 2125 \text{ N}$$

$$M_{max} = \frac{1}{2}(2125)(2.5) = 2656 \text{ N} \cdot \text{m}$$

The minimum section modulus needed to satisfy the allowable value of the flexural stress is given by Eq. (8-12) as

$$S = \frac{M_{max}}{\sigma_{all}} = \frac{2656}{8(10^6)} = 332.0(10^{-6}) \text{ m}^3 = 332.0(10^3) \text{ mm}^3$$

The lightest weight standard structural timber listed in Table B-16 with $S \geq 332.0(10^3)$ mm^3 is a timber with nominal dimensions of 51×254 mm. Some properties of this timber that will be needed later are

$$\text{Mass/unit length} = 6.38 \text{ kg/m}$$

$$\text{Area} = 9.88(10^3) \text{ mm}^2 = 9.88(10^{-3}) \text{ m}^2$$

$$I = 48.3(10^6) \text{ mm}^4 = 48.3(10^{-6}) \text{ m}^4$$

$$S = 400(10^3) \text{ mm}^3 = 400(10^{-6}) \text{ m}^3$$

$$E = 13 \text{ GPa (see Table B-18)}$$

Since the cross section is rectangular, the maximum shearing stress is

$$\tau_{max} = 1.5\frac{V_{max}}{A} = 1.5\frac{2125}{9.88(10^{-3})} = 0.3226(10^6) \text{ N/m}^2 = 0.3226 \text{ MPa} < 0.7 \text{ MPa}$$

Thus, the shearing stress requirement is satisfied.

For a simply supported beam with a uniformly distributed load, case 7 of Table B-19 gives the maximum deflection as

$$|y_{max}| = \frac{5wL^4}{384\,EI} = \frac{5(850)(5)^4}{384(13)(10^9)(48.3)(10^{-6})}$$

$$= 11.017(10^{-3}) \text{ m} = 11.017 \text{ mm} < 14 \text{ mm}$$

Therefore, the deflection requirement is satisfied. The 51 × 254-mm standard structural timber satisfies the requirements for flexural stress, shearing stress, and deflection; however, the analysis thus far neglected the weight of the beam. The timber beam weighs $(6.38)(9.81) = 62.59$ N/m. Adding this uniformly distributed load to the applied load gives a uniformly distributed load $w = 850 + 62.59 = 912.6$ N/m. For this loading, the maximum shear force is 2282 N and the maximum bending moment is 2852 N · m. The section modulus needed to satisfy the flexural stress requirement is

$$S = \frac{M_{max}}{\sigma_{all}} = \frac{2852}{8(10^6)} = 356.5(10^{-6}) \text{ m}^3 = 356.5(10^3) \text{ mm}^3 < 400(10^3) \text{ mm}^3$$

Thus, the 51 × 254-mm timber satisfies the flexural stress requirement. The maximum shearing stress and the maximum deflection with the beam weight included are

$$\tau_{max} = 1.5\frac{V_{max}}{A} = 1.5\frac{2282}{9.88(10^{-3})}$$

$$= 0.3465(10^6) \text{ N/m}^2 = 0.3465 \text{ MPa} < 0.7 \text{ MPa}$$

$$|y_{max}| = \frac{5wL^4}{384\,EI} = \frac{5(912.6)(5)^4}{384(13)(10^9)(48.3)(10^{-6})}$$

$$= 11.828(10^{-3}) \text{ m} = 11.828 \text{ mm} < 14 \text{ mm}$$

Thus, the 51 × 254-mm standard structural timber satisfies all requirements with the weight of the beam included. ■

Example Problem 9-16

A structural steel beam is loaded as shown in Fig. 9-24. If the allowable flexural stress is 24,000 psi, the allowable shearing stress is 14,000 psi, and the allowable deflection midway between supports A and B is 0.5 in., determine the lightest American Standard Section (S-shape) that can be used for the beam.

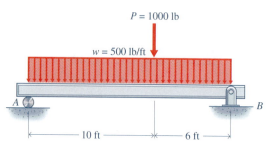

Figure 9-24

SOLUTION

Load (free-body), shear force, and bending moment diagrams for the beam of Fig. 9-24 are shown in Figs. 9-25a, b, and c, respectively. The moment equation $\Sigma M_B = 0$ yields

$$+\curvearrowright \Sigma M_B = R_A(16) - 500(16)(8) - 1000(6) = 0 \qquad R_A = 4375 \text{ lb}$$
$$+\uparrow \Sigma F_y = 4375 - 500(16) - 1000 + R_B = 0 \qquad R_B = 4625 \text{ lb}$$

From the shear and bending diagrams,

$$V_{max} = 4625 \text{ lb}$$

$$M_{max} = \frac{1}{2}(4375)(8.75) = 19{,}141 \text{ ft} \cdot \text{lb}$$

The minimum section modulus needed to satisfy the allowable value of the flexural stress is given by Eq. (8-12) as

$$S = \frac{M_{max}}{\sigma_{all}} = \frac{19{,}141(12)}{24{,}000} = 9.571 \text{ in.}^3$$

The lightest weight American Standard beam (S-shape) listed in Table B-3 with $S \geq 9.571$ in.^{3} is an S7 × 15.3 section. For this section

$$S = 10.5 \text{ in.}^3 \qquad\qquad I = 36.7 \text{ in.}^4$$

$$d = 7.00 \text{ in.} \qquad\qquad t_{web} = 0.252 \text{ in.}$$

$$E = 29{,}000 \text{ ksi}$$

The average value of the shearing stress in the web is

$$\tau_{avg} = \frac{V_{max}}{A_{web}} = \frac{4625}{7.00(0.252)} = 2622 \text{ psi} << 14{,}000 \text{ psi}$$

Thus, the shearing stress requirement is satisfied since the maximum shearing stress in the web of an American Standard beam is only slightly larger than the average shearing stress.

The deflection at midspan is found by using the method of superposition and Table B-19. The given loading is equivalent to the two loads, parts 1 and

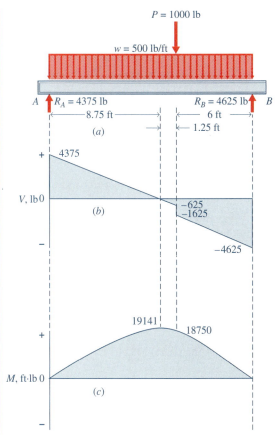

Figure 9-25

2, shown in Fig. 9-26b. For part 1, the midspan deflection is given as case 7 of Table B-19. It is

$$y_1 = -\frac{5wL^4}{384\,EI} = -\frac{5(500/12)[16(12)]^4}{384(29)(10^6)(36.7)} = -0.6927 \text{ in.}$$

For part 2, the midspan deflection is given as case 5 of Table B-19. It is

$$y_2 = -\frac{Pb(3L^2 - 4b^2)}{48\,EI} = -\frac{1000(72)[3(192)^2 - 4(72)^2]}{48(29)(10^6)(36.7)} = -0.12664 \text{ in.}$$

The midspan deflection is the algebraic sum of y_1 and y_2 and is

$$|y_{\text{midspan}}| = 0.6927 + 0.12664 = 0.8193 \text{ in.} > 0.50 \text{ in.}$$

Since $|y_{\text{midspan}}|$ is greater than $y_{\text{allowable}}$, a new section must be selected with sufficient I to satisfy the deflection requirement. Thus,

$$|y_{\text{midspan}}| = \frac{5wL^4}{384\,EI} + \frac{Pb(3L^2 - 4b^2)}{48\,EI}$$

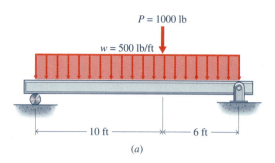

(a)

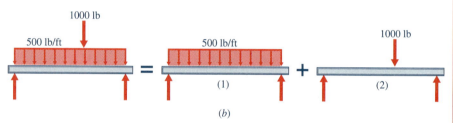

(b)

Figure 9-26

Solving for I yields

$$I = \frac{5wL^4}{384E|y_{mid}|} + \frac{Pb(3L^2 - 4b^2)}{48E|y_{mid}|}$$

$$= \frac{5(500/12)(192)^4}{384(29)(10^6)(0.5)} + \frac{1000(72)[3(192)^2 - 4(72)^2]}{48(29)(10^6)(0.5)} = 60.14 \text{ in.}^4$$

The lightest S-shape in Table B-3 with $I \geq 60.14$ in.^{4} is an S8 $\times$ 23 beam with $I = 64.9$ in.^{4} and $S = 16.2$ in.^{4}. The S8 $\times$ 23 beam satisfies the requirements for flexural stress, shearing stress, and deflection.

Consider now the effect of the weight of the beam on the deflection. The S8 $\times$ 23 beam weighs 23 lb/ft. Adding the weight of the beam to the 500 lb/ft applied distributed load gives a uniformly distributed load $w = 500 + 23 = 523$ lb/ft, resulting in maximum values of shear force and bending moment of 4809 lb and 19,870 ft · lb, respectively. The value of I required as a result of this increase in load is

$$I = \frac{5wL^4}{384E|y_{mid}|} + \frac{Pb(3L^2 - 4b^2)}{48E|y_{mid}|}$$

$$= \frac{5(523/12)(192)^4}{384(29)(10^6)(0.5)} + \frac{1000(72)[3(192)^2 - 4(72)^2]}{48(29)(10^6)(0.5)} = 62.49 \text{ in.}^4$$

For the S8 $\times$ 23 beam, $I = 64.9$ in.$^4 > 62.49$ in.^{4}; therefore, the addition of the weight of the beam does not change the selection of the beam. An S8 $\times$ 23 beam satisfies all requirements for flexural stress, shearing stress, and deflection. ■

PROBLEMS

Standard Problems

9-113 An air-dried Douglas fir beam is simply supported and has a span of 16 ft. The beam is subjected to a uniformly distributed load of 800 lb/ft over its entire length. If the allowable flexural stress is 1200 psi, the allowable shearing stress is 90 psi, and the allowable deflection at the middle of the span is $\frac{1}{2}$ in., select the lightest standard structural timber that can be used to support the load.

9-114 A 3-m simply supported beam is loaded with a uniformly distributed load of 2.6 kN/m over its entire length. The beam is made of air-dried Douglas fir with an allowable flexural stress of 8 MPa and an allowable shearing stress of 0.7 MPa. The maximum deflection at the center of the span must not exceed 10 mm. Select the lightest standard structural timber that can be used for the beam.

9-115 A portion of a pedestrian walkway along the side of a bridge is shown in Fig. P9-115. Cantilever beams support the loading, one of which is shown. The beams are select structural eastern Hemlock with allowable flexural and shearing stresses of 1300 psi and 80 psi, respectively. The allowable deflection is 0.2 in. The modulus of elasticity is $1.2(10^6)$ psi. Select the lightest standard structural timber that can be used for the beams.

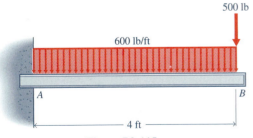

Figure P9-115

9-116 A standard structural steel pipe is to support the load shown in Fig. P9-116. The allowable flexural stress and deflection are 150 MPa and 5 mm, respectively. Select the lightest permissible standard steel pipe that can be used to support the load. Neglect the effects of shear.

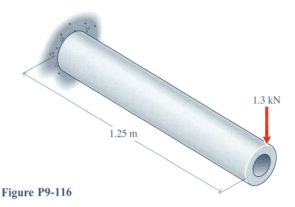

Figure P9-116

9-117 A simply supported beam has a span of 24 ft and carries a uniformly distributed load of 1200 lb/ft. The steel beam has an allowable flexural stress of 24 ksi, an allowable shearing stress of 14 ksi, and an allowable deflection of 1/360 of the span. Select the lightest American Standard beam that can be used to support the loading.

9-118 A solid circular shaft made of ASTM A36 steel is supported by bearings spaced 1.5 m apart. The shaft is to support a 4-kN load perpendicular to the shaft; the load may be placed at any point between the bearings. The allowable flexural stress is 152 MPa, the allowable shearing stress is 100 MPa, and the allowable deflection is 5 mm. If shafts are available with diameters in increments of 5 mm, determine the smallest diameter shaft that can be used to support the load. Neglect the weight of the shaft.

Challenging Problems

9-119 The simply supported beam shown in Fig. P9-119 is made of Douglas fir with an allowable flexural stress of 1900 psi, an allowable shearing stress of 85 psi, and an allowable deflection of 1/360 of the span. Select the lightest standard structural timber that can be used to support the loads shown in the figure.

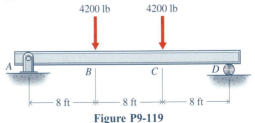

Figure P9-119

9-120 The simply supported steel beam shown in Fig. P9-120 has a modulus of elasticity of 200 GPa, an allowable flexural stress of 165 MPa, an allowable shearing stress of 100 MPa, and an allowable deflection of 1/360 of the span. Select the lightest wide-flange beam that can be used to support the loading shown in the figure.

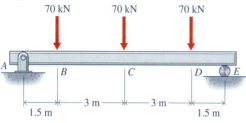

Figure P9-120

9-9 SUMMARY

A beam design is frequently not complete until the amount of deflection has been determined for the specified load. The deflection of a beam depends on the stiffness of the material and the dimensions of the beam as well as on the applied loads and type of supports.

When a straight beam is loaded and the action is elastic, the centroidal axis of the beam is a curve defined as the elastic curve. In regions of constant bending moment, the elastic curve is an arc of a circle of radius ρ. Since the beam is bent only with couples, plane sections remain plane and the deformation of the fibers (elongation and compression) is proportional to the distance from the neutral surface, which is unchanged in length. Thus

$$\frac{c}{\rho} = \frac{\delta}{L} = \epsilon = \frac{\sigma}{E} = \frac{Mc}{EI} \tag{9-1}$$

Therefore,

$$\frac{1}{\rho} = \frac{M}{EI} \tag{9-2}$$

which relates the radius of curvature of the neutral surface of the beam to the bending moment M, the stiffness of the material E, and the second moment I of the cross-sectional area of the beam with respect to the neutral axis.

Equation (9-2) for the curvature of the elastic curve is useful only when the bending moment is constant for the interval of the beam involved. For most beams, the bending moment is a function of position along the beam, and a more general expression is required. The curvature from the calculus (see any standard calculus textbook) is

$$\frac{1}{\rho} = \frac{d^2y/dx^2}{[1 + (dy/dx)^2]^{3/2}}$$

For most beams the slope dy/dx is very small, and its square can be neglected in comparison to unity. With this approximation,

$$\frac{1}{\rho} = \frac{d^2y}{dx^2}$$

and Eq. (9-2) becomes

$$EI\frac{d^2y}{dx^2} = M(x) \qquad\qquad (9\text{-}3)$$

which is the differential equation for the elastic curve of a beam where the moment M is a function of x.

    Whenever the bending moment can be readily expressed as an integrable function of x, Eq. (9-3) can be solved for the deflection y of the elastic curve of a beam at any point x along the beam. The constants of integration are evaluated from the applicable boundary conditions.

    The double integration method becomes tedious and time-consuming when several intervals and several sets of matching conditions are required. The labor involved in solving problems of this type, however, can be diminished by making use of singularity functions. Singularity functions are used to write one bending moment equation that applies in all intervals along a beam, thus eliminating the need for matching conditions.

    The method of superposition for determining beam deflections is based on the fact that the resultant effect of several loads acting simultaneously on a member is the sum of the contributions from each of the loads applied individually. The results for the separate loads are listed in tables in Appendix B.

    A beam, subjected only to transverse loads, with more than two reaction components is statically indeterminate because the equations of equilibrium are not sufficient to determine all the reactions. The additional relations needed for an evaluation of reactions (or other unknown forces) are obtained from deformation (slope or deflection) equations.

REVIEW PROBLEMS

9-121* A strip of hardened steel ($E = 30{,}000$ ksi) with a rectangular cross section $\frac{3}{4}$ in. wide by $\frac{1}{64}$ in. thick is bent to form a circular ring. If the maximum flexural stress in the strip must not exceed 40,000 psi, determine the minimum allowable diameter for the ring.

9-122* An aluminum alloy ($E = 73$ GPa) cantilever beam with a rectangular cross section has a moment of 2 kN · m applied at the free end. If the radius of curvature of the beam must be 12 m when the maximum flexural stress is 140 MPa, determine
(a) The required second moment of area for the cross section.
(b) The required width and thickness for the beam.

9-123 A beam is loaded and supported as shown in Fig. P9-123.
(a) Derive the equation of the elastic curve in terms of E, I, w, and L. Use the designated axes.
(b) Determine the deflection at the left end of the beam.

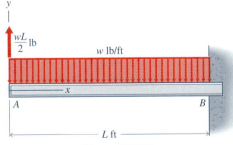

Figure P9-123

9-124 A beam is loaded and supported as shown in Fig. P9-124.
   (a) Derive the equation of the elastic curve in terms of E, I, w, and L for the interval of the beam between the supports. Use the designated axes.
   (b) Determine the deflection midway between the supports.

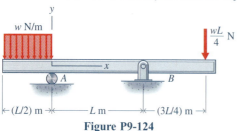

Figure P9-124

9-125* A timber beam 6 in. wide by 12 in. deep is loaded at midspan and supported as shown in Fig. P9-125. The modulus of elasticity of the timber is 1200 ksi. A pointer is attached to the right end of the beam. Determine
   (a) The deflection of the right end of the pointer.
   (b) The maximum deflection of the beam.

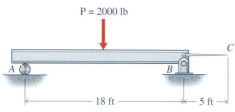

Figure P9-125

9-126* A beam is loaded and supported as shown in Fig. P9-126. The bending moment at the second support is $-wL^2/12$ and the vertical shear just to the right of the second support is $+5wL/12$. Determine
   (a) The equation of the elastic curve in terms of E, I, w, and L for the portion of the beam between the second and third supports. Use the designated axes.
   (b) The maximum deflection between the second and third supports.

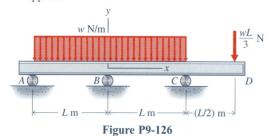

Figure P9-126

9-127 A beam is loaded and supported as shown in Fig. P9-127. Determine
   (a) The deflection at the left end of the distributed load in terms of w, L, E, and I.

(b) The deflection at the free end of the beam in terms of w, L, E, and I.

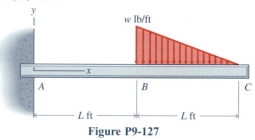

Figure P9-127

9-128 A beam is loaded and supported as shown in Fig. P9-128. Determine
   (a) The slope at the left support in terms of w, L, E, and I.
   (b) The deflection at a point midway between the supports in terms of w, L, E, and I.

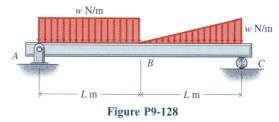

Figure P9-128

9-129* The left support of the beam shown in Fig. P9-129 settles under action of the load and the deflection becomes $wL^4/(24EI)$ downward. Determine
   (a) The reaction at the left end of the beam.
   (b) The deflection at a point midway between the supports in terms of w, L, E, and I.
   (c) The maximum bending moment in the beam.

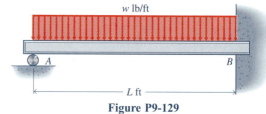

Figure P9-129

9-130* A beam is loaded and supported as shown in Fig. P9-130. Determine
   (a) All of the reactions in terms of w and L.
   (b) The deflection midway between the two right supports in terms of w, L, E, and I.

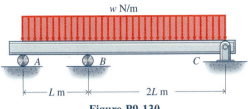

Figure P9-130

9-131 A beam is loaded and supported as shown in Fig. P9-131. Determine
(a) All of the reactions.
(b) The bending moment over the center support.

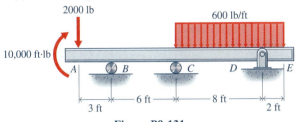

Figure P9-131

9-132 A beam is loaded and supported as shown in Fig. P9-132. When the loads are applied, the center support settles an amount equal to $wL^4/(12EI)$. Determine all of the reactions in terms of w, L, E, and I.

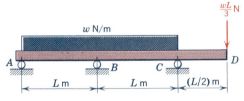

Figure P9-132

9-133* In Fig. P9-133, beam A is made of brass and beam B is made of steel. The second moment of the cross-sectional area of B with respect to the neutral axis is twice that of A, and the modulus of elasticity of the steel is twice that of the brass. The reaction at C is zero before the load w is applied. Determine the reaction at C on the beam B in terms of w and L.

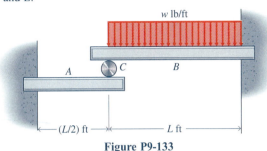

Figure P9-133

9-134 A uniformly distributed load w is supported by two beams arranged as shown in Fig. P9-134. Beam AB is fixed at the wall and beam CD is simply supported. Before the load is applied, the beams are in contact at B, but the reaction at B is zero. Both beams have the same cross-sectional area and are made of the same material. Determine the deflection at B when the load is applied in terms of w, L, E, and I.

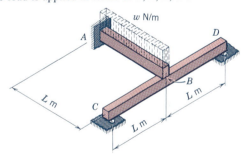

Figure P9-134

COMPUTER PROBLEMS

C9-135 A 160-lb diver walks slowly onto a diving board. The diving board is a wood (E = 1800 ksi) plank 10 ft long, 18 in. wide, and 2 in. thick, and is modeled as the cantilever beam shown in Fig. P9-135. For the diver at positions, $a = nL/5$ (n = 1, 2, . . ., 5), compute and plot the deflection curve for the diving board (plot y as a function of x for $0 \le x \le 10$ ft).

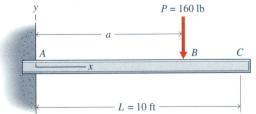

Figure P9-135

C9-136 A diving board consists of a wood (E = 12 GPa) plank that is pinned at the left end and rests on a movable support as shown in Fig. P9-136. The board is 3 m long, 500 mm wide, and 80 mm thick. If a 70-kg diver stands at the end of the board,
(a) Compute and plot the deflection curve for the beam (plot y as a function of x for $0 \le x \le 3$ m) for the right support at positions b = 0.5 m, 1.0 m, and 1.5 m.
(b) If the stiffness of the board is defined as the ratio of the diver's weight to the deflection at the end of the board ($k = W/y$), compute and plot the stiffness of the board as a function of b for $0 \le b \le 1.5$ m. Does the stiffness depend on the weight of the diver?

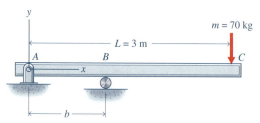

Figure P9-136

C9-137 The gangplank between a fishing boat and a dock consists of a wood ($E = 1800$ ksi) plank 10 ft long, 12 in. wide, and 2 in. thick. If the plank is modeled as the simply supported beam shown in Fig. P9-137, compute and plot the deflection curve for the gangplank (plot y as a function of x for $0 \le x \le 10$ ft) as a 170-lb man walks across the plank. Plot curves for the man at positions $a = nL/5$ ($n = 1, 2, \ldots 4$).

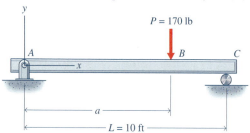

Figure P9-137

C9-138 A bridge over a small stream on a golf course consists of a wood deck (weighing 1000 newtons per meter) laid over two wood ($E = 12$ GPa) beams. Each of the beams are 5 m long, 100 mm wide, and h mm high; each beam carries half of the loading on the bridge. It is desired that the bridge support the weight of a loaded golf cart (total weight of 2400 N) with a maximum deflection of no more than 50 mm. Model the beams and loading as shown in Fig. P9-138 and

(a) Determine the minimum depth h_{min} of the beams that will support the given weight.

(b) If $h = 125$ mm, compute and plot the deflection curve for the beam (plot y as a function of x for $0 \le x \le L$) for the cart at positions $a = 1$ m, 2 m, and 3 m.

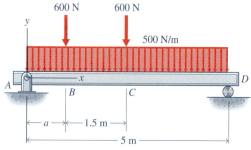

Figure P9-138

C9-139 A Cub Scout troop is marching across a small footbridge consisting of a wood ($E = 1800$ ksi) plank 8 ft long, 12 in. wide, and 1.5 in. thick. The scouts are separated by 2 ft and are modeled as static, concentrated loads (75 lb each) on a simply supported beam as shown in Fig. P9-139. Plot the deflection curve of the bridge (plot y as a function of x for $0 \le x \le 8$ ft) as the troop marches across the bridge. (Initially, only one scout is on the bridge—at position A. Then two scouts are on the bridge—at positions A and B. The troop continues to march across until finally only a single scout remains on the bridge—at position C.)

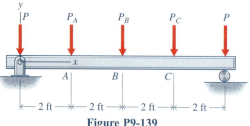

Figure P9-139

C9-140 A small weight W is suspended from the end of a meter stick that rests between two smooth pegs as shown in Fig. P9-140. The wide (30 mm) side of the wood ($E = 12$ GPa) stick is horizontal, and the thin (10 mm) side is vertical.

(a) Compute and plot the deflection curve (plot y as a function of x for $0 \le x \le 1$ m) for weights of $W = 25$ N, 50 N, and 75 N.

(b) Compute and plot the slope of the deflection curve [plot θ in radians as a function of x ($0 \le x \le 1$ m)] for weights of $W = 25$ N, 50 N, and 75 N.

(c) Which curves or what portion of the curves do you think are adequate approximations to the true deflection of the meter stick? Why?

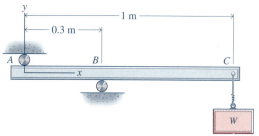

Figure P9-140

C9-141 A 30 ft–long beam carries a uniformly distributed load and is simply supported as shown in Fig. P9-141a. Modified designs of the beam consist of replacing the single long beam with a pair of shorter beams (Fig. P9-141b) or adding a center support to the long beam (Fig. P9-141c). If the beams are all constructed of wood ($EI = 12,000$ kip · ft^2), compute and plot the deflection curves for the three cases (plot y as a function of x for $0 \le x \le 30$ ft). Plot all three cases on the same graph. (Neglect horizontal reactions at the supports.)

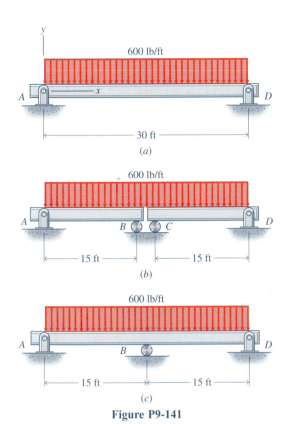

Figure P9-141

(a) The deflection curves for the beam (plot y as a function of x for $0 \le x \le 16$ ft) for load locations of $b = 3$ ft, 7 ft, and 11 ft.

(b) The bending moment distribution along the beam (plot M as a function of x for $0 \le x \le 16$ ft) for load locations of $b = 3$ ft, 7 ft, and 11 ft.

(c) The maximum flexural stresses in the beam as a function of b for $1 \le b \le 8$ ft. What range of b would be acceptable for this beam?

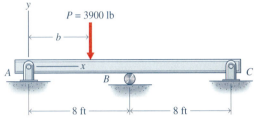

Figure P9-143

C9-142 A 10 m–long beam carries a uniformly distributed load and is supported as shown in Fig. P9-142. Horizontal reactions at the supports may be neglected. If the beam is constructed of wood ($EI = 1500$ kN · m²), compute and plot

(a) The deflection curves for the beam (plot y as a function of x for $0 \le x \le 10$ m) for center support locations of $b = 2$ m, 4 m, and 7 m.

(b) The bending moment distribution along the beam (plot M as a function of x for $0 \le x \le 10$ m) for center support locations of $b = 2$ m, 4 m, and 7 m.

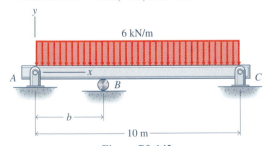

Figure P9-142

C9-143 A concentrated load $P = 3900$ lb moves slowly across the beam shown in Fig. P9-143. Horizontal reactions at the supports may be neglected. The 16 ft–long beam is a W4 × 13 structural steel section. Compute and plot

C9-144 A 6 m–long WT178 × 51 structural steel section is used for the cantilever beam shown in Fig. P9-144. The flange is at the top of the beam. If a roller support is added to the beam at B, compute and plot

(a) The deflection curves for the beam (plot y as a function of x for $0 \le x \le 6$ m) for support locations of $b = 3$ m, 4 m, and 5 m.

(b) The bending moment distribution along the beam (plot M as a function of x for $0 \le x \le 6$ m) for center support locations of $b = 3$ m, 4 m, and 5 m.

(c) The maximum tensile and compressive flexural stresses in the beam as a function of b ($1 \le b \le 6$ m). What range of b would be acceptable for this beam?

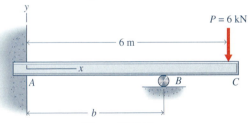

Figure P9-144

C9-145 An 18 ft–long C12 × 30 structural steel section is used for the cantilever beam shown in Fig. P9-145. If a roller support is added to the beam at B, compute and plot

(a) The deflection curves for the beam (plot y as a function of x for $0 \le x \le 18$ ft) for support locations of $b = 9$ ft, 12 ft, and 15 ft.

(b) The bending moment distribution along the beam (plot M as a function of x for $0 \le x \le 18$ ft) for center support locations of $b = 9$ ft, 12 ft, and 15 ft.

(c) The maximum tensile and compressive flexural stresses in the beam as a function of b ($3 \le b \le 18$ ft). What range of b would be acceptable for this beam?

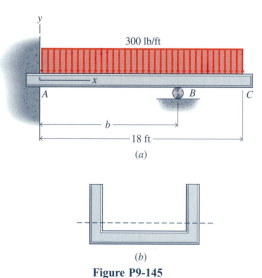

(a)

(b)

Figure P9-145

C9-146 A 6 m–long S457 × 104 structural steel section is used for the two-span beam shown in Fig. P9-146. The beam supports a uniformly distributed load w of 100 kN/m. The support at B settles with time until it provides no resistance to deflection of the beam.

(a) Determine b_{max}, the maximum amount that support B will settle.
(b) Compute and plot the deflection curves for the beam (plot y as a function of x for $0 \le x \le 6$ m) for support B initially ($b = 0$) and settled ($b = b_{max}/3$ and $b = 2b_{max}/3$).
(c) Compute and plot the bending moment distribution along the beam (plot M as a function of x for $0 \le x \le 6$ m) for support B initially ($b = 0$) and settled ($b = b_{max}/3$ and $b = 2b_{max}/3$).
(d) Compute and plot the maximum flexural stresses in the beam as a function of b ($0 \le b \le b_{max}$). What range of b would be acceptable for this beam?

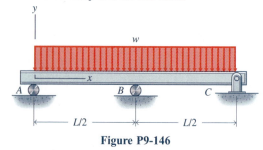

Figure P9-146

C9-147 A 20 ft–long S12 × 35 structural steel section is used for the two-span beam shown in Fig. P9-147. The beam supports a uniformly distributed load w of 1800 lb/ft. The support at A settles with time until it provides no resistance to deflection of the beam.

(a) Determine a_{max}, the maximum amount that support A will settle.

(b) Compute and plot the deflection curves for the beam (plot y as a function of x for $0 \le x \le 20$ ft) for support A initially ($a = 0$) and settled ($a = a_{max}/3$ and $a = 2a_{max}/3$).
(c) Compute and plot the bending moment distribution along the beam (plot M as a function of x for $0 \le x \le 20$ ft) for support A initially ($a = 0$) and settled ($a = a_{max}/3$ and $a = 2a_{max}/3$).
(d) Compute and plot the maximum flexural stresses in the beam as a function of a ($0 \le a \le a_{max}$). What range of a would be acceptable for this beam?

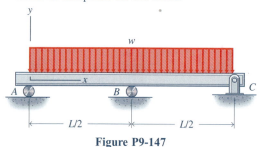

Figure P9-147

C9-148 A small weight $W = 100$ N is suspended from the end of a meter stick that rests between two smooth pegs as shown in Fig. P9-148. The wide (30 mm) side of the wood ($E = 12$ GPa) stick is horizontal, and the thin (10 mm) side is vertical. The approximation used in simple beam theory

$$EI \frac{d^2y}{dx^2} = M(x) \qquad (a)$$

is not very good here since neither the deflections nor slopes are small. Instead, the exact differential equation

$$EI \frac{d^2y/dx^2}{[1 + (dy/dx)^2]^{3/2}} = M(x) \qquad (b)$$

should be solved.

(a) Compute and plot the deflection curve (plot y as a function of x for $0 \le x \le 1$ m) using simple beam theory [Eq. (a)].
(b) On the same graph plot the deflection curve (plot y as a function of x) using the exact differential equation [Eq. (b)]. (*Note:* This curve will not go all the way to $x = 1$ m. It is the length of the stick that is 1 m, not the horizontal reach of the stick.)

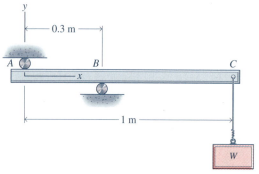

Figure P9-148

C9-149 One end of a yard stick is securely clamped to a table and a weight (W = 5 lb) is suspended from the other end as shown in Fig. P9-149. The wide (1 in.) side of the wood (E = 1800 ksi) stick is horizontal, and the thin ($\frac{1}{4}$ in.) side is vertical. The approximation used in simple beam theory

$$EI \frac{d^2y}{dx^2} = M(x) \qquad \text{(a)}$$

is not very good here since neither the deflections nor slopes are small. Instead, the exact differential equation

$$EI \frac{d^2y/dx^2}{[1 + (dy/dx)^2]^{3/2}} = M(x) \qquad \text{(b)}$$

should be solved.
(a) Compute and plot the deflection curve (plot y as a function of x for $0 \le x \le 32$ in.) using simple beam theory [Eq. (a)].

(b) On the same graph plot the deflection curve (plot y as a function of x) using the exact differential equation [Eq. (b)]. (*Note:* This curve will not go all the way to x = 32 in. It is the length of the stick that is 32 in., not the horizontal reach of the stick.)

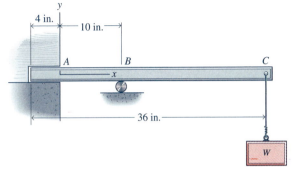

Figure P9-149

COMBINED STATIC LOADING

10-1 INTRODUCTION

In previous chapters, formulas were developed for determining normal and shearing stresses on specific planes in axially loaded bars, circular shafts, and beams. For example, the normal stress at a point on a transverse cross section of a beam can be determined by using the flexure formula (Eq. 8-11). The shearing stress at the same point on the cross section can be determined by using the shearing stress formula (Eq. 8-15). In this chapter, methods will be developed for determining the normal and shearing stresses on other planes through the point of interest.

For the case of an axially loaded bar, expressions were developed in Section 4-3 for determining the normal stress (Eq. 4-7) and shearing stress (Eq. 4-8) on inclined planes through the bar. This analysis indicated that maximum normal stresses occur on transverse planes and that maximum shearing stresses occur on planes inclined at 45° to the axis of the bar. Similarly, for the case of pure torsion in a circular shaft, maximum shearing stresses occur on transverse planes and maximum tensile and compressive stresses occur on planes inclined at 45° to the axis of the shaft. For the case of a beam subjected to arbitrary transverse loading, both normal and shearing stresses develop on a transverse cross section. The stresses on a specified inclined plane can be determined by using the free-body diagram method of Section 7-5; however, this method is not suitable for the determination of maximum normal and maximum shearing stresses, which are often required. A more general approach for determining stresses on inclined planes will be developed to handle this type of analysis.

10-2 STRESS AT A GENERAL POINT IN AN ARBITRARILY LOADED MEMBER

The nature of the internal force distribution at an arbitrary interior point O of a body of arbitrary shape that is in equilibrium under the action of a system of applied forces can be studied by exposing an interior plane through O as shown in Fig. 10-1a. The force distribution required on such an interior plane to maintain equilibrium of the isolated part of the body will not be uniform, in general; how-

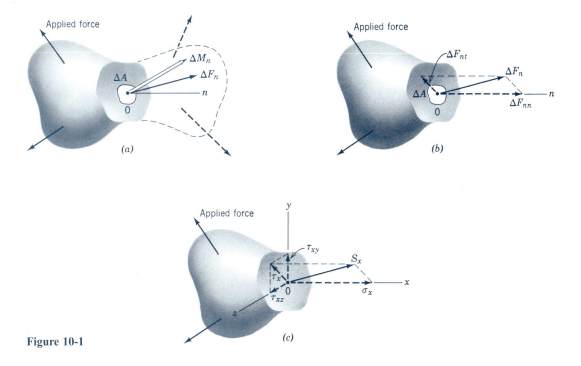

Figure 10-1

ever, any distributed force acting on a small area ΔA surrounding the point of interest O can be replaced by a statically equivalent resultant force ΔF_n through O and a couple ΔM_n. The subscript n indicates that the resultant force and couple are associated with a particular plane through O—namely, the one having an outward normal in the n-direction at O. For any other plane through O the values of ΔF and ΔM could be different. Note that the line of action of ΔF_n or ΔM_n may not coincide with the direction of n. If the resultant force ΔF_n is divided by the area ΔA, an average force per unit area (average resultant stress) is obtained. As the area ΔA is made smaller and smaller, the force distribution becomes more and more uniform and the couple ΔM_n vanishes. In the limit, a quantity known as the stress vector[1] or resultant stress is obtained. Thus,

$$S_n = \lim_{\Delta A \to 0} \frac{\Delta F_n}{\Delta A}$$

In Section 4-3 it was pointed out that materials respond to components of the stress vector rather than the stress vector itself. In particular, components normal and tangent to the internal plane are important. As shown in Fig. 10-1b, the resultant force ΔF_n can be resolved into components ΔF_{nn} normal to the plane and ΔF_{nt} tangent to the plane. A normal stress σ_n and a shearing stress τ_n are then defined as

$$\sigma_n = \lim_{\Delta A \to 0} \frac{\Delta F_{nn}}{\Delta A}$$

---

[1]The component of a tensor on a plane is a vector; therefore, on a particular plane, the stresses can be treated as vectors.

and

$$\tau_n = \underset{\Delta A \to 0}{\text{Lim}} \frac{\Delta F_{nt}}{\Delta A}$$

For purposes of analysis it is convenient to reference stresses to some co-ordinate system. In a Cartesian coordinate system, the stresses on planes having outward normals in the x-, y-, and z-directions are usually chosen. Consider the plane having an outward normal in the x-direction. In this case the normal and shearing stresses on the plane will be σ_x and τ_x, respectively. Since τ_x, in general, will not coincide with the y- or z-axes, it must be resolved into the components τ_{xy} and τ_{xz}, as shown in Fig. 10-1c.

Unfortunately, the state of stress at a point in a material is not completely defined by these three components of the stress vector, since the stress vector itself depends on the orientation of the plane with which it is associated. An infinite number of planes can be passed through the point, resulting in an infinite number of stress vectors being associated with the point. Fortunately it can be shown[2] that the specification of stresses on three mutually perpendicular planes is sufficient to completely describe the state of stress at the point. The rectangular components of stress vectors on planes having outward normals in the co-ordinate directions are shown in Fig. 10-2. The six faces of the small element are denoted by the directions of their outward normals so that the positive x-face is the one whose outward normal is in the direction of the positive x-axis. The coordinate axes x, y, and z are arranged as a right-hand system.

The sign convention for stresses is as follows.

Normal stresses are indicated by the symbol σ and a single subscript to indicate the plane (actually the outward normal to the plane) on which the stress acts. Normal stresses are positive if they point in the direction of the outward normal. Thus, normal stresses are positive if tensile and negative if compressive.

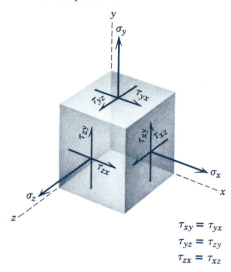

$$\tau_{xy} = \tau_{yx}$$
$$\tau_{yz} = \tau_{zy}$$
$$\tau_{zx} = \tau_{xz}$$

Figure 10-2

---

[2]See Section 1-9 of *Mechanics of Materials,* 4th Ed. by A. Higdon, E.H. Ohlsen, W.B. Stiles, J.A. Weese, and W.F. Riley, John Wiley & Sons, New York, 1985.

Shearing stresses are denoted by the symbol τ followed by two subscripts; the first subscript designates the normal to the plane on which the stress acts and the second the coordinate axis to which the stress is parallel. Thus, τ_{xz} is the shearing stress on a plane with outward normal in the x-direction. The stress acts parallel to the z-axis. A positive shearing stress points in the positive direction of the coordinate axis of the second subscript if it acts on a surface with an outward normal in the positive direction. Conversely, if the outward normal of the surface is in the negative direction, then the positive shearing stress points in the negative direction of the coordinate axis of the second subscript. The stresses shown on the element in Fig. 10-2 are all positive.

10-3 TWO-DIMENSIONAL OR PLANE STRESS

Considerable insight into the nature of stress distributions can be gained by considering a state of stress known as two-dimensional or plane stress. For this case, two parallel faces of the small element shown in Fig. 10-2 are assumed to be free of stress. For purposes of analysis, let these faces be perpendicular to the z-axis. Thus,

$$\sigma_z = \tau_{zx} = \tau_{zy} = 0$$

From Eq. (7-8), however, this also implies that

$$\tau_{xz} = \tau_{yz} = 0$$

Therefore, the components of stress present for plane stress analysis will be σ_x, σ_y, and $\tau_{xy} = \tau_{yx}$. For convenience this state of stress is represented by the two-dimensional sketch shown in Fig. 10-3. The three-dimensional element of which the two-dimensional sketch is a plane projection should be kept in mind at all times. Normal and shearing stresses on an arbitrary plane, such as plane A-A in Fig. 10-3, can be obtained by using the free-body diagram method discussed for axial loading in Section 4-3 and for torsional loading in Section 7-5. The solution to the following example problem illustrates this method of approach for plane stress problems.

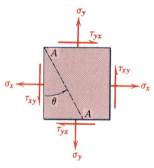

Figure 10-3

Example Problem 10-1

At a given point in a machine element, stresses are 8000 psi T and zero shear on a horizontal plane and 4000 psi C and zero shear on a vertical plane. Determine the stresses at this point on a plane having a slope of 3 vertical to 4 horizontal.

SOLUTION

As an aid to visualization of the data, the differential block of Fig. 10-4a is drawn (a stress picture, not a free-body diagram). A free-body diagram of a wedge-shaped element defined by the three given planes and subjected to or-

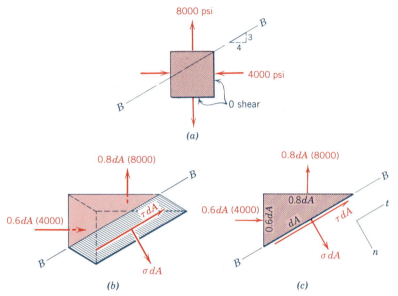

Figure 10-4

dinary force vectors is shown in Fig. 10-4*b*. The shaded area of Fig. 10-4*b* indicates the plane on which the stresses to be evaluated are acting. This area is arbitrarily assigned the magnitude *dA*: therefore, the corresponding areas of the horizontal and vertical faces of the element are 0.8 *dA* and 0.6 *dA*, respectively. For practical purposes, the two-dimensional free-body diagram of Fig. 10-4*c* is usually adequate.

Summing forces in the *n*-direction yields

$$+ \searrow \Sigma F_n = \sigma \, dA + 0.6 \, dA \, (4000)(0.6) - 0.8 \, dA \, (8000)(0.8) = 0$$

from which

$$\sigma = +3680 \text{ psi} = 3680 \text{ psi T} \qquad \textbf{Ans.}$$

Summing forces in the *t*-direction yields

$$+ \nearrow \Sigma F_t = \tau \, dA + 0.6 \, dA \, (4000)(0.8) + 0.8 \, dA \, (8000)(0.6) = 0$$

from which

$$\tau = -5760 \text{ psi} = 5760 \text{ psi} \swarrow \qquad \textbf{Ans.}$$

Normal stresses should always be designated as tension or compression. The presence of shearing stresses on the horizontal and vertical planes, had there been any, would merely have required two more forces on the free-body diagram: one parallel to the vertical face and one parallel to the horizontal face. The magnitudes of the shearing stresses (not the shearing forces) must be the same on any two orthogonal planes [Eq. (7-8)], and the vectors used to represent the shearing forces must be directed as in Fig. 10-3 or both vectors reversed, depending on the given data. ■

PROBLEMS

Standard Problems

10-1* At a point in a stressed body, there are normal stresses of 6250 psi T on a vertical plane and 12,500 psi T on a horizontal plane as shown in Fig. P10-1. Determine the normal and shearing stresses at this point on the inclined plane shown in the figure.

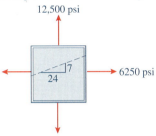

12,500 psi

6250 psi

7

24

Figure P10-1

10-2* At a point in a stressed body, there are normal stresses of 95 MPa T on a vertical plane and 150 MPa T on a horizontal plane as shown in Fig. P10-2. Determine the normal and shearing stresses at this point on the inclined plane shown in the figure.

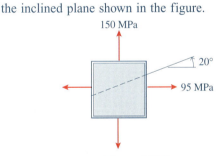

150 MPa

20°

95 MPa

Figure P10-2

10-3 The stresses shown in Fig. P10-3 act as a point in a stressed body. Determine the normal and shearing stresses at this point on the inclined plane shown in the figure.

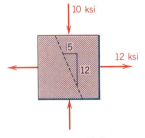

10 ksi

12 ksi

5

12

Figure P10-3

10-4 The stresses shown in Fig. P10-4 act at a point in a stressed body. Determine the normal and shearing stresses at this point on the inclined plane shown in the figure.

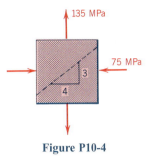

135 MPa

75 MPa

3

4

Figure P10-4

10-5* The stresses shown in Fig. P10-5 act at a point in a stressed body. Determine the normal and shearing stresses at this point on the inclined plane AB shown in the figure.

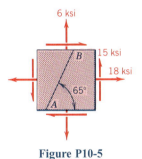

6 ksi

15 ksi

18 ksi

B

65°

A

Figure P10-5

10-6* The stresses shown in Fig. P10-6 act at a point in a stressed body. Determine the normal and shearing stresses at this point on the inclined plane AB shown in the figure.

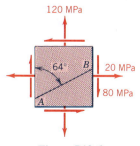

120 MPa

64°

B

20 MPa

80 MPa

A

Figure P10-6

10-7 The stresses shown in Fig. P10-7 act at a point in a stressed body. Determine the normal and shearing stresses at this point on the inclined plane *AB* shown in the figure.

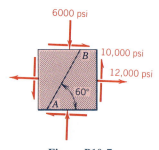

Figure P10-7

10-8 The stresses shown in Fig. P10-8 act at a point in a stressed body. Determine the normal and shearing stresses at this point on the inclined plane *AB* shown in the figure.

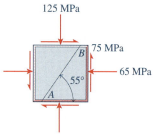

Figure P10-8

Challenging Problems

10-9* The stresses on horizontal and vertical planes at a point are shown in Fig. P10-9. The normal stress on plane *A-A* at this point is 8000 psi T. Determine
(a) The magnitude of the shearing stresses τ_h and τ_v.
(b) The magnitude and direction of the shearing stress on the inclined plane *A-A*.

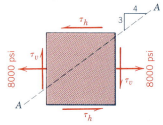

Figure P10-9

10-10* The stresses on horizontal and vertical planes at a point are shown in Fig. P10-10. The normal stress on plane *AB* is 15 MPa T. Determine

(a) The normal stress σ_x on the vertical plane.
(b) The magnitude and direction of the shearing stress on the inclined plane *AB*.

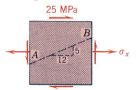

Figure P10-10

10-11 At a point in a structural member, there are stresses on horizontal and vertical planes as shown in Fig. P10-11. The magnitude of the compressive stress σ_c is three times the magnitude of the tensile stress σ_t. Specifications require that the shearing stress on plane *AB* not exceed 4000 psi, and the normal stress on plane *AB* not exceed 7800 psi. Determine the maximum value of stress σ_c that will satisfy the specifications.

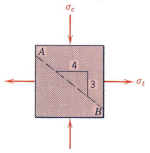

Figure P10-11

10-12 At a point in a structural member, there are stresses on horizontal and vertical planes as shown in Fig. P10-12. The normal stress on the inclined plane *AB* is 25 MPa T. Determine
(a) The normal stress σ_x on the vertical plane.
(b) The magnitude and direction of the shearing stress on the inclined plane *AB*.

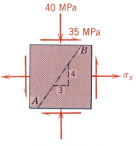

Figure P10-12

10-3-1 The Stress Transformation Equations for Plane Stress

Equations relating the normal and shearing stresses σ_n and τ_{nt} on an arbitrary plane (oriented at an angle θ with respect to a reference x-axis) through a point and the known stresses σ_x, σ_y, and $\tau_{xy} = \tau_{yx}$ on the reference planes can be developed using the free-body diagram method. Consider the plane stress situation indicated in Fig. 10-5a, where the dotted line A-A represents any plane through the point (all planes are perpendicular to the plane of zero stress—the plane of the paper). In the following derivation, a counterclockwise angle θ is positive.

Figure 10-5b is a free-body diagram of a wedge-shaped element in which the areas of the faces are dA for the inclined face (plane A-A), $dA \cos \theta$ for the vertical face, and $dA \sin \theta$ for the horizontal face. Summing forces in the n-direction gives

$$+ \nearrow \Sigma F_n = \sigma_n\, dA - \sigma_x(dA \cos \theta) \cos \theta - \sigma_y(dA \sin \theta) \sin \theta \\ - \tau_{yx}(dA \sin \theta) \cos \theta - \tau_{xy}(dA \cos \theta) \sin \theta = 0$$

from which, since $\tau_{yx} = \tau_{xy}$,

$$\sigma_n = \sigma_x \cos^2 \theta + \sigma_y \sin^2 \theta + 2\tau_{xy} \sin \theta \cos \theta \qquad (10\text{-}1a)$$

or, in terms of the double angle,

$$\sigma_n = \frac{\sigma_x(1 + \cos 2\theta)}{2} + \frac{\sigma_y(1 - \cos 2\theta)}{2} + \frac{2\tau_{xy} \sin 2\theta}{2}$$
$$= \frac{\sigma_x + \sigma_y}{2} + \frac{\sigma_x - \sigma_y}{2} \cos 2\theta + \tau_{xy} \sin 2\theta \qquad (10\text{-}1b)$$

Summing forces in the t-direction gives

$$+ \nwarrow \Sigma F_t = \tau_{nt}\, dA + \sigma_x(dA \cos \theta) \sin \theta - \sigma_y(dA \sin \theta) \cos \theta \\ - \tau_{xy}(dA \cos \theta) \cos \theta + \tau_{yx}(dA \sin \theta) \sin \theta = 0$$

from which

$$\tau_{nt} = -(\sigma_x - \sigma_y) \sin \theta \cos \theta + \tau_{xy} (\cos^2 \theta - \sin^2 \theta) \qquad (10\text{-}2a)$$

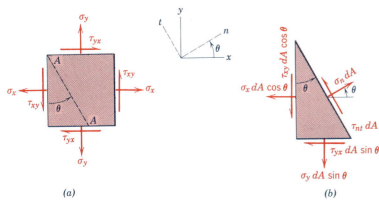

(a)   (b)

Figure 10-5

or, in terms of the double angle,

$$\tau_{nt} = -\frac{\sigma_x - \sigma_y}{2} \sin 2\theta + \tau_{xy} \cos 2\theta \qquad (10\text{-}2b)$$

Equations (10-1) and (10-2), the stress transformation equations for plane stress, provide a means for determining normal and shearing stresses on any plane whose outward normal is perpendicular to the z-axis and is oriented at an angle θ with respect to the reference x-axis. When these equations are used, the sign conventions used in their development must be rigorously followed; otherwise, erroneous results will be obtained. These sign conventions can be summarized as follows:

1. Tensile normal stresses are positive; compressive normal stresses are negative. All of the normal stresses shown on Fig. 10-5 are positive.
2. A shearing stress is positive if it points in the positive direction of the coordinate axis of the second subscript when it is acting on a surface whose outward normal is in a positive direction. Similarly, if the outward normal of the surface is in a negative direction, then a positive shearing stress points in the negative direction of the coordinate axis of the second subscript. All of the shearing stresses shown on Fig. 10-5 are positive. Shearing stresses pointing in the opposite directions would be negative.
3. An angle measured counterclockwise from the reference x-axis is positive. Conversely, angles measured clockwise from the reference x-axis are negative.
4. The (n, t, z) axes have the same order as the (x, y, z) axes. Both sets of axes form a right-hand coordinate system.

Example Problem 10-2

At a point in a structural member subjected to plane stress there are normal and shearing stresses on horizontal and vertical planes through the point, as shown in Fig. 10-6a. Use the stress transformation equations to determine

(a) The normal and shearing stresses on plane AB.
(b) The normal and shearing stresses on plane CD, which is perpendicular to plane AB.

SOLUTION

First define the x-y and n-t (if needed) directions if they have not been specified. On the basis of the axes shown in Fig. 10-6b and the established sign conventions, σ_x is positive, whereas σ_y and τ_{xy} are negative. Thus, the given values for use in Eqs. (10-1) and (10-2) are

$$\sigma_x = +80 \text{ MPa}, \ \sigma_y = -100 \text{ MPa}, \ \text{and} \ \tau_{xy} = -60 \text{ MPa}$$

(a) The angle θ for plane AB is $+42°$, as shown in Fig. 10-6c. Thus, from Eqs. (10-1) and (10-2),

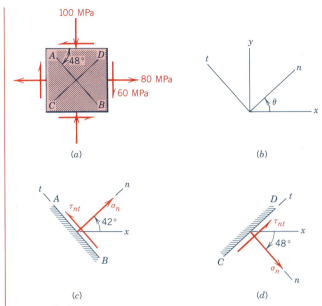

Figure 10-6

$$\sigma_n = 80 \cos^2 (+42°) + (-100) \sin^2 (+42°)$$
$$+ 2 (-60) \sin (+42°) \cos (+42°)$$
$$= -60.3 \text{ MPa} = 60.3 \text{ MPa C} \qquad \textbf{Ans.}$$

$$\tau_{nt} = -(80 + 100) \sin (+42°) \cos (+42°)$$
$$+ (-60)[\cos^2 (+42°) - \sin^2 (+42°)]$$
$$= -95.8 \text{ MPa} \qquad \textbf{Ans.}$$

(b) The angle θ for plane CD is $-48°$ (or $+132°$), as shown in Fig. 10-6d. Thus, from Eqs. (10-1) and (10-2),

$$\sigma_n = 80 \cos^2 (-48°) + (-100) \sin^2 (-48)$$
$$+ 2 (-60) \sin (-48°) \cos (-48°)$$
$$= +40.3 \text{ MPa} = 40.3 \text{ MPa T} \qquad \textbf{Ans.}$$

$$\tau_{nt} = -(80 + 100) \sin (-48°) \cos (-48°)$$
$$+ (-60)[\cos^2 (-48°) - \sin^2 (-48°)]$$
$$= +95.8 \text{ MPa} \qquad \textbf{Ans.}$$

Since planes AB and CD are orthogonal, the shearing stresses on the two planes must be equal in magnitude to satisfy Eq. (7-8). Also, one of the stresses must tend to produce a clockwise rotation of the element while the other tends to produce a counterclockwise rotation. The coordinate systems shown in Figs. 10-6c and 10-6d indicate that the shearing stresses calculated using Eq. (10-2) will have opposite signs on the two planes. ∎

PROBLEMS

Use the stress transformation equations for plane stress [Eqs. (10-1) and (10-2)] to solve the following problems.

Standard Problems

10-13* The stresses shown in Fig. P10-13 act at a point on the free surface of a stressed body. Determine the normal and shearing stresses at this point on the inclined plane *AB* shown in the figure.

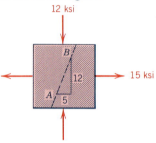

Figure P10-13

10-14* The stresses shown in Fig. P10-14 act at a point on the free surface of a stressed body. Determine the normal and shearing stresses at this point on the inclined plane *AB* shown in the figure.

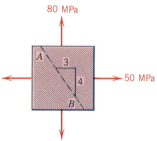

Figure P10-14

10-15 The stresses shown in Fig. P10-15 act at a point on the free surface of a stressed body. Determine the normal and shearing stresses at this point on the inclined plane *AB* shown in the figure.

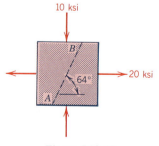

Figure P10-15

10-16 The stresses shown in Fig. P10-16 act at a point on the free surface of a stressed body. Determine the normal and shearing stresses at this point on the inclined plane *AB* shown in the figure.

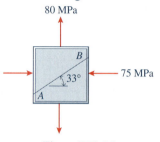

Figure P10-16

10-17* The stresses shown in Fig. P10-17 act at a point in a machine part. Determine the normal and shearing stresses at this point on the inclined plane *AB* shown in the figure.

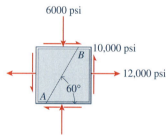

Figure P10-17

10-18* The stresses shown in Fig. P10-18 act at a point in a structural member. Determine the normal and shearing stresses at this point on the inclined plane *AB* shown in the figure.

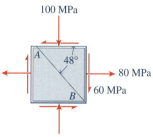

Figure P10-18

10-19 The stresses shown in Fig. P10-19 act at a point in a structural member. Determine the normal and shearing stresses at this point on the inclined plane AB shown in the figure.

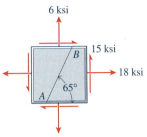

Figure P10-19

10-20 The stresses shown in Fig. P10-20 act at a point in a machine part. Determine the normal and shearing stresses at this point on the inclined plane AB shown in the figure.

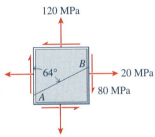

Figure P10-20

10-21* The stresses shown in Fig. P10-21 act at a point in a machine part. Determine the normal and shearing stresses at this point on the inclined plane AB shown in the figure.

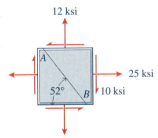

Figure P10-21

10-22* The stresses shown in Fig. P10-22 act at a point in a structural member. Determine the normal and shearing stresses at this point on the inclined plane AB shown in the figure.

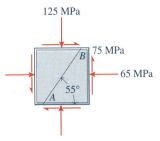

Figure P10-22

10-23 The stresses shown in Fig. P10-23a act at a point on the free surface of a stressed body. Determine the normal and shearing stresses at this point on the inclined plane AB shown in the figure.

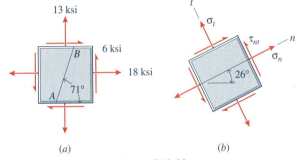

(a)   (b)

Figure P10-23

10-24 The stresses shown in Fig. P10-24a act at a point on the free surface of a stressed body. Determine the normal and shearing stresses at this point on the inclined plane AB shown in the figure.

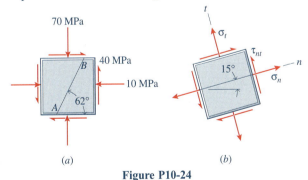

(a)   (b)

Figure P10-24

Challenging Problems

10-25* The stresses shown in Fig. P10-23a act at a point on the free surface of a stressed body. Determine the normal stresses σ_n and σ_t and the shearing stress τ_{nt} at this point if they act on the rotated stress element shown in Fig. P10-23b.

10-26* The stresses shown in Fig. P10-24a act at a point on the free surface of a stressed body. Determine the normal stresses σ_n and σ_t and the shearing stress τ_{nt} at this point if they act on the rotated stress element shown in Fig. P10-24b.

10-27 The stresses shown in Fig. P10-27 act at a point in a machine part. Determine the normal stresses σ_n and σ_t and the shearing stress τ_{nt} at the point.

10-28 The stresses shown in Fig. P10-28 act at a point in a structural member. Determine the normal stresses σ_n and σ_t and the shearing stress τ_{nt} at the point.

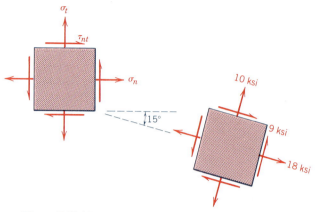

Figure P10-27

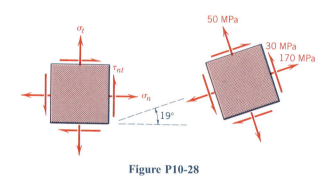

Figure P10-28

10-3-2 Principal Stresses and Maximum Shearing Stress

The transformation equations for plane stress [Eqs. (10-1) and (10-2)] provide a means for determining the normal stress σ_n and the shearing stress τ_{nt} on different planes through a point in a stressed body. As an example, consider the state of stress at a point on the free surface of a machine component or structural member shown in Fig. 10-7a. As the element is rotated through an angle θ about an axis perpendicular to the free surface, the normal stress σ_n and the shearing stress τ_{nt} on the different planes vary continuously, as shown in Fig. 10-7b. For design purposes, critical stresses at the point are usually the maximum tensile stress and the maximum shearing stress.

For a bar under axial load and for a shaft in pure torsion, the planes on which maximum normal stresses and maximum shearing stresses act are known from the results of Sections 4-3 and 7-5, respectively. For more complicated forms of loading, these stresses can be determined by plotting curves similar to those shown in Fig. 10-7b for each different state of stress encountered, but this process is time-consuming and inefficient. Therefore, more general methods for finding the critical stresses have been developed.

The transformation equations for plane stress developed previously are as follows:

For normal stress σ_n:

$$\sigma_n = \frac{\sigma_x + \sigma_y}{2} + \frac{\sigma_x - \sigma_y}{2} \cos 2\theta + \tau_{xy} \sin 2\theta \qquad (10\text{-}1b)$$

For shear stress τ_{nt}:

$$\tau_{nt} = -\frac{\sigma_x - \sigma_y}{2} \sin 2\theta + \tau_{xy} \cos 2\theta \qquad (10\text{-}2b)$$

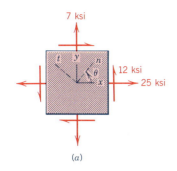

(a)

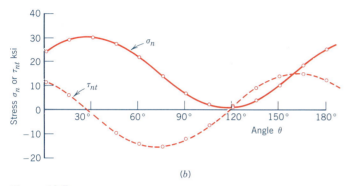

(b)

Figure 10-7

Maximum and minimum values of σ_n occur at values of θ for which $d\sigma_n/d\theta$ is equal to zero. Differentiation of σ_n with respect to θ yields

$$\frac{d\sigma_n}{d\theta} = -(\sigma_x - \sigma_y)\sin 2\theta + 2\tau_{xy}\cos 2\theta \qquad (a)$$

Setting Eq. (a) equal to zero and solving gives

$$\tan 2\theta_p = \frac{2\tau_{xy}}{\sigma_x - \sigma_y} \qquad (10\text{-}3)$$

where θ_p locates the planes of maximum and minimum values of σ_n. Note that the expression for $d\sigma_n/d\theta$ from Eq. (a) is numerically twice the value of the expression for τ_{nt} from Eq. (10-2b). Consequently, the shearing stress is zero on planes experiencing maximum and minimum values of normal stress. Planes free of shear stress are known as *principal planes*. Normal stresses occurring on principal planes are known as *principal stresses*. The values of θ_p from Eq. (10-3) give the orientations of two principal planes. A third principal plane for the plane stress state has an outward normal in the z-direction. For a given set of values of σ_x, σ_y, and τ_{xy}, there are two values of $2\theta_p$ differing by 180° and, consequently, two values of θ_p that are 90° apart. This proves that the principal planes are normal to each other.

When τ_{xy} and $(\sigma_x - \sigma_y)$ have the same sign, $\tan 2\theta_p$ is positive and one value of $2\theta_p$ is between 0° and 90°, with the other value 180° greater, as shown in Fig. 10-8. Consequently, one value of θ_p is between 0° and 45°, and the other

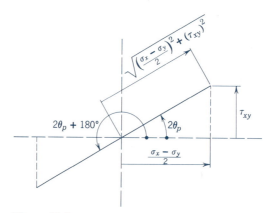

Figure 10-8

one is 90° greater. In the first case, both $\sin 2\theta_p$ and $\cos 2\theta_p$ are positive, and in the second case both are negative. When these functions of $2\theta_p$ are substituted into Eq. (10-1b), the two in-plane principal stresses σ_{p1} and σ_{p2} are found to be

$$\sigma_{p1, p2} = \frac{\sigma_x + \sigma_y}{2} \pm \sqrt{\left(\frac{\sigma_x - \sigma_y}{2}\right)^2 + \tau_{xy}^2} \tag{10-4}$$

Equation (10-4) gives the two principal stresses in the xy-plane, and the third principal stress is $\sigma_{p3} = \sigma_z = 0$. Equation (10-3) gives the angles θ_p and $\theta_p + 90°$ between the x- (or y-) plane and the mutually perpendicular planes on which the principal stresses act. When $\tan 2\theta_p$ is positive, θ_p is positive, and the rotation is counterclockwise from the x- and y-planes to the planes on which the two principal stresses act. When $\tan 2\theta_p$ is negative, the rotation is clockwise. Note that one value of θ_p will always be between positive and negative 45° (inclusive) with the other value 90° greater. The numerically greater principal stress will act on the plane that makes an angle of 45° or less with the plane of the numerically larger of the two given normal stresses. This statement can be confirmed by substituting the value of θ_p from Eq. (10-3) into Eq. (10-1b) to obtain the corresponding principal stress.

Note that if one or both of the principal stresses from Eq. (10-4) is negative, the algebraic maximum stress can have a smaller absolute value than the "minimum" stress.

The *maximum in-plane shearing stress* τ_p occurs on planes located by values of θ where $d\tau_{nt}/d\theta$ is equal to zero. Differentiation of Eq. (10-2b) yields

$$\frac{d\tau_{nt}}{d\theta} = -(\sigma_x - \sigma_y) \cos 2\theta - 2\tau_{xy} \sin 2\theta$$

When $d\tau_{nt}/d\theta$ is equated to zero, the value of θ_τ is given by the expression

$$\tan 2\theta_\tau = -\frac{(\sigma_x - \sigma_y)}{2\tau_{xy}} \tag{b}$$

where θ_τ locates the planes of maximum in-plane shearing stress. Comparison of Eqs. (b) and (10-3) reveals that the two tangents are negative reciprocals. Therefore, the two angles $2\theta_p$ and $2\theta_\tau$ differ by 90°, and θ_p and θ_τ are 45° apart. This means that the planes on which the maximum in-plane shearing stresses occur are 45° from the principal planes. The maximum in-plane shearing stresses are found by substituting values of angle functions obtained from Eq. (b) into Eq. (10-2b). The results are

$$\tau_p = \pm \sqrt{\left(\frac{\sigma_x - \sigma_y}{2}\right)^2 + \tau_{xy}^2} \qquad (10\text{-}5)$$

Equation (10-5) has the same magnitude as the second term of Eq. (10-4).

A useful relation between the principal stresses and the maximum in-plane shearing stress is obtained from Eqs. (10-4) and (10-5) by subtracting the values for the two in-plane principal stresses and substituting the value of the radical from Eq. (10-5). The result is

$$\tau_p = (\sigma_{p1} - \sigma_{p2})/2 \qquad (10\text{-}6)$$

or, in words, the maximum value of τ_{nt} is equal in magnitude to one half the difference between the two in-plane principal stresses.

In general, when stresses act in three directions it can be shown[3] that there are three orthogonal planes on which the shearing stress is zero. These planes are known as the *principal planes,* and the stresses acting on them (the *principal stresses*) will have three values: one maximum, one minimum, and a third stress between the other two. The maximum shearing stress τ_{max} on any plane that could be passed through the point is one half the difference between the maximum and minimum principal stresses and acts on planes that bisect the angles between the planes of the maximum and minimum normal stresses.

$$\tau_{max} = \frac{\sigma_{max} - \sigma_{min}}{2} \qquad (10\text{-}7)$$

When a state of plane stress exists, one of the principal stresses is zero. If the values of σ_{p1} and σ_{p2} from Eq. (10-4) have the same sign, then the third principal stress, σ_{p3} equals zero, will be either the maximum or the minimum normal stress. Thus, the maximum shearing stress may be

$$(\sigma_{p1} - \sigma_{p2})/2 \quad \text{or} \quad (\sigma_{p1} - 0)/2 \quad \text{or} \quad (0 - \sigma_{p2})/2$$

depending on the relative magnitudes and signs of the principal stresses. These three possibilities are illustrated in Fig. 10-9, in which one of the two orthogonal planes on which the maximum shearing stress acts is hatched for each example.

The direction of the maximum shearing stress can be determined by drawing a wedge-shaped block with two sides parallel to the planes having the maximum and minimum principal stresses, and with the third side at an angle of 45° with the other two sides. The direction of the maximum shearing stress must oppose the larger of the two principal stresses.

---

[3]See Section 1-9 of *Mechanics of Materials,* 4th Ed. by A. Higdon, E.H. Ohlsen, W.B. Stiles, J.A. Weese, and W.F. Riley, John Wiley & Sons, New York, 1985.

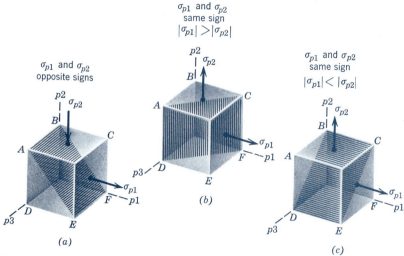

Figure 10-9

Another useful relation between the principal stresses and the normal stresses on the orthogonal planes shown in Fig. 10-10 is obtained by adding the values for the two principal stresses as given by Eq. (10-4). Thus,

$$\sigma_{p1} + \sigma_{p2} = \sigma_x + \sigma_y \qquad (10\text{-}8)$$

or, in words, for plane stress, the sum of the normal stresses on any two orthogonal planes through a point in a body is a constant or invariant.

In the preceding discussion, "maximum" and "minimum" stresses were considered as algebraic quantities, and it has already been pointed out that the minimum algebraic stress may have a larger magnitude than the maximum stress. However, in the application to engineering problems (which includes the problems in this book), the term "maximum" will always refer to the largest absolute value (largest magnitude).

Application of the formulas and procedures developed in this section are illustrated by the following examples.

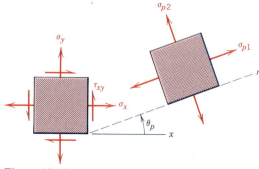

Figure 10-10

Example Problem 10-3

At a point in a structural member subjected to plane stress there are normal and shearing stresses on horizontal and vertical planes through the point, as shown in Fig. 10-11a.

(a) Determine the principal stresses and the maximum shearing stress at the point.

(b) Locate the planes on which these stresses act and show the stresses on a complete sketch.

SOLUTION

(a) On the basis of the axes shown in Fig. 10-11a and the established sign conventions, σ_x is positive, whereas σ_y and τ_{xy} are negative. For use in Eqs. (10-3) and (10-4), the given values are

$$\sigma_x = +10,000 \text{ psi} \qquad \sigma_y = -8000 \text{ psi} \qquad \tau_{xy} = -4000 \text{ psi}$$

When these values are substituted in Eq. (10-4), the principal stresses are found to be

$$\sigma_p = \frac{10,000 - 8000}{2} \pm \sqrt{\left(\frac{10,000 + 8000}{2}\right)^2 + (-4000)^2}$$

$$= 1000 \pm 9850$$

$$\sigma_{p1} = 10,850 \text{ psi} = 10,850 \text{ psi T} \qquad\qquad \textbf{Ans.}$$

$$\sigma_{p2} = -8850 \text{ psi} = 8850 \text{ psi C} \qquad\qquad \textbf{Ans.}$$

$$\sigma_{p3} = \sigma_z = 0 \qquad\qquad \textbf{Ans.}$$

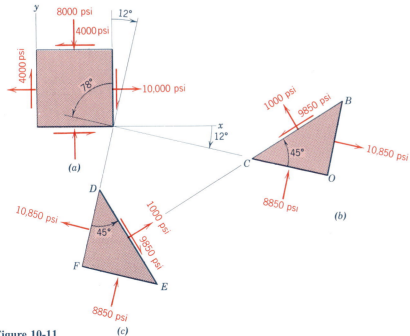

Figure 10-11

Since σ_{p1} and σ_{p2} are of opposite sign, the maximum shearing stress is

$$\tau_{max} = \frac{10{,}850 - (-8850)}{2} = 9850 \text{ psi} \qquad \textbf{Ans.}$$

(b) When the given data are substituted in Eq. (10-3), the results are

$$\tan 2\theta_p = \frac{2(-4000)}{10{,}000 - (-8000)} = -0.4444$$

from which

$$2\theta_p = -23.96°$$

and

$$\theta_p = -11.98° \cong 12° \,\swarrow$$

The required sketch is shown in Fig. 10-11b or c.

From Eq. (10-8), the sum of the normal stresses on any two orthogonal planes is a constant for plane stress. Planes CB and DE of Fig. 10-11b and c are orthogonal, and the normal stresses on them are obviously equal. Therefore,

$$2\sigma_n = \sigma_x + \sigma_y = 10{,}000 - 8000 = 2000$$

and

$$\sigma_n = 1000 \text{ psi on the planes of maximum shear} \quad \blacksquare$$

Example Problem 10-4

At a point in a structural member subjected to plane stress there are normal and shearing stresses on horizontal and vertical planes through the point, as shown in Fig. 10-12a.

(a) Determine the principal stresses and the maximum shearing stress at the point.
(b) Locate the planes on which these stresses act and show the stresses on a complete sketch.

SOLUTION

(a) On the basis of the axes shown in Fig. 10-12a and the established sign conventions, σ_x is positive, σ_y is positive, and τ_{xy} is negative. For use in Eqs. (10-3) and (10-4), the given values are

$$\sigma_x = +100 \text{ MPa} \qquad \sigma_y = +80 \text{ MPa} \qquad \tau_{xy} = -40 \text{ MPa}$$

When these values are substituted in Eq. (10-4), the principal stresses are found to be

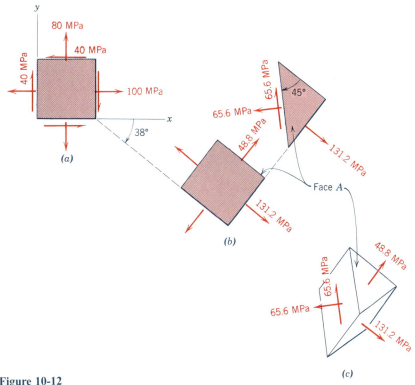

Figure 10-12

$$\sigma_p = \frac{100 + 80}{2} \pm \sqrt{\left(\frac{100 - 80}{2}\right)^2 + (-40)^2}$$

$$= 90 \pm 41.23$$

$$\sigma_{p1} = 131.2 \text{ MPa T} \qquad \qquad \textbf{Ans.}$$

$$\sigma_{p2} = 48.8 \text{ MPa T} \qquad \qquad \textbf{Ans.}$$

$$\sigma_{p3} = \sigma_z = 0 \qquad \qquad \textbf{Ans.}$$

Since σ_{p1} and σ_{p2} have the same sign, the maximum shearing stress is

$$\tau_{max} = \frac{131.2 - 0}{2} = 65.6 \text{ MPa} \qquad \qquad \textbf{Ans.}$$

(b) When the given data are substituted in Eq. (10-3), the results are

$$\tan 2\theta_p = \frac{2(-40)}{100 - 80} = -4.00$$

from which

$$2\theta_p = -75.97°$$

and

$$\theta_p = -37.985° \cong 38° \downarrow$$

The maximum shearing stress occurs on the plane making an angle of 45° with the planes of maximum and minimum normal stress—in this case, 131.2 MPa and zero. The complete sketch is given in Fig. 10-12b, where the upper (wedge-shaped) block is the orthographic projection of the lower block. The three-dimensional wedge (Fig. 10-12c) is presented as an aid to the visualization of Fig. 10-12b. ∎

PROBLEMS

Standard Problems

Normal and shearing stresses on horizontal and vertical planes through a point in a structural member subjected to plane stress are shown in the following figures. Determine and show on a sketch the principal and maximum shearing stresses.

10-29* At the point shown in Fig. P10-29.

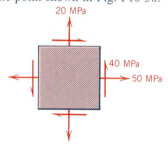

Figure P10-29

10-30* At the point shown in Fig. P10-30.

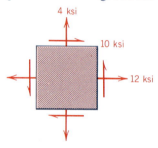

Figure P10-30

10-31 At the point shown in Fig. P10-31.

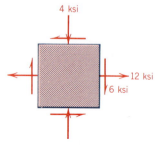

Figure P10-31

10-32 At the point shown in Fig. P10-32.

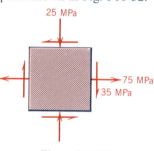

Figure P10-32

10-33* At the point shown in Fig. P10-33.

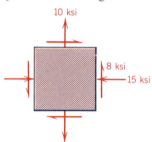

Figure P10-33

10-34* At the point shown in Fig. P10-34.

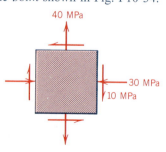

Figure P10-34

10-35 At the point shown in Fig. P10-35.

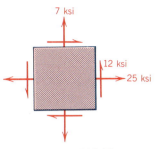

Figure P10-35

10-36 At the point shown in Fig. P10-36.

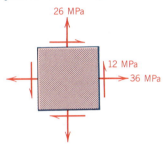

Figure P10-36

10-37* At the point shown in Fig. P10-37.

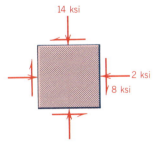

Figure P10-37

10-38* At the point shown in Fig. P10-38.

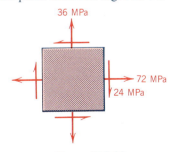

Figure P10-38

10-39 At the point shown in Fig. P10-39.

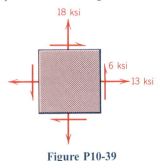

Figure P10-39

10-40 At the point shown in Fig. P10-40.

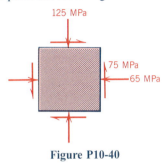

Figure P10-40

Challenging Problems

10-41* The stresses shown in Fig. P10-41 act at a point on the free surface of a stressed body. The principal stresses at the point are 20 ksi C and 12 ksi T. Determine the unknown normal stresses on the horizontal and vertical planes and the angle θ_p between the x-axis and the maximum tensile stress at the point.

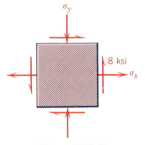

Figure P10-41

10-42* The stresses shown in Fig. P10-42 act at a point on the free surface of a stressed body. The magnitude of the maximum shearing stress at the point is 125 MPa. Determine the principal stresses, the magnitude of the unknown shearing stress on the vertical plane, and the angle θ_p between the x-axis and the maximum tensile stress at the point.

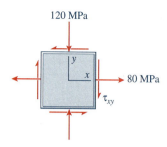

120 MPa

80 MPa

τ_{xy}

Figure P10-42

10-43 The principal compressive stress on a vertical plane through a point in a wooden block is equal to four times the principal compressive stress on a horizontal plane. The plane of the grain is 30° clockwise from the vertical plane. If the normal and shearing stresses on the plane of the grain must not exceed 300 psi C and 125 psi shear, determine the maximum allowable compressive stress on the horizontal plane.

10-44 At a point on the free surface of a stressed body, a normal stress of 64 MPa C and an unknown positive shearing stress exist on a horizontal plane. One princi-

pal stress at the point is 8 MPa C. The maximum shearing stress at the point has a magnitude of 95 MPa. Determine the unknown stresses on the horizontal and vertical planes and the unknown principal stress at the point.

10-45* At a point on the free surface of a stressed body, the normal stresses are 20 ksi T on a vertical plane and 30 ksi C on a horizontal plane. An unknown negative shearing stress acts on the vertical plane. The maximum shearing stress at the point has a magnitude of 32 ksi. Determine the principal stresses and the shearing stress on the vertical plane at the point.

10-46 At a point on the free surface of a stressed body, a normal stress of 75 MPa T and an unknown negative shearing stress exist on a horizontal plane. One principal stress at the point is 200 MPa T. The maximum in-plane shearing stress at the point has a magnitude of 85 MPa. Determine the unknown stresses on the vertical plane, the unknown principal stress, and the maximum shearing stress at the point.

10-3-3 Mohr's Circle for Plane Stress

The German engineer Otto Mohr (1835–1918) developed a useful pictorial or graphic interpretation of the transformation equations for plane stress. This representation, commonly called Mohr's circle, involves the construction of a circle in such a manner that the coordinates of each point on the circle represent the normal and shearing stresses on one plane through the stressed point, and the angular position of the radius to the point gives the orientation of the plane. The proof that normal and shearing components of stress on an arbitrary plane through a point can be represented as points on a circle follows from Eqs. (10-1) and (10-2). Recall Eqs. (10-1b) and (10-2b):

$$\sigma_n - \frac{\sigma_x + \sigma_y}{2} = \frac{\sigma_x - \sigma_y}{2} \cos 2\theta + \tau_{xy} \sin 2\theta$$

$$\tau_{nt} = -\frac{\sigma_x - \sigma_y}{2} \sin 2\theta + \tau_{xy} \cos 2\theta$$

Squaring both equations, adding, and simplifying yields

$$\left(\sigma_n - \frac{\sigma_x + \sigma_y}{2}\right)^2 + \tau_{nt}^2 = \left(\frac{\sigma_x - \sigma_y}{2}\right)^2 + \tau_{xy}^2$$

This is the equation of a circle in terms of the variables σ_n and τ_{nt}. The circle is centered on the σ axis at a distance $(\sigma_x + \sigma_y)/2$ from the τ axis, and the radius of the circle is given by

$$R = \sqrt{\left(\frac{\sigma_x - \sigma_y}{2}\right)^2 + \tau_{xy}^2}$$

Normal stresses are plotted as horizontal coordinates, with tensile stresses (positive) plotted to the right of the origin and compressive stresses (negative) plotted to the left. Shearing stresses are plotted as vertical coordinates, with those tending to produce a clockwise rotation of the stress element plotted above the σ-axis, and those tending to produce a counterclockwise rotation of the stress element plotted below the σ-axis. The method for interpreting the sign to be associated with a particular shear stress value obtained from a Mohr's circle analysis will be illustrated in the discussion and example problems that follow.

Mohr's circle for any point subjected to plane stress can be drawn when stresses on two mutually perpendicular planes through the point are known. Consider, for example, the stress situation of Fig. 10-13a with σ_x greater than σ_y and

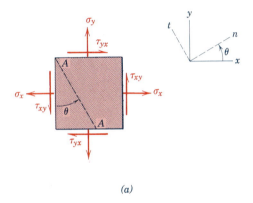

(a)

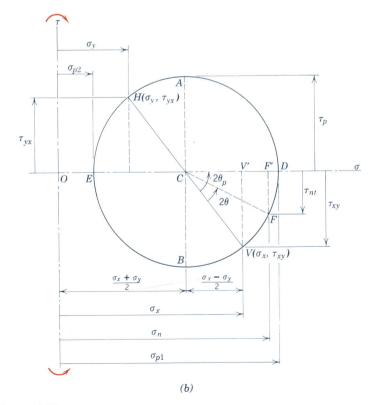

(b)

Figure 10-13

plot on Fig. 10-13b the points representing the given stresses. The coordinates of point V of Fig. 10-13b represent the stresses on the vertical plane through the stressed point of Fig. 10-13a, and the coordinates of point H represent the stresses on the horizontal plane through the point. Because $\tau_{yx} = \tau_{xy}$, point C, the center of the circle, is on the σ-axis. Line CV on Mohr's circle represents the plane (the vertical plane of Fig. 10-13a) through the stressed point from which the angle θ is measured. The coordinates of each point on the circle represent σ_n and τ_{nt} for one particular plane through the stressed point, the abscissa representing σ_n and the ordinate representing τ_{nt}. To demonstrate this statement, draw any radius CF in Fig. 10-13b at an angle 2θ counterclockwise from radius CV. From the figure, it is apparent that

$$OF' = OC + CF \cos (2\theta_p - 2\theta)$$

and since CF equals CV, the above equation reduces to

$$OF' = OC + CV \cos 2\theta_p \cos 2\theta + CV \sin 2\theta_p \sin 2\theta$$

Referring to Fig. 10-13b, note that

$$CV \cos 2\theta_p = CV' = (\sigma_x - \sigma_y)/2$$

$$CV \sin 2\theta_p = VV' = \tau_{xy}$$

$$OC = (\sigma_x + \sigma_y)/2 = \sigma_{avg}$$

Therefore,

$$OF' = OC + CV' \cos 2\theta + VV' \sin 2\theta$$
$$= \frac{\sigma_x + \sigma_y}{2} + \frac{\sigma_x - \sigma_y}{2} \cos 2\theta + \tau_{xy} \sin 2\theta$$

This expression is identical to Eq. (10-1b). Therefore, OF' is equal to σ_n. In a similar manner,

$$F'F = CF \sin (2\theta_p - 2\theta)$$
$$= CV \sin 2\theta_p \cos 2\theta - CV \cos 2\theta_p \sin 2\theta$$
$$= V'V \cos 2\theta - CV' \sin 2\theta$$
$$= \tau_{xy} \cos 2\theta - \frac{\sigma_x - \sigma_y}{2} \sin 2\theta$$

This expression is identical to Eq. (10-2b). Therefore $F'F$ is equal to τ_{nt}.

Since the horizontal coordinate of each point on the circle represents a particular value of σ_n, the maximum normal stress is represented by OD, and its value is

$$\sigma_{p1} = OD = OC + CD = OC + CV$$

$$= \frac{\sigma_x + \sigma_y}{2} + \sqrt{\left(\frac{\sigma_x - \sigma_y}{2}\right)^2 + \tau_{xy}^2}$$

which agrees with Eq. (10-4).

Likewise, the vertical coordinate of each point on the circle represents a particular value of τ_{nt} (called the in-plane shearing stress), which means that the maximum in-plane shearing stresses are represented by CA and CB, and their value is

$$\tau_p = CA = CV = \sqrt{\left(\frac{\sigma_x - \sigma_y}{2}\right)^2 + \tau_{xy}^2}$$

which agrees with Eq. (10-5). If the two nonzero principal stresses have the same sign, the maximum shearing stress at the point will not be in the plane of the applied stresses.

The angle $2\theta_p$ from CV to CD is counterclockwise or positive, and its tangent is

$$\tan 2\theta_p = \frac{\tau_{xy}}{(\sigma_x - \sigma_y)/2}$$

which is Eq. (10-3). From the derivation of Eq. (10-4), the angle between the vertical plane and one of the principal planes was θ_p. In obtaining the same equation from Mohr's circle, the angle between the radii representing these same two planes is $2\theta_p$. In other words, all angles on Mohr's circle are twice the corresponding angles for the actual stressed body. The angle from the vertical plane to the horizontal plane in Fig. 10-13a is 90°, but in Fig. 10-13b, the angle between line CV (which represents the vertical plane) and line CH (which represents the horizontal plane) on Mohr's circle is 180°.

The results obtained from Mohr's circle have been shown to be identical with the equations derived from the free-body diagram of Fig. 10-5. Thus, Mohr's circle provides an extremely useful aid for both the visualization of and the solution of stresses on various planes through a point in a stressed body in terms of the stresses on two mutually perpendicular planes through the point. Although Mohr's circle can be drawn to scale and used to obtain values of stresses and angles by direct measurements on the figure, it is probably more useful as a pictorial aid to the analyst who is performing analytical determinations of stresses and their directions at the point.

When the state of stress at a point is specified by means of a sketch of a small element, the procedure for drawing and using Mohr's circle to obtain specific stress information can be briefly summarized as follows:

1. Choose a set of x-y reference axes.
2. Identify the stresses σ_x, σ_y and $\tau_{xy} = \tau_{yx}$ and list them with the proper sign.
3. Draw a set of $\sigma\tau$-coordinate axes with σ and τ positive to the right and upward, respectively.
4. Plot the point $(\sigma_x, -\tau_{xy})$ and label it point V (vertical plane).
5. Plot the point (σ_y, τ_{yx}) and label it point H (horizontal plane).
6. Draw a line between V and H. This establishes the center C and the radius R of Mohr's circle.
7. Draw the circle.
8. An extension of the radius between C and V can be identified as the x-axis or the reference line for angle measurements (i.e., $\theta = 0°$).

By plotting points V and H as $(\sigma_x, -\tau_{xy})$ and (σ_y, τ_{yx}), respectively, shear stresses that tend to rotate the stress element clockwise will plot above the σ-axis, while those tending to rotate the element counterclockwise will plot below the σ-axis. The use of a negative sign with one of the shearing stresses (τ_{xy} or τ_{yx}) is required for plotting purposes, since for a given state of stress, the shearing stresses ($\tau_{xy} = \tau_{yx}$) have only one sign (both are positive or both are negative). The use of the negative sign at point V on Mohr's circle brings the direction of angular measurements 2θ on Mohr's circle into agreement with the direction of angular measurements θ on the stress element.

Once the circle has been drawn, the normal and shearing stresses on an arbitrary inclined plane AA having an outward normal n that is oriented at an angle θ with respect to the reference x-axis (see Fig. 10-13a) can be obtained from the coordinates of point F (see Fig. 10-13b) on the circle that is located at angular position 2θ from the reference axis through point V. The coordinates of point F must be interpreted as stresses σ_n and $-\tau_{nt}$. Other points on Mohr's circle that provide stresses of interest are

1. Point D that provides the principal stress σ_{p1}.
2. Point E that provides the principal stress σ_{p2}.
3. Point A that provides the maximum in-plane shearing stress $-\tau_p$ and the accompanying normal stress σ_{avg} that acts on the plane.

A negative sign must be used when interpreting shearing stresses τ_{nt} and τ_p obtained from the circle, since a shearing stress tending to produce a clockwise rotation of the stress element is a negative shearing stress when a right-hand nt-coordinate system is used.

Problems of the type presented in Section 10-3-1 and Section 10-3-2 can readily be solved by this semigraphic method as illustrated in the following examples.

Example Problem 10-5

At a point in a structural member subjected to plane stress there are normal and shearing stresses on horizontal and vertical planes through the point, as shown in Fig. 10-14a. Determine and show on a sketch:

(a) The principal and maximum shearing stresses at the point.

(b) The normal and shearing stresses on plane A-A through the point.

SOLUTION

Mohr's circle is constructed from the given data by plotting point V (representing the stresses on the vertical plane) at $(8, -4)$ because the stresses on the vertical plane are 8 ksi T and 4 ksi (counterclockwise) shear. Likewise, point H (representing the stresses on the horizontal plane) has the coordinates $(-6, 4)$. Draw line HV, which is a diameter of Mohr's circle, and note that the center of the circle is at $(1, 0)$. The radius of the circle is

$$CV = \sqrt{7^2 + 4^2} = 8.06 \text{ ksi}$$

(a) The principal stresses and the maximum shearing stress at the point are

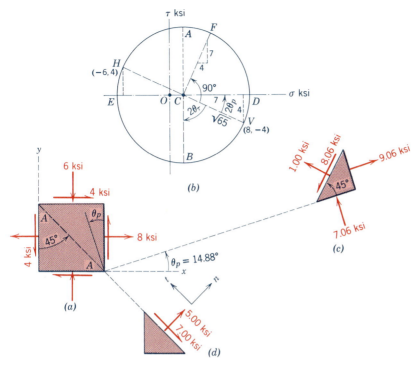

Figure 10-14

$$\sigma_{p1} = OD = 1 + 8.06 = +9.06 = 9.06 \text{ ksi T} \qquad \textbf{Ans.}$$

$$\sigma_{p2} = OE = 1 - 8.06 = -7.06 = 7.06 \text{ ksi C} \qquad \textbf{Ans.}$$

$$\sigma_{p3} = \sigma_z = 0 \qquad \textbf{Ans.}$$

Since σ_{p1} and σ_{p2} have opposite signs, the maximum shearing stress is

$$\tau_p = \tau_{\max} = CA = CB = 8.06 \text{ ksi} \qquad \textbf{Ans.}$$

The principal planes are represented by lines CD and CE, where

$$\tan 2\theta_p = 4/7 = 0.5714$$

which gives

$$2\theta_p = 29.75° \quad \text{or} \quad \theta_p = 14.88°$$

Since the angle $2\theta_p$ is counterclockwise, the principal planes are counterclockwise from the vertical and horizontal planes of the stress block, as shown in Fig. 10-14c. To determine which principal stress acts on which plane, note that as the radius of the circle rotates counterclockwise, the end of the radius CV moves from V to D, indicating that as the initially vertical plane rotates through 14.88°, the stresses change to 9.06 ksi T and zero. Note also that the end H of radius CH moves to E, indicating that as the initially horizontal plane rotates through 14.88°, the stresses change to 7.06 ksi C and zero. The required sketch is Fig. 10-14c.

(b) Plane *A-A* is 45° counterclockwise from the vertical plane; therefore, the corresponding radius of Mohr's circle is $2\theta = 90°$ counterclockwise from the line *CV*, and is shown as *CF* on Fig. 10-14*b*. The coordinates of the point *F* are seen to be (5, 7), which means that the stresses on plane *A-A* are 5.00 ksi T and 7.00 ksi shear in a direction to produce a clockwise moment on a stress element having plane *A-A* as one of its faces. This shear stress would be defined as a negative shear stress, since to produce a clockwise moment, it must be directed in the negative *t*-direction associated with plane *A-A* (see Fig. 10-14*d*). Note that Fig. 10-14*d* does not show all of the stresses acting on the element. ■

Example Problem 10-6

At a point in a structural member subjected to plane stress there are normal and shearing stresses on horizontal and vertical planes through the point, as shown in Fig. 10-15*a*. Determine, and show on a sketch, the principal stresses and the maximum shearing stress at the point.

SOLUTION

Mohr's circle is constructed from the given data by plotting point *V* (representing the stresses on the vertical plane) at (72, 24) and point *H* (representing the stresses on the horizontal plane) at (36, −24). Line *HV* (see Fig. 10-15*b*) between the two points is a diameter of Mohr's circle. The circle is centered at point *C* (54, 0) and has a radius *CV* equal to 30 MPa. The principal stresses at the point are

$$\sigma_{p1} = OD = 54 + 30 = 84 \text{ MPa T} \qquad \textbf{Ans.}$$

$$\sigma_{p2} = OE = 54 - 30 = 24 \text{ MPa T} \qquad \textbf{Ans.}$$

$$\sigma_{p3} = \sigma_z = 0 \qquad \textbf{Ans.}$$

The principal planes are represented by lines *CD* and *CE*, where

$$\tan 2\theta_p = -24/18 = -1.3333$$

which gives

$$2\theta_p = -53.13° = 53.13° \downarrow$$

$$\theta_p = -26.57° = 26.6° \downarrow$$

Since σ_{p1} and σ_{p2} have the same sign, the maximum in-plane shearing stress τ_p is not the maximum shearing stress $\tau_{\max}$ at the point. The maximum shearing stress at the point is represented on Mohr's circle by drawing an additional circle (shown dashed in Fig. 10-15*b*) that has line *OD* as a diameter. This circle is centered at point *F* (42, 0) and has a radius *FG* equal to 42 MPa. This circle represents the combinations of normal and shearing stresses existing on planes obtained by rotating the element about the principal axis associated with the principal stress σ_{p2}. A third Mohr's circle (shown dotted in Fig. 10-15*b*) has line *OE* as a diameter. This circle is centered at point *J* (12, 0) and has a radius *JK* equal to 12 MPa. This circle represents the combinations of normal and

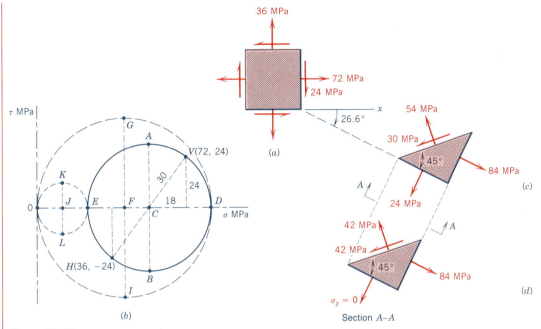

(a)

(b)

(c)

(d)

Section A–A

Figure 10-15

shearing stresses existing on planes obtained by rotating the element about the principal axis associated with the principal stress σ_{p1}. Thus

$$\tau_p = CA = 30 \text{ MPa}$$

$$\tau_{\max} = FG = 42 \text{ MPa} \qquad \textbf{Ans.}$$

The principal stresses σ_{p1}, σ_{p2}, and $\sigma_z = \sigma_{p3} = 0$, the maximum in-plane shearing stress τ_p, and the maximum shearing stress $\tau_{\max}$ at the point are all shown on Figs. 10-15c and d. ■

PROBLEMS

10-47* The stresses shown in Fig. P10-47 act at a point on the free surface of a stressed body. Use Mohr's circle to determine the normal and shearing stresses at this point on the inclined plane AB shown in the figure.

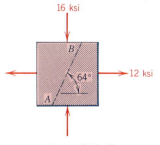

Figure P10-47

10-48* The stresses shown in Fig. P10-48 act at a point on the free surface of a stressed body. Use Mohr's circle to determine the normal and shearing stresses at this point on the inclined plane AB shown in the figure.

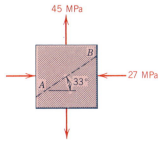

Figure P10-48

10-49 The stresses shown in Fig. P10-49 act at a point on the free surface of a stressed body. Use Mohr's circle to determine the normal and shearing stresses at this point on the inclined plane *AB* shown in the figure.

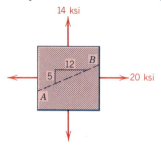

Figure P10-49

10-50 The stresses shown in Fig. P10-50 act at a point on the free surface of a stressed body. Use Mohr's circle to determine the normal and shearing stresses at this point on the inclined plane *AB* shown in the figure.

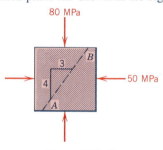

Figure P10-50

10-51* At a point in a structural member subjected to plane stress there are normal and shearing stresses on horizontal and vertical planes through the point, as shown in Fig. P10-51. Use Mohr's circle to determine
(a) The principal stresses and the maximum shearing stress at the point.
(b) The normal and shearing stresses on the inclined plane *AB* shown in the figure.

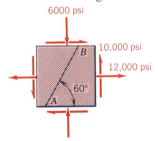

Figure P10-51

10-52* At a point in a structural member subjected to plane stress there are normal and shearing stresses on horizontal and vertical planes through the point, as

shown in Fig. P10-52. Use Mohr's circle to determine
(a) The principal stresses and the maximum shearing stress at the point.
(b) The normal and shearing stresses on the inclined plane *AB* shown in the figure.

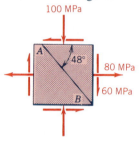

Figure P10-52

10-53 At a point in a structural member subjected to plane stress there are normal and shearing stresses on horizontal and vertical planes through the point, as shown in Fig. P10-53. Use Mohr's circle to determine
(a) The principal stresses and the maximum shearing stress at the point.
(b) The normal and shearing stresses on the inclined plane *AB* shown in the figure.

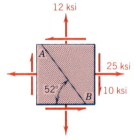

Figure P10-53

10-54 At a point in a structural member subjected to plane stress there are normal and shearing stresses on horizontal and vertical planes through the point, as shown in Fig. P10-54. Use Mohr's circle to determine
(a) The principal stresses and the maximum shearing stress at the point.
(b) The normal and shearing stresses on the inclined plane *AB* shown in the figure.

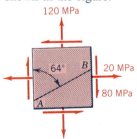

Figure P10-54

10-4 TWO-DIMENSIONAL OR PLANE STRAIN

The material of Section 4-4 serves to convey the concept of strain as a unit deformation, but it is inadequate for other than uniaxial loading. The extension of the concept to biaxial loading is essential because of the role played by strain in experimental methods of stress evaluation. In many practical problems involving the design of machine elements, the shape and loading are too complicated to permit stress determination solely by mathematical analysis; hence, this technique is supplemented by laboratory measurements.

Strains can be measured by several methods but, except for the simplest cases, stresses cannot be obtained directly. Consequently, the usual procedure in experimental stress analysis is to measure strains and calculate the state of stress by using stress–strain equations as illustrated later in Section 10-5.

The complete state of strain at an arbitrary point P in a body under load can be determined by considering the deformation associated with a small volume of material surrounding the point. For convenience the volume is normally assumed to have the shape of a rectangular parallelepiped with its faces oriented perpendicular to the reference x-, y-, and z-axes in the undeformed state, as shown in Fig. 10-16a. Since the element of volume is very small, deformations are assumed to be uniform; therefore, parallel planes remain plane and parallel and straight lines remain straight in the deformed element, as shown in Fig. 10-16b. The final size of the deformed element is determined by the lengths of the three edges dx', dy', and dz'. The distorted shape of the element is determined by the angles θ'_{xy}, θ'_{yz}, and θ'_{zx} between faces.

The Cartesian components of strain at the point can be expressed in terms of the deformations by using the definitions of normal and shearing strain presented in Section 4-4. These are the strain components associated with the Cartesian components of stress discussed in Section 10-3 and shown in Fig. 10-2. Thus,

$$\epsilon_x = \frac{dx' - dx}{dx} = \frac{d\delta_x}{dx} \qquad \gamma_{xy} = \frac{\pi}{2} - \theta'_{xy}$$

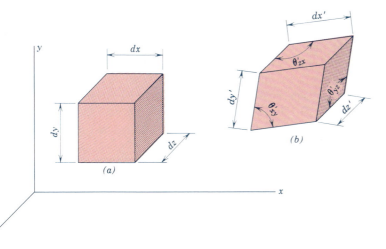

(a)

(b)

Figure 10-16

$$\epsilon_y = \frac{dy' - dy}{dy} = \frac{d\delta_y}{dy} \qquad \gamma_{yz} = \frac{\pi}{2} - \theta'_{yz} \qquad (10\text{-}9a)$$

$$\epsilon_z = \frac{dz' - dz}{dz} = \frac{d\delta_z}{dz} \qquad \gamma_{zx} = \frac{\pi}{2} - \theta'_{zx}$$

In a similar manner, the normal strain component associated with a line oriented in an arbitrary n-direction and the shearing strain component associated with two arbitrary orthogonal lines oriented in the n- and t-directions in the undeformed element are given by

$$\epsilon_n = \frac{dn' - dn}{dn} = \frac{d\delta_n}{dn} \qquad \gamma_{nt} = \frac{\pi}{2} - \theta'_{nt} \qquad (10\text{-}9b)$$

Alternative forms of Eq. (10-9), which will be useful in later developments, are

$$dx' = (1 + \epsilon_x)dx \qquad \theta'_{xy} = \frac{\pi}{2} - \gamma_{xy}$$

$$dy' = (1 + \epsilon_y)dy \qquad \theta'_{yz} = \frac{\pi}{2} - \gamma_{yz}$$

$$\qquad\qquad\qquad\qquad\qquad\qquad\qquad\qquad\qquad\qquad (10\text{-}10)$$

$$dz' = (1 + \epsilon_z)dz \qquad \theta'_{zx} = \frac{\pi}{2} - \gamma_{zx}$$

$$dn' = (1 + \epsilon_n)dn \qquad \theta'_{nt} = \frac{\pi}{2} - \gamma_{nt}$$

10-4-1 The Strain Transformation Equations for Plane Strain

The method of relating the components of strain associated with a Cartesian coordinate system to the normal and shearing strains associated with other orthogonal directions will be illustrated by considering the two-dimensional or plane strain case. If the xy-plane is taken as the reference plane, then for conditions of plane strain $\epsilon_z = \gamma_{zx} = \gamma_{zy} = 0$.[4]

Consider Fig. 10-17a in which the shaded rectangle represents a small unstrained element of material having the configuration of a rectangular parallelepiped. The sides of the rectangle are along directions for which the strains ϵ_x, ϵ_y, and γ_{xy} are known. The proportion of the rectangle is chosen so that the diagonal OB points in the direction n for which the strain ϵ_n is to be determined. When the body is subjected to a system of loads, the element assumes the shape indicated in Fig. 10-17b. The dimensions of the deformed element are given in terms of strains obtained from Eq. (10-10).

Normal Strain ϵ_n. An expression for the normal strain ϵ_n in the n-direction can be obtained by applying the law of cosines to the triangle $OC'B'$ shown in Fig. 10-17b. Thus,

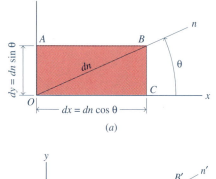

(a)

(b)

Figure 10-17

---

[4]Note that although there is no strain in the z-direction, $\epsilon_z = 0$, there must be a stress in the z-direction, $\sigma_z \neq 0$. That is, a stress is required in the z-direction to prevent deformation in the z-direction.

$$(OB')^2 = (OC')^2 + (C'B')^2 - 2(OC')(C'B') \cos\left(\frac{\pi}{2} + \gamma_{xy}\right)$$

or in terms of strains

$$[(1 + \epsilon_n)dn]^2 = [(1 + \epsilon_x)dx]^2 + [(1 + \epsilon_y)dy]^2$$
$$- 2[(1 + \epsilon_x)dx][(1 + \epsilon_y)dy][-\sin \gamma_{xy}] \qquad \text{(a)}$$

Substituting $dx = dn \cos \theta$ and $dy = dn \sin \theta$ (see Fig. 10-17a) into Eq. (a) yields

$$(1 + \epsilon_n)^2(dn)^2 = (1 + \epsilon_x)^2(dn)^2(\cos^2 \theta) + (1 + \epsilon_y)^2(dn)^2(\sin^2 \theta)$$
$$+ 2(dn)^2(\sin \theta)(\cos \theta)(1 + \epsilon_x)(1 + \epsilon_y)(\sin \gamma_{xy}) \qquad \text{(b)}$$

Since the strains are small, it follows that $\epsilon^2 << \epsilon$, $\sin \gamma \approx \gamma$, and so forth; hence, all second-degree terms such as ϵ^2, $\gamma\epsilon$, and the like, can be neglected as Eq. (b) is expanded to become

$$1 + 2\epsilon_n = (1 + 2\epsilon_x) \cos^2 \theta + (1 + 2\epsilon_y) \sin^2 \theta + 2\gamma_{xy} \sin \theta \cos \theta$$

which reduces to

$$\epsilon_n = \epsilon_x \cos^2 \theta + \epsilon_y \sin^2 \theta + \gamma_{xy} \sin \theta \cos \theta \qquad \text{(10-11a)}$$

or in terms of the double angle

$$\epsilon_n = \frac{\epsilon_x + \epsilon_y}{2} + \frac{\epsilon_x - \epsilon_y}{2} \cos 2\theta + \frac{\gamma_{xy}}{2} \sin 2\theta \qquad \text{(10-11b)}$$

Shearing Strain γ_{nt}.   The shearing strain γ_{nt} measures the amount by which the right angle between the n- and t-directions decreases as the material deforms. As the material deforms, the n-direction rotates counterclockwise through an angle ϕ_n as shown in Fig. 10-18. Applying the law of sines to triangle $OC'B'$

$$\frac{OB'}{\sin \angle OC'B'} = \frac{B'C'}{\sin \angle B'OC'}$$

which gives

$$B'C' \sin \angle OC'B' = OB' \sin \angle B'OC'$$

or in terms of the strains,

$$(1 + \epsilon_y)dy \sin\left(\frac{\pi}{2} + \gamma_{xy}\right) = (1 + \epsilon_n)dn \sin [\theta + (\phi_n - \psi)] \qquad \text{(c)}$$

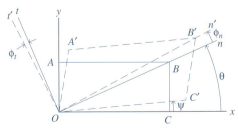

Figure 10-18

Since the strains (ϵ_x, ϵ_y, γ_{xy}, and ϵ_n) and the angles ϕ_n and ψ are all small,

$$\sin\left(\frac{\pi}{2} + \gamma_{xy}\right) = \cos\gamma_{xy} \cong 1$$

$$\sin[\theta + (\phi_n - \psi)] = \sin\theta\cos(\phi_n - \psi) + \cos\theta\sin(\phi_n - \psi)$$
$$\cong \sin\theta + (\phi_n - \psi)\cos\theta$$

and Eq. (c) can be written

$$(1 + \epsilon_y)dy \cong (1 + \epsilon_n)dn\,[\sin\theta + (\phi_n - \psi)\cos\theta] \qquad (d)$$

where $dy = dn\sin\theta$ (see Fig. 10-17a). Therefore, Eq. (d) can be reduced to

$$(\epsilon_y - \epsilon_n)\sin\theta \cong (\phi_n - \psi)\cos\theta + \epsilon_n(\phi_n - \psi)\cos\theta$$
$$\cong (\phi_n - \psi)\cos\theta \qquad (e)$$

since ϵ_n, ϕ_n, and ψ are all small. Substituting Eq. (10-11a) into Eq. (e) and solving for ϕ_n yields

$$\phi_n = -(\epsilon_x - \epsilon_y)\sin\theta\cos\theta - \gamma_{xy}\sin^2\theta + \psi \qquad (f)$$

Equation (f) gives the counterclockwise rotation of a line that makes an angle of θ with the x-axis initially. Thinking of ϕ_n as $\phi(\theta)$, the counterclockwise rotation of the t-axis can be written

$$\phi_t = \phi\left(\theta + \frac{\pi}{2}\right)$$
$$= -(\epsilon_x - \epsilon_y)\sin\left(\theta + \frac{\pi}{2}\right)\cos\left(\theta + \frac{\pi}{2}\right) - \gamma_{xy}\sin^2\left(\theta + \frac{\pi}{2}\right) + \psi$$
$$= (\epsilon_x - \epsilon_y)\cos\theta\sin\theta - \gamma_{xy}\cos^2\theta + \psi \qquad (g)$$

Finally, the shearing strain γ_{nt} is the decrease in the right angle between the n- and t-directions or the difference between the rotations ϕ_n and ϕ_t. Therefore,

$$\gamma_{nt} = \phi_n - \phi_t$$
$$= -2(\epsilon_x - \epsilon_y)\sin\theta\cos\theta + \gamma_{xy}(\cos^2\theta - \sin^2\theta) \qquad (10\text{-}12a)$$

or in terms of the double angle

$$\gamma_{nt} = -(\epsilon_x - \epsilon_y)\sin 2\theta + \gamma_{xy}\cos 2\theta \qquad (10\text{-}12b)$$

Equations (10-11) and (10-12) provide a means for determining the normal strain ϵ_n associated with a line oriented in an arbitrary n-direction in the xy-plane and the shearing strain γ_{nt} associated with any two orthogonal lines oriented in the n- and t-directions in the xy-plane when the strains ϵ_x, ϵ_y, and γ_{xy} associated with the coordinate directions are known. When these equations are used, the sign conventions used in their development must be rigorously followed. The sign conventions used are as follows:

1. Tensile strains are positive; compressive strains are negative.
2. Shearing strains that decrease the angle between the two lines at the origin of coordinates are positive.

3. Angles measured counterclockwise from the reference x-axis are positive.

4. The (n, t, z) axes have the same order as the (x, y, z) axes. Both sets of axes form a right-hand coordinate system.

PROBLEMS

10-55* The strain components at a point are $\epsilon_x = +800\mu$, $\epsilon_y = -1000\mu$, and $\gamma_{xy} = -600\mu$. Determine the strain components ϵ_n, ϵ_t, and γ_{nt} if the xy- and nt-axes are oriented as shown in Fig. P10-55.

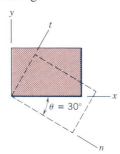

Figure P10-55

10-56* The strain components at a point are $\epsilon_x = +900\mu$, $\epsilon_y = +650\mu$, and $\gamma_{xy} = +300\mu$. Determine the strain components ϵ_n, ϵ_t, and γ_{nt} if the xy- and nt-axes are oriented as shown in Fig. P10-56.

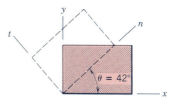

Figure P10-56

In Problems 10-57 through 10-62 the strain components ϵ_x, ϵ_y, and γ_{xy} are given for a point in a body subjected to plane strain. Determine the strain components ϵ_n, ϵ_t, and γ_{nt} at the point if the nt-axes are rotated with respect to the xy-axes by the amount and in the direction indicated by the angle θ.

Problem	ϵ_x	ϵ_y	γ_{xy}	θ
10-57	$+800\mu$	-950μ	-800μ	$-42°$
10-58	$+750\mu$	$+360\mu$	-300μ	$+52°$
10-59*	-100μ	-700μ	$+400\mu$	$-28°$
10-60*	$+900\mu$	$+650\mu$	-300μ	$+32°$
10-61	$+720\mu$	-480μ	$+360\mu$	$-30°$
10-62	$+850\mu$	$+450\mu$	$+600\mu$	$+20°$

10-4-2 Principal Strains and Maximum Shearing Strain

The similarity between Eqs. (10-11) and (10-12) for plane strain and Eqs. (10-1) and (10-2) for plane stress indicates that all the equations developed for plane stress can be applied to plane strain by substituting ϵ_x for σ_x, ϵ_y for σ_y, and $\gamma_{xy}/2$ for τ_{xy}. Thus, from Eqs. (10-3), (10-4), and (10-5), expressions are obtained for determining in-plane principal directions, in-plane principal strains, and the maximum in-plane shearing strain. Thus,

$$\tan 2\theta_p = \frac{\gamma_{xy}}{\epsilon_x - \epsilon_y} \qquad (10\text{-}13)$$

$$\epsilon_{p1}, \epsilon_{p2} = \frac{\epsilon_x + \epsilon_y}{2} \pm \sqrt{\left(\frac{\epsilon_x - \epsilon_y}{2}\right)^2 + \left(\frac{\gamma_{xy}}{2}\right)^2} \qquad (10\text{-}14)$$

$$\gamma_p = 2\sqrt{\left(\frac{\epsilon_x - \epsilon_y}{2}\right)^2 + \left(\frac{\gamma_{xy}}{2}\right)^2} \qquad (10\text{-}15)$$

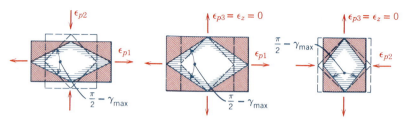

Figure 10-19

In the previous equations, normal strains that are tensile and shearing strains that decrease the angle between the faces of the element at the origin of coordinates (see Fig. 10-17) are positive.

When a state of plane strain exists, Eq. (10-14) gives the two in-plane principal strains while the third principal strain is $\epsilon_{p3} = \epsilon_z = 0$.[5] An examination of Eqs. (10-14) and (10-15) indicates that the maximum in-plane shearing strain is the difference between the in-plane principal strains, but this may not be the maximum shearing strain at the point. The maximum shearing strain at the point may be $(\epsilon_{p1} - \epsilon_{p2})$, $(\epsilon_{p1} - 0)$, or $(0 - \epsilon_{p2})$, depending on the relative magnitudes and signs of the principal strains. The lines associated with the maximum shearing strain bisect the angles between lines experiencing maximum and minimum normal strains. The three possibilities are illustrated in Fig. 10-19.

Example Problem 10-7

The strain components at a point in a body under a state of plane strain are $\epsilon_x = +1200\mu$, $\epsilon_y = -600\mu$, and $\gamma_{xy} = +900\mu$. Determine the principal strains and the maximum shearing strain at the point. Show the principal strain deformations and the maximum shearing strain distortion on a sketch.

SOLUTION
When the given data are substituted in Eqs. (10-13), (10-14), and (10-15), they yield the in-plane principal strains at the point, their orientations, and the maximum in-plane shearing strain. Thus,

$$\tan 2\theta_p = \frac{900}{1200 + 600}$$

$$(\epsilon_{p1}, \epsilon_{p2})(10^6) = \frac{1200 - 600}{2} \pm \sqrt{\left(\frac{1200 + 600}{2}\right)^2 + \left(\frac{900}{2}\right)^2}$$

$$\gamma_p(10^6) = 2\sqrt{\left(\frac{1200 + 600}{2}\right)^2 + \left(\frac{900}{2}\right)^2}$$

from which

$$\epsilon_{p1} = +1306\mu \qquad \qquad \textbf{Ans.}$$

$$\epsilon_{p2} = -706\mu \qquad \qquad \textbf{Ans.}$$

---

[5]Note that although there is no strain in the z-direction, $\epsilon_{p3} = \epsilon_z = 0$, there must be a stress in the z-direction, $\sigma_{p3} = \sigma_z \neq 0$. That is, a stress is required in the z-direction to prevent deformation in the z-direction.

$$\epsilon_{p3} = 0 \qquad \textbf{Ans.}$$

$$\theta_p = 13.3°$$

$$\gamma_p = +2012\mu = \gamma_{max} \qquad \textbf{Ans.}$$

The required sketch is given in Fig. 10-20. ■

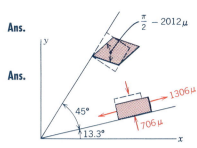

Figure 10-20

PROBLEMS

In Problems 10-63 through 10-74 the strain components ϵ_x, ϵ_y, and γ_{xy} are given for a point in a body subjected to plane strain. Determine the principal strains and the maximum shearing strain at the point. Show the principal strain deformations and the maximum shearing strain distortion on a sketch.

Problem	ϵ_x	ϵ_y	γ_{xy}
10-66	-900μ	$+600\mu$	$+480\mu$
10-67*	-750μ	$+1000\mu$	-250μ
10-68*	$+750\mu$	-410μ	$+360\mu$
10-69	$+720\mu$	$+520\mu$	$+480\mu$
10-70	-540μ	-980μ	-560μ
10-71*	$+864\mu$	$+432\mu$	-288μ
10-72*	$+650\mu$	$+900\mu$	$+300\mu$
10-73	-325μ	-625μ	$+380\mu$
10-74	$+900\mu$	$+650\mu$	$+600\mu$

Problem	ϵ_x	ϵ_y	γ_{xy}
10-63*	$+600\mu$	-200μ	$+500\mu$
10-64*	$+960\mu$	-320μ	-480μ
10-65	$+900\mu$	-300μ	-420μ

10-4-3 Mohr's Circle for Plane Strain

The pictorial or graphic representation of Eqs. (10-1) and (10-2), known as Mohr's circle for stress, can be used with Eqs. (10-11) and (10-12) to yield a Mohr's circle for strain. The equation for the strain circle obtained from the equation for the stress circle by using a change in variables is

$$\left(\epsilon_n - \frac{\epsilon_x + \epsilon_y}{2}\right)^2 + \left(\frac{\gamma_{nt}}{2}\right)^2 = \left(\frac{\epsilon_x - \epsilon_y}{2}\right)^2 + \left(\frac{\gamma_{xy}}{2}\right)^2$$

The variables in this equation are ϵ_n and $\gamma_{nt}/2$. The circle is centered on the ϵ axis at a distance $(\epsilon_x + \epsilon_y)/2$ from the origin and has a radius

$$R = \sqrt{\left(\frac{\epsilon_x - \epsilon_y}{2}\right)^2 + \left(\frac{\gamma_{xy}}{2}\right)^2}$$

Mohr's circle for the strains of Fig. 10-17 (with $\epsilon_x > \epsilon_y$) is given in Fig. 10-21. It is apparent that the sign convention for shearing strain needs to be extended to cover the construction of Mohr's circle. Observe that for positive shearing strain (indicated in Fig. 10-17), the edge of the element parallel to the x-axis tends to rotate counterclockwise while the edge parallel to the y-axis tends to rotate clockwise. For Mohr's circle construction, the clockwise rotation will be designated positive and the counterclockwise designated negative. This is consistent with the sign convention for shearing stresses given in Section 10-3. The Mohr's circle solution for Example Problem 10-7 is shown in Fig. 10-22.

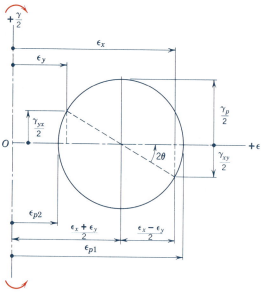

Figure 10-21

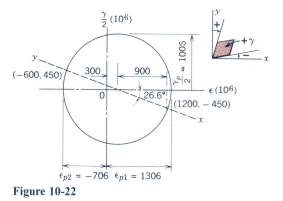

Figure 10-22

PROBLEMS

In Problems 10-75 through 10-82 certain strain components and angles are given for a point in a body subjected to plane strain. Use Mohr's circle to determine the unknown quantities for each problem and prepare a sketch showing the angle θ_p, the principal strain deformations, and the maximum shearing strain distortions. In some problems there may be more than one possible value of θ_p depending on the sign of γ_{xy}.

Problem	ϵ_x	ϵ_y	γ_{xy}	ϵ_{p1}	ϵ_{p2}	γ_p	γ_{max}	θ_p
10-75*	—	—	—	$+600\mu$	-400μ	—	—	$+18.43°$
10-76*	—	—	—	$+785\mu$	-945μ	—	—	$+16.85°$
10-77	—	—	—	$+708\mu$	-104μ	—	—	$-34.1°$
10-78	—	—	—	-114μ	-903μ	—	—	$+19.26°$
10-79*	$+950\mu$	-225μ	$+275\mu$	—	—	—	—	—
10-80*	$+900\mu$	-333μ	$+982\mu$	—	—	—	—	—
10-81	-750μ	-390μ	$+900\mu$	—	—	—	—	—
10-82	$+600\mu$	$+480\mu$	-480μ	—	—	—	—	—

10-5 GENERALIZED HOOKE'S LAW

Hooke's law (see Eq. 4-15) can be extended to include the biaxial (see Fig. 10-3) and triaxial (see Fig. 10-2) states of stress often encountered in engineering practice. Consider Fig. 10-23, which shows a differential element of material subjected to a biaxial state of normal stress. Shearing stresses have not been shown on the faces of the element since they produce distortion of the element (angle changes) but do not produce changes in the lengths of the sides of the element, which would contribute to the normal strains. The deformations of the element

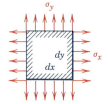

Figure 10-23

in the directions of the normal stresses, for a combined loading, can be determined by computing the deformations resulting from the individual stresses separately and adding the values obtained algebraically. This procedure is based on the principle of superposition, which states that the effects of separate loadings can be added algebraically if two conditions are satisfied:

1. Each effect is linearly related to the load that produced it.
2. The effect of the first load does not significantly change the effect of the second load.

Condition 1 is satisfied if the stresses do not exceed the proportional limit for the material. Condition 2 is satisfied if the deformations are small so that the small changes in the areas of the faces of the element do not produce significant changes in the stresses.

The deformations of the element of Fig. 10-23, associated with the stresses σ_x and σ_y, are shown in Fig. 10-24. The shaded square in Fig. 10-24a indicates the original or unstrained configuration of the element. Under the action of the tensile stress σ_x, the element experiences a tensile strain of σ_x/E in the x-direction and a compressive strain of $\nu\sigma_x/E$ in the y-direction. These strains cause the element to stretch an amount $(\sigma_x/E)dx$ in the x-direction and to contract an amount $(\nu\sigma_x/E)dy$ in the y-direction to the configuration indicated in Fig. 10-24b (the deformations are greatly exaggerated). Then, under the action of the tensile stress σ_y superimposed on the stress σ_x, the element experiences a tensile strain of σ_y/E

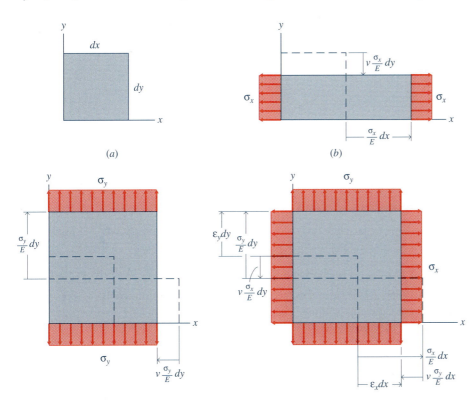

Figure 10-24

in the y-direction and a compressive strain of $\nu\sigma_y/E$ in the x-direction. These strains cause the element to stretch an amount $(\sigma_y/E)dy$ in the y-direction and to contract an amount $(\nu\sigma_y/E)dx$ in the x-direction, as shown in Fig. 10-24c. If the material is isotropic, Young's modulus has the same value for all directions and the final deformation in the x-direction is (Fig. 10-24d)

$$d\delta_x = \epsilon_x\, dx = \frac{\sigma_x}{E}\, dx - \nu\frac{\sigma_y}{E}\, dx$$

and the three normal strains are

$$\epsilon_x = \frac{1}{E}(\sigma_x - \nu\sigma_y)$$

$$\epsilon_y = \frac{1}{E}(\sigma_y - \nu\sigma_x) \tag{10-16}$$

$$\epsilon_z = -\frac{\nu}{E}(\sigma_x + \sigma_y)$$

The analysis above is readily extended to triaxial principal stresses, and the expressions for strain become

$$\epsilon_x = \frac{1}{E}[\sigma_x - \nu(\sigma_y + \sigma_x)]$$

$$\epsilon_y = \frac{1}{E}[\sigma_y - \nu(\sigma_x + \sigma_z)] \tag{10-17}$$

$$\epsilon_z = \frac{1}{E}[\sigma_z - \nu(\sigma_x + \sigma_y)]$$

In these expressions, tensile stresses and strains are considered positive, and compressive stresses and strains are considered negative.

When Eqs. (10-16) are solved for the stresses in terms of the strains, they give

$$\sigma_x = \frac{E}{1 - \nu^2}(\epsilon_x + \nu\epsilon_y)$$

$$\sigma_y = \frac{E}{1 - \nu^2}(\epsilon_y + \nu\epsilon_x) \tag{10-18}$$

Equations (10-18) can be used to calculate normal stresses from measured or computed normal strains. When Eqs. (10-17) are solved for stresses in terms of strains, they give

$$\sigma_x = \frac{E}{(1 + \nu)(1 - 2\nu)}[(1 - \nu)\,\epsilon_x + \nu\,(\epsilon_y + \epsilon_z)]$$

$$\sigma_y = \frac{E}{(1 + \nu)(1 - 2\nu)}[(1 - \nu)\,\epsilon_y + \nu\,(\epsilon_z + \epsilon_x)] \tag{10-19}$$

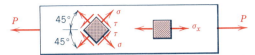

Figure 10-25

$$\sigma_z = \frac{E}{(1 + \nu)(1 - 2\nu)} [(1 - \nu)\,\epsilon_z + \nu\,(\epsilon_x + \epsilon_y)]$$

Torsion test specimens are used to study material behavior under pure shear, and it is observed that a shearing stress produces only a single corresponding shearing strain. Thus Hooke's law extended to shearing stresses is simply

$$\tau = G\gamma \qquad (10\text{-}20)$$

Equations (10-19) and (10-20) seem to indicate that three elastic constants—E, ν, and G—are required to determine the deformations (and strains) in a material resulting from an arbitrary state of stress. In fact, only two of these constants need be determined experimentally for a given material.

The relationship between the elastic constants E, ν, and G can be determined by considering the stresses and strains produced by an axial tensile load in a bar of the material. Equations (4-7) and (4-8) indicate that both normal and shearing stresses are produced on different inclined planes through the bar by the axial tensile load. For the discussion that follows, consider the stresses that develop on the faces of the two small square elements shown in Fig. 10-25. From Eqs. (4-7) and (4-8),

$$\sigma_x = \frac{P}{2A}(1 + \cos 0°) = \frac{P}{A}$$

$$\sigma = \frac{P}{2A}(1 + \cos 90°) = \frac{P}{2A} = \frac{1}{2}\sigma_x \qquad (a)$$

$$\tau = \frac{P}{2A}\sin 90° = \frac{P}{2A} = \frac{1}{2}\sigma_x$$

The strains produced by these stresses are given by Eqs. (10-16) and (10-20) as

$$\epsilon_x = \frac{\sigma_x}{E} \qquad \epsilon_y = -\frac{\nu\sigma_x}{E} \qquad \gamma = \frac{\tau}{G} = \frac{\sigma_x}{2G} \qquad (b)$$

Under the action of the axial tensile load, the two square elements shown in Fig. 10-25 deform into the shapes shown in Fig. 10-26. In this figure, the deformations of the two elements have been superimposed to show their relationships. The relationship between the shearing strain γ and the two normal strains ϵ_x and ϵ_y is determined from the geometry of Fig. 10-26. Thus,

$$\tan\left(\frac{\pi}{4} - \frac{\gamma}{2}\right) = \frac{(1 + \epsilon_y)(dL/2)}{(1 + \epsilon_x)(dL/2)} \qquad (c)$$

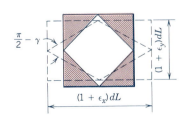

Figure 10-26

The tangent of the difference between two angles can be expressed as

$$\frac{\tan \dfrac{\pi}{4} - \tan \dfrac{\gamma}{2}}{1 + \tan \dfrac{\pi}{4} \tan \dfrac{\gamma}{2}} = \frac{1 + \epsilon_y}{1 + \epsilon_x} = \frac{1 - \nu\epsilon_x}{1 + \epsilon_x} \tag{d}$$

Since all of the strains are small, Eq. (d) reduces to

$$\frac{1 - \dfrac{\gamma}{2}}{1 + \dfrac{\gamma}{2}} = \frac{1 - \nu\epsilon_x}{1 + \epsilon_x} \tag{e}$$

Solving for the shearing strain γ yields

$$\gamma = \frac{(1 + \nu)\epsilon_x}{1 + (1/2)(1 - \nu)(\epsilon_x)} \approx (1 + \nu)\epsilon_x \tag{f}$$

Finally, if Eqs. (b) are substituted into Eq. (f),

$$\frac{\tau}{G} = \frac{\sigma_x}{2G} = (1 + \nu)\frac{\sigma_x}{E} \tag{g}$$

Solving Eq. (g) for G yields the desired relationship between G, E, and ν. Thus,

$$G = \frac{E}{2(1 + \nu)} \tag{10-21}$$

If Eq. (10-21) is substituted into Eq. (10-20), an alternate form of generalized Hooke's law for shearing stress and strain in isotropic materials is obtained. Thus

$$\tau_{xy} = G\gamma_{xy} = \frac{E}{2(1 + \nu)} \gamma_{xy}$$

$$\tau_{yz} = G\gamma_{yz} = \frac{E}{2(1 + \nu)} \gamma_{yz} \tag{10-22}$$

$$\tau_{zx} = G\gamma_{zx} = \frac{E}{2(1 + \nu)} \gamma_{zx}$$

Equations (10-16) thru (10-22) are widely used for experimental stress determinations. The following example illustrates the method of application.

Example Problem 10-8

At a point on the surface of an alloy steel ($E = 210$ GPa and $\nu = 0.30$) machine part subjected to a biaxial state of stress, the measured strains were

$\epsilon_x = +1394 \ \mu\text{m/m}$, $\epsilon_y = -660 \ \mu\text{m/m}$, and $\gamma_{xy} = 2054 \ \mu\text{rad}$. Determine the stresses σ_x, σ_y, and τ_{xy} at the point.

SOLUTION

The normal stresses σ_x and σ_y are obtained by using Eqs. (10-18). Thus,

$$\sigma_x = \frac{E}{1 - \nu^2} (\epsilon_x + \nu\epsilon_y)$$

$$= \frac{210(10^9)}{1 - (0.30)^2} [1394 + 0.3(-660)](10^{-6})$$

$$= +276(10^6) \ \text{N/m}^2 = 276 \ \text{MPa T} \qquad \text{Ans.}$$

$$\sigma_y = \frac{210(10^9)}{1 - (0.30)^2} [-660 + 0.3(1394)](10^{-6})$$

$$= -55.8(10^6) \ \text{N/m}^2 = 55.8 \ \text{MPa C} \qquad \text{Ans.}$$

The shearing stress τ_{xy} is obtained by using Eqs. (10-22). Thus,

$$\tau_{xy} = \frac{E}{2(1 + \nu)} \gamma_{xy}$$

$$= \frac{210(10^9)}{2(1 + 0.30)} (2054)(10^{-6})$$

$$= 165.9(10^6) \ \text{N/m}^2 = 165.9 \ \text{MPa} \qquad \text{Ans.}$$

Once the stresses σ_x, σ_y, and τ_{xy} are known, the methods presented in Section 10-3 can be used to determine the stresses on any other plane of interest through the point. ■

PROBLEMS

10-83* At a point on the surface of an aluminum alloy ($E = 10,000$ ksi and $G = 3800$ ksi) machine part subjected to a biaxial state of stress, the measured strains were $\epsilon_x = +900 \ \mu\text{in./in.}$, $\epsilon_y = -300 \ \mu\text{in./in.}$, and $\gamma_{xy} = -400 \ \mu\text{rad}$. Determine the stresses σ_x, σ_y, and τ_{xy} at the point.

10-84* At a point on the surface of a structural steel ($E = 200$ GPa and $G = 76$ GPa) machine part subjected to a biaxial state of stress, the measured strains were $\epsilon_x = +750 \ \mu\text{m/m}$, $\epsilon_y = +350 \ \mu\text{m/m}$, and $\gamma_{xy} = -560 \ \mu\text{rad}$. Determine the stresses σ_x, σ_y, and τ_{xy} at the point.

10-85 At a point on the surface of a titanium alloy ($E = 14,000$ ksi and $G = 5300$ ksi) aircraft part subjected to a biaxial state of stress, the measured strains were $\epsilon_x = +1250 \ \mu\text{in./in.}$, $\epsilon_y = +600 \ \mu\text{in./in.}$, and $\gamma_{xy} = 650 \ \mu\text{rad}$. Determine the stresses σ_x, σ_y, and τ_{xy} at the point.

10-86 At a point on the surface of a stainless steel ($E = 190$ GPa and $G = 76$ GPa) machine part subjected to a biaxial state of stress, the measured strains were $\epsilon_x = +1175 \ \mu\text{m/m}$, $\epsilon_y = -1250 \ \mu\text{m/m}$, and $\gamma_{xy} = +850 \ \mu\text{rad}$. Determine the stresses σ_x, σ_y, and τ_{xy} at the point.

10-87* Determine the state of strain that corresponds to the following state of stress at a point in a steel ($E = 30,000$ ksi and $\nu = 0.30$) machine part: $\sigma_x = 15,000$ psi, $\sigma_y = 5000$ psi, $\sigma_z = 7500$ psi, $\tau_{xy} = 5500$ psi, $\tau_{yz} = 4750$ psi, and $\tau_{zx} = 3200$ psi.

10-88 Determine the state of strain that corresponds to the following state of stress at a point in an aluminum alloy ($E = 73$ GPa and $\nu = 0.33$) machine part: $\sigma_x = 120$ MPa, $\sigma_y = -85$ MPa, $\sigma_z = 45$ MPa, $\tau_{xy} = 35$ MPa, $\tau_{yz} = 48$ MPa, and $\tau_{zx} = 76$ MPa.

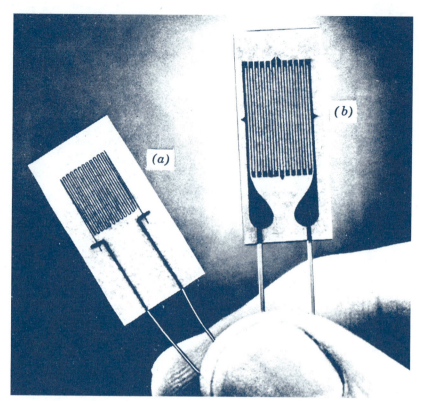

Figure 10-27

10-5-1 Strain Measurement and Rosette Analysis

Electrical resistance strain gages provide accurate measurements of normal strain. The gage may consist of a length of 0.001 in.–diameter wire arranged as shown in Fig. 10-27a and cemented between two pieces of paper, or it may be an etched foil conductor mounted on epoxy or polyimide backing (see Fig. 10-27b). The wire or foil gage is cemented to the material for which the strain is to be determined. As the material is strained, the wires are lengthened or shortened; this changes the electrical resistance of the gage. The change in resistance can be measured and calibrated to provide a normal strain. Shearing strains are more difficult to measure directly than normal strains and are often obtained by measuring normal strains in two or three different directions. The shearing strain γ_{xy} can be computed from normal strain data by using Eq. (10-11a). For example, consider the most general case of three arbitrary normal strain measurements as shown in Fig. 10-28. From Eq. (10-11a)

$$\epsilon_a = \epsilon_x \cos^2 \theta_a + \epsilon_y \sin^2 \theta_a + \gamma_{xy} \sin \theta_a \cos \theta_a$$

$$\epsilon_b = \epsilon_x \cos^2 \theta_b + \epsilon_y \sin^2 \theta_b + \gamma_{xy} \sin \theta_b \cos \theta_b$$

$$\epsilon_c = \epsilon_x \cos^2 \theta_c + \epsilon_y \sin^2 \theta_c + \gamma_{xy} \sin \theta_c \cos \theta_c$$

From the measured value of ϵ_a, ϵ_b, and ϵ_c and a knowledge of the gage orientations θ_a, θ_b, and θ_c with respect to the reference x-axis, the values of ϵ_x, ϵ_y, and γ_{xy} can be determined by simultaneous solution of the three equations. In practice the angles θ_a, θ_b, and θ_c are selected to simplify the calculations. Multiple-

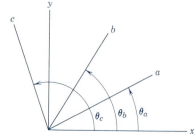

Figure 10-28

element strain gages used for this type of measurement are known as *strain rosettes*. Two rosette configurations marketed commercially are shown in Fig. 10-29. Once ϵ_x, ϵ_y, and γ_{xy} have been determined, Eqs. (10-13), (10-14), and (10-15) or the corresponding Mohr's circle can be used to determine the in-plane principal strains, their orientations, and the maximum in-plane shearing strain at the point.

The principal strain $\epsilon_z = \epsilon_{p3}$ can be determined from the measured data. From Eqs. (10-16) and (10-18), which represent Hooke's law for the case of plane stress,

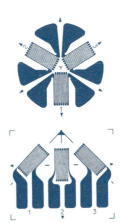

$$\epsilon_z = -\frac{\nu}{E}(\sigma_x + \sigma_y) = -\frac{\nu}{E}\left[\frac{E}{1-\nu^2}\right][(\epsilon_x + \nu\epsilon_y) + (\epsilon_y + \nu\epsilon_x)]$$

$$= -\frac{\nu}{1-\nu}(\epsilon_x + \epsilon_y) \qquad (10\text{-}23)$$

Figure 10-29

This out-of-plane principal strain is important since the maximum shearing strain at the point may be $(\epsilon_{p1} - \epsilon_{p2})$, $(\epsilon_{p1} - \epsilon_{p3})$, or $(\epsilon_{p3} - \epsilon_{p2})$, depending on the relative magnitudes and signs of the principal strains at the point.

The following example illustrates the application of Eqs. (10-11) through (10-23) and Mohr's circle to principal stress and strain and maximum shearing stress and strain determinations under conditions of plane stress.

Example Problem 10-9

A strain rosette, composed of three electrical resistance gages making angles of 0°, 60°, and 120° with the x-axis, was mounted on the free surface of a steel ($E = 30,000$ ksi and $\nu = 0.30$) machine component. Under load, the following strains were measured:

$$\epsilon_0 = \epsilon_x = +1000\mu \qquad \epsilon_{60} = -650\mu \qquad \epsilon_{120} = +750\mu$$

(a) Determine the principal strains and the maximum shearing strain at the point. Show the directions of the in-plane principal strains on a sketch.

(b) Determine the principal stresses and the maximum shearing stress at the point. Show the principal stresses and the maximum shearing stress on a sketch.

SOLUTION

(a) When the given data are substituted into Eq. (10-11b), the following equations (in terms of μ) are obtained:

$$\epsilon_x = 1000 \text{ (measured)}$$

$$-650 = \frac{1000 + \epsilon_y}{2} + \frac{1000 - \epsilon_y}{2}(\cos 120°) + \frac{\gamma_{xy}}{2}(\sin 120°)$$

$$750 = \frac{1000 + \epsilon_y}{2} + \frac{1000 - \epsilon_y}{2}(\cos 240°) + \frac{\gamma_{xy}}{2}(\sin 240°)$$

from which

$$\epsilon_y = -266.7\mu \qquad \gamma_{xy} = -1616.6\mu$$

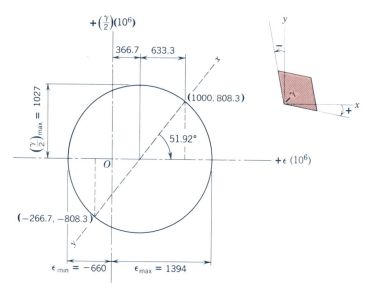

Figure 10-30

These values are used to construct the Mohr's circle of Fig. 10-30. The in-plane principal strains obtained from the circle are

$$\epsilon_{p1} = 1394\mu \qquad \epsilon_{p2} = -660\mu$$

The principal strain perpendicular to the surface, $\epsilon_z = \epsilon_{p3}$, is found from Eq. (10-23) as follows:

$$\epsilon_z = \epsilon_{p3} = -\frac{\nu}{1-\nu}(\epsilon_x + \epsilon_y)$$

$$= -\frac{0.30}{1-0.30}(1000\mu - 266.7\mu) = -314.3\mu$$

Since ϵ_{p3} is less compressive than the in-plane principal strain ϵ_{p2}, the maximum shearing strain at the point is the maximum in-plane shearing strain γ_p obtained from the circle. Thus,

$$\gamma_{max} = \gamma_p = \pm(1394\mu + 660\mu) = \pm2054\mu \cong \pm2050\mu \qquad \textbf{Ans.}$$

The principal strains are

$$\epsilon_{p1} = \epsilon_{max} = +1394\mu$$

$$\epsilon_{p3} = \epsilon_{int} = -314\mu \qquad \textbf{Ans.}$$

$$\epsilon_{p2} = \epsilon_{min} = -660\mu$$

(b) The principal stresses and the maximum shearing stress at the point are obtained from the strain results of part (a) by using Eqs. (10-18) and (10-22). Thus,

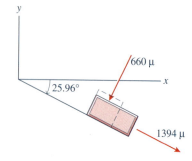

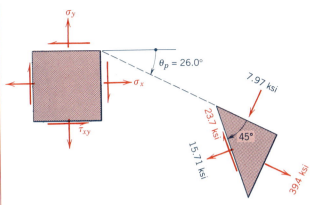

Figure 10-31

$$\sigma_{max} = \frac{30(10^6)}{1 - 0.09}[1394 + 0.30(-660)](10^{-6})$$
$$= +39{,}429 \text{ psi} \cong 39.4 \text{ ksi T} \qquad \textbf{Ans.}$$

$$\sigma_{min} = \frac{30(10^6)}{1 - 0.09}[-660 + 0.30(1394)](10^{-6})$$
$$= -7971 \text{ psi} \cong 7.97 \text{ ksi C} \qquad \textbf{Ans.}$$

$$\tau_{max} = \frac{30(10^6)}{2(1 + 0.30)}(2054)(10^{-6}) = 23{,}700 \text{ psi} = 23.7 \text{ ksi} \qquad \textbf{Ans.}$$

Note that the value for τ_{max} checks that given by Eq. (10-7); thus,

$$\tau_{max} = \frac{\sigma_{p1} - \sigma_{p2}}{2} = \frac{39{,}429 + 7971}{2} = 23{,}700 \text{ psi} = 23.7 \text{ ksi}$$

The directions of the principal stresses will be the same as those for the principal strains; hence, from Mohr's circle, $\theta_p = \frac{1}{2}(51.92) = 25.96° \cong 26.0°$ clockwise as shown in the sketch of Fig. 10-31.

ALTERNATE SOLUTION

The Cartesian components of strain obtained from the original rosette data were

$$\epsilon_x = 1000\mu \qquad \epsilon_y = -266.7\mu \qquad \gamma_{xy} = -1616.6\mu$$

The Cartesian components of stress at the point can be obtained from these strains by using Eqs. (10-18) and (10-22). Thus,

$$\sigma_x = \frac{E}{1 - \nu^2}(\epsilon_x + \nu\epsilon_y)$$
$$= \frac{30(10^6)}{1 - 0.09}[1000 + 0.30(-266.7)](10^{-6})$$
$$= 30{,}330 \text{ psi} \cong 30.3 \text{ ksi T}$$

$$\sigma_y = \frac{30(10^6)}{1 - 0.09}[-266.7 + 0.30(1000)](10^{-6})$$
$$= 1097.8 \text{ psi} \cong 1.098 \text{ ksi T}$$

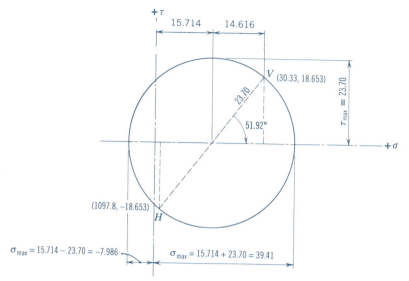

Figure 10-32

$$\tau_{xy} = \frac{E}{2(1 + \nu)}\, \gamma_{xy}$$
$$= \frac{30(10^6)}{2(1 + 0.30)}\,(-1616.6)(10^{-6})$$
$$= -18{,}653 \text{ psi} \cong -18.65 \text{ ksi}$$

The Mohr's circle of Fig. 10-32 is constructed from these data and the following results are obtained:

$$\sigma_{max} = 39.4 \text{ ksi T} \qquad\qquad\qquad \textbf{Ans.}$$
$$\sigma_{min} = 7.99 \text{ ksi C} \qquad\qquad\qquad \textbf{Ans.}$$
$$\tau_{max} = 23.7 \text{ ksi} \qquad\qquad\qquad\quad \textbf{Ans.}$$
$$\theta_p = 26.0° \qquad\qquad\qquad\qquad \textbf{Ans.}$$

The results are shown in the sketch of Fig. 10-31. Any differences in numerical values found in parts (a) and (b) of the Solution and Alternate Solution is due to round off. ■

PROBLEMS

In the following problems, when the material is named but the properties are not given, refer to Tables B-17 and B-18 in Appendix B for the properties. Use Eq. (10-21) to determine Poisson's ratio when data from Tables B-17 and B-18 are used.

Standard Problems

10-89* At a point on the free surface of a steel ($E = 30{,}000$ ksi and $\nu = 0.30$) machine part, the strain rosette shown in Fig. P10-89 was used to obtain the following normal strain data: $\epsilon_a = +750\mu$, $\epsilon_b = -125\mu$, and $\epsilon_c = -250\mu$. Determine

(a) The strain components ϵ_x, ϵ_y, and γ_{xy} at the point.
(b) The principal strains and the maximum shearing strain at the point.

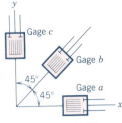

Figure P10-89

10-90* At a point on the free surface of an aluminum alloy ($E = 73$ GPa and $\nu = 0.33$) machine part, the strain rosette shown in Fig. P10-90 was used to obtain the following normal strain data: $\epsilon_a = +780\mu$, $\epsilon_b = +345\mu$, and $\epsilon_c = -332\mu$. Determine
(a) The strain components ϵ_x, ϵ_y, and γ_{xy} at the point.
(b) The principal strains and the maximum shearing strain at the point.

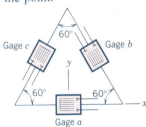

Figure P10-90

10-91 At a point on the free surface of an aluminum alloy ($E = 10,600$ ksi and $\nu = 0.33$) machine part, the strain rosette shown in Fig. P10-91 was used to obtain the following normal strain data: $\epsilon_a = +875\mu$, $\epsilon_b = +700\mu$, and $\epsilon_c = -350\mu$. Determine
(a) The strain components ϵ_x, ϵ_y, and γ_{xy} at the point.
(b) The principal strains and the maximum shearing strain at the point.

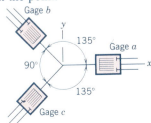

Figure P10-91

10-92 At a point on the free surface of a steel ($E = 200$ GPa and $\nu = 0.30$) machine part, the strain rosette shown in Fig. P10-92 was used to obtain the following normal strain data: $\epsilon_a = -555\mu$, $\epsilon_b = +925\mu$, and $\epsilon_c = +740\mu$. Determine
(a) The strain components ϵ_x, ϵ_y, and γ_{xy} at the point.
(b) The principal strains and the maximum shearing strain at the point.

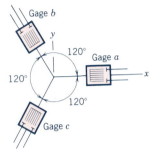

Figure P10-92

10-93* The strain rosette shown in Fig. P10-93 was used to obtain normal strain data at a point on the free surface of a structural steel machine part. The gage readings were $\epsilon_a = -350\mu$, $\epsilon_b = +1000\mu$, and $\epsilon_c = +550\mu$. Determine
(a) The strain components ϵ_x, ϵ_y, and γ_{xy} at the point.
(b) The stress components σ_x, σ_y, and τ_{xy} at the point.
(c) The principal stresses and the maximum shearing stress at the point. Express the stresses in U.S. customary units.

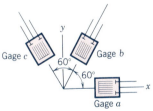

Figure P10-93

10-94* The strain rosette shown in Fig. P10-94 was used to obtain normal strain data at a point on the free surface of a 2024-T4 aluminum alloy machine part. The gage readings were $\epsilon_a = +525\mu$, $\epsilon_b = +450\mu$, and $\epsilon_c = +1425\mu$. Determine
(a) The strain components ϵ_x, ϵ_y, and γ_{xy} at the point.
(b) The stress components σ_x, σ_y, and τ_{xy} at the point.
(c) The principal stresses and the maximum shearing stress at the point. Express the stresses in SI units.

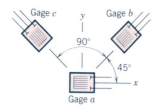

Figure P10-94

10-95 The strain rosette shown in Fig. P10-95 was used to obtain normal strain data at a point on the free surface of an 18-8 cold-rolled stainless steel machine part. The gage readings were $\epsilon_a = +665\mu$, $\epsilon_b = +390\mu$, and $\epsilon_c = +870\mu$. Determine

(a) The strain components ϵ_x, ϵ_y, and γ_{xy} at the point.
(b) The stress components σ_x, σ_y, and τ_{xy} at the point.
(c) The principal stresses and the maximum shearing stress at the point. Express the stresses in U.S. customary units.

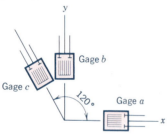

Figure P10-95

10-96 The strain rosette shown in Fig. P10-96 was used to obtain normal strain data at a point on the free surface of an annealed titanium alloy machine part. The gage readings were $\epsilon_a = -306\mu$, $\epsilon_b = -456\mu$, and $\epsilon_c = +906\mu$. Determine

(a) The strain components ϵ_x, ϵ_y, and γ_{xy} at the point.
(b) The stress components σ_x, σ_y, and τ_{xy} at the point.
(c) The principal stresses and the maximum shearing stress at the point. Express the stresses in SI units.

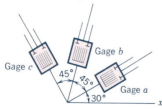

Figure P10-96

Challenging Problems

10-97* At a point on the free surface of a steel $(E = 30,000 \text{ ksi and } \nu = 0.30)$ machine part, the strain

rosette shown in Fig. P10-97 was used to obtain the following normal strain data: $\epsilon_a = +650\mu$, $\epsilon_b = +475\mu$, and $\epsilon_c = -250\mu$. Determine

(a) The stress components σ_x, σ_y, and τ_{xy} at the point.
(b) The principal strains and the maximum shearing strain at the point. Prepare a sketch showing all of these strains.
(c) The principal stresses and the maximum shearing stress at the point. Prepare a sketch showing all of these stresses.

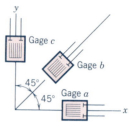

Figure P10-97

10-98* At a point on the free surface of an aluminum alloy $(E = 73 \text{ GPa and } \nu = 0.33)$ machine part, the strain rosette shown in Fig. P10-98 was used to obtain the following normal strain data: $\epsilon_a = +875\mu$, $\epsilon_b = +700\mu$, and $\epsilon_c = -650\mu$. Determine

(a) The stress components σ_x, σ_y, and τ_{xy} at the point.
(b) The principal strains and the maximum shearing strain at the point. Prepare a sketch showing all of these strains.
(c) The principal stresses and the maximum shearing stress at the point. Prepare a sketch showing all of these stresses.

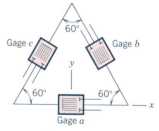

Figure P10-98

10-99 At a point on the free surface of an aluminum alloy $(E = 10,600 \text{ ksi and } \nu = 0.33)$ machine part, the strain rosette shown in Fig. P10-99 was used to obtain the following normal strain data: $\epsilon_a = +800\mu$, $\epsilon_b = +950\mu$, and $\epsilon_c = +600\mu$. Determine

(a) The stress components σ_x, σ_y, and τ_{xy} at the point.

(b) The principal strains and the maximum shearing strain at the point. Prepare a sketch showing all of these strains.

(c) The principal stresses and the maximum shearing stress at the point. Prepare a sketch showing all of these stresses.

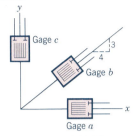

Figure P10-99

10-100 At a point on the free surface of a steel ($E = 200$ GPa and $\nu = 0.30$) machine part, the strain rosette shown in Fig. P10-100 was used to obtain the following normal strain data: $\epsilon_a = +875\mu$, $\epsilon_b = +700\mu$, and $\epsilon_c = +350\mu$. Determine

(a) The stress components σ_x, σ_y, and τ_{xy} at the point.

(b) The principal strains and the maximum shearing strain at the point. Prepare a sketch showing all of these strains.

(c) The principal stresses and the maximum shearing stress at the point. Prepare a sketch showing all of these stresses.

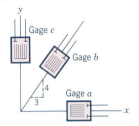

Figure P10-100

10-6 THIN-WALLED PRESSURE VESSELS

A pressure vessel is described as thin-walled when the ratio of the wall thickness to the radius of the vessel is so small that the distribution of normal stress on a plane perpendicular to the surface of the vessel is essentially uniform throughout the thickness of the vessel. Actually, this stress varies from a maximum value at the inside surface to a minimum value at the outside surface of the vessel, but it can be shown that if the ratio of the wall thickness to the inner radius of the vessel is less than 0.1, the maximum normal stress is less than 5 percent greater than the average stress. Boilers, gas storage tanks, pipelines, metal tires, and hoops are normally analyzed as thin-walled elements. Gun barrels, certain high-pressure vessels in the chemical processing industry, and cylinders and piping for heavy hydraulic presses need to be treated as thick-walled vessels.

Problems involving thin-walled vessels subjected to liquid (or gas) pressure p are readily solved with the aid of free-body diagrams of sections of the vessels together with the fluid contained therein. In the following subsections, spherical, cylindrical, and other thin shells of revolution are considered.

Spherical Pressure Vessels. A typical thin-walled spherical pressure vessel used for gas storage is shown in Fig. 10-33. If the weights of the gas and vessel are negligible (a common situation), symmetry of loading and geometry requires that stresses on sections that pass through the center of the sphere be equal. Thus, on the small element shown in Fig. 10-34a,

$$\sigma_x = \sigma_y = \sigma_n$$

Furthermore, there are no shearing stresses on any of these planes since there are no loads to induce them. The normal stress component in a sphere is known as a *meridional* or *axial stress* and is commonly denoted as σ_m or σ_a.

Figure 10-33

The free-body diagram shown in Fig. 10-34b can be used to evaluate the stress $\sigma_x = \sigma_y = \sigma_n = \sigma_a$ in terms of the pressure p, and the inside radius r and thickness t of the spherical vessel. The force R is the resultant of the internal forces that act on the cross-sectional area of the sphere that is exposed by passing a plane through the center of the sphere. The force P is the resultant of the fluid forces acting on the fluid remaining within the hemisphere. From a summation of forces in the x-direction,

$$R - P = 0$$

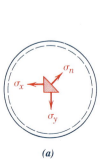

(a)

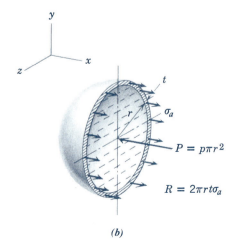

(b)

Figure 10-34

Figure 10-35

or

$$2\pi r t \sigma_a = p\pi r^2$$

from which

$$\sigma_a = \frac{pr}{2t} \qquad (10\text{-}24)$$

Cylindrical Pressure Vessels. A typical thin-walled cylindrical pressure vessel used for gas storage is shown in Fig. 10-35. Normal stresses, such as those shown on the small element of Fig. 10-36a, are easy to evaluate by using appropriate free-body diagrams. The normal stress component on a transverse plane is known as an *axial* or *meridional stress* and is commonly denoted as σ_a or σ_m. The normal stress component on a longitudinal plane is known as a *hoop, tangential,* or *circumferential stress* and is denoted as σ_h, σ_t, or σ_c. There are no shearing stresses on transverse or longitudinal planes.

The free-body diagram used for axial stress determinations is similar to Fig. 10-34b, which was used for the sphere, and the results are the same. The free-body diagram used for the hoop stress determination is shown in Fig. 10-36b.

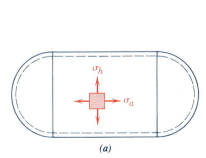

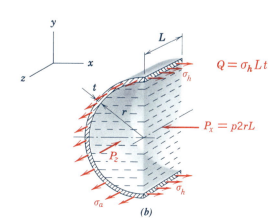

(a)

(b)

Figure 10-36

The force P_x is the resultant of the fluid forces acting on the fluid remaining within the portion of the cylinder isolated by the longitudinal plane and two transverse planes. The forces Q are the resultant of the internal forces on the cross-sectional area exposed by the longitudinal plane containing the axis of the cylinder. From a summation of forces in the x-direction,

$$2Q - P_x = 0$$

or

$$2\sigma_h L t = p2rL$$

from which

$$\sigma_h = \frac{pr}{t} \tag{10-25a}$$

Also [see Eq. (10-24)],

$$\sigma_a = \frac{pr}{2t} \tag{10-25b}$$

The previous analysis of the stresses in a cylindrical vessel subjected to uniform internal pressure indicates that the stress on a longitudinal plane is twice the stress on a transverse plane. Consequently, a longitudinal joint needs to be twice as strong as a transverse (or girth) joint.

Thin Shells of Revolution.   The discussion in the two previous subsections was limited to thin-walled cylindrical and spherical vessels under uniform internal pressure. The theory can be extended, however, to include other shapes and other loading conditions. Consider, for example, the thin shell of revolution shown in Fig. 10-37. Such shells are generated by rotating a plane curve, called the *meridian,* about an axis lying in the plane of the curve. Shapes that can be formed in this manner include the sphere, hemisphere, torus (doughnut), cylinder, cone, and ellipsoid. In shells of revolution, the two unknown principal stresses are a meridional stress σ_m that acts on a plane perpendicular to the meridian and a tangential stress σ_t that acts on a plane perpendicular to a parallel. Both of these stresses are shown on the small element of Fig. 10-37. The two stresses can be evaluated by using two equilibrium equations.

The small element of Fig. 10-37 is shown enlarged in Fig. 10-38. The element has a uniform thickness, is subjected to an internal pressure p, and has different curvatures in two orthogonal directions. The resultant forces on the various surfaces are shown on the diagram. Summing forces in the n-direction gives

$$P - 2F_m \sin d\theta_m - 2F_t \sin d\theta_t = 0$$

from which

$$p(2r_t d\theta_t)(2r_m d\theta_m) = 2\sigma_m(2tr_t d\theta_t) \sin d\theta_m + 2\sigma_t(2tr_m d\theta_m) \sin d\theta_t$$

and since, for small angles, $\sin d\theta \approx d\theta$, the above equation becomes

$$\frac{\sigma_m}{r_m} + \frac{\sigma_t}{r_t} = \frac{p}{t} \tag{10-26}$$

Since Eq. (10-26) contains two unknown stresses, an additional independent equation is needed. Such an equation can be obtained by considering equilibrium of

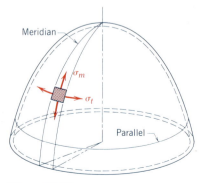

Figure 10-37

Meridian

σ_m

σ_t

Parallel

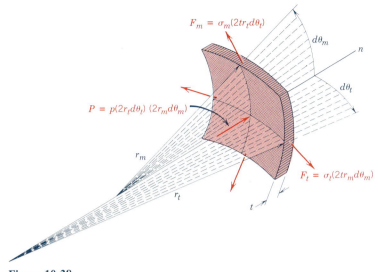

Figure 10-38

a portion of the vessel above or below the parallel that passes through the point of interest. Application of Eq. (10-26) is illustrated in Example Problem 10-11.

Example Problem 10-10

A cylindrical pressure vessel 1.50 m in diameter is constructed by wrapping a 15 mm–thick steel plate into a spiral and butt-welding the mating edges of the plate as shown in Fig. 10-39a. The butt-welded seams form an angle of 30° with a transverse plane through the cylinder. Determine the normal stress σ perpendicular to the weld and the shearing stress τ parallel to the weld when the internal pressure in the vessel is 1500 kPa.

SOLUTION

The hoop stress σ_h and the axial stress σ_a in the cylinder can be determined by using Eqs. (10-25a) and (10-25b). Thus,

$$\sigma_h = \frac{pr}{t} = \frac{1500(10^3)(0.75)}{0.015} = 75.0(10^6) \text{ N/m}^2 = 75.0 \text{ MPa}$$

$$\sigma_a = \frac{pr}{2t} = \frac{1500(10^3)(0.75)}{2(0.015)} = 37.5(10^6) \text{ N/m}^2 = 37.5 \text{ MPa}$$

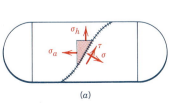

(a)

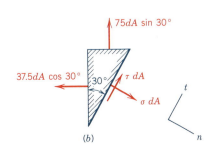

(b)

Figure 10-39

The normal stress σ perpendicular to the weld and the shearing stress τ parallel to the weld can be determined by using the free-body diagram shown in Fig. 10-39b.

From a summation of forces in the n-direction:

$$\sigma dA - 37.5\, dA \cos 30° \cos 30° - 75\, dA \sin 30° \sin 30° = 0$$

$$\sigma = 46.9 \text{ MPa} \qquad \textbf{Ans.}$$

From a summation of forces in the t-direction:

$$\tau dA - 37.5\, dA \cos 30° \sin 30° + 75\, dA \sin 30° \cos 30° = 0$$
$$\tau = -16.24 \text{ MPa} \qquad \textbf{Ans.}$$

The same results could have been obtained using the stress transformation equations, Eqs. (10-1) and (10-2). ■

Example Problem 10-11

A pressure vessel of $\frac{1}{4}$-in. steel plate has the shape of a paraboloid closed by a thick flat plate, as shown in Fig. 10-40a. The equation of the generating parabola is $y = x^2/4$, where x and y are in inches. Determine the meridional and tangential stresses σ_m and σ_t in the shell at a point 16 in. above the bottom of the vessel due to an internal gas pressure of 250 psi gage.

SOLUTION

Determining σ_m:

The meridional stress σ_m, which must be tangent to the shell, can be determined with the aid of the free-body diagram shown in Fig. 10-40b. This free-body diagram represents a thin slice ($h = 16$ in.) through the vessel and the gas. The slice is perpendicular to the end plate and contains the axis of the vessel. Forces perpendicular to the slice are omitted from the diagram since they are not needed for the determination of σ_m. From the equation of the parabola, the radius x and the slope of the shell dy/dx at $y = 16$ in. are determined to be 8 in. and 4/1, respectively. Summing forces in the y-direction gives

$$-\int_{A_p} dP + \int_{A_\sigma} dF \cos \alpha = 0$$
$$-p\pi x^2 + \sigma_m 2\pi x t \cos \alpha = 0$$

Substituting the given data yields

$$-250(\pi)(8^2) + \sigma_m(2\pi)(8)(1/4)(4/\sqrt{17}) = 0$$

from which

$$\sigma_m = 4123 \text{ psi} \cong 4120 \text{ psi T} \qquad \textbf{Ans.}$$

Determining σ_t:

To find σ_t from Eq. (10-26), the radii r_m and r_t at the point must be determined. The radius of curvature r_m of the shell in the xy-plane is determined from the expression

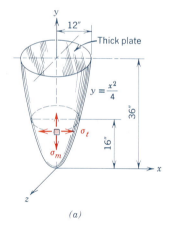

$y = \dfrac{x^2}{4}$

12″

Thick plate

36″

16″

σ_t

σ_m

(a)

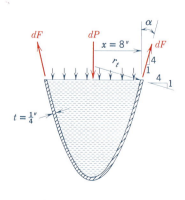

dF dP
$x = 8″$ dF
r_t

α

4
1

4
1

$t = \frac{1}{4}″$

(b)

Figure 10-40

$$r_m = \frac{[1 + (dy/dx)^2]^{1.5}}{d^2y/dx^2} = \frac{(1 + 4^2)^{1.5}}{1/2} = 140.19 \text{ in.}$$

and the perpendicular radius r_t is found from the geometry of Fig. 10-40*b* as

$$r_t = 8(\sqrt{17}/4) = 8.246 \text{ in.}$$

Then, from Eq. (10-26)

$$p/t = (\sigma_m/r_m) + (\sigma_t/r_t)$$

$$250(4) = (4123/140.19) + (\sigma_t/8.246)$$

from which

$$\sigma_t = 8003 \text{ psi} \cong 8000 \text{ psi T} \blacksquare \qquad \textbf{Ans.}$$

PROBLEMS

Standard Problems

10-101* Determine the maximum normal stress in a 12 in.–diameter basketball that has a $\frac{1}{16}$-in. wall thickness after it has been inflated to a pressure of 15 psi.

10-102* A spherical gas-storage tank 15 m in diameter is being constructed to store gas under an internal pressure of 1400 kPa. Determine the thickness of structural steel (see Appendix B for properties) plate required if a factor of safety of 3 with respect to failure by yielding is specified.

10-103 A steel pipe with an inside diameter of 12 in. will be used to transmit steam under a pressure of 1000 psi. If the hoop stress in the pipe must be limited to 10 ksi because of a longitudinal weld in the pipe, determine the minimum satisfactory thickness for the pipe.

10-104 A cylindrical propane tank, similar to the ones shown in Fig. 10-35, has an outside diameter of 3.25 m and a wall thickness of 22 mm. If the allowable hoop stress is 100 MPa and the allowable axial stress is 45 MPa, determine the maximum internal pressure that can be applied to the tank.

10-105* The force transmitted across a 10-in. length of the joint uniting the two halves of a spherical pressure vessel 20 ft in diameter must be limited to 3000 lb tension. Determine the maximum internal pressure that can be safely applied to the vessel.

10-106* A steel boiler 1 m in diameter is welded using a spiral seam that makes an angle of 30° with the longitudinal direction (axis of the boiler). Determine the

magnitudes of the normal and shearing forces transmitted across a 150-mm length of the seam due to an internal pressure of 950 kPa.

10-107 A standpipe 12 ft in diameter and 50 ft tall is being constructed for use as a storage tank for water ($\gamma = 62.4 \text{ lb/ft}^3$). Determine the minimum thickness of steel plate that can be used if the hoop stress in the standpipe must be limited to 5000 psi.

10-108 The cylindrical pressure tank, shown in Fig. P10-108, is made of 10-mm steel plate. The magnitude of the shearing stress at point A on plane B-B (which is perpendicular to the surface of the plate at A) is 45 MPa. Determine the air pressure in the tank.

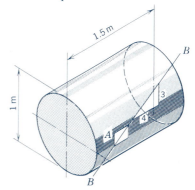

Figure P10-108

Challenging Problems

10-109* The strains measured on the outside surface of the cylindrical pressure vessel shown in Fig. P10-109 are $\epsilon_1 = +619$ μin./in. and $\epsilon_2 = +330$ μin./in. The angle $\theta = 30°$. The outside diameter of the vessel is 20 in. and the wall thickness is $\frac{1}{8}$ in. The vessel is made of 0.4 percent carbon hot-rolled steel (see Appendix B for properties). Determine

(a) The stresses σ_1 and σ_2 in the vessel.
(b) The internal pressure applied to the vessel.
(c) The factor of safety with respect to failure by fracture based on the hoop stress.

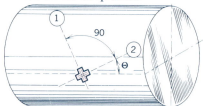

Figure P10-109

10-110* The thin-walled spherical water tank shown in Fig. P10-110 is completely full of water and open to the atmosphere at the top. Determine the meridional and tangential stresses σ_m and σ_t at a point on the equator of the sphere in terms of γ, r, and t, where γ is the specific weight of the water, r is the radius of the sphere, and t is the wall thickness of the sphere.

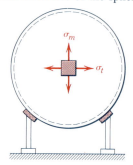

Figure P10-110

10-111 A hemispherical tank of radius r and thickness t is supported by a flange, as shown in the cross section of Fig. P10-111. The tank is filled with a fluid having a specific weight γ. Determine, in terms of γ, r, and t, the meridional and tangential stresses σ_m and σ_t at a depth $y = r/2$.

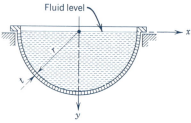

Figure P10-111

10-112 The conical water tank shown in Fig. P10-112 was fabricated from $\frac{1}{8}$-in. steel plate. When the tank is completely full of water (specific weight $\gamma = 62.4$ lb/ft^3), determine the axial and hoop stresses σ_a and σ_h at a point in the wall 8 ft below the apex of the cone.

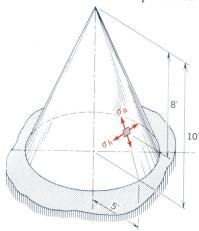

Figure P10-112

10-7 COMBINED AXIAL, TORSIONAL, AND FLEXURAL LOADS

In numerous industrial situations, machine members are subjected to combinations of the three general types of loads studied so far. When a flexural load is combined with torsional and axial loads, it is frequently difficult to locate the point or points where the most severe stresses occur. The following review of summary statements may be helpful in this regard.

The longitudinal and transverse shearing stresses in a beam are maximum where Q is maximum—usually at the centroidal axis of the section where V is maximum [see Eq. (8-15)]. The flexural stress is maximum at the greatest distance from the centroidal axis on the section where M is maximum [see Eq. (8-12)]. The torsional shearing stress is maximum at the surface of a shaft at the section where T is maximum [see Eq. (7-6)]. With these facts in mind, it is normally possible to locate one or more possible points of high stress. Even so, it may be necessary to calculate the various stresses at more than one point of a member before locating the most severely stressed point. For stresses below the proportional limit of the material, the superposition method can be used to combine stresses on any given plane at any specific point of a loaded member.

After the stresses on a pair of mutually perpendicular planes at a specific point are determined, the methods of Sections 10-3 and 10-4 can be used to find the principal stresses and the maximum shearing stress at the point.

The following Example Problems illustrate the procedure for the solution of elastic combined stress problems.

Example Problem 10-12

A hollow circular shaft of outer diameter 100 mm and inner diameter 50 mm is loaded as shown in Fig. 10-41a.

(a) Determine the principal stresses and the maximum shearing stress at the point (or points) where the stress situation is most severe.

(b) On a sketch show the approximate directions of these stresses.

SOLUTION

The axial stress is constant throughout the shaft and is equal to

$$\sigma = \frac{P}{A} = \frac{150(10^3)\pi}{\pi(0.05^2 - 0.025^2)} = 80.0(10^6) \text{ N/m}^2 = 80.0 \text{ MPa C}$$

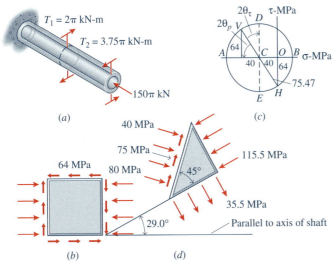

Figure 10-41

The resisting torque is highest to the right of the torque T_1, and the torsional stress is maximum on the outer surface of the shaft; therefore, the magnitude of the maximum torsional shearing stress is

$$\tau = \frac{Tc}{J} = \frac{3.75(10^3)(\pi)(0.05)}{(\pi/2)(0.05^4 - 0.025^4)} = 64.0(10^6) \text{ N/m}^2 = 64.0 \text{ MPa}$$

The stresses are

$$\sigma_x = -80.0 \text{ MPa} \qquad \sigma_y = 0 \qquad \tau_{xy} = -64.0 \text{ MPa}$$

These stresses are shown acting on mutually perpendicular planes through a point on the surface of the right portion of the shaft in Fig. 10-41b. Mohr's circle for the stresses at any point on the surface of the right portion of the shaft is shown in Fig. 10-41c.

(a) The principal stresses, principal directions, and the maximum shearing stress from Mohr's circle are

$$\sigma_{p1} = OA = -40.00 + 75.47 = +35.47 \text{ MPa} \cong 35.5 \text{ MPa T} \qquad \textbf{Ans.}$$

$$\sigma_{p2} = OB = -40.00 - 75.47 = -115.47 \text{ MPa} \cong 115.5 \text{ MPa C} \qquad \textbf{Ans.}$$

$$\theta_p = \frac{1}{2} \tan^{-1} \frac{64}{40} = 29.0° \text{⌐}$$

$$\tau_{max} = CD = 75.47 \text{ MPa} \cong 75.5 \text{ MPa} \qquad \textbf{Ans.}$$

(b) The stresses are shown in their proper directions in Fig. 10-41d. ■

Example Problem 10-13

The cast-iron frame of a small press is shown in Fig. 10-42a. The cross section a-a is shown in Fig. 10-42b. For a load P of 16 kip, and assuming linearly elastic action, determine

(a) The normal stress distribution on section a-a.

(b) The principal stresses for the critical points on section a-a.

SOLUTION

(a) A free-body diagram of the top portion of the frame is shown in Fig. 10-42c. The applied load P is equivalent to a centric load P and a couple of magnitude $18P$ as shown in Figs. 10-42d and e, respectively. The axial stress due to P is uniformly distributed over section a-a, as shown in Fig. 10-42d, and is

$$\sigma_1 = \frac{P}{A} = \frac{16,000}{40} = +400 \text{ psi} = 400 \text{ psi T}$$

The second moment of the cross-sectional area with respect to the centroidal axis c-c is

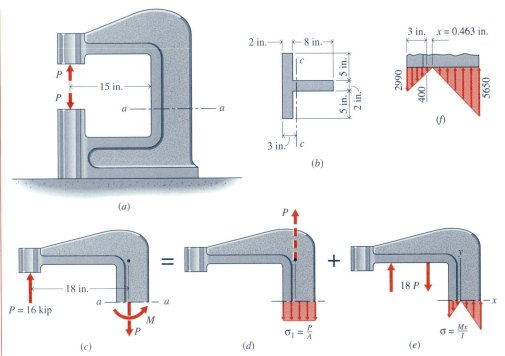

Figure 10-42

$$I = \frac{1}{3}(2)(7)^3 + \frac{1}{3}(12)(3)^3 - \frac{1}{3}(10)(1)^3 = 333.3 \text{ in.}^4$$

At the left edge of section *a-a* the flexural stress is

$$\sigma_2 = \frac{Mc}{I} = \frac{16,000(18)(3)}{333.3} = 2592 \text{ psi T}$$

At the right edge of section *a-a* the flexural stress is

$$\sigma_3 = \frac{Mc}{I} = \frac{16,000(18)(7)}{333.3} = 6049 \text{ psi C}$$

The distribution of flexural stresses is shown in Fig. 10-42*e*. The total normal stress at any point on section *a-a* is the algebraic sum of the component stresses at the point. When the stress distributions of Figs. 10-42*d* and *e* are added, the resultant stress distribution for section *a-a* is as shown in Fig. 10-42*f*, which is the answer for part (a).

(b) Since there are no shearing stresses on section *a-a*, the normal stresses are principal stresses, and the critical points, as observed from the stress distribution of Fig. 10-42*f*, are the left and right edges of the section.

The principal stresses for the right edge are

$$\sigma_{p2} = -6049 + 400 = -5649 \text{ psi} \cong 5650 \text{ psi C} \qquad \textbf{Ans.}$$

$$\sigma_{p1} = 0 \qquad \textbf{Ans.}$$

The principal stresses for the left edge are

$$\sigma_{p1} = +2592 + 400 = +2992 \text{ psi} \cong 2990 \text{ psi T} \qquad \textbf{Ans.}$$

$$\sigma_{p2} = 0 \qquad \textbf{Ans.}$$

Although not requested in this example, the location of the neutral axis (line of zero stress) can be determined as the place where the compressive flexural stress is 400 psi, because this stress will just balance the axial tensile stress of 400 psi. The distance on Fig. 10-42f can be obtained by writing

$$400 = \frac{16,000(18)x}{333.3}$$

from which

$$x = 0.463 \text{ in.}$$

to the right of the centroidal axis of the cross section. Note that the neutral axis for the combined loading ($x = 0.463$ in.) is not the same as the neutral axis for the flexural component of loading M, which is the centroidal axis c-c. ∎

Example Problem 10-14

A cantilever I-beam [$I = 1243(10^6)$ mm^4, $S = 4079(10^3)$ mm^3], supported at the left end, carries a uniformly distributed load of 160 kN/m on a span of 2.5 m. Determine the maximum normal and shearing stresses in the beam.

SOLUTION

For a cantilever beam with a uniformly distributed load, the maximum bending moment and the maximum transverse shear occur on the section at the support. In this case they are

$$M = \frac{wL^2}{2} = \frac{160(2.5^2)}{2} = 500 \text{ kN} \cdot \text{m}$$

$$V = wL = 160(2.5) = 400 \text{ kN}$$

The upper half of the cross section of the beam is shown in Fig. 10-43a. Stress values are calculated for three points—namely, at the neutral axis, in the web at the junction of the web and the top flange, and at the top surface. Both the fillets and the stress concentrations at the junction of the web and flange are neglected. At the neutral axis, the flexural stress is zero; however,

$$Q = 304.8(19.7)(295) + 285.1(11.9)(142.6) = 2.255(10^6) \text{ mm}^3$$

$$\tau = \frac{VQ}{It} = \frac{400(10^3)(2.255)(10^{-3})}{1243(10^{-6})(0.0119)} = 61.0(10^6) \text{ N/m}^2 = 61.0 \text{ MPa}$$

In the web at the junction with the top flange, the flexural stress is

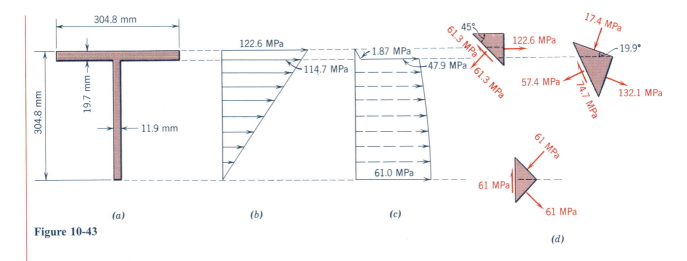

Figure 10-43

$$\sigma = \frac{My}{I} = \frac{500(10^3)(0.2851)}{1243(10^{-6})} = 114.7(10^6) \text{ N/m}^2 = 114.7 \text{ MPa T}$$

Also,

$$Q = 304.8(19.7)(295) = 1.771(10^6) \text{ mm}^3$$

$$\tau = \frac{VQ}{It} = \frac{400(10^3)(1.771)(10^{-3})}{1243(10^{-6})(0.0119)} = 47.9(10^6) \text{ N/m}^2 = 47.9 \text{ MPa}$$

At the top surface, the transverse shearing stress is zero, and the flexural stress is

$$\sigma = \frac{M}{S} = \frac{500(10^3)}{4079(10^{-6})} = 122.6(10^6) \text{ N/m}^2 = 122.6 \text{ MPa T}$$

The distribution of flexural stress for the top half of the section is shown in Fig. 10-43b. The distribution of the average transverse shearing stress is as shown in Fig. 10-43c.

The principal stresses and maximum shearing stresses for each of the three selected points are shown in Fig. 10-43d. The calculations, from the equations of Section 10-3, for a point at the junction of web and flange are

$$\sigma_p = \frac{114.7}{2} \pm \sqrt{\left(\frac{114.7}{2}\right)^2 + 47.9^2}$$

$$= 57.35 \pm 74.72$$

which gives

$$\sigma_{p1} = 132.1 \text{ MPa T} \qquad \sigma_{p2} = 17.4 \text{ MPa C} \qquad \tau_{max} = 74.7 \text{ MPa}$$

The angle is not particularly important in this case, but it can be obtained from the equation

$$\theta_p = \frac{1}{2} \tan^{-1}\left(-\frac{47.90}{57.35}\right) = -19.9^\circ = 19.9^\circ \downarrow$$

The other stresses of Fig. 10-43*d* are obtained by inspection. Note that the maximum tensile stress of 132.1 MPa is 7.75 percent above the maximum tensile flexural stress of 122.6 MPa and that the maximum shearing stress of 74.7 MPa is 22.5 percent above the maximum transverse shearing stress of 61.0 MPa. ∎

Example Problem 10-15

A 4 in.–diameter shaft is loaded and supported as shown in Fig. 10-44*a*. Determine the maximum permissible value of *P* for allowable stresses of 16,000 psi T and 9000 psi shear.

SOLUTION

The shaft acts as a cantilever beam with a concentrated load *P* lb at the right end, a torsion member subjected to a 20*P* in. · lb torque, and an axially loaded member with load of 48*P* lb. The couple produces a torsional shearing stress at all surface points of the shaft of

$$\tau_{xt} = \frac{Tc}{J} = \frac{20P(2)}{(\pi/2)(2)^4} = \frac{5P}{\pi}$$

as shown in Fig. 10-44*c*.

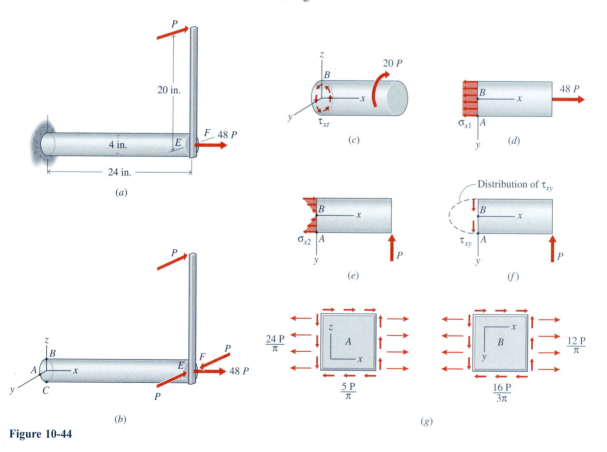

(a)

(b)

(c)

(d)

(e)

(f)

(g)

Distribution of τ_{xy}

Figure 10-44

The axial tensile stress at all points of the shaft is

$$\sigma_{x1} = \frac{P}{A} = \frac{48P}{\pi(2)^2} = \frac{12P}{\pi}$$

as shown in Fig. 10-44d.

The cantilever beam load P produces a maximum tensile flexural stress at point A and an equal compressive stress on the opposite side of the shaft. The axial stress is tensile; therefore, the compressive flexural stress is not critical. The tensile flexural stress at A is (see Fig. 10-44e)

$$\sigma_{x2} = \frac{Mc}{I} = \frac{24P(2)}{(\pi/4)(2)^4} = \frac{12P}{\pi}$$

The transverse shearing stress is equal to the longitudinal shearing stress, which is maximum on the xz-plane, including such surface points as B and C where the torsional shearing stress is also maximum. The maximum transverse shearing stress is (see Fig. 10-44f)

$$\tau_{xy} = \frac{VQ}{It} = \frac{P(\pi/2)(2)^2[4(2)/3\pi]}{(\pi/4)(2)^4(4)} = \frac{P}{3\pi}$$

The torsional and transverse shearing stresses are in the same direction along the top of the shaft and in opposite directions along the bottom of the shaft; therefore, any critical situation due to these two stresses will occur along the top of the shaft at some point such as B.

All four stresses are maximum at some point in the section of the shaft at the wall. The flexural stress is maximum only at this section; therefore, a complete analysis for the member can be made on the section at the wall. A complete analysis including interior points of this section is beyond the scope of this book; however, a fairly good solution can be obtained by investigating the combined stresses at two surface points A and B. At A, the flexural stress, axial stress, and torsional shearing stress are all maximum, whereas the transverse shearing stress is zero. At B, the axial stress, torsional shearing stress, and transverse shearing stress are all maximum, whereas the flexural stress is zero. The flexural stress and transverse shearing stress are never maximum at the same point. Therefore, the chance of finding a more severe stress at an interior point is unlikely.

The stresses on orthogonal planes through A and B are as shown in Fig. 10-44g. When these values are substituted in Eq. (10-4) of Section 10-3 and the resulting expressions compared, it is evident that the combination of stresses at A is more severe than that at B. The maximum stresses at A are

$$\sigma_A = \frac{12P}{\pi} + \sqrt{\left(\frac{12P}{\pi}\right)^2 + \left(\frac{5P}{\pi}\right)^2} = \frac{25P}{\pi}$$

$$\tau_A = \sqrt{\left(\frac{12P}{\pi}\right)^2 + \left(\frac{5P}{\pi}\right)^2} = \frac{13P}{\pi}$$

The allowable values of P from the allowable stresses and these relations are

$$16,000 = 25P_\sigma/\pi \qquad P_\sigma = 2010 \text{ lb}$$
$$9000 = 13P_\tau/\pi \qquad P_\tau = 2170 \text{ lb}$$

The smaller value controls and the maximum permissible value is

$$P = 2010 \text{ lb} \quad\blacksquare \qquad\qquad \textbf{Ans.}$$

PROBLEMS

Standard Problems

10-113* A 4 in.–diameter shaft is subjected to both a torque of 30 in. · kip and an axial tensile load of 50 kip as shown in Fig. P10-113.
  (a) Determine the principal stresses and the maximum shearing stress at point A on the surface of the shaft.
  (b) Show the stresses of part (a) and their approximate directions on a sketch.

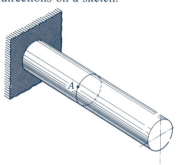

Figure P10-113

10-114* A hollow shaft with an outside diameter of 400 mm and an inside diameter of 300 mm is subjected to both a torque of 350 kN · m and an axial tensile load of 1500 kN as shown in Fig. P10-114.
  (a) Determine the principal stresses and the maximum shearing stress at a point on the outside surface of the shaft.

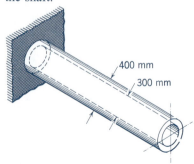

Figure P10-114

(b) Show the stresses of part (a) and their approximate directions on a sketch.

10-115 A 2 in.–diameter shaft is used in an aircraft engine to transmit 360 hp at 1500 rpm to a propeller that develops a thrust of 2800 lb. Determine the principal stresses and the maximum shearing stress produced at any point on the outside surface of the shaft.

10-116 A 60 mm–diameter shaft must transmit a torque of unknown magnitude while it is supporting an axial tensile load of 150 kN. Determine the maximum allowable value for the torque if the tensile principal stress on the outside surface of the shaft must not exceed 125 MPa.

10-117* A 6 in.–diameter shaft will be used to support the axial load and torques shown in Fig. P10-117.
  (a) Determine the principal stresses and the maximum shearing stress at point A on the surface of the shaft.
  (b) Show the stresses of part (a) and their approximate directions on a sketch.

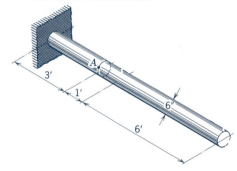

Figure P10-117

10-118 A steel shaft is loaded and supported as shown in Fig. P10-118. Determine the maximum allowable value for the axial load P if the maximum shearing stress in the shaft is not to exceed 75 MPa and the maximum compressive stress in the shaft is not to exceed 120 MPa.

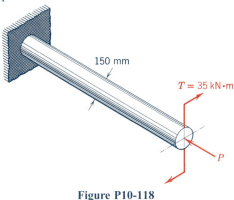

Figure P10-118

10-119* A solid circular bar with a diameter of 4 in. is used to support a tensile load of 10 kip. The line of action of the load is located 1.75 in. from the axis of the bar. Determine the maximum tensile and compressive stresses produced in the bar.

10-120* A hollow circular tube with an outside diameter of 100 mm and an inside diameter of 75 mm is used to support a compressive load of 50 kN. The line of action of the load is located 40 mm from the axis of the tube. Determine the maximum tensile and compressive stresses produced in the tube.

10-121 The T-section shown in Fig. P10-121 is used as a short post to support a compressive load P of 150 kip. The load is applied on the centerline of the stem at a distance $e = 2$ in. from the centroid of the cross section. Determine the normal stresses at points A and B on a transverse plane C-C near the base of the post.

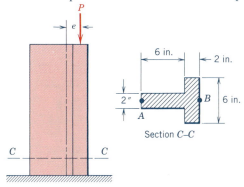

Figure P10-121

10-122 A W305 $\times$ 143 wide-flange section (see Appendix B) is used as a short post to support a load $P = 400$ kN as shown in Fig. P10-122. The load is applied

on the centerline of the web at a distance $e = 100$ mm from the centroid of the cross section. Determine the normal stresses at points A and B on a transverse plane C-C near the base of the post.

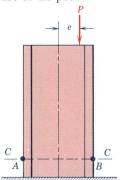

Figure P10-122

10-123* Determine the distribution of normal stress on section AB of the structural member shown in Fig. P10-123.

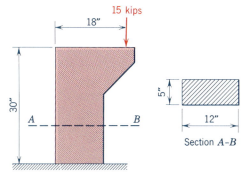

Figure P10-123

10-124* The cross section of the aluminum beam ($E = 70$ GPa) shown in Fig. P10-124 is a rectangle 100 mm wide by 150 mm deep. Determine the maximum normal stress on a vertical section at the wall.

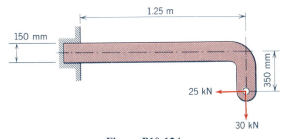

Figure P10-124

10-125 The straight portion *AB* of the cast steel machine part shown in Fig. P10-125 has a hollow rectangular cross section with outside dimensions 12 × 10 in. and walls 2-in. thick. Determine the maximum normal stress on a transverse plane within the portion *AB*.

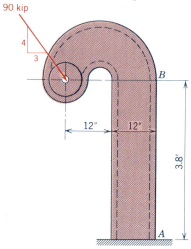

Figure P10-125

10-126 Determine the maximum normal stress on a transverse plane in the straight portion *AB* of the structure shown in Fig. P10-126 and clearly indicate where it occurs. The member is braced in the plane perpendicular to the plane of symmetry.

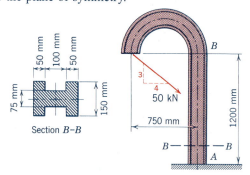

Figure P10-126

10-127* The beam shown in Fig. P10-127 has a 3 × 8-in. rectangular cross section and is loaded in a plane of symmetry. Determine and show on a sketch the principal and maximum shearing stresses at point *A*.

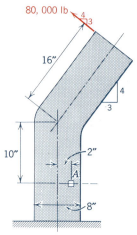

Figure P10-127

10-128 The machine element shown in Fig. P10-128 is loaded in a plane of symmetry. Determine and show on a sketch the principal and maximum shearing stresses at point *A* that is in the web just above the junction between the flange and the web.

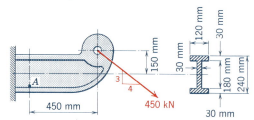

Figure P10-128

10-129* The cantilever beam shown in Fig. P10-129*a* has the cross section shown in Fig. P10-129*b*. If the allowable stresses are 9000 psi shear and 15,000 psi tension at point *A* (just below the flange), determine the maximum allowable load *P*.

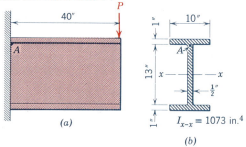

Figure P10-129

10-130* The simply supported beam shown in Fig. P10-130*a* has the cross section shown in Fig. 10-130*b*. If the allowable stresses are 75 MPa shear and 120 MPa tension at point A (just above the flange), determine the maximum allowable load P.

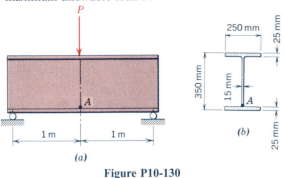

(a)

(b)

Figure P10-130

10-131 A W12 × 65 steel beam (see Appendix B) is loaded and supported as shown in Fig. P10-131. Determine the principal and maximum shearing stresses in the beam if the load P is 36 kip and the length L is 16 ft.

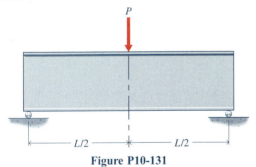

Figure P10-131

10-132 A W610 × 155 cantilever beam (see Appendix B) with a span of 3 m carries a uniformly distributed load of 160 kN/m. Determine the principal and maximum shearing stresses in the beam.

10-133* The cantilever beam shown in Fig. P10-133a has the cross section shown in Fig. P10-133b. If the allowable stresses are 10,000 psi shear and 16,000 psi tension and compression at point A (just below the flange), determine the maximum allowable load P.

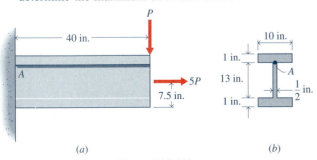

(a)

(b)

Figure P10-133

10-134* The cantilever beam shown in Fig. P10-134a has the cross section shown in Fig. P10-134b. If the stresses at point A (in the web) are $\sigma_x = 35$ MPa C and $\tau_{xy} = 7$ MPa, determine
(a) The magnitudes of loads P and T.
(b) The principal and maximum shearing stresses at point A.

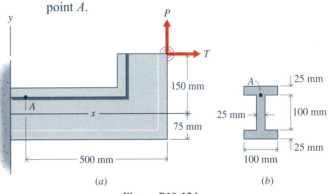

(a)

(b)

Figure P10-134

10-135 A W24 × 62 simply supported beam (see Appendix B) with a span of 18 ft carries a concentrated load P at the middle of the span. If the principal and maximum shearing stresses in the beam must not exceed 18 ksi and 10 ksi, respectively, determine the maximum allowable load P.

10-136 A W305 × 97 cantilever beam (see Appendix B) with a span of 3 m carries a concentrated load P at the free end of the beam. If the principal and maximum shearing stresses in the beam must not exceed 125 MPa and 75 MPa, respectively, determine the maximum allowable load P.

Challenging Problems

10-137* A 2 in.–diameter steel rod is loaded and supported as shown in Fig. P10-137. Determine, and show on a sketch, the principal stresses and the maximum shearing stress at the top surface of the rod adjacent to the support.

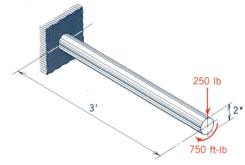

Figure P10-137

10-138* A 50 mm–diameter steel rod is loaded and supported as shown in Fig. P10-138. Determine, and show on a sketch, the principal stresses and the maximum shearing stress at the top surface of the rod adjacent to the support.

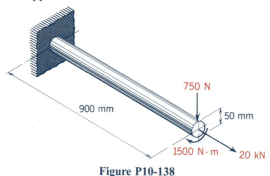

Figure P10-138

10-139 A 4 in.–diameter steel rod is loaded and supported as shown in Fig. P10-139. Determine the principal stresses and the maximum shearing stress
(a) At point A on a section adjacent to the support.
(b) At point B on a section adjacent to the support.

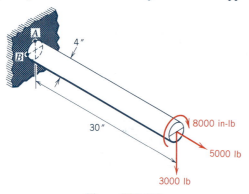

Figure P10-139

10-140 A 100 mm–diameter steel rod is loaded and supported as shown in Fig. P10-140. Determine the principal stresses and the maximum shearing stress

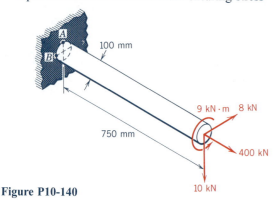

Figure P10-140

(a) At point A on a section adjacent to the support.
(b) At point B on a section adjacent to the support.

10-141* A steel shaft 4 in. in diameter is supported in flexible bearings at its ends. Two pulleys, each 2 ft in diameter, are keyed to the shaft. The pulleys carry belts with tensile forces as shown in Fig. P10-141. Determine the principal stresses and the maximum shearing stress at point A on the surface of the shaft.

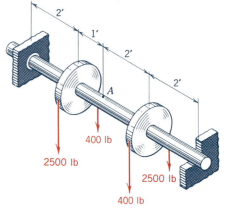

Figure P10-141

10-142* A steel shaft 120 mm in diameter is supported in flexible bearings at its ends. Two pulleys, each 500 mm in diameter, are keyed to the shaft. The pulleys carry belts with tensile forces as shown in Fig. P10-142. Determine the principal stresses and the maximum shearing stress at point A on the surface of the shaft.

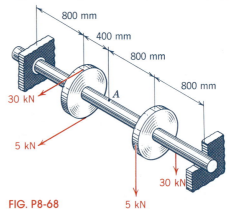

FIG. P8-68

Figure P10-142

10-143* The thin-walled cylindrical pressure vessel shown in Fig. P10-143 has an inside diameter of 24 in. and a wall thickness of $\frac{1}{2}$ in. The vessel is subjected to an internal pressure of 250 psi. In addition, an axial load P of 75 kip and a torque T of 150 ft · kip are applied to the vessel through rigid plates on the ends as shown in Fig. P10-143. Determine the maximum normal and shearing stresses at a point on the outside surface of the vessel.

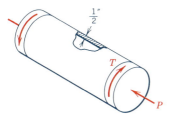

Figure P10-143

10-144 A thin-walled cylindrical pressure tank is fabricated by butt-welding 15-mm plate with a spiral seam as shown in Fig. P10-144. The pressure in the tank is 2500 kPa. Additional loads are applied to the cylinder through a rigid end plate as shown in Fig. P10-144. Determine

(a) The normal and shearing stresses on the plane of the weld.

(b) The principal stresses and the maximum shearing stress at a point on the inside surface of the tank.

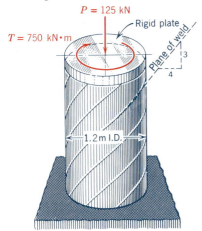

Figure P10-144

10-145 A 4 in.–diameter solid circular steel shaft is loaded and supported as shown in Fig. P10-145. If the maximum normal and shearing stresses at point A must be limited to 7500 psi T and 5000 psi, respectively, determine the maximum permissible value for the transverse load V.

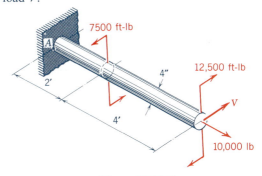

Figure P10-145

10-146 A 100 mm–diameter solid circular steel shaft is loaded and supported as shown in Fig. P10-146. If the maximum normal and shearing stresses at point A must be limited to 90 MPa T and 60 MPa, respectively, determine the maximum permissible value for the transverse load V.

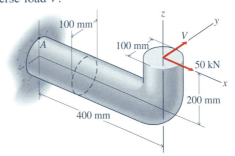

Figure P10-146

10-8 SUMMARY

In previous chapters, formulas were developed for determining normal and shearing stresses on specific planes in axially loaded bars, circular shafts, and beams. In this chapter, methods were developed for determining normal and shearing stresses on other planes through a point of interest.

Equations relating the normal and shearing stresses σ_n and τ_{nt} on an arbitrary plane (oriented at an angle θ with respect to a reference x-axis) through a point and the known stresses σ_x, σ_y, and $\tau_{xy} = \tau_{yx}$ on the reference planes were developed using the free-body diagram method introduced earlier for the axially loaded bars, circular shafts, and beams. The results are

$$\sigma_n = \sigma_x \cos^2 \theta + \sigma_y \sin^2 \theta + 2\tau_{xy} \sin \theta \cos \theta \qquad (10\text{-}1a)$$

$$\tau_{nt} = -(\sigma_x - \sigma_y) \sin \theta \cos \theta + \tau_{xy} (\cos^2 \theta - \sin^2 \theta) \qquad (10\text{-}2a)$$

Equations (10-1) and (10-2) provide a means for determining the normal stress σ_n and the shearing stress τ_{nt} on different planes through a point in a stressed body. For design purposes, critical stresses at the point are usually the maximum tensile stress and/or the maximum shearing stress.

Maximum and minimum values of σ_n occur at values of θ given by the expression

$$\tan 2\theta_p = \frac{2\tau_{xy}}{\sigma_x - \sigma_y} \qquad (10\text{-}3)$$

When these values of θ are substituted into Eq. (10-1), maximum and minimum normal stresses σ_{p1} and σ_{p2} at the point are found to be

$$\sigma_{p1,\,p2} = \frac{\sigma_x + \sigma_y}{2} \pm \sqrt{\left(\frac{\sigma_x - \sigma_y}{2}\right)^2 + \tau_{xy}^2} \qquad (10\text{-}4)$$

It is also found that the shearing stress τ_{nt} is zero on planes experiencing maximum and minimum values of normal stress. Planes free of shear stress are known as principal planes. Normal stresses occurring on principal planes are known as principal stresses.

In a similar manner, the maximum in-plane shearing stresses are found to be

$$\tau_p = \pm \sqrt{\left(\frac{\sigma_x - \sigma_y}{2}\right)^2 + \tau_{xy}^2} \qquad (10\text{-}5)$$

A useful relation between the principal stresses σ_{p1} and σ_{p2} and the maximum in-plane shearing stress is

$$\tau_p = (\sigma_{p1} - \sigma_{p2})/2 \qquad (10\text{-}6)$$

When stresses act in three directions it can be shown that there are, in general, three orthogonal planes on which the shearing stress is zero. These planes are principal planes, and the stresses acting on them (the principal stresses) will have three values: one maximum, one minimum, and a third stress between the other two. The maximum shearing stress τ_{max} on any plane that could be passed through the point is

$$\tau_{max} = \frac{\sigma_{max} - \sigma_{min}}{2} \qquad (10\text{-}7)$$

Another useful relation between the principal stresses and the normal stresses on the orthogonal reference planes is

$$\sigma_{p1} + \sigma_{p2} = \sigma_x + \sigma_y \qquad (10\text{-}8)$$

When a body is subjected to a system of loads, the body deforms. The dimensions of the deformed body can be expressed in terms of strains ϵ_x, ϵ_y, and γ_{xy} that can be measured with strain gages. Relationships between strains, similar to Eqs. (10-1) and (10-2), are

$$\epsilon_n = \epsilon_x \cos^2 \theta + \epsilon_y \sin^2 \theta + \gamma_{xy} \sin \theta \cos \theta \qquad (10\text{-}11a)$$

$$\gamma_{nt} = -2(\epsilon_x - \epsilon_y) \sin \theta \cos \theta + \gamma_{xy} (\cos^2 \theta - \sin^2 \theta) \qquad (10\text{-}12a)$$

The similarity between Eqs. (10-11) and (10-12) for plane strain and Eqs. (10-1) and (10-2) for plane stress indicates that all of the equations developed for plane stress can be applied to plane strain by substituting ϵ_x for σ_x, ϵ_y for σ_y, and $\gamma_{xy}/2$ for τ_{xy}. Thus,

$$\tan 2\theta_p = \frac{\gamma_{xy}}{\epsilon_x - \epsilon_y} \qquad (10\text{-}13)$$

$$\epsilon_{p1}, \epsilon_{p2} = \frac{\epsilon_x + \epsilon_y}{2} \pm \sqrt{\left(\frac{\epsilon_x - \epsilon_y}{2}\right)^2 + \left(\frac{\gamma_{xy}}{2}\right)^2} \qquad (10\text{-}14)$$

$$\gamma_p = 2\sqrt{\left(\frac{\epsilon_x - \epsilon_y}{2}\right)^2 + \left(\frac{\gamma_{xy}}{2}\right)^2} \qquad (10\text{-}15)$$

Hooke's law, Eq. (4-15), can be extended to biaxial and triaxial states of stress often encountered in engineering practice by using the principal of superposition. The results for a biaxial state of stress are

$$\epsilon_x = \frac{1}{E}(\sigma_x - \nu\sigma_y)$$

$$\epsilon_y = \frac{1}{E}(\sigma_y - \nu\sigma_x) \qquad (10\text{-}16)$$

$$\epsilon_z = -\frac{\nu}{E}(\sigma_x + \sigma_y)$$

When Eqs. (10-16) are solved for the stresses in terms of the strains,

$$\sigma_x = \frac{E}{1 - \nu^2}(\epsilon_x + \nu\epsilon_y)$$

$$\qquad (10\text{-}18)$$

$$\sigma_y = \frac{E}{1 - \nu^2}(\epsilon_y + \nu\epsilon_x)$$

Equations (10-18) can be used to calculate normal stresses from measured or computed normal strains.

Torsion test specimens are used to study material behavior under pure shear, and it is observed that a shearing stress produces only a single corresponding shear strain. Thus, Hooke's law extended to shearing stresses is simply

$$\tau = G\gamma \qquad (10\text{-}20)$$

The three elastic constants E, ν, and G for a material are not independent but are related by the equation

$$G = \frac{E}{2(1 + \nu)} \qquad (10\text{-}21)$$

REVIEW PROBLEMS

10-147* At a point in a structural member, there are stresses on horizontal and vertical planes, as shown in Fig. P10-147. Determine the normal and shearing stresses on inclined plane AB.

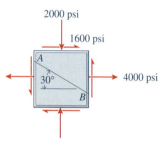

Figure P10-147

10-148* At a point in a structural member subjected to plane stress there are normal and shearing stresses on horizontal and vertical planes through the point, as shown in Fig. P10-148. Determine

(a) The normal and shearing stresses on the inclined plane AB.

(b) The principal stresses and the maximum shearing stress at the point.

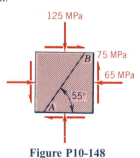

Figure P10-148

10-149 At a point in a machine component, the stresses on an inclined plane are $\sigma = 2400$ psi T and $\tau = 600$ psi, as shown in Fig. P10-149. The normal stress on a vertical plane through the point is zero. Determine

(a) The shearing stresses on horizontal and vertical planes.

(b) The normal stress on a horizontal plane through the point.

(c) The principal stresses and the maximum shearing stress at the point.

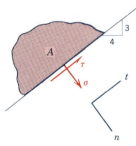

Figure P10-149

10-150 At a point in a stressed body, the nonzero principal stresses are oriented as shown in Fig. P10-150. Use Mohr's circle to determine

(a) The stresses on plane a-a.

(b) The stresses on horizontal and vertical planes at the point.

(c) The maximum shearing stress at the point.

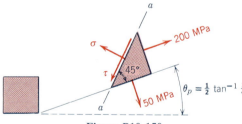

Figure P10-150

10-151* The strain components at a point in a body under a state of plane strain are $\epsilon_x = +1000\mu$, $\epsilon_y = -800\mu$, and $\gamma_{xy} = -800\mu$. Determine the principal strains and the maximum shearing strain at the point. Show the principal strain deformations and the maximum shearing strain distortion on a sketch.

10-152* The strain components at a point in a body under a state of plane strain are $\epsilon_x = +1200\mu$, $\epsilon_y = +960\mu$, and $\gamma_{xy} = -960\mu$. Determine the principal strains and the maximum shearing strain at the point. Show the principal strain deformations and the maximum shearing strain distortion on a sketch.

10-153 The strain rosette shown in Fig. P10-153 was used to obtain normal strain data at a point on the free surface of a machine part ($E = 29,000$ ksi and $\nu = 0.30$). The strain values obtained from the gages were $\epsilon_a = +1100\mu$, $\epsilon_b = +1000\mu$, and $\epsilon_c = -100\mu$. Determine

(a) The strain components ϵ_x, ϵ_y, and γ_{xy}.
(b) The principal strains and the maximum shearing strain at the point.

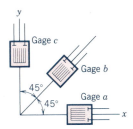

Figure P10-153

10-154 The strain rosette shown in Fig. P10-154 was used to obtain normal strain data at a point on the free surface of a machine part ($E = 200$ GPa and $\nu = 0.30$). The strain values obtained from the gages were $\epsilon_a = +900\mu$, $\epsilon_b = -450\mu$, and $\epsilon_c = +400\mu$. Determine

(a) The strain components ϵ_x, ϵ_y, and γ_{xy}.
(b) The principal strains and the maximum shearing strain at the point.

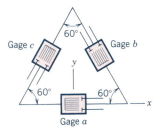

Figure P10-154

10-155* The strain rosette shown in Fig. P10-153 was used to obtain the following strain data at a point on the free surface of a steel ($E = 30,000$ ksi and $\nu = 0.30$) machine part: $\epsilon_a = +600\mu$, $\epsilon_b = +500\mu$, and $\epsilon_c = -200\mu$. Determine the principal stresses and the maximum shearing stress at the point. Show all of the stresses on a sketch.

10-156* The strain rosette shown in Fig. P10-154 was used to obtain the following strain data at a point on the free surface of an aluminum alloy ($E = 70$ GPa and $\nu = 0.33$) machine part: $\epsilon_a = +2000\mu$, $\epsilon_b = +1500\mu$, and $\epsilon_c = -1300\mu$. Determine the principal stresses and the maximum shearing stress at the point. Show all of the stresses on a sketch.

10-157 A cylindrical pressure tank is fabricated by butt-welding $\frac{3}{4}$-in. plate with a spiral seam, as shown in Fig. P10-157. The pressure in the tank is 400 psi, and an axial load of 30,000 lb is applied to the end of the tank through a rigid bearing plate. Determine the normal stress perpendicular to the weld and the shearing stress parallel to the weld.

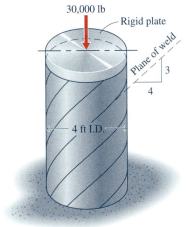

Figure P10-157

10-158 A short post is loaded by a force P of 25 kN and a force H of 5 kN as shown in Fig. P10-158. Determine the vertical normal stresses at corners A, B, C, and D of the post.

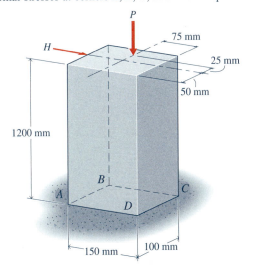

Figure P10-158

10-159* A steel shaft is loaded and supported as shown in Fig. P10-159. The allowable shearing stress is 8 ksi and the allowable tensile stress is 12 ksi. With T_2 equal to P in. · kip and T_1 equal to $P/2$ in. · kip, determine the maximum allowable axial load P.

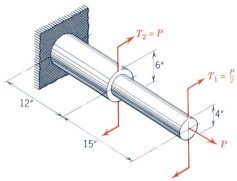

Figure P10-159

10-160* A solid shaft 100 mm in diameter is acted on by forces P and Q as shown in Fig. P10-160. Determine the principal stresses and the maximum shearing stress at point A on the surface of the shaft.

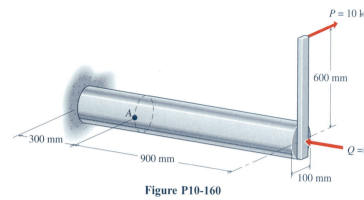

Figure P10-160

COMPUTER PROBLEMS

C10-161 At a point in a structural member subjected to plane stress, there are normal and shearing stresses on horizontal and vertical planes through the point as shown in Fig. P10-161.
(a) Determine the normal stress σ and the shearing stress τ at this point on the inclined plane AB as shown in the figure for various angles θ ($-90° < \theta < 90°$).
(b) Plot σ and τ as functions of θ ($-90° < \theta < 90°$).
(c) Plot $-\tau$ (vertical axis) as a function of σ (horizontal axis) for various angles θ ($-90° < \theta < 90°$). On the graph, clearly label the points corresponding to $\theta = 0°$, $30°$, $45°$, $60°$, $26.565°$, $-63.435°$, and $-18.435°$.

Note that the last three angles correspond to the axes for the principal stresses and the maximum shearing stress at the point.

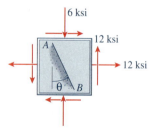

Figure P10-161

C10-162 A cylindrical pressure vessel 1.50 m in diameter is constructed by wrapping a 15 mm–thick steel plate into a spiral and butt-welding the meeting edges of the plate as shown in Fig. P10-162. The butt-welded seams form an angle

of θ with a transverse plane through the cylinder. If the internal pressure in the tank is $p = 800$ kPa,
(a) Compute and plot the normal stress σ perpendicular to the weld and the shearing stress τ parallel to the weld as functions of θ ($0° < \theta < 90°$).
(b) Plot $-\tau$ (vertical axis) as a function of σ (horizontal axis) for various angles θ ($0° < \theta < 90°$). On the graph, clearly label the points corresponding to $\theta = 0°$, $30°$, $45°$, $60°$, and $90°$.

Figure P10-162

C10-163 A strain gage rosette is used to obtain the strains ϵ_a, ϵ_b, and ϵ_c at the angles θ_a, θ_b, and θ_c (Fig. P10-163). For a state of plane stress, write a program to
(a) Compute and print the Cartesian strains, ϵ_x, ϵ_y, and γ_{xy}.
(b) Compute and print the principal strains ϵ_1, ϵ_2, and ϵ_3.
(c) Compute and print the angles for the in-plane principal strains, θ_{p1} and θ_{p2}.
(d) Compute and print the maximum shearing strains γ_{12}, γ_{23}, and γ_{31}.
(e) Compute (from ϵ_1, ϵ_2, and ϵ_3) and print the principal stresses σ_1, σ_2, and σ_3.
(f) Compute (from γ_{12}, γ_{23}, and γ_{31}) and print the maximum shearing stresses τ_{12}, τ_{23}, and τ_{31}.

(g) Use the program on the following eight sets of data:

Case	ϵ_a	ϵ_b	ϵ_c	θ_a	θ_b	θ_c	E	ν
1	750μ	-125μ	-250μ	$0°$	$45°$	$90°$	30,000 ksi	0.3
2	780μ	345μ	-332μ	$0°$	$-60°$	$60°$	10,000 ksi	1/3
3	875μ	700μ	-350μ	$30°$	$165°$	$-105°$	10,600 ksi	0.33
4	-555μ	925μ	740μ	$0°$	$120°$	$-120°$	200 GPa	0.3
5	525μ	450μ	1425μ	$0°$	$45°$	$135°$	73 GPa	0.304
6	-306μ	-456μ	906μ	$30°$	$75°$	$120°$	96 GPa	1/3
7	0μ	1000μ	0μ	$0°$	$45°$	$90°$	10,000 ksi	1/3
8	0μ	1000μ	0μ	$60°$	$105°$	$150°$	10,000 ksi	1/3

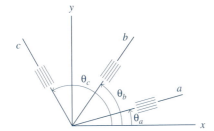

Figure P10-163

C10-164 A strain gage rosette is used to obtain the strains ϵ_a, ϵ_b, and ϵ_c at the angles θ_a, θ_b, and θ_c (Fig. P10-163). For a state of plane stress, write a program to

(a) Compute and print the Cartesian strains, ϵ_x, ϵ_y, γ_{xy}, and ϵ_z.

(b) Compute and print the Cartesian stresses σ_x, σ_y, τ_{xy}, and σ_z.

(c) Compute and print the principal stresses, σ_1, σ_2, and σ_3.

(d) Compute and print the angles for the in-plane principal stresses θ_{p1} and θ_{p2}.

(e) Compute and print the maximum shearing stresses τ_{12}, τ_{23}, and τ_{31}.

(f) Use the program on the following eight sets of data:

Case	ϵ_a	ϵ_b	ϵ_c	θ_a	θ_b	θ_c	E	ν
1	750μ	-125μ	-250μ	$0°$	$45°$	$90°$	30,000 ksi	0.3
2	780μ	345μ	-332μ	$0°$	$-60°$	$60°$	10,000 ksi	1/3
3	875μ	700μ	-350μ	$30°$	$165°$	$-105°$	10,600 ksi	0.33
4	-555μ	925μ	740μ	$0°$	$120°$	$-120°$	200 GPa	0.3
5	525μ	450μ	1425μ	$0°$	$45°$	$135°$	73 GPa	0.304
6	-306μ	-456μ	906μ	$30°$	$75°$	$120°$	96 GPa	1/3
7	0μ	1000μ	0μ	$0°$	$45°$	$90°$	10,000 ksi	1/3
8	0μ	1000μ	0μ	$60°$	$105°$	$150°$	10,000 ksi	1/3

C10-165 A $\frac{3}{8}$ in.–diameter by 2.5 in.–long steel bolt is used to fasten two machine parts as shown in Fig. P10-165. Because of friction in the threads, the shank of the bolt is subjected to torsional stresses as well as axial tensile stresses. As the bolt is tightened, the 2-in. shank of the bolt is stretched 0.05-in. per turn of the head of the bolt. The torque (in. · lb) is about 6 percent of the tension (lb) in the bolt. For a point on the surface of the bolt, compute and plot for $0° < \theta < 30°$.
(a) The axial stress σ_x as a function of the angle of twist θ of the head of the bolt.
(b) The shearing stress τ_{xy} as a function of θ.
(c) The principal stresses, σ_{p1} and σ_{p2}, as functions of θ.
(d) The maximum shearing stress τ_{max} as a function of θ.

Figure P10-165

C10-166 A hollow circular steel shaft 3 m long must be designed to withstand a torque of 300 kN · m and an axial load of 900 kN. If the angle of twist θ in the 3-m length must not exceed 0.22 deg and the maximum shearing stress τ_{max} in the shaft must not exceed 45 MPa, compute and plot the range of allowable outside diameters d_o as a function of the inside diameter d_i (0 mm $< d_i <$ 600 mm) of the shaft. Limit the wall thickness to values greater than 10 mm.

C10-167 As the C-clamp shown in Fig. P10-167 is tightened, the web of the C-section is subjected to flexural stresses as well as axial tensile stresses. If the web of the section has a width of 1 in. and a thickness of $\frac{1}{4}$ in., and the clamp has a capacity of 3 in.,
(a) Compute and plot the maximum tensile and compressive stresses on section A-A as a function of the clamp force P (0 lb $< P <$ 500 lb).
(b) Determine c, the distance from the inside edge of the web to the point where the normal stress is zero. Is c a function of P?

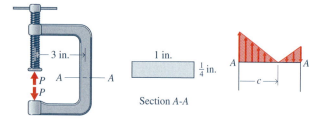

Figure P10-167

C10-168 A W356 × 179 wide-flange section is used for a simply supported beam that carries a concentrated load of 700 kN at the middle of a 3-m span (Fig. P10-168). Compute and plot
(a) The maximum normal stress σ_{max} at A, the bottom surface of the beam, and at B, in the web just above the lower flange of the beam, as functions of x (0 m $< x <$ 1.5 m).
(b) The maximum shearing stress τ_{max} at C, the neutral surface of the beam, and at B, in the web just above the lower flange of the beam, as functions of x (0 m $< x <$ 1.5 m).

Figure P10-168

C10-169 A WT4 × 29 T-section is used for a simply supported beam that carries a uniformly distributed load of 70 lb/ft for the full 25-ft span of the beam (Fig. P10-169). If the flange is placed at the bottom of the beam,
(a) Compute the principal stress σ_{p1} and σ_{p2} and the maximum shearing stress τ_{max} as functions of y, the distance from the neutral axis of the beam, for $x = 1$ ft.
(b) Show the variation of the principal stresses on a graph with the y-distance as the vertical axis and with stress as the horizontal axis.

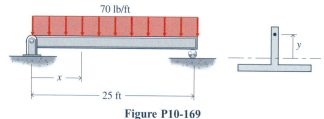

Figure P10-169

COLUMNS

11-1 INTRODUCTION

In their simplest form, columns are long, straight, prismatic bars subjected to compressive, axial loads. As long as a column remains straight, it can be analyzed by the methods of Chapter 4; however, if a column begins to deform laterally, the deflection may become large and lead to catastrophic failure. This situation, called *buckling,* can be defined as the sudden large deformation of a structure due to a slight increase of an existing load under which the structure had exhibited little, if any, deformation before the load was increased. For example, a yardstick will support a compressive load of several pounds without discernible lateral deformation, but once the load becomes large enough to cause the yardstick to "bow out" a slight amount, any further increase of load produces large lateral deflections.

Buckling of such a column is caused not by failure of the material of which the column is composed but by deterioration of what was a stable state of equilibrium to an unstable one. The three states of equilibrium can be illustrated with a ball at rest on a surface, as shown in Fig. 11-1. The ball in Fig. 11-1a is in a stable equilibrium position at the bottom of the pit because gravity will cause it to return to its equilibrium position if perturbed. The ball in Fig. 11-1b is in a neutral equilibrium position on the horizontal plane because it will remain at any new position to which it is displaced, tending neither to return to nor move farther from its original position. The ball in Fig. 11-1c, however, is in an unstable equilibrium position at the top of a hill because, if it is perturbed, gravity will cause it to move even farther from its original location until it eventually finds a stable equilibrium position at the bottom of another pit.

As the compressive load on a column is gradually increased from zero, the column is at first in a state of stable equilibrium. During this state, if the column is perturbed by small lateral loads, it will return to its straight configuration when

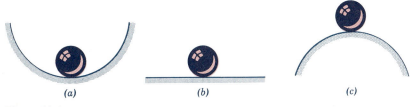

(a)   (b)   (c)

Figure 11-1

the lateral loads are removed. As the compressive load is increased further, a critical value is reached at which the column is on the verge of experiencing a lateral deflection so, if it is perturbed, it will not return to its straight configuration. The load cannot be increased beyond this value unless the column is restrained laterally; should the lateral restraints by removed, the slightest perturbation will trigger large lateral deflections. For long, slender columns, the critical buckling load (the maximum load for which the column is in stable equilibrium) occurs at stress levels much less than the proportional limit for the material. This indicates that this type of buckling is an elastic phenomenon.

11-2 BUCKLING OF LONG STRAIGHT COLUMNS

The first solution for buckling of long slender columns was published in 1757 by the Swiss mathematician Leonhard Euler (1707–1783). Although the results of this section can be used only for long slender columns, the analysis, similar to that used by Euler, is mathematically revealing and helps explain the behavior of columns.

The purpose of this analysis is to determine the minimum axial compressive load for which a column will experience lateral deflections. A straight, slender, pivot-ended column of length L centrically loaded by axial compressive forces P at each end is shown in Fig. 11-2a. A pivot-ended column is supported such that the bending moment and lateral movement are zero at the ends. In Fig. 11-2b, the load P has been increased sufficiently to cause a lateral deflection δ at the midpoint of the span. Axes are selected with the origin at the center of the span for convenience. If the column is sectioned at an arbitrary position x, a free-body diagram of the portion to the right of the section will appear as shown in Fig. 11-2c. The two forces constitute a couple of magnitude $P(\delta - y)$ that must

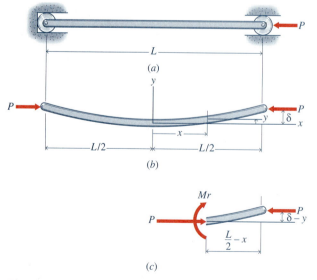

Figure 11-2

equal the resisting moment M_r; thus, $M_r = P(\delta - y)$. The differential equation for the elastic curve, as given by Eq. (9-3), becomes

$$EI \frac{d^2y}{dx^2} = M_r = P(\delta - y)$$

or

$$\frac{d^2y}{dx^2} + \frac{P}{EI} y = \frac{P\delta}{EI} \qquad \text{(a)}$$

Equation (a) is an ordinary second-order linear differential equation with constant coefficients and a constant right-hand side. Established methods for the solution of such equations show it to be of the form

$$y = A \sin px + B \cos px + C \qquad \text{(b)}$$

where A, B, C, and p are constants. By differentiating Eq. (b), substituting the results into Eq. (a), and collecting like terms, the following expression is obtained for evaluating p and C:

$$(-p^2 + P/EI)(A \sin px + B \cos px) + (P/EI)C = (P/EI)\delta$$

from which it follows that

$$p^2 = P/EI \qquad \text{and} \qquad C = \delta$$

The constants A and B can be obtained from the two boundary conditions that the slope and deflection of the elastic curve are zero at the origin; that is, at $x = 0$, $y = 0$, and at $x = 0$, $dy/dx = 0$. The purpose of this analysis is to determine the minimum load for which lateral deflections occur; therefore, it is necessary to have a third boundary condition; namely, at $x = L/2$, $y = \delta > 0$. The deflection δ is required to be nonzero; otherwise, $\delta = 0$, $y \equiv 0$ could be a solution of Eq. (a). When the boundary conditions are substituted into Eq. (b) and its first derivative, they give

$$0 = B + C$$

$$0 = Ap$$

from which

$$A = 0 \qquad \text{and} \qquad B = -C = -\delta$$

The solution to Eq. (a) thus becomes

$$y = \delta \left(1 - \cos \sqrt{\frac{P}{EI}} \, x \right) \qquad \text{(11-1)}$$

To satisfy the condition that at $x = L/2$, $y = \delta$, the cosine term in Eq. (11-1) must vanish. Thus,

$$\cos \sqrt{\frac{P}{EI}} \frac{L}{2} = 0$$

This equation is satisfied when the argument of the cosine is an odd multiple of $\pi/2$ or

$$\sqrt{\frac{P}{EI}} \frac{L}{2} = \frac{\pi}{2}, \frac{3\pi}{2}, \frac{5\pi}{2}, \cdots$$

Only the first value has physical significance since it determines the minimum value of P for a nontrivial solution. This value for P is called the *critical buckling load,* is designated by P_{cr}, and has the magnitude

$$P_{cr} = \frac{\pi^2 EI}{L^2} \tag{11-2}$$

The term P_{cr} is usually called the Euler buckling load[1] in honor of Leonhard Euler.

The second moment of the cross-sectional area I in Eq. (11-2) refers to the axis about which bending occurs. When I is replaced by Ar^2, where r is the radius of gyration (see Section 8-4-1) about the axis of bending, Eq. (11-2) becomes

$$\frac{P_{cr}}{A} = \frac{\pi^2 E}{(L/r)^2} \tag{11-3}$$

The quantity L/r is the slenderness ratio and is determined for the axis about which bending tends to occur. The slenderness ratio is dimensionless. For a pivot-ended, centrically loaded column with no intermediate bracing to restrain lateral motion, bending occurs about the axis of minimum second moment of area (minimum radius of gyration).

The Euler buckling load as given by Eq. (11-2) or (11-3) agrees well with experiment if the slenderness ratio is large ($L/r > 140$ for steel columns). Short compression members ($L/r < 40$ for steel columns) can be treated as compression blocks where yielding occurs before buckling. Many columns lie between these extremes in which neither solution is applicable. These intermediate-length columns are analyzed by using empirical formulas described in Section 11-4.

Example Problem 11-1

An 8-ft pivot-ended, timber [$E = 1.9(10^6)$ psi and $\sigma_e = 6400$ psi] column has a 2 × 4-in. rectangular cross section. Determine

(a) The slenderness ratio.
(b) The Euler buckling load.

---

[1]While the analysis predicts the buckling load, it does not determine the corresponding lateral deflection δ. This deflection can assume any nonzero value small enough that the nonlinear factor $[1 + (dy/dx)^2]^{3/2}$ in the curvature expression is approximately unity.

(c) The ratio of the axial stress under the action of the buckling load to the elastic strength σ_e of the material.

SOLUTION

(a) To determine the slenderness ratio, the minimum radius of gyration must be calculated. The moment of inertia of a rectangular cross section is $bh^3/12$ and the area is bh; therefore, the radius of gyration is $\sqrt{I/A}$ or $h/2\sqrt{3}$. The minimum radius of gyration is found by using the centroidal axis parallel to the longer side of the rectangle. Thus, $b = 4$ in. and $h = 2$ in. so that

$$r = h/2\sqrt{3} = 2/2\sqrt{3} = 0.5774 \text{ in.}$$

The slenderness ratio is then found to be

$$L/r = 8(12)/0.5774 = 166.3 \qquad \textbf{Ans.}$$

(b) The Euler buckling load is found by using Eq. (11-3). Thus,

$$P_{cr} = \pi^2 EA/(L/r)^2 = (\pi^2)(1.9)(10^6)(2)(4)/(166.3)^2$$
$$= 5425 \text{ lb} \cong 5430 \text{ lb} \qquad \textbf{Ans.}$$

(c) The axial stress under the action of the buckling load is

$$\sigma_{cr} = P_{cr}/A = 5425/8 = 678 \text{ psi}$$

$$\sigma_{cr}/\sigma_e = 678/6400 = 0.1059 = 10.59 \text{ percent} \qquad \textbf{Ans.}$$

(This demonstrates that buckling can occur at stresses well below the linearly elastic limit of a material for sufficiently slender columns.) ■

Example Problem 11-2

Two 51 × 51 × 3.2-mm structural steel angles 3 m long will be used as a pivot-ended column. Determine the slenderness ratio and the Euler buckling load if

(a) The two angles are not connected and each acts as an independent member.
(b) The two angles are fastened together as shown in Fig. 11-3 to act as a unit.

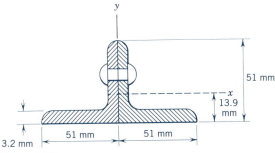

51 mm

x
13.9 mm

51 mm

51 mm   51 mm

3.2 mm

Figure 11-3

SOLUTION

(a) If the angles are not connected and each acts independently, the slenderness ratio is determined by using the minimum radius of gyration of the individual cross sections. From Appendix B, the minimum radius of gyration r_{min} for this angle is 10.1 mm about the Z-Z axis. Thus, the slenderness ratio is

$$\frac{L}{r} = \frac{L}{r_{min}} = \frac{3000}{10.1} = 297 \qquad \text{Ans.}$$

The area for each angle is 312 mm²; therefore, the cross-sectional area for the column is 624 mm². The modulus of elasticity for structural steel is 200 GPa; therefore, the buckling load is

$$P_{cr} = \frac{\pi^2 EA}{(L/r)^2} = \frac{\pi^2(200)(10^9)(624)(10^{-6})}{(297)^2}$$
$$= 13.964(10^3) \text{ N} \cong 13.96 \text{ kN} \qquad \text{Ans.}$$

(b) With the two angles connected as shown in Fig. 11-3, both I_x and I_y or r_x and r_y must be known to determine the minimum radius of gyration. Cross-sectional properties for the angles are given in Appendix B. The value of I_y for the two angles is obtained by using the parallel axis theorem; thus,

$$I_y = 2(I_C + Ad^2) = 2(Ar_C^2 + Ad^2)$$

and

$$r_y = \sqrt{I_y/2A} = \sqrt{\frac{2A(r_C^2 + d^2)}{2A}} = \sqrt{r_C^2 + d^2}$$

The above expression indicates that the radius of gyration for the two angles is the same as that for one angle, a fact obtained directly from the definition of radius of gyration. Then,

$$r_y = \sqrt{(15.9)^2 + (13.9)^2} = 21.1 \text{ mm}$$

In a similar fashion, the radius of gyration about the x-axis is the radius of gyration $r_x = 15.9$ mm for a single angle, since $d = 0$ for this axis. This means that the column tends to buckle about the x-axis where the slenderness ratio is

$$L/r = \frac{L}{r_{min}} = \frac{3000}{15.9} = 188.68 \cong 188.7 \qquad \text{Ans.}$$

The corresponding Euler buckling load is

$$P_{cr} = \frac{\pi^2 EA}{(L/r)^2} = \frac{\pi^2(200)(10^9)(624)(10^{-6})}{(188.68)^2}$$
$$= 34.599(10^3) \text{ N} \cong 34.6 \text{ kN} \quad \blacksquare \qquad \text{Ans.}$$

PROBLEMS

In the following problems, assume that all columns are pivot-ended and that Euler's formula [Eq. (11-2)] is applicable for steel with L/r greater than 140 and for aluminum or timber with L/r greater than 80.

Standard Problems

11-1* A steel ($E = 30,000$ ksi) rod 1 in. in diameter and 50 in. long will be used to support an axial compressive load P. Determine
(a) The slenderness ratio.
(b) The Euler buckling load.
(c) The axial stress in the column when the Euler load is applied.

11-2* A hollow circular steel ($E = 200$ GPa) column 6 m long has an outside diameter of 125 mm and an inside diameter of 100 mm. Determine
(a) The slenderness ratio.
(b) The Euler buckling load.
(c) The axial stress in the column when the Euler load is applied.

11-3 A 16-ft column with the cross section shown in Fig. P11-3 is constructed from four pieces of timber. The timbers are nailed together so that they act as a unit. Determine
(a) The slenderness ratio.
(b) The Euler buckling load. Use $E = 2000$ ksi for the timber.
(c) The axial stress in the column when the Euler load is applied.

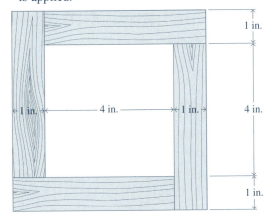

Figure P11-3

11-4 A 3-m column with the cross section shown in Fig. P11-4 is constructed from two pieces of timber. The timbers are nailed together so that they act as a unit. Determine
(a) The slenderness ratio.
(b) The Euler buckling load. Use $E = 13$ GPa for the timber.
(c) The axial stress in the column when the Euler load is applied.

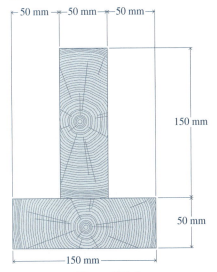

Figure P11-4

11-5* A WT6 × 36 structural steel section (see Appendix B for cross-sectional properties) is used for an 18-ft column. Determine
(a) The slenderness ratio.
(b) The Euler buckling load. Use $E = 29,000$ ksi for the steel.
(c) The axial stress in the column when the Euler load is applied.

11-6* A WT152 × 89 structural steel section (see Appendix B for cross-sectional properties) is used for a 6-m column. Determine
(a) The slenderness ratio
(b) The Euler buckling load. Use $E = 200$ GPa for the steel.
(c) The axial stress in the column when the Euler load is applied.

11-7 Determine the maximum allowable compressive load for a 12-ft aluminum alloy ($E = 10,600$ ksi) column having the cross section shown in Fig. P11-7 if a factor of safety of 2.50 is specified.

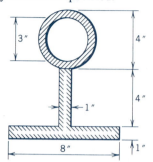

Figure P11-7

11-8 Determine the maximum allowable compressive load for a 6.5-m steel ($E = 200$ GPa) column having the cross section shown in Fig. P11-8 if a factor of safety of 1.92 is specified.

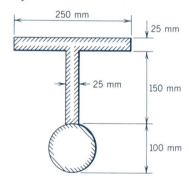

Figure P11-8

Challenging Problems

11-9* Determine the maximum total compressive load that three rectangular aluminum alloy ($E = 10,600$ ksi) bars $1 \times 4 \times 100$ in. long will carry as columns if
(a) Each bar acts independently of the other two.
(b) The three bars are welded together to form an H-column.

11-10* Two C229 $\times$ 30 structural steel channels (see Appendix B for cross-sectional properties) are used for a column that is 12 m long. Determine the total compressive load required to buckle the two members if
(a) They act independently of each other. Use $E = 200$ GPa.
(b) They are laced 150 mm back to back as shown in Fig. P11-10.

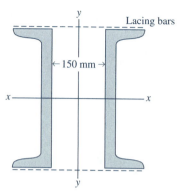

Figure P11-10

11-11 A column 40 ft long is made by riveting three S10 $\times$ 25.4 structural steel sections (see Appendix B for cross-sectional properties) together as shown in Fig. P11-11. Determine
(a) The maximum compressive load that this column can support. Use $E = 29,000$ ksi.
(b) The maximum compressive load that this column can support if the rivets are removed and the sections act as separate units.

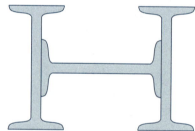

Figure P11-11

11-12 A 60-kN load is supported by a tie rod AB and a pipe strut BC as shown in Fig. P11-12. The tie rod has a diameter of 30 mm and is made of steel with a modulus of elasticity of 210 GPa and a yield strength of 360

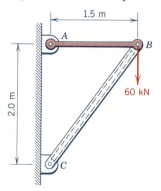

Figure P11-12

MPa. The pipe strut has an inside diameter of 50 mm and a wall thickness of 15 mm and is made of an aluminum alloy with a modulus of elasticity of 73 GPa and a yield strength of 280 MPa. Determine the factor of safety with respect to failure by slip or buckling for the structure.

11-13* A simple pin-connected truss is loaded and supported as shown in Fig. P11-13. The members of the truss were fabricated by bolting two C10 × 30 channel sections (see Appendix B for cross-sectional properties) back-to-back to form an H-section. The channels are made of structural steel with a modulus of elasticity of 29,000 ksi and a yield strength of 36 ksi. Determine the maximum load P that can be applied to the truss if a factor of safety of 1.75 with respect to failure by slip and a factor of safety of 4 with respect to failure by buckling is specified.

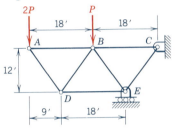

Figure P11-13

11-14 A simple pin-connected truss is loaded and supported as shown in Fig. P11-14. All members of the truss are WT102 × 43 sections (see Appendix B for cross-sectional properties) made of structural steel with a modulus of elasticity of 200 GPa and a yield strength of 250 MPa. Determine
(a) The factor of safety with respect to failure by slip.
(b) The factor of safety with respect to failure by buckling.

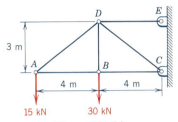

Figure P11-14

11-3 EFFECTS OF DIFFERENT IDEALIZED END CONDITIONS

The Euler buckling formula, as expressed by either Eq. (11-2) or Eq. (11-3), was derived for a column with pivoted ends. The Euler equation changes for columns with different end conditions, such as the four common ones shown in Fig. 11-4.

While it is possible to set up the differential equation with the appropriate boundary conditions to determine the Euler equation for each new case, a more common approach makes use of the concept of an effective length. The pivot-ended column, by definition, has zero bending moments at each end. The length L in the Euler equation, therefore, is the distance between successive points of zero bending moment. All that is needed to modify the Euler column formula for use with other end conditions is to replace L by L', where L' is defined as the effective length of the column (the distance between two successive inflection points or points of zero moment).

The ends of the column in Fig. 11-4b are built in or fixed. Since the deflection curve is symmetrical, the distance between successive points of zero moment (inflection points) is half the length of the column. Thus, the effective length

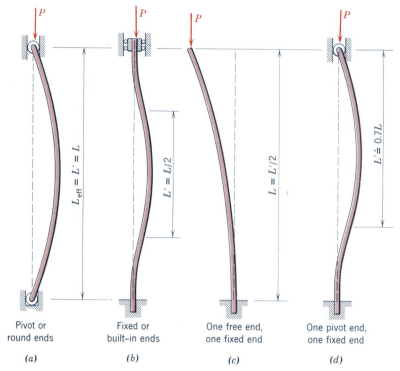

Figure 11-4

L' of a fixed-end column for use in the Euler column formula is half the true length ($L' = 0.5L$). The column in Fig. 11-4c, being fixed at one end and free at the other end, has zero moment only at the free end. If a mirror image of this column is visualized below the fixed end, however, the effective length between points of zero moment is seen to be twice the actual length of the column ($L' = 2L$). The column in Fig. 11-4d is fixed at one end and pinned at the other end. The effective length of this column cannot be determined by inspection, as could be done in the previous two cases; therefore, it is necessary to solve the differential equation to determine the effective length. This procedure yields $L' = 0.7L$.

A pivot-ended column is usually loaded through a pin that, as a result of friction, is not completely free to rotate; hence, there will always be an indeterminate moment at the ends of this type of column that will reduce the distance between the inflection points to a value less than L. Also, it is impossible to support a column so that all rotation is eliminated, and so the effective length of the column in Fig 11-4b will be somewhat greater than $L/2$. As a result, it is usually necessary to modify the effective column lengths indicated by the ideal end conditions. The amount of the corrections will depend on the individual application. In summary, the term L/r in all column formulas in this book is interpreted to mean the effective slenderness ratio L'/r. In the problems in this book, the length given for a member is assumed to be the effective length unless otherwise noted.

PROBLEMS

In the following problems, the Euler formula [Eq. (11-2)] is applicable for steel columns with an effective slenderness ratio L'/r greater than 140 and for aluminum or timber columns with L'/r greater than 80.

Standard Problems

11-15* A W8 × 15 structural steel section (see Appendix B for cross-sectional properties) is used for a fixed-end, free-end column having an actual length of 14 ft. Determine the maximum safe load for the column if a factor of safety of 2 with respect to failure by buckling is specified. Use $E = 29{,}000$ ksi.

11-16* An L102 × 76 × 6.4-mm aluminum alloy ($E = 70$ GPa) angle is used for a fixed-end, pivot-end column having an actual length of 2.5 m. Determine the maximum safe load for the column if a factor of safety of 1.75 with respect to failure by buckling is specified. See Appendix B for cross-sectional properties; they are the same as those for a steel angle of the same size.

11-17 A 6-in. × 6-in. × 20-ft long timber ($E = 1900$ ksi) is used as a fixed-end, pivot-end column to support a 40,000-lb load. Determine the factor of safety based on the Euler buckling load.

11-18 Determine the maximum load that a 50-mm × 75-mm × 2.5-m long aluminum alloy bar ($E = 73$ GPa) can support with a factor of safety of 3 with respect to failure by buckling if it is used as a fixed-end, pivot-end column.

11-19* A structural steel ($E = 29{,}000$ ksi) column 20 ft long must support an axial compressive load of 200 kip. The column can be considered pivoted at one end and fixed at the other end for bending about one axis and fixed at both ends for bending about the other axis. Select the lightest wide-flange or American standard section (see Appendix B) that can be used for the column.

11-20 A structural steel ($E = 200$ GPa) column 6 m long must support an axial compressive load of 500 kN. The column can be considered pivoted at both ends for bending about one axis and fixed at both ends for bending about the other axis. Select the lightest wide-flange or American standard section (see Appendix B) that can be used for the column.

Challenging Problems

11-21* Verify the effective length for the fixed-end, fixed-end column shown in Fig. 11-4b by solving the differential equation of the elastic curve and applying the appropriate boundary conditions.

11-22* Verify the effective length for the fixed-end, free-end column shown in Fig. 11-4c by solving the differential equation of the elastic curve and applying the appropriate boundary conditions.

11-23 A rigid block is supported by two fixed-end, fixed-end columns as shown in Fig. P11-23a. Determine the effective lengths of the columns by solving the differential equation of the elastic curve and applying the appropriate boundary conditions. Assume that buckling occurs as shown in Fig. P11-23b.

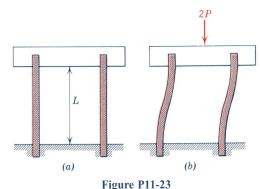

Figure P11-23

11-24 A free-body diagram for a fixed-end, pivot-end column is shown in Fig. P11-24. Use the diagram to develop the differential equation of the elastic curve. Verify the effective length for the fixed-end, pivot-end column shown in Fig. 11-4d by solving the differential equation of the elastic curve and applying the appropriate boundary conditions.

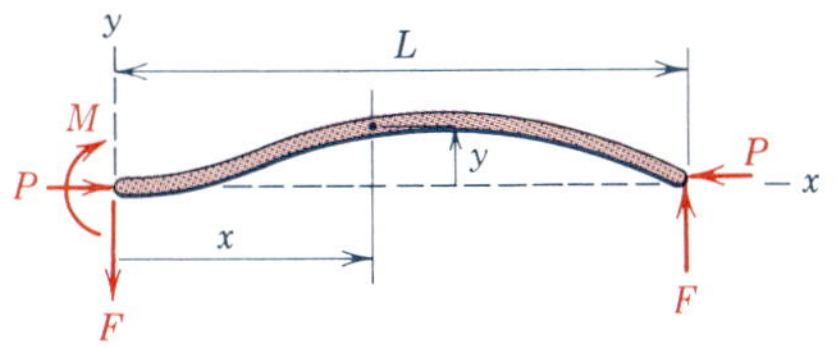

Figure P11-24

11-4 EMPIRICAL COLUMN FORMULAS—CENTRIC LOADING

Euler's formula, Eq. (11-2) or (11-3), is valid when the axial compressive stress for the column is less than the yield strength. The range of usefulness of the Euler formula is seen in Fig. 11-5, where the critical stress ($\sigma_{cr} = P_{cr}/A$) is plotted versus the slenderness ratio (L/r). The plot is for structural steel for which $E = 29 \times 10^6$ psi (200 GPa) and $\sigma_{yield} = 36,000$ psi (250 MPa). The Euler curve is truncated at 36,000 psi since the critical stress cannot exceed the yield strength. For structural steel, this occurs when $L/r = 89$. The dashed portion of the Euler curve indicates that the calculation of the critical stress is no longer useful since the stress exceeds the yield strength. For practical purposes, columns with large values of the slenderness ratio are useless since they support only small loads. At the low end of the slenderness scale, the column would behave essentially as a short compression block, and the critical stress would be the compressive strength of the material (for metals, usually the yield strength). The extent of this range—the compression block range—is a matter of judgment, or is dictated by specifications. The range between the compression block and the slender ranges is known as the *intermediate range*. Neither the compression-block theory nor the Euler formula give results in the intermediate range that agree with test results.

Experimental results of many tests on axially loaded columns are shown in Fig. 11-6. The Euler curve agrees well with experimental data for slenderness ratios in the slender range. Experimental data and compression-block theory agree in the compression-block range. Test results are not in agreement with the compression-block theory nor the Euler theory in the intermediate range. These intermediate-length columns are analyzed by empirical formulas.

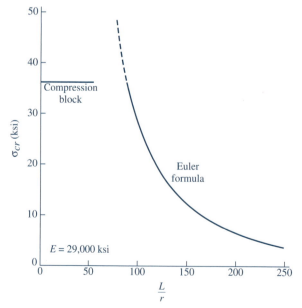

Figure 11-5

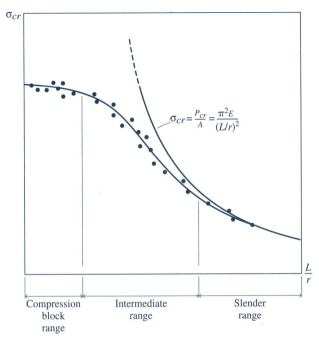

$$\sigma_{cr} = \frac{P_{cr}}{A} = \frac{\pi^2 E}{(L/r)^2}$$

Compression block range   Intermediate range   Slender range

$\frac{L}{r}$

Figure 11-6

For design purposes, the entire range of stresses for a given material is covered by an appropriate set of specifications known as a *column code*. Depending upon the code, the empirical formula may be specified for the intermediate range along with the limits of the intermediate range; the code may also specify the Euler formula in the slender range, and the limits of the slender range; or the code may specify the critical stress and limits of the range for the compression-block range. If the code is written for allowable loads or stresses, the factor of safety will either be specified or included in the constants for the empirical formula. A few representative codes are listed in Table 11-1. Each code is written for allowable (or safe) stresses. A few representative codes for steel, aluminum, and timber will be discussed. The codes are taken from references listed in the footnotes of Table 11-1.

The formulas in Table 11-1 are representative of column equations that have been incorporated in various design codes. Slenderness ratios in this table are always the effective slenderness ratio L'/r. If no end conditions are specified in the problems presented later, the stated length is the effective length. Note that the use of high-strength materials will increase the allowable (P/A) value for short columns but will have little effect on the strength of long columns, since the critical load (Euler load) depends on Young's modulus, not on the elastic strength of the material. Note also that the use of fixed or restrained ends, which has the effect of reducing the length of the column, materially increases the strength of slender columns but has much less influence on short compression members.

The discussion so far has been concerned with primary instability, in which the column deflects as a whole into a smooth curve. No discussion of compression loading is complete without reference to local instability, in which the member fails locally by crippling of thin sections. Thin open sections such as angles, channels, and H-sections are particularly sensitive to crippling failure. The de-

TABLE 11-1 Some Representative Column Codes for Centric Loading

Code No.	Source	Material	Compression-Block and/or Intermediate-Range Formulas and Limitations (L/r is the effective ratio L'/r)	Slender Range
1	a	Structural steel with a yield point σ_y	$0 \leq \dfrac{L}{r} \leq C_c \quad \sigma_{\text{all}} = \dfrac{\sigma_y}{FS}\left[1 - \dfrac{1}{2}\left(\dfrac{L/r}{C_c}\right)^2\right]$ $C_c^2 = \dfrac{2\pi^2 E}{\sigma_y}$ $FS = \dfrac{5}{3} + \dfrac{3}{8}\left(\dfrac{L/r}{C_c}\right) - \dfrac{1}{8}\left(\dfrac{L/r}{C_c}\right)^3$	$\dfrac{L}{r} \geq C_c$ $\sigma_{\text{all}} = \dfrac{\pi^2 E}{1.92(L/r)^2}$
2	b	2014-T6 (Alclad) Aluminum alloy	$\dfrac{L}{r} \leq 12 \quad \sigma_{\text{all}} = 28 \text{ ksi}$ $\qquad\qquad\quad = 193 \text{ MPa}$ $12 \leq \dfrac{L}{r} \leq 55 \quad \sigma_{\text{all}} = \left[30.7 - 0.23\left(\dfrac{L}{r}\right)\right] \text{ksi}$ $\qquad\qquad\qquad = \left[212 - 1.585\left(\dfrac{L}{r}\right)\right] \text{MPa}$	$\dfrac{L}{r} \geq 55$ $\sigma_{\text{all}} = \dfrac{54{,}000}{(L/r)^2} \text{ksi}$ $= \dfrac{372(10^3)}{(L/r)^2} \text{MPa}$
3	b	6061-T6 Aluminum alloy	$\dfrac{L}{r} \leq 9.5 \quad \sigma_{\text{all}} = 19 \text{ ksi}$ $\qquad\qquad\quad = 131 \text{ MPa}$ $9.5 \leq \dfrac{L}{r} \leq 66 \quad \sigma_{\text{all}} = \left[20.2 - 0.126\left(\dfrac{L}{r}\right)\right] \text{ksi}$ $\qquad\qquad\qquad = \left[139 - 0.868\left(\dfrac{L}{r}\right)\right] \text{MPa}$	$\dfrac{L}{r} \geq 66$ $\sigma_{\text{all}} = \dfrac{51{,}000}{(L/r)^2} \text{ksi}$ $= \dfrac{351(10^3)}{(L/r)^2} \text{MPa}$
4	c	Timber with a rectangular cross section $b \times d$ where $d < b$	$\dfrac{L}{d} \leq 11 \quad \sigma_{\text{all}} = F_c*$ $11 \leq \dfrac{L}{d} \leq k \quad \sigma_{\text{all}} = F_c\left[1 - \dfrac{1}{3}\left(\dfrac{L/d}{k}\right)^4\right]$ $k = 0.671\sqrt{E/F_c}$	$k \leq \dfrac{L}{d} \leq 50$ $\sigma_{\text{all}} = \dfrac{0.30E}{(L/d)^2}$

a. *Manual of Steel Construction,* 9th ed., American Institute of Steel Construction, New York, 1959.

b. *Specifications for Aluminum Structures,* Aluminum Association, Inc., Washington, D.C., 1986.

c. *Timber Construction Manual,* 3rd ed., American Institute of Timber Construction, John Wiley & Sons, Inc., New York, 1985.

*F_c is the allowable stress for a short block in compression parallel to the grain.

sign of such members to avoid crippling failure is usually governed by specifications controlling the width–thickness ratios of outstanding flanges. Closed-section members of thin material (thin-walled tubes, for example) must also be examined for crippling failure when the members are short.

Example Problem 11-3

Two structural steel C10 $\times$ 25 channels are laced 5 in. back to back, as shown in Figs. 11-7a and b, to form a column. Determine the maximum allowable axial load for effective lengths of 25 ft and 40 ft. Use Code 1 for structural steel (see Appendix B for properties).

SOLUTION

Both I_x and I_y (see Fig. 11-7b) or r_x and r_y must be known to determine the minimum radius of gyration. Properties of the channel section are given in Appendix B. The value of I_y for two channels is obtained by using the parallel axis theorem; thus,

$$I_y = 2(I_C + Ad^2) = 2(Ar_C^2 + Ad^2)$$

and

$$r_y = \sqrt{I_y/2A} = \sqrt{\frac{2A(r_C^2 + d^2)}{2A}} = \sqrt{r_C^2 + d^2}$$

The above expression indicates that the radius of gyration for the two channels is the same as that for one channel, a fact obtained directly from the definition of radius of gyration. Then,

$$r_y = \sqrt{(0.676)^2 + (2.50 + 0.617)^2} = 3.19 \text{ in.}$$

which is less than the tabular value of 3.52 in. for r_x. This means that the column tends to buckle with respect to the y axis, and the slenderness ratios are

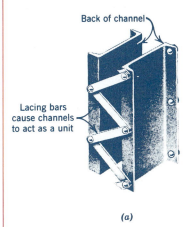

Back of channel

Lacing bars cause channels to act as a unit

(a)

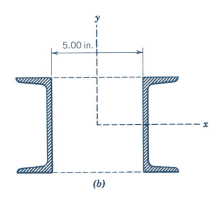

y

5.00 in.

x

(b)

Figure 11-7

$$\frac{L}{r} = \frac{12(25)}{3.19} = 94.0 \quad \text{and} \quad \frac{12(40)}{3.19} = 150.5$$

From Code 1,

$$C_c^2 = \frac{2\pi^2 E}{\sigma_y} = \frac{2\pi^2(29,000)}{36} = 15,901 \quad C_c = 126.1$$

The slenderness ratios above indicate that the 25-ft column is in the intermediate range (94.0 < 126.1); hence, the factor of safety is

$$FS = \frac{5}{3} + \frac{3}{8}\left(\frac{94.0}{126.1}\right) - \frac{1}{8}\left(\frac{94.0}{126.1}\right)^3 = 1.89$$

The allowable stress is

$$\sigma_{all} = \frac{\sigma_y}{FS}\left[1 - \frac{1}{2}\left(\frac{L/r}{C_c}\right)^2\right] = \frac{36(10^3)}{1.89}\left[1 - \frac{1}{2}\left(\frac{94.0}{126.1}\right)^2\right] = 13,755 \text{ psi}$$

Hence, the safe load for the 25-ft column is

$$P = \sigma_{all}(A) = 13,755(2)(7.35) = 202.2(10^3) \text{ lb} \cong 202 \text{ kip} \qquad \textbf{Ans.}$$

The 40-ft column has a slenderness ratio of 150.5; therefore, it is in the slender range (150.5 > 126.1). Hence,

$$\sigma_{all} = \frac{\pi^2 E}{1.92(L/r)^2} = \frac{\pi^2(29)(10^6)}{1.92(150.5)^2} = 6581 \text{ psi}$$

Hence, the safe load for the 40-ft column is

$$P = \sigma_{all}(A) = 6581(2)(7.35) = 96.74(10^3) \text{ lb} \cong 96.7 \text{ kip} \qquad\blacksquare \qquad \textbf{Ans.}$$

▌PROBLEMS

Refer to Appendix B for material properties and for cross-sectional properties of rolled structural shapes.

Standard Problems

11-25* Three structural steel bars with a 1×4-in. rectangular cross section will be used for a 10-ft. fixed-ended column. Determine the maximum compressive load permitted by Code 1 if
(a) The three bars act as independent axially loaded members.
(b) The three bars are welded together to form an H-column.

11-26* Three hollow circular structural steel tubes with inside diameters of 50 mm and outside diameters of 80 mm will be used for a 3.5-m pivot-ended column. Determine the maximum compressive load permitted by Code 1 if
(a) The three tubes act as independent axially loaded members.
(b) The three tubes are welded together as shown in Fig. P11-26.

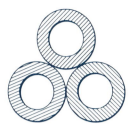

Figure P11-26

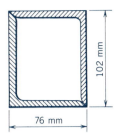

Figure P11-30

11-27 An L6 × 6 × 1-in. structural steel angle will be used as a pivot-ended column to support an axial compressive load P. Determine the maximum length permitted by Code 1
(a) If $P = 75$ kip.   (b) If $P = 120$ kip.

11-28 A WT178 × 36 structural steel T-section will be used as a pivot-ended column to support an axial compressive load P. Determine the maximum length permitted by Code 1
(a) If $P = 250$ kN.   (b) If $P = 500$ kN.

11-29* Two C10 × 15.3 structural steel channels 12 ft long are used as a fixed-ended, pivot-ended column. Determine the maximum load permitted by Code 1
(a) If the channels are not connected and each acts as an independent axially loaded member.
(b) If the channels are welded together to form a 10 × 5.2-in. box section as shown in Fig. P11-29.

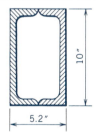

Figure P11-29

11-30* Two L102 × 76 × 9.5-mm structural steel angles 7 m long are used as a pivot-ended column. Determine the maximum load permitted by Code 1
(a) If the angles are not connected and each acts as an independent axially loaded member.
(b) If the angles are welded together to form a 102 × 76-mm box section as shown in Fig. P11-30.

11-31 Four L3 × 3 × $\frac{1}{2}$-in. structural steel angles 20 ft long are used as a fixed-ended column. Determine the maximum load permitted by Code 1
(a) If the angles are not connected and each acts as an independent axially loaded member.
(b) If the angles are fastened together as shown in Fig. P11-31.

Figure P11-31

11-32 Four L102 × 76 × 9.5-mm structural steel angles 4 m long are used as a pivot-ended column. Determine the maximum load permitted by Code 1
(a) If the angles are not connected and each acts as an independent axially loaded member.
(b) If the angles are welded together to form a 204 × 152-mm box section as shown in Fig. P11-32.

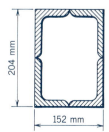

Figure P11-32

11-33* An L5 × 5 × $\frac{3}{4}$-in. aluminum alloy 2014-T6 angle will be used as a fixed-ended, pivot-ended column to support a load of 120 kip. The cross-sectional properties of steel and aluminum angles are the same (see Appendix B). Determine the maximum permissible length permitted by Code 2.

11-34* A column of aluminum alloy 2014-T6 is composed of two L127 × 127 × 19.1-mm angles riveted together as shown in Fig. P11-34. The length between end connections is 3 m, and the end connections are such that there is no restraint to bending about the y-axis; but restraint to bending about the x-axis reduces the effective length to 2.1 m. Determine the maximum axial compressive load permitted by Code 2. The cross-sectional properties of aluminum and steel angles are the same (see Appendix B).

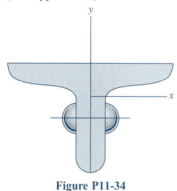

Figure P11-34

11-35 A sand bin is supported by four 8 × 10-in. fixed-ended rectangular columns 20 ft long. Assume that the load is equally divided among the columns and that the column load is axial. If the columns are made of air-dried Douglas fir ($E = 1900$ ksi and $F_c = 2500$ psi) determine the maximum load permitted by Code 4.

11-36 Two L152 × 89 × 12.7-mm angles of aluminum alloy 2014-T6 are welded together as shown in Fig. P11-36 to form a 4.75-m pivot-ended column. The pins provide no restraint to bending about the x-axis but reduce the effective length to 3.25 m for bending about the y-axis. The sectional properties of one angle are given on the figure where C is the centroid of one angle. Determine the maximum axial compressive load permitted by Code 2.

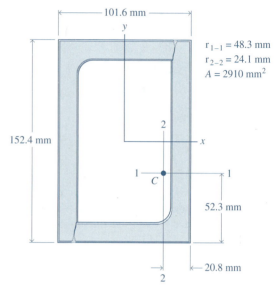

$r_{1-1} = 48.3$ mm
$r_{2-2} = 24.1$ mm
$A = 2910$ mm^2

Figure P11-36

11-37* A strut of aluminum alloy 6061-T6 having the cross section shown in Fig. P11-37 is to carry an axial compressive load of 175 kip. The strut is 4-ft long and is fixed at the bottom and pivoted at the top. Determine the dimension d of the cross section by using Code 3.

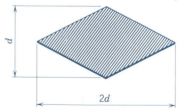

Figure P11-37

11-38 Douglas fir ($E = 13$ GPa and $F_c = 17.5$ MPa) timber columns with 200 × 300-mm rectangular cross sections will be used to support axial compressive loads. Determine the maximum loads permitted by Code 4 if
(a) The effective length of the column is 2 m.
(b) The effective length of the column is 4 m.
(c) The effective length of the column is 6 m.

Challenging Problems

11-39* A 25-ft plate and angle column consists of four L5 × $3\frac{1}{2}$ × $\frac{1}{2}$-in. structural steel angles riveted to a 10 × $\frac{1}{2}$-in. structural steel plate as shown in Fig. P11-39. Determine the maximum safe load permitted by Code 1

(a) If the column is pivoted at both ends.

(b) If the column is fixed at the base and pivoted at the top.

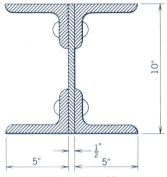

Figure P11-39

11-40* Four C178 × 22 structural steel channels 12 m long are used to fabricate a column with the cross section shown in Fig. P11-40. The column is fixed at the base and pivoted at the top. The pin at the top offers no restraint to bending about the y-axis; but for bending about the x-axis, the pin provides restraint sufficient to reduce the effective length to 7.5 m. Determine the maximum axial compressive load permitted by Code 1.

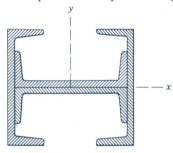

Figure P11-40

11-41 The machine part shown in Fig. P11-41 is made of SAE 4340 heat-treated steel and carries an axial compressive load. The pins at the ends offer no restraint to bending about the axis of the pin, but restraint about the perpendicular axis reduces the effective length to 36 in. Determine the maximum load permitted by Code 1.

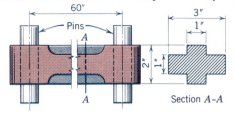

Figure P11-41

11-42 A connecting rod made of SAE 4340 heat-treated steel has the cross section shown in Fig. P11-42. The pins at the ends of the rod are parallel to the x-axis and are 1250 mm apart. Assume that the pins offer no restraint to bending about the x-axis but provide essentially complete fixity for bending about the y-axis. Determine the maximum axial compressive load permitted by Code 1.

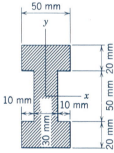

Figure P11-42

11-5 DESIGN PROBLEMS

In previous chapters design usually involved strength as a controlling parameter. Since buckling is an elastic phenomenon, the modulus of elasticity (stiffness) is a more significant parameter than is strength (such as yield strength if failure is by slip), if the column length is in the slender range. If the column is in the intermediate range, both yield strength and stiffness may be important parameters. When designing columns using the representative codes listed in Table 11-1, a designer must be aware of several factors. The codes are for specific materials, that is, materials with a specific value of yield strength and modulus of elasticity. In addition, some codes include a factor of safety, while others require that a factor of safety be introduced. Each of the codes has a range of applicability for the slenderness ratio. Finally, all codes in Table 11-1 are for axially loaded members.

For example, Code 2 is limited to a specific material, 2014-T6 aluminum alloy. The factor of safety (FS) is included in the code. If the slenderness ratio lies between 12 and 55, the column is in the intermediate range and the empirical column formula is valid. If the slenderness ratio exceeds 55, a form of the Euler formula is used with a factor of safety of approximately 1.94 included. For a slenderness ratio less than 12, the axially loaded member is in the compression-block range where buckling does not occur.

The following examples illustrate the use of the codes in Table 11-1.

Example Problem 11-4

Select the lightest structural steel wide-flange section listed in Appendix B to support an axial compressive load of 150 kip as a 15-ft column. Use Code 1.

SOLUTION

When a rolled section is to be selected to support a specified load, it is usually necessary to make several trial solutions since there is no direct relationship between areas and radii of gyration for different structural shapes. The best section is usually the section with the least area (smallest mass) that will support the load. A minimum area can be obtained by assuming $L/r = 0$. The load-carrying capacity of various sections with areas larger than this minimum can then be calculated, using the proper column formula, to determine the lightest one that will carry the specified load.

If L/r is small:

$$FS \cong \frac{5}{3} = 1.667$$

$$\sigma_{all} = \frac{\sigma_y}{FS} = \frac{36,000}{1.667} = 21,596 \text{ psi}$$

$$A_{min} = \frac{P}{\sigma_{all}} = \frac{150,000}{21,596} = 6.946 \text{ in.}^2$$

A column should be selected from Appendix B with an area greater than 6.95 in.^{2} for the first trial. In this case, try a W8 $\times$ 24 section for which A is 7.08 in.^{2} and r_{min} is 1.61 in. The value of L/r for this column is

$$\frac{L}{r} = \frac{15(12)}{1.61} = 111.8$$

To determine which of the equations for the allowable stress is applicable, first determine the value of C_c:

$$C_c = \sqrt{\frac{2\pi^2 E}{\sigma_y}} = \sqrt{\frac{2\pi^2 (29)(10^6)}{36,000}} = 126.1$$

Since $L/r = 111.8 < 126.1$, the column is in the intermediate range where the factor of safety is

$$FS = \frac{5}{3} + \frac{3}{8}\left(\frac{L/r}{C_c}\right) - \frac{1}{8}\left(\frac{L/r}{C_c}\right)^3 = \frac{5}{3} + \frac{3}{8}\left(\frac{111.8}{126.1}\right) - \frac{1}{8}\left(\frac{111.8}{126.1}\right)^3 = 1.912$$

and the allowable stress is

$$\sigma_{all} = \frac{\sigma_y}{FS}\left[1 - \frac{1}{2}\left(\frac{L/r}{C_c}\right)^2\right] = \frac{36,000}{1.912}\left[1 - \left(\frac{111.8}{126.1}\right)^2\right] = 11,428 \text{ psi}$$

The allowable load is

$$P_{all} = \sigma_{all}(A) = 11,428(7.08) = 80,910 \text{ lb} \cong 80.9 \text{ kip}$$

This load is less than the design load; therefore, a column with either a larger area, a larger radius of gyration, or both must be investigated. As a second trial value, use a W12 $\times$ 30 section for which A is 8.79 in.^{2} and r is 1.52 in. For this section L/r is 118.4 (intermediate range), the factor of safety is 1.92, the allowable stress is 10,485 psi, and the load this column can support is

$$P_{all} = \sigma_{all}(A) = 10,485(8.79) = 92,200 \text{ lb} = 92.2 \text{ kip}$$

This load is also less than the design load of 150 kip. For the third trial, use a W8 $\times$ 40 section for which A is 11.7 in.^{2} and r is 2.04 in. For this section L/r is 88.2 (intermediate range), the factor of safety is 1.89, the allowable stress is 14,542 psi, and the load this column can support is

$$P_{all} = \sigma_{all}(A) = 14,542(11.7) = 170,140 \text{ lb} \cong 170.1 \text{ kip}$$

Since the 40 lb/ft column (W8 $\times$ 40) is stronger than necessary and the 30 lb/ft column (W12 $\times$ 30) is not strong enough, any other section investigated should weigh between 30 and 40 lb/ft. The only other wide-flange section in Appendix B that might satisfy the requirements is a W8 $\times$ 31 section for which A is 9.13 in.^{2} and r is 2.02 in. For this section L/r is 89.1 (intermediate range), the factor of safety is 1.89, the allowable stress is 14,293 psi, and the load this column can support is 130,500 lb, which is less than the design load of 150 kip.

Thus, a W8 $\times$ 40 section should be used.   **Ans.**

The above is not necessarily the best procedure. Different designers have different approaches to the trial-and-error procedure, and for certain problems one approach may be better than another. The important point to make here is that the problem of design involving rolled shapes (other than simple geometric shapes) is, in general, solved by trial and error. ■

Example Problem 11-5

Determine the dimensions necessary for a 500-mm rectangular strut to carry an axial load of 6.75 kN. The material is aluminum alloy 2014-T6, and the width of the strut is to be twice the thickness. Use Code 2.

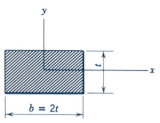

Figure 11-8

SOLUTION

The code is represented by three different equations that depend on the value of L/r, which in turn will depend on the equation used. Thus, it will be necessary to assume that one of the equations applies and use it to obtain the dimension of the column, after which the value of L/r must be calculated and used to check the validity of the equation used. Assume L/r is less than 55 but greater than 12, in which case the straight-line equation is valid. The cross section is shown in Fig. 11-8, and the least moment of inertia I_x is equal to $bt^3/12$, the area A is equal to bt, and the least radius of gyration is

$$r = \sqrt{\frac{bt^3/12}{bt}} = 0.2887t$$

The slenderness ratio is

$$\frac{L}{r} = \frac{0.500}{0.2887t} = \frac{1.7319}{t}$$

and, when this value and the expression for the area are substituted in the straight-line formula of Code 2, it becomes

$$\sigma_{all} = \frac{P_{all}}{A} = [212 - 1.585(L/r)](10^6)$$

$$\frac{6.75(10^3)}{2t^2} = \left[212 - 1.585\left(\frac{1.7319}{t}\right)\right](10^6)$$

from which

$$t = 14.08(10^{-3})\,\text{m} = 14.08\,\text{mm}$$

The value of L/r for this thickness is

$$\frac{L}{r} = \frac{1.7319}{0.01408} = 123.0$$

which is greater than 55 and indicates that the straight-line formula is not valid. The problem must be solved again using the Euler equation. Thus,

$$\frac{P}{A} = \frac{6.75(10^3)}{2t^2} = \frac{372(10^9)}{(1.7319/t)^2}$$

from which

$$t^4 = 27.22(10^{-9})\,\text{m}^4 \qquad \text{and} \qquad t = 12.84(10^{-3})\,\text{m} = 12.84\,\text{mm}$$

The value of L/r is 134.9 for this thickness, which confirms the use of the Euler formula. The dimensions of the cross section are

$$t = 12.84\,\text{mm} \qquad \text{and} \qquad b = 25.7\,\text{mm} \qquad \blacksquare \qquad \textbf{Ans.}$$

PROBLEMS

Standard Problems

11-43* A column 12 ft long must support an axial compressive load of 70,000 lb. Select the lightest standard-weight structural steel pipe that can be used. Use Code 1.

11-44* Select the lightest standard-weight structural steel pipe that can be used to support an axial compressive load of 200 kN as a 4-m column. Use Code 1.

11-45 A square 2014-T6 aluminum alloy member must support a 20,000-lb axial compressive load as a 12-ft column. Use Code 2 to determine the minimum cross-sectional area required.

11-46 A 2014-T6 aluminum alloy strut with a length of 4 m will be used to support an axial compressive load of 15 kN. Determine the minimum dimensions required if the width of the strut is to be twice the thickness. Use Code 2.

11-47* Select the lightest structural steel wide-flange section that can be used to support an axial compressive load of 200 kip as a 16-ft column. Use Code 1.

11-48* A 7 m–long structural steel column will be used to support an axial compressive load of 400 kN. Select the lightest wide-flange section that can be used. Use Code 1.

11-49 A Douglas fir ($E = 1800$ ksi and $F_c = 1350$ psi) timber column 16 ft long will be used to support an axial compressive load of 60 kip. Use Code 4 to determine the lightest structural timber that can be used.

11-50 A Douglas fir ($E = 12$ GPa and $F_c = 9.3$ MPa) timber column 4 m long will be used to support an axial compressive load of 100 kN. Use Code 4 to determine the lightest structural timber that can be used.

Challenging Problems

11-51* The structure shown in Fig. P11-51 consists of a solid steel tie rod BC and a standard-weight structural steel pipe AB. The tie rod has been adequately designed. Using Code 1, determine the lightest pipe that can be used to support the load. The effective length of pipe AB is 9 ft. Neglect the weight of the structure.

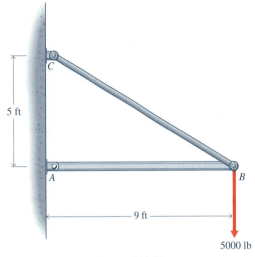

Figure P11-51

11-52* A structural steel standard-weight pipe is used for a spreader bar, as shown in Fig. P11-52. If the structure is to support a load $P = 45$ kN, and the cables have been adequately designed, determine the minimum size pipe needed to support the load using Code 1. The effective length of the spreader bar is 1.5 m. Neglect the weight of the structure.

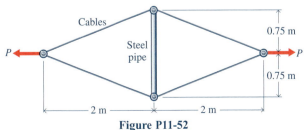

Figure P11-52

11-53 Select the lightest structural steel wide-flange section that can be used for the compression members of the truss shown in Fig. P11-53. Assume that buckling is limited to the plane of the structure and that the tension members have been adequately designed. Use Code 1.

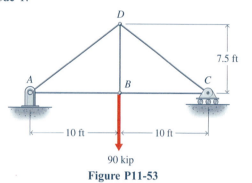

Figure P11-53

11-54 Member *ABC* of the structure shown in Fig. P11-54 supports a uniformly distributed load of 30 kN/m. Select the lightest standard-weight structural steel pipe that can be used for member *BD*. Use Code 1 and consider *BD* to be a pivot-ended member. Neglect the weight of the structure.

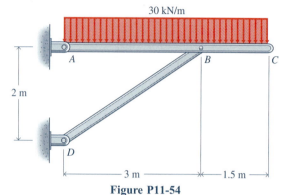

Figure P11-54

11-6 SUMMARY

Columns are long, straight, prismatic bars subjected to compressive, axial loads. As long as a column remains straight, it can be analyzed as an axially loaded member; however, if a column begins to deform laterally, the deflection may become large and lead to catastrophic failure called buckling. Buckling of a column is caused by deterioration of what was a stable state of equilibrium to an unstable one, not by failure of the material of which the column is composed. For long, slender columns the maximum load for which the column is in stable equilibrium (the critical buckling load) occurs at stress levels much less than the proportional limit for the material.

For a straight, slender, pivot-ended column that is centrically loaded by axial compressive forces P at the ends and that has experienced a small lateral deflection δ at the midpoint of the span the differential equation for the elastic curve is given by Eq. (9-3) as

$$EI \frac{d^2y}{dx^2} = M_r = P(\delta - y)$$

which has the solution

$$y = \delta \left(1 - \cos \sqrt{\frac{P}{EI}} x \right) \tag{11-1}$$

The minimum value of load P for a nontrivial solution of Eq. (11-1) is

$$P_{cr} = \frac{\pi^2 EI}{L^2} \tag{11-2}$$

Equation (11-2) is used to calculate the critical buckling load or the Euler load. The second moment of the cross-sectional area I in Eq. (11-2) refers to the axis about which bending occurs. When I is replaced by Ar^2, where r is the radius of gyration about the axis of bending, Eq. (11-2) becomes

$$\frac{P_{cr}}{A} = \frac{\pi^2 E}{(L/r)^2} \qquad (11\text{-}3)$$

The quantity L/r is called the slenderness ratio and is determined for the axis about which bending tends to occur. For a pivot-ended, centrically loaded column, bending occurs about the axis of minimum second moment of area (minimum radius of gyration).

Equation (11-2) agrees well with experiment if the slenderness ratio is large. Short compression members can be treated as compression blocks, in which yielding occurs before buckling. Columns that lie between these extremes are analyzed by using empirical formulas (column design codes).

REVIEW PROBLEMS

11-55* A 25-ft timber column has the cross section shown in Fig. P11-55. The timbers are nailed together so that they act as a unit. Determine
(a) The slenderness ratio.
(b) The Euler buckling load. Use $E = 1200$ ksi for the timber.
(c) The axial stress in the column when the Euler load is applied.

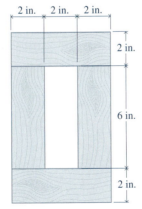

Figure P11-55

11-56* A 3-m column with the cross section shown in Fig. P11-56 is fabricated from three pieces of timber. The timbers are nailed together so that they act as a unit. Determine
(a) The slenderness ratio.
(b) The Euler buckling load. Use $E = 13$ GPa for the timber.
(c) The axial stress in the column when the Euler load is applied.

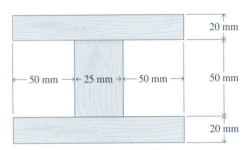

Figure P11-56

11-57 Determine the maximum compressive load that a W36 × 160 structural steel column (see Appendix B for cross-sectional properties) can support if it is 36 ft long and a factor of safety of 2.24 is specified.

11-58 Determine the maximum compressive load that a WT178 × 51 structural steel column (see Appendix B for cross-sectional properties) can support if it is 8 m long and a factor of safety of 1.92 is specified.

11-59* Determine the maximum allowable compressive load for a 12-ft aluminum alloy ($E = 10,600$ ksi) column having the cross section shown in Fig. P11-59 if a factor of safety of 2.25 is specified.

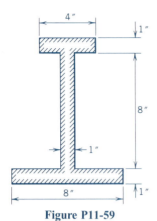

Figure P11-59

11-60* A 25 mm–diameter tie rod *AB* and a pipe strut *AC* with an inside diameter of 100 mm and a wall thickness of 25 mm are used to support a 100-kN load as shown in Fig. P11-60. Both the tie rod and the pipe strut are made of structural steel with a modulus of elasticity of 200 GPa and a yield strength of 250 MPa. Determine

(a) The factor of safety with respect to failure by slip.
(b) The factor of safety with respect to failure by buckling.

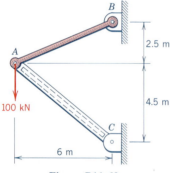

Figure P11-60

11-61 A WT7 × 24 structural steel section (see Appendix B for cross-sectional properties) is used for a column with an actual length of 45 ft. If the modulus of elasticity for the steel is 29,000 ksi and a factor of safety of 2 with respect to failure by buckling is specified, determine the maximum safe load for the column under the following support conditions.

(a) Pivot-pivot.
(b) Fixed-free.
(c) Pivot-fixed.
(d) Fixed-fixed.

11-62* A structural steel W356 × 122 wide-flange section will be used for a 9-m pivot-ended column. Determine the maximum axial compressive load permitted by Code 1 if

(a) The column is unbraced throughout its total length.
(b) The column is braced such that the effective length for bending about the *y*-axis is reduced to 6 m.

11-63 Three S10 × 25.4 structural steel sections 40 ft long are used to fabricate a column with the cross section shown in Fig. P11-63. The column is fixed at the base and pivoted at the top. The pin at the top offers no restraint to bending about the *y*-axis; but for bending about the *x*-axis, the pin provides restraint sufficient to reduce the effective length to 24 ft. Determine the maximum axial compressive load permitted by Code 1.

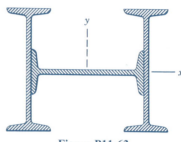

Figure P11-63

11-64 The compression member *AB* of the truss shown in Fig. P11-64 is a structural steel W254 × 67 wide-flange section with the *x-x* axis lying in the plane of the truss. The member is continuous from *A* to *B*. Consider all connections to be the equivalent of pivot ends. If Code 1 applies, determine

(a) The maximum safe load for member *AB*.
(b) The maximum safe load for member *AB* if the bracing member *CD* is removed.

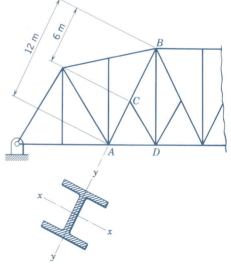

Figure P11-64

VECTOR OPERATIONS

APPENDIX

A-1 INTRODUCTION

Most of the physical quantities of interest in mechanics such as mass, force, time, and distance can be represented either as scalars or vectors. The mathematical operations required for their use in mechanics are presented in this Appendix.

A-1-1 Scalar Quantities

Scalar quantities can be completely described by their magnitudes. Mass, density, length, area, volume, speed, time, energy, and temperature are examples of scalar quantities in mechanics. Scalars follow the rules of elementary algebra in mathematical operations. Symbols representing scalar quantities are printed in lightface italic form (A) in this book.

A-1-2 Vector Quantities

A vector quantity has both a magnitude and a direction (line of action and sense) and obeys the parallelogram law of addition that will be described in Section A-2 of this Appendix. Displacement, velocity, acceleration, and force are examples of vector quantities in mechanics. A vector quantity can be represented graphically by using a directed line segment (arrow), as shown in Fig. A-1. The magnitude of the quantity is represented by the length of the arrow. The direction is specified by using the angle θ between the arrow and some known reference direction. The sense is indicated by the arrowhead at the tip of the line segment. Symbols representing vector quantities are printed in boldface type (**A**) to distinguish them from scalar quantities. Symbols representing magnitudes of vector quantities $|\mathbf{A}|$ will be printed in lightface italic type (A). In all handwritten work, it is important to distinguish between scalar and vector quantities since, in mathematical operations, vector quantities do not follow the rules of elementary algebra. A small arrow over the symbol for a vector quantity ($\vec{A}$) is often used in handwritten work to take the place of boldface type in printed material.

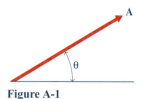

Figure A-1

Vectors can be classified into three types—free, sliding, or fixed.

1. A free vector has a specific magnitude, slope, and sense but its line of action does not pass through a unique point in space.
2. A sliding vector has a specific magnitude, slope, and sense and its line of action passes through a unique point in space. The point of application of a sliding vector can be anywhere along its line of action.
3. A fixed vector has a specific magnitude, slope, and sense and its line of action passes through a unique point in space. The point of application of a fixed vector is confined to a fixed point on its line of action.

A-2 ELEMENTARY VECTOR OPERATIONS

A-2-1 Addition of Vectors

Two free vectors **A** and **B,** can be translated so that their tails meet at a common point, as shown in Fig. A-2. The parallelogram law of vector addition states that the two vectors **A** and **B** are equivalent to a vector **R** that is the diagonal of a parallelogram constructed by using vectors **A** and **B** as the two adjacent sides (see Fig. A-2). The vector **R** is the diagonal that passes through the point of intersection of vectors **A** and **B**. Vector **R** is known as the resultant of the two vectors **A** and **B**. Vector quantities must obey this parallelogram law of addition. The resultant **R** can be represented mathematically by the vector equation

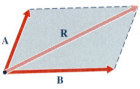

Figure A-2

$$\mathbf{R} = \mathbf{A} + \mathbf{B} \qquad\qquad \text{(A-1)}$$

The plus sign used in conjunction with vector quantities **A** and **B** (boldface type) indicates vector (parallelogram law) addition, not scalar (algebraic) addition.

It is obvious from Fig. A-2 that the two vectors **A** and **B** can be added in a head-to-tail fashion, as shown in the top part of Fig. A-3, to obtain the vector sum **R**. Similarly, the two vectors can be added in the manner shown in the bottom part of Fig. A-3 to obtain **R** since the two triangles shown in Fig. A-3 are the two different halves of the parallelogram shown in Fig. A-2. The procedure of adding vectors in a head-to-tail fashion is known as the *triangle law.* Note from Fig. A-3 that the order of addition of the vectors does not affect their sum; therefore,

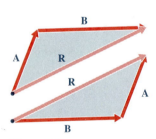

Figure A-3

$$\mathbf{A} + \mathbf{B} = \mathbf{B} + \mathbf{A} \qquad\qquad \text{(A-2)}$$

Equation (A-2) establishes the fact that vector addition is commutative.

The sum of three or more vectors **A** + **B** + **C** + . . . is obtained by first adding the vectors **A** and **B**, and then adding the vector **C** to the vector sum (**A** + **B**). The procedure, which is illustrated in Fig. A-4, can be expressed mathematically as

$$\mathbf{A} + \mathbf{B} + \mathbf{C} = (\mathbf{A} + \mathbf{B}) + \mathbf{C} \qquad\qquad \text{(A-3)}$$

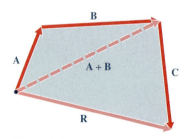

Figure A-4

Similarly, the sum of four vectors is obtained by adding the fourth vector to the sum of the first three. Thus, the sum of any number of vectors can be obtained by applying the parallelogram law, as expressed mathematically by Eq. (A-3), to successive pairs of vectors until all the given vectors are replaced by a single resultant vector **R**.

The sum of three or more vectors **A** + **B** + **C** + . . . can also be obtained by repeated application of the triangle law. In Fig. A-5, the sum of three vectors is obtained directly by connecting the three vectors, in sequence, in a tip-to-tail fashion. The resultant **R** is the vector that connects the tail of the first vector with the tip of the last one. The procedure, illustrated in Fig. A-5, can be applied to any number of vectors and is known as the *polygon rule for vector addition.* The sketch of the vectors connected in a tip-to-tail fashion is known as a *vector polygon.* As illustrated in Fig. A-6, the resultant **R** is independent of the order in which the vectors **A**, **B**, and **C** are introduced into the vector polygon. This fact, which can be expressed mathematically as

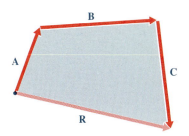

Figure A-5

$$\mathbf{A} + \mathbf{B} + \mathbf{C} = (\mathbf{A} + \mathbf{B}) + \mathbf{C} = \mathbf{A} + (\mathbf{B} + \mathbf{C}) \qquad \text{(A-4)}$$

shows that vector addition is associative.

The resultant **R** of two or more vectors can be determined graphically by drawing either the parallelogram or the vector polygon to scale. In practice, the magnitude R of the resultant is determined algebraically by applying the cosine law for a general triangle, as illustrated in Fig. A-7. Thus,

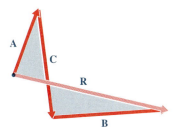

Figure A-6

$$R = \sqrt{A^2 + B^2 - 2AB \cos \theta} \qquad \text{(A-5)}$$

The angles between resultant **R** and vectors **A** and **B** can be determined by using the law of sines for a general triangle. Thus, for the triangle shown in Fig. A-7,

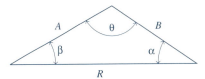

Figure A-7

$$\frac{\sin \alpha}{A} = \frac{\sin \beta}{B} = \frac{\sin \theta}{R} \qquad \text{(A-6)}$$

A-2-2 Subtraction of Vectors

The difference between vectors **A** and **B** is defined by the relation

$$\mathbf{A} - \mathbf{B} = \mathbf{A} + (-\mathbf{B}) \qquad \text{(A-7)}$$

The vector $-\mathbf{B}$ has the same magnitude and direction angle θ as vector **B** but is of opposite sense. Two vectors **A** and **B** are shown in Fig. A-8a. Their sum and difference are illustrated in Figs. A-8b and c, respectively. A zero or null vector is obtained when a vector is subtracted from itself; that is,

$$\mathbf{A} - \mathbf{A} = \mathbf{0} \qquad \text{(A-8)}$$

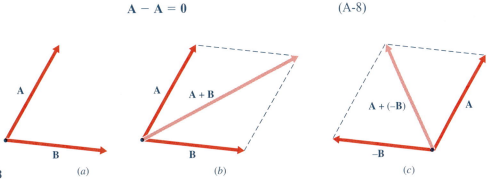

Figure A-8   (a)   (b)   (c)

It is important to realize that the zero that occurs on the right-hand side of this vector equation must be a vector.

The minus sign used in conjunction with vectors **A** and **B** (boldface type) indicates vector (parallelogram law) subtraction and not scalar subtraction. Since vector subtraction is defined as a special case of vector addition, all of the rules for vector addition discussed in Section A-2-1 apply to vector subtraction.

A-2-3 Multiplication of Vectors by Scalars

The product of a vector **A** and a scalar m is a vector m**A**, whose magnitude is mA (a positive number) and whose direction has the same sense as **A** if m is positive and a sense opposite to **A** if m is negative. Operations involving the products of scalars m and n and vectors **A** and **B** include the following:

$$(m + n)\mathbf{A} = m\mathbf{A} + n\mathbf{A}$$

$$m(\mathbf{A} + \mathbf{B}) = m\mathbf{A} + m\mathbf{B} \qquad \text{(A-9)}$$

$$m(n\mathbf{A}) = (mn)\mathbf{A} = n(m\mathbf{A})$$

A-3 CARTESIAN VECTORS

A-3-1 Unit Vectors

A vector of unit magnitude is called a *unit vector*. Unit vectors directed along the x-, y-, and z-axes of a Cartesian coordinate system are normally given in the symbols **i**, **j**, and **k**, respectively. The sense of these unit vectors is indicated analytically by using a plus sign if the unit vector points in a positive x-, y-, or z-direction and a minus sign if the unit vector points in a negative x-, y-, or z-direction. The unit vectors shown in Fig. A-9 are positive. In this book, the symbol **e** with a subscript is used to denote a unit vector in a direction other than a coordinate direction.

Any vector quantity can be written as the product of its magnitude (a positive number) and a unit vector in the direction of the given vector. Thus, for a vector **A** in the positive n-direction

$$\mathbf{A} = |\mathbf{A}|\mathbf{e}_n = A\mathbf{e}_n \qquad \text{(A-10)}$$

For a vector **B** in the negative n-direction

$$\mathbf{B} = |\mathbf{B}|(-\mathbf{e}_n) = -B\mathbf{e}_n \qquad \text{(A-11)}$$

It is obvious from Eq. (A-10) that the unit vector $\mathbf{e}_n$ can be expressed as

$$\mathbf{e}_n = \frac{\mathbf{A}}{|\mathbf{A}|} = \frac{\mathbf{A}}{A} \qquad \text{(A-12)}$$

The Cartesian coordinate axes shown in Fig. A-9 are arranged as a right-hand system. In a right-hand system, if the fingers of the right hand are curled about the z-axis in a direction from the positive x-axis toward the positive y-axis, then the thumb points in the positive z-direction, as shown in Fig. A-10. All of the vector relationships developed in the remainder of this appendix will utilize a right-hand coordinate system.

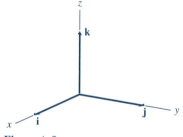

Figure A-9

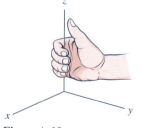

Figure A-10

A-3-2 Cartesian Components of a Vector

The process of adding two or more vectors to obtain a resultant vector **R** was discussed in Section A-2-1. The reverse process of resolving a given vector **A** into two or more components will now be discussed. The vector **A** can be resolved into any number of components provided the components sum to the given vector **A** by the parallelogram law. Mutually perpendicular components, called *rectangular components,* along the x-, y-, and z-coordinate axes are most useful. Any vector **A** can be resolved into rectangular components $\mathbf{A}_x$, $\mathbf{A}_y$, and $\mathbf{A}_z$ by constructing a rectangular parallelepiped (see Fig. A-11) with vector **A** as the diagonal and rectangular components $\mathbf{A}_x$, $\mathbf{A}_y$, and $\mathbf{A}_z$ along the x-, y-, and z-axes, respectively, as the edges. The three components $\mathbf{A}_x$, $\mathbf{A}_y$, and $\mathbf{A}_z$ can be written in Cartesian vector form, as illustrated in Fig. A-12, by using Eq. (A-10). Thus

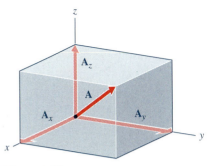

Figure A-11

$$\mathbf{A}_x = A_x\mathbf{i} \qquad \mathbf{A}_y = A_y\mathbf{j} \qquad \mathbf{A}_z = A_z\mathbf{k} \qquad \text{(A-13)}$$

Furthermore, from Eqs. (A-1), (A-10), and (A-13),

$$\mathbf{A} = \mathbf{A}_x + \mathbf{A}_y + \mathbf{A}_z = A_x\mathbf{i} + A_y\mathbf{j} + A_z\mathbf{k} = A\mathbf{e}_n \qquad \text{(A-14)}$$

The magnitudes of the rectangular components are related to the magnitude of vector **A** by the expressions

$$A_x = A \cos \theta_x \qquad A_y = A \cos \theta_y \qquad A_z = A \cos \theta_z \qquad \text{(A-15)}$$

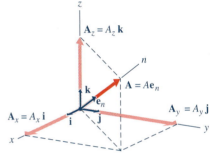

Figure A-12

The terms $\cos \theta_x$, $\cos \theta_y$, and $\cos \theta_z$ are called the *direction cosines* of the vector **A**. These cosines are also the direction cosines of the unit vector $\mathbf{e}_n$ since its line of action, as illustrated in Fig. A-13, coincides with the line of action of vector **A**.

The magnitude of vector **A** can be expressed in terms of the magnitudes of its rectangular components by the Pythagorean theorem as

$$A = \sqrt{A_x^2 + A_y^2 + A_z^2} \qquad \text{(A-16)}$$

Also, from Eq. (A-15), the direction cosines of a vector **A** can be expressed in terms of the magnitude of vector **A** and the magnitudes of its rectangular components as

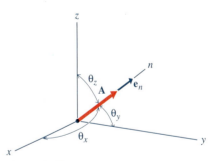

Figure A-13

$$\cos \theta_x = \frac{A_x}{A} \qquad \cos \theta_y = \frac{A_y}{A} \qquad \cos \theta_z = \frac{A_z}{A} \qquad \text{(A-17)}$$

Finally, the unit vector $\mathbf{e}_n$ can be written in terms of vector **A** and its rectangular components by using Eqs. (A-14) and (A-15). Thus,

$$\mathbf{e}_n = \frac{\mathbf{A}}{A} = \frac{A_x}{A}\mathbf{i} + \frac{A_y}{A}\mathbf{j} + \frac{A_z}{A}\mathbf{k}$$

$$= \cos \theta_x\mathbf{i} + \cos \theta_y\mathbf{j} + \cos \theta_z\mathbf{k} \qquad \text{(A-18)}$$

Since the magnitude of $\mathbf{e}_n = 1$,

$$\cos^2 \theta_x + \cos^2 \theta_y + \cos^2 \theta_z = 1 \qquad \text{(A-19)}$$

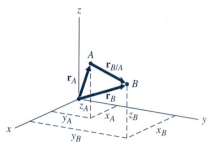

Figure A-14

A-3-3 Position Vectors

A directed line segment $\mathbf{r}$, known as a *position vector,* can be used to locate a point in space relative to the origin of a coordinate system or relative to another point in space. For example, consider the two points A and B shown in Fig. A-14. The location of point A can be specified by using the position vector $\mathbf{r}_A$ drawn from the origin of coordinates to point A. Position vector $\mathbf{r}_A$ can be written in Cartesian vector form as

$$\mathbf{r}_A = x_A\mathbf{i} + y_A\mathbf{j} + z_A\mathbf{k}$$

Similarly, the location of point B with respect to the origin of coordinates can be specified by using the position vector $\mathbf{r}_B$, which can be written in Cartesian vector form as

$$\mathbf{r}_B = x_B\mathbf{i} + y_B\mathbf{j} + z_B\mathbf{k}$$

Finally, the position of point B with respect to point A can be specified by using the position vector $\mathbf{r}_{B/A}$ where the subscript B/A indicates B with respect to A. Observe in Fig. A-14 that

$$\mathbf{r}_B = \mathbf{r}_A + \mathbf{r}_{B/A}$$

Therefore,

$$
\begin{aligned}
\mathbf{r}_{B/A} &= \mathbf{r}_B - \mathbf{r}_A \\
&= (x_B\mathbf{i} + y_B\mathbf{j} + z_B\mathbf{k}) - (x_A\mathbf{i} + y_A\mathbf{j} + z_A\mathbf{k}) \\
&= (x_B - x_A)\mathbf{i} + (y_B - y_A)\mathbf{j} + (z_B - z_A)\mathbf{k}
\end{aligned}
\tag{A-20}
$$

Example Problem A-1

Determine the magnitude and direction of the position vector from point A to B if the coordinates of points A and B are (3, 4, 5) and (6, −3, −2), respectively.

SOLUTION

The position vector $\mathbf{r}_{B/A}$ is given by Eq. (A-20) as

$$
\begin{aligned}
\mathbf{r}_{B/A} &= (x_B - x_A)\mathbf{i} + (y_B - y_A)\mathbf{j} + (z_B - z_A)\mathbf{k} \\
&= (6 - 3)\mathbf{i} + (-3 - 4)\mathbf{j} + (-2 - 5)\mathbf{k} \\
&= 3\mathbf{i} - 7\mathbf{j} - 7\mathbf{k}
\end{aligned}
$$

The magnitude of $\mathbf{r}_{B/A}$ is determined by using Eq. (A-16). Thus,

$$
\begin{aligned}
r_{B/A} &= \sqrt{(x_B - x_A)^2 + (y_B - y_A)^2 + (z_B - z_A)^2} \\
&= \sqrt{(3)^2 + (-7)^2 + (-7)^2} = 10.34
\end{aligned}
$$
Ans.

The direction is determined by using Eqs. (A-17). Thus,

$$\theta_x = \cos^{-1}\frac{x_B - x_A}{r_{B/A}} = \cos^{-1}\frac{3}{10.34} = 73.1°$$
Ans.

$$\theta_y = \cos^{-1}\frac{y_B - y_A}{r_{B/A}} = \cos^{-1}\frac{-7}{10.34} = 132.6° \qquad \text{Ans.}$$

$$\theta_z = \cos^{-1}\frac{z_B - z_A}{r_{B/A}} = \cos^{-1}\frac{-7}{10.34} = 132.6° \ \blacksquare \qquad \text{Ans.}$$

A-4 ADDITION OF CARTESIAN VECTORS

The vector operations of addition and subtraction are greatly simplified when more than two vectors are involved if the vectors are expressed in Cartesian vector form. For example, consider the two vectors **A** and **B**, which can be written as

$$\mathbf{A} = A_x\mathbf{i} + A_y\mathbf{j} + A_z\mathbf{k}$$

$$\mathbf{B} = B_x\mathbf{i} + B_y\mathbf{j} + B_z\mathbf{k}$$

The sum of the two vectors is

$$\mathbf{R} = \mathbf{A} + \mathbf{B} = (A_x\mathbf{i} + A_y\mathbf{j} + A_z\mathbf{k}) + (B_x\mathbf{i} + B_y\mathbf{j} + B_z\mathbf{k})$$

$$= (A_x + B_x)\mathbf{i} + (A_y + B_y)\mathbf{j} + (A_z + B_z)\mathbf{k} \qquad \text{(A-21)}$$

Thus, the resultant vector **R** has components that represent the scalar sums of the rectangular components of **A** and **B**. The process represented by Eq. (A-21) can be extended to any number of vectors. Thus,

$$\mathbf{R}_x = (A_x + B_x + C_x + \dots)\mathbf{i} = R_x\mathbf{i}$$

$$\mathbf{R}_y = (A_y + B_y + C_y + \dots)\mathbf{j} = R_y\mathbf{j}$$

$$\mathbf{R}_z = (A_z + B_z + C_z + \dots)\mathbf{k} = R_z\mathbf{k}$$

$$\mathbf{R} = \mathbf{R}_x + \mathbf{R}_y + \mathbf{R}_z = R_x\mathbf{i} + R_y\mathbf{j} + R_z\mathbf{k}$$

The magnitude and direction of the resultant **R** are then obtained by using Eqs. (A-16) and (A-17). Thus,

$$R = \sqrt{(R_x)^2 + (R_y)^2 + (R_z)^2} \qquad \text{(A-22)}$$

$$\theta_x = \cos^{-1}\frac{R_x}{R} \qquad \theta_y = \cos^{-1}\frac{R_y}{R} \qquad \theta_z = \cos^{-1}\frac{R_z}{R} \qquad \text{(A-23)}$$

Example Problem A-2

Determine the magnitude and direction of the resultant of the following three vectors: $\mathbf{A} = 3\mathbf{i} + 7\mathbf{j} + 8\mathbf{k}$, $\mathbf{B} = 4\mathbf{i} - 5\mathbf{j} + 3\mathbf{k}$, and $\mathbf{C} = 2\mathbf{i} + 3\mathbf{j} - 4\mathbf{k}$. Express the resultant **R** and the unit vector $\mathbf{e}_R$ associated with the resultant in Cartesian vector form.

SOLUTION

The x-, y-, and z-components of the resultant $\mathbf{R}$ are

$$R_x = A_x + B_x + C_x = 3 + 4 + 2 = 9$$

$$R_y = A_y + B_y + C_y = 7 - 5 + 3 = 5$$

$$R_z = A_z + B_z + C_z = 8 + 3 - 4 = 7$$

The magnitude of the resultant is determined by using Eq. (A-22). Thus,

$$R = \sqrt{(R_x)^2 + (R_y)^2 + (R_z)^2} = \sqrt{(9)^2 + (5)^2 + (7)^2} = 12.45 \qquad \textbf{Ans.}$$

The direction is determined by using Eqs. (A-23). Thus,

$$\theta_x = \cos^{-1}\frac{R_x}{R} = \cos^{-1}\frac{9}{12.45} = 43.7° \qquad \textbf{Ans.}$$

$$\theta_y = \cos^{-1}\frac{R_y}{R} = \cos^{-1}\frac{5}{12.45} = 66.3° \qquad \textbf{Ans.}$$

$$\theta_z = \cos^{-1}\frac{R_z}{R} = \cos^{-1}\frac{7}{12.45} = 55.8° \qquad \textbf{Ans.}$$

The resultant can be expressed in Cartesian vector form as

$$\mathbf{R} = R_x\mathbf{i} + R_y\mathbf{j} + R_z\mathbf{k} = 9\mathbf{i} + 5\mathbf{j} + 7\mathbf{k} \qquad \textbf{Ans.}$$

The unit vector $\mathbf{e}_R$ associated with the resultant $\mathbf{R}$ is given by Eq. (A-18) as

$$\mathbf{e}_R = \frac{\mathbf{R}}{R} = \frac{9\mathbf{i} + 5\mathbf{j} + 7\mathbf{k}}{12.45} = 0.723\mathbf{i} + 0.402\mathbf{j} + 0.562\mathbf{k} \quad ■ \qquad \textbf{Ans.}$$

A-5 MULTIPLICATION OF CARTESIAN VECTORS

Two types of multiplication involving vector quantities are performed; namely, the scalar or dot product (written as $\mathbf{A} \cdot \mathbf{B}$) and the vector or cross product (written as $\mathbf{A} \times \mathbf{B}$). These two types of multiplication have entirely different properties and are used for different purposes.

A-5-1 Dot or Scalar Product

The dot or scalar product of two intersecting vectors is defined as the product of the magnitudes of the vectors and the cosine of the angle between them. Thus, by definition, for the vectors $\mathbf{A}$ and $\mathbf{B}$ shown in Fig. A-15,

$$\mathbf{A} \cdot \mathbf{B} = \mathbf{B} \cdot \mathbf{A} = AB \cos \theta \qquad (A\text{-}24)$$

where $0° \leq \theta \leq 180°$. This type of vector multiplication yields a scalar, not a vector. For $0° \leq \theta < 90°$, the scalar is positive. For $90° < \theta \leq 180°$, the scalar is negative. When $\theta = 90°$, the two vectors are perpendicular and the scalar is zero.

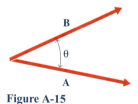

Figure A-15

The dot product can be used to obtain the rectangular scalar component of a vector in a specified direction. For example, the rectangular scalar component of vector $\mathbf{A}$ along the x-axis is

$$A_x = \mathbf{A} \cdot \mathbf{i} = A(1) \cos \theta_x = A \cos \theta_x$$

Similarly, the rectangular scalar component of vector $\mathbf{A}$ in a direction n (see Fig. A-16a) is

$$A_n = \mathbf{A} \cdot \mathbf{e}_n = A \cos \theta_n \qquad \text{(A-25)}$$

where $\mathbf{e}_n$ is the unit vector associated with the direction n. The vector component of $\mathbf{A}$ in the direction n (see Fig. A-16b) is then given by Eq. (A-10) as

$$\mathbf{A}_n = (\mathbf{A} \cdot \mathbf{e}_n)\mathbf{e}_n \qquad \text{(A-26)}$$

The component of vector $\mathbf{A}$ perpendicular to the direction n lies in the plane containing $\mathbf{A}$ and n, as shown in Fig. A-16b, and can be obtained from the expression

$$\mathbf{A}_t = \mathbf{A} - \mathbf{A}_n \qquad \text{(a)}$$

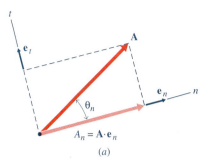

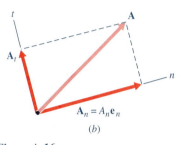

(a)

(b)

Figure A-16

Once an expression for $\mathbf{A}_t$ has been obtained by using Eq. (a), the magnitude and direction of $\mathbf{A}_t$ can be determined by using Eqs. (A-22) and (A-18).

If two vectors $\mathbf{A}$ and $\mathbf{B}$ are written in Cartesian vector form, the scalar product becomes

$$\begin{aligned}
\mathbf{A} \cdot \mathbf{B} &= (A_x\mathbf{i} + A_y\mathbf{j} + A_z\mathbf{k}) \cdot (B_x\mathbf{i} + B_y\mathbf{j} + B_z\mathbf{k}) \\
&= A_xB_x(\mathbf{i} \cdot \mathbf{i}) + A_xB_y(\mathbf{i} \cdot \mathbf{j}) + A_xB_z(\mathbf{i} \cdot \mathbf{k}) \\
&+ A_yB_x(\mathbf{j} \cdot \mathbf{i}) + A_yB_y(\mathbf{j} \cdot \mathbf{j}) + A_yB_z(\mathbf{j} \cdot \mathbf{k}) \\
&+ A_zB_x(\mathbf{k} \cdot \mathbf{i}) + A_zB_y(\mathbf{k} \cdot \mathbf{j}) + A_zB_z(\mathbf{k} \cdot \mathbf{k})
\end{aligned}$$

Since $\mathbf{i}$, $\mathbf{j}$, and $\mathbf{k}$ are orthogonal,

$$\mathbf{i} \cdot \mathbf{j} = \mathbf{j} \cdot \mathbf{k} = \mathbf{k} \cdot \mathbf{i} = (1)(1) \cos 90° = 0$$
$$\mathbf{i} \cdot \mathbf{i} = \mathbf{j} \cdot \mathbf{j} = \mathbf{k} \cdot \mathbf{k} = (1)(1) \cos 0° = 1$$

Therefore,

$$\mathbf{A} \cdot \mathbf{B} = A_xB_x + A_yB_y + A_zB_z \qquad \text{(A-27)}$$

Note also the special case

$$\mathbf{A} \cdot \mathbf{A} = A^2 \cos 0° = A^2 = A_x^2 + A_y^2 + A_z^2$$

which verifies Eq. (A-16) obtained by using the Pythagorean theorem.

By combining Eqs. (A-24) and (A-27), an expression for the angle between two vectors is obtained. Thus,

$$\mathbf{A} \cdot \mathbf{B} = A_xB_x + A_yB_y + A_zB_z = AB \cos \theta$$

or

$$\cos \theta = \frac{\mathbf{A} \cdot \mathbf{B}}{AB} = \frac{A_xB_x + A_yB_y + A_zB_z}{AB} \qquad \text{(A-28)}$$

In mechanics, the scalar product is used to determine the rectangular component of a vector (force, moment, velocity, acceleration, etc.) along a line and to find the angle between two vectors (two forces, a force and a line, a force and an acceleration, etc.).

Example Problem A-3

Determine the rectangular components of vector $\mathbf{A} = -312\mathbf{i} + 72\mathbf{j} - 228\mathbf{k}$ parallel and perpendicular to a line whose direction is given by the unit vector $\mathbf{e}_n = -0.60\mathbf{i} + 0.80\mathbf{j}$. Express the unit vector $\mathbf{e}_t$ associated with the perpendicular components of vector $\mathbf{A}$ in Cartesian vector form.

SOLUTION

The magnitude of the component of vector $\mathbf{A}$ parallel to the line is given by Eq. (A-27) as

$$A_n = \mathbf{A} \cdot \mathbf{e}_n = (-312\mathbf{i} + 72\mathbf{j} - 228\mathbf{k}) \cdot (-0.60\mathbf{i} + 0.80\mathbf{j})$$
$$= -312(-0.60) + 72(0.80) - 228(0) = 244.8$$

Thus, from Eq. (A-26),

$$\mathbf{A}_n = (\mathbf{A} \cdot \mathbf{e}_n)\,\mathbf{e}_n = 244.8(-0.60\mathbf{i} + 0.80\mathbf{j})$$
$$= -146.9\mathbf{i} + 195.8\mathbf{j} \qquad \textbf{Ans.}$$

From Eq. (A-21):

$$\mathbf{A} = \mathbf{A}_n + \mathbf{A}_t$$

Therefore,

$$\mathbf{A}_t = \mathbf{A} - \mathbf{A}_n = (-312\mathbf{i} + 72\mathbf{j} - 228\mathbf{k}) - (-146.9\mathbf{i} + 195.8\mathbf{j})$$
$$= -165.1\mathbf{i} - 123.8\mathbf{j} - 228\mathbf{k} \qquad \textbf{Ans.}$$

The magnitude of vector $\mathbf{A}_t$ is determined by using Eq. (A-22). Thus,

$$A_t = |\mathbf{A}_t| = \sqrt{(-165.1)^2 + (-123.8)^2 + (-228)^2} = 307.5$$

The unit vector $\mathbf{e}_t$ is given by Eq. (A-18) as

$$\mathbf{e}_t = \frac{\mathbf{A}_t}{A_t} = \frac{-165.1\mathbf{i} - 123.8\mathbf{j} - 228\mathbf{k}}{307.5}$$
$$= -0.537\mathbf{i} - 0.403\mathbf{j} - 0.741\mathbf{k} \quad \blacksquare \qquad \textbf{Ans.}$$

Example Problem A-4

Determine the angle θ between the vectors

$$\mathbf{A} = 8\mathbf{i} + 9\mathbf{j} + 7\mathbf{k} \qquad \text{and} \qquad \mathbf{B} = 6\mathbf{i} - 5\mathbf{j} + 3\mathbf{k}$$

SOLUTION

The magnitudes of vectors **A** and **B** are determined by using Eq. (A-22). Thus,

$$A = \sqrt{(A_x)^2 + (A_y)^2 + (A_z)^2} = \sqrt{(8)^2 + (9)^2 + (7)^2} = 13.93$$
$$B = \sqrt{(B_x)^2 + (B_y)^2 + (B_z)^2} = \sqrt{(6)^2 + (-5)^2 + (3)^2} = 8.37$$

The dot product of vectors **A** and **B** is given by Eq. (A-27) as

$$\mathbf{A} \cdot \mathbf{B} = A_x B_x + A_y B_y + A_z B_z = 8(6) + 9(-5) + 7(3) = 24$$

The angle θ between the two vectors is given by Eq. (A-28) as

$$\theta = \cos^{-1} \frac{\mathbf{A} \cdot \mathbf{B}}{AB} = \cos^{-1} \frac{24}{13.93(8.37)} = 78.1° \quad\blacksquare \qquad \text{Ans.}$$

A-5-2 Cross or Vector Product

The cross or vector product of two intersecting vectors **A** and **B**, by definition, yields a vector **C** that has a magnitude that is the product of the magnitudes of vectors **A** and **B** and the sine of the angle θ between them and a direction that is perpendicular to the plane containing the vectors **A** and **B**. Thus, by definition, for the vectors **A** and **B** shown in Fig. A-17,

$$\mathbf{C} = \mathbf{A} \times \mathbf{B} = (AB \sin \theta)\mathbf{e}_C \qquad (A\text{-}29)$$

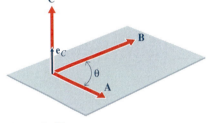

Figure A-17

where $0 \le \theta \le 180°$ and $\mathbf{e}_C$ is a unit vector in a direction perpendicular to the plane containing the vectors **A** and **B**. The sense of $\mathbf{e}_C$ is obtained by using the right-hand rule; in other words, if the fingers of the right hand are curled from **A** toward **B** about an axis perpendicular to the plane containing vectors **A** and **B** then the thumb points in the direction of vector **C**. Because of this definition of the unit vector $\mathbf{e}_C$, the cross product is not commutative. In fact,

$$\mathbf{A} \times \mathbf{B} = -\mathbf{B} \times \mathbf{A}$$

It is apparent from Eq. (A-29), however, that

$$\mathbf{A} \times s\mathbf{B} = s(\mathbf{A} \times \mathbf{B})$$

and by combining Eqs. (A-29) and the law of addition that

$$\mathbf{A} \times (\mathbf{B} + \mathbf{C}) = (\mathbf{A} \times \mathbf{B}) + (\mathbf{A} \times \mathbf{C})$$

If two vectors **A** and **B** are written in Cartesian vector form, the cross product becomes

$$
\begin{aligned}
\mathbf{A} \times \mathbf{B} &= (A_x\mathbf{i} + A_y\mathbf{j} + A_z\mathbf{k}) \times (B_x\mathbf{i} + B_y\mathbf{j} + B_z\mathbf{k}) \\
&= A_x B_x(\mathbf{i} \times \mathbf{i}) + A_x B_y(\mathbf{i} \times \mathbf{j}) + A_x B_z(\mathbf{i} \times \mathbf{k}) \\
&\quad + A_y B_x(\mathbf{j} \times \mathbf{i}) + A_y B_y(\mathbf{j} \times \mathbf{j}) + A_y B_z(\mathbf{j} \times \mathbf{k}) \\
&\quad + A_z B_x(\mathbf{k} \times \mathbf{i}) + A_z B_y(\mathbf{k} \times \mathbf{j}) + A_z B_z(\mathbf{k} \times \mathbf{k})
\end{aligned}
\qquad (a)
$$

Since **i**, **j**, and **k** are orthogonal,

$$\mathbf{i} \times \mathbf{i} = [(1)(1) \sin 0°]\mathbf{k} = \mathbf{0}$$

$$\mathbf{i} \times \mathbf{j} = [(1)(1) \sin 90°] \mathbf{k} = \mathbf{k}$$

Similarly,

$$
\begin{array}{lll}
\mathbf{i} \times \mathbf{i} = \mathbf{0} & \mathbf{j} \times \mathbf{i} = -\mathbf{k} & \mathbf{k} \times \mathbf{i} = \mathbf{j} \\
\mathbf{i} \times \mathbf{j} = \mathbf{k} & \mathbf{j} \times \mathbf{j} = \mathbf{0} & \mathbf{k} \times \mathbf{j} = -\mathbf{i} \qquad \text{(A-30)} \\
\mathbf{i} \times \mathbf{k} = -\mathbf{j} & \mathbf{j} \times \mathbf{k} = \mathbf{i} & \mathbf{k} \times \mathbf{k} = \mathbf{0}
\end{array}
$$

Equations (A-30) can be represented graphically by arranging the unit vectors **i**, **j**, and **k** in a circle in a counterclockwise order, as shown in Fig. A-18. The product of two unit vectors will be positive if they follow each other in counterclockwise order, and negative if they follow each other in clockwise order. Substituting Eqs. (A-30) in to Eq. (a) yields

$$
\begin{aligned}
\mathbf{A} \times \mathbf{B} &= A_x B_y \mathbf{k} + A_x B_z(-\mathbf{j}) + A_y B_x(-\mathbf{k}) + A_y B_z \mathbf{i} + A_z B_x \mathbf{j} + A_z B_y(-\mathbf{i}) \\
&= (A_y B_z - A_z B_y)\mathbf{i} + (A_z B_x - A_x B_z)\mathbf{j} + (A_x B_y - A_y B_x)\mathbf{k} \qquad \text{(A-31)}
\end{aligned}
$$

which is the expanded form of the determinant

$$
\mathbf{A} \times \mathbf{B} = \begin{vmatrix} \mathbf{i} & \mathbf{j} & \mathbf{k} \\ A_x & A_y & A_z \\ B_x & B_y & B_z \end{vmatrix} = \mathbf{C} \qquad \text{(A-32)}
$$

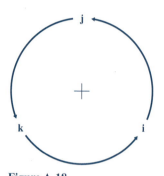

Figure A-18

Note carefully the arrangement of terms in the determinant, which places the unit vectors **i**, **j**, **k** in the first row, the components A_x, A_y, A_z of **A** in the second row, and the components B_x, B_y, B_z of **B** in the third row. If the two bottom rows of the determinant are interchanged, the sign of the determinant will change. Thus,

$$\mathbf{A} \times \mathbf{B} = -\mathbf{B} \times \mathbf{A}$$

In mechanics, the cross product is used to find the moment of a force about a point. The direction of the moment vector is the axis about which the force tends to rotate a body, and the magnitude of the moment is the strength of the tendency to rotate the body.

Example Problem A-5

If $\mathbf{A} = -3.75\mathbf{i} - 2.50\mathbf{j} + 1.50\mathbf{k}$ and $\mathbf{B} = 32\mathbf{i} + 44\mathbf{j} + 64\mathbf{k}$, determine the magnitude and direction of the vector $\mathbf{C} = \mathbf{A} \times \mathbf{B}$. Express the unit vector $\mathbf{e}_C$ associated with the vector **C** in Cartesian vector form.

SOLUTION

The vector cross product $\mathbf{A} \times \mathbf{B}$ is obtained by using Eq. (A-32). Thus,

$$
\mathbf{C} = \mathbf{A} \times \mathbf{B} = \begin{vmatrix} \mathbf{i} & \mathbf{j} & \mathbf{k} \\ A_x & A_y & A_z \\ B_x & B_y & B_z \end{vmatrix} = \begin{vmatrix} \mathbf{i} & \mathbf{j} & \mathbf{k} \\ -3.75 & -2.50 & 1.50 \\ 32 & 44 & 64 \end{vmatrix}
$$

$$= -226\mathbf{i} + 288\mathbf{j} - 85.0\mathbf{k} \qquad \textbf{Ans.}$$

The magnitude of vector **C** is determined by using Eq. (A-22). Thus,

$$C = |\mathbf{C}| = \sqrt{C_x^2 + C_y^2 + C_z^2} = \sqrt{(-226)^2 + (288)^2 + (-85.0)^2} = 376 \quad \textbf{Ans.}$$

The direction is determined by using Eqs. (A-23). Thus,

$$\theta_x = \cos^{-1}\frac{C_x}{|\mathbf{C}|} = \cos^{-1}\frac{-226}{376} = 126.9° \qquad \textbf{Ans.}$$

$$\theta_y = \cos^{-1}\frac{C_y}{|\mathbf{C}|} = \cos^{-1}\frac{288}{376} = 40.0° \qquad \textbf{Ans.}$$

$$\theta_z = \cos^{-1}\frac{C_z}{|\mathbf{C}|} = \cos^{-1}\frac{-85.0}{376} = 103.1° \qquad \textbf{Ans.}$$

The unit vector $\mathbf{e}_C$ is given by Eq. (A-18) as

$$\mathbf{e}_C = \frac{\mathbf{C}}{|\mathbf{C}|} = \frac{-226\mathbf{i} + 288\mathbf{j} - 85.0\mathbf{k}}{376} = -0.601\mathbf{i} + 0.766\mathbf{j} - 0.226\mathbf{k} \quad \blacksquare \quad \textbf{Ans.}$$

A-5-3 Triple Scalar Product

The triple scalar product involves the dot product of vector **A** and the cross product of vectors **B** and **C**. The triple scalar product is written as

$$\mathbf{A} \cdot (\mathbf{B} \times \mathbf{C}) \qquad \text{or} \qquad (\mathbf{B} \times \mathbf{C}) \cdot \mathbf{A} \qquad \text{(A-33)}$$

Expressing vectors **A**, **B**, and **C** in Cartesian vector form and expanding yields

$$\mathbf{A} \cdot (\mathbf{B} \times \mathbf{C}) = (A_x\mathbf{i} + A_y\mathbf{j} + A_z\mathbf{k}) \cdot [(B_x\mathbf{i} + B_y\mathbf{j} + B_z\mathbf{k}) \times (C_x\mathbf{i} + C_y\mathbf{j} + C_z\mathbf{k})]$$
$$= A_x(B_yC_z - B_zC_y) + A_y(B_zC_x - B_xC_z) + A_z(B_xC_y - B_yC_x) \quad \text{(A-34)}$$

which is the expanded form of the determinant

$$\mathbf{A} \cdot (\mathbf{B} \times \mathbf{C}) = \begin{vmatrix} A_x & A_y & A_z \\ B_x & B_y & B_z \\ C_x & C_y & C_z \end{vmatrix} \qquad \text{(A-35)}$$

The result of this product is a scalar; hence the name triple scalar product. However, the vectors cannot be indiscriminately interchanged since vector cross products are not commutative. It is true, however, that

$$\mathbf{A} \cdot (\mathbf{B} \times \mathbf{C}) = (\mathbf{A} \times \mathbf{B}) \cdot \mathbf{C} = (\mathbf{B} \times \mathbf{C}) \cdot \mathbf{A} = \mathbf{B} \cdot (\mathbf{C} \times \mathbf{A})$$

Since the cross product is used to find the moment of a force about a point and the scalar product is used to find the component of a vector along a line, the triple scalar product represents the component of the moment of a force along a line. That is, the triple scalar product represents the tendency of a force to rotate a body about a line.

A-5-4 Triple Vector Product

The triple vector product involves the cross product of a vector **A** with the result of a cross product of vectors **B** and **C**. Thus, the triple vector product is written as

$$\mathbf{A} \times (\mathbf{B} \times \mathbf{C})$$

If the vectors are written in Cartesian vector form, Eq. (A-31) gives the cross product $(\mathbf{B} \times \mathbf{C})$ as

$$\mathbf{B} \times \mathbf{C} = (B_x\mathbf{i} + B_y\mathbf{j} + B_z\mathbf{k}) \times (C_x\mathbf{i} + C_y\mathbf{j} + C_z\mathbf{k})$$
$$= (B_yC_z - B_zC_y)\mathbf{i} + (B_zC_x - B_xC_z)\mathbf{j} + (B_xC_y - B_yC_x)\mathbf{k}$$

Similarly, Eq. (A-31) gives the cross product $\mathbf{A} \times (\mathbf{B} \times \mathbf{C})$ as

$$\mathbf{A} \times (\mathbf{B} \times \mathbf{C}) = (A_x\mathbf{i} + A_y\mathbf{j} + A_z\mathbf{k}) \times [(B_yC_z - B_zC_y)\mathbf{i}$$
$$+ (B_zC_x - B_xC_z)\mathbf{j} + (B_xC_y - B_yC_x)\mathbf{k}]$$
$$= [A_y(B_xC_y - B_yC_x) - A_z(B_zC_x - B_xC_z)]\mathbf{i}$$
$$+ [A_z(B_yC_z - B_zC_y) - A_x(B_xC_y - B_yC_x)]\mathbf{j}$$
$$+ [A_x(B_zC_x - B_xC_z) - A_y(B_yC_z - B_zC_y)]\mathbf{k}$$
$$= (A_yC_y + A_zC_z)B_x\mathbf{i} + (A_zC_z + A_xC_x)B_y\mathbf{j} + (A_xC_x + A_yC_y)B_z\mathbf{k}$$
$$- (A_yB_y + A_zB_z)C_x\mathbf{i} - (A_zB_z + A_xB_x)C_y\mathbf{j} - (A_xB_x + A_yB_y)C_z\mathbf{k}$$

Adding and subtracting the vector

$$A_xB_xC_x\mathbf{i} + A_yB_yC_y\mathbf{j} + A_zB_zC_z\mathbf{k}$$

and regrouping yields

$$\mathbf{A} \times (\mathbf{B} \times \mathbf{C}) = (A_xC_x + A_yC_y + A_zC_z)(B_x\mathbf{i} + B_y\mathbf{j} + B_z\mathbf{k})$$
$$- (A_xB_x + A_yB_y + A_zB_z)(C_x\mathbf{i} + C_y\mathbf{j} + C_z\mathbf{k})$$

However, from Eq. (A-27),

$$A_xC_x + A_yC_y + A_zC_z = \mathbf{A} \cdot \mathbf{C}$$
$$A_xB_x + A_yB_y + A_zB_z = \mathbf{A} \cdot \mathbf{B}$$

Therefore

$$\mathbf{A} \times (\mathbf{B} \times \mathbf{C}) = (\mathbf{A} \cdot \mathbf{C})\mathbf{B} - (\mathbf{A} \cdot \mathbf{B})\mathbf{C} \qquad \text{(A-36)}$$

In a similar fashion it can be established that

$$(\mathbf{A} \times \mathbf{B}) \times \mathbf{C} = (\mathbf{A} \cdot \mathbf{C})\mathbf{B} - (\mathbf{B} \cdot \mathbf{C})\mathbf{A} \qquad \text{(A-37)}$$

The triple vector product is used in dynamics to determine the acceleration of points in a rotating body.

Example Problem A-6

Determine the triple scalar product $\mathbf{A} \cdot (\mathbf{B} \times \mathbf{C})$ and the triple vector product $\mathbf{A} \times (\mathbf{B} \times \mathbf{C})$ for the vectors

$$\mathbf{A} = 3\mathbf{i} + 5\mathbf{j} + 8\mathbf{k}$$
$$\mathbf{B} = 4\mathbf{i} - 5\mathbf{j} + 3\mathbf{k}$$
$$\mathbf{C} = 2\mathbf{i} + 3\mathbf{j} - 4\mathbf{k}$$

SOLUTION

The triple scalar product $\mathbf{A} \cdot (\mathbf{B} \times \mathbf{C})$ can be expressed in determinate form as given by Eq. (A-35). Thus,

$$\mathbf{A} \cdot (\mathbf{B} \times \mathbf{C}) = \begin{vmatrix} A_x & A_y & A_z \\ B_x & B_y & B_z \\ C_x & C_y & C_z \end{vmatrix} = \begin{vmatrix} 3 & 5 & 8 \\ 4 & -5 & 3 \\ 2 & 3 & -4 \end{vmatrix}$$

$$= 3(20 - 9) - 5(-16 - 6) + 8(12 + 10)$$
$$= 319 \qquad \qquad \textbf{Ans.}$$

The triple vector product $\mathbf{A} \times (\mathbf{B} \times \mathbf{C})$ is given by Eq. (A-36) as

$$\begin{aligned}
\mathbf{A} \times (\mathbf{B} \times \mathbf{C}) &= (\mathbf{A} \cdot \mathbf{C})\mathbf{B} - (\mathbf{A} \cdot \mathbf{B})\mathbf{C} \\
&= [(3\mathbf{i} + 5\mathbf{j} + 8\mathbf{k}) \cdot (2\mathbf{i} + 3\mathbf{j} - 4\mathbf{k})](4\mathbf{i} - 5\mathbf{j} + 3\mathbf{k}) \\
&\quad - [(3\mathbf{i} + 5\mathbf{j} + 8\mathbf{k}) \cdot (4\mathbf{i} - 5\mathbf{j} + 3\mathbf{k})](2\mathbf{i} + 3\mathbf{j} - 4\mathbf{k}) \\
&= [3(2) + 5(3) + 8(-4)](4\mathbf{i} - 5\mathbf{j} + 3\mathbf{k}) \\
&\quad - [3(4) + 5(-5) + 8(3)](2\mathbf{i} + 3\mathbf{j} - 4\mathbf{k}) \\
&= -11(4\mathbf{i} - 5\mathbf{j} + 3\mathbf{k}) - 11(2\mathbf{i} + 3\mathbf{j} - 4\mathbf{k}) \\
&= -66\mathbf{i} + 22\mathbf{j} + 11\mathbf{k} \qquad \qquad \textbf{Ans.}
\end{aligned}$$

B
APPENDIX

TABLES OF PROPERTIES

TABLE B-1   Wide-Flange Beams (U.S. Customary Units)

Desig-nation*	Area (in.^{2})	Depth (in.)	FLANGE Width (in.)	Thick-ness (in.)	Web Thick-ness (in.)	AXIS X–X I (in.^{4})	S (in.^{3})	r (in.)	AXIS Y–Y I (in.^{4})	S (in.^{3})	r (in.)
W36 × 230	67.6	35.90	16.470	1.260	0.760	15000	837	14.9	940	114	3.73
× 160	47.0	36.01	12.000	1.020	0.650	9750	542	14.4	295	49.1	2.50
W33 × 201	59.1	33.68	15.745	1.150	0.715	11500	684	14.0	749	95.2	3.56
× 152	44.7	33.49	11.565	1.055	0.635	8160	487	13.5	273	47.2	2.47
× 130	38.3	33.09	11.510	0.855	0.580	6710	406	13.2	218	37.9	2.39
W30 × 132	38.9	30.31	10.545	1.000	0.615	5770	380	12.2	196	37.2	2.25
× 108	31.7	29.83	10.475	0.760	0.545	4470	299	11.9	146	27.9	2.15
W27 × 146	42.9	27.38	13.965	0.975	0.605	5630	411	11.4	443	63.5	3.21
× 94	27.7	26.92	9.990	0.745	0.490	3270	243	10.9	124	24.8	2.12
W24 × 104	30.6	24.06	12.750	0.750	0.500	3100	258	10.1	259	40.7	2.91
× 84	24.7	24.10	9.020	0.770	0.470	2370	196	9.79	94.4	20.9	1.95
× 62	18.2	23.74	7.040	0.590	0.430	1550	131	9.23	34.5	9.80	1.38
W21 × 101	29.8	21.36	12.290	0.800	0.500	2420	227	9.02	248	40.3	2.89
× 83	24.3	21.43	8.355	0.835	0.515	1830	171	8.67	81.4	19.5	1.83
× 62	18.3	20.99	8.240	0.615	0.400	1330	127	8.54	57.5	13.9	1.77
W18 × 97	28.5	18.59	11.145	0.870	0.535	1750	188	7.82	201	36.1	2.65
× 76	22.3	18.21	11.035	0.680	0.425	1330	146	7.73	152	27.6	2.61
× 60	17.6	18.24	7.555	0.695	0.415	984	108	7.47	50.1	13.3	1.69
W16 × 100	29.4	16.97	10.425	0.985	0.585	1490	175	7.10	186	35.7	2.52
× 67	19.7	16.33	10.235	0.665	0.395	954	117	6.96	119	23.2	2.46
× 40	11.8	16.01	6.995	0.505	0.305	518	64.7	6.63	28.9	8.25	1.57
× 26	7.68	15.69	5.500	0.345	0.250	301	38.4	6.26	9.59	3.49	1.12
W14 × 120	35.3	14.48	14.670	0.940	0.590	1380	190	6.24	495	67.5	3.74
× 82	24.1	14.31	10.130	0.855	0.510	882	123	6.05	148	29.3	2.48
× 43	12.6	13.66	7.995	0.530	0.305	428	62.7	5.82	45.2	11.3	1.89
× 30	8.85	13.84	6.730	0.385	0.270	291	42.0	5.73	19.6	5.82	1.49
W12 × 96	28.2	12.71	12.160	0.900	0.550	833	131	5.44	270	44.4	3.09
× 65	19.1	12.12	12.000	0.605	0.390	533	87.9	5.28	174	29.1	3.02
× 50	14.7	12.19	8.080	0.640	0.370	394	64.7	5.18	56.3	13.9	1.96
× 30	8.79	12.34	6.520	0.440	0.260	238	38.6	5.21	20.3	6.24	1.52
W10 × 60	17.6	10.22	10.080	0.680	0.420	341	66.7	4.39	116	23.0	2.57
× 45	13.3	10.10	8.020	0.620	0.350	248	49.1	4.33	53.4	13.3	2.01
× 30	8.84	10.47	5.810	0.510	0.300	170	32.4	4.38	16.7	5.75	1.37
× 22	6.49	10.17	5.750	0.360	0.240	118	23.2	4.27	11.4	3.97	1.33
W8 × 40	11.7	8.25	8.070	0.560	0.360	146	35.5	3.53	49.1	12.2	2.04
× 31	9.13	8.00	7.995	0.435	0.285	110	27.5	3.47	37.1	9.27	2.02
× 24	7.08	7.93	6.495	0.400	0.245	82.8	20.9	3.42	18.3	5.63	1.61
× 15	4.44	8.11	4.015	0.315	0.245	48.0	11.8	3.29	3.41	1.70	0.876
W6 × 25	7.34	6.38	6.080	0.455	0.320	53.4	16.7	2.70	17.1	5.61	1.52
× 16	4.74	6.28	4.030	0.405	0.260	32.1	10.2	2.60	4.43	2.20	0.967
W5 × 16	4.68	5.01	5.000	0.360	0.240	21.3	8.51	2.13	7.51	3.00	1.27
W4 × 13	3.83	4.16	4.060	0.345	0.280	11.3	5.46	1.72	3.86	1.90	1.00

Courtesy of the American Institute of Steel Construction.

*W means wide-flange beam, followed by the nominal depth in inches, then the weight in pounds per foot of length.

TABLE B-2 — Wide-Flange Beams (SI Units)

Designation*	Area (mm²)	Depth (mm)	FLANGE Width (mm)	FLANGE Thickness (mm)	Web Thickness (mm)	AXIS X–X I (10⁶ mm⁴)	AXIS X–X S (10³ mm³)	r (mm)	AXIS Y–Y I (10⁶ mm⁴)	AXIS Y–Y S (10³ mm³)	r (mm)
W914 × 342	43610	912	418	32.0	19.3	6245	13715	378	391	1870	94.7
× 238	30325	915	305	25.9	16.5	4060	8880	366	123	805	63.5
W838 × 299	38130	855	400	29.2	18.2	4785	11210	356	312	1560	90.4
× 226	28850	851	294	26.8	16.1	3395	7980	343	114	775	62.7
× 193	24710	840	292	21.7	14.7	2795	6655	335	90.7	620	60.7
W762 × 196	25100	770	268	25.4	15.6	2400	6225	310	81.6	610	57.2
× 161	20450	758	266	19.3	13.8	1860	4900	302	60.8	457	54.6
W686 × 217	27675	695	355	24.8	15.4	2345	6735	290	184	1040	81.5
× 140	17870	684	254	18.9	12.4	1360	3980	277	51.6	406	53.8
W610 × 155	19740	611	324	19.1	12.7	1290	4230	257	108	667	73.9
× 125	15935	612	229	19.6	11.9	985	3210	249	39.3	342	49.5
× 92	11750	603	179	15.0	10.9	645	2145	234	14.4	161	35.1
W533 × 150	19225	543	312	20.3	12.7	1005	3720	229	103	660	73.4
× 124	15675	544	212	21.2	13.1	762	2800	220	33.9	320	46.5
× 92	11805	533	209	15.6	10.2	554	2080	217	23.9	228	45.0
W457 × 144	18365	472	283	22.1	13.6	728	3080	199	83.7	592	67.3
× 113	14385	463	280	17.3	10.8	554	2395	196	63.3	452	66.3
× 89	11355	463	192	17.7	10.5	410	1770	190	20.9	218	42.9
W406 × 149	18970	431	265	25.0	14.9	620	2870	180	77.4	585	64.0
× 100	12710	415	260	16.9	10.0	397	1915	177	49.5	380	62.5
× 60	7615	407	178	12.8	7.7	216	1060	168	12.0	135	39.9
× 39	4950	399	140	8.8	6.4	125	629	159	3.99	57.2	28.4
W356 × 179	22775	368	373	23.9	15.0	574	3115	158	206	1105	95.0
× 122	15550	363	257	21.7	13.0	367	2015	154	61.6	480	63.0
× 64	8130	347	203	13.5	7.7	178	1025	148	18.8	185	48.0
× 45	5710	352	171	9.8	6.9	121	688	146	8.16	95.4	37.8
W305 × 143	18195	323	309	22.9	14.0	347	2145	138	112	728	78.5
× 97	12325	308	305	15.4	9.9	222	1440	134	72.4	477	76.7
× 74	9485	310	205	16.3	9.4	164	1060	132	23.4	228	49.8
× 45	5670	313	166	11.2	6.6	99.1	633	132	8.45	102	38.6
W254 × 89	11355	260	256	17.3	10.7	142	1095	112	48.3	377	65.3
× 67	8580	257	204	15.7	8.9	103	805	110	22.2	218	51.1
× 45	5705	266	148	13.0	7.6	70.8	531	111	6.95	94.2	34.8
× 33	4185	258	146	9.1	6.1	49.1	380	108	4.75	65.1	33.8
W203 × 60	7550	210	205	14.2	9.1	60.8	582	89.7	20.4	200	51.8
× 46	5890	203	203	11.0	7.2	45.8	451	88.1	15.4	152	51.3
× 36	4570	201	165	10.2	6.2	34.5	342	86.7	7.61	92.3	40.9
× 22	2865	206	102	8.0	6.2	20.0	193	83.6	1.42	27.9	22.3
W152 × 37	4735	162	154	11.6	8.1	22.2	274	68.6	7.12	91.9	38.6
× 24	3060	160	102	10.3	6.6	13.4	167	66.0	1.84	36.1	24.6
W127 × 24	3020	127	127	9.1	6.1	8.87	139	54.1	3.13	49.2	32.3
W102 × 19	2470	106	103	8.8	7.1	4.70	89.5	43.7	1.61	31.1	25.4

*W means wide-flange beam, followed by the nominal depth in mm, then the mass in kg per meter of length.

American Standard Beams (U.S. Customary Units)

Desig-nation*	Area (in.²)	Depth (in.)	FLANGE Width (in.)	FLANGE Thick-ness (in.)	Web Thick-ness (in.)	AXIS X–X I (in.⁴)	AXIS X–X S (in.³)	AXIS X–X r (in.)	AXIS Y–Y I (in.⁴)	AXIS Y–Y S (in.³)	AXIS Y–Y r (in.)
S24 × 121	35.6	24.50	8.050	1.090	0.800	3160	258	9.43	83.3	20.7	1.53
× 106	31.2	24.50	7.870	1.090	0.620	2940	240	9.71	77.1	19.6	1.57
× 100	29.3	24.00	7.245	0.870	0.745	2390	199	9.02	47.7	13.2	1.27
× 90	26.5	24.00	7.125	0.870	0.625	2250	187	9.21	44.9	12.6	1.30
× 80	23.5	24.00	7.000	0.870	0.500	2100	175	9.47	42.2	12.1	1.34
S20 × 96	28.2	20.30	7.200	0.920	0.800	1670	165	7.71	50.2	13.9	1.33
× 86	25.3	20.30	7.060	0.920	0.660	1580	155	7.89	46.8	13.3	1.36
× 75	22.0	20.00	6.385	0.795	0.635	1280	128	7.62	29.8	9.32	1.16
× 66	19.4	20.00	6.255	0.795	0.505	1190	119	7.83	27.7	8.85	1.19
S18 × 70	20.6	18.00	6.251	0.691	0.711	926	103	6.71	24.1	7.72	1.08
× 54.7	16.1	18.00	6.001	0.691	0.461	804	89.4	7.07	20.8	6.94	1.14
S15 × 50	14.7	15.00	5.640	0.622	0.550	486	64.8	5.75	15.7	5.57	1.03
× 42.9	12.6	15.00	5.501	0.622	0.411	447	59.6	5.95	14.4	5.23	1.07
S12 × 50	14.7	12.00	5.477	0.659	0.687	305	50.8	4.55	15.7	5.74	1.03
× 40.8	12.0	12.00	5.252	0.659	0.462	272	45.4	4.77	13.6	5.16	1.06
× 35	10.3	12.00	5.078	0.544	0.428	229	38.2	4.72	9.87	3.89	0.980
× 31.8	9.35	12.00	5.000	0.544	0.350	218	36.4	4.83	9.36	3.74	1.00
S10 × 35	10.3	10.00	4.944	0.491	0.594	147	29.4	3.78	8.36	3.38	0.901
× 25.4	7.46	10.00	4.661	0.491	0.311	124	24.7	4.07	6.79	2.91	0.954
S8 × 23	6.77	8.00	4.171	0.426	0.441	64.9	16.2	3.10	4.31	2.07	0.798
× 18.4	5.41	8.00	4.001	0.426	0.271	57.6	14.4	3.26	3.73	1.86	0.831
S7 × 20	5.88	7.00	3.860	0.392	0.450	42.4	12.1	2.69	3.17	1.64	0.734
× 15.3	4.50	7.00	3.662	0.392	0.252	36.7	10.5	2.86	2.64	1.44	0.766
S6 × 17.25	5.07	6.00	3.565	0.359	0.465	26.3	8.77	2.28	2.31	1.30	0.675
× 12.5	3.67	6.00	3.332	0.359	0.232	22.1	7.37	2.45	1.82	1.09	0.705
S5 × 14.75	4.34	5.00	3.284	0.326	0.494	15.2	6.09	1.87	1.67	1.01	0.620
× 10	2.94	5.00	3.004	0.326	0.214	12.3	4.92	2.05	1.22	0.809	0.643
S4 × 9.5	2.79	4.00	2.796	0.293	0.326	6.79	3.39	1.56	0.903	0.646	0.569
× 7.7	2.26	4.00	2.663	0.293	0.193	6.08	3.04	1.64	0.764	0.574	0.581
S3 × 7.5	2.21	3.00	2.509	0.260	0.349	2.93	1.95	1.15	0.586	0.468	0.516
× 5.7	1.67	3.00	2.330	0.260	0.170	2.52	1.68	1.23	0.455	0.390	0.522

Courtesy of The American Institute of Steel Construction.

*S means standard beam, followed by the nominal depth in inches, then the weight in pounds per foot of length.

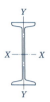

TABLE B-4

American Standard Beams (SI Units)

Desig-nation[*]	Area (mm²)	Depth (mm)	FLANGE Width (mm)	FLANGE Thick-ness (mm)	Web Thick-ness (mm)	AXIS X–X I (10^6 mm⁴)	AXIS X–X S (10^3 mm³)	r (mm)	AXIS Y–Y I (10^6 mm⁴)	AXIS Y–Y S (10^3 mm³)	r (mm)
S610 × 180	22970	622.3	204.5	27.7	20.3	1315	4225	240	34.7	339	38.9
× 158	20130	622.3	199.9	27.7	15.7	1225	3935	247	32.1	321	39.9
× 149	18900	609.6	184.0	22.1	18.9	995	3260	229	19.9	216	32.3
× 134	17100	609.6	181.0	22.1	15.9	937	3065	234	18.7	206	33.0
× 119	15160	609.6	177.8	22.1	12.7	874	2870	241	17.6	198	34.0
S508 × 143	18190	515.6	182.9	23.4	20.3	695	2705	196	20.9	228	33.8
× 128	16320	515.6	179.3	23.4	16.8	658	2540	200	19.5	218	34.5
× 112	14190	508.0	162.2	20.2	16.1	533	2100	194	12.4	153	29.5
× 98	12520	508.0	158.9	20.2	12.8	495	1950	199	11.5	145	30.2
S457 × 104	13290	457.2	158.8	17.6	18.1	358	1690	170	10.0	127	27.4
× 81	10390	457.2	152.4	17.6	11.7	335	1465	180	8.66	114	29.0
S381 × 74	9485	381.0	143.3	15.8	14.0	202	1060	146	6.53	91.3	26.2
× 64	8130	381.0	139.7	15.8	10.4	186	977	151	5.99	85.7	27.2
S305 × 74	9485	304.8	139.1	16.7	17.4	127	832	116	6.53	94.1	26.2
× 61	7740	304.8	133.4	16.7	11.7	113	744	121	5.66	84.6	26.9
× 52	6645	304.8	129.0	13.8	10.9	95.3	626	120	4.11	63.7	24.1
× 47	6030	304.8	127.0	13.8	8.9	90.7	596	123	3.90	61.3	25.4
S254 × 52	6645	254.0	125.6	12.5	15.1	61.2	482	96.0	3.48	55.4	22.9
× 38	4815	254.0	118.4	12.5	7.9	51.6	408	103	2.83	47.7	24.2
S203 × 34	4370	203.2	105.9	10.8	11.2	27.0	265	78.7	1.79	33.9	20.3
× 27	3490	203.2	101.6	10.8	6.9	24.0	236	82.8	1.55	30.5	21.1
S178 × 30	3795	177.8	98.0	10.0	11.4	17.6	198	68.3	1.32	26.9	18.6
× 23	2905	177.8	93.0	10.0	6.4	15.3	172	72.6	1.10	23.6	19.5
S152 × 26	3270	152.4	90.6	9.1	11.8	10.9	144	57.9	0.961	21.3	17.1
× 19	2370	152.4	84.6	9.1	5.9	9.20	121	62.2	0.758	17.9	17.9
S127 × 22	2800	127.0	83.4	8.3	12.5	6.33	99.8	47.5	0.695	16.6	15.7
× 15	1895	127.0	76.3	8.3	5.4	5.12	80.6	52.1	0.508	13.3	16.3
S102 × 14	1800	101.6	71.0	7.4	8.3	2.83	55.6	39.6	0.376	10.6	14.5
× 11	1460	101.6	67.6	7.4	4.9	2.53	49.8	41.7	0.318	9.41	14.8
S76 × 11	1425	76.2	63.7	6.6	8.9	1.22	32.0	29.2	0.244	7.67	13.1
× 8.5	1075	76.2	59.2	6.6	4.3	1.05	27.5	31.2	0.189	6.39	13.3

[*]S means standard beam, followed by the nominal depth in mm, then the mass in kg per meter of length.

TABLE B-5   Standard Channels (U.S. Customary Units)

Desig-nation[†]	Area (in.²)	Depth (in.)	FLANGE Width (in.)	FLANGE Average Thick-ness (in.)	Web Thick-ness (in.)	AXIS X–X I (in.⁴)	AXIS X–X S (in.³)	AXIS X–X r (in.)	AXIS Y–Y I (in.⁴)	AXIS Y–Y S (in.³)	AXIS Y–Y r (in.)	x_C (in.)
*C18 × 58	17.1	18.00	4.200	0.625	0.700	676	75.1	6.29	17.8	5.32	1.02	0.862
× 51.9	15.3	18.00	4.100	0.625	0.600	627	69.7	6.41	16.4	5.07	1.04	0.858
× 45.8	13.5	18.00	4.000	0.625	0.500	578	64.3	6.56	15.1	4.82	1.06	0.866
× 42.7	12.6	18.00	3.950	0.625	0.450	554	61.6	6.64	14.4	4.69	1.07	0.877
C15 × 50	14.7	15.00	3.716	0.650	0.716	404	53.8	5.24	11.0	3.78	0.867	0.798
× 40	11.8	15.00	3.520	0.650	0.520	349	46.5	5.44	9.23	3.37	0.886	0.777
× 33.9	9.96	15.00	3.400	0.650	0.400	315	42.0	5.62	8.13	3.11	0.904	0.787
C12 × 30	8.82	12.00	3.170	0.501	0.510	162	27.0	4.29	5.14	2.06	0.763	0.674
× 25	7.35	12.00	3.047	0.501	0.387	144	24.1	4.43	4.47	1.88	0.780	0.674
× 20.7	6.09	12.00	2.942	0.501	0.282	129	21.5	4.61	3.88	1.73	0.799	0.698
C10 × 30	8.82	10.00	3.033	0.436	0.673	103	20.7	3.42	3.94	1.65	0.669	0.649
× 25	7.35	10.00	2.886	0.436	0.526	91.2	18.2	3.52	3.36	1.48	0.676	0.617
× 20	5.88	10.00	2.739	0.436	0.379	78.9	15.8	3.66	2.81	1.32	0.692	0.606
× 15.3	4.49	10.00	2.600	0.436	0.240	67.4	13.5	3.87	2.28	1.16	0.713	0.634
C9 × 20	5.88	9.00	2.648	0.413	0.448	60.9	13.5	3.22	2.42	1.17	0.642	0.583
× 15	4.41	9.00	2.485	0.413	0.285	51.0	11.3	3.40	1.93	1.01	0.661	0.586
× 13.4	3.94	9.00	2.433	0.413	0.233	47.9	10.6	3.48	1.76	0.962	0.669	0.601
C8 × 18.75	5.51	8.00	2.527	0.390	0.487	44.0	11.0	2.82	1.98	1.01	0.599	0.565
× 13.75	4.04	8.00	2.343	0.390	0.303	36.1	9.03	2.99	1.53	0.854	0.615	0.553
× 11.5	3.38	8.00	2.260	0.390	0.220	32.6	8.14	3.11	1.32	0.781	0.625	0.571
C7 × 14.75	4.33	7.00	2.299	0.366	0.419	27.2	7.78	2.51	1.38	0.779	0.564	0.532
× 12.25	3.60	7.00	2.194	0.366	0.314	24.2	6.93	2.60	1.17	0.703	0.571	0.525
× 9.8	2.87	7.00	2.090	0.366	0.210	21.3	6.08	2.72	0.968	0.625	0.581	0.540
C6 × 13	3.83	6.00	2.157	0.343	0.437	17.4	5.80	2.13	1.05	0.642	0.525	0.514
× 10.5	3.09	6.00	2.034	0.343	0.314	15.2	5.06	2.22	0.866	0.564	0.529	0.499
× 8.2	2.40	6.00	1.920	0.343	0.200	13.1	4.38	2.34	0.693	0.492	0.537	0.511
C5 × 9	2.64	5.00	1.885	0.320	0.325	8.90	3.56	1.83	0.632	0.450	0.489	0.478
× 6.7	1.97	5.00	1.750	0.320	0.190	7.49	3.00	1.95	0.479	0.378	0.493	0.484
C4 × 7.25	2.13	4.00	1.721	0.296	0.321	4.59	2.29	1.47	0.433	0.343	0.450	0.459
× 5.4	1.59	4.00	1.584	0.296	0.184	3.85	1.93	1.56	0.319	0.283	0.449	0.457
C3 × 6	1.76	3.00	1.596	0.273	0.356	2.07	1.38	1.08	0.305	0.268	0.416	0.455
× 5	1.47	3.00	1.498	0.273	0.258	1.85	1.24	1.12	0.247	0.233	0.410	0.438
× 4.1	1.21	3.00	1.410	0.273	0.170	1.66	1.10	1.17	0.197	0.202	0.404	0.436

Courtesy of The American Institute of Steel Construction.

*Not part of the American Standard Series.

[†]C means channel, followed by the nominal depth in inches, then the weight in pounds per foot of length.

TABLE B-6 — Standard Channels (SI Units)

Desig-nation*	Area (mm²)	Depth (mm)	FLANGE Width (mm)	FLANGE Thick-ness (mm)	Web Thick-ness (mm)	AXIS X–X I (10^6 mm⁴)	AXIS X–X S (10^3 mm³)	AXIS X–X r (mm)	AXIS Y–Y I (10^6 mm⁴)	AXIS Y–Y S (10^3 mm³)	AXIS Y–Y r (mm)	x_C (mm)
C457 × 86	11030	457.2	106.7	15.9	17.8	281	1230	160	7.41	87.2	25.9	21.9
× 77	9870	457.2	104.1	15.9	15.2	261	1140	163	6.83	83.1	26.4	21.8
× 68	8710	457.2	101.6	15.9	12.7	241	1055	167	6.29	79.0	26.9	22.0
× 64	8130	457.2	100.3	15.9	11.4	231	1010	169	5.99	76.9	27.2	22.3
C381 × 74	9485	381.0	94.4	16.5	18.2	168	882	133	4.58	61.9	22.0	20.3
× 60	7615	381.0	89.4	16.5	13.2	145	762	138	3.84	55.2	22.5	19.7
× 50	6425	381.0	86.4	16.5	10.2	131	688	143	3.38	51.0	23.0	20.0
C305 × 45	5690	304.8	80.5	12.7	13.0	67.4	442	109	2.14	33.8	19.4	17.1
× 37	4740	304.8	77.4	12.7	9.8	59.9	395	113	1.86	30.8	19.8	17.1
× 31	3930	304.8	74.7	12.7	7.2	53.7	352	117	1.61	28.3	20.3	17.7
C254 × 45	5690	254.0	77.0	11.1	17.1	42.9	339	86.9	1.64	27.0	17.0	16.5
× 37	4740	254.0	73.3	11.1	13.4	38.0	298	89.4	1.40	24.3	17.2	15.7
× 30	3795	254.0	69.6	11.1	9.6	32.8	259	93.0	1.17	21.6	17.6	15.4
× 23	2895	254.0	66.0	11.1	6.1	28.1	221	98.3	0.949	19.0	18.1	16.1
C229 × 30	3795	228.6	67.3	10.5	11.4	25.3	221	81.8	1.01	19.2	16.3	14.8
× 22	2845	228.6	63.1	10.5	7.2	21.2	185	86.4	0.803	16.6	16.8	14.9
× 20	2540	228.6	61.8	10.5	5.9	19.9	174	88.4	0.733	15.7	17.0	15.3
C203 × 28	3555	203.2	64.2	9.9	12.4	18.3	180	71.6	0.824	16.6	15.2	14.4
× 20	2605	203.2	59.5	9.9	7.7	15.0	148	75.9	0.637	14.0	15.6	14.0
× 17	2180	203.2	57.4	9.9	5.6	13.6	133	79.0	0.549	12.8	15.9	14.5
C178 × 22	2795	177.8	58.4	9.3	10.6	11.3	127	63.8	0.574	12.8	14.3	13.5
× 18	2320	177.8	55.7	9.3	8.0	10.1	114	66.0	0.487	11.5	14.5	13.3
× 15	1850	177.8	53.1	9.3	5.3	8.87	99.6	69.1	0.403	10.2	14.8	13.7
C152 × 19	2470	152.4	54.8	8.7	11.1	7.24	95.0	54.1	0.437	10.5	13.3	13.1
× 16	1995	152.4	51.7	8.7	8.0	6.33	82.9	56.4	0.360	9.24	13.4	12.7
× 12	1550	152.4	48.8	8.7	5.1	5.45	71.8	59.4	0.288	8.06	13.6	13.0
C127 × 13	1705	127.0	47.9	8.1	8.3	3.70	58.3	46.5	0.263	7.37	12.4	12.1
× 10	1270	127.0	44.5	8.1	4.8	3.12	49.2	49.5	0.199	6.19	12.5	12.3
C102 × 11	1375	101.6	43.7	7.5	8.2	1.91	37.5	37.3	0.180	5.62	11.4	11.7
× 8	1025	101.6	40.2	7.5	4.7	1.60	31.6	39.6	0.133	4.64	11.4	11.6
C76 × 9	1135	76.2	40.5	6.9	9.0	0.862	22.6	27.4	0.127	4.39	10.6	11.6
× 7	948	76.2	38.0	6.9	6.6	0.770	20.3	28.4	0.103	3.82	10.4	11.1
× 6	781	76.2	35.8	6.9	4.6	0.691	18.0	29.7	0.082	3.31	10.3	11.1

*C means channel, followed by the nominal depth in mm, then the mass in kg per meter of length.

Equal Leg Angles (U.S. Customary Units)

Size and Thickness (in.)	Weight (lb/ft)	Area (in.^{2})	AXIS X–X or Y–Y				AXIS Z–Z
			I (in.^{4})	S (in.^{3})	r (in.)	x_C or y_C (in.)	r (in.)
L8 × 8 × 1	51.0	15.0	89.0	15.8	2.44	2.37	1.56
× 7/8	45.0	13.2	79.6	14.0	2.45	2.32	1.57
× 3/4	38.9	11.4	69.7	12.2	2.47	2.28	1.58
× 5/8	32.7	9.61	59.4	10.3	2.49	2.23	1.58
× 1/2	26.4	7.75	48.6	8.36	2.50	2.19	1.59
L6 × 6 × 1	37.4	11.0	35.5	8.57	1.80	1.86	1.17
× 7/8	33.1	9.73	31.9	7.63	1.81	1.82	1.17
× 3/4	28.7	8.44	28.2	6.66	1.83	1.78	1.17
× 5/8	24.2	7.11	24.2	5.66	1.84	1.73	1.18
× 1/2	19.6	5.75	19.9	4.61	1.86	1.68	1.18
× 3/8	14.9	4.36	15.4	3.53	1.88	1.64	1.19
L5 × 5 × 7/8	27.2	7.98	17.8	5.17	1.49	1.57	0.973
× 3/4	23.6	6.94	15.7	4.53	1.51	1.52	0.975
× 5/8	20.0	5.86	13.6	3.86	1.52	1.48	0.978
× 1/2	16.2	4.75	11.3	3.16	1.54	1.43	0.983
× 3/8	12.3	3.61	8.74	2.42	1.56	1.39	0.990
L4 × 4 × 3/4	18.5	5.44	7.67	2.81	1.19	1.27	0.778
× 5/8	15.7	4.61	6.66	2.40	1.20	1.23	0.779
× 1/2	12.8	3.75	5.56	1.97	1.22	1.18	0.782
× 3/8	9.8	2.86	4.36	1.52	1.23	1.14	0.788
× 1/4	6.6	1.94	3.04	1.05	1.25	1.09	0.795
L3½ × 3½ × 1/2	11.1	3.25	3.64	1.49	1.06	1.06	0.683
× 3/8	8.5	2.48	2.87	1.15	1.07	1.01	0.687
× 1/4	5.8	1.69	2.01	0.794	1.09	0.968	0.694
L3 × 3 × 1/2	9.4	2.75	2.22	1.07	0.898	0.932	0.584
× 3/8	7.2	2.11	1.76	0.833	0.913	0.888	0.58
× 1/4	4.9	1.44	1.24	0.577	0.930	0.84	.592
L2½ × 2½ × 1/2	7.7	2.25	1.23	0.724	0.739	0.806	0.487
× 3/8	5.9	1.73	0.984	0.566	0.753	0.762	0.487
× 1/4	4.1	1.19	0.703	0.394	0.769	0.717	0.491
L2 × 2 × 3/8	4.7	1.36	0.479	0.351	0.594	0.636	0.389
× 1/4	3.19	0.938	0.348	0.247	0.609	0.592	0.391
× 1/8	1.65	0.484	0.190	0.131	0.626	0.546	0.398

Courtesy of The American Institute of Steel Construction.

Equal Leg Angles (SI Units)

Size and Thickness (mm)	Mass (kg/m)	Area (mm²)	AXIS X–X or Y–Y				AXIS Z–Z
			I (10^6 mm⁴)	S (10^3 mm³)	r (mm)	x_C or y_C (mm)	r (mm)
L203 × 203 × 25.4	75.9	9675	37.0	259	62.0	60.2	39.6
× 22.2	67.0	8515	33.1	229	62.2	58.9	39.9
× 19.1	57.9	7355	29.0	200	62.7	57.9	40.1
× 15.9	48.7	6200	24.7	169	63.2	56.6	40.1
× 12.7	39.3	5000	20.2	137	63.5	55.6	40.4
L152 × 152 × 25.4	55.7	7095	14.8	140	45.7	47.2	29.7
× 22.2	49.3	6275	13.3	125	46.0	46.2	29.7
× 19.1	42.7	5445	11.7	109	46.5	45.2	29.7
× 15.9	36.0	4585	10.1	92.8	46.7	43.9	30.0
× 12.7	29.2	3710	8.28	75.5	47.2	42.7	30.0
× 9.5	22.2	2815	6.61	57.8	47.8	41.7	30.2
L127 × 127 × 22.2	40.5	5150	7.41	84.7	37.8	39.9	24.7
× 19.1	35.1	4475	6.53	74.2	38.4	38.6	24.8
× 15.9	29.8	3780	5.66	63.3	38.6	37.6	24.8
× 12.7	24.1	3065	4.70	51.8	39.1	36.3	25.0
× 9.5	18.3	2330	3.64	39.7	39.6	35.3	25.1
L102 × 102 × 19.1	27.5	3510	3.19	46.0	30.2	32.3	19.8
× 15.9	23.4	2975	2.77	39.3	30.5	31.2	19.8
× 12.7	19.0	2420	2.31	32.3	31.0	30.0	19.9
× 9.5	14.6	1845	1.81	24.9	31.2	29.0	20.0
× 6.4	9.8	1250	1.27	17.2	31.8	27.7	20.2
L89 × 89 × 12.7	16.5	2095	1.52	24.4	26.9	26.9	17.3
× 9.5	12.6	1600	1.19	18.8	27.2	25.7	17.4
× 6.4	8.6	1090	0.837	13.0	27.7	24.6	17.6
L76 × 76 × 12.7	14.0	1775	0.924	17.5	22.8	23.7	14.8
× 9.5	10.7	1360	0.732	13.7	23.2	22.6	14.9
× 6.4	7.3	929	0.516	9.46	23.6	21.4	15.0
L64 × 64 × 12.7	11.5	1450	0.512	11.9	18.8	20.5	12.4
× 9.5	8.8	1115	0.410	9.28	19.1	19.4	12.4
× 6.4	6.1	768	0.293	6.46	19.5	18.2	12.5
L51 × 51 × 9.5	7.0	877	0.199	5.75	15.1	16.2	9.88
× 6.4	4.75	605	0.145	4.05	15.5	15.0	9.93
× 3.2	2.46	312	0.079	2.15	15.9	13.9	10.1

TABLE B-9   Unequal Leg Angles (U.S. Customary Units)

Size and Thickness	Weight (lb/ft)	Area (in.^{2})	AXIS X–X				AXIS Y–Y				AXIS Z–Z	
			I (in.^{4})	S (in.^{3})	r (in.)	y_C (in.)	I (in.^{4})	S (in.^{3})	r (in.)	x_C (in.)	r (in.)	Tan α
L9 × 4 × 5/8	26.3	7.73	64.9	11.5	2.90	3.36	8.32	2.65	1.04	0.858	0.847	0.216
× 1/2	21.3	6.25	53.2	9.34	2.92	3.31	6.92	2.17	1.05	0.810	0.854	0.220
L8 × 6 × 1	44.2	13.0	80.8	15.1	2.49	2.65	38.8	8.92	1.73	1.65	1.28	0.543
× 3/4	33.8	9.94	63.4	11.7	2.53	2.56	30.7	6.92	1.76	1.56	1.29	0.551
× 1/2	23.0	6.75	44.3	8.02	2.56	2.47	21.7	4.79	1.79	1.47	1.30	0.558
L8 × 4 × 1	37.4	11.0	69.6	14.1	2.52	3.05	11.6	3.94	1.03	1.05	0.846	0.247
× 3/4	28.7	8.44	54.9	10.9	2.55	2.95	9.36	3.07	1.05	0.953	0.852	0.258
× 1/2	19.6	5.75	38.5	7.49	2.59	2.86	6.74	2.15	1.08	0.859	0.865	0.267
L7 × 4 × 3/4	26.2	7.69	37.8	8.42	2.22	2.51	9.05	3.03	1.09	1.01	0.860	0.324
× 1/2	17.9	5.25	26.7	5.81	2.25	2.42	6.53	2.12	1.11	0.917	0.872	0.335
× 3/8	13.6	3.98	20.6	4.44	2.27	2.37	5.10	1.63	1.13	0.870	0.880	0.340
L6 × 4 × 3/4	23.6	6.94	24.5	6.25	1.88	2.08	8.68	2.97	1.12	1.08	0.860	0.428
× 1/2	16.2	4.75	17.4	4.33	1.91	1.99	6.27	2.08	1.15	0.987	0.870	0.440
× 3/8	12.3	3.61	13.5	3.32	1.93	1.94	4.90	1.60	1.17	0.941	0.877	0.446
L6 × 3$\frac{1}{2}$ × 1/2	15.3	4.50	16.6	4.24	1.92	2.08	4.25	1.59	0.972	0.833	0.759	0.344
× 3/8	11.7	3.42	12.9	3.24	1.94	2.04	3.34	1.23	0.988	0.787	0.767	0.350
L5 × 3$\frac{1}{2}$ × 3/4	19.8	5.81	13.9	4.28	1.55	1.75	5.55	2.22	0.977	0.996	0.748	0.464
× 1/2	13.6	4.00	9.99	2.99	1.58	1.66	4.05	1.56	1.01	0.906	0.755	0.479
× 3/8	10.4	3.05	7.78	2.29	1.60	1.61	3.18	1.21	1.02	0.861	0.762	0.486
× 1/4	7.0	2.06	5.39	1.57	1.62	1.56	2.23	0.830	1.04	0.814	0.770	0.492
L5 × 3 × 1/2	12.8	3.75	9.45	2.91	1.59	1.75	2.58	1.15	0.829	0.750	0.648	0.357
× 3/8	9.8	2.86	7.37	2.24	1.61	1.70	2.04	0.888	0.845	0.704	0.654	0.364
× 1/4	6.6	1.94	5.11	1.53	1.62	1.66	1.44	0.614	0.861	0.657	0.663	0.371
L4 × 3$\frac{1}{2}$ × 1/2	11.9	3.50	5.32	1.94	1.23	1.25	3.79	1.52	1.04	1.00	0.722	0.750
× 3/8	9.1	2.67	4.18	1.49	1.25	1.21	2.95	1.17	1.06	0.955	0.727	0.755
× 1/4	6.2	1.81	2.91	1.03	1.27	1.16	2.09	0.808	1.07	0.909	0.734	0.759
L4 × 3 × 1/2	11.1	3.25	5.05	1.89	1.25	1.33	2.42	1.12	0.864	0.827	0.639	0.543
× 3/8	8.5	2.48	3.96	1.46	1.26	1.28	1.92	0.866	0.879	0.782	0.644	0.551
× 1/4	5.8	1.69	2.77	1.00	1.28	1.24	1.36	0.599	0.896	0.736	0.651	0.558
L3$\frac{1}{2}$ × 3 × 1/2	10.2	3.00	3.45	1.45	1.07	1.13	2.33	1.10	0.881	0.875	0.621	0.714
× 3/8	7.9	2.30	2.72	1.13	1.09	1.08	1.85	0.851	0.897	0.830	0.625	0.721
× 1/4	5.4	1.56	1.91	0.776	1.11	1.04	1.30	0.589	0.914	0.785	0.631	0.727
L3$\frac{1}{2}$ × 2$\frac{1}{2}$ × 1/2	9.4	2.75	3.24	1.41	1.09	1.20	1.36	0.760	0.704	0.705	0.534	0.486
× 3/8	7.2	2.11	2.56	1.09	1.10	1.16	1.09	0.592	0.719	0.660	0.537	0.496
× 1/4	4.9	1.44	1.80	0.755	1.12	1.11	0.777	0.412	0.735	0.614	0.544	0.506
L3 × 2$\frac{1}{2}$ × 1/2	8.5	2.50	2.08	1.04	0.913	1.00	1.30	0.744	0.722	0.750	0.520	0.667
× 3/8	6.6	1.92	1.66	0.810	0.928	0.956	1.04	0.581	0.736	0.706	0.522	0.676
× 1/4	4.5	1.31	1.17	0.561	0.945	0.911	0.743	0.404	0.753	0.661	0.528	0.684
L3 × 2 × 1/2	7.7	2.25	1.92	1.00	0.924	1.08	0.672	0.474	0.546	0.583	0.428	0.414
× 3/8	5.9	1.73	1.53	0.781	0.940	1.04	0.543	0.371	0.559	0.539	0.430	0.428
× 1/4	4.1	1.19	1.09	0.542	0.957	0.993	0.392	0.260	0.574	0.493	0.435	0.440
L2$\frac{1}{2}$ × 2 × 3/8	5.3	1.55	0.912	0.547	0.768	0.813	0.514	0.363	0.577	0.581	0.420	0.614
× 1/4	3.62	1.06	0.654	0.381	0.784	0.787	0.372	0.254	0.592	0.537	0.424	0.626

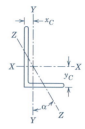

TABLE B-10   Unequal Leg Angles (SI Units)

Size and Thickness	Mass (kg/m)	Area (mm²)	AXIS X–X I (10^6 mm⁴)	S (10^3 mm³)	r (mm)	y_C (mm)	AXIS Y–Y I (10^6 mm⁴)	S (10^3 mm³)	r (mm)	x_C (mm)	AXIS Z–Z r (mm)	Tan α
L229 × 102 × 15.9	39.1	4985	27.0	188	73.7	85.3	3.46	43.4	26.4	21.8	21.5	0.216
× 12.7	31.7	4030	22.1	153	74.2	84.1	2.88	35.6	26.7	20.6	21.7	0.220
L203 × 152 × 25.4	65.8	8385	33.6	247	63.2	67.3	16.1	146	43.9	41.9	32.5	0.543
× 19.1	50.3	6415	26.4	192	64.3	65.0	12.8	113	44.7	39.6	32.8	0.551
× 12.7	34.2	4355	18.4	131	65.0	62.7	9.03	78.5	45.5	37.3	33.0	0.558
L203 × 102 × 25.4	55.7	7095	29.0	231	64.0	77.5	4.83	64.6	26.2	26.7	21.5	0.247
× 19.1	42.7	5445	22.9	179	64.8	74.9	3.90	50.3	26.7	24.2	21.6	0.258
× 12.7	29.2	3710	16.0	123	65.8	72.6	2.81	35.2	27.4	21.8	22.0	0.267
L178 × 102 × 19.1	39.0	4960	15.7	138	56.4	63.8	3.77	49.7	27.7	25.7	21.8	0.324
× 12.7	26.6	3385	11.1	95.2	57.2	61.5	2.72	34.7	28.2	23.3	22.1	0.335
× 9.5	20.2	2570	8.57	72.8	57.7	60.2	2.12	26.7	28.7	22.1	22.4	0.340
L152 × 102 × 19.1	35.1	4475	10.2	102	47.8	52.8	3.61	48.7	28.4	27.4	21.8	0.428
× 12.7	24.1	3065	7.24	71.0	48.5	50.5	2.61	34.1	29.2	25.1	22.1	0.440
× 9.5	18.3	3230	5.62	54.4	49.0	49.3	2.04	26.2	29.7	23.9	22.3	0.446
L152 × 89 × 12.7	22.8	2905	6.91	69.5	48.8	52.8	1.77	26.1	24.7	21.2	19.3	0.344
× 9.5	17.4	2205	5.37	53.1	49.3	51.8	1.39	20.2	25.1	20.0	19.5	0.350
L127 × 89 × 19.1	29.5	3750	5.79	70.1	39.4	44.5	2.31	36.4	24.8	25.3	19.0	0.464
× 12.7	20.2	2580	4.16	49.0	40.1	42.2	1.69	25.6	25.7	23.0	19.2	0.479
× 9.5	15.5	1970	3.24	37.5	40.6	40.9	1.32	19.8	25.9	21.9	19.4	0.486
× 6.4	10.4	1330	2.24	25.7	41.1	39.6	0.928	13.6	26.4	20.7	19.6	0.492
L127 × 76 × 12.7	19.0	2420	3.93	47.7	40.4	44.5	1.07	18.8	21.1	19.1	16.5	0.357
× 9.5	14.6	1845	3.07	36.7	40.9	43.2	0.849	14.6	21.5	17.9	16.6	0.364
× 6.4	9.82	1250	2.13	25.1	41.1	42.2	0.599	10.1	21.9	16.7	16.8	0.371
L102 × 89 × 12.7	17.7	2260	2.21	31.8	31.2	31.8	1.58	24.9	26.4	25.4	18.3	0.750
× 9.5	13.5	1725	1.74	24.4	31.8	30.7	1.23	19.2	26.9	24.3	18.5	0.755
× 6.4	9.22	1170	1.21	16.9	32.3	29.5	0.870	13.2	27.2	23.1	18.6	0.759
L102 × 76 × 12.7	16.5	2095	2.10	31.0	31.8	33.8	1.01	18.4	21.9	21.0	16.2	0.543
× 9.5	12.6	1600	1.65	23.9	32.0	32.5	0.799	14.2	22.3	19.9	16.4	0.551
× 6.4	8.63	1090	1.15	16.4	32.5	31.5	0.566	9.82	22.8	18.7	16.5	0.558
L89 × 76 × 12.7	15.2	1935	1.44	23.8	27.2	28.7	0.970	18.0	22.4	22.2	15.8	0.714
× 9.5	11.8	1485	1.13	18.5	27.7	27.4	0.770	13.9	22.8	21.1	15.9	0.721
× 6.4	8.04	1005	0.795	12.7	28.2	26.4	0.541	9.65	23.2	19.9	16.0	0.727
L89 × 64 × 12.7	14.0	1775	1.35	23.1	27.7	30.5	0.566	12.5	17.9	17.9	13.6	0.486
× 9.5	10.7	1360	1.07	17.9	27.9	29.5	0.454	8.70	18.3	16.8	13.6	0.496
× 6.4	7.29	929	0.749	12.4	28.4	28.2	0.323	6.75	18.7	15.6	13.8	0.506
L76 × 64 × 12.7	12.6	1615	0.866	17.0	23.2	25.4	0.541	12.2	18.3	19.1	13.2	0.667
× 9.5	9.82	1240	0.691	13.3	23.6	24.3	0.433	9.52	18.7	17.9	13.3	0.676
× 6.4	6.70	845	0.487	9.19	24.0	23.1	0.309	6.62	19.1	16.8	13.4	0.684
L76 × 51 × 12.7	11.5	1450	0.799	16.4	23.5	27.4	0.280	7.77	13.9	14.8	10.9	0.414
× 9.5	8.78	1115	0.637	12.8	23.9	26.4	0.226	6.08	14.2	13.7	10.9	0.428
× 6.4	6.10	768	0.454	8.88	24.3	25.2	0.163	4.26	14.6	12.5	11.0	0.440
L64 × 51 × 9.5	7.89	1000	0.380	8.96	19.5	20.7	0.214	5.95	14.7	14.8	10.7	0.614
× 6.4	5.39	684	0.272	6.24	19.9	20.0	0.155	4.16	15.0	13.6	10.8	0.626

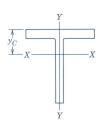

TABLE B-11   Structural Tees (U.S. Customary Units)

Desig-nation*	Area (in.²)	Depth of Tee (in.)	FLANGE Width (in.)	Thick-ness (in.)	Stem Thick-ness (in.)	AXIS X–X I (in.⁴)	S (in.³)	r (in.)	y_C (in.)	AXIS Y–Y I (in.⁴)	S (in.³)	r (in.)
WT18 × 115	33.8	17.950	16.470	1.260	0.760	934	67.0	5.25	4.01	470	57.1	3.73
× 80	23.5	18.005	12.000	1.020	0.650	740	55.8	5.61	4.74	147	24.6	2.50
WT15 × 66	19.4	15.155	10.545	1.000	0.615	421	37.4	4.66	3.90	98.0	18.6	2.25
× 54	15.9	14.915	10.475	0.760	0.545	349	32.0	4.69	4.01	73.0	13.9	2.15
WT12 × 52	15.3	12.030	12.750	0.750	0.500	189	20.0	3.51	2.59	130	20.3	2.91
× 47	13.8	12.155	9.065	0.875	0.515	186	20.3	3.67	2.99	54.5	12.0	1.98
× 42	12.4	12.050	9.020	0.770	0.470	166	18.3	3.67	2.97	47.2	10.5	1.95
× 31	9.11	11.870	7.040	0.590	0.430	131	15.6	3.79	3.46	17.2	4.90	1.38
WT9 × 38	11.2	9.105	11.035	0.680	0.425	71.8	9.83	2.54	1.80	76.2	13.8	2.61
× 30	8.82	9.120	7.555	0.695	0.415	64.7	9.29	2.71	2.16	25.0	6.63	1.69
× 25	7.33	8.995	7.495	0.570	0.355	53.5	7.79	2.70	2.12	20.0	5.35	1.65
× 20	5.88	8.950	6.015	0.525	0.315	44.8	6.73	2.76	2.29	9.55	3.17	1.27
WT8 × 50	14.7	8.485	10.425	0.985	0.585	76.8	11.4	2.28	1.76	93.1	17.9	2.51
× 25	7.37	8.130	7.070	0.630	0.380	42.3	6.78	2.40	1.89	18.6	5.26	1.59
× 20	5.89	8.005	6.995	0.505	0.305	33.1	5.35	2.37	1.81	14.4	4.12	1.57
× 13	3.84	7.845	5.500	0.345	0.250	23.5	4.09	2.47	2.09	4.80	1.74	1.12
WT7 × 60	17.7	7.240	14.670	0.940	0.590	51.7	8.61	1.71	1.24	247	33.7	3.74
× 41	12.0	7.155	10.130	0.855	0.510	41.2	7.14	1.85	1.39	74.2	14.6	2.48
× 34	9.99	7.020	10.035	0.720	0.415	32.6	5.69	1.81	1.29	60.7	12.1	2.46
× 24	7.07	6.985	8.030	0.595	0.340	24.9	4.48	1.87	1.35	25.7	6.40	1.91
× 15	4.42	6.920	6.730	0.385	0.270	19.0	3.55	2.07	1.58	9.79	2.91	1.49
× 11	3.25	6.870	5.000	0.335	0.230	14.8	2.91	2.14	1.76	3.50	1.40	1.04
WT6 × 60	17.6	6.560	12.320	1.105	0.710	43.4	8.22	1.57	1.28	172	28.0	3.13
× 48	14.1	6.355	12.160	0.900	0.550	32.0	6.12	1.51	1.13	135	22.2	3.09
× 36	10.6	6.125	12.040	0.670	0.430	23.2	4.54	1.48	1.02	97.5	16.2	3.04
× 25	7.34	6.095	8.080	0.640	0.370	18.7	3.79	1.60	1.17	28.2	6.97	1.96
× 15	4.40	6.170	6.520	0.440	0.260	13.5	2.75	1.75	1.27	10.2	3.12	1.52
× 8	2.36	5.995	3.990	0.265	0.220	8.70	2.04	1.92	1.74	1.41	0.706	0.773
WT5 × 56	16.5	5.680	10.415	1.250	0.755	28.6	6.40	1.32	1.21	118	22.6	2.68
× 44	12.9	5.420	10.265	0.990	0.605	20.8	4.77	1.27	1.06	89.3	17.4	2.63
× 30	8.82	5.110	10.080	0.680	0.420	12.9	3.04	1.21	0.884	58.1	11.5	2.57
× 15	4.42	5.235	5.810	0.510	0.300	9.28	2.24	1.45	1.10	8.35	2.87	1.37
× 6	1.77	4.935	3.960	0.210	0.190	4.35	1.22	1.57	1.36	1.09	0.551	0.785
WT4 × 29	8.55	4.375	8.220	0.810	0.510	9.12	2.61	1.03	0.874	37.5	9.13	2.10
× 20	5.87	4.125	8.070	0.560	0.360	5.73	1.69	0.988	0.735	24.5	6.08	2.04
× 12	3.54	3.965	6.495	0.400	0.245	3.53	1.08	0.999	0.695	9.14	2.81	1.61
× 9	2.63	4.070	5.250	0.330	0.230	3.41	1.05	1.14	0.834	3.98	1.52	1.23
× 5	1.48	3.945	3.940	0.205	0.170	2.15	0.717	1.20	0.953	1.05	0.532	0.841
WT3 × 10	2.94	3.100	6.020	0.365	0.260	1.76	0.693	0.774	0.560	6.64	2.21	1.50
× 6	1.78	3.015	4.000	0.280	0.230	1.32	0.564	0.861	0.677	1.50	0.748	0.918
WT2 × 6.5	1.91	2.080	4.060	0.345	0.280	0.526	0.321	0.524	0.440	1.93	0.950	1.00

Courtesy of The American Institute of Steel Construction.
*WT means structural T-section (cut from a W-section), followed by the nominal depth in inches, then the weight in pounds per foot of length.

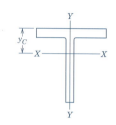

TABLE B-12 — Structural Tees (SI Units)

Desig-nation*	Area (mm²)	Depth of Tee (mm)	FLANGE Width (mm)	FLANGE Thick-ness (mm)	Stem Thick-ness (mm)	AXIS X–X I (10^6 mm⁴)	AXIS X–X S (10^3 mm³)	r (mm)	y_C (mm)	AXIS Y–Y I (10^6 mm⁴)	AXIS Y–Y S (10^3 mm³)	r (mm)
WT457 × 171	21805	455.9	418.3	32.0	19.3	389	1098	133	102	196	936	94.7
× 119	15160	457.3	304.8	25.9	16.5	308	914	142	120	61.2	403	63.5
WT381 × 98	12515	384.9	267.8	25.4	15.6	175	613	118	99.1	40.8	305	57.2
× 80	10260	378.8	266.1	19.3	13.8	145	524	119	102	30.4	228	54.6
WT305 × 77	9870	305.6	323.9	19.1	12.7	78.7	328	89.2	65.8	54.1	333	73.9
× 70	8905	308.7	230.3	22.2	13.1	77.4	333	93.2	75.9	22.7	197	50.3
× 63	8000	306.1	229.1	19.6	11.9	69.1	300	93.2	75.4	19.6	172	49.5
× 46	5875	301.5	178.8	15.0	10.9	54.5	256	96.3	87.9	7.16	80.3	35.1
WT229 × 57	7225	231.3	280.3	17.3	10.8	29.9	161	64.5	45.7	31.7	226	66.3
× 45	5690	231.6	191.9	17.7	10.5	26.9	152	68.8	54.9	10.4	109	42.9
× 37	4730	228.5	190.4	14.5	9.0	22.3	128	68.6	53.8	8.32	87.7	41.9
× 30	3795	227.3	152.8	13.3	8.0	18.6	110	70.1	58.2	3.98	51.9	32.3
WT203 × 74	9485	215.5	264.8	25.0	14.9	32.0	187	57.9	44.7	38.8	293	63.8
× 37	4755	206.5	179.6	16.0	9.7	17.6	111	61.0	48.0	7.74	86.2	40.4
× 30	3800	203.3	177.7	12.8	7.7	13.8	87.7	60.2	46.0	5.99	67.5	39.9
× 19	2475	199.3	139.7	8.8	6.4	9.78	67.0	62.7	53.1	2.00	28.5	28.4
WT178 × 89	11420	183.9	372.6	23.9	15.0	21.5	141	43.4	31.5	103	552	95.0
× 61	7740	181.7	257.3	21.7	13.0	17.1	117	47.0	35.3	30.9	239	63.0
× 51	6445	178.3	254.9	18.3	10.5	13.6	93.2	46.0	32.8	25.3	198	62.5
× 36	4560	177.4	204.0	15.1	8.6	10.4	73.4	47.5	34.3	10.7	105	48.5
× 22	2850	175.8	170.9	9.8	6.9	7.91	58.2	52.6	40.1	4.07	47.7	37.8
× 16	2095	174.5	127.0	8.5	5.8	6.16	47.7	54.4	44.7	1.46	22.9	26.4
WT152 × 89	11355	166.6	312.9	28.1	18.0	18.1	135	39.9	32.5	71.6	459	79.5
× 71	9095	161.4	308.9	22.9	14.0	13.3	100	38.4	28.7	56.2	364	78.5
× 54	6840	155.6	305.8	17.0	10.9	9.66	74.4	37.6	25.9	40.6	265	77.2
× 37	4735	154.8	205.2	16.2	9.4	7.78	62.1	40.6	29.7	11.7	114	49.8
× 22	2840	156.7	165.6	11.2	6.6	5.62	45.1	44.5	32.3	4.25	51.1	38.6
× 12	1525	152.3	101.3	6.7	5.6	3.62	33.4	48.8	44.2	0.587	11.6	19.6
WT127 × 83	10645	144.3	264.5	31.8	19.2	11.9	105	33.5	30.7	49.1	370	68.1
× 65	8325	137.7	260.7	25.1	15.4	8.66	78.2	32.3	26.9	37.2	285	66.8
× 45	5690	129.8	256.0	17.3	10.7	5.37	49.8	30.7	22.5	24.2	188	65.3
× 22	2850	133.0	147.6	13.0	7.6	3.86	36.7	36.8	27.9	3.48	47.0	34.8
× 9	1140	125.3	100.6	5.3	4.8	1.81	20.0	39.9	34.5	0.454	9.03	19.9
WT102 × 43	5515	111.1	208.8	20.6	13.0	3.80	42.8	26.2	22.2	15.6	150	53.3
× 30	3785	104.8	205.0	14.2	9.1	2.39	27.7	25.1	18.7	10.2	99.6	51.8
× 18	2285	100.7	165.0	10.2	6.2	1.47	17.7	25.4	17.7	3.80	46.0	40.9
× 13	1695	103.4	133.4	8.4	5.8	1.42	17.2	29.0	21.2	1.66	24.9	31.2
× 7	955	100.2	100.1	5.2	4.3	0.895	11.7	30.5	24.2	0.437	8.72	21.4
WT76 × 15	1895	78.7	152.9	9.3	6.6	0.733	11.4	19.7	14.2	2.76	36.2	38.1
× 9	1150	76.6	101.6	7.1	5.8	0.549	9.24	21.9	17.2	0.624	12.3	23.3
WT51 × 10	1230	52.8	103.1	8.8	7.1	0.219	5.26	13.3	11.2	0.803	15.6	25.4

*WT means structural T-section (cut from a W-section), followed by the nominal depth in mm, then the mass in kg per meter of length.

TABLE B-13 Properties of Standard Steel Pipe (U.S. Customary Units)

Nominal Diam. d (in.)	Outside Diam. d_o (in.)	Inside Diam. d_i (in.)	Wall Thickness t (in.)	Weight w (lb/ft)	A (in.^{2})	I (in.^{4})	S (in.^{3})	r (in.)
Standard Weight								
$\frac{1}{2}$	0.840	0.622	0.109	0.85	0.250	0.017	0.041	0.26
$\frac{3}{4}$	1.050	0.824	0.113	0.13	0.333	0.037	0.071	0.33
1	1.315	1.049	0.133	1.68	0.494	0.087	0.133	0.42
$1\frac{1}{4}$	1.660	1.380	0.140	2.27	0.669	0.195	0.235	0.54
$1\frac{1}{2}$	1.900	1.610	0.145	2.72	0.799	0.310	0.326	0.62
2	2.375	2.067	0.154	3.65	1.075	0.666	0.561	0.79
$2\frac{1}{2}$	2.875	2.469	0.203	5.79	1.704	1.530	1.064	0.95
3	3.500	3.068	0.216	7.58	2.228	3.017	1.724	1.16
$3\frac{1}{2}$	4.000	3.548	0.226	9.11	2.680	4.787	2.39	1.34
4	4.500	4.026	0.237	10.79	3.174	7.233	3.21	1.51
5	5.563	5.047	0.258	14.62	4.300	15.16	5.45	1.88
6	6.625	6.065	0.280	18.97	5.581	28.14	8.50	2.25
8	8.625	7.981	0.322	28.55	8.399	72.49	16.81	2.94
10	10.750	10.020	0.365	40.48	11.91	160.7	29.9	3.67
12	12.750	12.000	0.375	49.56	14.58	279.3	43.8	4.38
Extra Strong								
$1\frac{1}{2}$	1.900	1.500	0.200	3.63	1.068	0.391	0.412	0.61
2	2.375	1.939	0.218	5.02	1.477	0.868	0.731	0.77
$2\frac{1}{2}$	2.875	2.323	0.276	7.66	2.254	1.924	1.338	0.92
3	3.500	2.900	0.300	10.25	3.016	3.894	2.23	1.14
4	4.500	3.826	0.337	14.98	4.407	9.610	4.27	1.48
6	6.625	5.761	0.432	28.57	8.405	40.49	12.22	2.20
Double Extra Strong								
$1\frac{1}{2}$	1.900	1.100	0.400	6.41	1.885	0.568	0.564	0.55
2	2.375	1.503	0.436	9.03	2.656	1.311	1.104	0.70
$2\frac{1}{2}$	2.875	1.771	0.552	13.69	4.028	2.871	1.997	0.84
3	3.500	2.300	0.600	18.58	5.466	5.993	3.42	1.05
4	4.500	3.152	0.674	27.54	8.101	15.28	6.79	1.37
6	6.625	4.897	0.864	53.16	15.64	66.33	20.0	2.06

TABLE B-14   Properties of Standard Steel Pipe (SI Units)

Nominal Diam. d (mm)	Outside Diam. d_o (mm)	Inside Diam. d_i (mm)	Wall Thickness t (mm)	Mass m (kg/m)	A (mm²)	I (10^6) (mm⁴)	S (10^3) (mm³)	r (mm)
	DIMENSIONS				**PROPERTIES**			
Standard Weight								
13	21.3	15.8	2.77	1.264	161.3	0.007	0.672	6.6
19	26.7	20.9	2.87	1.681	214.8	0.015	1.163	8.5
25	33.4	26.6	3.38	2.499	318.7	0.036	2.179	10.7
32	42.2	35.1	3.56	3.376	431.6	0.081	3.851	13.7
38	48.1	40.9	3.68	4.045	515.5	0.129	5.342	15.8
51	60.3	52.5	3.91	5.428	693.5	0.277	9.193	20.0
64	73.0	62.7	5.16	8.611	1099	0.637	17.44	24.1
76	88.9	77.9	5.49	11.27	1437	1.256	28.25	29.5
89	101.6	90.1	5.74	13.55	1729	1.992	39.17	34.0
102	114.3	102.3	6.02	16.05	2048	3.011	52.60	38.4
127	141.3	128.2	6.55	21.74	2774	6.310	89.31	47.8
152	168.3	154.1	7.11	28.21	3600	11.71	139.3	57.2
203	219.1	202.7	8.18	42.46	5419	30.2	275.5	74.7
254	273.1	254.5	9.27	60.20	7684	66.9	490	93.2
305	323.9	304.8	9.53	73.71	9406	116.3	718	111.3
Extra Strong								
38	48.3	38.1	5.08	5.399	689	0.163	6.75	15.4
51	60.3	49.3	5.54	7.466	953	0.361	11.98	19.5
64	70.0	59.0	7.01	11.39	1454	0.801	21.93	23.5
76	88.9	73.7	7.62	15.24	1946	1.621	36.54	29.0
102	114.3	97.2	8.56	22.28	2843	4.000	69.67	37.6
152	168.3	146.3	10.97	42.49	5423	16.85	200	55.9
Double Extra Strong								
38	48.3	27.9	10.16	9.53	1216	0.236	0.564	13.9
51	60.3	38.2	11.07	13.43	1714	0.546	1.104	17.9
64	70.0	45.0	14.02	20.36	2600	1.195	1.997	21.4
76	88.9	58.4	15.24	27.63	3526	2.494	3.42	26.7
102	114.3	80.1	17.12	40.96	5226	6.360	6.79	34.8
152	168.3	124.4	21.95	79.06	10090	27.61	20.0	52.3

TABLE B-15   Properties of Standard Structural Timber (U.S. Customary Units)

DIMENSIONS*		PROPERTIES			
Nominal Size $b \times h$ (in.)	Dressed Size (in.)	Weight w (lb/ft)	Area A (in.²)	Moment of Inertia I (in.⁴)	Section Modulus S (in.³)
2×4	$1\frac{5}{8} \times 3\frac{5}{8}$	1.64	5.89	6.45	3.56
6	$5\frac{5}{8}$	2.54	9.14	24.1	8.57
8	$7\frac{1}{2}$	3.39	12.2	57.1	15.3
10	$9\frac{1}{2}$	4.29	15.4	116	24.4
12	$11\frac{1}{2}$	5.19	18.7	206	35.8
4×4	$3\frac{5}{8} \times 3\frac{5}{8}$	3.65	13.1	14.4	7.94
6	$5\frac{5}{8}$	5.66	20.4	53.8	19.1
8	$7\frac{1}{2}$	7.55	27.2	127	34.0
10	$9\frac{1}{2}$	9.57	34.4	259	54.5
12	$11\frac{1}{2}$	11.6	41.7	459	79.9
6×6	$5\frac{1}{2} \times 5\frac{1}{2}$	8.40	30.3	76.3	27.7
8	$7\frac{1}{2}$	11.4	41.3	193	51.6
10	$9\frac{1}{2}$	14.5	52.3	393	82.7
12	$11\frac{1}{2}$	17.5	63.3	697	121
14	$13\frac{1}{2}$	20.6	74.3	1128	167
8×8	$7\frac{1}{2} \times 7\frac{1}{2}$	15.6	56.3	264	70.3
10	$9\frac{1}{2}$	19.8	71.3	536	113
12	$11\frac{1}{2}$	23.9	86.3	951	165
14	$13\frac{1}{2}$	28.0	101	1538	228
16	$15\frac{1}{2}$	32.0	116	2327	300
10×10	$9\frac{1}{2} \times 9\frac{1}{2}$	25.0	90.3	679	143
12	$11\frac{1}{2}$	30.3	109	1204	209
14	$13\frac{1}{2}$	35.6	128	1948	289
16	$15\frac{1}{2}$	40.9	147	2948	380
18	$17\frac{1}{2}$	46.1	166	4243	485
12×12	$11\frac{1}{2} \times 11\frac{1}{2}$	36.7	132	1458	253
14	$13\frac{1}{2}$	43.1	155	2358	349
16	$15\frac{1}{2}$	49.5	178	3569	460
18	$17\frac{1}{2}$	55.9	201	5136	587
20	$19\frac{1}{2}$	62.3	224	7106	729

*Properties and weights are for dressed sizes.

TABLE B-16   Properties of Standard Structural Timber (SI Units)

DIMENSIONS*		PROPERTIES			
Nominal Size $b \times h$ (mm)	Dressed Size (mm)	Mass m (kg/m)	Area A (10^3 mm^2)	Moment of Inertia I (10^6 mm^4)	Section Modulus S (10^3 mm^3)
51 × 102	41 × 92	2.44	3.77	2.68	58.3
152	140	3.78	5.86	10.03	140
203	191	5.04	7.83	23.8	251
254	241	6.38	9.88	48.3	400
305	292	7.72	12.0	85.7	587
102 × 102	92 × 92	5.43	8.46	5.99	130
152	140	8.42	13.2	22.4	313
203	191	11.2	17.6	52.9	557
254	241	14.2	22.2	107.8	893
305	292	17.3	26.9	191.1	1310
152 × 152	140 × 140	12.5	19.6	31.8	454
203	191	17.0	26.7	80.3	846
254	241	21.6	33.7	163.6	1350
305	292	26.0	40.9	290	1980
356	343	30.6	48.0	470	2740
203 × 203	191 × 191	23.2	36.5	110	1150
254	241	29.4	46.0	223	1850
305	292	35.5	55.8	396	2700
356	343	41.6	65.5	640	3740
406	394	47.6	75.3	969	4920
254 × 254	241 × 241	37.2	58.1	283	2340
305	292	45.1	70.4	501	3420
356	343	52.9	82.7	810	4740
406	394	60.8	95.0	1227	6230
457	445	68.6	107	1766	7950
305 × 305	292 × 292	54.6	85.3	607	4150
356	343	64.1	100	981	5720
406	394	73.6	115	1486	7540
457	445	83.1	130	2138	5620
508	495	92.7	145	2958	11950

*Properties and masses are for dressed sizes.

TABLE B-17 Properties of Selected Engineering Materials (U.S. Customary Units)

Exact values may vary widely with changes in composition, heat treatment, and mechanical working. More precise information can be obtained from manufacturers.

Materials	Specific Weight (lb/in.^{3})	ELASTIC STRENGTH[a]			ULTIMATE STRENGTH			Endurance Limit[c] (ksi)	Modulus of Elasticity (1000 ksi)	Modulus of Rigidity (1000 ksi)	Percent Elongation in 2 in.	Coefficient of Thermal Expansion (10^{-6}/°F)
		Tension (ksi)	Comp. (ksi)	Shear (ksi)	Tension (ksi)	Comp. (ksi)	Shear (ksi)					
Ferrous metals												
Wrought iron	0.278	30	b		48	b	25	23	28		30[d]	6.7
Structural steel	0.284	36	b		66	b		28	29	11.0	28[d]	6.6
Steel, 0.2% C hardened	0.284	62	b		90	b			30	11.6	22	6.6
Steel, 0.4% C hot-rolled	0.284	53	b		84	b		38	30	11.6	29	
Steel, 0.8% C hot-rolled	0.284	76	b		122	b			30	11.6	8	
Cast iron—gray	0.260				25	100		12	15		0.5	6.7
Cast iron—malleable	0.266	32	b		50	b			25		20	6.6
Cast iron—nodular	0.266	70			100				25		4	6.6
Stainless steel (18-8) annealed	0.286	36	b		85	b		40	28	12.5	55	9.6
Stainless steel (18-8) cold-rolled	0.286	165	b		190	b		90	28	12.5	8	9.6
Steel, SAE 4340, heat-treated	0.283	132	145		150	b	95	76	29	11.0	19	
Nonferrous metal alloys												
Aluminum, cast, 195-T6	0.100	24	25		36		30	7	10.3	3.8	5	
Aluminum, wrought, 2014-T4	0.101	41	41	24	62	b	38	18	10.6	4.0	20	12.5
Aluminum, wrought, 2024-T4	0.100	48	48	28	68	b	41	18	10.6	4.0	19	12.5
Aluminum, wrought, 6061-T6	0.098	40	40	26	45	b	30	13.5	10.0	3.8	17	12.5
Magnesium, extrusion, AZ80X	0.066	35	26		49	b	21	19	6.5	2.4	12	14.4
Magnesium, sand cast, AZ63-HT	0.066	14	14		40	b	19	14	6.5	2.4	12	14.4
Monel, wrought, hot-rolled	0.319	50	b		90	b		40	26	9.5	35	7.8
Red brass, cold-rolled	0.316	60			75				15	5.6	4	9.8
Red brass, annealed	0.316	15	b		40	b			15	5.6	50	9.8
Bronze, cold-rolled	0.320	75			100				15	6.5	3	9.4
Bronze, annealed	0.320	20	b		50	b			15	6.5	50	9.4
Titanium alloy, annealed	0.167	135	b		155	b			14	5.3	13	
Invar, annealed	0.292	42	b		70	b			21	8.1	41	0.6
Nonmetallic materials												
Douglas fir, green[e]	0.022	4.8	3.4			3.9	0.9		1.6			
Douglas fir, air dry[e]	0.020	8.1	6.4			7.4	1.1		1.9			
Red oak, green[e]	0.037	4.4	2.6			3.5	1.2		1.4			1.9
Red oak, air dry[e]	0.025	8.4	4.6			6.9	1.8		1.8			
Concrete, medium strength	0.087		1.2			3.0			3.0			6.0
Concrete, fairly high strength	0.087		2.0			5.0			4.5			6.0

[a]Elastic strength may be represented by proportional limit, yield point, or yield strength at a specified offset (usually 0.2 percent for ductile metals).

[b]For ductile metals (those with an appreciable ultimate elongation), it is customary to assume the properties in compression have the same values as those in tension.

[c]Rotating beam.

[d]Elongation in 8 in.

[e]All timber properties are parallel to the grain.

TABLE B-18 Properties of Selected Engineering Materials (SI Units)

Exact values may vary widely with changes in composition, heat treatment, and mechanical working. More precise information can be obtained from manufacturers.

Materials	Density (Mg/m³)	ELASTIC STRENGTH[a] Tension (MPa)	Comp. (MPa)	Shear (MPa)	ULTIMATE STRENGTH Tension (MPa)	Comp. (MPa)	Shear (MPa)	Endurance Limit[c] (MPa)	Modulus of Elasticity (GPa)	Modulus of Rigidity (GPa)	Percent Elongation in 50 mm	Coefficient of Thermal Expansion (10^{-6}/°F)
Ferrous metals												
Wrought iron	7.70	210	[b]		330	[b]	170	160	190		30[d]	12.1
Structural steel	7.87	250	[b]		450	[b]		190	200	76	28[d]	11.9
Steel, 0.2% C hardened	7.87	430	[b]		620	[b]			210	80	22	11.9
Steel, 0.4% C hot-rolled	7.87	360	[b]		580	[b]	260		210	80	29	
Steel, 0.8% C hot-rolled	7.87	520	[b]		840	[b]			210	80	8	
Cast iron—gray	7.20				170	690		80	100		0.5	12.1
Cast iron—malleable	7.37	220	[b]		340	[b]			170		20	11.9
Cast iron—nodular	7.37	480			690				170		4	11.9
Stainless steel (18-8) annealed	7.92	250	[b]		590	[b]		270	190	86	55	17.3
Stainless steel (18-8) cold-rolled	7.92	1140	[b]		1310	[b]		620	190	86	8	17.3
Steel, SAE 4340, heat-treated	7.84	910	1000		1030	[b]	650	520	200	76	19	
Nonferrous metal alloys												
Aluminum, cast, 195-T6	2.77	160	170		250		210	50	71	26	5	
Aluminum, wrought, 2014-T4	2.80	280	280	160	430	[b]	260	120	73	28	20	22.5
Aluminum, wrought, 2024-T4	2.77	330	330	190	470	[b]	280	120	73	28	19	22.5
Aluminum, wrought, 6061-T6	2.71	270	270	180	310	[b]	210	93	70	26	17	22.5
Magnesium, extrusion, AZ80X	1.83	240	180		340	[b]	140	130	45	16	12	25.9
Magnesium, sand cast, AZ63-HT	1.83	100	96		270	[b]	130	100	45	16	12	25.9
Monel, wrought, hot-rolled	8.84	340	[b]		620	[b]		270	180	65	35	14.0
Red brass, cold-rolled	8.75	410			520				100	39	4	17.6
Red brass, annealed	8.75	100	[b]		270	[b]			100	39	50	17.6
Bronze, cold-rolled	8.86	520			690				100	45	3	16.9
Bronze, annealed	8.86	140	[b]		340	[b]			100	45	50	16.9
Titanium alloy, annealed	4.63	930	[b]		1070	[b]			96	36	13	
Invar, annealed	8.09	290	[b]		480	[b]			140	56	41	1.1
Nonmetallic materials												
Douglas fir, green[e]	0.61	33	23			27	6.2		11			
Douglas fir, air dry[e]	0.55	56	44			51	7.6		13			
Red oak, green[e]	1.02	30	18			24	8.3		10			3.4
Red oak, air dry[e]	0.69	58	32			48	12.4		12			
Concrete, medium strength	2.41		8			21			21			10.8
Concrete, fairly high strength	2.41		14			34			31			10.8

[a]Elastic strength may be represented by proportional limit, yield point, or yield strength at a specified offset (usually 0.2 percent for ductile metals).
[b]For ductile metals (those with an appreciable ultimate elongation), it is customary to assume the properties in compression have the same values as those in tension.
[c]Rotating beam.
[d]Elongation in 200 mm.
[e]All timber properties are parallel to the grain.

TABLE B-19   Beam Deflections and Slopes

Case	Load and Support (Length L)	Slope at End $(+\triangle)$	Maximum Deflection (+ upward)
1		$\theta = -\dfrac{PL^2}{2EI}$ at $x = L$	$y_{max} = -\dfrac{PL^3}{3EI}$ at $x = L$
2		$\theta = -\dfrac{wL^3}{6EI}$ at $x = L$	$y_{max} = -\dfrac{wL^4}{8EI}$ at $x = L$
3		$\theta = -\dfrac{wL^3}{24EI}$ at $x = L$	$y_{max} = -\dfrac{wL^4}{30EI}$ at $x = L$
4		$\theta = +\dfrac{ML}{EI}$ at $x = L$	$y_{max} = +\dfrac{ML^2}{2EI}$ at $x = L$
5	$b > a$	$\theta_1 = -\dfrac{Pb(L^2 - b^2)}{6LEI}$ at $x = 0$ $\theta_2 = +\dfrac{Pa(L^2 - a^2)}{6LEI}$ at $x = L$	$y_{max} = -\dfrac{Pb(L^2-b^2)^{3/2}}{9\sqrt{3}\,LEI}$ at $x = \sqrt{(L^2-b^2)/3}$ $y_{\substack{center\\ not\ max}} = -\dfrac{Pb(3L^2 - 4b^2)}{48EI}$
6		$\theta_1 = -\dfrac{PL^2}{16EI}$ at $x = 0$ $\theta_2 = +\dfrac{PL^2}{16EI}$ at $x = L$	$y_{max} = -\dfrac{PL^2}{48EI}$ at $x = L/2$
7		$\theta_1 = -\dfrac{wL^3}{24EI}$ at $x = 0$ $\theta_2 = +\dfrac{wL^3}{24EI}$ at $x = L$	$y_{max} = -\dfrac{5wL^4}{384EI}$ at $x = L/2$
8		$\theta_1 = -\dfrac{ML}{6EI}$ at $x = 0$ $\theta_2 = +\dfrac{ML}{3EI}$ at $x = L$	$y_{max} = -\dfrac{ML^2}{9\sqrt{3}EI}$ at $x = L/\sqrt{3}$ $y_{\substack{center\\ not\ max}} = -\dfrac{ML^2}{16EI}$

TABLE B-20   Astronomical Data

Universal Gravitational Constant
$G = 6.673(10^{-11}) \ m^3/(kg \cdot s^2) = 3.439(10^{-8}) \ ft^4/(lb \cdot s^4)$

The Sun

Mass	$1.990(10^{30})$ kg	$1.364(10^{29})$ lb $\cdot$ s^2/ft
Mean radius	696,000 km	432,000 mi

The Earth

Mass	$5.976(10^{24})$ kg	$4.095(10^{23})$ lb $\cdot$ s^2/ft
Mean radius	6370 km	3960 mi
Rotation rate	23.93 hr	

The Moon

Mass	$7.350(10^{22})$ kg	$5.037(10^{21})$ lb $\cdot$ s^2/ft
Mean radius	1740 km	1080 mi
Mean distance to the Earth (center to center)	384,000 km	239,000 mi
Eccentricity (e)	0.055	

The Solar System

Planet	Mean Distance to Sun A.U.[a]	e	Mean Diameter (relative to Earth)	Mass (relative to Earth)
Mercury	0.387	0.206	0.380	0.05
Venus	0.723	0.007	0.975	0.81
Earth	1.000	0.017	1.000	1.00
Mars	1.524	0.093	0.532	0.11
Jupiter	5.203	0.048	11.27	317.8
Saturn	9.539	0.056	9.49	95.2

[a]Astronomical Unit (A.U.) is equal to the mean distance from the Earth to the sun = $149.6(10^6)$ km = $92.96(10^6)$ mi.

ANSWERS TO PROBLEMS

$\theta_z = 118.0°$

3-15 $F_4 = 39.5$ kip

$\theta_x = 101.9°$

$\theta_y = 81.1°$

$\theta_z = 165.1°$

3-19 $F_1 = 21.7$ kip

$F_2 = 17.24$ kip

3-20 $\theta = 1.894°$

$F_4 = 1377$ N

3-21 $F_B = 340$ lb

$F_C = 879$ lb

$F_A = 539$ lb

3-22 $F_A = 467$ N

$F_B = 902$ N

$\mathbf{F}_C = 234$ N $\measuredangle$ 2.90°

3-25 $T_A = 400$ lb

$T_B = 0$

$T_C = 400$ lb

3-26 $F_1 = 2.34$ kN

$F_2 = 43.1$ kN

$F_3 = 44.2$ kN

CHAPTER 4

4-1 (a) $\sigma_{AB} = 23.1$ ksi C

(b) $\sigma_{BC} = 30.0$ ksi C

(c) $\sigma_{CD} = 20.0$ ksi C

4-2 (a) $A_{AB} = 4770$ mm²

(b) $A_{BC} = 2600$ mm²

(c) $A_{CD} = 1167$ mm²

4-5 $L_1 = 2.55$ in.

$L_2 = 3.40$ in.

4-6 $L = 175.4$ mm

4-9 (a) $d_o = 8.16$ in.

(b) $t = 0.750$ in.

4-10 $P_{max} = 589$ kN

4-13 $P_{AB} = 4.88$ kip

4-14 (a) $P_{mid} = 113.8$ kN

(b) $P_b = 156.0$ kN

(c) $P_s = 54.3$ kN

(d) $P_s = 216$ kN

4-17 $d_{min} = 1.596$ in.

4-18 $\sigma_n = 5.37$ MPa C

$\tau_n = 8.27$ MPa

4-21 $P_{max} = 9.38$ kip

4-22 $P_{max} = 291$ kN

4-25 $P_{max} = 5.44$ kip

4-26 (a) $\phi = 59.0°$

(b) $P_{max} = 34.0$ kN

4-29 $\epsilon = 2080$ μ

4-30 $\Delta L = 0.480$ mm

4-33 $\gamma = 1396$ μ rad

4-34 $\gamma = 0.0279$ rad

4-37 (a) $\delta = \gamma L^2/6E$

(b) $\epsilon_{avg} = \gamma L/6E$

(c) $\epsilon_{max} = \gamma L/3E$

4-38 (a) $\delta = 5.00$ mm

(b) $\epsilon_{avg} = 1667$ μm/m

(c) $\epsilon_{max} = 5000$ μm/m

4-41 (a) $E = 16,300$ ksi

(b) $v = 0.333$

(c) $\sigma_{PL} = 24.4$ ksi

4-42 (a) $E = 78.7$ GPa

(b) $v = 0.326$

(c) $\sigma_{PL} = 354$ MPa

4-47 $P_{max} = 39.3$ kip

4-48 $P_{max} = 187.5$ kN

4-51 $\delta = 0.1214$ in.

4-52 (a) $\delta_{a-a} = 3.31$ mm

(b) $\delta_{b-b} = 5.63$ mm

4-53 (a) $\delta_{a-a} = 0.01061$ in.

(b) $\delta_{b-b} = 0.0615$ in.

4-57 $\delta = 0.0617$ in.

4-61 (a) $\sigma_c = 1222$ psi C

$\sigma_s = 12,220$ psi C

(b) $\delta = 0.00978$ in.

4-62 (a) $P = 245$ kN

(b) $\delta = 0.375$ mm

4-65 (a) $\sigma_A = 18.93$ ksi

$\sigma_B = 19.02$ ksi

(b) $\delta_C = 0.0454$ in.

4-66 (a) $\sigma_P = 1.074$ MPa T

$\sigma_b = 11.93$ MPa T

(b) $a = 200.1023$ mm

4-71 $\Delta L = 48.7$ in.

4-72 $\Delta L = -72.0$ mm

4-75 $\delta_P = 0.059$ in. $\uparrow$

4-76 $\delta_P = 1.270$ mm

4-79 (a) $\sigma_{AB} = 12.65$ ksi C

$\sigma_{BC} = 7.12$ ksi C

(b) $\delta_{BC} = 0.00581$ in.

4-81 $\sigma_B = 11.05$ ksi T

$\sigma_C = 2.10$ ksi C

4-82 $\sigma = 52.5$ MPa C

4-85 (a) $FS = 3.53$

(b) $FS = 6.48$

4-86 (a) $FS = 2.50$

(b) $FS = 4.50$

4-89 $W_{max} = 1.561$ kip

4-105 $d = 0.230$ in.

$L = 88.3$ in.

4-106 (a) $\delta = 3.89$ mm

(b) $FS = 4.61$

4-109 (a) $\epsilon_D = 0.1000$ in./in.

(b) $\epsilon_C = 0.1000$ in./in.

4-110 (a) $\epsilon_a = 1920$ μm/m

(b) $\epsilon_t = -213.3$ μm/m

(c) $\Delta V = 186.6$ mm³

4-113 (a) $\gamma_o = 1326$ μin./in.

(b) $\gamma_i = 2650$ μin./in.

(c) $\delta = 0.001839$ in.

4-114 (a) $\sigma_A = 56.6$ MPa C

$\sigma_B = 84.9$ MPa T

(b) $\delta_C = 0.849$ mm $\downarrow$

CHAPTER 5

5-1 (a) $\mathbf{M}_E = 1000$ in. · lb $\curvearrowright$

(b) $\mathbf{M}_A = 2400$ in. · lb $\curvearrowright$

(c) $\mathbf{M}_B = 4200$ in. · lb $\curvearrowleft$

5-2 (a) $\mathbf{M}_A = 2.65$ kN · m $\curvearrowright$

(b) $\mathbf{M}_B = 6.93$ kN · m $\curvearrowright$

(c) $\mathbf{M}_A = 4.43$ kN · m $\curvearrowright$

5-5 (a) $\mathbf{M}_o = 600$ in. · lb $\curvearrowleft$

(b) $\mathbf{M}_o = 600$ in. · lb $\curvearrowright$

(c) $\mathbf{M}_A = 1358$ in. · lb $\curvearrowright$

5-6 (a) $\mathbf{M}_B = 25.2$ N · m $\curvearrowright$

(b) $\mathbf{M}_A = 66.0$ N · m $\curvearrowright$

(c) $\mathbf{M}_C = 36.4$ N · m $\curvearrowleft$

(d) $\mathbf{M}_E = 41.1$ N · m $\curvearrowleft$

5-9 (a) $\mathbf{M}_A = 6.00$ in. · kip $\curvearrowright$

(b) $\mathbf{M}_B = 12.60$ in. · kip $\curvearrowleft$

5-10 (a) $\mathbf{M}_{A1} = 63.0$ N · m $\curvearrowright$

(b) $\mathbf{M}_{A2} = 92.5$ N · m $\curvearrowleft$

5-13 $\mathbf{M}_B = 2.23$ in. · kip $\curvearrowright$

5-14 (a) $\mathbf{M}_B = 94.4$ N · m $\curvearrowleft$

(b) $\mathbf{M}_C = 184.4$ N · m $\curvearrowleft$

5-17 (a) $\mathbf{M}_A = 5.06$ in. · kip $\curvearrowright$

(b) $\mathbf{M}_B = 5.22$ in. · kip $\curvearrowleft$

5-18 (a) $\mathbf{M}_o = 0.707$ kN · m $\curvearrowleft$

(b) $\mathbf{M}_A = 1.358$ kN · m $\curvearrowleft$

5-21 (a) $\mathbf{M}_A = 375$ in. · lb $\curvearrowright$

(b) $\mathbf{M}_B = 990$ in. · lb $\curvearrowright$

5-22 (a) $\mathbf{M}_A = 435$ N · m $\curvearrowright$

(b) $\mathbf{M}_B = 251$ N · m $\curvearrowright$

5-25 $\mathbf{M}_B = -3.36\ \mathbf{i} - 0.450\ \mathbf{j} - 2.88\ \mathbf{k}$ in. · kip

5-26 $\mathbf{M}_B = 0\ \mathbf{i} - 140\ \mathbf{j} - 210\ \mathbf{k}$ N · m

5-29 (a) $\mathbf{M}_B = 2.40\ \mathbf{i} + 4.20\ \mathbf{j} + 7.98\ \mathbf{k}$ in. · kip

$M_B = 9.33$ in. · kip

(b) $\theta_x = 75.1°$

$\theta_y = 63.2°$

$\theta_z = 31.2°$

5-30 (a) $\mathbf{M}_B = -89.9\ \mathbf{i} + 112.3\ \mathbf{j} - 179.7\ \mathbf{k}$ N · m

$M_B = 230$ N · m

(b) $\theta_x = 113.0°$

$\theta_y = 60.8°$

$\theta_z = 141.3°$

5-33 $\mathbf{M}_B = -4.35\ \mathbf{i} + 2.37\ \mathbf{j} - 0.510\ \mathbf{k}$ in. · kip

5-34 $\mathbf{M}_B = -190.4\ \mathbf{i} + 220\ \mathbf{j} + 214\ \mathbf{k}$ N · m

5-37 $M_{BC} = 1216$ in. · lb

5-38 $\mathbf{M}_{CD} = 0$

5-41 (a) $\mathbf{M}_{oC} = 10.35$ in. · kip

(b) $\mathbf{M}_{DE} = -3.63$ in. · kip

5-43 $\mathbf{M}_{oC} = -2.19$ in. · kip

5-44 $\mathbf{M}_{oB} = 260\ \mathbf{j} + 94.5\ \mathbf{k}$ N · m

5-47 $M = 1469$ ft · lb

$d = 9.17$ ft

5-48 $M = 64.0$ N · m

$d = 186.3$ mm

5-51 $C = 887$ in. · lb

$\theta_x = 105.7°$

$\theta_y = 109.8°$

$\theta_z = 154.3°$

5-52 $C = 11.61$ N · m

$\theta_x = 74.4°$

$\theta_y = 64.5°$

$\theta_z = 30.5°$

5-53 $\mathbf{F} = 125.0\ \mathbf{i} + 217\ \mathbf{j}$ lb

$\mathbf{C} = 650\ \mathbf{k}$ ft · lb

5-54 $\mathbf{F} = 410\ \mathbf{i} + 287\ \mathbf{j}$ N

5-57
$\mathbf{C} = -124.9 \text{ k N} \cdot \text{m}$
$\mathbf{R} = 95.0 \text{ j lb}$
$\mathbf{C} = -380 \text{ k in.} \cdot \text{lb}$

5-58
$\mathbf{R} = 585 \text{ N} \measuredangle 70.0°$
$d = 170.9 \text{ mm}$

5-61
(a) $\mathbf{R} = 2.13 \text{ kip} \nearrow 63.3°$
(b) $d_R = 6.28 \text{ ft}$

5-62
(a) $\mathbf{R} = 12.37 \text{ kN} \searrow 76.0°$
(b) $d_R = 3.23 \text{ m}$

5-65
$\mathbf{R} = 40 \text{ i lb}$
$y_R = -8.13 \text{ ft}$
$z_R = -1.50 \text{ ft}$

5-66
$\mathbf{R} = -50 \text{ j}$
$x_R = 13.5 \text{ m}$
$z_R = 6.00 \text{ m}$

5-67
$\mathbf{R} = 238 \text{ i} - 47.0 \text{ j} + 238 \text{ k lb}$
$\mathbf{C} = -953 \text{ j} - 1071 \text{ k ft} \cdot \text{lb}$

5-68
$\mathbf{R} = 465 \text{ i} + 61.6 \text{ j} + 474 \text{ k N}$
$\mathbf{C} = 600 \text{ i} - 41.1 \text{ j} - 123.2 \text{ k N} \cdot \text{m}$

5-71
$\mathbf{R} = 75.0 \text{ i} + 40.0 \text{ j} + 50.0 \text{ k lb}$
$\mathbf{C} = 900 \text{ i} + 720 \text{ j} - 1350 \text{ k ft} \cdot \text{lb}$

5-72
$\mathbf{R} = -20.6 \text{ i} + 195.7 \text{ j} - 241 \text{ k N}$
$\mathbf{C} = -168.8 \text{ i} + 80.0 \text{ j} + 33.7 \text{ k N} \cdot \text{m}$

5-75
$x_G = 2.09 \text{ in.}$
$y_G = 7.83 \text{ in.}$
$z_G = 1.739 \text{ in.}$

5-76
$x_G = 53.3 \text{ mm}$
$y_G = 77.8 \text{ mm}$
$z_G = 160.0 \text{ mm}$

5-79
$x_G = 4.83 \text{ in.}$
$y_G = 9.42 \text{ in.}$
$z_G = 4.10 \text{ in.}$

5-80
$x_G = 218 \text{ mm}$
$y_G = 253 \text{ mm}$
$z_G = 139.4 \text{ mm}$

5-83
$x_C = 4.00 \text{ in.}$
$y_C = 2.67 \text{ in.}$

5-84
$x_C = 133.3 \text{ mm}$
$y_C = 100.0 \text{ mm}$

5-87
$x_C = L/2$
$y_C = \pi a/8$

5-88
$x_C = (\pi - 2)L/\pi$
$y_C = \pi a/8$

5-91
$x_C = 2a/5$
$y_C = a/2$

5-92
$y_C = (10 - 3\pi) r/(12 - 3\pi)$
$x_C = (10 - 3\pi) r/(12 - 3\pi)$

5-95
$x_C = a/4$
$y_C = b/4$
$z_C = c/4$

5-96
$x_C = r/\pi$
$y_C = r/\pi$
$z_C = h/4$

5-99
$x_C = 3.18 \text{ in.}$
$y_C = 6.68 \text{ in.}$

5-100
$x_C = 66.5 \text{ mm}$
$y_C = 190.2 \text{ mm}$

5-103
$x_C = 0$
$y_C = 4.00 \text{ in.}$

5-104
$x_C = 101.5 \text{ mm}$

$y_C = 173.8 \text{ mm}$

5-107
$x_C = 3.75 \text{ in.}$
$y_C = 7.73 \text{ in.}$

5-108
$x_C = 38.0 \text{ mm}$
$y_C = 156.2 \text{ mm}$

5-109
(a) $x_C = 2.40 \text{ in.}$
$y_C = 3.55 \text{ in.}$
$z_C = 1.494 \text{ in.}$
(b) $x_G = 3.22 \text{ in.}$
$y_G = 3.76 \text{ in.}$
$z_G = 0.794 \text{ in.}$

5-110
(a) $x_C = 0$
$y_C = 0$
$z_C = 229 \text{ mm}$
(b) $x_G = 0$
$y_G = 0$
$z_G = 197.0 \text{ mm}$

5-113
$x_G = 10.86 \text{ in.}$
$y_G = 9.95 \text{ in.}$
$z_G = 3.50 \text{ in.}$

5-114
$x_G = 0$
$y_G = 165.0 \text{ mm}$
$z_G = 0$

5-115
$R = 1350 \text{ lb}$
$d = 3.83 \text{ ft}$

5-116
$R = 4.50 \text{ kN}$
$d = 4.35 \text{ m}$

5-117
$R = 3775 \text{ lb} \downarrow$
$d = 8.06 \text{ ft}$

5-118
$R = 2.25 \text{ kN}$
$d = 2.67 \text{ m}$

5-121
$R = 11,750 \text{ lb} \downarrow$
$d = 15.71 \text{ ft} \rightarrow$

5-122
$R = 2390 \text{ N} \downarrow$
$d = 3.17 \text{ m} \rightarrow$

5-127
(a) $\mathbf{M} = 11.47 \text{ k ft} \cdot \text{kip}$
(b) $\mathbf{M} = -86.8 \text{ k ft} \cdot \text{kip}$

5-128
(a) $\mathbf{M}_o = -177.4 \text{ i} + 35.5 \text{ j} - 71.0 \text{ k N} \cdot \text{m}$
(b) $M_{oD} = 169.3 \text{ N} \cdot \text{m}$

5-131
$\mathbf{R} = -130.0 \text{ k lb}$
$x_R = 4.15 \text{ ft}$
$y_R = 29.2 \text{ ft}$

5-132
(a) $\mathbf{R} = 450 \text{ k N}$
$\mathbf{C} = 67.5 \text{ i} + 202 \text{ j N} \cdot \text{m}$
(b) $M_{oA} = 202 \text{ N} \cdot \text{m}$
$M_{BA} = 47.7 \text{ N} \cdot \text{m}$
$M_{BC} = 67.5 \text{ N} \cdot \text{m}$

5-135
$x_C = 0.900 \text{ in.}$
$y_C = 0.900 \text{ in.}$

5-136
$x_C = 0$
$y_C = 173.3 \text{ mm}$

5-139
$R = 2500 \text{ lb}$
$d = 10.50 \text{ ft}$

CHAPTER 6

6-13
$\mathbf{A} = 1240 \text{ lb} \uparrow$
$\mathbf{B} = 1160 \text{ lb} \uparrow$

6-14
$\mathbf{A} = 2.00 \text{ kN} \uparrow$
$\mathbf{M}_A = 11.00 \text{ kN} \cdot \text{m} \downarrow$

6-15
$\mathbf{A} = 250 \text{ lb} \uparrow$

$\mathbf{M}_A = 750 \text{ ft} \cdot \text{lb} \uparrow$

6-16
$\mathbf{B} = 424 \text{ N} \searrow 45°$
$\mathbf{A} = 300 \text{ N} \rightarrow$

6-19
$\mathbf{A} = 856 \text{ lb} \searrow 83.3°$
$\mathbf{B} = 950 \text{ lb} \uparrow$

6-20
$\mathbf{B} = 60 \text{ kN} \uparrow$
$\mathbf{A} = 100 \text{ kN} \uparrow$

6-23
$\mathbf{B} = 1950 \text{ lb} \uparrow$
$\mathbf{A} = 1800 \text{ lb} \uparrow$

6-24
(a) $\mathbf{B} = 600 \text{ N} \leftarrow$
$\mathbf{C} = 960 \text{ N} \measuredangle 51.3°$
(b) $\tau = 6.11 \text{ MPa}$
(c) $\delta = 0.465 \text{ mm}$

6-27
$T = 100 \text{ lb}$

6-28
$T = 429 \text{ N}$

6-31
(a) $\mathbf{B} = 250 \text{ lb} \measuredangle 60.0°$
$\mathbf{A} = 150.3 \text{ lb} \searrow 33.8°$
(b) $\tau = 1361 \text{ psi}$

6-32
(a) $\mathbf{B} = 3.42 \text{ kN} \searrow 60°$
$\mathbf{A} = 2.54 \text{ kN} \searrow 9.01°$
(b) $\tau = 16.18 \text{ MPa}$

6-35
(a) $\mathbf{T} = 104.1 \text{ lb} \searrow 20.6°$
$\mathbf{C} = 134.3 \text{ lb} \searrow 83.6°$
(b) $\delta = 0.00966 \text{ in.}$

6-36
(a) $\mathbf{N}_1 = 392 \text{ N} \measuredangle 53.1°$
$\mathbf{N}_2 = 294 \text{ N} \searrow 36.9°$
(b) $\mathbf{C} = 147.2 \text{ N} \searrow 73.7°$
$\mathbf{A} = 197.4 \text{ N} \searrow 10.3°$
(c) $\tau = 5.21 \text{ MPa}$

6-39
(a) $\mathbf{D} = 2610 \text{ lb} \uparrow$
$\mathbf{A} = 2250 \text{ lb} \searrow 77.2°$
(b) $\tau = 6.65 \text{ ksi}$
(c) $\sigma_b = 8.99 \text{ ksi}$

6-40
(a) $\mathbf{E} = 197.6 \text{ N} \searrow 30.0°$
$\mathbf{A} = 312 \text{ N} \measuredangle 56.8°$
(b) $\tau = 11.05 \text{ MPa}$
(c) $\sigma_b = 4.12 \text{ MPa}$

6-43
(a) $\mathbf{E} = 588 \text{ lb} \leftarrow$
$\mathbf{A} = 914 \text{ lb} \measuredangle 50.0°$
(b) $\sigma_{CD} = 5.31 \text{ ksi}$
(c) $\delta_{CD} = 0.1201 \text{ in.}$
(d) $\tau = 4.66 \text{ ksi}$

6-45
$T_{AB} = 1000 \text{ lb (T)}$
$T_{AC} = 1500 \text{ lb (T)}$
$T_{BC} = 1732 \text{ lb (C)}$

6-46
$T_{AB} = 0.373 \text{ kN (C)}$
$T_{AC} = 0.557 \text{ kN (T)}$
$T_{BC} = 0.928 \text{ kN (C)}$

6-47
$T_{AB} = 1128 \text{ lb (C)}$
$T_{AC} = 205 \text{ lb (T)}$
$T_{BC} = 564 \text{ lb (T)}$

6-48
$T_{AB} = 2.50 \text{ kN (C)}$
$T_{AD} = 2.17 \text{ kN (T)}$
$T_{BD} = 5.00 \text{ kN (T)}$
$T_{BC} = 4.33 \text{ kN (C)}$
$T_{CD} = 2.17 \text{ kN (T)}$

6-51
$T_{AB} = 1186 \text{ lb (C)}$
$T_{AH} = 893 \text{ lb (T)}$
$T_{BH} = 0 \text{ (zero force member)}$
$T_{GH} = 893 \text{ lb (T)}$
$T_{BG} = 0 \text{ (zero force member)}$

$T_{BC} = 1186$ lb (C)
$T_{CD} = 1186$ lb (C)
$T_{CG} = 1100$ lb (T)
$T_{DE} = 1712$ lb (C)
$T_{EF} = 1290$ lb (T)
$T_{DF} = 0$ (zero force member)
$T_{FG} = 1290$ lb (T)
$T_{DG} = 343$ lb (C)
$T_{BC} = 5.71$ kN (C)

6-52
$T_{AB} = 3.46$ kN (T)
$T_{AF} = 1.732$ kN (C)
$T_{BF} = 4.62$ kN (C)
$T_{EF} = 8.08$ kN (C)
$T_{BC} = 1.155$ kN (C)
$T_{BE} = 4.62$ kN (T)
$T_{CD} = 1.732$ kN (T)
$T_{CE} = 4.62$ kN (C)
$T_{DE} = 3.46$ kN (C)

6-55
$T_{CD} = 750$ lb (T)
$T_{CF} = 2500$ lb (C)
$T_{FG} = 750$ lb (T)

6-56
$T_{BC} = 10.00$ kN (C)
$T_{CG} = 2.08$ kN (C)
$T_{FG} = 11.25$ kN (T)

6-59
$T_{DE} = 1732$ lb (C)
$T_{DF} = 1400$ lb (T)
$T_{EF} = 1000$ lb (T)

6-60
$T_{CD} = 16.00$ kN (C)
$T_{CI} = 11.67$ kN (T)
$T_{CJ} = 0$ (zero force member)

6-63
$T_{CD} = 2730$ lb (C)
$T_{EF} = 2660$ lb (T)

6-64
$T_{CD} = 8.63$ kN (C)
$T_{EF} = 7.00$ kN (T)

6-67
$T_{CD} = 1949$ lb (C)
$T_{DF} = 486$ lb (T)
$T_{EF} = 1699$ lb (T)

6-68
$T_{EJ} = 3.89$ kN (C)
$T_{HJ} = 9.20$ kN (T)

6-71
(a) $\sigma_{FG} = 3.84$ ksi (C)
(b) $A_{CD} = 0.693$ in.^{2}
(c) $\delta_{CF} = 0.01468$ in.

6-72
(a) $\sigma_{AB} = 117.7$ MPa (T)
(b) $A_{EF} = 589$ mm^2
(c) $A_{AG} = 117.7$ mm^2
(d) $\delta_{DE} = -3.68$ mm

6-75
$T_{CD} = 5750$ lb (C)
$T_{FG} = 5750$ lb (T)

6-76
$T_{AB} = 5.83$ kN (T)
$T_{FG} = 5.83$ kN (C)

6-79
$T_{DE} = 10,000$ lb (C)
$T_{DJ} = 8000$ lb (C)
$T_{JK} = 7000$ lb (T)

6-80
(a) $T_{CD} = 9.66$ kN (C)
$T_{EF} = 10.83$ kN (T)
$T_{DF} = 6.50$ kN (C)
(b) $\sigma_{DG} = 13.89$ MPa (C)
(c) $\delta_{FG} = 0.577$ mm

6-83
$B = 48.2$ lb ∠ 57.4°
$C = 85.7$ lb ⤢ 45.0°

6-84
$A = 1269$ N ⤡ 23.2°

$B = N$ ⤢ 40.6°

6-87
$F_{block} = 150$ lb ↓
$A = 50$ lb ↓
$B = 100$ lb ↑
$C = 50$ lb ↓

6-88
$A = 35.1$ N →
$B = 53.5$ N ⤢ 49.1°
$C = 40.5$ N ↓
$F_{block} = 35.1$ N →

6-91
$A = 70.5$ lb ⤡ 7.13°
$B = 61.3$ lb ↓
$C = 99.0$ lb ⤢ 45.0°
$M_C = 245$ ft · lb ↺

6-92
(a) $\sigma_{AC} = 87.2$ MPa (T)
(b) $\delta_{BC} = 0.812$ mm
(c) $\tau_B = 82.7$ MPa

6-95
$A = 25.3$ lb ∠ 81.9°
$B = 10.85$ lb ⤢ 39.8°
$C = 18.67$ lb ⤡ 75.2°

6-96
$B = 13.75$ N ↑
$D = 8.75$ N ↓

6-98
(a) $A = 15.35$ kN ⤡ 26.6°
$B = 15.41$ kN ∠ 82.1°
$D = 17.17$ kN ⤢ 22.6°
(b) $\tau_A = 43.4$ MPa
(c) $\tau_E = 17.35$ MPa

6-99
$P = 138.6$ lb
$A = 72.6$ lb ⤡ 77.3°
$B = 138.6$ lb ⤢ 60.0°
$C = 72.6$ lb ⤡ 42.7°

6-101
(a) $\sigma_{CD} = 1.597$ ksi (T)
(b) $\tau_F = 6.87$ ksi
(c) $\delta_{BE} = -0.00307$ in.

6-102
(a) $\sigma_{BD} = 3.89$ MPa (T)
(b) $\tau_C = 8.31$ MPa
(c) $\delta_{BD} = 0.01945$ mm

6-105
$R_A = 50$ lb ↑
$C_A = 1150\,\mathbf{i} + 350\,\mathbf{j}$ in. · lb

6-106
$A = -1900\,\mathbf{i} + 500\,\mathbf{j}$ N
$B = 1580\,\mathbf{i} - 800\,\mathbf{j}$ N
$C = 720\,\mathbf{i} + 200\,\mathbf{k}$ N

6-109
(a) $T_B = 643$ lb
$T_C = 573$ lb
(b) $R_A = 1036\,\mathbf{j} + 364\,\mathbf{k}$ lb
(c) $\sigma_B = 13.09$ ksi (T)
$\sigma_C = 11.68$ ksi (T)
$\delta_B = 0.0656$ in.
$\delta_C = 0.0625$ in.

6-110
$T_A = 1561$ k N
$T_B = 1394$ k N
$T_C = 1951$ k N

6-113
(a) System is in equilibrium
(b) $P_{max} = 800$ lb

6-114
(a) System is not in equilibrium
(b) $P_{max} = 1583$ N

6-117
(a) $P = 197.8$ lb
(b) $P = 25.8$ lb

6-118
$d = 2.33$ m

6-121
$x = 3.21$ ft

6-122
$h = 1.071$ m

6-125
$W_C(max) = 12.00$ lb (tipping)

6-126
(a) $W_{min} = 100$ N
$W_{max} = 220$ N
(b) $h = 2.05$ m

6-129
(a) $P_{min} = 1063$ lb
(b) $\tau_B = 1654$ psi

6-130
(a) $P_{max} = 196.2$ N
(b) $\theta = 9.46°$

6-131
(a) $W_{max} = 75.0$ lb
(b) $\tau_C = 1080$ psi

6-135
$P_{min} = 16.63$ lb

6-136
(a) $P_{min} = 1347$ N
(b) $P_{min} = 3460$ N

6-139
$W_{min} = 9.37$ lb
$W_{max} = 144.0$ lb

6-140
$P_{max} = 392$ N

6-143
(a) $T_{max} = 3990$ in. · lb
(b) $\tau_B = 2.04$ ksi

6-160
$A = 803$ N ⤢ 61.6°
$B = 855$ N ∠ 63.4°

6-161
$A = 600$ lb ↑
$B = 1008$ lb ⤡ 30°
$C = 1008$ lb ⤢ 30.0°

6-164
$A = 5.48$ kN ⤢ 47.3°
$B = 5.26$ kN ↑

6-165
(a) $W = 4.93$ kip
(b) $T = 6.28$ kip
(c) $A = 11.38$ kip ∠ 31.3°

6-168
$T_C = 756$ N
$A = 185.8\,\mathbf{j} - 107.3\,\mathbf{k}$ N
$B = 307\,\mathbf{i} + 185.9\,\mathbf{j} + 261\,\mathbf{k}$ N

6-169
$A = -375\,\mathbf{j} + 30\,\mathbf{k}$ lb
$B = 255$ lb ↑
$\mathbf{T}_C = 510$ lb ↑

6-172
$T_{CD} = 9.19$ kN (C)
$T_{CG} = 0.750$ kN (C)

6-173
$T_{BC} = 1170$ lb (C)
$T_{CF} = 752$ lb (T)
$T_{FG} = 1111$ lb (T)
$T_{GE} = 1288$ lb (T)

6-174
(a) $D = 6.91$ kN ⤡ 49.4°
(b) $A = 2.65$ kN ⤡ 8.13°
$E = 4.28$ kN ⤢ 52.1°

6-175
$\mu_{min} = 1/3$

6-178
$T = 21.1$ kN
$C_A = 1.460$ kN · m
$C_B = 0.450$ kN · m

6-179
$a = 4.52$ in.

CHAPTER 7

7-1
(a) $T_{AB} = 80$ ft · kip ↓
$T_{BC} = 20$ ft · kip ↗
$T_{CD} = 20$ ft · kip ↓
$T_{DE} = 45$ ft · kip ↓

7-2
(a) $T_{BC} = 500$ N · m ↓
$T_{CD} = 400$ N · m ↓
$T_{DE} = 250$ N · m ↓

7-5
(a) $\tau = 11.46$ ksi
(b) $\theta = 0.0688$ rad

7-6
(a) $\tau = 65.8$ MPa
(b) $\tau = 43.9$ MPa
(c) $\theta = 0.0274$ rad

7-9
(a) $\tau_{AB} = 9.43$ ksi
$\tau_{BC} = 3.77$ ksi
(b) $\theta_{B/A} = 0.0252$ rad
$\theta_{C/A} = 0.0327$ rad

7-10
(a) $d_{AB} = 76.1$ mm
$d_{AC} = 61.9$ mm
(b) $\theta_{B/A} = 0.0458$ rad
(c) $\theta_{C/D} = -0.0218$ rad

7-13
(a) $\tau_{max} = 11.21$ ksi
(b) $\theta_{D/B} = 0.00764$ rad —↗—

7-14
(a) $d_{min} = 99.3$ mm
(b) $\theta_{D/A} = 0.0288$ rad —↗—

7-17
(a) $\tau_{max} = 4.75$ ksi
(b) $\theta_{C/A} = 0.00335$ rad —↗—

7-18
(a) $d_i = 70.5$ mm
(b) $d_i = 70.0$ mm

7-21
(a) $T_3 = 1068$ in. · kip
(b) $\theta_B = 0.0600$ rad —↙—
(c) $\theta_D = 0.0727$ rad —↙—

7-22
$T_1 = 15.02$ kN · m
$T_2 = 111.5$ kN · m

7-25
(a) $d_{AB} = 0.546$ in.
(b) $d_{CD} = 0.402$ in.
(c) $L = 2.12$ ft

7-26
(a) $d_{AB} = 68.8$ mm
$d_{BC} = 93.4$ mm
(b) $d = 74.9$ mm

7-29
(a) $T_{max} = 1091$ in. · lb
(b) $\theta = 0.1112$ rad

7-30
$T_{max} = 11.96$ kN · m

7-33
(a) $T_1 = 32.0$ in. · kip
(b) $\sigma_{max} = 20.4$ ksi (T)
(c) $\sigma_{max} = 6.05$ ksi (C)

7-34
(a) $\sigma_{max} = 83.0$ MPa (T)
(b) $\sigma_{max} = 76.4$ MPa (C)
(c) $\theta_{C/A} = 0.01179$ rad —↗—

7-35
(a) $U = 2100$ ft · lb
(b) $U = -2100$ ft · lb

7-36
(a) $U = 30.0$ kJ
(b) $U = -23.5$ kJ

7-39
(a) $U = 243$ ft · lb
(b) $U = 0$
(c) $U = -243$ ft · lb

7-41 $U = 28$ ft · lb
7-42 $U = -10$ J
7-45 $P = 9340$ hp
7-46 $d = 85.9$ mm
7-49 $P = 170.3$ hp
7-50
(a) $P = 286$ kW
(b) $\theta = 0.0600$ rad

7-53
(a) $d = 1.399$ in.
(b) $d_i = 2.92$ in.
(c) % = 76.5

7-54 $d_{min} = 84.4$ mm
7-57
(a) $\sigma_{max} = 19.26$ ksi
(b) $\sigma_{max} = 3.80$ ksi
(c) $\theta_{C/A} = -0.0355$ rad

7-58
(a) $d_1 = 80.4$ mm
(b) $d_2 = 57.9$ mm
(c) $\theta_{C/A} = 0.0538$ rad

7-61
(a) $\tau_{max} = 16.04$ ksi

(b) $\theta_B = 0.0321$ rad —↗—
7-62
(a) $T = 15.94$ kN · m
(b) $\theta_{B/A} = 0.0556$ rad —↗—

7-65 $\tau_b = 8.91$ ksi
7-66
(a) $\tau_s = 13.49$ MPa
$\tau_m = 15.35$ MPa
(b) $\theta = 0.01349$ rad

7-69
(a) $\tau_b = 6.11$ ksi
$\tau_s = 18.99$ ksi
$\tau_a = 7.64$ ksi
(b) $\theta_b = 0.1048$ rad

7-70
(a) $T = 28.1$ kN · m
(b) $\theta = 0.0438$ rad

7-73
(a) $\tau_{max} = 16.25$ ksi
(b) $\theta_{B/A} = 0.0406$ rad —↗—

7-74 $d = 77.5$ mm
7-77
$d_C = 0.0564$ in.
$d_D = 0.219$ in.

7-78
$\tau_s = 61.1$ MPa
$\tau_b = 18.46$ MPa

7-81 $\tau_A = 1778$ psi
7-82
$\tau_b = 24.7$ MPa
$\tau_s = 67.7$ MPa

7-97
(a) $\tau_{max} = 6.82$ ksi
(b) $\tau_i = 5.12$ ksi
(c) $\theta = 0.0341$ rad

7-98 $d_{nom} = 102$ mm
7-101 $T_2 = 228$ in. · kip
7-102
(a) $\tau_{max} = 41.0$ MPa
(b) $\theta_{B/A} = 0.0256$ rad —↗—
(c) $\theta_{C/A} = 0.0409$ rad —↗—

7-105 $P = 175.8$ hp
7-106
(a) $P = 4.83$ MW
(b) $\theta = 0.0281$ rad

7-109 $d = 2.50$ in.
7-110
(a) $\tau_A = 73.3$ MPa
$\tau_B = 67.8$ MPa
(b) $\theta_A = 0.01878$ rad —↗—

CHAPTER 8

8-1
$I_{xC} = 56.0$ in.^{4}
$I_{yC} = 11.00$ in.^{4}

8-2
$I_{xC} = 116.5(10^6)$ mm^4
$I_{yC} = 318(10^6)$ mm^4

8-5
$I_{xC} = 647$ in.^{4}
$I_{yC} = 134.9$ in.^{4}

8-6
$I_{xC} = 390(10^6)$ mm^4
$I_{yC} = 142.4(10^6)$ mm^4

8-9
$I_x = 4810$ in.^{4}
$I_y = 20,200$ in.^{4}

8-10
$I_x = 127.4(10^6)$ mm^4
$I_y = 343(10^6)$ mm^4

8-13 $M = 64.0$ in. · kip
8-14 $M = 33.8$ kN · m
8-17 $M = 102.7$ in. · kip
8-18
(a) $\sigma_B = 5.00$ MPa (T)
(b) $\sigma_C = 2.50$ MPa (C)
(c) $\sigma_D = 6.25$ MPa (T)

8-21 $h = 0.0249$ in.
8-22 $R = 4.55$ m
8-25 (a) $\sigma_{bot} = 8.51$ ksi (T)

(b) $\sigma_{top} = 5.41$ ksi (C)
8-26 $M_r = 8.08$ kN · m
8-29 $M_F = 78.1\%$
8-30 $M_F = 80.0\%$
8-33
(a) $\sigma = 13.91$ ksi (T)
(b) $\sigma = 7.36$ ksi (C)

8-37
$V_x = 1000$ lb
$M_x = 1000x + 8000$ ft · lb.

8-38
$V_x = -15x + 50$ kN
$M_x = -7.5x^2 + 50x$ kN · m

8-41
$V = -1000x + 6000$ lb
$M = -500x^2 + 6000x - 8000$ ft · lb.

8-42
(a) $V = w(L + x)$ N
$M = \frac{w}{2}(x^2 + 2Lx - 2L^2)$ N · m
(b) $V = 2w(2L - x)$ N
$M = -w(x^2 - 4Lx + 4L^2)$ N · m

8-45
$V = 1200 - 200x + 1200$ lb
$M = -100x^2 + 1200x - 3000$ ft · lb

8-46
$V = -4x + 9$ kN
$M = -2x^2 + 9x - 3$ kN · m

8-49 $P_{max} = 3.67$ kip
8-50 $w_{max} = 3.26$ kN/m
8-53
(a) $\sigma_A = 354$ psi (C)
(b) $\sigma_B = 591$ psi (T)

8-54
(a) $\sigma_A = 2.87$ MPa (C)
(b) $\sigma_B = 12.45$ MPa (T)

8-57 $L = 43.2$ ft
8-59
(a) $V = (10,000/\pi) \cos (\pi x/10)$
$M = (100,000/\pi^2) \sin (\pi x/10)$
(b) $V_{max} = V_{x=0} = V_{x=10} = 3.18$ kip
$M_{max} = M_{x=5} = 10.13$ ft · kip

8-60
(a) $V = (400/\pi^2) - (200/\pi) \sin (\pi x/8)$ kN
$M = (400/\pi^2)(x - 4) + (1600/\pi^2) \cos (\pi x/8)$ kN · m
(b) $V_{max} = V_{x=0} = 40.5$ kN
$M_{max} = M_{x=1.7573} = 34.1$ kN · m

8-63 $P = 9.40$ kip
8-65
$V_0 = 5.75$ kip
$V_5 = 5.75$ kip and 0.750 kip
$V_{15} = 0.750$ kip and -7.25 kip
$M_0 = 0$
$M_5 = 28.75$ ft · kip
$M_{15} = 36.25$ ft · kip

8-66
$V_0 = -15$ kN
$V_{2.5} = -15$ kN and 15.5 kN
$V_{5.5} = 15.5$ kN and -4.5 kN
$M_0 = 0$
$M_{2.5} = -37.5$ kN · m
$M_{5.5} = 9$ kN · m

8-69
$V_0 = 11$ kip
$V_{12} = -13$ kip
$V_{16} = -13$ kip and 8 kip
$M_0 = 0$
$M_{5.5} = 30.25$ ft · kip
$M_{12} = -12$ ft · kip
$M_{16} = -64$ ft · kip

8-70 $V_0 = 45$ kN

$V_2 = 45$ kN and 25 kN
$V_4 = 25$ kN
$V_8 = -95$ kN
$M_0 = 0$
$M_2 = 90$ kN $\cdot$ m
$M_4 = 140$ kN $\cdot$ m
$M_{4.83} = 100.4$ kN $\cdot$ m

8-73 $V_0 = 1800$ lb
$V_5 = 1800$ lb and 300 lb
$V_{10} = 300$ lb
$V_{20} = -2200$ lb
$M_0 = 0$
$M_5 = 9$ ft $\cdot$ kip and 8 ft $\cdot$ kip
$M_{10} = 9.5$ ft $\cdot$ kip
$M_{11.2} = 9.68$ ft $\cdot$ kip

8-74 $V_0 = 0$
$V_2 = -36$ kN and 76.5 kN
$V_6 = 4.5$ kN
$V_8 = 4.5$ kN and -31.5 kN
$M_0 = -9$ kN $\cdot$ m
$M_2 = -45$ kN $\cdot$ m
$M_6 = 117$ kN $\cdot$ m
$M_8 = 126$ kN $\cdot$ m

8-77 $V_0 = -1000$ lb and 1750 lb
$V_6 = -250$ lb and -2250 lb
$V_8 = -2250$ lb and 1000 lb
$M_0 = 0$
$M_6 = 6500$ ft $\cdot$ lb and 2500 ft $\cdot$ lb
$M_8 = -2000$ ft $\cdot$ lb

8-78 $V_0 = -6$ kN and 9.33 kN
$V_2 = 9.33$ kN
$V_4 = -0.67$ kN and -3.67 kN
$M_0 = -4$ kN $\cdot$ m
$M_2 = 14.67$ kN $\cdot$ m and 8.67 kN $\cdot$ m
$M_4 = 17.33$ kN $\cdot$ m

8-81 Use a W8 $\times$ 15 Section
8-82 Use a W254 $\times$ 33 Section
8-85 (a) $\sigma_T = 5.44$ ksi (T)
(b) $\sigma_C = 8.16$ ksi (C)
8-87 $V_0 = 0$
$V_3 = -3000$ lb and 5000 lb
$V_6 = 2000$ lb
$V_9 = 2000$ lb and -3000 lb
$V_{12} = -3000$ lb and 6000 lb
$M_0 = 0$
$M_3 = -4500$ ft $\cdot$ lb
$M_6 = 6000$ ft $\cdot$ lb
$M_9 = 12{,}000$ ft $\cdot$ lb and 3000 ft $\cdot$ lb
$M_{12} = -6000$ ft $\cdot$ lb
8-88 $V_0 = 0$
$V_1 = -7.5$ kN and 16.75 kN
$V_3 = 1.75$ kN
$V_4 = 1.75$ kN and -13.25 kN
$V_5 = -13.25$ kN and 6.75 kN
$M_0 = 0$
$M_1 = -3.75$ kN $\cdot$ m
$M_3 = 14.75$ kN $\cdot$ m
$M_4 = 16.50$ kN $\cdot$ m and 6.50 kN $\cdot$ m
$M_5 = -6.75$ kN $\cdot$ m
8-91 $V_0 = -1000$ lb
$V_2 = -1000$ lb and 2333 lb

$V_6 = 333$ lb and -1667 lb
$V_8 = -1667$ lb and 1000 lb
$M_0 = 2000$ ft $\cdot$ lb
$M_2 = 0$
$M_6 = 5333$ ft $\cdot$ lb and 1333 ft $\cdot$ lb
$M_8 = -2000$ ft $\cdot$ lb
8-93 $w = 2.43$ kip/ft
8-94 $w = 8.28$ kN/m
8-95 $w = 10.43$ kip/ft
8-99 (a) $\tau_V = 143.7$ psi
(b) $\tau_{max} = 171.2$ psi
8-100 (a) $\tau_H = 2.31$ MPa
(b) $\tau_{max} = 2.60$ MPa
8-103 (a) $\tau_H = 140.6$ psi
(b) $\tau_{max} = 187.5$ psi
(c) $\sigma_{max} = 4.50$ ksi (T)
8-104 (a) $\tau_{max} = 1800$ kPa
(b) $\tau_V = 645$ kPa
(c) $\sigma_{max} = 21.6$ MPa (C)
8-107 (a) $w = 768$ lb/ft
(b) $\tau_H = 48.9$ psi
(c) $\sigma_{max} = 1800$ psi (T)
8-108 (a) $\tau_H = 1957$ kPa
(b) $\tau_H = 2120$ kPa
(c) $\tau_{max} = 3.17$ MPa
(d) $\sigma_{max} = 30.8$ MPa (T)
8-111 (a) $\sigma_{max} = 1600$ psi (T)
(b) $\tau_{max} = 83.3$ psi
8-112 $P_{max} = 2.19$ kN
8-115 (a) $F_J = 519$ lb
(b) Use 6 nails spaced 2 in. apart
8-116 (a) $F_B = 28.5$ kN
(b) $d_{min} = 24.6$ mm
8-119 $s = 10.00$ in.
8-133 (a) $\sigma_{max} = 10.00$ ksi (C)
(b) $M_r = 544$ in. $\cdot$ kip
8-134 $\sigma_b = 206$ MPa (T)
$\sigma_t = 137.1$ MPa (C)
8-137 (a) $V_0 = 0$
$V_6 = 1200$ lb and 2100 lb
$V_{16} = -900$ lb and -1500 lb
$M_0 = -5600$ ft $\cdot$ lb
$M_6 = -2000$ ft $\cdot$ lb
$M_{13} = 5350$ ft $\cdot$ lb
$M_{15} = 4000$ ft $\cdot$ lb
$M_{21} = -3500$ ft $\cdot$ lb
(b) $V = -300x + 2100$ lb
$M = -150x^2 + 2100x - 2000$ ft $\cdot$ lb
8-138 (a) $V_0 = 0$
$V_3 = -60$ N
$V_5 = 80$ N
$V_7 = 80$ N and -45 N
$M_0 = 0$
$M_3 = -90$ N $\cdot$ m
$M_{3.86} = -115.7$ N $\cdot$ m
$M_5 = -70$ N $\cdot$ m
$M_7 = 90$ N $\cdot$ m
(b) $V = 70x - 60$ N
$M = 35x^2 - 60x - 90$ N $\cdot$ m
8-141 (a) $\sigma_{maxT} = 8.47$ ksi (T)

$\sigma_{maxC} = 6.62$ ksi (C)
(b) $\tau_{max} = 496$ psi
8-142 (a) $\sigma_{maxT} = 32.3$ MPa (T)
$\sigma_{maxC} = 50.7$ MPa (C)
(b) $\tau_{max} = 1.991$ MPa

CHAPTER 9

9-1 (a) $y = \dfrac{P}{6EI}(-x^3 + 3L^2x - 2L^3)$

(b) $\theta = y'_{x=0} = \dfrac{PL^2}{2EI}$ ⟋

(c) $y_{x=0} = -\dfrac{PL^3}{3EI} = \dfrac{PL^3}{3EI}$ ↓

9-2 (a) $y = \dfrac{w}{24EI}(-x^4 + 4L^3x - 3L^4)$

(b) $\theta = y'_{x=0} = \dfrac{wL^3}{6EI}$ ⟋

(c) $y_{x=0} = -\dfrac{wL^4}{8EI} = \dfrac{wL^4}{8EI}$ ↓

9-5 (a) $\rho = 892$ ft
(b) $y_{x\,L/2} = 0.242$ in. ↓
9-6 (a) $\rho = 640$ m
(b) $y_{x=L/2} = 0.781$ mm ↑
9-9 (a) $y = \dfrac{w}{72EI}(-3x^4 + 2Lx^3 + 12L^2x^2 - 11L^3x)$

(b) $y_{x=L/2} = \dfrac{13wL^4}{384EI}$ ↓

9-10 (a) $y = \dfrac{w}{48EI}(-2x^4 + 3Lx^3 - 3L^2x^2 + 2L^3x)$

(b) $y_{x=L/2} = \dfrac{wL^4}{96EI}$ ↑

9-13 (a) $y = \dfrac{w}{120EIL}(-x^5 + 5L^4x - 4L^5)$

(b) $\theta_A = \dfrac{wL^3}{24EI}$ ⟋

(c) $y_{x=0} = \dfrac{wL^4}{30EI}$ ↓

9-14 (a) $y = \dfrac{w}{120EIL}(x^5 - 5Lx^4 + 10L^2x^3 - 10L^3x^2)$

(b) $\theta_B = \dfrac{wL^3}{24EI}$ ⟍

(c) $y_{x=L} = \dfrac{wL^4}{30EI}$ ↓

9-17 (a) $\theta_A = \dfrac{5wL^3}{24EI}$ ⟍

(b) $y_{x=L} = \dfrac{2wL^4}{15EI}$ ↓

9-18 (a) $\theta_A = \dfrac{ML}{12EI}$ ⟋

(b) $y_{max} = \dfrac{\sqrt{3}ML^2}{54EI}$ ↑

9-21 $y_{x=3a/2} = \dfrac{205wa^4}{384EI}$ ↓

9-22 $y_{x=L} = \dfrac{49wL^4}{240EI}$ ↓

9-25 (a) $y = \dfrac{W}{12EI}(2x^3 - 3Lx^2)$

(b) $\delta_C = 0.1333$ in. $\downarrow$

9-27 (a) $y_{x=L} = \dfrac{PL^3}{2EI} \downarrow$

(b) $y_{x=2L} = \dfrac{11PL^3}{6EI} \downarrow$

9-28 (a) $y_{x=L} = \dfrac{2PL^3}{3EI} \downarrow$

(b) $y_{x=2L} = \dfrac{2PL^3}{EI} \downarrow$

9-31 $y_{x=2L} = \dfrac{wL^4}{24EI} \uparrow$

9-32 $y_{x=0} = \dfrac{29wL^4}{24EI} \downarrow$

9-35 (a) $y_{x=L} = \dfrac{2wL^4}{3EI} \downarrow$

(b) $y_{x=2L} = \dfrac{23wL^4}{12EI} \downarrow$

9-36 (a) $y_{x=L} = \dfrac{7wL^4}{24EI} \downarrow$

(b) $y_{x=2L} = \dfrac{7wL^4}{8EI} \downarrow$

9-39 (a) $y_{x=5L/4} = \dfrac{101wL^4}{128EI} \downarrow$

(b) $y_{max} = \dfrac{101.08wL^4}{128EI} \downarrow$

9-40 (a) $y_{x=L} = \dfrac{147wL^4}{720EI} \downarrow$

(b) $y_{max} = \dfrac{147.06wL^4}{720EI} \downarrow$

9-43 (a) $y = \dfrac{w}{840EIL^3}$
$[-x^7 + 7L^6x - 6L^7]$

(b) $y_{x=0} = \dfrac{wL^4}{140EI} \downarrow$

(c) $V_B = \dfrac{wL}{4} \uparrow$

$M_B = \dfrac{wL^2}{20} \downarrow$

9-44 (a) $y = \dfrac{w}{840EIL^3}$
$[-x^7 + 7L^4x^3 - 6L^6x]$

(b) $y_{x=L/2} = \dfrac{13wL^4}{5120EI} \downarrow$

(c) $y_{max} = \dfrac{13.11wL^4}{5120EI} \downarrow$

(d) $R_A = \dfrac{wL}{20} \uparrow$

$R_B = \dfrac{wL}{5} \uparrow$

9-47 (a) $y = -\dfrac{wL^4}{\pi^4EI} \sin \dfrac{\pi x}{L}$

(b) $y_{x=L/2} = \dfrac{wL^4}{\pi^4EI} \downarrow$

(c) $y'_{x=0} = \dfrac{wL^3}{\pi^3EI} \searrow$

(d) $R_A = \dfrac{wL}{\pi} \uparrow$

$R_B = \dfrac{wL}{\pi} \uparrow$

9-49 $\delta_A = 2.50$ in. $\downarrow$

9-50 $\delta_C = \dfrac{PL^3}{2EI} \uparrow$

9-53 $\delta = 12.58$ mm $\downarrow$
9-54 $\delta = 11.14$ mm $\downarrow$
9-57 $\delta = \dfrac{7wL^4}{768EI} \downarrow$

9-58 $\delta_A = \dfrac{5wL^4}{32EI} \downarrow$

9-61 $\delta_C = 0.243$ in. $\downarrow$

9-63 $\delta_A = \dfrac{29wL^4}{48EI} \downarrow$

9-64 $\delta_D = \dfrac{23wL^4}{8EI} \downarrow$

9-67 $\delta_D = \dfrac{wL^4}{4EI} \downarrow$

9-68 $\delta_C = \dfrac{7wL^4}{10EI} \downarrow$

9-71 (a) $M = PL/2$

(b) $y_{x=0} = \dfrac{PL^3}{12EI} \uparrow$

9-72 (a) $M = wL^2/6$

(b) $M = wL^2/4$

9-75 (a) $R_A = 3M/2L \uparrow$
$V_B = 3M/2L \downarrow$
$M_B = M/2 \downarrow$

(b) $y_{x=L/2} = \dfrac{ML^2}{32EI} \uparrow$

9-76 (a) $R_A = 3wL/8 \uparrow$
$V_B = 5wL/8 \uparrow$
$M_B = wL^2/8 \downarrow$

(b) $y_{x=L/2} = \dfrac{wL^4}{192EI} \downarrow$

9-79 (a) $R_A = wL/24 \uparrow$
$V_B = 7wL/24 \uparrow$
$M_B = wL^2/24 \downarrow$

(b) $y_{max} = 0.001267 \dfrac{wL^4}{EI} \downarrow$

9-80 (a) $V_A = wL/2 \uparrow$
$V_B = wL/2 \uparrow$
$M_A = 5wL^2/24 \downarrow$
$M_B = 5wL^2/24 \downarrow$

(b) $y_{x=L} = \dfrac{7wL^4}{240EI} \downarrow$

(c) $M_{x=L} = wL^2/8 \;\ulcorner\!\!-\!\!\urcorner$

9-83 (a) $R_A = 7wL/16 \uparrow$
$R_B = 5wL/8 \uparrow$
$R_C = wL/16 \downarrow$

(b) $M_B = wL^2/16 \downarrow\!\!-\!\!\downarrow$

(c) $y_{x=3L/2} = \dfrac{wL^4}{256EI} \uparrow$

9-84 (a) $R_A = 3wL/7 \uparrow$
$R_B = 19wL/28 \uparrow$
$V_C = 3wL/28 \downarrow$

$M_C = wL^2/28 \;\urcorner$

(b) $M_B = wL^2/14 \;\downarrow\!\!-\!\!\downarrow$

(c) $y_{x=L/2} = \dfrac{23wL^4}{2688EI} \downarrow$

9-87 (a) $R_A = 11wL/24 \uparrow$
$V_B = 13wL/24 \uparrow$
$M_B = wL^2/24 \downarrow$

(b) $y_{max} = 0.01043 \dfrac{wL^4}{EI} \downarrow$

(c) $M_{max} = 121wL^2/1152$

9-88 (a) $P = wL/4$

(b) $y_{max} = 0.00542 \dfrac{wL^4}{EI} \downarrow$

9-91 (a) $R_A = 1960$ lb $\uparrow$
$R_B = 4467$ lb $\uparrow$
$R_D = 3373$ lb $\uparrow$

(b) $y_{x=6} = 0.000784$ in. $\uparrow$

9-93 (a) $M = wL^2/6$

(b) $M = wL^2/4$

9-94 (a) $P = wL/3$

(b) $P = 3wL/8$

9-97 (a) $R_A = 109wL/128 \uparrow$
$V_D = 275wL/128 \uparrow$
$M_D = 51wL^2/32 \downarrow$

(b) $\delta_C = \dfrac{95wL^4}{96EI} \downarrow$

9-98 (a) $R_A = 38wL/27 \uparrow$
$V_B = 70wL/27 \uparrow$
$M_B = 16wL^2/9 \downarrow$

(b) $\delta_C = \dfrac{62wL^4}{81EI} \downarrow$

9-101 $R_A = 7P/4 \uparrow$
$V_B = 3P/4 \downarrow$
$M_B = PL/4 \;\urcorner$

9-102 $R_A = 17wL/16 \uparrow$
$V_B = 7wL/16 \uparrow$
$M_B = wL^2/16 \downarrow$

9-105 $V_A = 2P \uparrow$
$M_A = PL/2 \downarrow$
$R_C = 3P \uparrow$

9-106 $V_A = 139P/288 \uparrow$
$M_A = 47PL/432 \downarrow$
$R_B = 499P/216 \uparrow$
$R_C = 179P/864 \uparrow$

9-109 (a) $R_B = 2810$ lb

(b) $R_B = 2490$ lb

9-110 $R_B = 28.3$ kN
9-121 $d_{min} = 11.72$ in.
9-122 (a) $I = 329(10^3)$ mm^4

(b) $h = 46.0$ mm
$b = 40.5$ mm

9-125 (a) $\delta_c = 0.338$ in. $\uparrow$

(b) $y_{max} = 0.405$ in. $\downarrow$

9-126 (a) $y = \dfrac{w}{72EI}(-3x^4 + 5Lx^3 - 3L^2x^2 + L^3x)$

(b) $y_{max} = 0.00312 \dfrac{wL^4}{EI} \downarrow$

9-129 (a) $R_A = wL/4$

(b) $y_{x=L/2} = \dfrac{7wL^4}{384EI} \downarrow$

(c) $M_{max} = -wL^2/4$

9-130 (a) $R_A = wL/8 \uparrow$
$R_B = 33wL/16 \uparrow$
$R_C = 13wL/16 \uparrow$

(b) $y_{x=2L} = \dfrac{11wL^4}{96EI} \downarrow$

9-133 $R_C = wL/4 \uparrow$

CHAPTER 10

10-1 $\sigma = 12.01$ ksi (T)
$\tau = 1.680$ ksi

10-2 $\sigma = 143.6$ MPa (T)
$\tau = 17.68$ MPa

10-5 $\sigma = 4.37$ ksi (T)
$\tau = 14.24$ ksi

10-6 $\sigma = 163.4$ MPa (T)
$\tau = 9.85$ MPa

10-9 (a) $\tau_h = \tau_v = 5.33$ ksi
(b) $\tau = 5.33$ ksi

10-10 (a) $\sigma_x = 221$ MPa (T)
(b) $\tau = 61.0$ MPa

10-13 $\sigma_n = 11.01$ ksi (T)
$\tau_{nt} = 9.59$ ksi

10-14 $\sigma_n = 60.8$ MPa (T)
$\tau_{nt} = 14.40$ MPa

10-17 $\sigma_n = 1.160$ ksi (C)
$\tau_{nt} = 12.79$ ksi

10-18 $\sigma_n = 60.3$ MPa (C)
$\tau_{nt} = -95.8$ MPa

10-21 $\sigma_n = 10.37$ ksi (T)
$\tau_{nt} = -8.73$ ksi

10-22 $\sigma_n = 155.2$ MPa (C)
$\tau_{nt} = 53.8$ MPa

10-25 $\sigma_n = 21.8$ ksi (T)
$\sigma_t = 9.23$ ksi (T)
$\tau_{nt} = 1.724$ ksi

10-26 $\sigma_n = 5.98$ MPa (T)
$\sigma_n = 86.0$ MPa (C)
$\tau_{nt} = 19.64$ MPa

10-29 $\sigma_{p1} = 18.77$ ksi (T)
$\sigma_{p2} = 2.77$ ksi (C)
$\sigma_{p3} = 0$
$\tau_p = \tau_{max} = 10.77$ ksi
$\theta_p = 34.1°$

10-30 $\sigma_{p1} = 77.7$ MPa (T)
$\sigma_{p2} = 7.72$ MPa (C)
$\sigma_{p3} = 0$
$\tau_p = \tau_{max} = 42.7$ MPa
$\theta_p = 34.7°$

10-33 $\sigma_{p1} = 12.34$ ksi (T)
$\sigma_{p2} = 17.34$ ksi (C)
$\sigma_{p3} = 0$
$\tau_p = \tau_{max} = 14.84$ ksi
$\theta_p = -16.31°$

10-34 $\sigma_{p1} = 41.4$ MPa (T)
$\sigma_{p2} = 31.4$ MPa (C)
$\sigma_{p3} = 0$
$\tau_p = \tau_{max} = 36.4$ MPa

10-37 $\theta_p = 7.97°$
$\sigma_{p1} = 2.00$ ksi (T)
$\sigma_{p2} = 18.00$ ksi (C)
$\sigma_{p3} = 0$
$\tau_p = \tau_{max} = 10.00$ ksi
$\theta_p = -26.6°$

10-38 $\sigma_{p1} = 84.0$ MPa (T)
$\sigma_{p2} = 24.0$ MPa (T)
$\sigma_{p3} = 0$
$\tau_{max} \neq \tau_p = 30.00$ MPa
$\tau_{max} = 42.0$ MPa
$\theta_p = -26.6°$

10-41 $\sigma_x = 9.86$ ksi (T)
$\sigma_y = 17.86$ ksi (C)
$\theta_p = 15.00°$

10-42 $\sigma_{p1} = 105.0$ MPa (T)
$\sigma_{p2} = 145.0$ MPa (C)
$\sigma_{p3} = 0$
$\tau_{xy} = -75.0$ MPa
$\theta_p = 18.43°$

10-45 $\tau_{xy} = -19.97$ ksi
$\sigma_{p1} = 27.0$ ksi (T)
$\sigma_{p2} = 37.0$ ksi (C)
$\sigma_{p3} = 0$

10-47 $\sigma_n = 6.62$ ksi (T)
$\tau_{nt} = 11.03$ ksi

10-48 $\sigma_n = 23.6$ MPa (T)
$\tau_{nt} = -32.9$ MPa

10-51 (a) $\sigma_{p1} = 16.45$ ksi (T)
$\sigma_{p2} = 10.45$ ksi (C)
$\sigma_{p3} = 0$
$\tau_p = \tau_{max} = 13.45$ ksi
(b) $\sigma_n = 1.159$ ksi (C)
$\tau_{nt} = +12.79$ ksi

10-52 (a) $\sigma_{p1} = 98.2$ MPa (T)
$\sigma_{p2} = 118.2$ MPa (C)
$\sigma_{p3} = 0$
$\tau_p = \tau_{max} = 108.2$ MPa
(b) $\sigma_n = 60.3$ MPa (C)
$\tau_{nt} = -95.8$ MPa

10-55 $\epsilon_n = 610\mu$
$\epsilon_t = -810\mu$
$\gamma_{nt} = 1259$ μrad

10-56 $\epsilon_n = 937\mu$
$\epsilon_t = 613\mu$
$\gamma_{nt} = -217$ μrad

10-59 $\epsilon_n = -398\mu$
$\epsilon_t = -402\mu$
$\gamma_{nt} = 721$ μrad

10-60 $\epsilon_n = 695\mu$
$\epsilon_t = 855\mu$
$\gamma_{nt} = -356$ μrad

10-63 $\epsilon_{p1} = 672\mu$
$\epsilon_{p2} = -272\mu$
$\epsilon_{p3} = 0$
$\gamma_p = \gamma_{max} = 943$ μrad
$\theta_p = 16.00°$

10-64 $\epsilon_{p1} = 1004\mu$
$\epsilon_{p2} = -364\mu$
$\epsilon_{p3} = 0$
$\gamma_p = \gamma_{max} = 1367$ μrad

10-67 $\theta_p = -10.28°$
$\epsilon_{p1} = 1009\mu$
$\epsilon_{p2} = -759\mu$
$\epsilon_{p3} = 0$
$\gamma_p = \gamma_{max} = 1768$ μrad
$\theta_p = 4.07°$

10-68 $\epsilon_{p1} = 777\mu$
$\epsilon_{p2} = -437\mu$
$\epsilon_{p3} = 0$
$\gamma_p = \gamma_{max} = 1215$ μrad
$\theta_p = 8.62°$

10-71 $\epsilon_{p1} = 908\mu$
$\epsilon_{p2} = 388\mu$
$\epsilon_{p3} = 0$
$\gamma_{max} \neq \gamma_p = 519$ μrad
$\gamma_{max} = 908$ μrad
$\theta_p = -16.85°$

10-72 $\epsilon_{p1} = 970\mu$
$\epsilon_{p2} = 580\mu$
$\epsilon_{p3} = 0$
$\gamma_{max} \neq \gamma_p = 391$ μrad
$\gamma_{max} = 970$ μrad
$\theta_p = -25.1°$

10-75 $\epsilon_x = 500$ μ
$\epsilon_y = -300$ μ
$\gamma_{xy} = +600$ μrad
$\gamma_p = \gamma_{max} = 1000$ μrad

10-76 $\epsilon_x = 640$ μ
$\epsilon_y = -800$ μ
$\gamma_{xy} = +960$ μrad
$\gamma_p = \gamma_{max} = +1730$ μrad

10-79 $\epsilon_{p1} = 966\mu$
$\epsilon_{p2} = -241\mu$
$\epsilon_{p3} = 0$
$\gamma_{pp} = \gamma_{max} = 1207$ μrad
$\theta_p = 6.59°$

10-80 $\epsilon_{p1} = 1072\mu$
$\epsilon_{p2} = -505\mu$
$\epsilon_{p3} = 0$
$\gamma_p = \gamma_{max} = 1576$ μrad
$\theta_p = 19.27°$

10-83 $\sigma_x = 8.94$ ksi (T)
$\sigma_y = 0.1753$ ksi (C)
$\tau_{xy} = -1.520$ ksi

10-84 $\sigma_x = 191.2$ MPa (T)
$\sigma_y = 130.4$ MPa (T)
$\tau_{xy} = -42.6$ MPa

10-87 $\epsilon_x = 375$ μin./in.
$\epsilon_y = -58.3$ μin./in.
$\epsilon_z = 50.0$ μin./in.
$\gamma_{xy} = 477$ μrad
$\gamma_{yz} = 412$ μrad
$\gamma_{zx} = 277$ μrad

10-89 (a) $\epsilon_x = +750\mu$
$\epsilon_y = -250\mu$
$\gamma_{xy} = -750$ μrad
(b) $\epsilon_{p1} = 875\mu$
$\epsilon_{p2} = -375\mu$
$\epsilon_{p3} = -214\mu$
$\gamma_p = \gamma_{max} = 1250$ μrad
$\theta_p = -18.43°$

10-90 (a) $\epsilon_x = +780\mu$
$\epsilon_y = -251\mu$
$\gamma_{xy} = -782\ \mu\text{rad}$
(b) $\epsilon_{p1} = 912\mu$
$\epsilon_{p2} = -383\mu$
$\epsilon_{p3} = -261\mu$
$\gamma_p = \gamma_{max} = 1294\ \mu\text{rad}$
$\theta_p = -18.59°$

10-93 (a) $\epsilon_x = -350\mu$
$\epsilon_y = +1150\mu$
$\gamma_{xy} = 520\ \mu\text{rad}$
(b) $\sigma_x = 0.514$ ksi (T)
$\sigma_y = 33.5$ ksi (T)
$\tau_{xy} = 5.72$ ksi
(c) $\sigma_{p1} = 34.5$ ksi (T)
$\sigma_{p2} = 0.451$ ksi (C)
$\sigma_{p3} = 0$
$\tau_p = \tau_{max} = 17.46$ ksi
$\theta_p = -9.56°$

10-94 (a) $\epsilon_x = +525\mu$
$\epsilon_y = +1350\mu$
$\gamma_{xy} = -975\ \mu\text{rad}$
(b) $\sigma_x = 71.2$ MPa (T)
$\sigma_y = 121.4$ MPa (T)
$\tau_{xy} = -27.3$ MPa
(c) $\sigma_{p1} = 134.0$ MPa (T)
$\sigma_{p2} = 62.5$ MPa (T)
$\sigma_{p3} = 0$
$\tau_p = 35.8$ MPa
$\tau_{max} = 67.0$ MPa
$\theta_p = 24.9°$

10-97 (a) $\sigma_x = 18.96$ ksi (T)
$\sigma_y = 1.813$ ksi (C)
$\tau_{xy} = 6.35$ ksi
(b) $\epsilon_{p1} = 727\mu$
$\epsilon_{p2} = -327\mu$
$\epsilon_{p3} = -171.4\mu$
$\gamma_p = \gamma_{max} = 1055\ \mu\text{rad}$
$\theta_p = 15.71°$
(c) $\sigma_{p1} = 20.7$ ksi (T)
$\sigma_{p2} = 3.60$ ksi (C)
$\tau_{max} = 12.17$ ksi

10-98 (a) $\sigma_x = 64.7$ MPa (T)
$\sigma_y = 2.52$ MPa (T)
$\tau_{xy} = -42.8$ MPa
(b) $\epsilon_{p1} = 1272\mu$
$\epsilon_{p2} = -655\mu$
$\epsilon_{p3} = -304\mu$
$\gamma_p = \gamma_{max} = 1927\ \mu\text{rad}$
$\theta_p = -27.0°$
(c) $\sigma_{p1} = 86.5$ MPa (T)
$\sigma_{p2} = 19.28$ MPa (C)
$\sigma_{p3} = 0$
$\tau_{max} = 52.9$ MPa

10-101 $\sigma = 720$ psi (T)
10-102 $t = 63.0$ mm
10-105 $p = 5.00$ psi
10-106 $N = 62.3$ kN
$V = 15.42$ kN
10-109 (a) $\sigma_1 = 23.5$ ksi (T)
$\sigma_2 = 16.78$ ksi (T)

(b) $pu = 340$ psi
(c) $FS = 3.13$
10-110 $\sigma_m = \gamma r^2/6t$ (T)
$\sigma_t = 5\gamma r^2/6t$ (T)
10-113 $\sigma_{p1} = 5.10$ ksi (T)
$\sigma_{p2} = 1.119$ ksi (C)
$\sigma_{p3} = 0$
$\tau_p = \tau_{max} = 3.11$ ksi
$\theta_p = 25.1°$
10-114 $\sigma_{p1} = 56.6$ MPa (T)
$\sigma_{p2} = 29.3$ MPa (C)
$\sigma_{p3} = 0$
$\tau_p = \tau_{max} = 43.0$ MPa
$\theta_p = 35.7°$
10-117 $\sigma_{p1} = 8.25$ ksi (T)
$\sigma_{p2} = 2.06$ ksi (C)
$\sigma_{p3} = 0$
$\tau_p = \tau_{max} = 5.16$ ksi
$\theta_p = 26.6°$
10-119 $\sigma_T = 3.58$ ksi (T)
$\sigma_C = 1.989$ ksi (C)
10-120 $\sigma_T = 15.25$ MPa (T)
$\sigma_C = 44.3$ MPa (C)
10-123 $\sigma_A = 1000$ psi (T)
$\sigma_B = 1400$ psi (C)
10-124 $\sigma_{max} = 125.0$ MPa (C)
10-127 $\sigma_{p1} = 32.3$ ksi (T)
$\sigma_{p2} = 0.279$ ksi (C)
$\sigma_{p3} = 0$
$\tau_{max} = \tau_p = 16.28$ ksi
$\theta_P = 5.31°$ ⭜
10-129 $P_{max} = 50.1$ kip
10-130 $P_{max} = 497$ kN
10-133 $P_{max} = 34.2$ kip
10-134 (a) $P = 24.5$ kN
$T = 14.70$ kN (C)
(b) $\sigma_{p1} = 1.348$ MPa (T)
$\sigma_{p2} = 36.3$ MPa (C)
$\sigma_{p3} = 0$
$\tau_{max} = \tau_p = 18.85$ MPa
10-137 $\sigma_{p1} = 13.83$ ksi (T)
$\sigma_{p2} = 2.37$ ksi (C)
$\sigma_{p3} = 0$
$\tau_p = \tau_{max} = 8.10$ ksi
$\theta_p = 22.5°$ ⭜
10-138 $\sigma_{p1} = 101.9$ MPa (T)
$\sigma_{p2} = 36.7$ MPa (C)
$\sigma_{p3} = 0$
$\tau_{max} = \tau_p = 69.3$ MPa
$\theta_p = 31.0°$ ⭜
10-141 $\sigma_{p1} = 0.352$ ksi (T)
$\sigma_{p2} = 11.43$ ksi (C)
$\sigma_{p3} = 0$
$\tau_{max} = \tau_p = 5.89$ ksi
$\theta_p = 9.95°$ ⭛
10-142 $\sigma_{p1} = 5.07$ MPa (T)
$\sigma_{p2} = 75.8$ MPa (C)
$\sigma_{p3} = 0$
$\tau_p = \tau_{max} = 40.4$ MPa
$\theta_p = 14.50°$ ⭛
10-143 $\sigma_{p1} = 8.14$ ksi (T)

$\sigma_{p2} = 1.089$ ksi (C)
$\sigma_{p3} = 0$
$\tau_p = \tau_{max} = 4.61$ ksi
$\theta_p = 28.8°$ ⭛
10-147 $\sigma = 886$ psi (T)
$\tau = 3400$ psi
10-148 (a) $\sigma_n = 155.2$ MPa (C)
$\tau_{nt} = 53.8$ MPa
(b) $\sigma_{p1} = 14.22$ MPa (C)
$\sigma_{p2} = 175.8$ MPa (C)
$\sigma_{p3} = 0$
$\tau_{max} \neq \tau_p = 80.8$ MPa
$\tau_{max} = 87.9$ MPa
$\theta_p = 34.1°$
10-151 $\epsilon_{p1} = 1085\mu$
$\epsilon_{p2} = -885\mu$
$\epsilon_{p3} = 0$
$\gamma_p = \gamma_{max} = 1970\ \mu\text{rad}$
$\theta_p = 11.98°$ ⭛
10-152 $\epsilon_{p1} = 1575\mu$
$\epsilon_{p2} = 585\mu$
$\epsilon_{p3} = 0$
$\gamma_{max} \neq \gamma_p = 990\ \mu\text{rad}$
$\gamma_{max} = 1575\ \mu\text{rad}$
$\theta_p = 38.0°$ ⭛
10-155 $\sigma_{p1} = 20.1$ ksi (T)
$\sigma_{p2} = 2.97$ ksi (C)
$\sigma_{p3} = 0$
$\tau_{max} = 11.54$ ksi
10-156 $\sigma_{p1} = 184.8$ MPa (T)
$\sigma_{p2} = 31.5$ MPa (C)
$\sigma_{p3} = 0$
$\tau_{max} = 108.1$ MPa
10-159 $P_{max} = 125.0$ kip
10-160 $\sigma_{p1} = 91.7$ MPa (T)
$\sigma_{p2} = 10.19$ MPa (C)
$\sigma_{p3} = 0$
$\tau_{max} = \tau_P = 50.9$ MPa
$\theta_p = 18.44°$ ⭜

CHAPTER 11
11-1 (a) $L/r = 200$
(b) $P_{CR} = 5.81$ kip
(c) $\sigma = 7.40$ ksi (C)
11-2 (a) $L/r = 149.9$
(b) $P_{CR} = 388$ kN
(c) $\sigma = 87.8$ MPa (C)
11-5 (a) $L/r = 145.9$
(b) $P_{CR} = 142.4$ kip
(c) $\sigma = 13.44$ ksi (C)
11-6 (a) $L/r = 150.4$
(b) $P_{CR} = 991$ kN
(c) $\sigma = 87.3$ MPa (C)
11-9 (a) $P_{max} = 10.46$ kip
(b) $P_{max} = 115.1$ kip
11-10 (a) $P = 27.6$ kN
(b) $P = 696$ kN
11-13 $P_{max} = 3.78$ kip
11-15 $P_{max} = 4.32$ kip
11-16 $P_{max} = 38.3$ kN
11-19 Use a W10 × 22 section

11-21 $L' = L/2$

11-22 $L' = 2L$

11-25 (a) $P_{max} = 41.4$ kip

(b) $P_{max} = 206$ kip

11-26 (a) $P_{max} = 429$ kN

(b) $P_{max} = 928$ kN

11-29 (a) $P_{max} = 67.0$ kip

(b) $P_{max} = 166.3$ kip

11-30 (a) $P_{max} = 18.06$ kN

(b) $P_{max} = 55.4$ kN

11-33 $L = 6.49$ ft

11-34 $P_{max} = 1097$ kN

11-37 $d = 3.50$ in.

11-39 (a) $P_{max} = 163.0$ kip

(b) $P_{max} = 280$ kip

11-40 $P_{max} = 455$ kN

11-55 (a) $L/r = 156.7$

(b) $P_{CR} = 23.2$ kip

(c) $\sigma = 483$ psi (C)

11-56 (a) $L/r = 92.6$

(b) $P_{CR} = 93.4$ kN

(c) $\sigma = 14.95$ MPa (C)

11-59 $P_{max} = 109.1$ kip

11-60 (a) $FS_{AB} = 1.322$

$FS_{AC} = 22.9$

(b) $FS_{AC} = 6.53$

11-62 (a) $P_{max} = 783$ kN

(b) $P_{max} = 1461$ kN

11-65 $P_{max} = 80.6$ kip

INDEX

SECOND MOMENTS OF PLANE AREAS

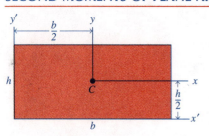

$$I_x = \frac{bh^3}{12}$$

$$I_{x'} = \frac{bh^3}{3}$$

$$A = bh$$

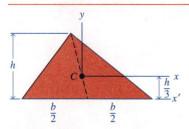

$$I_x = \frac{bh^3}{36}$$

$$I_{x'} = \frac{bh^3}{12}$$

$$A = \frac{1}{2}bh$$

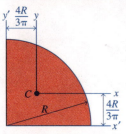

$$I_x = \frac{\pi R^4}{4}$$

$$I_{x'} = \frac{5\pi R^4}{4}$$

$$A = \pi R^2$$

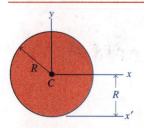

$$I_x = \frac{\pi R^4}{8} - \frac{8R^4}{9\pi}$$

$$I_y = \frac{\pi R^4}{8}$$

$$I_{x'} = \frac{\pi R^4}{8}$$

$$A = \frac{1}{2}\pi R^2$$

$$I_x = \frac{\pi R^4}{16} - \frac{4R^4}{9\pi}$$

$$I_{x'} = \frac{\pi R^4}{16}$$

$$A = \frac{1}{4}\pi R^2$$

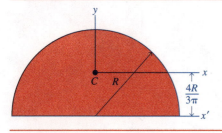

$$I_x = \frac{R^4}{4}\left(\theta - \frac{1}{2}\sin 2\theta\right)$$

$$x_C = \frac{2}{3}\frac{R\sin\theta}{\theta}$$

$$I_y = \frac{R^4}{4}\left(\theta + \frac{1}{2}\sin 2\theta\right)$$

$$A = \theta R^2$$